W9-CCI-155

Springer Texts in Statistics

Series Editors:

R. DeVeaux
S.E. Fienberg
I. Olkin

More information about this series at http://www.springer.com/series/417

David Ruppert • David S. Matteson

Statistics and Data Analysis for Financial Engineering

with R examples

Second Edition

 Springer

David Ruppert
Department of Statistical
  Science and School of ORIE
Cornell University
Ithaca, NY, USA

David S. Matteson
Department of Statistical Science
Department of Social Statistics
Cornell University
Ithaca, NY, USA

ISSN 1431-875X
Springer Texts in Statistics
ISBN 978-1-4939-2613-8
DOI 10.1007/978-1-4939-2614-5

ISSN 2197-4136   (electronic)

ISBN 978-1-4939-2614-5   (eBook)

Library of Congress Control Number: 2015935333

Springer New York Heidelberg Dordrecht London

Printed on acid-free paper

Springer Science+Business Media LLC New York is part of Springer Science+Business Media (www.
springer.com)

To Susan

David Ruppert

To my grandparents

David S. Matteson

Preface

The first edition of this book has received a very warm reception. A number of instructors have adopted this work as a textbook in their courses. Moreover, both novices and seasoned professionals have been using the book for self-study. The enthusiastic response to the book motivated a new edition. One major change is that there are now two authors. The second edition improves the book in several ways: all known errors have been corrected and changes in R have been addressed. Considerably more R code is now included. The GARCH chapter now uses the `rugarch` package, and in the Bayes chapter we now use `JAGS` in place of `OpenBUGS`.

The first edition was designed primarily as a textbook for use in university courses. Although there is an Instructor's Manual with solutions to all exercises and all problems in the R labs, this manual has been available only to instructors. No solutions have been available for readers engaged in self-study. To address this problem, the number of exercises and R lab problems has increased and the solutions to many of them are being placed on the book's web site.

Some data sets in the first edition were in R packages that are no longer available. These data sets are also on the web site. The web site also contains R scripts with the code used in the book.

We would like to thank Peter Dalgaard, Guy Yollin, and Aaron Fox for many helpful suggestions. We also thank numerous readers for pointing out errors in the first edition.

The book's web site is http://people.orie.cornell.edu/davidr/SDAFE2/index.html.

Ithaca, NY, USA
Ithaca, NY, USA
January 2015

David Ruppert
David S. Matteson

Preface to the First Edition

I developed this textbook while teaching the course *Statistics for Financial Engineering* to master's students in the financial engineering program at Cornell University. These students have already taken courses in portfolio management, fixed income securities, options, and stochastic calculus, so I concentrate on teaching statistics, data analysis, and the use of R, and I cover most sections of Chaps. 4–12 and 18–20. These chapters alone are more than enough to fill a one-semester course. I do not cover regression (Chaps. 9–11 and 21) or the more advanced time series topics in Chap. 13, since these topics are covered in other courses. In the past, I have not covered cointegration (Chap. 15), but I will in the future. The master's students spend much of the third semester working on projects with investment banks or hedge funds. As a faculty adviser for several projects, I have seen the importance of cointegration.

A number of different courses might be based on this book. A two-semester sequence could cover most of the material. A one-semester course with more emphasis on finance would include Chaps. 16 and 17 on portfolios and the CAPM and omit some of the chapters on statistics, for instance, Chaps. 8, 14, and 20 on copulas, GARCH models, and Bayesian statistics. The book could be used for courses at both the master's and Ph.D. levels.

Readers familiar with my textbook *Statistics and Finance: An Introduction* may wonder how that volume differs from this book. This book is at a somewhat more advanced level and has much broader coverage of topics in statistics compared to the earlier book. As the title of this volume suggests, there is more emphasis on data analysis and this book is intended to be more than just "an introduction." Chapters 8, 15, and 20 on copulas, cointegration, and Bayesian statistics are new. Except for some figures borrowed from *Statistics and Finance*, in this book R is used exclusively for computations, data analysis, and graphing, whereas the earlier book used SAS and MATLAB. Nearly all of the examples in this book use data sets that are available in R, so readers can reproduce the results. In Chap. 20 on Bayesian statistics,

WinBUGS is used for Markov chain Monte Carlo and is called from R using the R2WinBUGS package. There is some overlap between the two books, and, in particular, a substantial amount of the material in Chaps. 2, 3, 9, 11–13, and 16 has been taken from the earlier book. Unlike *Statistics and Finance*, this volume does not cover options pricing and behavioral finance.

The prerequisites for reading this book are knowledge of calculus, vectors, and matrices; probability including stochastic processes; and statistics typical of third- or fourth-year undergraduates in engineering, mathematics, statistics, and related disciplines. There is an appendix that reviews probability and statistics, but it is intended for reference and is certainly not an introduction for readers with little or no prior exposure to these topics. Also, the reader should have some knowledge of computer programming. Some familiarity with the basic ideas of finance is helpful.

This book does not teach R programming, but each chapter has an "R lab" with data analysis and simulations. Students can learn R from these labs and by using R's help or the manual *An Introduction to R* (available at the CRAN web site and R's online help) to learn more about the functions used in the labs. Also, the text does indicate which R functions are used in the examples. Occasionally, R code is given to illustrate some process, for example, in Chap. 16 finding the tangency portfolio by quadratic programming. For readers wishing to use R, the bibliographical notes at the end of each chapter mention books that cover R programming and the book's web site contains examples of the R and WinBUGS code used to produce this book. Students enter my course *Statistics for Financial Engineering* with quite disparate knowledge of R. Some are very accomplished R programmers, while others have no experience with R, although all have experience with some programming language. Students with no previous experience with R generally need assistance from the instructor to get started on the R labs. Readers using this book for self-study should learn R first before attempting the R labs.

Ithaca, NY, USA   David Ruppert
July 2010

Contents

Notation

The following conventions are observed as much as possible:

- Lowercase letters, e.g., a and b, are used for nonrandom scalars.
- Lowercase boldface letters, e.g., $\boldsymbol{a}$, $\boldsymbol{b}$, and $\boldsymbol{\theta}$, are used for nonrandom vectors.
- Uppercase letters, e.g., X and Y, are used for random variables.
- Uppercase bold letters either early in the Roman alphabet or in Greek without a "hat," e.g., $\boldsymbol{A}$, $\boldsymbol{B}$, and $\boldsymbol{\Omega}$, are used for nonrandom matrices.
- A hat over a parameter or parameter vector, e.g., $\widehat{\theta}$ and $\widehat{\boldsymbol{\theta}}$, denotes an estimator of the corresponding parameter or parameter vector.
- $\boldsymbol{I}$ denotes the identity matrix with dimension appropriate for the context.
- $\mathrm{diag}(d_1, \ldots, d_p)$ is a diagonal matrix with diagonal elements $d_1, \ldots, d_p$.
- Greek letters with a "hat" or uppercase bold letters later in the Roman alphabet, e.g., $\boldsymbol{X}$, $\boldsymbol{Y}$, and $\widehat{\boldsymbol{\theta}}$, will be used for random vectors.
- $\log(x)$ is the natural logarithm of x and $\log_{10}(x)$ is the base-10 logarithm.
- $E(X)$ is the expected value of a random variable X.
- $\mathrm{Var}(X)$ and σ_X^2 are used to denote the variance of a random variable X.
- $\mathrm{Cov}(X, Y)$ and σ_{XY} are used to denote the covariance between the random variables X and Y.
- $\mathrm{Corr}(X, Y)$ and ρ_{XY} are used to denote the correlation between the random variables X and Y.
- $\mathrm{COV}(\boldsymbol{X})$ is the covariance matrix of a random vector $\boldsymbol{X}$.
- $\mathrm{CORR}(\boldsymbol{X})$ is the correlation matrix of a random vector $\boldsymbol{X}$.
- A Greek letter denotes a parameter, e.g., θ.
- A boldface Greek letter, e.g., $\boldsymbol{\theta}$, denotes a vector of parameters.
- $\Re$ is the set of real numbers and $\Re^p$ is the p-dimensional Euclidean space, the set of all real p-dimensional vectors.
- $A \cap B$ and $A \cup B$ are, respectively, the intersection and union of the sets A and B.
- $\emptyset$ is the empty set.

- If A is some statement, then $I\{A\}$ is called the indicator function of A and is equal to 1 if A is true and equal to 0 if A is false.
- If f_1 and f_2 are two functions of a variable x, then

$$f_1(x) \sim f_2(x) \text{ as } x \to x_0$$

means that

$$\lim_{x \to x_0} \frac{f_1(x)}{f_2(x)} = 1.$$

Similarly,

$$a_n \sim b_n$$

means that the sequences $\{a_n\}$ and $\{b_n\}$ are such that

$$\frac{a_n}{b_n} \to 1 \text{ as } n \to \infty.$$

- Vectors are column vectors and transposed vectors are rows, e.g.,

$$\boldsymbol{x} = \begin{pmatrix} x_1 \\ \vdots \\ x_n \end{pmatrix}$$

and

$$\boldsymbol{x}^\mathsf{T} = (\, x_1 \quad \cdots \quad x_n \,).$$

- $|\boldsymbol{A}|$ is the determinant of a square matrix $\boldsymbol{A}$.
- $\mathrm{tr}(\boldsymbol{A})$ is the trace (sum of the diagonal elements) of a square matrix $\boldsymbol{A}$.
- $f(x) \propto g(x)$ means that $f(x)$ is proportional to $g(x)$, that is, $f(x) = ag(x)$ for some nonzero constant a.
- A word appearing in italic font is being defined or introduced in the text.

1

Introduction

This book is about the analysis of financial markets data. After this brief introductory chapter, we turn immediately in Chaps. 2 and 3 to the sources of the data, returns on equities and prices and yields on bonds. Chapter 4 develops methods for informal, often graphical, analysis of data. More formal methods based on statistical inference, that is, estimation and testing, are introduced in Chap. 5. The chapters that follow Chap. 5 cover a variety of more advanced statistical techniques: ARIMA models, regression, multivariate models, copulas, GARCH models, factor models, cointegration, Bayesian statistics, and nonparametric regression.

Much of finance is concerned with financial risk. The *return* on an investment is its revenue expressed as a fraction of the initial investment. If one invests at time t_1 in an asset with price P_{t_1} and the price later at time t_2 is P_{t_2}, then the net return for the holding period from t_1 to t_2 is $(P_{t_2} - P_{t_1})/P_{t_1}$. For most assets, future returns cannot be known exactly and therefore are random variables. *Risk* means uncertainty in future returns from an investment, in particular, that the investment could earn less than the expected return and even result in a loss, that is, a negative return. Risk is often measured by the standard deviation of the return, which we also call the volatility. Recently there has been a trend toward measuring risk by value-at-risk (VaR) and expected shortfall (ES). These focus on large losses and are more direct indications of financial risk than the standard deviation of the return. Because risk depends upon the probability distribution of a return, probability and statistics are fundamental tools for finance. Probability is needed for risk calculations, and statistics is needed to estimate parameters such as the standard deviation of a return or to test hypotheses such as the so-called random walk hypothesis which states that future returns are independent of the past.

© Springer Science+Business Media New York 2015
D. Ruppert, D.S. Matteson, *Statistics and Data Analysis for Financial Engineering*, Springer Texts in Statistics,
DOI 10.1007/978-1-4939-2614-5_1

In financial engineering there are two kinds of probability distributions that can be estimated. Objective probabilities are the true probabilities of events. Risk-neutral or pricing probabilities give model outputs that agree with market prices and reflect the market's beliefs about the probabilities of future events. The statistical techniques in this book can be used to estimate both types of probabilities. Objective probabilities are usually estimated from historical data, whereas risk-neutral probabilities are estimated from the prices of options and other financial instruments.

Finance makes extensive use of probability models, for example, those used to derive the famous Black–Scholes formula. Use of these models raises important questions of a statistical nature such as: Are these models supported by financial markets data? How are the parameters in these models estimated? Can the models be simplified or, conversely, should they be elaborated?

After Chaps. 4–8 develop a foundation in probability, statistics, and exploratory data analysis, Chaps. 12 and 13 look at ARIMA models for time series. Time series are sequences of data sampled over time, so much of the data from financial markets are time series. ARIMA models are stochastic processes, that is, probability models for sequences of random variables. In Chap. 16 we study optimal portfolios of risky assets (e.g., stocks) and of risky assets and risk-free assets (e.g., short-term U.S. Treasury bills). Chapters 9–11 cover one of the most important areas of applied statistics, regression. Chapter 15 introduces cointegration analysis. In Chap. 17 portfolio theory and regression are applied to the CAPM. Chapter 18 introduces factor models, which generalize the CAPM. Chapters 14–21 cover other areas of statistics and finance such as GARCH models of nonconstant volatility, Bayesian statistics, risk management, and nonparametric regression.

Several related themes will be emphasized in this book:

Always look at the data According to a famous philosopher and baseball player, Yogi Berra, "You can see a lot by just looking." This is certainly true in statistics. The first step in data analysis should be plotting the data in several ways. Graphical analysis is emphasized in Chap. 4 and used throughout the book. Problems such as bad data, outliers, mislabeling of variables, missing data, and an unsuitable model can often be detected by visual inspection. *Bad data* refers to data that are outlying because of errors, e.g., recording errors. Bad data should be corrected when possible and otherwise deleted. Outliers due, for example, to a stock market crash are "good data" and should be retained, though the model may need to be expanded to accommodate them. It is important to detect both bad data and outliers, and to understand which is which, so that appropriate action can be taken.

All models are false Many statisticians are familiar with the observation of George Box that "all models are false but some models are useful." This fact should be kept in mind whenever one wonders whether a statistical, economic, or financial model is "true." Only computer-simulated data have a "true model." No model can be as complex as the real world, and even if such a model did exist, it would be too complex to be useful.

Bias-variance tradeoff If useful models exist, how do we find them? The answer to this question depends ultimately on the intended uses of the model. One very useful principle is *parsimony* of parameters, which means that we should use only as many parameters as necessary. Complex models with unnecessary parameters increase estimation error and make interpretation of the model more difficult. However, a model that is too simple will not capture important features of the data and will lead to serious biases. Simple models have large biases but small variances of the estimators. Complex models have small biases but large variances. Therefore, model choice involves finding a good tradeoff between bias and variance.

Uncertainty analysis It is essential that the uncertainty due to estimation and modeling errors be quantified. For example, portfolio optimization methods that assume that return means, variances, and correlations are known exactly are suboptimal when these parameters are only estimated (as is always the case). Taking uncertainty into account leads to other techniques for portfolio selection—see Chap. 16. With complex models, uncertainty analysis could be challenging in the past, but no longer is so because of modern statistical techniques such as resampling (Chap. 6) and Bayesian MCMC (Chap. 20).

Financial markets data are not normally distributed Introductory statistics textbooks model continuously distributed data with the normal distribution. This is fine in many domains of application where data are well approximated by a normal distribution. However, in finance, stock returns, changes in interest rates, changes in foreign exchange rates, and other data of interest have many more outliers than would occur under normality. For modeling financial markets data, heavy-tailed distributions such as the t-distributions are much more suitable than normal distributions—see Chap. 5. *Remember:* In finance, the normal distribution is not normal.

Variances are not constant Introductory textbooks also assume constant variability. This is another assumption that is rarely true for financial markets data. For example, the daily return on the market on Black Monday, October 19, 1987, was -23%, that is, the market lost 23% of its value in a single day! A return of this magnitude is virtually impossible under a normal model with a constant variance, and it is still quite unlikely under a t-distribution with constant variance, but much more likely under a t-distribution model with conditional heteroskedasticity, e.g., a GARCH model (Chap. 14).

1.1 Bibliographic Notes

The dictum that "All models are false but some models are useful" is from Box (1976).

References

Box, G. E. P. (1976) Science and statistics, *Journal of the American Statistical Association*, 71, 791–799.

2

Returns

2.1 Introduction

The goal of investing is, of course, to make a profit. The revenue from investing, or the loss in the case of negative revenue, depends upon both the change in prices and the amounts of the assets being held. Investors are interested in revenues that are high relative to the size of the initial investments. Returns measure this, because returns on an asset, e.g., a stock, a bond, a portfolio of stocks and bonds, are changes in price expressed as a fraction of the initial price.

2.1.1 Net Returns

Let P_t be the price of an asset at time t. Assuming no dividends, the *net return* over the holding period from time $t-1$ to time t is

$$R_t = \frac{P_t}{P_{t-1}} - 1 = \frac{P_t - P_{t-1}}{P_{t-1}}.$$

The numerator $P_t - P_{t-1}$ is the revenue or profit during the holding period, with a negative profit meaning a loss. The denominator, P_{t-1}, was the initial investment at the start of the holding period. Therefore, the net return can be viewed as the relative revenue or profit rate.

The revenue from holding an asset is

$$\text{revenue} = \text{initial investment} \times \text{net return}.$$

For example, an initial investment of \$10,000 and a net return of 6 % earns a revenue of \$600. Because $P_t \geq 0$,

© Springer Science+Business Media New York 2015
D. Ruppert, D.S. Matteson, *Statistics and Data Analysis for Financial Engineering*, Springer Texts in Statistics,
DOI 10.1007/978-1-4939-2614-5_2

$$R_t \geq -1, \tag{2.1}$$

so the worst possible return is -1, that is, a 100% loss, and occurs if the asset becomes worthless.

2.1.2 Gross Returns

The simple *gross return* is

$$\frac{P_t}{P_{t-1}} = 1 + R_t.$$

For example, if $P_t = 2$ and $P_{t+1} = 2.1$, then $1 + R_{t+1} = 1.05$, or 105%, and $R_{t+1} = 0.05$, or 5%. One's final wealth at time t is one's initial wealth at time $t-1$ times the gross return. Stated differently, if X_0 is the initial at time $t-1$, then $X_0(1 + R_t)$ is one's wealth at time t.

Returns are scale-free, meaning that they do not depend on units (dollars, cents, etc.). Returns are *not* unitless. Their unit is time; they depend on the units of t (hour, day, etc.). In this example, if t is measured in years, then, stated more precisely, the net return is 5% per year.

The *gross return over the most recent k periods* is the product of the k single-period gross returns (from time $t - k$ to time t):

$$1 + R_t(k) = \frac{P_t}{P_{t-k}} = \left(\frac{P_t}{P_{t-1}}\right)\left(\frac{P_{t-1}}{P_{t-2}}\right) \cdots \left(\frac{P_{t-k+1}}{P_{t-k}}\right)$$
$$= (1 + R_t) \cdots (1 + R_{t-k+1}).$$

The k-period net return is $R_t(k)$.

2.1.3 Log Returns

Log returns, also called *continuously compounded returns*, are denoted by r_t and defined as

$$r_t = \log(1 + R_t) = \log\left(\frac{P_t}{P_{t-1}}\right) = p_t - p_{t-1},$$

where $p_t = \log(P_t)$ is called the *log price*.

Log returns are approximately equal to returns because if x is small, then $\log(1+x) \approx x$, as can been seen in Fig. 2.1, where $\log(1+x)$ is plotted. Notice in that figure that $\log(1 + x)$ is very close to x if $|x| < 0.1$, e.g., for returns that are less than 10%.

For example, a 5% return equals a 4.88% log return since $\log(1 + 0.05) = 0.0488$. Also, a -5% return equals a -5.13% log return since $\log(1 - 0.05) = -0.0513$. In both cases, $r_t = \log(1 + R_t) \approx R_t$. Also, $\log(1 + 0.01) = 0.00995$ and $\log(1 - 0.01) = -0.01005$, so log returns of $\pm 1\%$ are very close to the corresponding net returns. Since returns are smaller in magnitude over shorter periods, we can expect returns and log returns to be similar for daily returns, less similar for yearly returns, and not necessarily similar for longer periods such as 10 years.

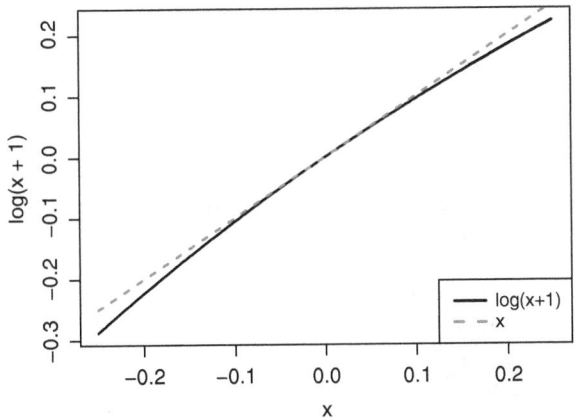

Fig. 2.1. *Comparison of functions* $\log(1 + x)$ *and* x.

The return and log return have the same sign. The magnitude of the log return is smaller (larger) than that of the return if they are both positive (negative). The difference between a return and a log return is most pronounced when both are very negative. Returns close to the lower bound of -1, that is complete losses, correspond to log return close to $-\infty$.

One advantage of using log returns is simplicity of multiperiod returns. A k-period log return is simply the sum of the single-period log returns, rather than the product as for gross returns. To see this, note that the k-period log return is

$$
\begin{aligned}
r_t(k) &= \log\{1 + R_t(k)\} \\
&= \log\left\{(1 + R_t) \cdots (1 + R_{t-k+1})\right\} \\
&= \log(1 + R_t) + \cdots + \log(1 + R_{t-k+1}) \\
&= r_t + r_{t-1} + \cdots + r_{t-k+1}.
\end{aligned}
$$

2.1.4 Adjustment for Dividends

Many stocks, especially those of mature companies, pay dividends that must be accounted for when computing returns. Similarly, bonds pay interest. If a dividend (or interest) D_t is paid prior to time t, then the gross return at time t is defined as

$$1 + R_t = \frac{P_t + D_t}{P_{t-1}}, \tag{2.2}$$

and so the net return is $R_t = (P_t + D_t)/P_{t-1} - 1$ and the log return is $r_t = \log(1 + R_t) = \log(P_t + D_t) - \log(P_{t-1})$. Multiple-period gross returns are products of single-period gross returns so that

$$1 + R_t(k) = \left(\frac{P_t + D_t}{P_{t-1}}\right) \left(\frac{P_{t-1} + D_{t-1}}{P_{t-2}}\right) \cdots \left(\frac{P_{t-k+1} + D_{t-k+1}}{P_{t-k}}\right)$$
$$= (1 + R_t)(1 + R_{t-1}) \cdots (1 + R_{t-k+1}), \tag{2.3}$$

where, for any time s, $D_s = 0$ if there is no dividend between $s - 1$ and s. Similarly, a k-period log return is

$$r_t(k) = \log\{1 + R_t(k)\} = \log(1 + R_t) + \cdots + \log(1 + R_{t-k+1})$$
$$= \log\left(\frac{P_t + D_t}{P_{t-1}}\right) + \cdots + \log\left(\frac{P_{t-k+1} + D_{t-k+1}}{P_{t-k}}\right).$$

2.2 The Random Walk Model

The *random walk hypothesis* states that the single-period log returns, $r_t = \log(1 + R_t)$, are independent. Because

$$1 + R_t(k) = (1 + R_t) \cdots (1 + R_{t-k+1})$$
$$= \exp(r_t) \cdots \exp(r_{t-k+1})$$
$$= \exp(r_t + \cdots + r_{t-k+1}),$$

we have

$$\log\{1 + R_t(k)\} = r_t + \cdots + r_{t-k+1}. \tag{2.4}$$

It is sometimes assumed further that the log returns are $N(\mu, \sigma^2)$ for some constant mean and variance. Since sums of normal random variables are themselves normal, normality of single-period log returns implies normality of multiple-period log returns. Under these assumptions, $\log\{1 + R_t(k)\}$ is $N(k\mu, k\sigma^2)$.

2.2.1 Random Walks

Model (2.4) is an example of a random walk model. Let $Z_1, Z_2, \ldots$ be i.i.d. (independent and identically distributed) with mean μ and standard deviation σ. Let S_0 be an arbitrary starting point and

$$S_t = S_0 + Z_1 + \cdots + Z_t, \quad t \geq 1. \tag{2.5}$$

From (2.5), S_t is the position of the random walker after t steps starting at S_0.

The process $S_0, S_1, \ldots$ is called a *random walk* and $Z_1, Z_2, \ldots$ are its steps. If the steps are normally distributed, then the process is called a *normal random walk*. The expectation and variance of S_t, conditional given S_0, are $E(S_t|S_0) = S_0 + \mu t$ and $\text{Var}(S_t|S_0) = \sigma^2 t$. The parameter μ is called the *drift* and determines the general direction of the random walk. The parameter σ is the *volatility* and determines how much the random walk fluctuates about the conditional mean $S_0 + \mu t$. Since the standard deviation of S_t given S_0 is $\sigma\sqrt{t}$, $(S_0 + \mu t) \pm \sigma\sqrt{t}$ gives the mean plus and minus one standard deviation, which, for a normal random walk, gives a range containing 68 % probability. The width of this range grows proportionally to $\sqrt{t}$, as is illustrated in Fig. 2.2, showing that at time $t = 0$ we know far less about where the random walk will be in the distant future compared to where it will be in the immediate future.

2.2.2 Geometric Random Walks

Recall that $\log\{1 + R_t(k)\} = r_t + \cdots + r_{t-k+1}$. Therefore,

$$\frac{P_t}{P_{t-k}} = 1 + R_t(k) = \exp(r_t + \cdots + r_{t-k+1}), \qquad (2.6)$$

so taking $k = t$, we have

$$P_t = P_0 \exp(r_t + r_{t-1} + \cdots + r_1). \qquad (2.7)$$

We call such a process whose logarithm is a random walk a *geometric random walk* or an *exponential random walk*. If $r_1, r_2, \ldots$ are i.i.d. $N(\mu, \sigma^2)$, then P_t is lognormal for all t and the process is called a *lognormal geometric random walk with parameters* (μ, σ^2). As discussed in Appendix A.9.4, μ is called the log-mean and σ is called the log-standard deviation of the log-normal distribution of $\exp(r_t)$. Also, μ is sometimes called the log-drift of the lognormal geometric random walk.

2.2.3 Are Log Prices a Lognormal Geometric Random Walk?

Much work in mathematical finance assumes that prices follow a lognormal geometric random walk or its continuous-time analog, geometric Brownian motion. So a natural question is whether this assumption is usually true. The quick answer is "no." The lognormal geometric random walk makes two assumptions: (1) the log returns are normally distributed and (2) the log returns are mutually independent.

In Chaps. 4 and 5, we will investigate the marginal distributions of several series of log returns. The conclusion will be that, though the return density has a bell shape somewhat like that of normal densities, the tails of the log return distributions are generally much heavier than normal tails. Typically, a

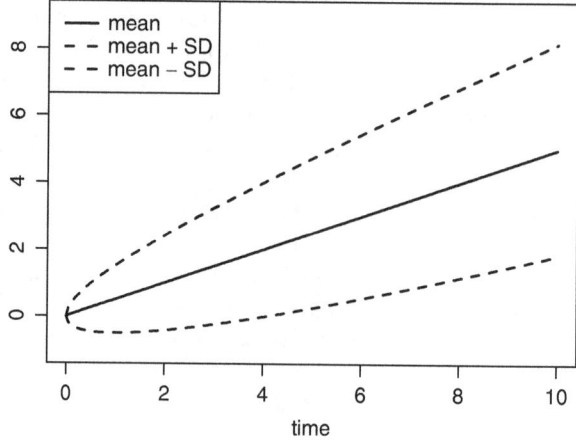

Fig. 2.2. *Mean and bounds (mean plus and minus one standard deviation) on a random walk with $S_0 = 0$, $\mu = 0.5$, and $\sigma = 1$. At any given time, the probability of being between the bounds (dashed curves) is 68 % if the distribution of the steps is normal. Since $\mu > 0$, there is an overall positive trend that would be reversed if μ were negative.*

t-distribution with a small degrees-of-freedom parameter, say 4–6, is a much better fit than the normal model. However, the log-return distributions do appear to be symmetric, or at least nearly so.

The independence assumption is also violated. First, there is some correlation between returns. The correlations, however, are generally small. More seriously, returns exhibit *volatility clustering*, which means that if we see high volatility in current returns then we can expect this higher volatility to continue, at least for a while. Volatility clustering can be detected by checking for correlations between the *squared* returns.

Before discarding the assumption that the prices of an asset are a lognormal geometric random walk, it is worth remembering Box's dictum that "all models are false, but some models are useful." This assumption is sometimes useful, e.g., for deriving the famous Black–Scholes formula.

2.3 Bibliographic Notes

The random walk hypothesis is related to the so-called efficient market hypothesis; see Ruppert et al. (2003) for discussion and further references. Bodie et al. (1999) and Sharpe et al. (1995) are good introductions to the random walk hypothesis and market efficiency. A more advanced discussion of the random walk hypothesis is found in Chap. 2 of Campbell et al. (1997) and Lo and MacKinlay (1999). Much empirical evidence about the behavior of returns is reviewed by Fama (1965, 1970, 1991, 1998). Evidence against the efficient market hypothesis can be found in the field of behavioral finance which uses the study of human behavior to understand market behavior; see Shefrin (2000), Shleifer (2000), and Thaler (1993). One indication of market inefficiency is excess volatility of market prices; see Shiller (1992) or Shiller (2000) for a less technical discussion.

R will be used extensively in what follows. Dalgaard (2008) and Zuur et al. (2009) are good places to start learning R.

2.4 R Lab

2.4.1 Data Analysis

Obtain the data set `Stock_bond.csv` from the book's website and put it in your working directory. Start R[1] and you should see a console window open up. Use **Change Dir** in the "File" menu to change to the working directory. Read the data with the following command:

```
dat = read.csv("Stock_bond.csv", header = TRUE)
```

The data set `Stock_bond.csv` contains daily volumes and adjusted closing (AC) prices of stocks and the S&P 500 (columns B–W) and yields on bonds (columns X–AD) from 2-Jan-1987 to 1-Sep-2006.

This book does not give detailed information about R functions since this information is readily available elsewhere. For example, you can use R's help to obtain more information about the `read.csv()` function by typing "?read.csv" in your R console and then hitting the Enter key. You should also use the manual *An Introduction to R* that is available on R's help file and also on CRAN. Another resource for those starting to learn R is Zuur et al. (2009).

An alternative to typing commands in the console is to start a new script from the "file" menu, put code into the editor, highlight the lines, and then press Ctrl-R to run the code that has been highlighted.[2] This technique is useful for debugging. You can save the script file and then reuse or modify it.

Once a file is saved, the entire file can be run by "sourcing" it. You can use the "file" menu in R to source a file or use the `source()` function. If the file is in the editor, then it can be run by hitting Ctrl-A to highlight the entire file and then Ctrl-R.

The next lines of code print the names of the variables in the data set, attach the data, and plot the adjusted closing prices of GM and Ford.

---

[1] You can also run R from **Rstudio** and, in fact, **Rstudio** is highly recommended. The authors switched from R to **Rstudio** while the second edition of this book was being written.

[2] Or click the "run" button in Rstudio.

```
1 names(dat)
2 attach(dat)
3 par(mfrow = c(1, 2))
4 plot(GM_AC)
5 plot(F_AC)
```

Here and elsewhere in this book, line numbers are often added when listing R code. The line numbers are not part of the code.

By default, as in lines 4 and 5, points are plotted with the character "o". To plot a line instead, use, for example `plot(GM_AC, type = "l")`. Similarly, `plot(GM_AC, type = "b")` plots both points and a line.

The R function `attach()` puts a database into the R search path. This means that the database is searched by R when evaluating a variable, so objects in the database can be accessed by simply giving their names. If `dat` was not attached, then line 4 would be replaced by `plot(dat$GM_AC)` and similarly for line 5.

The function `par()` specifies plotting parameters and `mfrow=c(n1,n2)` specifies "make a figure, fill by rows, n1 rows and n2 columns." Thus, the first n1 plots fill the first row and so forth. `mfcol(n1,n2)` fills by columns and so would put the first n2 plots in the first column. As mentioned before, more information about these and other R functions can be obtained from R's online help or the manual *An Introduction to R*.

Run the code below to find the sample size (n), compute GM and Ford returns, and plot GM net returns versus the Ford returns.

```
1 n = dim(dat)[1]
2 GMReturn = GM_AC[-1] / GM_AC[-n] - 1
3 FReturn = F_AC[-1] / F_AC[-n] - 1
4 par(mfrow = c(1, 1))
5 plot(GMReturn,FReturn)
```

On lines 2 and 3, the index `-1` means all indices except the first and similarly `-n` means all indices except the last.

Problem 1 *Do the GM and Ford returns seem positively correlated? Do you notice any outlying returns? If "yes," do outlying GM returns seem to occur with outlying Ford returns?*

Problem 2 *Compute the log returns for GM and plot the returns versus the log returns. How highly correlated are the two types of returns? (The R function cor() computes correlations.)*

Problem 3 *Repeat Problem 1 with Microsoft (MSFT) and Merck (MRK).*

When you exit R, you can "Save workspace image," which will create an R workspace file in your working directory. Later, you can restart R and load this workspace image into memory by right-clicking on the R workspace file. When R starts, your working directory will be the folder containing the R workspace that was opened. A useful trick when starting a project in a new folder is to put an empty saved workspace into this folder. Double-clicking on the workspace starts R with the folder as the working directory.

2.4.2 Simulations

Hedge funds can earn high profits through the use of leverage, but leverage also creates high risk. The simulations in this section explore the effects of leverage in a simplified setting.

Suppose a hedge fund owns $1,000,000 of stock and used $50,000 of its own capital and $950,000 in borrowed money for the purchase. Suppose that if the value of the stock falls below $950,000 at the end of any trading day, then the hedge fund will sell all the stock and repay the loan. This will wipe out its $50,000 investment. The hedge fund is said to be leveraged 20:1 since its position is 20 times the amount of its own capital invested.

Suppose that the daily log returns on the stock have a mean of 0.05/year and a standard deviation of 0.23/year. These can be converted to rates per trading day by dividing by 253 and $\sqrt{253}$, respectively.

Problem 4 *What is the probability that the value of the stock will be below $950,000 at the close of at least one of the next 45 trading days? To answer this question, run the code below.*

```
1  niter = 1e5          #  number of iterations
2  below = rep(0, niter)  # set up storage
3  set.seed(2009)
4  for (i in 1:niter)
5  {
6     r = rnorm(45, mean = 0.05/253,
7        sd = 0.23/sqrt(253)) # generate random numbers
8     logPrice = log(1e6) + cumsum(r)
9     minlogP = min(logPrice) # minimum price over next 45 days
10    below[i] = as.numeric(minlogP < log(950000))
11 }
12 mean(below)
```

On line 10, `below[i]` equals 1 if, for the ith simulation, the minimum price over 45 days is less that 950,000. Therefore, on line 12, `mean(below)` is the proportion of simulations where the minimum price is less than 950,000.

If you are unfamiliar with any of the R functions used here, then use R's help to learn about them; e.g., type `?rnorm` to learn that `rnorm()` generates normally distributed random numbers. You should study each line of code, understand what it is doing, and convince yourself that the code estimates the probability being requested. Note that anything that follows a pound sign is a comment and is used only to annotate the code.

Suppose the hedge fund will sell the stock for a profit of at least $100,000 if the value of the stock rises to at least $1,100,000 at the end of one of the first 100 trading days, sell it for a loss if the value falls below $950,000 at the end of one of the first 100 trading days, or sell after 100 trading days if the closing price has stayed between $950,000 and $1,100,000.

The following questions can be answered by simulations much like the one above. Ignore trading costs and interest when answering these questions.

Problem 5 *What is the probability that the hedge fund will make a profit of at least $100,000?*

Problem 6 *What is the probability the hedge fund will suffer a loss?*

Problem 7 *What is the expected profit from this trading strategy?*

Problem 8 *What is the expected return? When answering this question, remember that only $50,000 was invested. Also, the units of return are time, e.g., one can express a return as a daily return or a weekly return. Therefore, one must keep track of how long the hedge fund holds its position before selling.*

2.4.3 Simulating a Geometric Random Walk

In this section you will use simulations to see how stock prices evolve when the log-returns are i.i.d. normal, which implies that the price series is a geometric random walk.

Run the following R code. The set.seed() command insures that everyone using this code will have the same random numbers and will obtain the same price series. There are 253 trading days per year, so you are simulating 1 year of daily returns nine times. The price starts at 120.

The code par(mfrow=c(3,3)) on line 3 opens a graphics window with three rows and three columns and rnorm() on line 6 generates normally distributed random numbers.

```
1  set.seed(2012)
2  n = 253
3  par(mfrow=c(3,3))
4  for (i in (1:9))
5  {
6      logr = rnorm(n, 0.05 / 253, 0.2 / sqrt(253))
```

```
7    price = c(120, 120 * exp(cumsum(logr)))
8    plot(price, type = "b")
9 }
```

Problem 9 *In this simulation, what are the mean and standard deviation of the log-returns for 1 year?*

Problem 10 *Discuss how the price series appear to have momentum. Is the appearance of momentum real or an illusion?*

Problem 11 *Explain what the code* c(120,120*exp(cumsum(logr))) *does.*

2.4.4 Let's Look at McDonald's Stock

In this section we will be looking at daily returns on McDonald's stock over the period 2010–2014. To start the lab, run the following commands to get daily adjusted prices over this period:

```
1 data = read.csv('MCD_PriceDaily.csv')
2 head(data)
3 adjPrice = data[, 7]
```

Problem 12 *Compute the returns and log returns and plot them against each other. As discussed in Sect. 2.1.3, does it seem reasonable that the two types of daily returns are approximately equal?*

Problem 13 *Compute the mean and standard deviation for both the returns and the log returns. Comment on the similarities and differences you perceive in the first two moments of each random variable. Does it seem reasonable that they are the same?*

Problem 14 *Perform a t-test to compare the means of the returns and the log returns. Comment on your findings. Do you reject the null hypothesis that they are the same mean at 5 % significance? Or do you accept it? [Hint: Should you be using an independent samples t-test or a paired-samples t-test?]*
What are the assumptions behind the t-test? Do you think that they are met in this example? If the assumptions made by the t-test are not met, how would this affect your interpretation of the results of the test?

Problem 15 *After looking at return and log return data for McDonald's, are you satisfied that for small values, log returns and returns are interchangeable?*

Problem 16 *Assume that McDonald's log returns are normally distributed with mean and standard deviation equal to their estimates and that you have been made the following proposition by a friend: If at any point within the next 20 trading days, the price of McDonald's falls below 85 dollars, you will be paid $100, but if it does not, you have to pay him $1. The current price of McDonald's is at the end of the sample data, $93.07. Are you willing to make the bet? (Use 10,000 iterations in your simulation and use the command* set.seed(2015) *to ensure your results are the same as the answer key)*

Problem 17 *After coming back to your friend with an unwillingness to make the bet, he asks you if you are willing to try a slightly different deal. This time the offer stays the same as before, except he would pay an additional $25 if the price ever fell below $84.50. You still only pay him $1 for losing. Do you now make the bet?*

2.5 Exercises

1. Suppose that the daily log returns on a stock are independent and normally distributed with mean 0.001 and standard deviation 0.015. Suppose you buy $1,000 worth of this stock.
   (a) What is the probability that after one trading day your investment is worth less than $990? (**Note:** The R function pnorm() will compute a normal CDF, so, for example, pnorm(0.3, mean = 0.1, sd = 0.2) is the normal CDF with mean 0.1 and standard deviation 0.2 evaluated at 0.3.)
   (b) What is the probability that after five trading days your investment is worth less than $990?
2. The yearly log returns on a stock are normally distributed with mean 0.1 and standard deviation 0.2. The stock is selling at $100 today. What is the probability that 1 year from now it is selling at $110 or more?
3. The yearly log returns on a stock are normally distributed with mean 0.08 and standard deviation 0.15. The stock is selling at $80 today. What is the probability that 2 years from now it is selling at $90 or more?
4. Suppose the prices of a stock at times 1, 2, and 3 are $P_1 = 95$, $P_2 = 103$, and $P_3 = 98$. Find $r_3(2)$.
5. The prices and dividends of a stock are given in the table below.
   (a) What is R_2?
   (b) What is $R_4(3)$?
   (c) What is r_3?

t	P_t	D_t
1	52	0.2
2	54	0.2
3	53	0.2
4	59	0.25

6. The prices and dividends of a stock are given in the table below.
   (a) Find $R_3(2)$,
   (b) Find $r_4(3)$.

t	P_t	D_t
1	82	0.1
2	85	0.1
3	83	0.1
4	87	0.125

7. Let r_t be a log return. Suppose that $r_1, r_2, \ldots$ are i.i.d. $N(0.06, 0.47)$.
   (a) What is the distribution of $r_t(4) = r_t + r_{t-1} + r_{t-2} + r_{t-3}$?
   (b) What is $P\{r_1(4) < 2\}$?
   (c) What is the covariance between $r_2(1)$ and $r_2(2)$?
   (d) What is the conditional distribution of $r_t(3)$ given $r_{t-2} = 0.6$?
8. Suppose that $X_1, X_2, \ldots$ is a lognormal geometric random walk with parameters (μ, σ^2). More specifically, suppose that $X_k = X_0 \exp(r_1 + \cdots + r_k)$, where X_0 is a fixed constant and $r_1, r_2, \ldots$ are i.i.d. $N(\mu, \sigma^2)$.
   (a) Find $P(X_2 > 1.3 X_0)$.
   (b) Use (A.4) to find the density of X_1.
   (c) Find a formula for the 0.9 quantile of X_k for all k.
   (d) What is the expected value of X_k^2 for any k? (Find a formula giving the expected value as a function of k.)
   (e) Find the variance of X_k for any k.
9. Suppose that $X_1, X_2, \ldots$ is a lognormal geometric random walk with parameters $\mu = 0.1, \sigma = 0.2$.
   (a) Find $P(X_3 > 1.2X_0)$.
   (b) Find the conditional variance of X_k/k given X_0 for any k.
   (c) Find the minimum number of days before the probability is at least 0.9 of doubling one's money, that is, find the small value of t such that $P(P_t/P_0 \geq 2) \geq 0.9$.
10. The daily log returns on a stock are normally distributed with mean 0.0002 and standard deviation 0.03. The stock price is now \$97. What is the probability that it will exceed \$100 after 20 trading days?
11. Suppose that daily log-returns are $N(0.0005, 0.012)$. Find the smallest value of t such that $P(P_t/P_0 \geq 2) \geq 0.9$, that is, that after t days the probability the price has doubled is at least 90 %.

References

Bodie, Z., Kane, A., and Marcus, A. (1999) *Investments*, 4th ed., Irwin/ McGraw-Hill, Boston.

Campbell, J., Lo, A., and MacKinlay, A. (1997) *The Econometrics of Financial Markets*, Princeton University Press, Princeton, NJ.

Dalgaard, P. (2008) *Introductory Statistics with* R, *2nd ed.*, Springer.

Fama, E. (1965) The behavior of stock market prices. *Journal of Business*, **38**, 34–105.

Fama, E. (1970) Efficient capital markets: A review of theory and empirical work. *Journal of Finance*, **25**, 383–417.

Fama, E. (1991) Efficient Capital Markets: II. *Journal of Finance.* **46**, 1575–1618.

Fama, E. (1998) Market efficiency, long-term returns, and behavioral finance. *Journal of Financial Economics*, **49**, 283–306.

Lo, A. W., and MacKinlay, A. C. (1999) *A Non-Random Walk Down Wall Street*, Princeton University Press, Princeton and Oxford.

Ruppert, D. (2003) *Statistics and Finance: An Introduction*, Springer, New York.

Sharpe, W. F., Alexander, G. J., and Bailey, J. V. (1995) *Investments,* 6th ed., Simon and Schuster, Upper Saddle River, NJ.

Shefrin, H. (2000) *Beyond Greed and Fear: Understanding Behavioral Finance and the Psychology of Investing*, Harvard Business School Press, Boston.

Shiller, R. (1992) *Market Volatility,* Reprint ed., MIT Press, Cambridge, MA.

Shiller, R. (2000) *Irrational Exuberance*, Broadway, New York.

Shleifer, A. (2000) *Inefficient Markets: An Introduction to Behavioral Finance*, Oxford University Press, Oxford.

Thaler, R. H. (1993) *Advances in Behavioral Finance*, Russell Sage Foundation, New York.

Zuur, A., Ieno, E., Meesters, E., and Burg, D. (2009) *A Beginner's Guide to R*, Springer, New York.

3

Fixed Income Securities

3.1 Introduction

Corporations finance their operations by selling stock and bonds. Owning a share of stock means partial ownership of the company. Stockholders share in both the profits and losses of the company. Owning a bond is different. When you buy a bond you are loaning money to the corporation, though bonds, unlike loans, are tradeable. The corporation is obligated to pay back the principal and to pay interest as stipulated by the bond. The bond owner receives a fixed stream of income, unless the corporation defaults on the bond. For this reason, bonds are called "fixed income" securities.

It might appear that bonds are risk-free, almost stodgy, but this is not the case. Many bonds are long-term, e.g., 5, 10, 20, or even 30 years. Even if the corporation stays solvent or if you buy a U.S. Treasury bond, where default is for all intents and purposes impossible, your income from the bond is guaranteed only if you keep the bond to maturity. If you sell the bond before maturity, your return will depend on changes in the price of the bond. Bond prices move in opposite direction to interest rates, so a decrease in interest rates will cause a bond "rally," where bond prices increase. Long-term bonds are more sensitive to interest-rate changes than short-term bonds. The interest rate on your bond is fixed, but in the market interest rates fluctuate. Therefore, the market value of your bond fluctuates too. For example, if you buy a bond paying 5 % and the rate of interest increases to 6 %, then your bond is inferior to new bonds offering 6 %. Consequently, the price of your bond will decrease. If you sell the bond, you could lose money.

© Springer Science+Business Media New York 2015
D. Ruppert, D.S. Matteson, *Statistics and Data Analysis for Financial Engineering*, Springer Texts in Statistics,
DOI 10.1007/978-1-4939-2614-5_3

The interest rate of a bond depends on its maturity. For example, on March 28, 2001, the interest rate of Treasury bills[1] was 4.23 % for 3-month bills. The yields on Treasury notes and bonds were 4.41 %, 5.01 %, and 5.46 % for 2-, 10-, and 30-year maturities, respectively. The *term structure* of interest rates describes how rates change with maturity.

3.2 Zero-Coupon Bonds

Zero-coupon bonds, also called *pure discount bonds* and sometimes known as "zeros," pay no principal or interest until maturity. A "zero" has a *par value* or *face value*, which is the payment made to the bondholder at maturity. The zero sells for less than the par value, which is the reason it is a discount bond.

For example, consider a 20-year zero with a par value of $1,000 and 6 % interest compounded annually. The market price is the present value of $1,000 with an annual interest rate of 6 % with annual discounting. That is, the market price is

$$\frac{\$1,000}{(1.06)^{20}} = \$311.80.$$

If the annual interest rate is 6 % but compounded every 6 months, then the price is

$$\frac{\$1,000}{(1.03)^{40}} = \$306.56,$$

and if the annual rate is 6 % compounded continuously, then the price is

$$\frac{\$1,000}{\exp\{(0.06)(20)\}} = \$301.19.$$

3.2.1 Price and Returns Fluctuate with the Interest Rate

For concreteness, assume semiannual compounding. Suppose you bought the zero for $306.56 and then 6 months later the interest rate increased to 7 %. The market price would now be

$$\frac{\$1,000}{(1.035)^{39}} = \$261.41,$$

so the value of your investment would drop by ($306.56 − $261.41) = $45.15. You will still get your $1,000 if you keep the bond for 20 years, but if you sold it now, you would lose $45.15. This is a return of

---

[1] Treasury bills have maturities of 1 year or less, Treasury notes have maturities from 1 to 10 years, and Treasury bonds have maturities from 10 to 30 years.

$$\frac{-45.15}{306.56} = -14.73\%$$

for a half-year, or $-29.46,\%$ per year. And the interest rate only changed from 6% to 7%![2] Notice that the interest rate went up and the bond price went down. This is a general phenomenon. Bond prices always move in the opposite direction of interest rates.

If the interest rate dropped to 5% after 6 months, then your bond would be worth

$$\frac{\$1,000}{(1.025)^{39}} = \$381.74.$$

This would be an annual rate of return of

$$2\left(\frac{381.74 - 306.56}{306.56}\right) = 49.05\%.$$

If the interest rate remained unchanged at 6%, then the price of the bond would be

$$\frac{\$1,000}{(1.03)^{39}} = \$315.75.$$

The annual rate of return would be

$$2\left(\frac{315.75 - 306.56}{306.56}\right) = 6\%.$$

Thus, if the interest rate does not change, you can earn a 6% annual rate of return, the same return rate as the interest rate, by selling the bond before maturity. If the interest rate does change, however, the 6% annual rate of return is guaranteed only if you keep the bond until maturity.

General Formula

The price of a zero-coupon bond is given by

$$\text{PRICE} = \text{PAR}(1+r)^{-T}$$

if T is the time to maturity in years and the annual rate of interest is r with annual compounding. If we assume semiannual compounding, then the price is

$$\text{PRICE} = \text{PAR}(1+r/2)^{-2T}. \tag{3.1}$$

---

[2] Fortunately for investors, a rate change as large as going from 6% to 7% is rare on a 20-year bond.

3.3 Coupon Bonds

Coupon bonds make regular interest payments. Coupon bonds generally sell at or near the par value when issued. At maturity, one receives a principal payment equal to the par value of the bond and the final interest payment.

As an example, consider a 20-year coupon bond with a par value of $1,000 and 6 % annual coupon rate with semiannual coupon payments, so effectively the 6 % is compounded semiannually. Each coupon payment will be $30. Thus, the bondholder receives 40 payments of $30, one every 6 months plus a principal payment of $1,000 after 20 years. One can check that the present value of all payments, with discounting at the 6 % annual rate (3 % semiannual), equals $1,000:

$$\sum_{t=1}^{40} \frac{30}{(1.03)^t} + \frac{1000}{(1.03)^{40}} = 1000.$$

After 6 months, if the interest rate is unchanged, then the bond (including the first coupon payment, which is now due) is worth

$$\sum_{t=0}^{39} \frac{30}{(1.03)^t} + \frac{1000}{(1.03)^{39}} = (1.03)\left(\sum_{t=1}^{40} \frac{30}{(1.03)^t} + \frac{1000}{(1.03)^{40}}\right) = 1030,$$

which is a semiannually compounded 6 % annual return as expected. If the interest rate increases to 7 %, then after 6 months the bond (plus the interest due) is only worth

$$\sum_{t=0}^{39} \frac{30}{(1.035)^t} + \frac{1000}{(1.035)^{39}} = (1.035)\left(\sum_{t=1}^{40} \frac{30}{(1.035)^t} + \frac{1000}{(1.035)^{40}}\right) = 924.49.$$

This is an annual return of

$$2\left(\frac{924.49 - 1000}{1000}\right) = -15.1\,\%.$$

If the interest rate drops to 5 % after 6 months, then the investment is worth

$$\sum_{t=0}^{39} \frac{30}{(1.025)^t} + \frac{1000}{(1.025)^{39}} = (1.025)\left(\sum_{t=1}^{40} \frac{30}{(1.025)^t} + \frac{1000}{(1.025)^{40}}\right) = 1{,}153.70,$$

$$(3.2)$$

and the annual return is

$$2\left(\frac{1153.7 - 1000}{1000}\right) = 30.72\,\%.$$

3.3.1 A General Formula

Let's derive some useful formulas. If a bond with a par value of PAR matures in T years and makes semiannual coupon payments of C and the yield (rate of interest) is r per half-year, then the value of the bond when it is issued is

$$
\sum_{t=1}^{2T} \frac{C}{(1+r)^t} + \frac{\text{PAR}}{(1+r)^{2T}} = \frac{C}{r}\left\{1 - (1+r)^{-2T}\right\} + \frac{\text{PAR}}{(1+r)^{2T}}
$$

$$
= \frac{C}{r} + \left\{\text{PAR} - \frac{C}{r}\right\}(1+r)^{-2T}. \qquad (3.3)
$$

Derivation of (3.3)

The summation formula for a finite geometric series is

$$
\sum_{i=0}^{T} r^i = \frac{1 - r^{T+1}}{1 - r}, \qquad (3.4)
$$

provided that $r \neq 1$. Therefore,

$$
\sum_{t=1}^{2T} \frac{C}{(1+r)^t} = \frac{C}{1+r}\sum_{t=0}^{2T-1}\left(\frac{1}{1+r}\right)^t = \frac{C\{1-(1+r)^{-2T}\}}{(1+r)\{1-(1+r)^{-1}\}}
$$

$$
= \frac{C}{r}\{1 - (1+r)^{-2T}\}. \qquad (3.5)
$$

The remainder of the derivation is straightforward algebra.

3.4 Yield to Maturity

Suppose a bond with $T = 30$ and $C = 40$ is selling for \$1,200, \$200 above par value. If the bond were selling at par value, then the interest rate would be 0.04/half-year $(= 0.08/\text{year})$. The 4 %/half-year rate is called the *coupon rate*.

But the bond is *not* selling at par value. If you purchase the bond at \$1,200, you will make *less* than 8 % per year interest. There are two reasons that the rate of interest is less than 8 %. First, the coupon payments are \$40 or $40/1200 = 3.333$ %/half-year (or 6.67 %/year) for the \$1,200 investment; 6.67 %/year is called the *current yield*. Second, at maturity you only get back \$1,000, not the entire \$1,200 investment. The current yield of 6.67 %/year, though less than the coupon rate of 8 %/year, overestimates the return since it does not account for this loss of capital.

The *yield to maturity*, often shortened to simply *yield*, is the average rate of
return, including the loss (or gain) of capital because the bond was purchased
above (or below) par. For this bond, the yield to maturity is the value of r
that solves

$$1200 = \frac{40}{r} + \left\{ 1000 - \frac{40}{r} \right\} (1 + r)^{-60}. \tag{3.6}$$

The right-hand side of (3.6) is (3.3) with $C = 40$, $T = 30$, and PAR $= 1000$.
It is easy to solve equation (3.6) numerically. The R program in Sect. 3.10.1
does the following:

- computes the bond price for each r value on a grid;
- graphs bond price versus r (this is not necessary, but it is fun to see the
  graph); and
- interpolates to find the value of r such that the bond value equals $1,200.

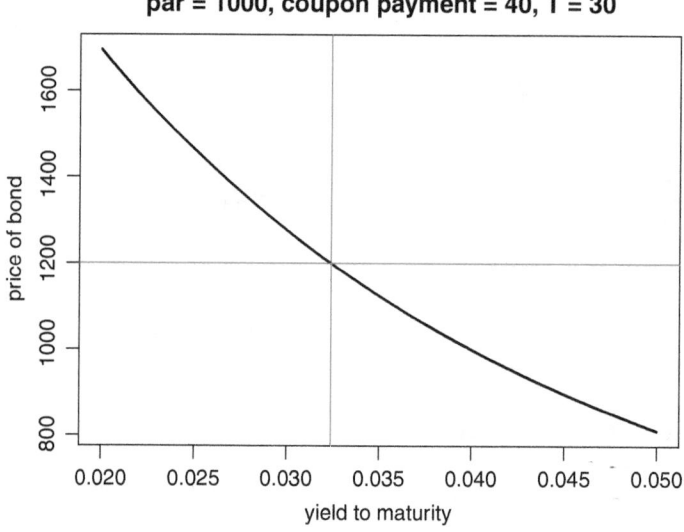

par = 1000, coupon payment = 40, T = 30

Fig. 3.1. *Bond price versus yield to maturity. The horizontal red line is at the bond
price of $1,200. The price/yield curve intersects this line at 0.0324 as indicated by
the vertical red line. Therefore, 0.0324 is the bond's yield.*

One finds that the yield to maturity is 0.0324, that is, 3.24 %/half-year. Figure 3.1 shows the graph of bond price versus the yield (r) and shows that
$r = 0.0324$ maps to a bond price of $1,200.

The yield to maturity of 0.0324 is less than the current yield of 0.0333,
which is less than the coupon rate of $40/1000 = 0.04$. (All three rates are rates
per half-year.) Whenever, as in this example, the bond is selling above par

value, then the coupon rate is greater than the current yield because the bond sells above par value, and the current yield is greater than the yield to maturity because the yield to maturity accounts for the loss of capital when at the maturity date you get back only the par value, not the entire investment. In summary,

$$\text{price} > \text{par} \Rightarrow \text{coupon rate} > \text{current yield} > \text{yield to maturity}.$$

Everything is reversed if the bond is selling below par value. For example, if the price of the bond were only $900, then the yield to maturity would be 0.0448 (as before, this value can be determined by interpolation), the current yield would be $40/900 = 0.0444$, and the coupon rate would still be $40/1000 = 0.04$. In general,

$$\text{price} < \text{par} \Rightarrow \text{coupon rate} < \text{current yield} < \text{yield to maturity}.$$

3.4.1 General Method for Yield to Maturity

The yield to maturity (on a semiannual basis) of a coupon bond is the value of r that solves

$$\text{PRICE} = \frac{C}{r} + \left\{ \text{PAR} - \frac{C}{r} \right\} (1+r)^{-2T}. \tag{3.7}$$

Here PRICE is the market price of the bond, PAR is the par value, C is the semiannual coupon payment, and T is the time to maturity in years and assumed to be a multiple of $1/2$.

For a zero-coupon bond, $C = 0$ and (3.7) becomes

$$\text{PRICE} = \text{PAR}(1+r)^{-2T}. \tag{3.8}$$

3.4.2 Spot Rates

The yield to maturity of a zero-coupon bond of maturity n years is called the n-year *spot rate* and is denoted by y_n. One uses the n-year spot rate to discount a payment n years from now, so a payment of $1 to be made n years from now has a net present value (NPV) of $\$1/(1+y_n)^n$ if y_n is the spot rate per annum or $\$1/(1+y_n)^{2n}$ if y_n is a semiannual rate.

A coupon bond is a bundle of zero-coupon bonds, one for each coupon payment and a final one for the principal payment. The component zeros have different maturity dates and therefore different spot rates. The yield to maturity of the coupon bond is, thus, a complex "average" of the spot rates of the zeros in this bundle.

Example 3.1. Finding the price and yield to maturity of a coupon bond using spot rates

Consider the simple example of 1-year coupon bond with semiannual coupon payments of $40 and a par value of $1,000. Suppose that the one-half-year spot rate is 2.5 %/half-year and the 1-year spot rate is 3 %/half-year. Think of the coupon bond as being composed of two zero-coupon bonds, one with $T = 1/2$ and a par value of $40 and the second with $T = 1$ and a par value of $1,040. The price of the bond is the sum of the prices of these two zeros. Applying (3.8) twice to obtain the prices of these zeros and summing, we obtain the price of the zero-coupon bond:

$$\frac{40}{1.025} + \frac{1040}{(1.03)^2} = 1019.32.$$

The yield to maturity on the coupon bond is the value of y that solves

$$\frac{40}{1+y} + \frac{1040}{(1+y)^2} = 1019.32.$$

The solution is $y = 0.0299$/half-year. Thus, the annual yield to maturity is twice 0.0299, or 5.98 %/year.   □

General Formula

In this section we will find a formula that generalizes Example 3.1. Suppose that a coupon bond pays semiannual coupon payments of C, has a par value of PAR, and has T years until maturity. Let $y_1, y_2, \ldots, y_{2T}$ be the half-year spot rates for zero-coupon bonds of maturities $1/2, 1, 3/2, \ldots, T$ years. Then the yield to maturity (on a half-year basis) of the coupon bond is the value of y that solves

$$\frac{C}{1+y_1} + \frac{C}{(1+y_2)^2} + \cdots + \frac{C}{(1+y_{2T-1})^{2T-1}} + \frac{\text{PAR}+C}{(1+y_n)^{2T}}$$
$$= \frac{C}{1+y} + \frac{C}{(1+y)^2} + \cdots + \frac{C}{(1+y)^{2T-1}} + \frac{\text{PAR}+C}{(1+y)^{2T}}. \quad (3.9)$$

The left-hand side of Eq. (3.9) is the price of the coupon bond, and the yield to maturity is the value of y that makes the right-hand side of (3.9) equal to the price.

Methods for solving (3.9) are explored in the R lab in Sect. 3.10.

3.5 Term Structure

3.5.1 Introduction: Interest Rates Depend Upon Maturity

On January 26, 2001, the 1-year T-bill rate was 4.83 % and the 30-year Treasury bond rate was 6.11 %. This is typical. Short- and long-term rates usually differ. Often short-term rates are lower than long-term rates. This makes

sense since long-term bonds are riskier, because long-term bond prices fluctuate more with interest-rate changes. However, during periods of very high
short-term rates, the short-term rates may be higher than the long-term rates.
The reason is that the market believes that rates will return to historic levels and no one will commit to the high interest rate for, say, 20 or 30 years.
Figure 3.2 shows weekly values of the 90-day, 10-year, and 30-year Treasury
rates from 1970 to 1993, inclusive. Notice that the 90-day rate is more volatile
than the longer-term rates and is usually less than them. However, in the early
1980s, when interest rates were very high, the short-term rates were higher
than the long-term rates. These data were taken from the Federal Reserve
Bank of Chicago's website.

The *term structure* of interest rates is a description of how, *at a given
time*, yield to maturity depends on maturity.

3.5.2 Describing the Term Structure

Term structure for all maturities up to n years can be described by any one
of the following:

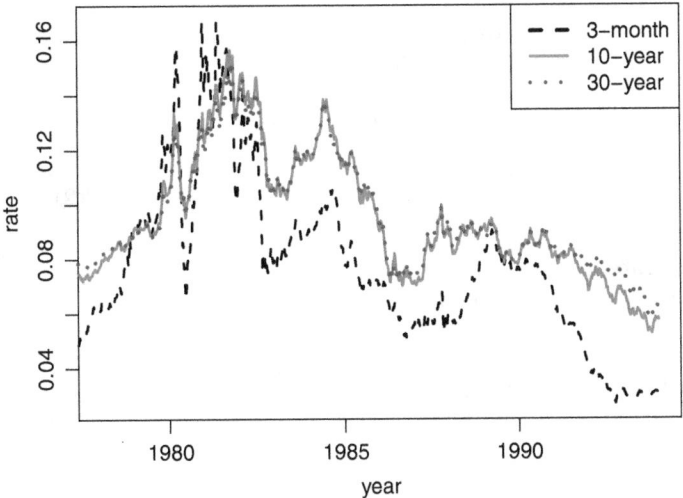

Fig. 3.2. *Treasury rates of three maturities. Weekly time series. The data were
taken from the website of the Federal Reserve Bank of Chicago.*

- prices of zero-coupon bonds of maturities 1-year, 2-years, ..., n-years are
  denoted here by $P(1), P(2), \ldots, P(n)$;
- spot rates (yields of maturity of zero-coupon bonds) of maturities 1-year,
  2-years, ..., n-years are denoted by $y_1, \ldots, y_n$;

- forward rates $r_1, \ldots, r_n$, where r_i is the forward rate that can be locked in now for borrowing in the ith future year ($i = 1$ for next year, and so on).

As discussed in this section, each of the sets $\{P(1), \ldots, P(n)\}$, $\{y_1, \ldots, y_n\}$, and $\{r_1, \ldots, r_n\}$ can be computed from either of the other sets. For example, equation (3.11) ahead gives $\{P(1), \ldots, P(n)\}$ in terms of $\{r_1, \ldots, r_n\}$, and equations (3.12) and (3.13) ahead give $\{y_1, \ldots, y_n\}$ in terms of $\{P(1), \ldots, P(n)\}$ or $\{r_1, \ldots, r_n\}$, respectively.

Term structure can be described by breaking down the time interval between the present time and the maturity time of a bond into short time segments with a constant interest rate within each segment, but with interest rates varying between segments. For example, a 3-year loan can be considered as three consecutive 1-year loans, or six consecutive half-year loans, and so forth.

Example 3.2. Finding prices from forward rates

As an illustration, suppose that loans have the forward interest rates listed in Table 3.1. Using the forward rates in the table, we see that a par \$1,000 1-year zero would sell for

$$\frac{1000}{1 + r_1} = \frac{1000}{1.06} = \$943.40 = P(1).$$

A par \$1,000 2-year zero would sell for

$$\frac{1000}{(1 + r_1)(1 + r_2)} = \frac{1000}{(1.06)(1.07)} = \$881.68 = P(2),$$

since the rate r_1 is paid the first year and r_2 the following year. Similarly, a par \$1,000 3-year zero would sell for

$$\frac{1000}{(1 + r_1)(1 + r_2)(1 + r_3)} = \frac{1000}{(1.06)(1.07)(1.08)} = 816.37 = P(3).$$

Table 3.1. *Forward interest rates used in Examples 3.2 and 3.3*

Year (i)	Interest rate (r_i)(%)
1	6
2	7
3	8

□

The general formula for the present value of $1 paid n periods from now is

$$\frac{1}{(1+r_1)(1+r_2)\cdots(1+r_n)}. \tag{3.10}$$

Here r_i is the *forward interest rate* during the ith period. If the periods are years, then the price of an n-year par $1,000 zero-coupon bond $P(n)$ is $1,000 times the discount factor in (3.10); that is,

$$P(n) = \frac{1000}{(1+r_1)\cdots(1+r_n)}. \tag{3.11}$$

Example 3.3. Back to Example 3.2: Finding yields to maturity from prices and from the forward rates

In this example, we first find the yields to maturity from the prices derived in Example 3.2 using the interest rates from Table 3.1. For a 1-year zero, the yield to maturity y_1 solves

$$\frac{1000}{(1+y_1)} = 943.40,$$

which implies that $y_1 = 0.06$. For a 2-year zero, the yield to maturity y_2 solves

$$\frac{1000}{(1+y_2)^2} = 881.68,$$

so that

$$y_2 = \sqrt{\frac{1000}{881.68}} - 1 = 0.0650.$$

For a 3-year zero, the yield to maturity y_3 solves

$$\frac{1000}{(1+y_3)^3} = 816.37,$$

and equals 0.070.

The yields can also be found from the forward rates. First, trivially, $y_1 = r_1 = 0.06$. Next, y_2 is given by

$$y_2 = \sqrt{(1+r_1)(1+r_2)} - 1 = \sqrt{(1.06)(1.07)} - 1 = 0.0650.$$

Also,

$$y_3 = \{(1+r_1)(1+r_2)(1+r_3)\}^{1/3} - 1$$
$$= \{(1.06)(1.07)(1.08)\}^{1/3} - 1 = 0.0700,$$

or, more precisely, 0.06997. Thus, $(1 + y_3)$ is the geometric average of 1.06, 1.07, and 1.08 and very nearly equal to their arithmetic average, which is 1.07.

$\square$

Recall that $P(n)$ is the price of a par $1,000 n-year zero-coupon bond. The general formulas for the yield to maturity y_n of an n-year zero are

$$y_n = \left\{ \frac{1000}{P(n)} \right\}^{1/n} - 1, \tag{3.12}$$

to calculate the yield from the price, and

$$y_n = \{(1+r_1)\cdots(1+r_n)\}^{1/n} - 1 \tag{3.13}$$

to obtain the yield from the forward rate.

Equations (3.12) and (3.13) give the yields to maturity in terms of the bond prices and forward rates, respectively. Also, inverting (3.12) gives the formula

$$P(n) = \frac{1000}{(1+y_n)^n} \tag{3.14}$$

for $P(n)$ as a function of the yield to maturity.

As mentioned before, interest rates for future years are called *forward rates*. A forward contract is an agreement to buy or sell an asset at some fixed future date at a fixed price. Since $r_2, r_3, \ldots$ are rates that can be locked in now for future borrowing, they are forward rates.

The general formulas for determining forward rates from yields to maturity are

$$r_1 = y_1, \tag{3.15}$$

and

$$r_n = \frac{(1+y_n)^n}{(1+y_{n-1})^{n-1}} - 1, \quad n = 2, 3, \ldots. \tag{3.16}$$

Now suppose that we only observed bond prices. Then we can calculate yields to maturity and forward rates using (3.12) and then (3.16).

Table 3.2. *Bond prices used in Example 3.4*

Maturity	Price
1 Year	$920
2 Years	$830
3 Years	$760

Example 3.4. Finding yields and forward rates from prices

Suppose that one-, two-, and three-year par $1,000 zeros are priced as given in Table 3.2. Using (3.12), the yields to maturity are

$$y_1 = \frac{1000}{920} - 1 = 0.087,$$

$$y_2 = \left\{ \frac{1000}{830} \right\}^{1/2} - 1 = 0.0976,$$

$$y_3 = \left\{ \frac{1000}{760} \right\}^{1/3} - 1 = 0.096.$$

Then, using (3.15) and (3.16),

$$r_1 = y_1 = 0.087,$$

$$r_2 = \frac{(1+y_2)^2}{(1+y_1)} - 1 = \frac{(1.0976)^2}{1.0876} - 1 = 0.108, \text{ and}$$

$$r_3 = \frac{(1+y_3)^3}{(1+y_2)^2} - 1 = \frac{(1.096)^3}{(1.0976)^2} - 1 = 0.092.$$

The formula for finding r_n from the prices of zero-coupon bonds is

$$r_n = \frac{P(n-1)}{P(n)} - 1, \tag{3.17}$$

which can be derived from

$$P(n) = \frac{1000}{(1+r_1)(1+r_2)\cdots(1+r_n)},$$

and

$$P(n-1) = \frac{1000}{(1+r_1)(1+r_2)\cdots(1+r_{n-1})}.$$

To calculate r_1 using (3.17), we need $P(0)$, the price of a 0-year bond, but $P(0)$ is simply the par value.[3]

Example 3.5. Forward rates from prices

Thus, using (3.17) and the prices in Table 3.2, the forward rates are

$$r_1 = \frac{1000}{920} - 1 = 0.087,$$

$$r_2 = \frac{920}{830} - 1 = 0.108,$$

and

$$r_3 = \frac{830}{760} - 1 = 0.092.$$

---

[3] Trivially, a bond that must be paid back immediately is worth exactly its par value.

3.6 Continuous Compounding

Now assume continuous compounding with forward rates $r_1, \ldots, r_n$. Using continuously compounded rates simplifies the relationships among the forward rates, the yields to maturity, and the prices of zero-coupon bonds.

If $P(n)$ is the price of a $1,000 par value n-year zero-coupon bond, then

$$P(n) = \frac{1000}{\exp(r_1 + r_2 + \cdots + r_n)}. \tag{3.18}$$

Therefore,

$$\frac{P(n-1)}{P(n)} = \frac{\exp(r_1 + \cdots + r_n)}{\exp(r_1 + \cdots + r_{n-1})} = \exp(r_n), \tag{3.19}$$

and

$$\log\left\{\frac{P(n-1)}{P(n)}\right\} = r_n. \tag{3.20}$$

The yield to maturity of an n-year zero-coupon bond solves the equation

$$P(n) = \frac{1000}{\exp(ny_n)},$$

and is easily seen to be

$$y_n = (r_1 + \cdots + r_n)/n. \tag{3.21}$$

Therefore, $\{r_1, \ldots, r_n\}$ is easily found from $\{y_1, \ldots, y_n\}$ by the relationship

$$r_1 = y_n,$$

and

$$r_n = ny_n - (n-1)y_{n-1} \text{ for } n > 1.$$

Example 3.6. Continuously compounded forward rates and yields from prices

Using the prices in Table 3.2, we have $P(1) = 920$, $P(2) = 830$, and $P(3) = 760$. Therefore, using (3.20),

$$r_1 = \log\left\{\frac{1000}{920}\right\} = 0.083,$$

$$r_2 = \log\left\{\frac{920}{830}\right\} = 0.103,$$

and

$$r_3 = \log\left\{\frac{830}{760}\right\} = 0.088.$$

Also, $y_1 = r_1 = 0.083$, $y_2 = (r_1 + r_2)/2 = 0.093$, and $y_3 = (r_1 + r_2 + r_3)/3 = 0.091$. □

3.7 Continuous Forward Rates

So far, we have assumed that forward interest rates vary from year to year but are constant within each year. This assumption is, of course, unrealistic and was made only to simplify the introduction of forward rates. Forward rates should be modeled as a function varying continuously in time.

To specify the term structure in a realistic way, we assume that there is a function $r(t)$ called the *forward-rate function* such that the current price of a zero-coupon bond of maturity T and with par value equal to 1 is given by

$$D(T) = \exp\left\{-\int_0^T r(t)dt\right\}. \tag{3.22}$$

$D(T)$ is called the discount function and the price of any zero-coupon bond is given by discounting its par value by multiplication with the discount function; that is,

$$P(T) = \text{PAR} \times D(T), \tag{3.23}$$

where $P(T)$ is the price of a zero-coupon bond of maturity T with par value equal to PAR. Also,

$$\log P(T) = \log(\text{PAR}) - \int_0^T r(t)dt,$$

so that

$$-\frac{d}{dT}\log P(T) = r(T) \text{ for all } T. \tag{3.24}$$

Formula (3.22) is a generalization of formula (3.18). To appreciate this, suppose that $r(t)$ is the piecewise constant function

$$r(t) = r_k \text{ for } k - 1 < t \leq k.$$

With this piecewise constant r, for any integer T, we have

$$\int_0^T r(t)dr = r_1 + r_2 + \cdots + r_T,$$

so that

$$\exp\left\{-\int_0^T r(t)dt\right\} = \exp\{-(r_1 + \cdots + r_T)\}$$

and therefore (3.18) agrees with (3.22) in this special situation. However, (3.22) is a more general formula since it applies to noninteger T and to arbitrary $r(t)$, not only to piecewise constant functions.

The yield to maturity of a zero-coupon bond with maturity date T is defined to be

$$y_T = \frac{1}{T} \int_0^T r(t)\, dt. \tag{3.25}$$

inking of the right-hand side of (3.25) as the average of $r(t)$ over the interval $\leq t \leq T$, we see that (3.25) is the analog of (3.21). From (3.22) and (3.25) it follows that the discount function can be obtained from the yield to maturity by the formula

$$D(T) = \exp\{-Ty_T\}, \tag{3.26}$$

so that the price of a zero-coupon bond maturing at time T is the same as it would be if there were a constant forward interest rate equal to y_T. It follows from (3.26) that

$$y_T = -\log\{D(T)\}/T. \tag{3.27}$$

Example 3.7. Finding continuous yield and discount functions from forward rates

Suppose the forward rate is the linear function $r(t) = 0.03 + 0.0005\,t$. Find $r(15)$, y_{15}, and $D(15)$.

Answer: $r(15) = 0.03 + (0.0005)(15) = 0.0375$,

$$y_{15} = (15)^{-1} \int_0^{15} (0.03 + 0.0005\,t)dt$$

$$= (15)^{-1}(0.03\,t + 0.0005\,t^2/2)\Big|_0^{15} = 0.03375,$$

and $D(15) = \exp(-15y_{15}) = \exp\{-(15)(0.03375)\} = \exp(-0.5055) = 0.603$.
□

The linear forward rate in Example 3.7 was chosen for simplicity and is not realistic. The Nelson-Siegel and Svensson parametric families of curves introduced in Sect. 11.3 are used in practice to model forward rates and yield curves. The European Community Bank uses the Svensson family. Nonparametric estimation of a forward rate by local polynomial and spline estimation is discussed in Examples 21.1 and 21.3, respectively. The Federal Reserve, the Bank of England, and the Bank of Canada use splines. The European Central Bank uses the Svensson family.

The discount function $D(T)$ and forward-rate function $r(t)$ in formula (3.22) depend on the current time, which is taken to be zero in that formula. However, we could be interested in how the discount function and forward rate function change over time. In that case we define the discount function $D(s, T)$ to be the price at time s of a zero-coupon bond, with a par value of \$1, maturing at time T. Also, if the forward-rate curve at time s is $r(s, t)$, $t \geq s$, then

$$D(s,T) = \exp\left\{ -\int_s^T r(s,t)dt \right\}. \tag{3.28}$$

The yield at time s of a bond maturing at time $T > s$ is

$$y(s,T) = (T-s)^{-1} \int_s^T r(s,u)du.$$

Since $r(t)$ and $D(t)$ in (3.22) are $r(0,t)$ and $D(0,t)$ in our new notation, (3.22) is the special case of (3.28) with $s = 0$. Similarly, y_T is equal to $y(0,T)$ in the new notation. However, for the remainder of this chapter we assume that $s = 0$ and return to the simpler notation of $r(t)$ and $D(t)$.

3.8 Sensitivity of Price to Yield

As we have seen, bonds are risky because bond prices are sensitive to interest rates. This problem is called *interest-rate risk*. This section describes a traditional method of quantifying interest-rate risk.

Using Eq. (3.26), we can approximate how the price of a zero-coupon bond changes if there is a small change in yield. Suppose that y_T changes to $y_T + \delta$, where the change in yield δ is small. Then the change in $D(T)$ is approximately δ times

$$\frac{d}{dy_T} \exp\{-Ty_T\} \approx -T \exp\{-Ty_T\} = -TD(T). \tag{3.29}$$

Therefore, by Eq. (3.23), for a zero-coupon bond of maturity T,

$$\frac{\text{change bond price}}{\text{bond price}} \approx -T \times \text{change in yield}. \tag{3.30}$$

In this equation "$\approx$" means that the ratio of the right- to left-hand sides converges to 1 as $\delta \to 0$.

Equation (3.30) is worth examining. The minus sign on the right-hand side shows us something we already knew, that bond prices move in the opposite direction to interest rates. Also, the relative change in the bond price, which is the left-hand side of the equation, is proportional to T, which quantifies the principle that longer-term bonds have higher interest-rate risks than short-term bonds.

3.8.1 Duration of a Coupon Bond

Remember that a coupon bond can be considered a bundle of zero-coupon bonds of various maturities. The *duration* of a coupon bond, which we will denote by DUR, is the weighted average of these maturities with weights in proportion to the net present value of the cash flows (coupon payments and par value at maturity).

Now assume that all yields change by a constant amount δ, that is, y_T changes to $y_T + \delta$ for all T. This restrictive assumption is needed to define duration. Because of this assumption, Eq. (3.30) applies to each of these cash flows and averaging them with these weights gives us that for a coupon bond,

$$\frac{\text{change bond price}}{\text{bond price}} \approx -\text{DUR} \times \delta. \tag{3.31}$$

The details of the derivation of (3.31) are left as an exercise (Exercise 15). *Duration analysis* uses (3.31) to approximate the effect of a change in yield on bond prices.

We can rewrite (3.31) as

$$\text{DUR} \approx \frac{-1}{\text{price}} \times \frac{\text{change in price}}{\text{change in yield}} \tag{3.32}$$

and use (3.32) as a *definition* of duration. Notice that "bond price" has been replaced by "price." The reason for this is that (3.32) can define the durations of not only bonds but also of derivative securities whose prices depend on yield, for example, call options on bonds. When this definition is extended to derivatives, duration has nothing to do with maturities of the underlying securities. Instead, duration is solely a measure of sensitivity of price to yield. Tuckman (2002) gives an example of a 10-year coupon bond with a duration of 7.79 years and a call option on this bond with a duration of 120.82 years. These durations show that the call is much riskier than the bond since it is 15.5 ($= 129.82/7.79$) times more sensitive to changes in yield.

Unfortunately, the underlying assumption behind (3.31) that all yields change by the same amount is not realistic, so duration analysis is falling into disfavor and value-at-risk is replacing duration analysis as a method for evaluating interest-rate risk.[4] Value-at-risk and other risk measures are covered in Chap. 19.

3.9 Bibliographic Notes

Tuckman (2002) is an excellent comprehensive treatment of fixed income securities; it is written at an elementary mathematical level and is highly recommended for readers wishing to learn more about this topic. Bodie, Kane, and Marcus (1999), Sharpe, Alexander, and Bailey (1999), and Campbell, Lo, and MacKinlay (1997) provide good introductions to fixed income securities, with the last-named being at a more advanced level. James and Webber (2000) is an advanced book on interest rate modeling. Jarrow (2002) covers

---

[4] See Dowd (1998).

many advanced topics that are not included in this book, including modeling the evolution of term structure, bond trading strategies, options and futures on bonds, and interest-rate derivatives.

3.10 R Lab

3.10.1 Computing Yield to Maturity

The following R function computes the price of a bond given its coupon payment, maturity, yield to maturity, and par value.

```
bondvalue = function(c, T, r, par)
{
#       Computes bv = bond values (current prices) corresponding
#       to all values of yield to maturity in the
#       input vector r
#
#       INPUT
#       c = coupon payment (semiannual)
#       T = time to maturity (in years)
#       r = vector of yields to maturity (semiannual rates)
#       par = par value
#
bv = c / r + (par - c / r) * (1 + r)^(-2 * T)
bv
}
```

The R code that follows computes the price of a bond for 300 semiannual interest rates between 0.02 and 0.05 for a 30-year par $1,000 bond with coupon payments of $40. Then interpolation is used to find the yield to maturity if the current price is $1,200.

```
price = 1200    #   current price of the bond
C = 40          #   coupon payment
T= 30           #   time to maturity
par = 1000      #   par value of the bond

r = seq(0.02, 0.05, length = 300)
value = bondvalue(C, T, r, par)
yield2M = spline(value, r, xout = price) # spline interpolation
```

The final bit of R code below plots price as a function of yield to maturity and graphically interpolates to show the yield to maturity when the price is $1,200.

```
plot(r, value, xlab = 'yield to maturity', ylab = 'price of bond',
    type = "l", main = "par = 1000, coupon payment = 40,
```

```
    T = 30", lwd = 2)
  abline(h = 1200)
  abline(v = yield2M)
```

Problem 1 *Use the plot to estimate graphically the yield to maturity. Does this estimate agree with that from spline interpolation?*

As an alternative to interpolation, the yield to maturity can be found using a nonlinear root finder (equation solver) such as uniroot(), which is illustrated here:

```
uniroot(function(r) r^2 - .5, c(0.7, 0.8))
```

Problem 2 *What does the code*

```
uniroot(function(r) r^2 - 0.5, c(0.7, 0.8))
```

do?

Problem 3 *Use uniroot() to find the yield to maturity of the 30-year par $1,000 bond with coupon payments of $40 that is selling at $1,200.*

Problem 4 *Find the yield to maturity of a par $10,000 bond selling at $9,800 with semiannual coupon payments equal to $280 and maturing in 8 years.*

Problem 5 *Use uniroot() to find the yield to maturity of the 20-year par $1,000 bond with semiannual coupon payments of $35 that is selling at $1,050.*

Problem 6 *The yield to maturity is 0.035 on a par $1,000 bond selling at $950.10 and maturing in 5 years. What is the coupon payment?*

3.10.2 Graphing Yield Curves

R's fEcofin package had many interesting financial data sets but is no longer available. The data sets mk.maturity.csv and mk.zero2.csv used in this example were taken from this package and are now available on this book's webpage. The data set mk.zero2 has yield curves of U.S. zero coupon bonds recorded monthly at 55 maturities. These maturities are in the data set mk.maturity. The following code plots the yield curves on four consecutive months.

```
mk.maturity = read.csv("mk.maturity.csv", header = T)
mk.zero2 = read.csv("mk.zero2.csv", header = T)
plot(mk.maturity[,1], mk.zero2[5,2:56], type = "l",
   xlab = "maturity", ylab = "yield")
lines(mk.maturity[,1], mk.zero2[6,2:56], lty = 2, type = "l")
lines(mk.maturity[,1], mk.zero2[7,2:56], lty = 3, type = "l")
lines(mk.maturity[,1], mk.zero2[8,2:56], lty = 4, type = "l")
legend("bottomright", c("1985-12-01", "1986-01-01",
   "1986-02-01", "1986-03-01"), lty = 1:4)
```

Run the code above and then, to zoom in on the short end of the curves, rerun the code with maturities restricted to 0 to 3 years; to do that, use xlim in the plot function.

Problem 7 *Describe how the yield curve changes between December 1, 1985 and March 1, 1986. Describe the behavior of both the short and long ends of the yield curves.*

Problem 8 *Plot the yield curves from December 1, 1986 to March 1, 1987 and describe how the yield curve changes during this period.*

The next set of code estimates the forward rate for 1 month. Line 1 estimates the integrated forward rate, called intForward, which is $T_{yT} = \int_0^T r(t)dt$ where $r(t)$ is the forward rate. Line 3 interpolates the estimated integrated forward rate onto a grid of 200 points from 0 to 20. This grid is created on line 2.

If a function f is evaluated on a grid, $t_1, \ldots, t_L$, then $\{f(t_\ell) - f(t_{\ell-1})\}/(t_\ell - t_{\ell-1})$ approximates $f'((t_\ell + t_{\ell-1})/2)$ for $\ell = 2, \ldots, L$. Line 4 numerically differentiates the integrated forward rate to approximate the forward rate on the grid calculated at Line 5.

```
1 intForward = mk.maturity[, 1] * mk.zero2[6, 2:56]
2 xout = seq(0, 20, length = 200)
3 z1 = spline(mk.maturity[ ,1], intForward, xout = xout)
4 forward = diff(z1$y) / diff(z1$x)
5 T_grid = (xout[-1] + xout[-200]) / 2
6 plot(T_grid, forward, type = "l", lwd = 2, ylim = c(0.06, 0.11))
```

Problem 9 *Plot the forward rates on the same dates used before, 1985-12-01, 1986-01-01, 1986-02-01, and 1986-03-01. Describe how the forward rates changed from month to month.*

The approximate forward rates found by numerically differentiating a interpolating spline are "wiggly." The wiggles can be removed, or at least reduced, by using a penalized spline instead of an interpolating spline. See Chap. 21.

3.11 Exercises

1. Suppose that the forward rate is $r(t) = 0.028 + 0.00042t$.
   (a) What is the yield to maturity of a bond maturing in 20 years?
   (b) What is the price of a par $1,000 zero-coupon bond maturing in 15 years?

2. Suppose that the forward rate is $r(t) = 0.04 + 0.0002t - 0.00003t^2$.
   (a) What is the yield to maturity of a bond maturing in 8 years?
   (b) What is the price of a par $1,000 zero-coupon bond maturing in 5 years?
   (c) Plot the forward rate and the yield curve. Describe the two curves. Which are convex and which are concave? How do they differ?
   (d) Suppose you buy a 10-year zero-coupon bond and sell it after 1 year. What will be the return if the forward rate does not change during that year?

3. A coupon bond has a coupon rate of 3 % and a current yield of 2.8 %.
   (a) Is the bond selling above or below par? Why or why not?
   (b) Is the yield to maturity above or below 2.8 %? Why or why not?

4. Suppose that the forward rate is $r(t) = 0.032 + 0.001t + 0.0002t^2$.
   (a) What is the 5-year continuously compounded spot rate?
   (b) What is the price of a zero-coupon bond that matures in 5 years?

5. The 1/2-, 1-, 1.5-, and 2-year semiannually compounded spot rates are 0.025, 0.028, 0.032, and 0.033, respectively. A par $1,000 coupon bond matures in 2 years and has semiannual coupon payments of $35. What is the price of this bond?

6. Verify the following equality:

$$\sum_{t=1}^{2T} \frac{C}{(1+r)^t} + \frac{\text{PAR}}{(1+r)^{2T}} = \frac{C}{r} + \left\{ \text{PAR} - \frac{C}{r} \right\} (1+r)^{-2T}.$$

7. One year ago a par $1,000 20-year coupon bond with semiannual coupon payments was issued. The annual interest rate (that is, the coupon rate) at that time was 8.5 %. Now, a year later, the annual interest rate is 7.6 %.
   (a) What are the coupon payments?
   (b) What is the bond worth now? Assume that the second coupon payment was just received, so the bondholder receives an additional 38 coupon payments, the next one in 6 months.
   (c) What would the bond be worth if instead the second payment were just about to be received?

8. A par $1,000 zero-coupon bond that matures in 5 years sells for $828. Assume that there is a constant continuously compounded forward rate r.
   (a) What is r?
   (b) Suppose that 1 year later the forward rate r is still constant but has changed to be 0.042. Now what is the price of the bond?

(c) If you bought the bond for the original price of $828 and sold it 1 year later for the price computed in part (b), then what is the net return?

9. A coupon bond with a par value of $1,000 and a 10-year maturity pays semiannual coupons of $21.
   (a) Suppose the yield for this bond is 4 % per year compounded semiannually. What is the price of the bond?
   (b) Is the bond selling above or below par value? Why?

10. Suppose that a coupon bond with a par value of $1,000 and a maturity of 7 years is selling for $1,040. The semiannual coupon payments are $23.
    (a) Find the yield to maturity of this bond.
    (b) What is the current yield on this bond?
    (c) Is the yield to maturity less or greater than the current yield? Why?

11. Suppose that the continuous forward rate is $r(t) = 0.033 + 0.0012t$. What is the current value of a par $100 zero-coupon bond with a maturity of 15 years?

12. Suppose the continuous forward rate is $r(t) = 0.04 + 0.001t$ when a 8-year zero coupon bond is purchased. Six months later the forward rate is $r(t) = 0.03 + 0.0013t$ and bond is sold. What is the return?

13. Suppose that the continuous forward rate is $r(t) = 0.03 + 0.001t - 0.00021(t - 10)_+$. What is the yield to maturity on a 20-year zero-coupon bond? Here x_+ is the *positive part function* defined by

$$ x_+ = \begin{cases} x, & x > 0, \\ 0, & x \le 0. \end{cases} $$

14. An investor is considering the purchase of zero-coupon bonds with maturities of one, three, or 5 years. Currently the spot rates for 1-, 2-, 3-, 4-, and 5-year zero-coupon bonds are, respectively, 0.031, 0.035, 0.04, 0.042, and 0.043 per year with semiannual compounding. A financial analyst has advised this investor that interest rates will increase during the next year and the analyst expects all spot rates to increase by the amount 0.005, so that the 1-year spot rate will become 0.036, and so forth. The investor plans to sell the bond at the end of 1 year and wants the greatest return for the year. This problem does the bond math to see which maturity, 1, 3, or 5 years, will give the best return under two scenarios: interest rates are unchanged and interest rates increase as forecast by the analyst.
    (a) What are the current prices of 1-, 3-, and 5-year zero-coupon bonds with par values of $1,000?
    (b) What will be the prices of these bonds 1 year from now if spot rates remain unchanged?
    (c) What will be the prices of these bonds 1 year from now if spot rates each increase by 0.005?
    (d) If the analyst is correct that spot rates will increase by 0.005 in 1 year, which maturity, 1, 3, or 5 years, will give the investor the greatest return when the bond is sold after 1 year? Justify your answer.

(e) If instead the analyst is incorrect and spot rates remain unchanged, then which maturity, 1, 3, or 5 years, earns the highest return when the bond is sold after 1 year? Justify your answer.

(f) The analyst also said that if the spot rates remain unchanged, then the bond with the highest spot rate will earn the greatest 1-year return. Is this correct? Why?

(*Hint:* Be aware that a bond will not have the same maturity in 1 year as it has now, so the spot rate that applies to that bond will change.)

15. Suppose that a bond pays a cash flow C_i at time T_i for $i = 1, \ldots, N$. Then the net present value (NPV) of cash flow C_i is

$$\text{NPV}_i = C_i \exp(-T_i \, y_{T_i}).$$

Define the weights

$$\omega_i = \frac{\text{NPV}_i}{\sum_{j=1}^{N} \text{NPV}_j}$$

and define the duration of the bond to be

$$\text{DUR} = \sum_{i=1}^{N} \omega_i T_i,$$

which is the weighted average of the times of the cash flows. Show that

$$\frac{d}{d\delta} \sum_{i=1}^{N} C_i \exp\{-T_i(y_{T_i} + \delta)\} \bigg|_{\delta=0} = -\text{DUR} \sum_{i=1}^{N} C_i \exp\{-T_i \, y_{T_i}\}$$

and use this result to verify Eq. (3.31).

16. Assume that the yield curve is $Y_T = 0.04 + 0.001\,T$.

(a) What is the price of a par-$1,000 zero-coupon bond with a maturity of 10 years?

(b) Suppose you buy this bond. If 1 year later the yield curve is $Y_T = 0.042 + 0.001\,T$, then what will be the net return on the bond?

17. A coupon bond has a coupon rate of 3 % and a current yield of 2.8 %.

(a) Is the bond selling above or below par? Why or why not?

(b) Is the yield to maturity above or below 2.8 %? Why or why not?

18. Suppose that the forward rate is $r(t) = 0.03 + 0.001t + 0.0002t^2$

(a) What is the 5-year spot rate?

(b) What is the price of a zero-coupon bond that matures in 5 years?

19. The 1/2-, 1-, 1.5-, and 2-year spot rates are 0.025, 0.029, 0.031, and 0.035, respectively. A par $1,000 coupon bond matures in 2 years and has semi-annual coupon payments of $35. What is the price of this bond?

20. Par $1,000 zero-coupon bonds of maturities of 0.5-, 1-, 1.5-, and 2-years are selling at $980.39, $957.41, $923.18, and $888.489, respectively.

(a) Find the 0.5-, 1-, 1.5-, and 2-year semiannual spot rates.

(b) A par $1,000 coupon bond has a maturity of 2 years. The semiannual coupon payment is $21. What is the price of this bond?

21. A par $1,000 bond matures in 4 years and pays semiannual coupon payments of $25. The price of the bond is $1,015. What is the semiannual yield to maturity of this bond?

22. A coupon bond matures in 4 years. Its par is $1,000 and it makes eight coupon payments of $21, one every one-half year. The continuously compounded forward rate is

$$r(t) = 0.022 + 0.005\,t - 0.004\,t^2 + 0.0003\,t^3.$$

(a) Find the price of the bond.

(b) Find the duration of this bond.

References

Bodie, Z., Kane, A., and Marcus, A. (1999) *Investments,* 4th ed., Irwin/McGraw-Hill, Boston.

Campbell, J. Y., Lo, A. W., and MacKinlay, A. C. (1997) *Econometrics of Financial Markets,* Princeton University Press, Princeton, NJ.

Dowd, K. (1998) *Beyond Value at Risk,* Wiley, Chichester.

James, J., and Webber, N. (2000) *Interest Rate Modeling,* Wiley, Chichester.

Jarrow, R. (2002) *Modeling Fixed-Income Securities and Interest Rate Options,* 2nd ed., Stanford University Press, Stanford, CA.

Sharpe, W., Alexander, G., and Bailey, J. (1999) *Investments,* 6th ed., Prentice-Hall, Englewood Cliffs, NJ.

Tuckman, B. (2002) *Fixed Income Securities,* 2nd ed., Wiley, Hoboken, NJ.

4

Exploratory Data Analysis

4.1 Introduction

This book is about the statistical analysis of financial markets data such as equity prices, foreign exchange rates, and interest rates. These quantities vary randomly thereby causing financial risk as well as the opportunity for profit. Figures 4.1, 4.2, and 4.3 show, respectively, time series plots of daily log returns on the S&P 500 index, daily changes in the Deutsch Mark (DM) to U.S. dollar exchange rate, and changes in the monthly risk-free return, which is 1/12th the annual risk-free interest rate. A *time series* is a sequence of observations

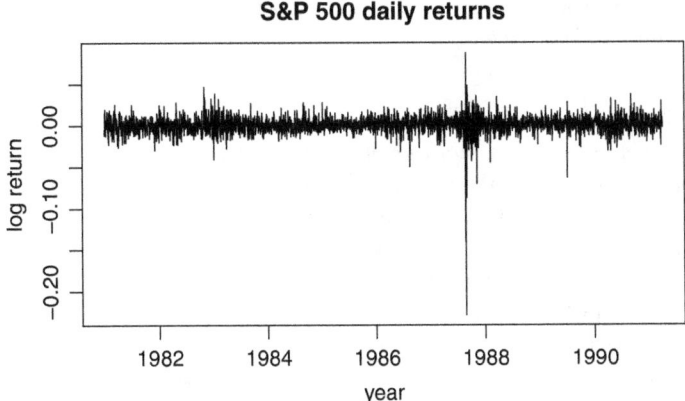

S&P 500 daily returns

Fig. 4.1. *Daily log returns on the S&P 500 index from January 1981 to April 1991. This data set is the variable* r500 *in the* SP500 *series in the* Ecdat *package in* R. *Notice the extreme volatility in October 1987.*

© Springer Science+Business Media New York 2015
D. Ruppert, D.S. Matteson, *Statistics and Data Analysis for Financial Engineering*, Springer Texts in Statistics,
DOI 10.1007/978-1-4939-2614-5_4

of some quantity or quantities, e.g., equity prices, taken over time, and a *time series plot* is a plot of a time series in chronological order. Figure 4.1 was produced by the following code:

```
data(SP500, package = "Ecdat")
SPreturn = SP500$r500
n = length(SPreturn)
year_SP = 1981 + (1:n) * (1991.25 - 1981) / n
plot(year_SP, SPreturn, main = "S&P 500 daily returns",
     xlab = "year", type = "l", ylab = "log return")
```

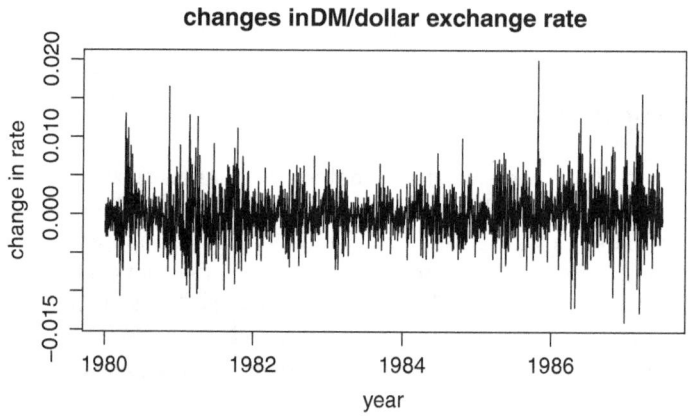

Fig. 4.2. *Daily changes in the DM/dollar exchange rate, January 2, 1980, to May 21, 1987. The data come from the* Garch *series in the* Ecdat *package in* R. *The DM/dollar exchange rate is the variable* dm.

Despite the large random fluctuations in all three time series, we can see that each series appears *stationary*, meaning that the nature of its random variation is constant over time. In particular, the series fluctuate about means that are constant, or nearly so. We also see *volatility* clustering, because there are periods of higher, and of lower, variation within each series. Volatility clustering does *not* indicate a lack of stationarity but rather can be viewed as a type of dependence in the conditional variance of each series. This point will be discussed in detail in Chap. 14.

Each of these time series will be modeled as a sequence $Y_1, Y_2, \ldots$ of random variables, each with a CDF that we will call F.[1] F will vary between series but, because of stationarity, is assumed to be constant within each series. F is also called the marginal distribution function. By the *marginal distribution* of a stationary time series, we mean the distribution of Y_t given no knowledge

---

[1] See Appendix A.2.1 for definitions of CDF, PDF, and other terms in probability theory.

changes in risk–free interest return

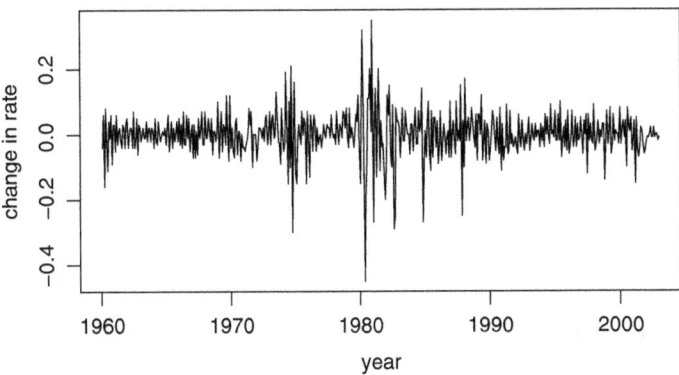

Fig. 4.3. *Monthly changes in the risk-free rate, January* 1960 *to December* 2002. *The rates are the variable* rf *in the* Capm *series in the* Ecdat *package in* R.

of the other observations, that is, no knowledge of Y_s for any $s \neq t$. Thus, when modeling a marginal distribution, we disregard dependencies in the time series. For this reason, a marginal distribution is also called an *unconditional distribution*. Dependencies such as autocorrelation and volatility clustering will be discussed in later chapters.

In this chapter, we explore various methods for modeling and estimating marginal distributions, in particular, graphical methods such as histograms, density estimates, sample quantiles, and probability plots.

4.2 Histograms and Kernel Density Estimation

Assume that the marginal CDF F has a probability density function f. The histogram is a simple and well-known estimator of probability density functions. Panel (a) of Fig. 4.4 is a histogram of the S&P 500 log returns using 30 cells (or bins). There are some outliers in this series, especially a return near -0.23 that occurred on Black Monday, October 19, 1987. Note that a return of this size means that the market lost 23 % of its value in a single day. The outliers are difficult, or perhaps impossible, to see in the histogram, except that they have caused the x-axis to expand. The reason that the outliers are difficult to see is the large sample size. When the sample size is in the thousands, a cell with a small frequency is essentially invisible. Panel (b) of Fig. 4.4 zooms in on the high-probability region. Note that only a few of the 30 cells are in this area.

The histogram is a fairly crude density estimator. A typical histogram looks more like a big city skyline than a density function and its appearance is sensitive to the number and locations of its cells—see Fig. 4.4, where panels (b), (c), and (d) differ only in the number of cells. A much better estimator is

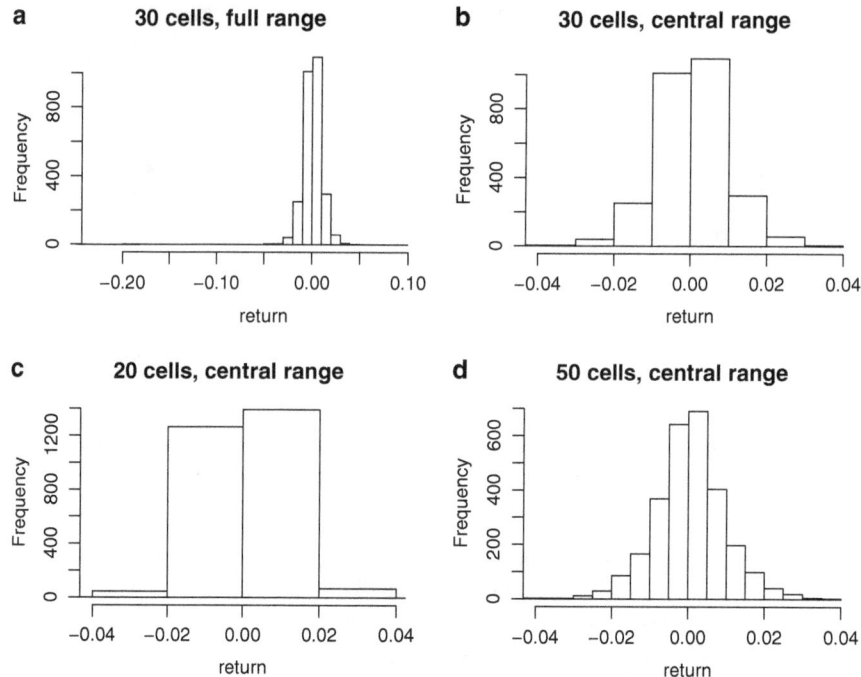

Fig. 4.4. *Histograms of the daily log returns on the S&P 500 index from January 1981 to April 1991. This data set is the* **SP500** *series in the* `Ecdat` *package in* R.

the *kernel density estimator* (KDE). The estimator takes its name from the so-called kernel function, denoted here by K, which is a probability density function that is symmetric about 0. The standard[2] normal density function is a common choice for K and will be used here. The kernel density estimator based on $Y_1, \ldots, Y_n$ is

$$\widehat{f}(y) = \frac{1}{nb} \sum_{i=1}^{n} K\left(\frac{y - Y_i}{b}\right) \tag{4.1}$$

where b, which is called the bandwidth, determines the resolution of the estimator.

Figure 4.5 illustrates the construction of kernel density estimates using a small simulated data set of six observations from a standard normal distribution. The small sample size is needed for visual clarity but, of course, does not lead to an accurate estimate of the underlying normal density. The six data points are shown at the bottom of the figure as short vertical lines called a "rug." The bandwidth in the top plot is 0.4, and so each of the six dashed lines is 1/6 times a normal density with standard deviation equal to 0.4 and

---

[2] "Standard" means having expectation 0 and variance 1.

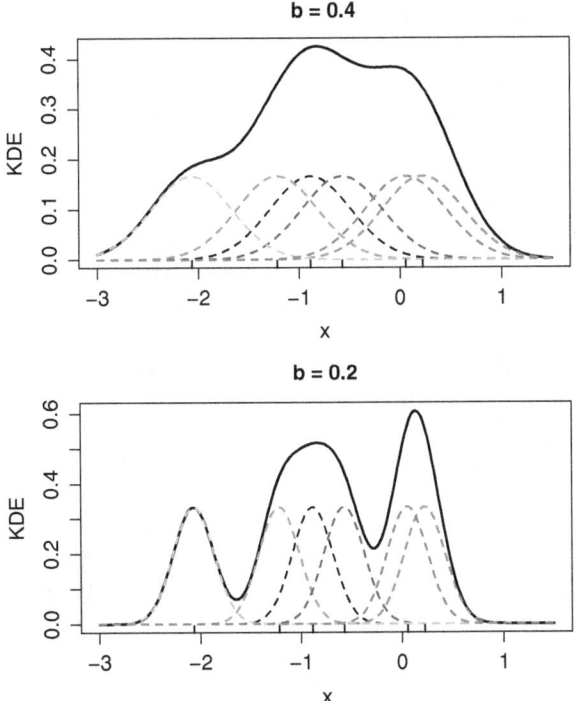

Fig. 4.5. *Illustration of kernel density estimates using a sample of size* 6 *and two bandwidths. The six dashed curves are the kernels centered at the data points, which are indicated by vertical lines at the bottom. The solid curve is the kernel density estimate created by adding together the six kernels. Although the same data are used in the top and bottom panels, the density estimates are different because of the different bandwidths.*

centered at one of the data points. The solid curve is the superposition, that is, the sum as in Eq. (4.1), of the six dashed curves and estimates the density of the data.

A small value of b allows the density estimator to detect fine features in the true density, but it also permits a high degree of random variation. This can be seen in the plot in the bottom of Fig. 4.5 where the bandwidth is only half as large as in the plot on the top. Conversely, a large value of b dampens random variation but obscures fine detail in the true density. Stated differently, a small value of b causes the kernel density estimator to have high variance and low bias, and a large value of b results in low variance and high bias.

Choosing b requires one to make a tradeoff between bias and variance. Appropriate values of b depend on both the sample size n and the true density and, of course, the latter is unknown, though it can be estimated. Roughly speaking, nonsmooth or "wiggly" densities require a smaller bandwidth.

Fortunately, a large amount of research has been devoted to automatic selection of b, which, in effect, estimates the roughness of the true density. As a result of this research, modern statistical software can select the bandwidth automatically. However, automatic bandwidth selectors are not foolproof and density estimates should be checked visually and, if necessary, adjusted as described below.

The solid curve in Fig. 4.6 has the default bandwidth from the density() function in R. The dashed and dotted curves have the default bandwidth multiplied by 1/3 and 3, respectively. The tuning parameter adjust in R is the multiplier of the default bandwidth, so that adjust is 1, 1/3, and 3 in the three curves. The solid curve with adjust equal to 1 appears to have a proper amount of smoothness. The dashed curve corresponding to adjust = 1/3 is wiggly, indicating too much random variability; such a curve is called undersmoothed and overfit. The dotted curve is very smooth but underestimates the peak near 0, a sign of bias. Such a curve is called oversmoothed or underfit. Here *overfit* means that the density estimate adheres too closely to the data and so is unduly influenced by random variation. Conversely, *underfit* means that the density estimate does not adhere closely enough to the data and misses features in the true density. Stated differently, over- and underfitting means a poor bias–variance tradeoff with an overfitted curve having too much variance and an underfitted curve having too much bias.

Automatic bandwidth selectors are very useful, but there is nothing magical about them, and often one will use an automatic selector as a starting

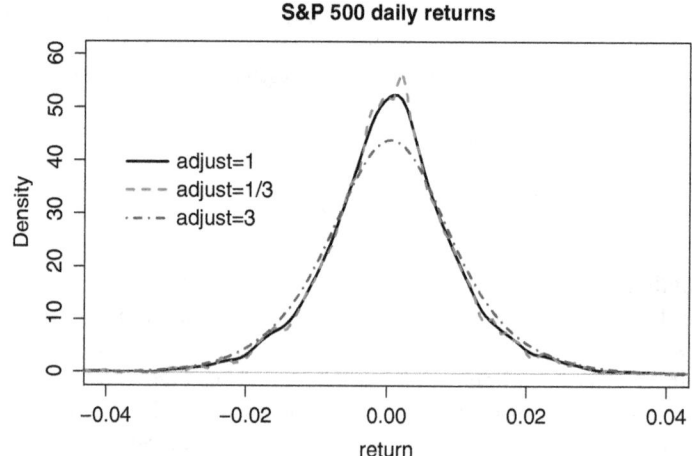

Fig. 4.6. *Kernel density estimates of the daily log returns on the S&P 500 index using three bandwidths. Each bandwidth is the default bandwidth times* adjust *and* adjust *is 1/3, 1, and 3. This data set is the* SP500 *series in the* Ecdat *package in* R. *The KDE is plotted only for a limited range of returns to show detail in the middle of the distribution.*

point and then "fine-tune" the bandwidth; this is the point of the adjust parameter. Generally, adjust will be much closer to 1 than the values, 1/3 and 3, used above. The reason for using 1/3 and 3 in Fig. 4.6 was to emphasize the effects of under- and oversmoothing.

Often a kernel density estimate is used to suggest a parametric statistical model. The density estimates in Fig. 4.6 are bell-shaped, suggesting that a normal distribution might be a suitable model. To further investigate the suitability of the normal model, Fig. 4.7 compares the kernel density estimate with adjust = 1 with normal densities. In panel (a), the normal density has mean and standard deviation equal to the sample mean and standard deviation of the returns. We see that the kernel estimate and the normal density are somewhat dissimilar. The reason is that the outlying returns inflate the sample standard deviation and cause the fitted normal density to be too dispersed in the middle of the data. Panel (b) shows a normal density that is much closer to the kernel estimator. This normal density uses robust estimators which are less sensitive to outliers—the mean is estimated by the sample median and the MAD estimator is used for the standard deviation. The MAD estimator is the median absolute deviation from the median but scaled so that it estimates the standard deviation of a normal population.[3] The sample

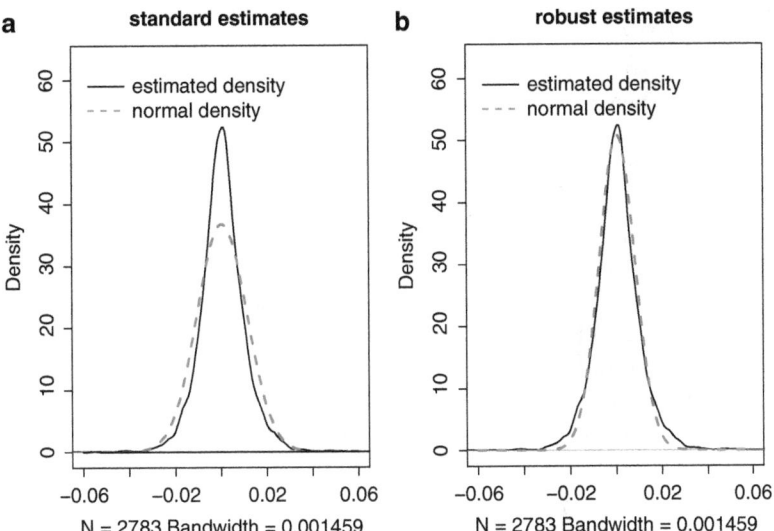

Fig. 4.7. *Kernel density estimates (solid) of the daily log returns on the S&P 500 index compared with normal densities (dashed). (**a**) The normal density uses the sample mean and standard deviation. (**b**) The normal density uses the sample median and MAD estimate of standard deviation. This data set is the* SP500 *series in the* Ecdat *package in* R.

---

[3] See Sect. 5.16 for more discussion of robust estimation and the precise definition of MAD.

standard deviation is 0.011, but the MAD is smaller, 0.0079; these values were computed using the R functions sd() and mad(). Even the normal density in panel (b) shows some deviation from the kernel estimator, and, as we will soon see, the t-distribution provides a better model for the return distribution than does the normal distribution. The need for robust estimators is itself a sign of nonnormality.

We have just seen a problem with using a KDE to suggest a good model for the distribution of the data in a sample—the parameters in the model must be estimated properly. Normal probability plots and, more generally, quantile–quantile plots, which will be discussed in Sects. 4.3.2 and 4.3.4, are better methods for comparing a sample with a theoretical distribution.

Though simple to compute, the KDE has some problems. In particular, it is often too bumpy in the tails. An improvement to the KDE is discussed in Sect. 4.8.

4.3 Order Statistics, the Sample CDF, and Sample Quantiles

Suppose that $Y_1, \ldots, Y_n$ is a random sample from a probability distribution with CDF F. In this section we estimate F and its quantiles. The *sample* or *empirical CDF* $F_n(y)$ is defined to be the proportion of the sample that is less than or equal to y. For example, if 10 out of 40 $(= n)$ elements of a sample are 3 or less, then $F_n(3) = 0.25$. More generally,

$$F_n(y) = \frac{\sum_{i=1}^{n} I\{Y_i \le y\}}{n}, \tag{4.2}$$

where $I\{\cdot\}$ is the indicator function so that $I\{Y_i \le y\}$ is 1 if $Y_i \le y$ and is 0 otherwise. Therefore, the sum in the numerator in (4.2) counts the number of Y_i that are less than or equal to y. Figure 4.8 shows F_n for a sample of size 150 from an $N(0, 1)$ distribution. The true CDF (Φ) is shown as well. The sample CDF differs from the true CDF because of random variation. The sample CDF is also called the empirical distribution function, or EDF.

The function ecdf() computes a sample CDF. The code to produce Fig. 4.8 is:

```
1  set.seed("991155")
2  edf_norm = ecdf(rnorm(150))
3  pdf("normalcdfplot.pdf", width = 6, height = 5)  ## Figure 4.8
4  par(mfrow = c(1, 1))
5  plot(edf_norm, verticals = TRUE, do.p = FALSE, main = "EDF and CDF")
6  tt = seq(from = -3, to = 3, by = 0.01)
7  lines(tt, pnorm(tt), lty = 2, lwd = 2, col = "red")
8  legend(1.5, 0.2, c("EDF", "CDF"), lty = c(1, 2),
9     lwd = c(1.5, 2), col = c("black", "red"))
10 graphics.off()
```

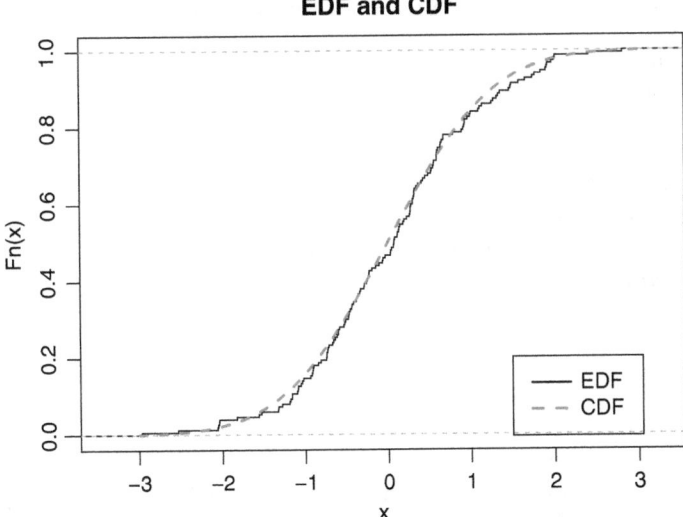

Fig. 4.8. *The EDF F_n (solid) and the true CDF (dashed) for a simulated random sample from an $N(0,1)$ population. The sample size is 150.*

The *order statistics* $Y_{(1)}, Y_{(2)}, \ldots, Y_{(n)}$ are the values $Y_1, \ldots, Y_n$ ordered from smallest to largest. The subscripts of the order statistics are in parentheses to distinguish them from the unordered sample. For example, Y_1 is simply the first observation in the original sample while $Y_{(1)}$ is the smallest observation in that sample. The *sample quantiles* are defined in slightly different ways by different authors, but roughly the q-sample quantile, $0 < q < 1$, is $Y_{(k)}$, where k is qn rounded to an integer. Some authors round up, others round to the nearest integer, and still others interpolate. The function `quantile()` in R has nine different types of sample quantiles, the three used by SAS[TM], S-PLUS[TM], and SPSS[TM] and Minitab[TM], plus six others. With the large sample sizes typical of financial markets data, the different choices lead to nearly identical estimates, but for small samples they can be somewhat different.

The qth quantile is also called the $100q$th *percentile*. Certain quantiles have special names. The 0.5 sample quantile is the 50th percentile and is usually called the *median*. The 0.25 and 0.75 sample quantiles are called the first and third *quartiles*, and the median is also called the second quartile. The 0.2, 0.4, 0.6, and 0.8 quantiles are the *quintiles* since they divide the data into five equal-size subsets, and the 0.1, 0.2, ..., 0.9 quantiles are the *deciles*.[4]

---

[4] Somewhat confusingly, the bottom 10 % of the data is also called the first decile and similarly for the second 10 %, and so forth. Thus, the first decile could refer to the 10th percentile of the data or to all of the data at or below this percentile. In like fashion, the bottom 20 % of the sample is called the first quintile and the second to fifth quantiles are defined analogously.

4.3.1 The Central Limit Theorem for Sample Quantiles

Many estimators have an approximate normal distribution if the sample size is sufficiently large. This is true of sample quantiles by the following central limit theorem.

Result 4.1 *Let $Y_1, \ldots, Y_n$ be an i.i.d. sample with a CDF F. Suppose that F has a density f that is continuous and positive at $F^{-1}(q)$, $0 < q < 1$. Then for large n, the qth sample quantile is approximately normally distributed with mean equal to the population quantile $F^{-1}(q)$ and variance equal to*

$$\frac{q(1-q)}{n\left[f\{F^{-1}(q)\}\right]^2}. \tag{4.3}$$

This result is not immediately applicable, for example, for constructing a confidence interval for a population quantile, because $\left[f\{F^{-1}(q)\}\right]^2$ is unknown. However, f can be estimated by kernel density estimation (Sect. 4.2) and $F^{-1}(q)$ can be estimated by the qth sample quantile. Alternatively, a confidence interval can be constructed by resampling. Resampling is introduced in Chap. 6.

4.3.2 Normal Probability Plots

Many statistical models assume that a random sample comes from a normal distribution. *Normal probability* plots are used to check this assumption, and, if the normality assumption seems false, to investigate how the distribution of the data differs from a normal distribution. If the normality assumption is true, then the qth sample quantile will be approximately equal to $\mu + \sigma\,\Phi^{-1}(q)$, which is the population quantile. Therefore, except for sampling variation, a plot of the sample quantiles versus Φ^{-1} will be linear. One version of the normal probability plot is a plot of $Y_{(i)}$ versus $\Phi^{-1}\{(i-1/2)/n\}$. These are the $(i-1/2)/n$ sample and population quantiles, respectively. The subtraction of $1/2$ from i in the numerator is used to avoid $\Phi^{-1}(1) = +\infty$ when $i = n$.

Systematic deviation of the plot from a straight line is evidence of non-normality. There are other versions of the normal plot, e.g., a plot of the order statistics versus their expectations under normality, but for large samples these will all be similar, except perhaps in the extreme tails.

Statistical software differs about whether the data are on the x-axis (horizontal axis) and the theoretical quantiles on the y-axis (vertical axis) or vice versa. The qqnorm() function in R allows the data to be on either axis depending on the choice of the parameter datax. When interpreting a normal plot with a nonlinear pattern, it is essential to know which axis contains the data. In this book, the data will always be plotted on the x-axis and the theoretical quantiles on the y-axis, so in R, datax = TRUE was used to construct the plots rather than the default, which is datax = FALSE.

If the pattern in a normal plot is nonlinear, then to interpret the pattern one checks where the plot is convex and where it is concave. A convex curve is one such that as one moves from left to right, the slope of the tangent line increases; see Fig. 4.9a. Conversely, if the slope decreases as one moves from left to right, then the curve is concave; see Fig. 4.9b. A convex-concave curve is convex on the left and concave on the right and, similarly, a concave-convex curve is concave on the left and convex on the right; see Fig. 4.9c and d.

A convex, concave, convex-concave, or concave-convex normal plot indicates, respectively, left skewness, right skewness, heavy tails (compared to the normal distribution), or light tails (compared to the normal distribution)—these interpretations require that the sample quantiles are on the horizontal axis and need to be changed if the sample quantiles are plotted on the vertical axis. *Tails* of a distribution are the regions far from the center. Reasonable definitions of the "tails" would be that the left tail is the region from $-\infty$ to $\mu - 2\sigma$ and the right tail is the region from $\mu + 2\sigma$ to $+\infty$, though the choices of $\mu - 2\sigma$ and $\mu + 2\sigma$ are somewhat arbitrary. Here μ and σ are the mean and standard deviation, though they might be replaced by the median and MAD estimator, which are less sensitive to tail weight.

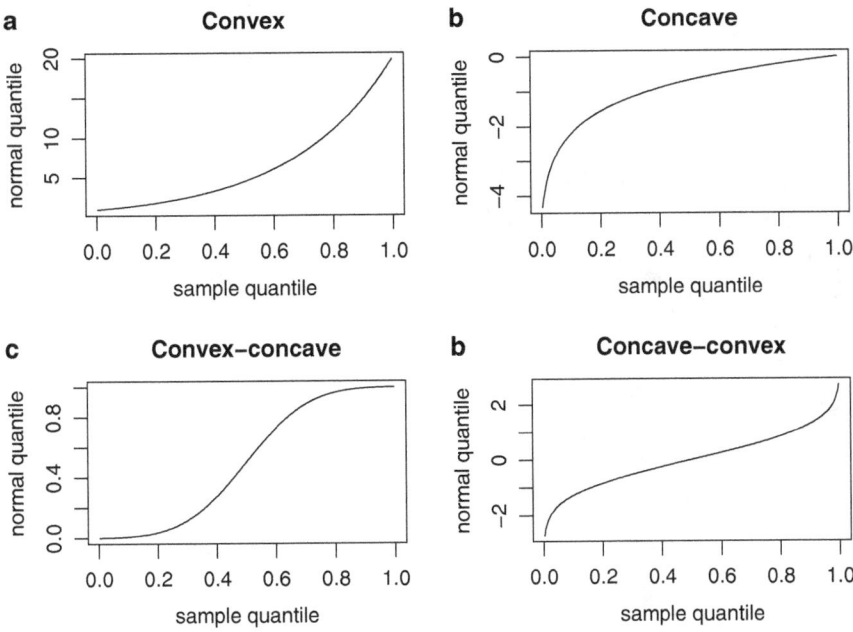

Fig. 4.9. *As one moves from (**a**) to (**d**), the curves are convex, concave, convex-concave, and concave-convex. Normal plots with these patterns indicate left skewness, right skewness, heavier tails than a normal distribution, and lighter tails than a normal distribution, respectively, assuming that the data are on the x-axis and the normal quantiles on the y-axis, as will always be the case in this textbook.*

Figure 4.10 contains normal plots of samples of size 20, 150, and 1000 from a normal distribution. To show the typical amount of random variation in normal plots, two independent samples are shown for each sample size. The plots are only close to linear because of random variation. Even for normally distributed data, some deviation from linearity is to be expected, especially for smaller sample sizes. With larger sample sizes, the only deviations from linearity are in the extreme left and right tails, where the plots are more variable.

Often, a reference line is added to the normal plot to help the viewer determine whether the plot is reasonably linear. One choice for the reference line goes through the pair of first quartiles and the pair of third quartiles; this is what R's qqline() function uses. Other possibilities would be a least-squares fit to all of the quantiles or, to avoid the influence of outliers, some subset of the quantiles, e.g., all between the 0.1 and 0.9-quantiles.

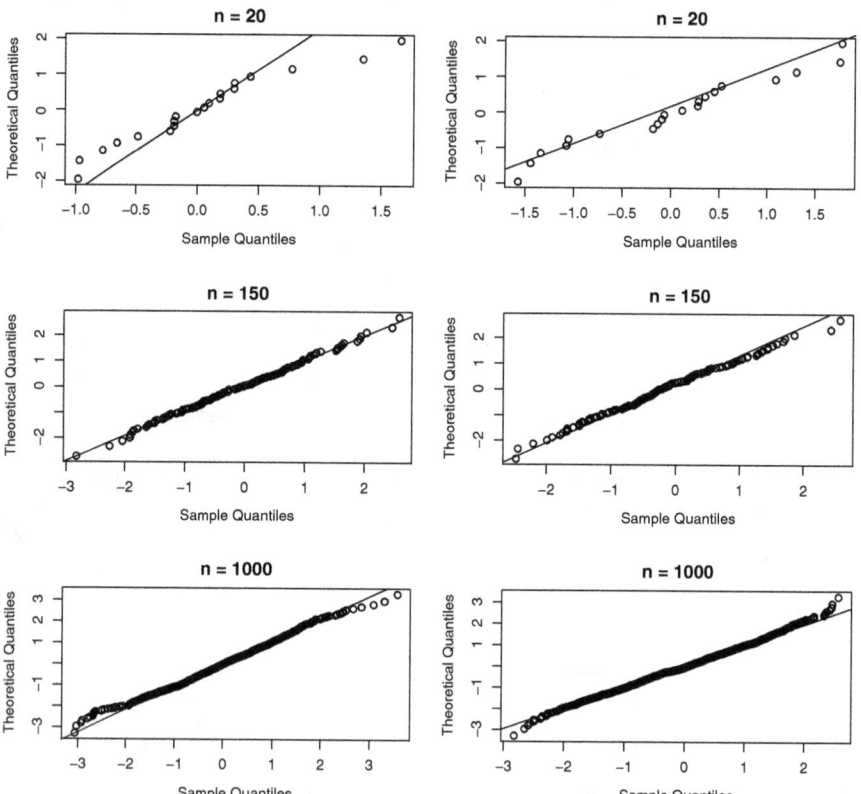

Fig. 4.10. *Normal probability plots of random samples of size 20, 150, and 1000 from an $N(0,1)$ population. The plots were produced by the R function* qqnorm(). *The reference lines pass through the first and third quartiles and were produced by R's* qqline() *function.*

Figure 4.11 contains normal probability plots of samples of size 150 from lognormal $(0, \sigma^2)$ distributions,[5] with the log-standard deviation $\sigma = 1, 1/2$, and $1/5$. The concave shapes in Fig. 4.11 indicate right skewness. The skewness when $\sigma = 1$ is quite strong, and when $\sigma = 1/2$, the skewness is still very noticeable. With σ reduced to $1/5$, the right skewness is much less pronounced and might not be discernable with smaller sample sizes.

Figure 4.12 contains normal plots of samples of size 150 from t-distributions with 4, 10, and 30 degrees of freedom. The first two distributions have heavy tails or, stated differently, are outlier-prone, meaning that the extreme observations on both the left and right sides are significantly more extreme than would be expected for a normal distribution. One can see that the tails are heavier in the sample with 4 degrees of freedom compared to the sample with

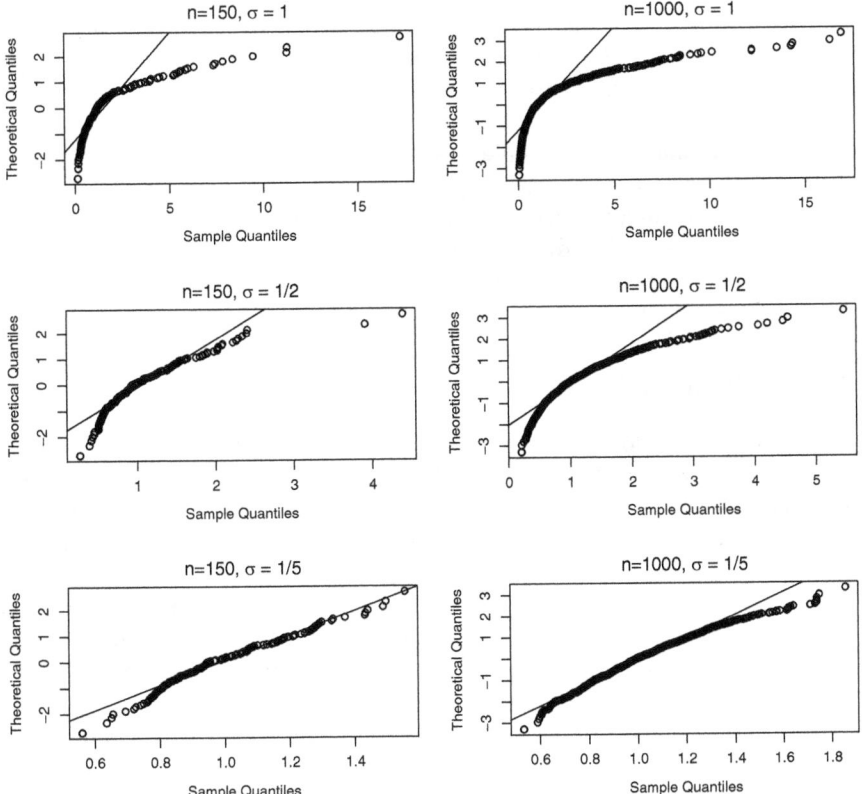

Fig. 4.11. *Normal probability plots of random samples of sizes 150 and 1000 from lognormal populations with $\mu = 0$ and $\sigma = 1, 1/2$, or $1/5$. The reference lines pass through the first and third quartiles.*

---

[5] See Appendix A.9.4 for an introduction to the lognormal distribution and the definition of the log-standard deviation.

10 degrees of freedom, and the tails of the t-distribution with 30 degrees of freedom are not much different from the tails of a normal distribution. It is a general property of the t-distribution that the tails become heavier as the degrees of freedom parameter decreases and the distribution approaches the normal distribution as the degrees of freedom approaches infinity. Any t-distribution is symmetric,[6] so none of the samples is skewed. Heavy-tailed distributions with little or no skewness are common in finance and, as we will see, the t-distribution is a reasonable model for stock returns and other financial markets data.

Sometimes, a normal plot will not have any of the patterns discussed here but instead will have more complex behavior. An example is shown in Fig. 4.13, which uses a simulated sample from a trimodal density. The alternation of the QQ plot between concavity and convexity indicates complex behavior which should be investigated by a KDE. Here, the KDE reveals the trimodality. Multimodality is somewhat rare in practice and often indicates a mixture of several distinct groups of data.

It is often rather difficult to decide whether a normal plot is close enough to linear to conclude that the data are normally distributed, especially when the sample size is small. For example, even though the plots in Fig. 4.10 are close to linear, there is some nonlinearity. Is this nonlinearity due to nonnormality or just due to random variation? If one did not know that the data were simulated from a normal distribution, then it would be difficult to tell, unless one were very experienced with normal plots. In such situations, a test of normality is very helpful. These tests are discussed in Sect. 4.4.

4.3.3 Half-Normal Plots

The half-normal plot is a variation of the normal plot used for detecting outlying data rather than checking for a normal distribution. For example, suppose one has data $Y_1, \ldots, Y_n$ and wants to see whether any of the absolute deviations $|Y_1 - \overline{Y}|, \ldots, |Y_n - \overline{Y}|$ from the mean are unusual. In a half-normal plot, these deviation are plotted against the quantiles of $|Z|$, where Z is $N(0, 1)$ distributed. More precisely, a half-normal plot is a scatterplot of the order statistics of the absolute values of the data against $\Phi^{-1}\{(n + i)/(2n + 1)\}$, $i = 1, \ldots, n$, where n is the sample size. The function `halfnorm()` in R's `faraway` package creates a half-normal plot and labels the `nlab` most outlying observations, where `nlab` is an argument of this function with a default value of 2.

Example 4.1. DM/dollar exchange rate—Half-normal plot

Figure 4.14 is a half-normal plot of changes in the DM/dollar exchange rate. The plot shows that case #1447 is the most outlying, with case #217

---

[6] However, t-distributions have been generalized in at least two different ways to the so-called skewed-t-distributions, which need not be symmetric. See Sect. 5.7.

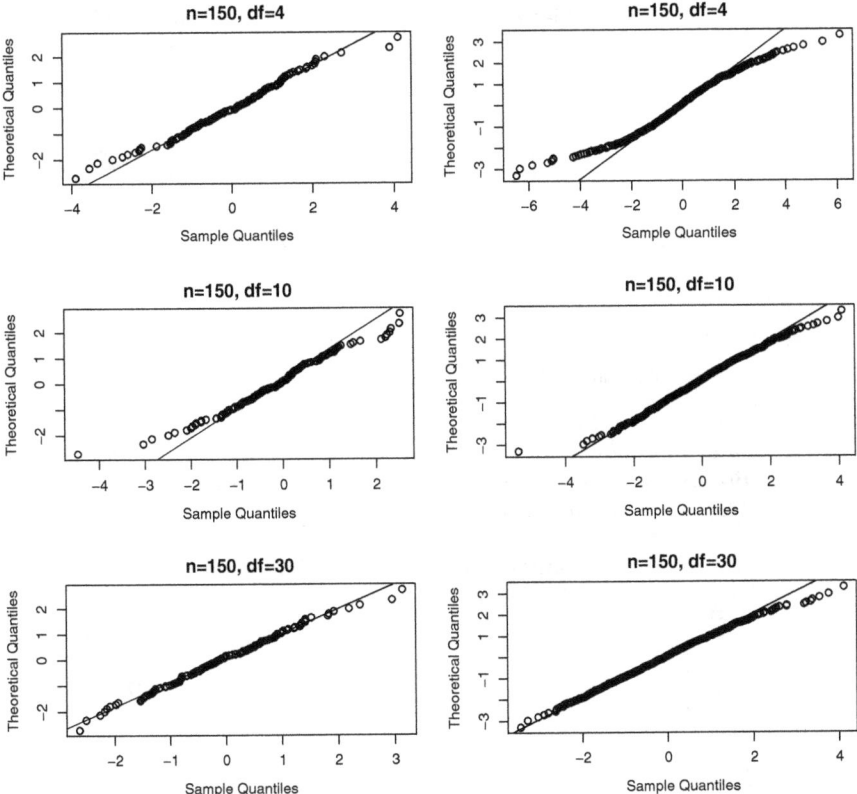

Fig. 4.12. *Normal probability plot of a random sample of size* 150 *and* 1000 *from a t-distribution with* 4, 10, *and* 30 *degrees of freedom. The reference lines pass through the first and third quartiles.*

the next most outlying. Only the two most outlying cases are labeled because the default value of `nlab` was used. The code to produce this figure is below.

```
1 data(Garch, package = "Ecdat")
2 diffdm = diff(dm)  # Deutsch mar
3 pdf("dm_halfnormal.pdf" ,width = 7, height = 6)  # Figure 4.14
4 halfnorm(abs(diffdm), main = "changes in DM/dollar exchange rate",
5    ylab = "Sorted data")
6 graphics.off()
```

□

Another application of half-normal plotting can be found in Sect. 10.1.3.

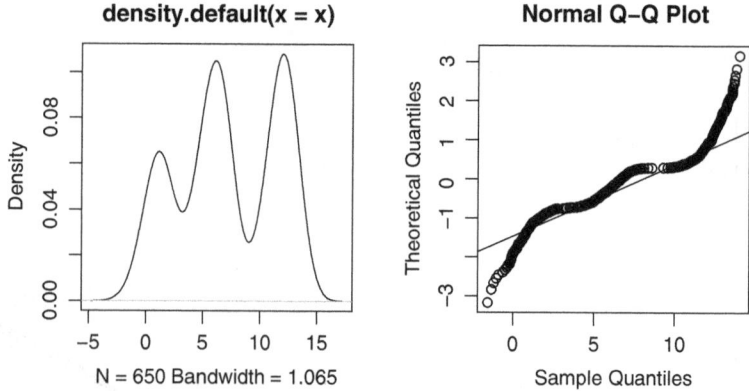

Fig. 4.13. *Kernel density estimate (left) and normal plot (right) of a simulated sample from a trimodal density. The reference lines pass through the first and third quartiles. Because of the three modes, the normal plot changes convexity three times, concave to convex to concave to convex, going from left to right.*

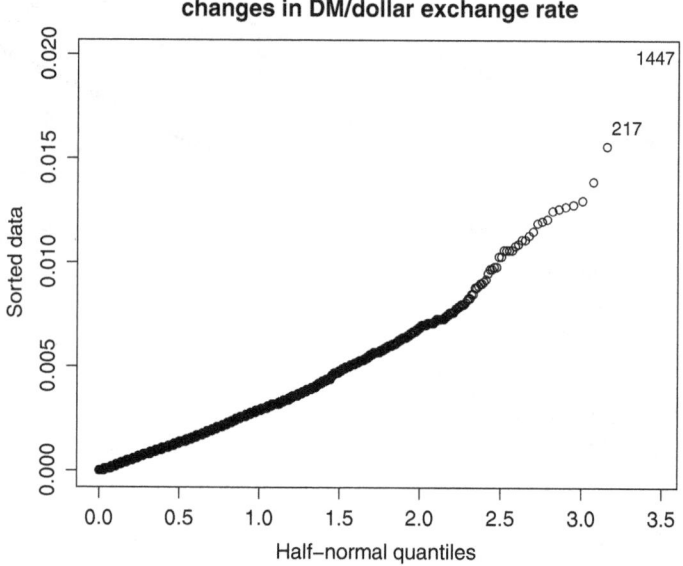

Fig. 4.14. *Half-normal plot of changes in DM/dollar exchange rate.*

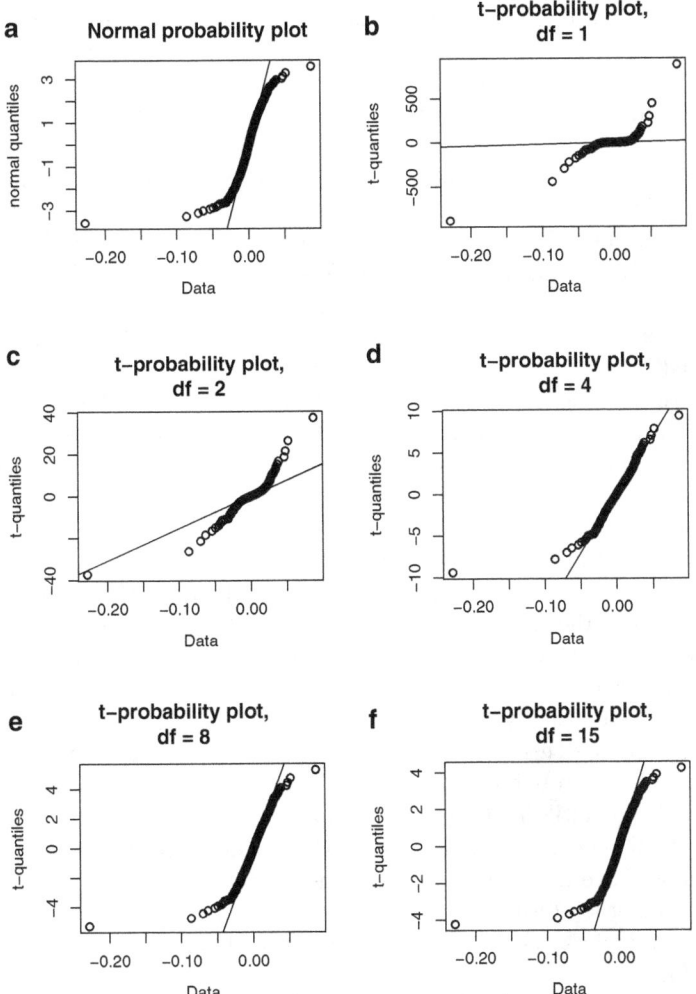

Fig. 4.15. *Normal and t probability plots of the daily returns on the S&P* 500 *index from January* 1981 *to April* 1991. *This data set is the* SP500 *series in the* Ecdat *package in* R. *The reference lines pass through the first and third quartiles.*

4.3.4 Quantile–Quantile Plots

Normal probability plots are special cases of *quantile-quantile plots*, also known as QQ plots. A *QQ plot* is a plot of the quantiles of one sample or distribution against the quantiles of a second sample or distribution.

For example, suppose that we wish to model a sample using the $t_\nu(\mu, \sigma^2)$ distribution defined in Sect. 5.5.2. The parameter ν is called the "degrees of freedom," or simply "df." Suppose, initially, that we have a hypothesized value of ν, say $\nu = 6$ to be concrete. Then we plot the sample quantiles against the quantiles of the $t_6(0, 1)$ distribution. If the data are from a $t_6(\mu, \sigma^2)$ distribution, then, apart from random variation, the plot will be linear with intercept and slope depending on μ and σ.

Figure 4.15 contains a normal plot of the S&P 500 log returns in panel (a) and t-plots with 1, 2, 4, 8, and 15 df in panels (b) through (f). None of the plots looks exactly linear, but the t-plot with 4 df is rather straight through the bulk of the data. There are approximately nine returns in the left tail and four in the right tail that deviate from a line through the remaining data, but these are small numbers compared to the sample size of 2783. Nonetheless, it is worthwhile to keep in mind that the historical data have more extreme outliers than a t-distribution. The t-model with 4 df and mean and standard deviation estimated by maximum likelihood[7] implies that a daily log return of -0.228, the return on Black Monday, or less has probability 3.2×10^{-6}. This means approximately 3 such returns every 1,000,000 days or 40,000 years, assuming 250 trading days per year. Thus, the t-model implies that Black Monday was extremely unlikely, and anyone using that model should be mindful that it did happen.

There are two reasons why the t-model does not give a credible probability of a negative return as extreme as on Black Monday. First, the t-model is symmetric, but the return distribution appears to have some skewness in the extreme left tail, which makes extreme negative returns more likely than under the t-model. Second, the t-model assumes constant conditional volatility, but volatility was unusually high in October 1987. GARCH models (Chap. 14) can accommodate this type of volatility clustering and provide more realistic estimates of the probability of an extreme event such as Black Monday.

Quantile–quantile plots are useful not only for comparing a sample with a theoretical model, as above, but also for comparing two samples. If the two samples have the same sizes, then one need only plot their order statistics against each other. Otherwise, one computes the same sets of sample quantiles for each and plots them. This is done automatically with the R command qqplot().

The interpretation of convex, concave, convex-concave, and concave-convex QQ plots is similar to that with QQ plots of theoretical quantiles versus sample quantiles. A concave plot implies that the sample on the x-axis is more right-skewed, or less left-skewed, than the sample on the y-axis. A convex plot implies that the sample on the x-axis is less right-skewed, or more left-skewed, than the sample on the y-axis. A convex-concave (concave-convex) plot implies that the sample on the x-axis is more (less) heavy-tailed than the sample on the y-axis. As before, a straight line, e.g., through the first and third quartiles, is often added for reference.

Figure 4.16 contains sample QQ plots for all three pairs of the three time series, S&P 500 returns, changes in the DM/dollar rate, and changes in the risk-free return, used as examples in this chapter. One sees that the S&P 500

---

[7] See Sect. 5.14.

returns have more extreme outliers than the other two series. The changes in DM/dollar and risk-free returns have somewhat similar shapes, but the changes in the risk-free rate have slightly more extreme outliers in the left tail. To avoid any possible confusion, it should be mentioned that the plots in Fig. 4.16 only compare the marginal distributions of the three time series. They tell us nothing about dependencies between the series and, in fact, the three series were observed on different time intervals so correlations between these time series cannot be estimated from these data.

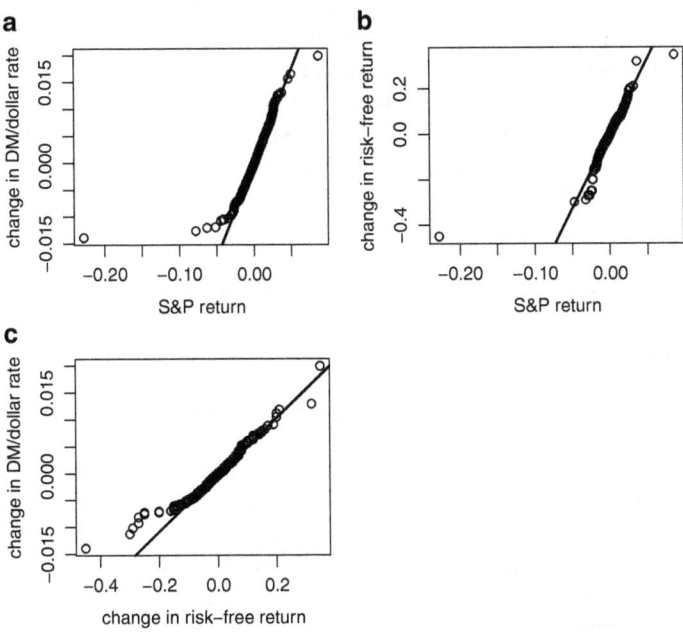

Fig. 4.16. *Sample QQ plots. The straight lines pass through the first and third sample quantiles. (**a**) Change in DM/dollar rate versus S&P return. (**b**) Change in risk-free rate versus S&P return. (**c**) Change in DM/dollar rate versus change in risk-free rate.*

The code for panel (a) of Fig. 4.16 is below. The code for the other panels is similar and so is omitted.

```
1 qqplot(SPreturn, diffdm, xlab = "S&P return",
2   ylab = "change in DM/dollar rate", main = "(a)")
3 xx = quantile(SPreturn, c(0.25, 0.75))
4 yy = quantile(diffdm, c(0.25, 0.75))
5 slope = (yy[2] - yy[1]) / (xx[2] - xx[1])
6 inter = yy[1] - slope*xx[1]
7 abline(inter, slope, lwd = 2 )
```

4.4 Tests of Normality

When viewing a normal probability plot, it is often difficult to judge whether any deviation from linearity is systematic or instead merely due to sampling variation, so a statistical test of normality is useful. The null hypothesis is that the sample comes from a normal distribution and the alternative is that the sample is from a nonnormal distribution.

The Shapiro–Wilk test of these hypotheses uses something similar to a normal plot. Specifically, the Shapiro–Wilk test is based on the association between sample order statistics $Y_{(i)}$ and the expected normal order statistics which, for large samples, are close to $\Phi^{-1}\{i/(n+1)\}$, the quantiles of the standard normal distribution. The vector of expected order statistics is multiplied by the inverse of its covariance matrix. Then the correlation between this product and the sample order statistics is used as the test statistic. Correlation and covariance matrices will be discussed in greater detail in Chap. 7. For now, only a few facts will be mentioned. The *covariance* between two random variables X and Y is

$$\mathrm{Cov}(X,Y) = \sigma_{XY} = E\Big[\{X - E(X)\}\{Y - E(Y)\}\Big],$$

and the *Pearson correlation coefficient* between X and Y is

$$\mathrm{Corr}(X,Y) = \rho_{XY} = \sigma_{XY}/\sigma_X\,\sigma_Y. \qquad (4.4)$$

A correlation equal to 1 indicates a perfect positive linear relationship, where $Y = \beta_0 + \beta_1 X$ with $\beta_1 > 0$. Under normality, the correlation between sample order statistics and the expected normal order statistics should be close to 1 and the null hypothesis of normality is rejected for small values of the correlation coefficient. In R, the Shapiro–Wilk test can be implemented using the `shapiro.test()` function.

The Jarque–Bera test uses the sample skewness and kurtosis coefficients and is discussed in Sect 5.4 where skewness and kurtosis are introduced. Other tests of normality in common use are the Anderson–Darling, Cramér–von Mises, and Kolmogorov–Smirnov tests. These tests compare the sample CDF to the normal CDF with mean equal to $\overline{Y}$ and variance equal to s_Y^2. The Kolmogorov–Smirnov test statistic is the maximum absolute difference between these two functions, while the Anderson–Darling and Cramér–von Mises tests are based on a weighted integral of the squared difference. The p-values of the Shapiro–Wilk, Anderson–Darling, Cramér–von Mises, and Kolmogorov–Smirnov tests are routinely part of the output of statistical software. A small p-value is interpreted as evidence that the sample is not from a normal distribution.

A recent comparison of eight tests of normality (Yap and Sim 2011) found that the Shapiro-Wilk test was as powerful as its competitors for both short- and long-tailed symmetric alternatives and was the most powerful test for asymmetric alternatives. The tests in this study were: Shapiro-Wilk, Kolmogorov-Smirnov, Lilliefors, Cramér-vo Mises, Anderson-Darling, D'Agostino-Pearson, Jarque-Bera, and chi-squared.

For the S&P 500 returns, the Shapiro–Wilk test rejects the null hypothesis of normality with a p-value less than 2.2×10^{-16}. The Shapiro–Wilk also strongly rejects normality for the changes in DM/dollar rate and for the changes in risk-free return. With large sample sizes, e.g., 2783, 1866, and 515, for the S&P 500 returns, changes in DM/dollar rate, and changes in risk-free return, respectively, it is quite likely that normality will be rejected, since any real population will deviate to some extent from normality and any deviation, no matter how small, will be detected with a large enough sample. When the sample size is large, it is important to look at normal plots to see whether the deviation from normality is of practical importance. For financial time series, the deviation from normality in the tails is often large enough to be important.[8]

4.5 Boxplots

The boxplot is a useful graphical tool for comparing several samples. The appearance of a boxplot depends somewhat on the specific software used. In this section, we will describe boxplots produced by the R function boxplot(). The three boxplots in Fig. 4.17 were created by boxplot() with default choice of tuning parameters. The "box" in the middle of each plot extends from the first to the third quartile and thus gives the range of the middle half of the data, often called the *interquartile range*, or IQR. The line in the middle of the box is at the median. The "whiskers" are the vertical dashed lines extending from the top and bottom of each box. The whiskers extend to the smallest and largest data points whose distance from the bottom or top of the box is at most 1.5 times the IQR.[9] The ends of the whiskers are indicated by horizontal lines. All observations beyond the whiskers are plotted with an "o". The most obvious differences among the three boxplots in Fig. 4.17 are differences in scale, with the monthly risk-free return changes being the most variable and the daily DM/dollar changes being the least variable. It is not surprising that the changes in the risk-free return are most variable, since these are changes over months, not days as with the other series.

These scale differences obscure differences in shape. To remedy this problem, in Fig. 4.18 the three series have been standardized by subtracting the median and then dividing by the MAD. Now, differences in shape are clearer. One can see that the S&P 500 returns have heavier tails because the "o"s are farther from the whiskers. The return of the S&P 500 on Black Monday

---

[8] See Chap. 19 for a discussion on how tail weight can greatly affect risk measures such as VaR and expected shortfall.

[9] The factor 1.5 is the default value of the range parameter and can be changed.

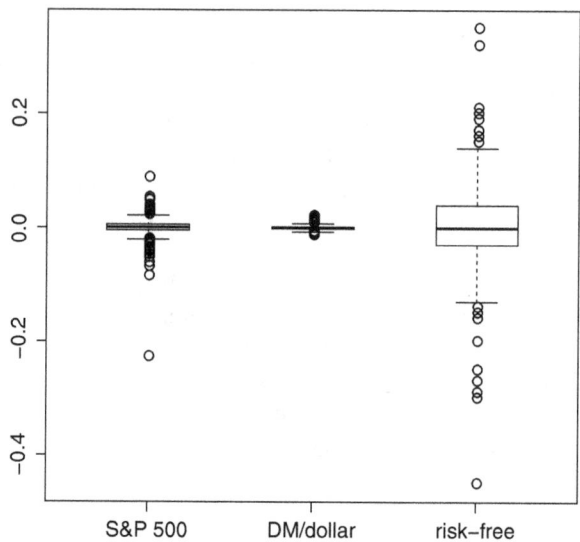

Fig. 4.17. *Boxplots of the S&P 500 daily log returns, daily changes in the DM/dollar exchange rate, and monthly changes in the risk-free returns.*

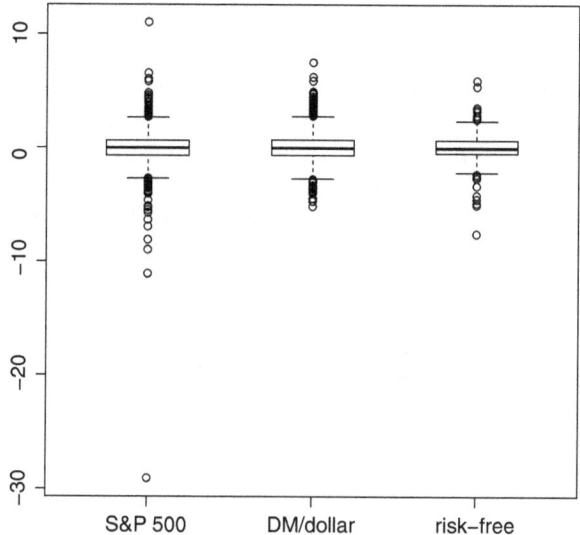

Fig. 4.18. *Boxplots of the standardized S&P 500 daily log returns, daily changes in the DM/dollar exchange rate, and monthly changes in the risk-free returns.*

is quite detached from the remaining data. Of course, one should be aware of differences in scale, so it is worthwhile to look at boxplots of the variables both without and with standardization.

When comparing several samples, boxplots and QQ plots provide different views of the data. It is best to use both. However, if there are N samples, then the number of QQ plots is $N(N-1)/2$ or $N(N-1)$ if, by interchanging axes, one includes two plots for each pair of samples. This number can get out of hand quickly, so, for large values of N, one might use boxplots augmented with a few selected QQ plots.

4.6 Data Transformation

There are a number of reasons why data analysts often work not with the original variables, but rather with transformations of the variables such as logs, square roots, or other power transformations. Many statistical methods work best when the data are normally distributed or at least symmetrically distributed and have a constant variance, and the transformed data will often exhibit less skewness and a more constant variance compared to the original variables, especially if the transformation is selected to induce these features.

A transformation is called *variance stabilizing* if it removes a dependence between the conditional variance and the conditional mean of a variable. For example, if Y is Poisson distributed with a conditional mean depending on X, then its conditional variance is equal to the conditional mean. A transformation h would be variance-stabilizing for Y if the conditional variance of $h(Y)$ did not depend on the conditional mean of $h(Y)$.

The logarithm transformation is probably the most widely used transformation in data analysis, though the square root is a close second. The log stabilizes the variance of a variable whose conditional standard deviation is proportional to its conditional mean. This is illustrated in Fig. 4.19, which plots monthly changes in the risk-free return (top row) and changes in the log of the return (bottom row) against the lagged risk-free return (left column) or year (right column). Notice that the changes in the return are more variable when the lagged return is higher. This behavior is called nonconstant conditional variance or conditional heteroskedasticity. We see in the bottom row that the changes in the log return have relatively constant variability, at least compared to changes in the return.

The log transformation is sometimes embedded into the power transformation family by using the so-called Box–Cox power transformation

$$y^{(\alpha)} = \begin{cases} \frac{y^\alpha - 1}{\alpha}, & \alpha \neq 0 \\ \log(y), & \alpha = 0. \end{cases} \tag{4.5}$$

In (4.5), the subtraction of 1 from y^α and the division by α are not essential, but they make the transformation continuous in α at 0 since

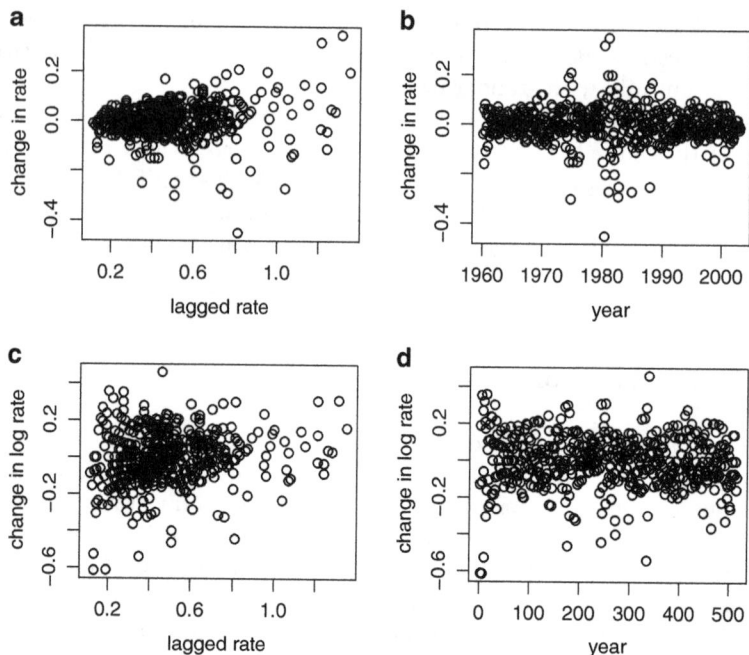

Fig. 4.19. *Changes in risk-free rate (top) and changes in the logarithm of the risk-free rate (bottom) plotted against time and against lagged rate. The risk-free returns are the variable* rf *of the* Capm *data set in R's* Ecdat *package. (a) Change in risk-free rate versus change in lagged rate. (b) Change in rate versus year. (c) Change in log(rate) versus lagged rate. (d) Change in log(rate) versus year.*

$$\lim_{\alpha \to 0} \frac{y^{\alpha} - 1}{\alpha} = \log(y).$$

Note that division by α ensures that the transformation is increasing even when $\alpha < 0$. This is convenient though not essential. For the purposes of inducing symmetry and a constant variance, y^{α} and $y^{(\alpha)}$ work equally well and can be used interchangeably, especially if, when $\alpha < 0$, y^{α} replaced by $-y^{\alpha}$ to ensure that the transformation is monotonically increasing for all values of α. The use of a monotonically decreasing, rather than increasing, transformation is inconvenient since decreasing transformations reverse ordering and, for example, transform the pth quantile to the $(1 - p)$th quantile.

It is commonly the case that the response is right-skewed and the conditional response variance is an increasing function of the conditional response mean. In such cases, a concave transformation, e.g., a Box–Cox transformation with $\alpha < 1$, will remove skewness and stabilize the variance. If a Box–Cox transformation with $\alpha < 1$ is used, then the smaller the value of α, the greater the effect of the transformation. One can go too far—if the transformed response is *left*-skewed or has a conditional variance that is decreasing as a function of the conditional mean, then α has been chosen too small. Instances of this type of overtransformation are given in Examples 4.2, 4.4, and 13.2.

Typically, the value of α that is best for symmetrizing the data is not the same value of α that is best for stabilizing the variance. Then, a compromise is needed so that the transformation is somewhat too weak for one purpose and somewhat too strong for the other. Often, however, the compromise is not severe, and near symmetry and homoskedasticity can both be achieved.

Example 4.2. Gas flows in pipelines

In this example, we will use a data set of daily flows of natural gas in three pipelines. These data are part of a larger data set used in an investigation of the relationships between flows in the pipelines and prices. Figure 4.20 contains histograms of the daily flows. Notice that all three distributions are left-skewed. For left-skewed data, a Box–Cox transformation should use $\alpha > 1$.

Figure 4.21 shows KDEs of the flows in pipeline 1 after a Box–Cox transformation using $\alpha = 1, 2, 3, 4, 5, 6$. One sees that α between 3 and 4 removes most

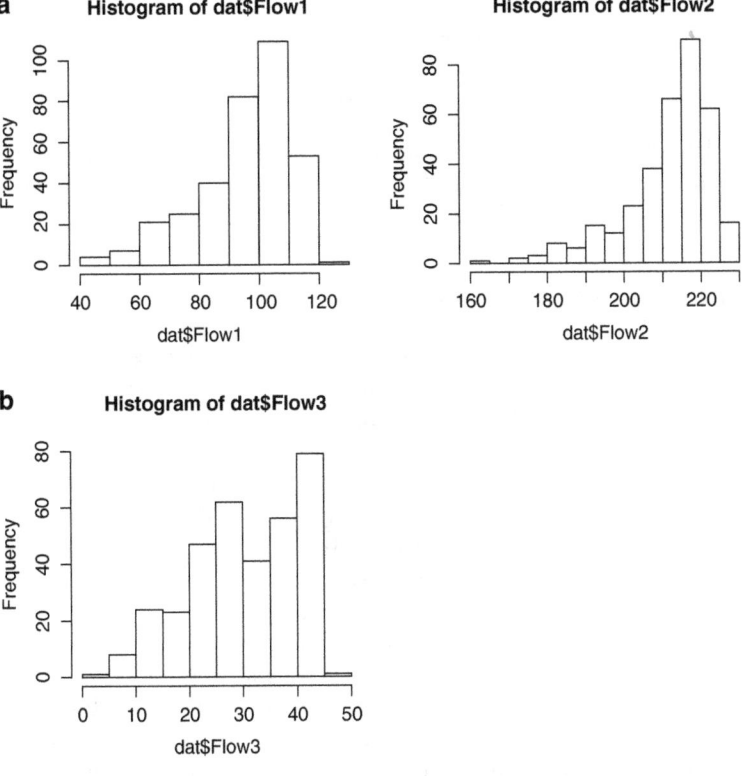

Fig. 4.20. *Histograms of daily flows in three pipelines.*

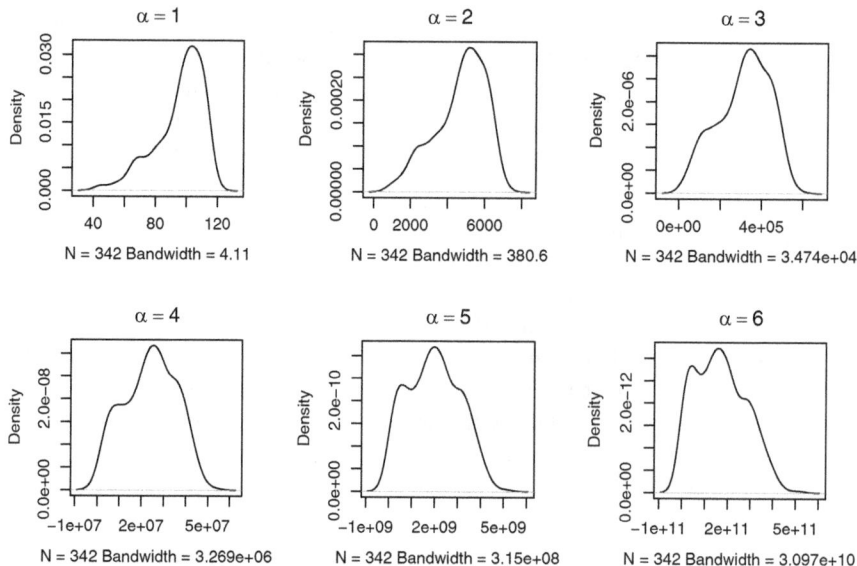

Fig. 4.21. *Kernel density estimates for gas flows in pipeline 1 with Box–Cox transformations.*

of the left-skewness and $\alpha = 5$ or greater overtransforms to right-skewness. Later, in Example 5.7, we will illustrate an automatic method for selecting α and find that $\alpha = 3.5$ is chosen. □

Example 4.3. t-Tests and transformations

This example shows the deleterious effect of skewness and nonconstant variance on hypothesis testing and how a proper data transformation can remedy this problem. The boxplots on the panel (a) in Fig. 4.22 are of independent samples of size 15 from lognormal(1,4) (left) and lognormal(3,4) distributions. Panel (b) shows boxplots of the log-transformed data.

Suppose one wants to test the null hypothesis that the two populations have the same means against a two-sided alternative. The transformed data satisfy the assumptions of the t-test that the two populations are normally distributed with the same variance, but of course the original data do not meet these assumptions. Two-sided independent-samples t-tests have p-values of 0.105 and 0.00467 using the original data and the log-transformed data, respectively. These two p-values lead to rather different conclusions, for the first test that the means are not significantly different at the usual $\alpha = 0.05$, and not quite significant even at $\alpha = 0.1$, and for the second test that the difference is highly significant. The first test reaches an incorrect conclusion because its assumptions are not met. □

The previous example illustrates some general principles to keep in mind. All statistical estimators and tests make certain assumptions about the distribution of the data. One should check these assumptions, and graphical methods are often the most convenient way to diagnose problems. If the assumptions are not met, then one needs to know how sensitive the estimator or test is to violations of the assumptions. If the estimator or test is likely to be seriously degraded by violations of the assumptions, which is called *nonrobustness*, then there are two recourses. The first is to find a new estimator or test that is suitable for the data. The second is to transform the data so that the transformed data satisfy the assumptions of the original test or estimator.

4.7 The Geometry of Transformations

Response transformations induce normality of a distribution and stabilize variances because they can stretch apart data in one region and push observations together in other regions. Figure 4.23 illustrates this behavior. On the horizontal axis is a sample of data from a right-skewed lognormal distribution. The transformation $h(y)$ is the logarithm. The transformed data are plotted on the vertical axis. The dashed lines show the transformation of y to $h(y)$ as one moves from a y-value on the x-axis upward to the curve and then to $h(y)$ on the y-axis. Notice the near symmetry of the transformed data. This symmetry is achieved because the log transformation stretches apart data with small values and shrinks together data with large values. This can be seen by observing the derivative of the log function. The derivative of $\log(y)$ is $1/y$, which is a decreasing function of y. The derivative is, of course, the slope of the tangent line and the tangent lines at $y = 1$ and $y = 5$ are plotted to show the decrease in the derivative as y increases.

Consider an arbitrary increasing transformation, $h(y)$. If x and x' are two nearby data points that are transformed to $h(x)$ and $h(x')$, respectively, then

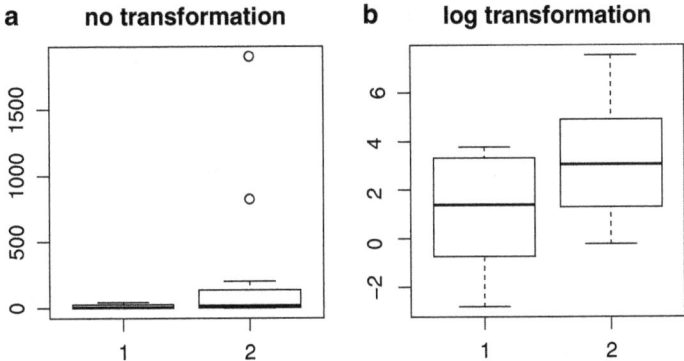

Fig. 4.22. *Boxplots of samples from two lognormal distributions without (**a**) and with (**b**) log transformation.*

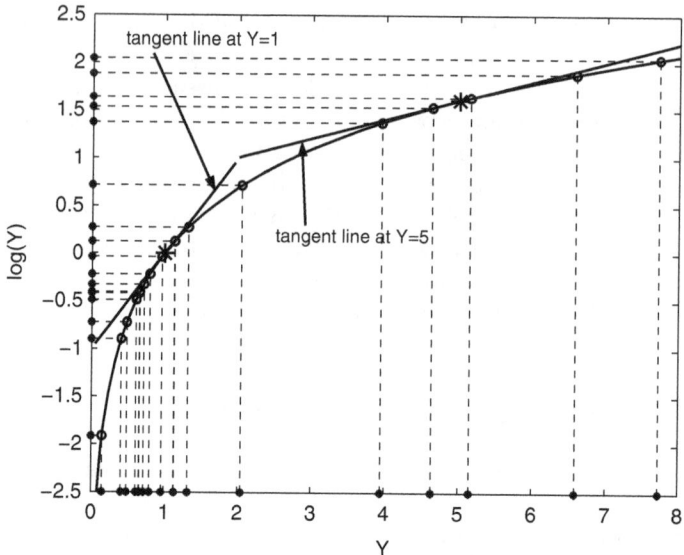

Fig. 4.23. *A symmetrizing transformation. The skewed lognormal data on the horizontal axis are transformed to symmetry by the log transformation.*

the distance between transformed values is $|h(x) - h(x')| \approx h^{(1)}(x)|x - x'|$. Therefore, $h(x)$ and $h(x')$ are stretched apart where $h^{(1)}$ is large and pushed together where $h^{(1)}$ is small. A function h is called concave if $h^{(1)}(y)$ is a decreasing function of y. As can be seen in Fig. 4.23, concave transformations can remove right skewness.

Concave transformations can also stabilize the variance when the untransformed data are such that small observations are less variable than large observations. This is illustrated in Fig. 4.24. There are two groups of responses, one with a mean of 1 and a relatively small variance and another with a mean of 5 and a relatively large variance. If the expected value of the response Y_i, conditional on $\boldsymbol{X}_i$, followed a regression model $m(\boldsymbol{X}_i; \boldsymbol{\beta})$, then two groups like these would occur if there were two possible values of $\boldsymbol{X}_i$, one with a small value of $m(\boldsymbol{X}_i; \boldsymbol{\beta})$ and the other with a large value. Because of the concavity of the transformation h, the variance of the group with a mean of 5 is reduced by transformation. After the transformation, the groups have nearly the same variance, as can be seen by observing the scatter of the two groups on the y-axis.

The strength of a transformation can be measured by how much its derivative changes over some interval, say a to b. More precisely, for $a < b$, the strength of an increasing transformation h is the derivative ratio $h'(b)/h'(a)$. If the transformation is concave, then the derivative ratio is less than 1 and the smaller the ratio the stronger the concavity. Conversely, if the transformation is convex, then the derivative ratio is greater than 1 and the larger the ratio,

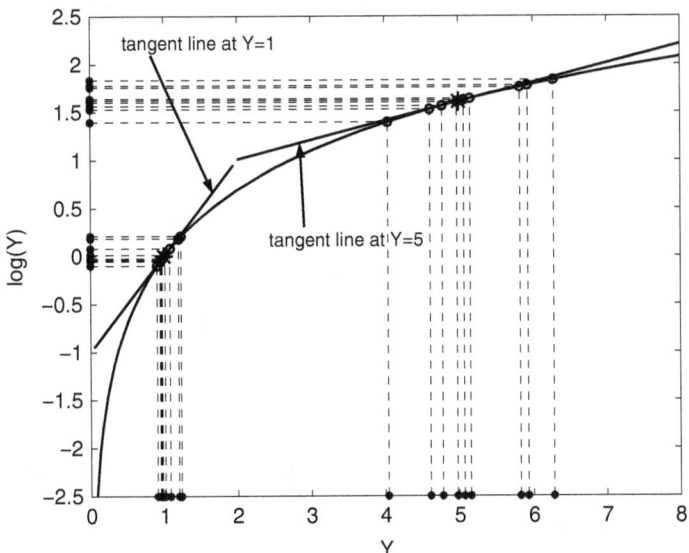

Fig. 4.24. *A variance-stabilizing transformation.*

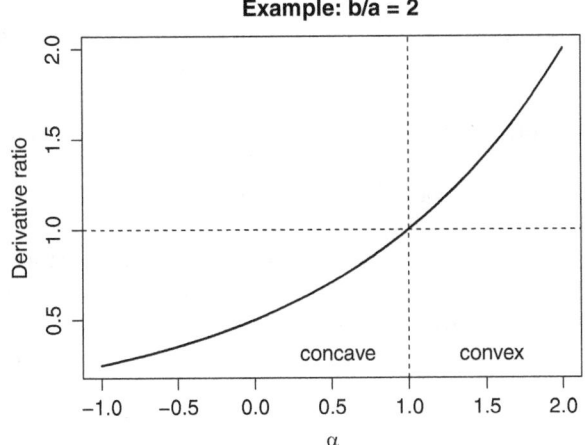

Fig. 4.25. *Derivative ratio for Box–Cox transformations.*

the greater the convexity. For a Box–Cox transformation, the derivative ratio is $(b/a)^{\alpha-1}$ and so depends on a and b only through the ratio b/a. Figure 4.25 shows the derivative ratio of Box–Cox transformations when $b/a = 2$. One can see that the Box–Cox transformation is concave when $\alpha < 1$, with the concavity becoming stronger as α decreases. Similarly, the transformation is convex for $\alpha > 1$, with increasing convexity as α increases.

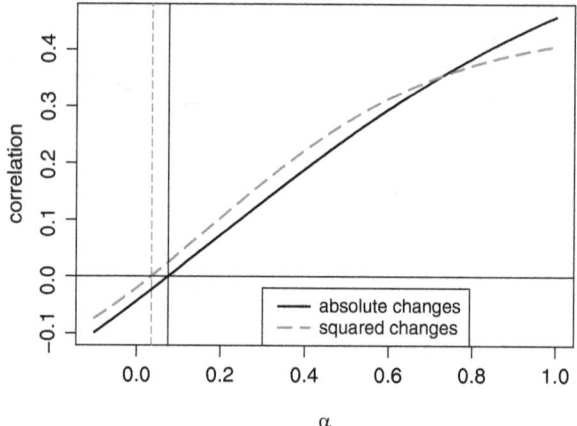

Fig. 4.26. *Correlations between the lagged risk-free returns and absolute (solid) and squared (dashed) changes in the Box–Cox transformed returns. A zero correlation indicates a constant conditional variance. Zero correlations are achieved with the transformation parameter α equal to 0.036 and 0.076 for the absolute and squared changes, respectively, as indicated by the vertical lines. If $\alpha \approx 0$, then the data are conditionally homoskedastic, or at least nearly so.*

Example 4.4. Risk-free returns—Strength of the Box–Cox transformation for variance stabilization

In this example, we return to the changes in the risk-free interest rates. In Fig. 4.19, it was seen that there is noticeable conditional heteroskedasticity in the changes in the untransformed rate but little or no heteroskedasticity in the changes in the logarithms of the rate. We will see that for a Box–Cox transformation intermediate in strength between the identity transformation ($\alpha = 1$) and the log transformation ($\alpha = 0$), some but not all of the heteroskedasticity is removed, and that a transformation with $\alpha < 0$ is too strong for this application so that a new type of heteroskedasticity is induced.

The strength of a Box–Cox transformation for this example is illustrated in Fig. 4.26. In that figure, the correlations between the lagged risk-free interest returns, r_{t-1}, and absolute and squared changes, $|r_t^{(\alpha)} - r_{t-1}^{(\alpha)}|$ and $\{r_t^{(\alpha)} - r_{t-1}^{(\alpha)}\}^2$, in the transformed rate are plotted against α. The two correlations are similar, especially when they are near zero. Any deviations of the correlations from zero indicate conditional heteroskedasticity where the standard deviation of the change in the transformed rate depends on the previous value of the rate. We see that the correlations decrease as α decreases from 1 so that the concavity of the transformation increases. The correlations are equal to zero when α is very close to 0, that is, the log transformation. If α is much below 0, then the transformation is too strong and the overtransformation induces a negative correlation, which indicates that the conditional standard deviation is a decreasing function of the lagged rate.   □

4.8 Transformation Kernel Density Estimation

The kernel density estimator (KDE) discussed in Sect. 4.2 is popular because of its simplicity and because it is available on most software platforms. However, the KDE has some drawbacks. One disadvantage of the KDE is that it undersmooths densities with long tails. For example, the solid curve in Fig. 4.27 is a KDE of annual earnings in 1988–1989 for 1109 individuals. The data are in the Earnings data set in R's Ecdat package. The long right tail of the density estimate exhibits bumps, which seem due solely to random variation in the data, not to bumps in the true density. The problem is that there is no single bandwidth that works well both in the center of the data and in the right tail. The automatic bandwidth selector chose a bandwidth that is a compromise, undersmoothing in the tails and perhaps oversmoothing in the center. The latter problem can cause the height of the density at the mode(s) to be underestimated.

A better density estimate can be obtained by the *transformation kernel density estimator* (TKDE). The idea is to transform the data so that the density of the transformed data is easier to estimate by the KDE. For the earnings data, the square roots of the earnings are closer to being symmetric and have a shorter right tail than the original data; see Fig. 4.28, which compares histograms of the original data and the data transformed by the square root. The KDE should work well for the square roots of the earnings.

Of course, we are interested in the density of the earnings, not the density of their square roots. However, it is easy to convert an estimate of the latter to

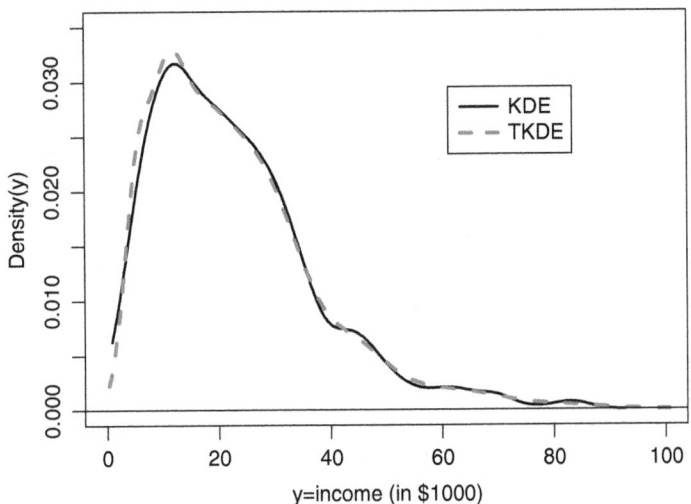

Fig. 4.27. *Kernel density and transformation kernel density estimates of annual earnings in 1988–1989 expressed in thousands of 1982 dollars. These data are the same as in Fig. 4.28.*

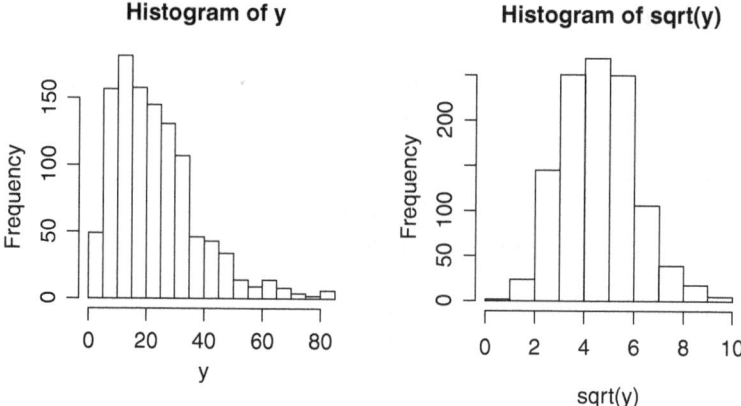

Fig. 4.28. *Histograms of earnings (y) and the square roots of earnings. The data are from the* Earnings *data set in R's* Ecdat *package and use only age group* g1.

one of the former. To do that, one uses the change-of-variables formula (A.4). For convenience, we repeat the result here—if $X = g(Y)$, where g is monotonic and f_X and f_Y are the densities of X and Y, respectively, then

$$f_Y(y) = f_X\{g(y)\}\,|g'(y)|. \qquad (4.6)$$

For example, if $x = g(y) = \sqrt{y}$, then $g'(y) = y^{-1/2}/2$ and

$$f_Y(y) = \{f_X(\sqrt{y})y^{-1/2}\}/2.$$

Putting $y = g^{-1}(x)$ into Eq. (4.6), we obtain

$$f_Y\{g^{-1}(x)\} = f_X(x)\,|g'\{g^{-1}(x)\}|. \qquad (4.7)$$

Equation (4.7) suggests a convenient method for computing the TKDE:

1. start with data $Y_1, \ldots, Y_n$;
2. transform the data to $X_1 = g(Y_1), \ldots, X_n = g(Y_n)$;
3. let $\widehat{f}_X$ be the usual KDE calculated on a grid $x_1, \ldots, x_m$ using $X_1, \ldots, X_n$;
4. plot the pairs $\left[g^{-1}(x_j),\ \widehat{f}_X(x_j)\,|g'\{g^{-1}(x_j)\}|\right]$, $j = 1, \ldots, m$.

The red dashed curve in Fig. 4.27 is a plot of the TKDE of the earnings data using the square-root transformation. Notice the smoother right tail, the faster decrease to 0 at the left boundary, and the somewhat sharper peak at the mode compared to the KDE (solid curve).

When using a TKDE, it is important to choose a good transformation. For positive, right-skewed variables such as the earnings data, a concave transformation is needed. A power transformation, y^α, for some $\alpha < 1$ is a common choice. Although there are automatic methods for choosing α (see Sect. 4.9), trial-and-error is often good enough.

4.9 Bibliographic Notes

Exploratory data analysis was popularized by Tukey (1977). Hoaglin, Mosteller, and Tukey (1983,1985) are collections of early articles on exploratory data analysis, data transformations, and robust estimation. Kleiber and Zeileis (2008) is an introduction to econometric modeling with R and covers exploratory data analysis as well as material in latter chapters of this book including regression and time series analysis. The R package AER accompanies Kleiber and Zeileis's book.

The central limit theorem for sample quantiles is stated precisely and proved in textbooks on asymptotic theory such as Serfling (1980); Lehmann (1999), and van der Vaart (1998).

Silverman (1986) is an early book on nonparametric density estimation and is still well worth reading. Scott (1992) covers both univariate and multivariate density estimation. Wand and Jones (1995) has an excellent treatment of kernel density estimation as well as nonparametric regression, which we cover in Chap. 21. Wand and Jones cover more recent developments such as transformation kernel density estimation. An alternative to the TKDE is variable-bandwidth KDE; see Sect. 2.10 of Wand and Jones (1995) as well as Abramson (1982) and Jones (1990).

Atkinson (1985) and Carroll and Ruppert (1988) are good sources of information about data transformations.

Wand, Marron, and Ruppert (1991) is an introduction to the TKDE and discusses methods for automatic selection of the transformation to minimize the expected squared error of the estimator. Applications of TKDE to losses can be found in Bolance, Guillén, and Nielsen (2003).

4.10 R Lab

4.10.1 European Stock Indices

This lab uses four European stock indices in R's EuStockMarkets database. Run the following code to access the database, learn its mode and class, and plot the four time series. The plot() function will produce a plot tailored to the class of the object on which it is acting. Here four time series plots are produced because the class of EuStockMarkets is mts, multivariate time series.

```
data(EuStockMarkets)
mode(EuStockMarkets)
class(EuStockMarkets)
plot(EuStockMarkets)
```

If you right-click on the plot, a menu for printing or saving will open. There are alternative methods for printing graphs. For example,

```
pdf("EuStocks.pdf", width = 6, height = 5)
plot(EuStockMarkets)
graphics.off()
```

will send a pdf file to the working directory and the `width` and `height` parameters allow one to control the size and aspect ratio of the plot.

Problem 1 *Write a brief description of the time series plots of the four indices. Do the series look stationary? Do the fluctuations in the series seem to be of constant size? If not, describe how the volatility fluctuates.*

Next, run the following R code to compute and plot the log returns on the indices.

```
logR = diff(log(EuStockMarkets))
plot(logR)
```

Problem 2 *Write a brief description of the time series plots of the four series of log returns. Do the series look stationary? Do the fluctuations in the series seem to be of constant size? If not, describe how the volatility fluctuates.*

In R, data can be stored as a data frame, which does not assume that the data are in time order and would be appropriate, for example, with cross-sectional data. To appreciate how `plot()` works on a data frame rather than on a multivariate time series, run the following code. You will be plotting the same data as before, but they will be plotted in a different way.

```
plot(as.data.frame(logR))
```

Run the code that follows to create normal plots of the four indices and to test each for normality using the Shapiro–Wilk test. You should understand what each line of code does.

```
par(mfrow=c(2, 2))
for(i in colnames(logR))
{
  qqnorm(logR[ ,i], datax = T, main = i)
  qqline(logR[ ,i], datax = T)
  print(shapiro.test(logR[ ,i]))
}
```

Problem 3 *Briefly describe the shape of each of the four normal plots and state whether the marginal distribution of each series is skewed or symmetric and whether its tails appear normal. If the tails do not appear normal, do they appear heavier or lighter than normal? What conclusions can be made from the Shapiro–Wilk tests? Include the plots with your work.*

The next set of R code creates *t*-plots with 1, 4, 6, 10, 20, and 30 degrees of freedom and all four indices. However, for the remainder of this lab, only the DAX index will be analyzed. Notice how the reference line is created by the abline() function, which adds lines to a plot, and the lm() function, which fits a line to the quantiles. The lm() function is discussed in Chap. 9.

```
1  n=dim(logR)[1]
2  q_grid = (1:n) / (n + 1)
3  df_grid = c(1, 4, 6, 10, 20, 30)
4  index.names = dimnames(logR)[[2]]
5  for(i in 1:4)
6  {
7    # dev.new()
8    par(mfrow = c(3, 2))
9    for(df in df_grid)
10   {
11     qqplot(logR[,i], qt(q_grid,df),
12        main = paste(index.names[i], ", df = ", df) )
13     abline(lm(qt(c(0.25, 0.75), df = df) ~
14        quantile(logR[,i], c(0.25, 0.75))))
15   }
16 }
```

If you are running R from Rstudio, then line 7 should be left as it is. If you are working directly in R, then remove the "#" in this line to open a new window for each plot.

Problem 4 *What does the code* q.grid = (1:n) / (n + 1) *do? What does* qt(q.grid, df = df[j]) *do? What does* paste *do?*

Problem 5 *For the DAX index, state which choice of the degrees of freedom parameter gives the best-fitting t-distribution and explain why.*

Run the next set of code to create a kernel density estimate and two parametric density estimates, *t* with df degrees of freedom and normal, for the DAX index. Here df equals 5, but you should vary df so that the *t* density agrees as closely as possible with the kernel density estimate.

At lines 5–6, a robust estimator of the standard deviation of the *t*-distribution is calculated using the mad() function. The default value of the argument constant is 1.4826, which is calibrated to the normal distribution since $1/\Phi^{-1}(3/4) = 1.4826$. To calibrate to the *t*-distribution, the normal quantile is replaced by the corresponding *t*-quantile and multiplied by df/(df − 2) to convert from the scale parameter to the standard deviation.

```
1  library("fGarch")
2  x=seq(-0.1, 0.1,by = 0.001)
```

```
 3  par(mfrow = c(1, 1))
 4  df = 5
 5  mad_t = mad(logR[ , 1],
 6      constant = sqrt(df / (df - 2)) / qt(0.75, df))
 7  plot(density(logR[ , 1]), lwd = 2, ylim = c(0, 60))
 8  lines(x, dstd(x, mean = mean(logR[,1]), sd = mad_t, nu = df),
 9      lty = 5, lwd = 2, col = "red")
10  lines(x, dnorm(x, mean = mean(logR[ ,1]), sd = sd(logR[ ,1])),
11      lty = 3, lwd = 4, col = "blue")
12  legend("topleft", c("KDE", paste("t: df = ",df), "normal"),
13      lwd = c(2, 2, 4), lty = c(1, 5, 3),
14      col = c("black", "red", "blue"))
```

To examine the left and right tails, plot the density estimate two more times, once zooming in on the left tail and then zooming in on the right tail. You can do this by using the xlim parameter of the plot() function and changing ylim appropriately. You can also use the adjust parameter in density() to smooth the tail estimate more than is done with the default value of adjust.

Problem 6 *Do either of the parametric models provide a reasonably good fit to the first index? Explain.*

Problem 7 *Which bandwidth selector is used as the default by* density? *What is the default kernel?*

Problem 8 *For the CAC index, state which choice of the degrees of freedom parameter gives the best-fitting t-distribution and explain why.*

4.10.2 McDonald's Prices and Returns

This section analyzes daily stock prices and returns of the McDonald's Corporation (MCD) over the period Jan-4-10 to Sep-5-14. The data set is in the file MCD_PriceDail.csv. Run the following commands to load the data and plot the adjusted closing prices:

```
data = read.csv('MCD_PriceDaily.csv')
head(data)
adjPrice = data[ , 7]
plot(adjPrice, type = "l", lwd = 2)
```

Problem 9 *Does the price series appear stationary? Explain your answer.*

Problem 10 *Transform the prices into log returns and call that series* LogRet. *Create a time series plot of* LogRet *and discuss whether or not this series appears stationary.*

The following code produces a histogram of the McDonald's log returns. The histogram will have 80 evenly spaced bins, and the argument freq = FALSE specifies the density scale.

```
hist(LogRet, 80, freq = FALSE)
```

Also, make a QQ plot of LogRet.

Problem 11 *Discuss any features you see in the histogram and QQ plot, and, specifically, address the following questions: Do the log returns appear to be normally distributed? If not, in what ways do they appear non-normal? Are the log returns symmetrically distributed? If not, how are they skewed? Do the log returns seems heavy tailed compared to a normal distribution? How do the left and right tails compare; is one tail heavier than the other?*

4.11 Exercises

1. This problem uses the data set ford.csv on the book's web site. The data were taken from the ford.s data set in R's fEcofin package. This package is no longer on CRAN. This data set contains 2000 daily Ford returns from January 2, 1984, to December 31, 1991.
   (a) Find the sample mean, sample median, and standard deviation of the Ford returns.
   (b) Create a normal plot of the Ford returns. Do the returns look normally distributed? If not, how do they differ from being normally distributed?
   (c) Test for normality using the Shapiro–Wilk test? What is the p-value? Can you reject the null hypothesis of a normal distribution at 0.01?
   (d) Create several t-plots of the Ford returns using a number of choices of the degrees of freedom parameter (df). What value of df gives a plot that is as linear as possible? The returns include the return on Black Monday, October 19, 1987. Discuss whether or not to ignore that return when looking for the best choices of df.
   (e) Find the standard error of the sample median using formula (4.3) with the sample median as the estimate of $F^{-1}(0.5)$ and a KDE to estimate f. Is the standard error of the sample median larger or smaller than the standard error of the sample mean?
2. Column seven of the data set RecentFord.csv on the book's web site contains Ford daily closing prices, adjusted for splits and dividends, for the years 2009–2013. Repeat Problem 1 using these more recent returns. One of returns is approximately −0.175. For part (d), use that return in place of Black Monday. (Black Monday, of course, is not in this data set.) On what date did this return occur? Search the Internet for news about Ford that day. Why did the Ford price drop so precipitously that day?

3. This problems uses the `Garch` data set in R's `Ecdat` package.

   (a) Using a solid curve, plot a kernel density estimate of the first differences of the variable `dy`, which is the U.S. dollar/Japanese yen exchange rate. Using a dashed curve, superimpose a normal density with the same mean and standard deviation as the sample. Do the two estimated densities look similar? Describe how they differ.

   (b) Repeat part (a), but with the mean and standard deviation equal to the median and MAD. Do the two densities appear more or less similar compared to the two densities in part (a)?

4. Suppose in a normal plot that the sample quantiles are plotted on the vertical axis, rather than on the horizontal axis as in this book.

   (a) What is the interpretation of a convex pattern?

   (b) What is the interpretation of a concave pattern?

   (c) What is the interpretation of a convex-concave pattern?

   (d) What is the interpretation of a concave-convex pattern?

5. Let `diffbp` be the changes (that is, differences) in the variable `bp`, the U.S. dollar to British pound exchange rate, which is in the `Garch` data set of R's `Ecdat` package.

   (a) Create a 3×2 matrix of normal plots of `diffbp` and in each plot add a reference line that goes through the p- and $(1 - p)$-quantiles, where $p = 0.25, 0.1, 0.05, 0.025, 0.01$, and 0.0025, respectively, for the six plots. Create a second set of six normal plots using n simulated $N(0, 1)$ random variables, where n is the number of changes in `bp` plotted in the first figure. Discuss how the reference lines change with the value of p and how the set of six different reference lines can help detect nonnormality.

   (b) Create a third set of six normal plots using changes in the logarithm of `bp`. Do the changes in `log(bp)` look closer to being normally distributed than the changes in `bp`?

6. Use the following fact about the standard normal cumulative distribution function $\Phi(\cdot)$:
$$\Phi^{-1}(0.025) = -1.96.$$

   (a) What value is $\Phi^{-1}(0.975)$? Why?

   (b) What is the 0.975-quantile of the normal distribution with mean -1 and variance 2?

7. Suppose that $Y_1, \ldots, Y_n$ are i.i.d. with a uniform distribution on the interval (0,1), with density function f and distribution function F defined as

$$f(x) = \begin{cases} 1 & \text{if } x \in (0, 1), \\ 0 & \text{otherwise,} \end{cases} \quad and \quad F(x) = \begin{cases} 0 & \text{if } x \le 0, \\ x & \text{if } x \in (0, 1), \\ 1 & \text{if } x \ge 1. \end{cases}$$

Use Result 4.1 to conclude which sample quantile q will have the smallest variance?

References

Abramson, I. (1982) On bandwidth variation in kernel estimates—a square root law. *Annals of Statistics*, **9**, 168–176.

Atkinson, A. C. (1985) *Plots, transformations, and regression: An introduction to graphical methods of diagnostic regression analysis*, Clarendon Press, Oxford.

Bolance, C., Guillén, M., and Nielsen, J. P. (2003) Kernel density estimation of actuarial loss functions. *Insurance: Mathematics and Economics*, **32**, 19–36.

Carroll, R. J., and Ruppert, D. (1988) *Transformation and Weighting in Regression*, Chapman & Hall, New York.

Hoaglin, D. C., Mosteller, F., and Tukey, J. W., Eds. (1983) *Understanding Robust and Exploratory Data Analysis*, Wiley, New York.

Hoaglin, D. C., Mosteller, F., and Tukey, J. W., Eds. (1985) *Exploring Data Tables, Trends, and Shapes*, Wiley, New York.

Jones, M. C. (1990) Variable kernel density estimates and variable kernel density estimates. *Australian Journal of Statistics*, **32**, 361–371. (Note: The title is intended to be ironic and is not a misprint.)

Kleiber, C., and Zeileis, A. (2008) *Applied Econometrics with R*, Springer, New York.

Lehmann, E. L. (1999) *Elements of Large-Sample Theory*, Springer-Verlag, New York.

Scott, D. W. (1992) *Multivariate Density Estimation: Theory, Practice, and Visualization*, Wiley-Interscience, New York.

Serfling, R. J. (1980) *Approximation Theorems of Mathematical Statistics*, Wiley, New York.

Silverman, B. W. (1986) *Density Estimation for Statistics and Data Analysis*, Chapman & Hall, London.

Tukey, J. W. (1977) *Exploratory Data Analysis*, Addison-Wesley, Reading, MA.

van der Vaart, A. W. (1998) *Asymptotic Statistics*, Cambridge University Press, Cambridge.

Wand, M. P., and Jones, M. C. (1995) *Kernel Smoothing*, Chapman & Hall, London.

Wand, M. P., Marron, J. S., and Ruppert, D. (1991) Transformations in density estimation, *Journal of the American Statistical Association*, **86**, 343–366.

Yap, B. W., and Sim, C. H. (2011) Comparisons of various types of normality tests. *Journal of Statistical Computation and Simulation*, **81**, 2141–2155.

5

Modeling Univariate Distributions

5.1 Introduction

As seen in Chap. 4, usually the marginal distributions of financial time series are not well fit by normal distributions. Fortunately, there are a number of suitable alternative models, such as t-distributions, generalized error distributions, and skewed versions of t- and generalized error distributions. All of these will be introduced in this chapter. Typically, the parameters in these distributions are estimated by maximum likelihood. Sections 5.9 and 5.14 provide an introduction to the maximum likelihood estimator (MLE), and Sect. 5.18 provides references for further study on this topic.

Software for maximum likelihood is readily available for standard models, and a reader interested only in data analysis and modeling often need not be greatly concerned with the technical details of maximum likelihood. However, when performing a statistical analysis, it is always worthwhile to understand the underlying theory, at least at a conceptual level, since doing so can prevent misapplications. Moreover, when using a nonstandard model, often there is no software available for automatic computation of the MLE and one needs to understand enough theory to write a program to compute the MLE.

5.2 Parametric Models and Parsimony

In a parametric statistical model, the distribution of the data is completely specified except for a finite number of unknown parameters. For example, assume that $Y_1, \ldots, Y_n$ are i.i.d. from a t-distribution[1] with mean μ, variance

---

[1] The reader who is unfamiliar with t-distributions should look ahead to Sect. 5.5.2.

© Springer Science+Business Media New York 2015
D. Ruppert, D.S. Matteson, *Statistics and Data Analysis for Financial Engineering*, Springer Texts in Statistics,
DOI 10.1007/978-1-4939-2614-5_5

σ^2, and degrees of freedom ν. Then this is a parametric model provided that, as is usually the case, one or more of μ, σ^2, and ν are unknown.

A model should have only as many parameters as needed to capture the important features of the data. Each unknown parameter is another quantity to estimate and another source of estimation error. Estimation error, among other things, increases the uncertainty when one forecasts future observations. On the other hand, a statistical model must have enough parameters to adequately describe the behavior of the data. A model with too few parameters can create biases because the model does not fit the data well.

A statistical model with little bias, but without excess parameters, is called *parsimonious* and achieves a good tradeoff between bias and variance. Finding one or a few parsimonious models is an important part of data analysis.

5.3 Location, Scale, and Shape Parameters

Parameters are often classified as location, scale, or shape parameters depending upon which properties of a distribution they determine. A *location parameter* is a parameter that shifts a distribution to the right or left without changing the distribution's shape or variability. Scale parameters quantify dispersion. A parameter is a *scale parameter* for a univariate sample if the parameter is increased by the amount $|a|$ when the data are multiplied by a. Thus, if $\sigma(X)$ is a scale parameter for a random variable X, then $\sigma(aX) = |a|\sigma(X)$. A scale parameter is a constant multiple of the standard deviation provided that the latter is finite. Many examples of location and scale parameters can be found in the following sections. If λ is a scale parameter, then λ^{-1} is called an inverse-scale parameter. Since scale parameters quantify dispersion, inverse-scale parameters quantify precision.

If $f(y)$ is any fixed density, then $f(y - \mu)$ is a family of distributions with location parameter μ; $\theta^{-1}f(y/\theta)$, $\theta > 0$, is a family of distributions with a scale parameter θ; and $\theta^{-1}f\{\theta^{-1}(y - \mu)\}$ is a family of distributions with location parameter μ and scale parameter θ. These facts can be derived by noting that if Y has density $f(y)$ and $\theta > 0$, then, by Result A.1, $Y + \mu$ has density $f(y - \mu)$, θY has density $\theta^{-1}f(\theta^{-1}y)$, and $\theta Y + \mu$ has density $\theta^{-1}f\{\theta^{-1}(y - \mu)\}$.

A *shape* parameter is defined as any parameter that is not changed by location and scale changes. More precisely, for any $f(y)$, μ, and $\theta > 0$, the value of a shape parameter for the density $f(y)$ will equal the value of that shape parameter for $\theta^{-1}f\{\theta^{-1}(y - \mu)\}$. The degrees-of-freedom parameters of t-distributions and the log-standard deviations of lognormal distributions are shape parameters. Other shape parameters will be encountered later in this chapter. Shape parameters are often used to specify the skewness or tail weight of a distribution.

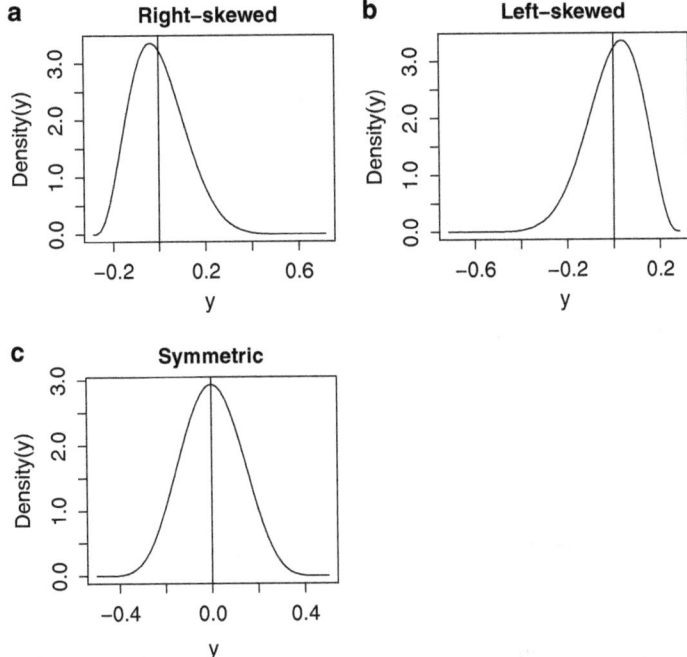

Fig. 5.1. *Skewed and symmetric densities. In each case, the mean is zero and is indicated by a vertical line. The distributions in panels (**a**)–(**c**) are beta(4,10), beta(10,4), and beta(7,7), respectively. The R function* dbeta() *was used to calculate these densities.*

5.4 Skewness, Kurtosis, and Moments

Skewness and kurtosis help characterize the shape of a probability distribution. *Skewness* measures the degree of asymmetry, with symmetry implying zero skewness, positive skewness indicating a relatively long right tail compared to the left tail, and negative skewness indicating the opposite. Figure 5.1 shows three densities, all with an expectation equal to 0. The densities are right-skewed, left-skewed, and symmetric about 0, respectively, in panels (a)–(c).

Kurtosis indicates the extent to which probability is concentrated in the center and especially the tails of the distribution rather than in the "shoulders," which are the regions between the center and the tails.

In Sect. 4.3.2, the left tail was defined as the region from $-\infty$ to $\mu - 2\sigma$ and the right tail as the region from $\mu + 2\sigma$ to $+\infty$. Here μ and σ could be the mean and standard deviation or the median and MAD. Admittedly, these definitions are somewhat arbitrary. Reasonable definitions of *center* and *shoulder* would be that the center is the region from $\mu - \sigma$ to $\mu + \sigma$, the left

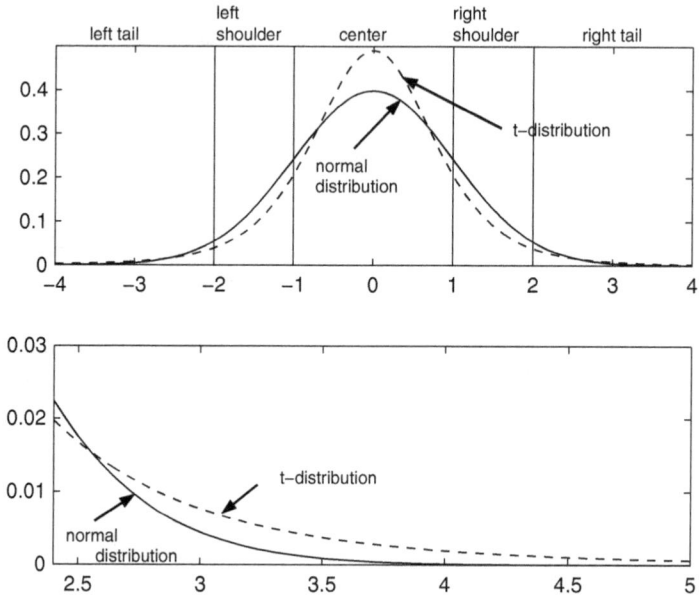

Fig. 5.2. *Comparison of a normal density and a t-density with 5 degrees of freedom. Both densities have mean 0 and standard deviation 1. The upper plot also indicates the locations of the center, shoulders, and tail regions. The lower plot zooms in on the right tail region.*

shoulder is from $\mu - 2\sigma$ to $\mu - \sigma$, and the right shoulder is from $\mu + \sigma$ to $\mu + 2\sigma$. See the upper plot in Fig. 5.2. Because skewness and kurtosis measure shape, they do not depend on the values of location and scale parameters.

The skewness of a random variable Y is

$$\mathrm{Sk} = E\left\{\left(\frac{Y - E(Y)}{\sigma}\right)^3\right\} = \frac{E\{Y - E(Y)\}^3}{\sigma^3}.$$

To appreciate the meaning of the skewness, it is helpful to look at an example; the binomial distribution is convenient for that purpose. The skewness of the Binomial(n, p) distribution is

$$\mathrm{Sk}(n, p) = \frac{1 - 2p}{\sqrt{np(1 - p)}}, \quad 0 < p < 1.$$

Figure 5.3 shows the binomial probability distribution and its skewness for $n = 10$ and four values of p. Notice that

1. the skewness is positive if $p < 0.5$, negative if $p > 0.5$, and 0 if $p = 0.5$;
2. the absolute skewness becomes larger as p moves closer to either 0 or 1 with n fixed;
3. the absolute skewness decreases to 0 as n increases to ∞ with p fixed;

Positive skewness is also called right skewness and negative skewness is called left skewness. A distribution is *symmetric* about a point θ if $P(Y > \theta + y) = P(Y < \theta - y)$ for all $y > 0$. In this case, θ is a location parameter and equals $E(Y)$, provided that $E(Y)$ exists. The skewness of any symmetric distribution is 0. Property 3 is not surprising in light of the central limit theorem. We know that the binomial distribution converges to the symmetric normal distribution as $n \to \infty$ with p fixed and not equal to 0 or 1.

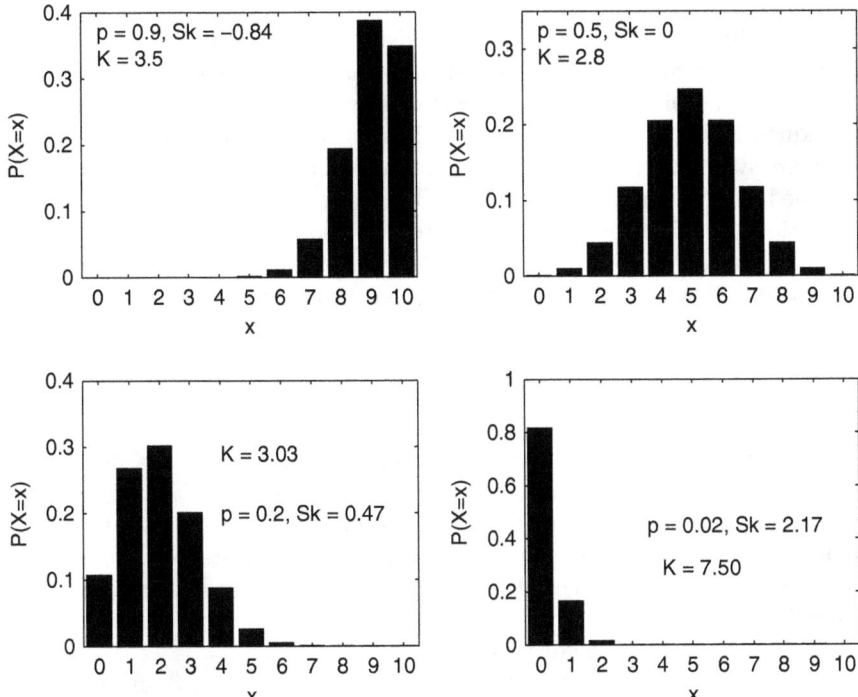

Fig. 5.3. *Several binomial probability distributions with $n = 10$ and their skewness determined by the shape parameter p. Sk = skewness coefficient and K = kurtosis coefficient. The top left plot has left-skewness (Sk = −0.84). The top right plot has no skewness (Sk = 0). The bottom left plot has moderate right-skewness (Sk = 0.47). The bottom-left plot has strong right skewness (Sk = 2.17).*

The kurtosis of a random variable Y is

$$\text{Kur} = E\left\{ \frac{Y - E(Y)}{\sigma} \right\}^4 = \frac{E\{Y - E(Y)\}^4}{\sigma^4}.$$

The kurtosis of a normal random variable is 3. The smallest possible value of the kurtosis is 1 and is achieved by any random variable taking exactly two distinct values, each with probability $1/2$. The kurtosis of a Binomial(n, p) distribution is

$$\text{Kur}^{\text{Bin}}(n, p) = 3 + \frac{1 - 6p(1 - p)}{np(1 - p)}.$$

Notice that $\text{Kur}^{\text{Bin}}(n, p) \to 3$, the value at the normal distribution, as $n \to \infty$ with p fixed, which is another sign of the central limit theorem at work. Figure 5.3 also gives the kurtosis of the distributions in that figure. $\text{Kur}^{\text{Bin}}(n, p)$ equals 1, the minimum value of kurtosis, when $n = 1$ and $p = 1/2$.

It is difficult to interpret the kurtosis of an asymmetric distribution because, for such distributions, kurtosis may measure both asymmetry and tail weight, so the binomial is not a particularly good example for understanding kurtosis. For that purpose we will look instead at t-distributions because they are symmetric. Figure 5.2 compares a normal density with the t_5-density rescaled to have variance equal to 1. Both have a mean of 0 and a standard deviation of 1. The mean and standard deviation are location and scale parameters, respectively, and do not affect kurtosis. The parameter ν of the t-distribution is a shape parameter. The kurtosis of a t_ν-distribution is finite if $\nu > 4$ and then the kurtosis is

$$\text{Kur}^{t}(\nu) = 3 + \frac{6}{\nu - 4}. \tag{5.1}$$

For example, the kurtosis is 9 for a t_5-distribution. Since the densities in Fig. 5.2 have the same mean and standard deviation, they also have the same tail, center, and shoulder regions, at least according to our somewhat arbitrary definitions of these regions, and these regions are indicated on the top plot. The bottom plot zooms in on the right tail. Notice that the t_5-density has more probability in the tails and center than the $N(0, 1)$ density. This behavior of t_5 is typical of symmetric distributions with high kurtosis.

Every normal distribution has a skewness coefficient of 0 and a kurtosis of 3. The skewness and kurtosis must be the same for all normal distributions, because the normal distribution has only location and scale parameters, no shape parameters. The kurtosis of 3 agrees with formula (5.1) since a normal distribution is a t-distribution with $\nu = \infty$. The "excess kurtosis" of a distribution is $(\text{Kur} - 3)$ and measures the deviation of that distribution's kurtosis from the kurtosis of a normal distribution. From (5.1) we see that the excess kurtosis of a t_ν-distribution is $6/(\nu - 4)$.

An exponential distribution[2] has a skewness equal to 2 and a kurtosis of 9. A double-exponential distribution has skewness 0 and kurtosis 6. Since the exponential distribution has only a scale parameter and the double-exponential has only a location and a scale parameter, their skewness and kurtosis must be constant since skewness and kurtosis depend only on shape parameters.

---

[2] The exponential and double-exponential distributions are defined in Appendix A.9.5.

The Lognormal(μ, σ^2) distribution, which is discussed in Appendix A.9.4, has the log-mean μ as a scale parameter and the log-standard deviation σ as a shape parameter—even though μ and σ are location and scale parameters for the normal distribution itself, they are scale and shape parameters for the lognormal. The effects of σ on lognormal shapes can be seen in Figs. 4.11 and A.1. The skewness coefficient of the lognormal(μ, σ^2) distribution is

$$\{\exp(\sigma^2) + 2\}\sqrt{\exp(\sigma^2) - 1}. \tag{5.2}$$

Since μ is a scale parameter, it has no effect on the skewness. The skewness is always positive and increases from 0 to ∞ as σ increases from 0 to ∞.

Estimation of the skewness and kurtosis of a distribution is relatively straightforward if we have a sample, $Y_1, \ldots, Y_n$, from that distribution. Let the sample mean and standard deviation be $\overline{Y}$ and s. Then the sample skewness, denoted by $\widehat{\text{Sk}}$, is

$$\widehat{\text{Sk}} = \frac{1}{n} \sum_{i=1}^{n} \left(\frac{Y_i - \overline{Y}}{s} \right)^3, \tag{5.3}$$

and the sample kurtosis, denoted by $\widehat{\text{Kur}}$, is

$$\widehat{\text{Kur}} = \frac{1}{n} \sum_{i=1}^{n} \left(\frac{Y_i - \overline{Y}}{s} \right)^4. \tag{5.4}$$

Often the factor $1/n$ in (5.3) and (5.4) is replaced by $1/(n-1)$, in analogy with the sample variance. Both the sample skewness and the excess kurtosis should be near 0 if a sample is from a normal distribution. Deviations of the sample skewness and kurtosis from these values are an indication of nonnormality.

A word of caution is in order. Skewness and kurtosis are highly sensitive to outliers. Sometimes outliers are due to *contaminants*, that is, bad data not from the population being sampled. An example would be a data recording error. A sample from a normal distribution with even a single contaminant that is sufficiently outlying will appear highly nonnormal according to the sample skewness and kurtosis. In such a case, a normal plot *will* look linear, except that the single contaminant will stick out. See Fig. 5.4, which is a normal plot of a sample of 999 $N(0,1)$ data points plus a contaminant equal to 30. This figure shows clearly that the sample is nearly normal but with an outlier. The sample skewness and kurtosis, however, are 10.85 and 243.04, which might give the false impression that the sample is far from normal. Also, even if there were no contaminants, a distribution could be extremely close to a normal distribution except in the extreme tails and yet have a skewness or excess kurtosis that is very different from 0.

5.4.1 The Jarque–Bera Test

The Jarque–Bera test of normality compares the sample skewness and kurtosis to 0 and 3, their values under normality. The test statistic is

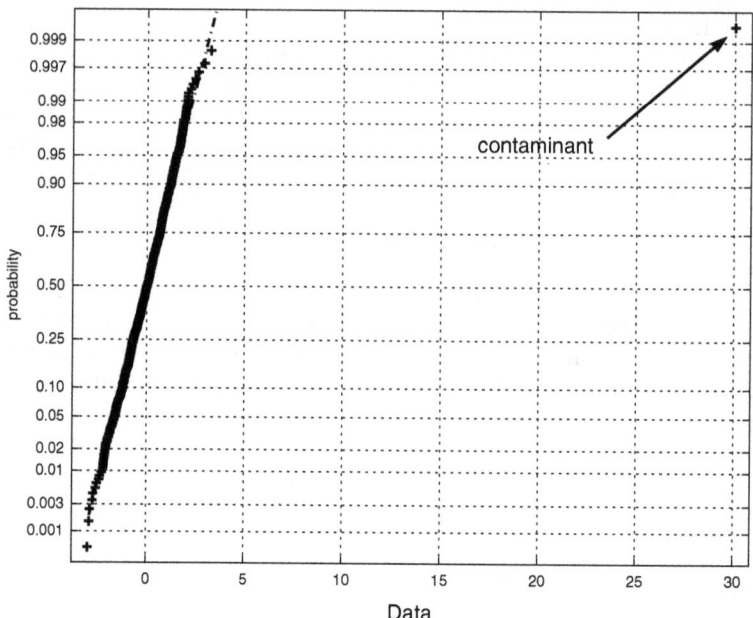

Fig. 5.4. *Normal plot of a sample of 999 $N(0,1)$ data plus a contaminant.*

$$\mathrm{JB} = n\{\widehat{\mathrm{Sk}}^2/6 + (\widehat{\mathrm{Kur}} - 3)^2/24\},$$

which, of course, is 0 when $\widehat{\mathrm{Sk}}$ and $\widehat{\mathrm{Kur}}$, respectively, have the values 0 and 3, the values expected under normality, and increases as $\widehat{\mathrm{Sk}}$ and $\widehat{\mathrm{Kur}}$ deviate from these values. In R, the test statistic and its p-value can be computed with the `jarque.bera.test()` function.

A large-sample approximation is used to compute a p-value. Under the null hypothesis, JB converges to the chi-square distribution with 2 degrees of freedom (χ_2^2) as the sample size becomes infinite, so the approximate p-value is $1 - F_{\chi_2^2}(\mathrm{JB})$, where $F_{\chi_2^2}$ is the CDF of the χ_2^2-distribution.

5.4.2 Moments

The expectation, variance, skewness coefficient, and kurtosis of a random variable are all special cases of moments, which will be defined in this section.

Let X be a random variable. The kth moment of X is $E(X^k)$, so in particular the first moment is the expectation of X. The kth absolute moment is $E|X|^k$.

The kth central moment is

$$\mu_k = E\left[\{X - E(X)\}^k\right], \tag{5.5}$$

so, for example, μ_2 is the variance of X. The skewness coefficient of X is

$$\text{Sk}(X) = \frac{\mu_3}{(\mu_2)^{3/2}}, \tag{5.6}$$

and the kurtosis of X is

$$\text{Kur}(X) = \frac{\mu_4}{(\mu_2)^2}. \tag{5.7}$$

5.5 Heavy-Tailed Distributions

As discussed in earlier chapters, distributions with higher tail probabilities compared to a normal distribution are called *heavy-tailed*. Because kurtosis is particularly sensitive to tail weight, high kurtosis is nearly synonymous with having a heavy tailed distribution. Heavy-tailed distributions are important models in finance, because equity returns and other changes in market prices usually have heavy tails. In finance applications, one is especially concerned when the return distribution has heavy tails because of the possibility of an extremely large negative return, which could, for example, entirely deplete the capital reserves of a firm. If one sells short,[3] then large positive returns are also worrisome.

5.5.1 Exponential and Polynomial Tails

Double-exponential distributions have slightly heavier tails than normal distributions. This fact can be appreciated by comparing their densities. The density of the double-exponential with scale parameter θ is proportional to $\exp(-|y/\theta|)$ and the density of the $N(0, \sigma^2)$ distribution is proportional to $\exp\{-0.5(y/\sigma)^2\}$. The term $-y^2$ converges to $-\infty$ much faster than $-|y|$ as $|y| \to \infty$. Therefore, the normal density converges to 0 much faster than the double-exponential density as $|y| \to \infty$. The generalized error distributions discussed soon in Sect. 5.6 have densities proportional to

$$\exp\left(-|y/\theta|^\alpha\right), \tag{5.8}$$

where $\alpha > 0$ is a shape parameter and θ is a scale parameter. The special cases of $\alpha = 1$ and 2 are, of course, the double-exponential and normal densities. If $\alpha < 2$, then a generalized error distribution will have heavier tails than a normal distribution, with smaller values of α implying heavier tails. In particular, $\alpha < 1$ implies a tail heavier than that of a double-exponential distribution.

However, no density of the form (5.8) will have truly heavy tails, and, in particular, $E(|Y|^k) < \infty$ for all k, so all moments are finite. To achieve a very heavy right tail, the density must be such that

$$f(y) \sim Ay^{-(a+1)} \text{ as } y \to \infty \tag{5.9}$$

---

[3] See Sect. 16.5 for a discussion of short selling.

for some $A > 0$ and $a > 0$, which will be called a *right polynomial tail*, in contrast to

$$f(y) \sim A \exp(-y/\theta) \text{ as } y \to \infty \qquad (5.10)$$

for some $A > 0$ and $\theta > 0$, which will be called an *exponential right tail*. Polynomial and exponential left tails are defined analogously.

A polynomial tail is also called a *Pareto tail* after the Pareto distribution defined in Appendix A.9.8. The parameter a of a polynomial tail is called the *tail index*. The smaller the value of a, the heavier the tail. The value of a must be greater than 0, because if $a \le 0$, then the density integrates to ∞, not 1. An exponential tail as in (5.8) is lighter than any polynomial tail, since

$$\frac{\exp(-|y/\theta|^{\alpha})}{|y|^{-(a+1)}} \to 0 \text{ as } |y| \to \infty$$

for all $\theta > 0$, $\alpha > 0$, and $a > 0$.

It is, of course, possible to have left and right tails that behave quite differently from each other. For example, one could be polynomial and the other exponential, or they could both be polynomial but with different indices.

A density with both tails polynomial will have a finite kth absolute moment only if the smaller of the two tail indices is larger than k. If both tails are exponential, then all moments are finite.

5.5.2 t-Distributions

The t-distributions have played an extremely important role in classical statistics because of their use in testing and confidence intervals when the data are modeled as having normal distributions. More recently, t-distributions have gained added importance as models for the distribution of heavy-tailed phenomena such as financial markets data.

We will start with some definitions. If Z is $N(0,1)$, W is chi-squared[4] with ν degrees of freedom, and Z and W are independent, then the distribution of

$$Z/\sqrt{W/\nu} \qquad (5.11)$$

is called the *t-distribution* with ν *degrees of freedom* and denoted t_{ν}. The α-upper quantile of the t_{ν}-distribution is denoted by $t_{\alpha,\nu}$ and is used in tests and confidence intervals about population means, regression coefficients, and parameters in time series models.[5] In testing and interval estimation, the parameter ν generally assumes only positive integer values, but when the t-distribution is used as a model for data, ν is restricted only to be positive.

The density of the t_{ν}-distribution is

$$f_{t,\nu}(y) = \left[\frac{\Gamma\{(\nu+1)/2\}}{(\pi\nu)^{1/2}\Gamma(\nu/2)}\right] \frac{1}{\{1+(y^2/\nu)\}^{(\nu+1)/2}}. \qquad (5.12)$$

---

[4] Chi-squared distributions are discussed in Appendix A.10.1.
[5] See Appendix A.17.1 for confidence intervals for the mean.

Here Γ is the *gamma function* defined by

$$\Gamma(t) = \int_0^\infty x^{t-1} \exp(-x)dx, \quad t > 0. \tag{5.13}$$

The quantity in large square brackets in (5.12) is just a constant, though a somewhat complicated one.

The variance of a t_ν is finite and equals $\nu/(\nu - 2)$ if $\nu > 2$. If $0 < \nu \le 1$, then the expected value of the t_ν-distribution does not exist and the variance is not defined.[6] If $1 < \nu \le 2$, then the expected value is 0 and the variance is infinite. If Y has a t_ν-distribution, then

$$\mu + \lambda Y$$

is said to have a $t_\nu(\mu, \lambda^2)$ distribution, and λ will be called *the scale parameter*. With this notation, the t_ν and $t_\nu(0, 1)$ distributions are the same. If $\nu > 1$, then the $t_\nu(\mu, \lambda^2)$ distribution has a mean equal to μ, and if $\nu > 2$, then it has a variance equal to $\lambda^2 \nu/(\nu - 2)$.

The t-distribution will also be called the *classical t-distribution* to distinguish it from the standardized t-distribution defined in the next section.

Standardized t-Distributions

Instead of the classical t-distribution just discussed, some software uses a "standardized" version of the t-distribution. The difference between the two versions is merely notational, but it is important to be aware of this difference.

The $t_\nu\{0, (\nu - 2)/\nu\}$ distribution with $\nu > 2$ has a mean equal to 0 and variance equal to 1 and is called a *standardized t-distribution*, and will be denoted by $t_\nu^{\mathrm{std}}(0, 1)$. More generally, for $\nu > 2$, define the $t_\nu^{\mathrm{std}}(\mu, \sigma^2)$ distribution to be equal to the $t_\nu[\mu, \{(\nu - 2)/\nu\}\sigma^2]$ distribution, so that μ and σ^2 are the mean and variance of the $t_\nu^{\mathrm{std}}(\mu, \sigma^2)$ distribution. For $\nu \le 2$, $t_\nu^{\mathrm{std}}(\mu, \sigma^2)$ cannot be defined since the t-distribution does not have a finite variance in this case. The advantage in using the $t_\nu^{\mathrm{std}}(\mu, \sigma^2)$ distribution is that σ^2 is the variance, whereas for the $t_\nu(\mu, \lambda^2)$ distribution, λ^2 is not the variance but instead λ^2 is the variance times $(\nu - 2)/\nu$.

Some software uses the standardized t-distribution while other software uses the classical t-distribution. It is, of course, important to understand which t-distribution is being used in any specific application. However, estimates from one model can be translated easily into the estimates one would obtain from the other model; see Sect. 5.14 for an example.

---

[6] See Appendix A.3 for discussion of when the mean and the variance exist and when they are finite.

t-Distributions Have Polynomial Tails

The t-distributions are a class of heavy-tailed distributions and can be used to model heavy-tail returns data. For t-distributions, both the kurtosis and the weight of the tails increase as ν gets smaller. When $\nu \leq 4$, the tail weight is so high that the kurtosis is infinite. For $\nu > 4$, the kurtosis is given by (5.1). By (5.12), the t-distribution's density is proportional to

$$\frac{1}{\{1 + (y^2/\nu)\}^{(\nu+1)/2}}$$

which for large values of $|y|$ is approximately

$$\frac{1}{(y^2/\nu)^{(\nu+1)/2}} \propto |y|^{-(\nu+1)}.$$

Therefore, the t-distribution has polynomial tails with tail index $a = \nu$. The smaller the value of ν, the heavier the tails.

5.5.3 Mixture Models

Discrete Mixtures

Another class of models containing heavy-tailed distributions is the set of *mixture models*. Consider a distribution that is 90 % $N(0,1)$ and 10 % $N(0,25)$. A random variable Y with this distribution can be obtained by generating a normal random variable X with mean 0 and variance 1 and a uniform(0,1) random variable U that is independent of X. If $U < 0.9$, then $Y = X$. If $U \geq 0.9$, then $Y = 5X$. If an independent sample from this distribution is generated, then the expected percentage of observations from the $N(0,1)$ component is 90 %. The actual percentage is random; in fact, it has a Binomial(n, 0.9) distribution, where n is a sample size. By the law of large numbers, the actual percentage converges to 90 % as $n \to \infty$. This distribution could be used to model a market that has two *regimes*, the first being "normal volatility" and second "high volatility," with the first regime occurring 90 % of the time.

This is an example of a *finite* or *discrete normal mixture distribution*, since it is a mixture of a finite number, here two, different normal distributions called the *components*. A random variable with this distribution has a variance equal to 1 with 90 % probability and equal to 25 with 10 % probability. Therefore, the variance of this distribution is $(0.9)(1) + (0.1)(25) = 3.4$, so its standard deviation is $\sqrt{3.4} = 1.84$. This distribution is much different from an $N(0, 3.4)$ distribution, even though the two distributions have the same mean and variance. To appreciate this, look at Fig. 5.5.

You can see in Fig. 5.5a that the two densities look quite different. The normal density looks much more dispersed than the normal mixture, but they actually have the same variances. What is happening? Look at the detail of

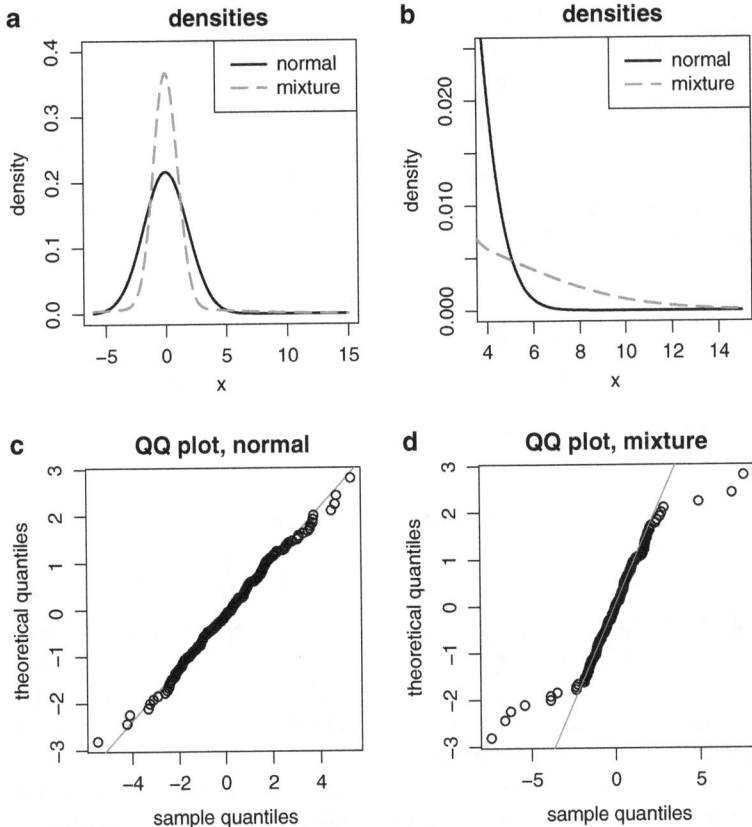

Fig. 5.5. *Comparison of $N(0, 3.4)$ distribution and heavy-tailed normal mixture distributions. These distributions have the same mean and variance. The normal mixture distribution is 90 % $N(0, 1)$ and 10 % $N(0, 25)$. In (**c**) and (**d**) the sample size is 200. In panel (**a**), the left tail is not shown fully to provide detail at the center and because the left tail is the mirror image of the right tail. (**b**) Detail of right tail.*

the right tails in panel (b). The normal mixture density is much higher than the normal density when x is greater than 6. This is the "outlier" region (along with $x < -6$).[7] The normal mixture has far more outliers than the normal distribution and the outliers come from the 10 % of the population with a variance of 25. Remember that ± 6 is only $6/5$ standard deviations from the mean, using the standard deviation 5 of the component from which they come. Thus, these observations are not outlying relative to their component's standard deviation of 5, only relative to the population standard deviation of

---

[7] There is nothing special about "6" to define the boundary of the outlier range, but a specific number was needed to make numerical comparisons. Clearly, $|x| > 7$ or $|x| > 8$, say, would have been just as appropriate as outlier ranges.

$\sqrt{3.4} = 1.84$ since $6/1.84 = 3.25$ and three or more standard deviations from the mean is generally considered rather outlying.

Outliers have a powerful effect on the variance and this small fraction of outliers inflates the variance from 1.0 (the variance of 90 % of the population) to 3.4.

Let's see how much more probability the normal mixture distribution has in the outlier range $|x| > 6$ compared to the normal distribution. For an $N(0, \sigma^2)$ random variable Y,

$$P\{|Y| > y\} = 2\{1 - \Phi(y/\sigma)\}.$$

Therefore, for the normal distribution with variance 3.4,

$$P\{|Y| > 6\} = 2\{1 - \Phi(6/\sqrt{3.4})\} = 0.0011.$$

For the normal mixture population that has variance 1 with probability 0.9 and variance 25 with probability 0.1, we have that

$$P\{|Y| > 6\} = 2\left[0.9\{1 - \Phi(6)\} + 0.1\{1 - \Phi(6/5)\}\right]$$
$$= 2\{(0.9)(0) + (0.1)(0.115)\} = 0.023.$$

Since $0.023/0.0011 \approx 21$, the normal mixture distribution is 21 times more likely to be in this outlier range than the $N(0, 3.4)$ population, even though both have a variance of 3.4. In summary, the normal mixture is much more prone to outliers than a normal distribution with the same mean and standard deviation. So, we should be much more concerned about very large negative returns if the return distribution is more like the normal mixture distribution than like a normal distribution. Large positive returns are also likely under a normal mixture distribution and would be of concern if an asset was sold short.

It is not difficult to compute the kurtosis of this normal mixture. Because a normal distribution has kurtosis equal to 3, if Z is $N(\mu, \sigma^2)$, then $E(Z - \mu)^4 = 3\sigma^4$. Therefore, if Y has this normal mixture distribution, then

$$E(Y^4) = 3\{0.9 + (0.1)25^2\} = 190.2$$

and the kurtosis of X is $190.2/3.4^2 = 16.45$.

Normal probability plots of samples of size 200 from the normal and normal mixture distributions are shown in panels (c) and (d) of Fig. 5.5. Notice how the outliers in the normal mixture sample give the probability plot a convex-concave pattern typical of heavy-tailed data. The deviation of the plot of the normal sample from linearity is small and is due entirely to randomness.

In this example, the conditional variance of any observation is 1 with probability 0.9 and 25 with probability 0.1. Because there are only two components, the conditional variance is discrete, in fact, with only two possible values, and the example was easy to analyze. This example is a normal *scale mixture* because only the scale parameter σ varies between components. It is also a *discrete mixture* because there are only a finite number of components.

Continuous Mixtures

The marginal distributions of the GARCH processes studied in Chap. 14 are also normal scale mixtures, but with infinitely many components and a continuous distribution of the conditional variance. Although GARCH processes are more complex than the simple mixture model in this section, the same theme applies—a nonconstant conditional variance of a mixture distribution induces heavy-tailed marginal distributions even though the conditional distributions are normal distributions and have relatively light tails.

   The general definition of a normal scale mixture is that it is the distribution of the random variable

$$\mu + \sqrt{U} Z \tag{5.14}$$

where μ is a constant equal to the mean, Z is $N(0,1)$, U is a positive random variable giving the variance of each component, and Z and U are independent. If U can assume only a finite number of values, then (5.14) is a *discrete* (or finite) scale mixture distribution. If U is continuously distributed, then we have a *continuous scale mixture distribution*. The distribution of U is called the *mixing distribution*. By (5.11), a t_ν-distribution is a continuous normal scale mixture with $\mu = 0$ and $U = \nu/W$, where ν and W are as defined above Eq. (5.11).

   Despite the apparent heavy tails of a *finite* normal mixture, the tails are exponential, not polynomial. A continuous normal mixture can have a polynomial tail if the mixture distribution's tail is heavy enough, e.g., as in t-distributions.

5.6 Generalized Error Distributions

Generalized error distributions mentioned briefly in Sect. 5.5.1 have exponential tails. This section provides more detailed information about them. The standardized generalized error distribution, or GED, with shape parameter ν has density

$$f_{\text{ged}}^{\text{std}}(y|\nu) = \kappa(\nu) \exp\left\{-\frac{1}{2}\left|\frac{y}{\lambda_\nu}\right|^\nu\right\}, \quad -\infty < y < \infty,$$

where $\kappa(\nu)$ and λ_ν are constants given by

$$\lambda_\nu = \left\{\frac{2^{-2/\nu}\Gamma(\nu^{-1})}{\Gamma(3/\nu)}\right\}^{1/2} \quad \text{and} \quad \kappa(\nu) = \frac{\nu}{\lambda_\nu 2^{1+1/\nu}\Gamma(\nu^{-1})}$$

and chosen so that the function integrates to 1, as it must to be a density, and the variance is 1. The latter property is not necessary but is often convenient.

   The shape parameter $\nu > 0$ determines the tail weight, with smaller values of ν giving greater tail weight. When $\nu = 2$, a GED is a normal distribution, and when $\nu = 1$, it is a double-exponential distribution. The generalized error distributions can give tail weights intermediate between the normal and double-exponential distributions by having $1 < \nu < 2$. They can also give tail weights more extreme than the double-exponential distribution by having $\nu < 1$.

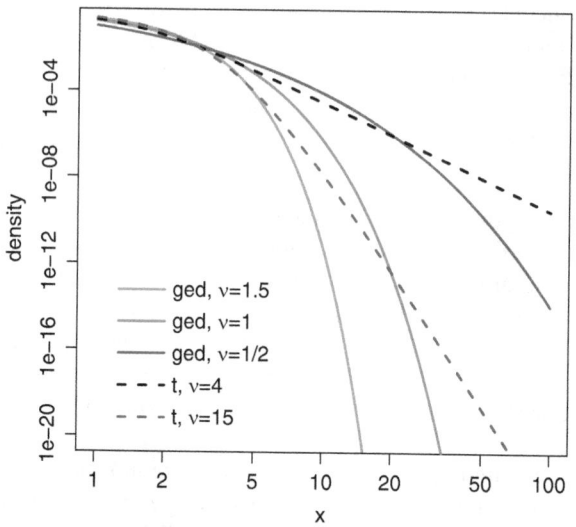

Fig. 5.6. *A comparison of the tails of several generalized error (solid curves) and t-distributions (dashed curves).*

Figure 5.6 shows the right tails of several t- and generalized error densities with mean 0 and variance 1.[8] Since they are standardized, the argument y is the number of standard deviations from the median of 0. Because t-distributions have polynomial tails, any t-distribution is heavier-tailed than any generalized error distribution. However, this is only an asymptotic result as $y \to \infty$. In the more practical range of y, tail weight depends as much on the tail weight parameter as it does on the choice between a t-distribution or a generalized error distribution.

The t-distributions and generalized error densities also differ in their shapes at the median. This can be seen in Fig. 5.7, where the generalized error densities have sharp peaks at the median with the sharpness increasing as ν decreases. In comparison, a t-density is smooth and rounded near the median, even with ν small. If a sample is better fit by a t-distribution than by a generalized error distribution, this may be due more to the sharp central peaks of generalized error densities than to differences between the tails of the two types of distributions.

---

[8] This plot and Fig. 5.7 used the R functions `dged()` and `dstd()` in the `fGarch` package.

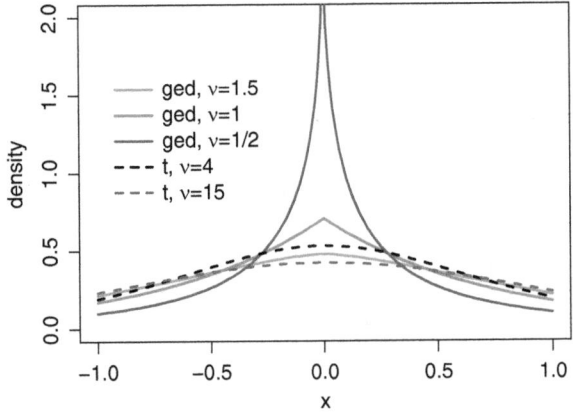

Fig. 5.7. *A comparison of the centers of several generalized error (solid) and t-densities (dashed) with mean 0 and variance 1.*

The $f_{\text{ged}}^{\text{std}}(y|\nu)$ density is symmetric about 0, which is its mean, median, and mode, and has a variance equal to 1. However, it can be shifted and rescaled to create a location-scale family. The GED distribution with mean μ, variance σ^2, and shape parameter ν has density

$$f_{\text{ged}}^{\text{std}}(y|\mu, \sigma^2, \nu) := f_{\text{ged}}^{\text{std}}\{(y - \mu)/\sigma|\nu\}/\sigma.$$

5.7 Creating Skewed from Symmetric Distributions

Returns and other financial markets data typically have no natural lower or upper bounds, so one would like to use models with support equal to $(-\infty, \infty)$. This is fine if the data are symmetric since then one can use, for example, normal, t, or generalized error distributions as models. What if the data are skewed? Unfortunately, many of the well-known skewed distributions, such as gamma and log-normal distributions, have support $[0, \infty)$ and so are not suitable for modeling the changes in many types of financial markets data. This section describes a remedy to this problem.

Fernandez and Steel (1998) have devised a clever way for inducing skewness in symmetric distributions such as normal and t-distributions. The fGarch package in R implements their idea. Let ξ be a positive constant and f a density that is symmetric about 0. Define

$$f^*(y|\xi) = \begin{cases} f(y\xi) & \text{if } y < 0, \\ f(y/\xi) & \text{if } y \geq 0. \end{cases} \tag{5.15}$$

Since $f^*(y|\xi)$ integrates to $(\xi^{-1} + \xi)/2$, $f^*(y|\xi)$ is divided by this constant to create a probability density. After this normalization, the density is given a

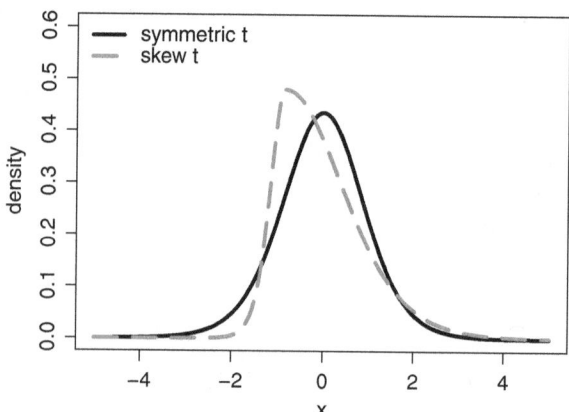

Fig. 5.8. *Symmetric (solid) and skewed (dashed) t-densities, both with mean 0, standard deviation 1, and $\nu = 10$. $\xi = 2$ in the skewed density. Notice that the mode of the skewed density lies to the left of its mean, a typical behavior of right-skewed densities.*

location shift and scale change to induce a mean equal to 0 and variance of 1. The final result is denoted by $f(y|\xi)$.

If $\xi > 1$, then the right half of $f(y|\xi)$ is elongated relative to the left half, which induces right skewness. Similarly, $\xi < 1$ induces left skewness. Figure 5.8 shows standardized symmetric and skewed t-distributions[9] with $\nu = 10$ in both cases and $\xi = 2$ for the skewed distribution. Similarly, if $\xi < 1$, then $f(y|\xi)$ is left skewed.

If f is a t-distribution, then $f(y|\xi)$ is called a skewed t-distribution. Skewed t-distributions include symmetric t-distributions as special cases where $\xi = 1$. In the same way, skewed generalized error distributions are created when f is a generalized error distribution. The skewed distributions just described will be called Fernandez–Steel or F-S skewed distributions.

Fernandez and Steel's technique is not the only method for creating skewed versions of the normal and t-distributions. Azzalini and Capitanio (2003) have created somewhat different skewed normal and t-distributions.[10] These distributions have a shape parameter α that determines the skewness; the distribution is left-skewed, symmetric, or right-skewed according to whether α is negative, zero, or positive. An example is given in Sect. 5.14 and multivariate versions are discussed in Sect. 7.9. We will refer to these as Azzalini–Capitanio or A-C skewed distributions.

---

[9] R's `dstd()` (for symmetric t) and `dsstd()` (for skewed t) functions in the `fGarch` package were used for to create this plot.

[10] Programs for fitting these distributions, computing their densities, quantile, and distribution functions, and generating random samples are available in R's `sn` package.

The A-C skewed normal density is $g(y|\xi, \omega, \alpha) = (2/\omega)\phi(z)\Phi(\alpha z)$ where $z = (y - \xi)/\omega$ and $\phi()$ and $\Phi()$ are the $N(0, 1)$ density and CDF, respectively. The parameters ξ, ω, and α determine location, scale, and skewness and are called the direct parameters or DP. The parameters ξ and ω are the mean and standard deviation of $\phi(z)$ and α determines the amount of skewness induced by $\Phi(\alpha z)$. The skewness of $g(y|\xi, \omega, \alpha)$ is positive if $\alpha > 0$ and negative if $\alpha < 0$.

The direct parameters do not have simple interpretations for the skew normal density $g(y|\xi, \omega, \alpha)$. Therefore, the so-called centered parameters (CP) are defined to be the mean, standard deviation, and skewness of $g(y|\xi, \omega, \alpha)$.

The A-C skew-t distribution has four parameters. The four DP are the mean, scale, and degrees of freedom of a t-density and α which measures the amount of skewness induced into that density. The CP are the mean, standard deviation, skewness, and kurtosis of the skew t.

5.8 Quantile-Based Location, Scale, and Shape Parameters

As has been seen, the mean, standard deviation, skewness coefficient, and kurtosis are moments-based location, scale, and shape parameters. Although they are widely used, they have the drawbacks that they are sensitive to outliers and may be undefined or infinite for distributions with heavy tails. An alternative is to use parameters based on quantiles.

Any quantile $F^{-1}(p)$, $0 < p < 1$, is a location parameter. A positive weighted average of quantiles, that is, $\sum_{\ell=1}^{L} w_\ell F^{-1}(p_\ell)$, where $w_\ell > 0$ for all ℓ and $\sum_{\ell=1}^{L} w_\ell = 1$, is also a location parameter. A simple example is $\{F^{-1}(1 - p) + F^{-1}(p)\}/2$ where $0 < p < 1/2$, which equals the mean and median if F is symmetric.

A scale parameter can be obtained from the difference between two quantiles:

$$s(p_1, p_2) = \frac{F^{-1}(p_2) - F^{-1}(p_1)}{a}$$

where $0 < p_1 < p_2 < 1$ and a is a positive constant. An obvious choice is $p_1 < 1/2$ and $p_2 = 1 - p_1$. If $a = \Phi^{-1}(p_2) - \Phi^{-1}(p_1)$, then $s(p_1, p_2)$ is equal to the standard deviation when F is a normal distribution. If $a = 1$, then $s(1/4, 3/4)$ is called the *interquartile range* or IQR.

A quantile-based shape parameter that quantifies skewness is a ratio with the numerator the difference between two scale parameters and the denominator a scale parameter:

$$\frac{s(1/2, p_2) - s(1/2, p_1)}{s(p_3, p_4)}. \tag{5.16}$$

where $p_1 < 1/2$, $p_2 > 1/2$, and $0 < p_3 < p_4 < 1$. For example, one could use $p_2 = 1 - p_1$, $p_4 = p_2$, and $p_3 = p_1$.

A quantile-based shape parameter that quantifies tail weight is the ratio of two scale parameters:

$$\frac{s(p_1, 1 - p_1)}{s(p_2, 1 - p_2)}, \tag{5.17}$$

where $0 < p_1 < p_2 < 1/2$. For example, one might have $p_1 = 0.01$ or 0.05 and $p_2 = 0.25$.

5.9 Maximum Likelihood Estimation

Maximum likelihood is the most important and widespread method of estimation. Many well-known estimators such as the sample mean, and the least-squares estimator in regression are maximum likelihood estimators if the data have a normal distribution. Maximum likelihood estimation generally provides more efficient (less variable) estimators than other techniques of estimation. As an example, for a t-distribution, the maximum likelihood estimator of the mean is more efficient than the sample mean.

Let $\boldsymbol{Y} = (Y_1, \ldots, Y_n)^\mathsf{T}$ be a vector of data and let $\boldsymbol{\theta} = (\theta_1, \ldots, \theta_p)^\mathsf{T}$ be a vector of parameters. Let $f(\boldsymbol{Y}|\boldsymbol{\theta})$ be the density of $\boldsymbol{Y}$, which depends on the parameters.

The function $L(\boldsymbol{\theta}) = f(\boldsymbol{Y}|\boldsymbol{\theta})$ viewed as a function of $\boldsymbol{\theta}$ with $\boldsymbol{Y}$ fixed at the observed data is called the *likelihood function*. It tells us the likelihood of the sample that was actually observed. The *maximum likelihood estimator* (MLE) is the value of $\boldsymbol{\theta}$ that maximizes the likelihood function. In other words, the MLE is the value of $\boldsymbol{\theta}$ at which the likelihood of the observed data is largest. We denote the MLE by $\widehat{\boldsymbol{\theta}}_{\mathrm{ML}}$. Often it is mathematically easier to maximize $\log\{L(\boldsymbol{\theta})\}$, which is called the log-likelihood. If the data are independent, then the likelihood is the product of the marginal densities and products are cumbersome to differentiate. Taking the logarithm converts the product into an easily differentiated sum. Also, in numerical computations, using the log-likelihood reduces the possibility of underflow or overflow. Since the log function is increasing, maximizing $\log\{L(\boldsymbol{\theta})\}$ is equivalent to maximizing $L(\boldsymbol{\theta})$.

In examples found in introductory statistics textbooks, it is possible to find an explicit formula for the MLE. With more complex models such as the ones we will mostly be using, there is no explicit formula for the MLE. Instead, one must write a program that computes $\log\{L(\boldsymbol{\theta})\}$ for any $\boldsymbol{\theta}$ and then use optimization software to maximize this function numerically; see Example 5.3. The R functions optim() and nlminb() minimize functions and can be applied to $-L(\theta)$.

For many important models, such as the examples in the Sect. 5.14 and the ARIMA and GARCH time series models discussed in Chap. 12, R and other software packages contain functions to find the MLE for these models.

5.10 Fisher Information and the Central Limit Theorem for the MLE

Standard errors are essential for measuring the accuracy of estimators. We have formulas for the standard errors of simple estimators such as $\overline{Y}$, but what about standard errors for other estimators? Fortunately, there is a simple method for calculating the standard error of a maximum likelihood estimator.

We assume for now that θ is one-dimensional. The *Fisher information* is defined to be minus the expected second derivative of the log-likelihood, so if $\mathcal{I}(\theta)$ denotes the Fisher information, then

$$\mathcal{I}(\theta) = -E\left[\frac{d^2}{d\theta^2}\log\{L(\theta)\}\right]. \tag{5.18}$$

The standard error of $\widehat{\theta}$ is simply the inverse square root of the Fisher information, with the unknown θ replaced by $\widehat{\theta}$:

$$s_{\widehat{\theta}} = \frac{1}{\sqrt{\mathcal{I}(\widehat{\theta})}}. \tag{5.19}$$

Example 5.1. Fisher information for a normal model mean

Suppose that $Y_1,\ldots,Y_n$ are i.i.d. $N(\mu,\sigma^2)$ with σ^2 known. The log-likelihood for the unknown parameter μ is

$$\log\{L(\mu)\} = -\frac{n}{2}\{\log(\sigma^2) + \log(2\pi)\} - \frac{1}{2\sigma^2}\sum_{i=1}^{n}(Y_i - \mu)^2.$$

Therefore,

$$\frac{d}{d\mu}\log\{L(\mu)\} = \frac{1}{\sigma^2}\sum_{i=1}^{n}(Y_i - \mu),$$

so that $\overline{Y}$ is the MLE of μ and

$$\frac{d^2}{d\mu^2}\log\{L(\mu)\} = -\frac{\sum_{i=1}^{n}1}{\sigma^2} = -\frac{n}{\sigma^2}.$$

It follows that $\mathcal{I}(\widehat{\mu}) = n/\sigma^2$ and $s_{\widehat{\mu}} = \sigma/\sqrt{n}$. Since the MLE of μ is $\overline{Y}$, this result is the familiar fact that when σ is known, then $s_{\overline{Y}} = \sigma/\sqrt{n}$ and when σ is unknown, then $s_{\overline{Y}} = s/\sqrt{n}$. $\square$

The theory justifying using these standard errors is the central limit theorem for the maximum likelihood estimator. This theorem can be stated in a mathematically precise manner that is difficult to understand without advanced probability theory. The following less precise statement is more easily understood:

Result 5.1 *Under suitable assumptions, for large enough sample sizes, the maximum likelihood estimator is approximately normally distributed with mean equal to the true parameter and with variance equal to the inverse of the Fisher information.*

The central limit theorem for the maximum likelihood estimator justifies the following large-sample confidence interval for the MLE of θ:

$$\widehat{\theta} \pm s_{\widehat{\theta}} \, z_{\alpha/2}, \tag{5.20}$$

where $z_{\alpha/2}$ is the $\alpha/2$-upper quantile of the normal distribution and $s_{\widehat{\theta}}$ is defined in (5.19).

The observed Fisher information is

$$\mathcal{I}^{\mathrm{obs}}(\theta) = -\frac{d^2}{d\theta^2} \log\{L(\theta)\}, \tag{5.21}$$

which differs from (5.18) in that there is no expectation taken. In many examples, (5.21) is a sum of many independent terms and, by the law of large numbers, will be close to (5.18). The expectation in (5.18) may be difficult to compute and using (5.21) instead is a convenient alternative.

The standard error of $\widehat{\theta}$ based on observed Fisher information is

$$s_{\widehat{\theta}}^{\mathrm{obs}} = \frac{1}{\sqrt{\mathcal{I}^{\mathrm{obs}}(\widehat{\theta})}}. \tag{5.22}$$

Often $s_{\widehat{\theta}}^{\mathrm{obs}}$ is used in place of $s_{\widehat{\theta}}$ in the confidence interval (5.20). There is theory suggesting that using the observed Fisher information will result in a more accurate confidence interval, that is, an interval with the true coverage probability closer to the nominal value of $1-\alpha$, so observed Fisher information can be justified by more than mere convenience; see Sect. 5.18.

So far, it has been assumed that θ is one-dimensional. In the multivariate case, the second derivative in (5.18) is replaced by the Hessian matrix of second derivatives,[11] and the result is called the *Fisher information matrix*. Analogously, the observed Fisher information matrix is the multivariate analog of (5.21). The covariance matrix of the MLE can be estimated by the inverse of the observed Fisher information matrix. If the negative of the log-likelihood is minimized by the R function optim(), then the observed Fisher information matrix is computed numerically and returned if hessian = TRUE

---

[11] The Hessian matrix of a function $f(x_1, \ldots, x_m)$ of m variables is the $m \times m$ matrix whose i, jth entry is the second partial derivative of f with respect to x_i and x_j.

in the call to this function. See Example 5.3 for an example where standard errors of the MLEs are computed numerically. Fisher information matrices are discussed in more detail in Sect. 7.10.

Bias and Standard Deviation of the MLE

In many examples, the MLE has a small bias that decreases to 0 at rate n^{-1} as the sample size n increases to ∞. More precisely,

$$\text{BIAS}(\widehat{\theta}_{\text{ML}}) = E(\widehat{\theta}_{\text{ML}}) - \theta \sim \frac{A}{n}, \text{ as } n \to \infty, \tag{5.23}$$

for some constant A. The bias of the MLE of a normal variance is an example and $A = -\sigma^2$ in this case.

Although this bias can be corrected in some special problems, such as estimation of a normal variance, usually the bias is ignored. There are two good reasons for this. First, the log-likelihood usually is the sum of n terms and so grows at rate n. The same is true of the Fisher information. Therefore, the variance of the MLE decreases at rate n^{-1}, that is,

$$\text{Var}(\widehat{\theta}_{\text{ML}}) \sim \frac{B}{n}, \text{ as } n \to \infty, \tag{5.24}$$

for some $B > 0$. Variability should be measured by the standard deviation, not the variance, and by (5.24),

$$\text{SD}(\widehat{\theta}_{\text{ML}}) \sim \frac{\sqrt{B}}{\sqrt{n}}, \text{ as } n \to \infty. \tag{5.25}$$

The convergence rate in (5.25) can also be obtained from the CLT for the MLE. Comparing (5.23) and (5.25), one sees that as n gets larger, the bias of the MLE becomes negligible compared to the standard deviation. This is especially important with financial markets data, where sample sizes tend to be large.

Second, even if the MLE of a parameter θ is unbiased, the same is not true for a nonlinear function of θ. For example, even if $\widehat{\sigma}^2$ is unbiased for σ^2, $\widehat{\sigma}$ is biased for σ. The reason for this is that for a nonlinear function g, in general,

$$E\{g(\widehat{\theta})\} \neq g\{E(\widehat{\theta})\}.$$

Therefore, it is impossible to correct for all biases.

5.11 Likelihood Ratio Tests

Some readers may wish to review hypothesis testing by reading Appendix A.18 before starting this section.

Likelihood ratio tests, like maximum likelihood estimation, are based upon the likelihood function. Both are convenient, all-purpose tools that are widely used in practice.

Suppose that $\boldsymbol{\theta}$ is a parameter vector and that the null hypothesis puts m equality constraints on $\boldsymbol{\theta}$. More precisely, there are m functions $g_1, \ldots, g_m$ and the null hypothesis is that $g_i(\boldsymbol{\theta}) = 0$ for $i = 1, \ldots, m$. The models without and with the constraints are called the full and reduced models, respectively.

It is also assumed that none of these constraints is redundant, that is, implied by the others. To illustrate redundancy, suppose that $\boldsymbol{\theta} = (\theta_1, \theta_2, \theta_3)$ and the constraints are $\theta_1 = \theta_2$, $\theta_2 = \theta_3$, and $\theta_1 = \theta_3$. Then the constraints have a redundancy since any two of them imply the third. Thus, $m = 2$, not 3.

Of course, redundancies need not be so easy to detect. One way to check is that the $m \times \dim(\boldsymbol{\theta})$ matrix

$$\begin{pmatrix} \nabla g_1(\boldsymbol{\theta}) \\ \vdots \\ \nabla g_m(\boldsymbol{\theta}) \end{pmatrix} \tag{5.26}$$

must have rank m. Here $\nabla g_i(\boldsymbol{\theta})$ is the gradient of g_i.

As an example, one might want to test that a population mean is zero; then $\boldsymbol{\theta} = (\mu, \sigma)^{\mathsf{T}}$ and $m = 1$ since the null hypothesis puts one constraint on $\boldsymbol{\theta}$, specifically that $\mu = 0$.

Let $\widehat{\boldsymbol{\theta}}_{\mathrm{ML}}$ be the maximum likelihood estimator without restrictions and let $\widehat{\boldsymbol{\theta}}_{0,\mathrm{ML}}$ be the value of $\boldsymbol{\theta}$ that maximizes $L(\boldsymbol{\theta})$ subject to the restrictions of the null hypothesis. If H_0 is true, then $\widehat{\boldsymbol{\theta}}_{0,\mathrm{ML}}$ and $\widehat{\boldsymbol{\theta}}_{\mathrm{ML}}$ should both be close to $\boldsymbol{\theta}$ and therefore $L(\widehat{\boldsymbol{\theta}}_{0,\mathrm{ML}})$ should be similar to $L(\widehat{\boldsymbol{\theta}})$. If H_0 is false, then the constraints will keep $\widehat{\boldsymbol{\theta}}_{0,\mathrm{ML}}$ far from $\widehat{\boldsymbol{\theta}}_{\mathrm{ML}}$ and so $L(\widehat{\boldsymbol{\theta}}_{0,\mathrm{ML}})$ should be noticeably *smaller* that $L(\widehat{\boldsymbol{\theta}})$.

The likelihood ratio test rejects H_0 if

$$2\left[\log\{L(\widehat{\boldsymbol{\theta}}_{\mathrm{ML}})\} - \log\{L(\widehat{\boldsymbol{\theta}}_{0,\mathrm{ML}})\} \right] \geq c, \tag{5.27}$$

where c is a critical value. The left-hand side of (5.27) is twice the log of the likelihood ratio $L(\widehat{\boldsymbol{\theta}}_{\mathrm{ML}})/L(\widehat{\boldsymbol{\theta}}_{0,\mathrm{ML}})$, hence the name likelihood ratio test. Often, an *exact critical value* can be found. A critical value is exact if it gives a level that is exactly equal to α. When an exact critical value is unknown, then the usual choice of the critical value is

$$c = \chi^2_{\alpha,m}, \tag{5.28}$$

where, as defined in Appendix A.10.1, $\chi^2_{\alpha,m}$ is the α-upper quantile value of the chi-squared distribution with m degrees of freedom.[12] The critical value (5.28)

---

[12] The reader should now appreciate why it is essential to calculate m correctly by eliminating redundant constraints. The wrong value of m will cause an incorrect critical value to be used.

is only approximate and uses the fact that under the null hypothesis, as the sample size increases the distribution of twice the log-likelihood ratio converges to the chi-squared distribution with m degrees of freedom if certain assumptions hold. One of these assumptions is that the null hypothesis is *not* on the boundary of the parameter space. For example, if the null hypothesis is that a variance parameter is zero, then the null hypothesis is on the boundary of the parameter space since a variance must be zero or greater. In this case (5.27) should not be used; see Self and Liang (1987). Also, if the sample size is small, then the large-sample approximation (5.27) is suspect and should be used with caution. An alternative is to use the bootstrap to determine the rejection region. The bootstrap is discussed in Chap. 6.

Computation of likelihood ratio tests is often very simple. In some cases, the test is computed automatically by statistical software. In other cases, software will compute the log-likelihood for each model (full and reduced) and these can be plugged into the left-hand side of (5.27).

5.12 AIC and BIC

An important practical problem is choosing between two or more statistical models that might be appropriate for a data set. The maximized value of the log-likelihood, denoted here by $\log\{L(\widehat{\boldsymbol{\theta}}_{\mathrm{ML}})\}$, can be used to measure how well a model fits the data or to compare the fits of two or more models. However, $\log\{L(\widehat{\boldsymbol{\theta}}_{\mathrm{ML}})\}$ can be increased simply by adding parameters to the model. The additional parameters do not necessarily mean that the model is a better description of the data-generating mechanism, because the additional model complexity due to added parameters may simply be fitting random noise in the data, a problem that is called *overfitting*. Therefore, models should be compared both by fit to the data and by model complexity. To find a parsimonious model one needs a good tradeoff between maximizing fit and minimizing model complexity.

AIC (Akaike's information criterion) and BIC (Bayesian information criterion) are two means for achieving a good tradeoff between fit and complexity. They differ slightly and BIC seeks a somewhat simpler model than AIC. They are defined by

$$\mathrm{AIC} = -2\log\{L(\widehat{\boldsymbol{\theta}}_{\mathrm{ML}})\} + 2p \qquad (5.29)$$

$$\mathrm{BIC} = -2\log\{L(\widehat{\boldsymbol{\theta}}_{\mathrm{ML}})\} + \log(n)p, \qquad (5.30)$$

where p equals the number of parameters in the model and n is the sample size. For both criteria, "smaller is better," since small values tend to maximize $L(\widehat{\boldsymbol{\theta}}_{\mathrm{ML}})$ (minimize $-\log\{L(\widehat{\boldsymbol{\theta}}_{\mathrm{ML}})\}$) and minimize p, which measures model complexity. The terms $2p$ and $\log(n)p$ are called "complexity penalties" since the penalize larger models.

The term *deviance* is often used for minus twice the log-likelihood, so AIC = deviance + $2p$ and BIC = deviance + $\log(n)p$. Deviance quantifies model fit, with smaller values implying better fit.

Generally, from a group of candidate models, one selects the model that minimizes whichever criterion, AIC or BIC, is being used. However, any model that is within 2 or 3 of the minimum value is a good candidate and might be selected instead, for example, because it is simpler or more convenient to use than the model achieving the absolute minimum. Since $\log(n) > 2$ provided, as is typical, that $n > 8$, BIC penalizes model complexity more than AIC does, and for this reason BIC tends to select simpler models than AIC. However, it is common for both criteria to select the same, or nearly the same, model. Of course, if several candidate models all have the same value of p, then AIC, BIC, and $-2\log\{L(\widehat{\boldsymbol{\theta}}_{\mathrm{ML}})\}$ are minimized by the same model.

5.13 Validation Data and Cross-Validation

When the same data are used both to estimate parameters and to assess fit, there is a strong tendency towards overfitting. Data contain both a *signal* and *noise*. The signal contains characteristics that are present in the population and therefore in each sample from the population, but the noise is random and varies from sample to sample. *Overfitting* means selecting an unnecessarily complex model to fit the noise. The obvious remedy to overfitting is to diagnose model fit using data that are independent of the data used for parameter estimation. We will call the data used for estimation the *training data* and the data used to assess fit the *validation data* or *test data*.

Example 5.2. Estimating the expected returns of midcap stocks

This example uses 500 daily returns on 20 midcap stocks in the file midcapD.ts.csv on the book's web site. The data were originally in the midcapD.ts data set in R's fEcofin package. The data are from 28-Feb-91 to 29-Dec-95. Suppose we need to estimate the 20 expected returns. Consider two estimators. The first, called "separate-means," is simply the 20 sample means. The second, "common-mean," uses the average of the 20 sample means as the common estimator of all 20 expected returns.

The rationale behind the common-mean estimator is that midcap stocks should have similar expected returns. The common-mean estimator pools data and greatly reduces the variance of the estimator. The common-mean estimator has some bias because the true expected returns will not be identical, which is the requirement for unbiasedness of the common-mean estimator. The separate-means estimator is unbiased but at the expense of a higher variance. This is a classic example of a bias–variance tradeoff.

Which estimator achieves the best tradeoff? To address this question, the data were divided into the returns for the first 250 days (training data) and for the last 250 days (validation data). The criterion for assessing goodness-of-fit was the sum of squared errors, which is

$$\sum_{k=1}^{20} \left(\widehat{\mu}_k^{\text{train}} - \overline{Y}_k^{\text{val}}\right)^2,$$

where $\widehat{\mu}_k^{\text{train}}$ is the estimator (using the training data) of the kth expected return and $\overline{Y}_k^{\text{val}}$ is the validation data sample mean of the returns on the kth stock. The sum of squared errors are 3.262 and 0.898, respectively, for the separate-means and common-mean estimators. The conclusion, of course, is that in this example the common-mean estimator is much more accurate than using separate means.

Suppose we had used the training data also for validation? The goodness-of-fit criterion would have been

$$\sum_{k=1}^{20} \left(\widehat{\mu}_k^{\text{train}} - \overline{Y}_k^{\text{train}}\right)^2,$$

where $\overline{Y}_k^{\text{train}}$ is the training data sample mean for the kth stock and is also the separate-means estimator for that stock. What would the results have been? Trivially, the sum of squared errors for the separate-means estimator would have been 0—each mean is estimated by itself with perfect accuracy! The common-mean estimator has a sum of squared errors equal to 0.920. The inappropriate use of the training data for validation would have led to the erroneous conclusion that the separate-means estimator is more accurate.

There are compromises between the two extremes of a common mean and separate means. These compromise estimators shrink the separate means toward the common mean. Bayesian estimation, discussed in Chap. 20, is an effective method for selecting the amount of shrinkage; see Example 20.12, where this set of returns is analyzed further. $\square$

A common criterion for judging fit is the deviance, which is -2 times the log-likelihood. The deviance of the validation data is

$$- 2\log f\left(\boldsymbol{Y}^{\text{val}}|\widehat{\boldsymbol{\theta}}^{\text{train}}\right), \tag{5.31}$$

where $\widehat{\boldsymbol{\theta}}^{\text{train}}$ is the MLE of the training data, $\boldsymbol{Y}^{\text{val}}$ is the validation data, and $f(\boldsymbol{y}^{\text{val}}|\boldsymbol{\theta})$ is the density of the validation data.

When the sample size is small, splitting the data once into training and validation data is wasteful. A better technique is *cross-validation*, often called simply CV, where each observation gets to play both roles, training and validation. K-fold cross-validation divides the data set into K subsets of roughly equal size. Validation is done K times. In the kth validation, $k = 1, \ldots, K$, the kth subset is the validation data and the other $K - 1$ subsets are combined to form the training data. The K estimates of goodness-of-fit are combined, for example, by averaging them. A common choice is n-fold cross-validation, also called *leave-one-out* cross-validation. With leave-one-out cross-validation, each observation takes a turn at being the validation data set, with the other $n - 1$ observations as the training data.

An alternative to actually using validation data is to calculate what would happen if new data could be obtained and used for validation. This is how AIC was derived. AIC is an approximation to the expected deviance of a hypothetical new sample that is independent of the actual data. More precisely, AIC approximates

$$E\left[-2\log f\left\{\boldsymbol{Y}^{\text{new}}\middle| \widehat{\boldsymbol{\theta}}(\boldsymbol{Y}^{\text{obs}})\right\}\right], \tag{5.32}$$

where $\boldsymbol{Y}^{\text{obs}}$ is the observed data, $\widehat{\boldsymbol{\theta}}(\boldsymbol{Y}^{\text{obs}})$ is the MLE computed from $\boldsymbol{Y}^{\text{obs}}$, and $\boldsymbol{Y}^{\text{new}}$ is a hypothetical new data set such that $\boldsymbol{Y}^{\text{obs}}$ and $\boldsymbol{Y}^{\text{new}}$ are i.i.d. Stated differently, $\boldsymbol{Y}^{\text{new}}$ is an unobserved independent replicate of $\boldsymbol{Y}^{\text{obs}}$. Since $\boldsymbol{Y}^{\text{new}}$ is not observed but has the same distribution as $\boldsymbol{Y}^{\text{obs}}$, to obtain AIC one substitutes $\boldsymbol{Y}^{\text{obs}}$ for $\boldsymbol{Y}^{\text{new}}$ in (5.32) and omits the expectation in (5.32). Then one calculates the effect of this substitution. The approximate effect is to reduce (5.32) by twice the number of parameters. Therefore, AIC compensates by adding $2p$ to the deviance, so that

$$\text{AIC} = -2\log f\left\{\boldsymbol{Y}^{\text{obs}}\middle| \widehat{\boldsymbol{\theta}}(\boldsymbol{Y}^{\text{obs}})\right\} + 2p, \tag{5.33}$$

which is a reexpression of (5.29).

The approximation used in AIC becomes more accurate when the sample size increases. A small-sample correction to AIC is

$$\text{AIC}_c = \text{AIC} + \frac{2p(p+1)}{n-p-1}. \tag{5.34}$$

Financial markets data sets are often large enough that the correction term $2p(p+1)/(n-p-1)$ is small, so that AIC is adequate and AIC_c is not needed. For example, if $n = 200$, then $2p(p+1)/(n-p-1)$ is 0.12, 0.21, 0.31, and 0.44 and for $p = 3$, 4, 5, and 6, respectively. Since a difference less than 1 in AIC values is usually considered inconsequential, the correction would have little effect when comparing models with 3 to 6 parameters when n is at least 200. Even more dramatically, when n is 500, then the corrections for 3, 4, 5, and 6 parameters are only 0.05, 0.08, 0.12, and 0.17.

Traders often develop trading strategies using a set of historical data and then test the strategies on new data. This is called *back-testing* and is a form of validation.

5.14 Fitting Distributions by Maximum Likelihood

As mentioned previously, one can find a formula for the MLE only for a few "textbook" examples. In most cases, the MLE must be found numerically. As an example, suppose that $Y_1, \ldots, Y_n$ is an i.i.d. sample from a t-distribution. Let

$$f_{t,\nu}^{\text{std}}(y \mid \mu, \sigma) \tag{5.35}$$

be the density of the standardized t-distribution with ν degrees of freedom and with mean μ and standard deviation σ. Then the parameters ν, μ, and σ are estimated by maximizing

$$\sum_{i=1}^{n} \log\left\{ f_{t,\nu}^{\text{std}}(Y_i \mid \mu, \sigma) \right\} \tag{5.36}$$

using any convenient optimization software. Estimation of other models is similar.

In the following examples, t-distributions and generalized error distributions are fit.

Example 5.3. Fitting a t-distribution to changes in risk-free returns

This example uses one of the time series in Chap. 4, the changes in the risk-free returns that has been called `diffrf`. This time series will be used to illustrate several methods for fitting a t-distribution. The simplest method uses the R function `fitdistr()`.

```
data(Capm, package = "Ecdat")
x = diff(Capm$rf)
fitdistr(x,"t")
```

The output is:

```
> fitdistr(x,"t")
      m              s             df
  0.0012243      0.0458549     3.3367036
 (0.0024539)    (0.0024580)   (0.5000096)
```

The parameters, in order, are the mean, the scale parameter, and the degrees of freedom. The numbers in parentheses are the standard errors.

Next, we fit the t-distribution by writing a function to return the negative log-likelihood and using R's `optim()` function to minimize the log-likelihood. We compute standard errors by using `solve()` to invert the Hessian and then taking the square roots of the diagonal elements of the inverted Hessian. We also compute AIC and BIC.

```
library(fGarch)
n = length(x)
start = c(mean(x), sd(x), 5)
loglik_t = function(beta) sum( - dt((x - beta[1]) / beta[2],
    beta[3], log = TRUE) + log(beta[2]) )
fit_t = optim(start, loglik_t, hessian = T,
    method = "L-BFGS-B", lower = c(-1, 0.001, 1))
AIC_t = 2 * fit_t$value + 2 * 3
BIC_t = 2 * fit_t$value + log(n) * 3
sd_t = sqrt(diag(solve(fit_t$hessian)))
fit_t$par
sd_t
AIC_t
BIC_t
```

The results are below. The estimates and the standard errors agree with those produced by `fitdistr()`, except for small numerical errors.

```
> fit_t$par
[1] 0.00122 0.04586 3.33655
> sd_t
[1] 0.00245 0.00246 0.49982
> AIC_t
[1] -1380.4
> BIC_t
[1] -1367.6
```

The standardized *t*-distribution can be fit by changing `dt()` to `dstd()`. Then the parameters are the mean, standard deviation, and degrees of freedom.

```
loglik_std = function(beta) sum(- dstd(x, mean = beta[1],
    sd = beta[2], nu = beta[3], log = TRUE))
fit_std = optim(start, loglik_std, hessian = T,
    method = "L-BFGS-B", lower = c(-0.1, 0.01, 2.1))
AIC_std = 2*fit_std$value + 2 * 3
BIC_std = 2*fit_std$value + log(n) * 3
sd_std = sqrt(diag(solve(fit_std$hessian)))
fit_std$par
sd_std
AIC_std
BIC_std
```

The results are below. The estimates agree with those when using `dt()` since $0.0725 = 0.0459\sqrt{3.33/(3.33 - 2)}$, aside from numerical error. Notice that AIC and BIC are unchanged, as expected since we are fitting the same model as before and only changing the parameterization.

```
> fit_std$par
[1] 0.0012144 0.0725088 3.3316132
> sd_std
[1] 0.0024538 0.0065504 0.4986456
> AIC_std
[1] -1380.4
> BIC_std
[1] -1367.6
```

□

Example 5.4. Fitting an F-S skewed t-distribution to changes in risk-free returns

Next, we fit the F-S skewed t-distribution.

```
loglik_sstd = function(beta) sum(- dsstd(x, mean = beta[1],
    sd = beta[2], nu = beta[3], xi = beta[4], log = TRUE))
start = c(mean(x), sd(x), 5, 1)
fit_sstd = optim(start, loglik_sstd, hessian = T,
    method = "L-BFGS-B", lower = c(-0.1, 0.01, 2.1, -2))
AIC_sstd = 2*fit_sstd$value + 2 * 4
BIC_sstd = 2*fit_sstd$value + log(n) * 4
sd_sstd = sqrt(diag(solve(fit_sstd$hessian)))
fit_sstd$par
sd_sstd
AIC_sstd
BIC_sstd
```

The results are below. The estimate of ξ (the fourth parameter) is very close to 1, which corresponds to the usual t-distribution. Both AIC and BIC increase since the extra skewness parameter does not improve the fit but adds 1 to the number of parameters.

```
> fit_sstd$par
[1] 0.0011811 0.0724833 3.3342759 0.9988491
> sd_sstd
[1] 0.0029956 0.0065790 0.5057846 0.0643003
> AIC_sstd
[1] -1378.4
> BIC_sstd
[1] -1361.4
```

□

Example 5.5. Fitting a generalized error distribution to changes in risk-free returns

The fit of the generalized error distribution to `diffrf` was obtained using `optim()` similarly to the previous example.

```
> fit_ged$par
[1] -0.00019493  0.06883004  1.00006805
> sd_ged
[1] 0.0011470 0.0033032 0.0761374
> AIC_ged
[1] -1361.4
> BIC_ged
[1] -1344.4
```

The three parameters are the estimates of the mean, standard deviation, and the shape parameter ν, respectively. The estimated shape parameter is extremely close to 1, implying a double-exponential distribution. Note that AIC and BIC are considerably larger than for the t-distribution. Therefore, t-distributions appear to be better models for these data compared to generalized error distributions. A possible reason for this is that, like the t-distributions, the density of the data seems to be rounded near the median; see the kernel density estimate in Fig. 5.9. QQ plots in Fig. 5.10 of `diffrf` versus the quantiles of the fitted t- and generalized error distributions are similar, indicating that neither model has a decidedly better fit than the other. However, the QQ plot of the t-distribution is slightly more linear.   □

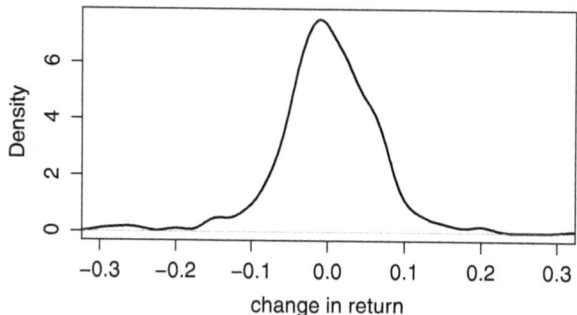

Fig. 5.9. *Kernel estimate of the probability density of* `diffrf`, *the changes in the risk-free returns.*

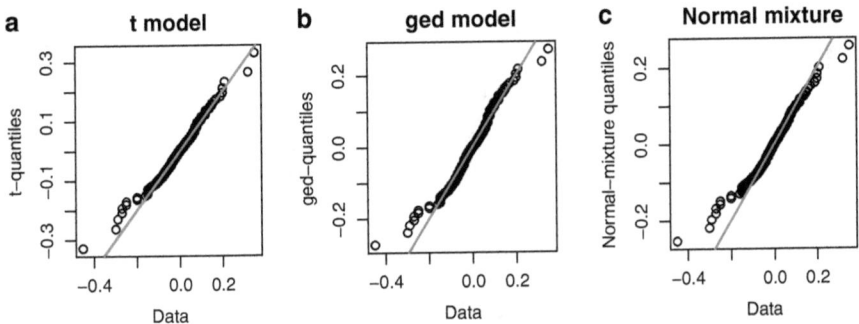

Fig. 5.10. *(a) QQ plot of* `diffrf` *versus the quantiles of a* $t_\nu^{\mathrm{std}}(\mu, s^2)$ *distribution with* μ, s^2, *and* ν *estimated by maximum likelihood. A 45° line through the origin has been added for reference.* *(b) A similar plot for the generalized error distribution.* *(c) A similar plot for a normal mixture model.*

Example 5.6. A-C skewed t-distribution fit to pipeline flows

This example uses the daily flows in natural gas pipelines introduced in Example 4.2. Recall that all three distributions are left-skewed. There are many well-known parametric families of right-skewed distributions, such as, the gamma and log-normal distributions, but there are not as many families of left-skewed distributions. The F-S skewed *t*- and A-C skewed *t*-distributions, which contain both right- and left-skewed distributions, are important exceptions. In this example, the A-C skewed normal distributions will be used.

Figure 5.11 has one row of plots for each variable. The left plots have two density estimates, an estimate using the Azzalini–Capitanio skewed normal distribution (solid) and a KDE (dashed). The right plots are QQ plots using the fitted skewed normal distributions.

The flows in pipelines 1 and, to a lesser extent, 2 are fit reasonably well by the A-C skewed normal distribution. This can be seen in the agreement between the parametric density estimates and the KDEs and in the nearly straight patterns in the QQ plots. The flows in pipeline 3 have a KDE with either a wide, flat mode or, perhaps, two modes. This pattern cannot be accommodated very well by the A-C skewed normal distributions. The result is less agreement between the parametric and KDE fits and a curved QQ plot. Nonetheless, a skewed normal distribution might be an adequate approximation for some purposes.

The following code produced the top row of Fig. 5.11. The code for the remaining rows is similar. The function `sn.mple()` at line 7 computed the MLEs using the CD parametrization and the function `cp2dp()` at line 8 converted the MLEs to the DP parametrization, which is used by the functions `dsn()` and `qsn()` at lines 9 and 18 that were needed in the plots. The red reference line through the quartiles in the QQ plot is created at lines 20–22.

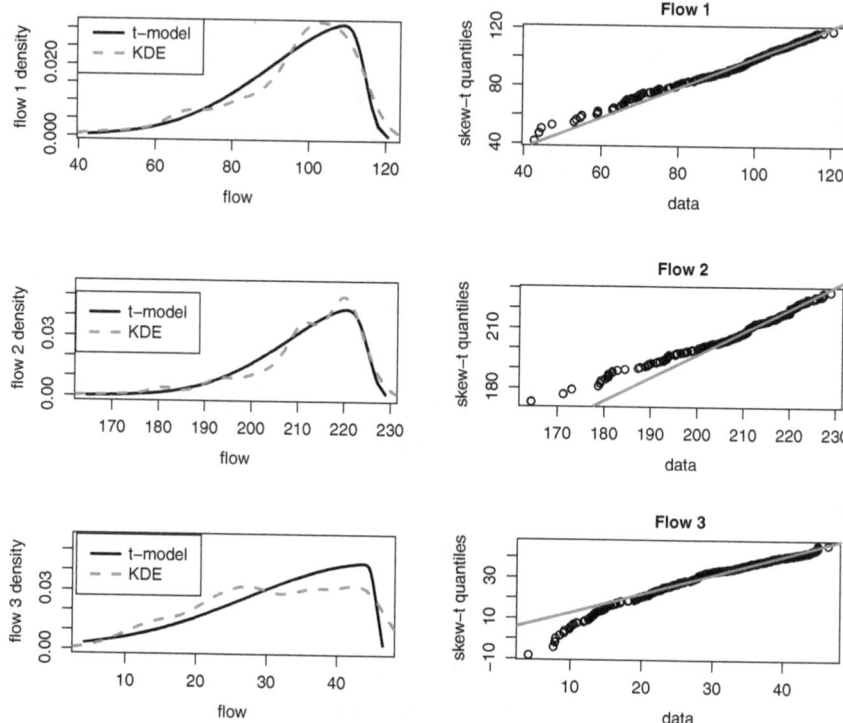

Fig. 5.11. *Parametric (solid) and nonparametric (dashed) density estimates for daily flows in three pipelines (left) and QQ plots for the parametric fits (right). The reference lines go through the first and third quartiles.*

```
1  library(sn)
2  dat = read.csv("FlowData.csv")
3  dat = dat/10000
4  par(mfrow = c(3, 2))
5  x = dat$Flow1
6  x1 = sort(x)
7  fit1 = sn.mple(y = x1, x = as.matrix(rep(1, length(x1))))
8  est1 = cp2dp(fit1$cp, family = "SN")
9  plot(x1, dsn(x1, dp = est1),
10     type = "l", lwd = 2, xlab = "flow",
11     ylab = "flow 1 density")
12 d = density(x1)
13 lines(d$x, d$y, lty = 2, lwd = 2)
14 legend(40, 0.034, c("t-model", "KDE"), lty = c(1, 2),
15    lwd = c(2, 2))
16 n = length(x1)
17 u=(1:n) / (n + 1)
18 plot(x1, qsn(u, dp = est1),xlab = "data",
```

```
19    ylab = "skew-t quantiles", main = "Flow 1")
20 lmfit = lm(qsn(c(0.25, 0.75), dp = est1) ~ quantile(x1,
21    c(0.25, 0.75)) )
22 abline(lmfit)
```

□

5.15 Profile Likelihood

Profile likelihood is a technique based on the likelihood ratio test introduced in Sect. 5.11. Profile likelihood is used to create confidence intervals and is often a convenient way to find a maximum likelihood estimator. Suppose the parameter vector is $\boldsymbol{\theta} = (\theta_1, \boldsymbol{\theta}_2)$, where θ_1 is a scalar parameter and the vector $\boldsymbol{\theta}_2$ contains the other parameters in the model. The profile log-likelihood for θ_1 is

$$L_{\max}(\theta_1) = \max_{\boldsymbol{\theta}_2} L(\theta_1, \boldsymbol{\theta}_2). \tag{5.37}$$

The right-hand side of (5.37) means the $L(\theta_1, \boldsymbol{\theta}_2)$ is maximized over $\boldsymbol{\theta}_2$ with θ_1 fixed to create a function of θ_1 only. Define $\widehat{\boldsymbol{\theta}}_2(\theta_1)$ as the value of $\boldsymbol{\theta}_2$ that maximizes the right-hand side of (5.37).

The MLE of θ_1 is the value, $\widehat{\theta}_1$, that maximizes $L_{\max}(\theta_1)$ and the MLE of $\boldsymbol{\theta}_2$ is $\widehat{\boldsymbol{\theta}}_2(\widehat{\theta}_1)$. Let $\theta_{0,1}$ be a hypothesized value of θ_1. By the theory of likelihood ratio tests in Sect. 5.11, one accepts the null hypothesis $\mathrm{H}_0 : \theta_1 = \theta_{0,1}$ if

$$L_{\max}(\theta_{0,1}) > L_{\max}(\widehat{\theta}_1) - \frac{1}{2}\chi^2_{\alpha,1}. \tag{5.38}$$

Here $\chi^2_{\alpha,1}$ is the α-upper quantile of the chi-squared distribution with one degree of freedom. The profile likelihood confidence interval (or, more properly, confidence region since it need not be an interval) for θ_1 is the set of all null values that would be accepted, that is,

$$\left\{ \theta_1 : L_{\max}(\theta_1) > L_{\max}(\widehat{\theta}_1) - \frac{1}{2}\chi^2_{\alpha,1} \right\}. \tag{5.39}$$

The profile likelihood can be defined for a subset of the parameters, rather than for just a single parameter, but this topic will not be pursued here.

Example 5.7. Estimating a Box–Cox transformation

An automatic method for estimating the transformation parameter for a Box–Cox transformation[13] assumes that for some values of α, μ, and σ, the transformed data $Y_1^{(\alpha)}, \ldots, Y_n^{(\alpha)}$ are i.i.d. $N(\mu, \sigma^2)$-distributed. All three

---

[13] See Eq. (4.5).

parameters can be estimated by maximum likelihood. For a fixed value of α, $\widehat{\mu}$ and $\widehat{\sigma}$ are the sample mean and variance of $Y_1^{(\alpha)}, \dots, Y_n^{(\alpha)}$ and these values can be plugged into the log-likelihood to obtain the profile log-likelihood for α. This can be done with the function boxcox() in R's MASS package, which plots the profile log-likelihood with confidence intervals.

Estimating α by the use of profile likelihood will be illustrated using the data on gas pipeline flows. Figure 5.12 shows the profile log-likelihoods and the KDEs and normal QQ plots of the flows transformed using the MLE of α.

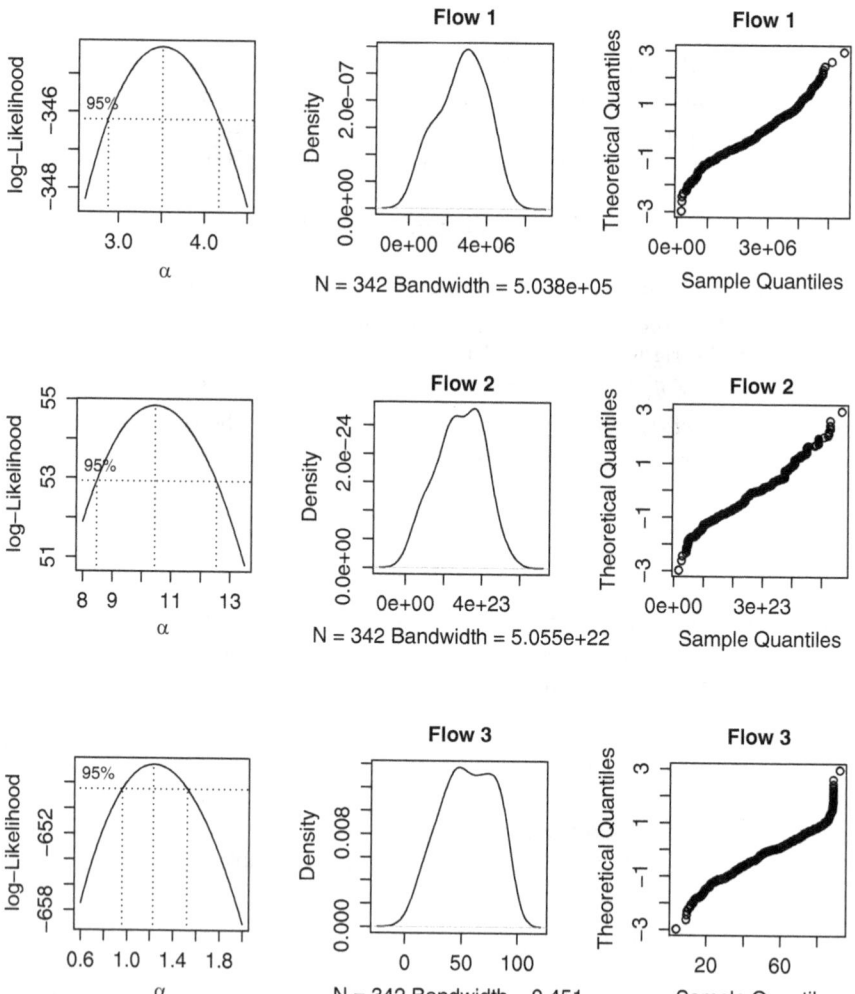

Fig. 5.12. *Profile log-likelihoods and 95 % confidence intervals for the parameter α of the Box–Cox transformation (left), KDEs of the transformed data (middle column), and normal plots of the transformed data (right).*

The KDE used `adjust` = 1.5 to smooth out local bumpiness seen with the default bandwidth. For the flows in pipeline 1, the MLE is $\widehat{\alpha} = 3.5$. Recall that in Example 4.2, we saw by trial-and-error that α between 3 and 4 was best for symmetrizing the data. It is gratifying to see that maximum likelihood corroborates this choice. The QQ plots show that the Box–Cox transformed flows have light tails. Light tails are not usually considered to be a problem and are to be expected here since the pipeline flows are bounded, below by 0 and above by the capacity of the pipeline.

The top row of Fig. 5.12 was produced by the following code. The function `boxcox()` at line 8 created the top-left plot containing the profile likelihood of the transformation parameter.

```
1 dat = read.csv("FlowData.csv")
2 dat = dat / 10000
3 library("MASS") ####  for boxcox()
4 adj = 1.5
5 par(mfrow = c(3, 3))
6 x = dat$Flow1
7 x1 = sort(x)
8 bcfit1 = boxcox(x1 ~ 1, lambda = seq(2.6, 4.5, 1 / 100),
9     xlab = expression(alpha))
10 text(3, -1898.75, "Flow 1")
11 plot(density((x1^3.5 - 1) / 3.5, adjust = adj), main = "Flow 1")
12 qqnorm((x1^3.5 - 1) / 3.5, datax = TRUE, main = "Flow 1")
```

□

It is worth pointing out that we have now seen two distinct methods for accommodating the left skewness in the pipeline flows, modeling the untransformed data by a skewed t-distribution (Example 5.6) and Box–Cox transformation to a normal distribution (Example 5.7). A third method would be to forego parametric modeling and use the kernel density estimation. This is not an atypical situation; often data can be analyzed in several different, but equally appropriate, ways.

5.16 Robust Estimation

Although maximum likelihood estimators have many attractive properties, they have one serious drawback of which anyone using them should be aware. Maximum likelihood estimators can be very sensitive to the assumptions of the statistical model. For example, the MLE of the mean of a normal population is the sample mean and the MLE of σ^2 is the sample variance, except with the minor change of a divisor of n rather than $n - 1$. The sample mean and variance are efficient estimators when the population is truly normally distributed, but these estimators are very sensitive to outliers, especially the sample standard deviation. Because these estimators are averages of the data and the squared deviations from the mean, respectively, a single outlier in the sample can drive the sample mean and variance to wildly absurd values if the outlier is far enough removed from the other data. Extreme outliers are nearly impossible with exactly normally distributed data, but if the data are only approximately normal with heavier tails than the normal distribution, then outliers are more probable and, when they do occur, more likely to be extreme. Therefore, the sample mean and variance can be very inefficient estimators. Statisticians say that the MLE is not *robust* to mild deviations from the assumed model. This is bad news and has led researchers to find estimators that are robust.

A robust alternative to the sample mean is the *trimmed mean*. An α-trimmed mean is computed by ordering the sample from smallest to largest, removing the fraction α of the smallest and the same fraction of the largest observations, and then taking the mean of the remaining observations. The idea behind trimming is simple and should be obvious: The sample is trimmed of extreme values before the mean is calculated. There is a mathematical formulation of the α-trimmed mean. Let $k = n\alpha$ rounded[14] to an integer; k is the number of observations removed from both ends of the sample. Then the α-trimmed mean is

$$\overline{X}_\alpha = \frac{\sum_{i=k+1}^{n-k} Y_{(i)}}{n - 2k},$$

where $Y_{(i)}$ is the ith order statistic. Typical values of α are 0.1, 0.15, 0.2, and 0.25. As α approaches 0.5, the α-trimmed mean approaches the sample median, which is the 0.5-sample quantile.

Dispersion refers to the variation in a distribution or sample. The sample standard deviation is the most common estimate of dispersion, but as stated it is nonrobust. In fact, the sample standard deviation is even more nonrobust than the sample mean, because squaring makes outliers more extreme. A robust estimator of dispersion is the *MAD* (*median absolute deviation*) estimator, defined as

$$\widehat{\sigma}^{\mathrm{MAD}} = 1.4826 \times \mathrm{median}\{|Y_i - \mathrm{median}(Y_i)|\}. \tag{5.40}$$

This formula should be interpreted as follows. The expression "median(Y_i)" is the sample median, $|Y_i - \mathrm{median}(Y_i)|$ is the absolute deviation of the observations from their median, and $\mathrm{median}\{|Y_i - \mathrm{median}(Y_i)|\}$ is the median of these absolute deviations. For normally distributed data, the median$\{|Y_i - \mathrm{median}(Y_i)|\}$ estimates not σ but rather $\Phi^{-1}(0.75)\sigma = \sigma/1.4826$, because for normally distributed data the median$\{|Y_i - \mathrm{median}(Y_i)|\}$ will converge to $\sigma/1.4826$ as the sample size increases. Thus, the factor 1.4826 in Eq. (5.40) calibrates $\widehat{\sigma}^{\mathrm{MAD}}$ so that it estimates σ when applied to normally distributed data.

---

[14] Definitions vary and the rounding could be either upward or to the nearest integer.

$\widehat{\sigma}^{\mathrm{MAD}}$ does not estimate σ for a nonnormal population. It does measure dispersion, but not dispersion as measured by the standard deviation. But this is just the point. For nonnormal populations the standard deviation can be very sensitive to the tails of the distribution and does not tell us much about the dispersion in the central range of the distribution, just in the tails.

In R, mad() computes (5.40). Some authors define MAD to be median$\{|Y_i - \mathrm{median}(Y_i)|\}$, that is, without 1.4826. Here the notation $\widehat{\sigma}^{\mathrm{MAD}}$ is used to emphasize the standardization by 1.4826 in order to estimate a normal standard deviation.

An alternative to using robust estimators is to assume a model where outliers are more probable. Then the MLE will automatically downweight outliers. For example, the MLE of the parameters of a t-distribution is much more robust to outliers than the MLE of the parameters of a normal distribution.

5.17 Transformation Kernel Density Estimation with a Parametric Transformation

We saw in Sect. 4.8 that the transformation kernel density estimator (TKDE) can avoid the bumps seen when the ordinary KDE is applied to skewed data. The KDE also can exhibit bumps in the tails when both tails are long, as is common with financial markets data. An example is the variable diffrf whose KDE is in Fig. 5.9. For such data, the TKDE needs a transformation that is convex to the right of the mode and concave to the left of the mode. There are many such transformations, and in this section we will use some facts from probability theory, as well as maximum likelihood estimation, to select a suitable one.

The key ideas used here are that (1) normally distributed data have light tails and are suitable for estimation with the KDE, (2) it is easy to transform data to normality if one knows the CDF, and (3) the CDF can be estimated by assuming a parametric model and using maximum likelihood. If a random variable has a continuous distribution F, then $F(X)$ has a uniform distribution and $\Phi^{-1}\{F(X)\}$ has an $N(0,1)$ distribution; here Φ is the standard normal CDF. Of course, in practice F is unknown, but one can estimate F parametrically, assuming, for example, that F is some t-distribution. It is not necessary that F actually be a t-distribution, only that a t-distribution can provide a reasonable enough fit to F in the tails so that an appropriate transformation is selected. If it was known that F was a t-distribution, then, of course, there would be no need to use a KDE or TKDE to estimate its density. The transformation to use in the TKDE is $g(y) = \Phi^{-1}\{F(y)\}$, which has inverse $g^{-1}(x) = F^{-1}\{\Phi(x)\}$. The derivative of g is needed to compute the TKDE and is

$$g'(y) = \frac{f(y)}{\phi[\Phi^{-1}\{F(y)\}]}. \tag{5.41}$$

Example 5.8. TKDE for risk-free returns

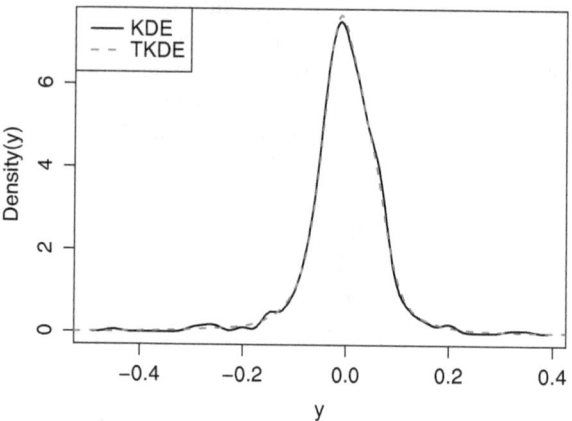

Fig. 5.13. *Kernel density and transformation kernel density estimates of monthly changes in the risk-free returns, January* 1960 *to December* 2002. *The data are in the* Capm *series in the* Ecdat *package in* R.

This example uses the changes in the risk-free returns in Fig. 4.3. We saw in Sect. 5.14 that these data are reasonably well fit by a t-distribution with mean, standard deviation, and ν equal to 0.00121, 0.0724, and 3.33, respectively. This distribution will be used as F. Figure 5.13 compares the ordinary KDE to the TKDE for this example. Notice that the TKDE is much smoother in the tails; this can be seen better in Fig. 5.14, which gives detail on the left tail.

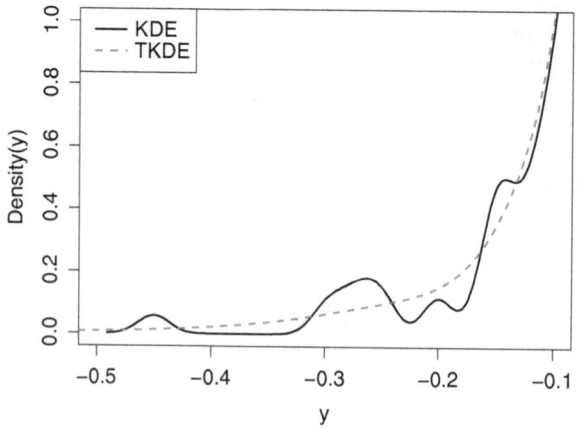

Fig. 5.14. *Kernel density and transformation kernel density estimates of monthly changes in the risk-free returns, January* 1960 *to December* 2002, *zooming in on left tail.*

The transformation used in this example is shown in Fig. 5.15. Notice the concave-convex shape that brings the left and right tails closer to the center and results in transformed data without the heavy tails seen in the original data. The removal of the heavy tails can be seen in Fig. 5.16, which is a normal plot of the transformed data.
The code to create Fig. 5.13 is below:

```
1 data(Capm, package = "Ecdat")
2 y = diff(Capm$rf)
3 diffrf = y
4 library(fGarch)
5 x1 = pstd(y, mean = 0.001, sd = .0725, nu = 3.34)
6 x = qnorm(x1)
```

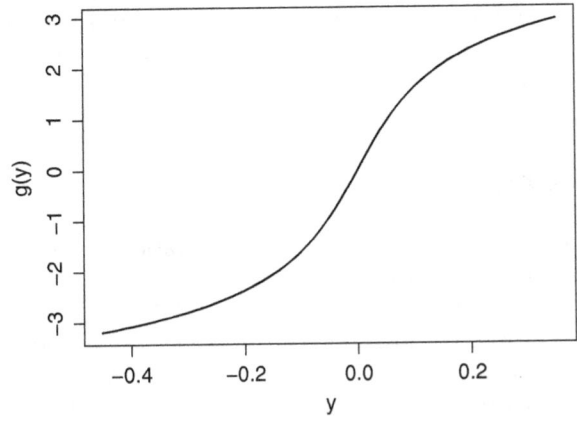

Fig. 5.15. *Plot of the transformation used in Example 5.8.*

Normal Q–Q Plot

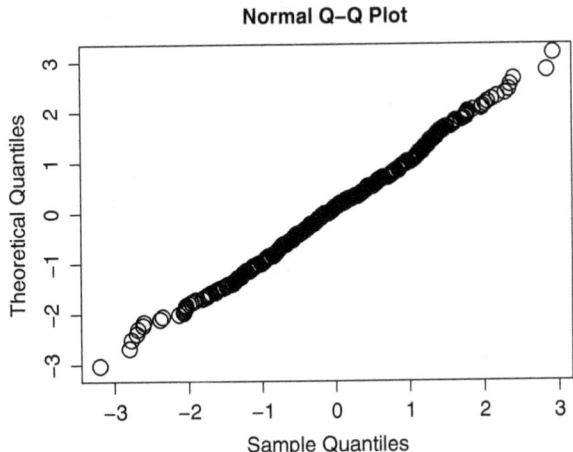

Fig. 5.16. *Normal plot of the transformed data used in Example 5.8.*

```
 7 par(mfrow = c(1, 1))
 8 d1 = density(diffrf)
 9 plot(d1$x, d1$y, type = "l", xlab = "y", ylab = "Density(y)",
10    lwd = 2)
11 d2 = density(x)
12 ginvx = qstd(pnorm(d2$x), mean = 0.001, sd = .0725, nu = 3.34)
13 gprime_num = dstd(ginvx, mean = 0.001, sd = .0725, nu = 3.34)
14 gprime_den = dnorm(qnorm(pstd(ginvx, mean = 0.001,
15    sd = .0725, nu = 3.34)))
16 gprime = gprime_num / gprime_den
17 lines(ginvx,d2$y * gprime, type = "l", lty = 2, col = "red", lwd = 2)
18 legend("topleft", c("KDE", "TKDE"), lty = c(1,2), lwd = 2,
19    col = c("black", "red"))
```

Lines 5–6 compute the transformation. Line 8 computes the KDE of the untransformed data and line 11 computes the KDE of the transformed data. Lines 12–16 compute g' in (5.41). At line 17 the KDE of the transformed data is multiplied by g' as in Eq. (4.6) to compute the TKDE.   □

5.18 Bibliographic Notes

Maximum likelihood estimation and likelihood ratio tests are discussed in all textbooks on mathematical statistics, including Boos and Stefanski (2013); Casella and Berger (2002), and Wasserman (2004).

Burnham and Anderson (2002) is a comprehensive introduction to model selection and is highly recommended for further reading. They also cover multimodel inference, a more advanced topic that includes *model averaging* where estimators or predictions are averaged across several models. Chapter 7 of Burnham and Anderson provides the statistical theory behind AIC as an approximate deviance of hypothetical validation data. The small-sample corrected AIC is due to Hurvich and Tsai (1989).

Buch-Larsen et al. (2005) and Ruppert and Wand (1992) discuss other methods for choosing the transformation when the TKDE is applied to heavy-tailed data.

The central limit theorem for the MLE is stated precisely and proved in textbooks on asymptotic theory such as Serfling (1980), van der Vaart (1998), and Lehmann (1999).

Observed and expected Fisher information are compared by Efron and Hinkley (1978), who argue that the observed Fisher information gives superior standard errors.

Box–Cox transformations were introduced by Box and Dox (1964). See Azzalini (2014); Azzalini and Capitanio (2003), and Arellano-Valle and Azzalini (2013) for discussion of the A-C skewed distributions.

5.19 R Lab

5.19.1 Earnings Data

Run the following R code to find a symmetrizing transformation for 1998 earnings data from the Current Population Survey. The code looks at the untransformed data and the square-root and log-transformed data. The transformed data are compared by normal plots, boxplots, and kernel density estimates.

```
library("Ecdat")
?CPSch3
data(CPSch3)
dimnames(CPSch3)[[2]]

male.earnings = CPSch3[CPSch3[ ,3] == "male", 2]
sqrt.male.earnings = sqrt(male.earnings)
log.male.earnings = log(male.earnings)

par(mfrow = c(2, 2))
qqnorm(male.earnings ,datax = TRUE, main = "untransformed")
qqnorm(sqrt.male.earnings, datax = TRUE,
    main = "square-root transformed")
qqnorm(log.male.earnings, datax = TRUE, main = "log-transformed")

par(mfrow = c(2, 2))
boxplot(male.earnings, main = "untransformed")
boxplot(sqrt.male.earnings, main = "square-root transformed")
boxplot(log.male.earnings, main = "log-transformed")

par(mfrow = c(2,2))
plot(density(male.earnings), main = "untransformed")
plot(density(sqrt.male.earnings), main = "square-root transformed")
plot(density(log.male.earnings), main = "log-transformed")
```

Problem 1 *Which of the three transformation provides the most symmetric distribution? Try other powers beside the square root. Which power do you think is best for symmetrization? You may include plots with your work if you find it helpful to do that.*

Next, you will estimate the Box–Cox transformation parameter by maximum likelihood. The model is that the data are $N(\mu, \sigma^2)$-distributed after being transformed by some λ. The unknown parameters are λ, μ, and σ.

Run the following R code to plot the profile likelihood for λ on the grid `seq(-2, 2, 1/10)` (this is the default and can be changed). The command `boxcox` takes an R formula as input. The left-hand side of the formula is the variable to be transformed. The right-hand side is a linear model (see Chap. 9). In this application, the model has only an intercept, which is indicated by

"1." "MASS" is an acronym for "Modern Applied Statistics with S-PLUS," a highly-regarded textbook whose fourth edition also covers R. The MASS library accompanies this book.

```
library("MASS")
par(mfrow = c(1, 1))
boxcox(male.earnings ~ 1)
```

The default grid of λ values is large, but you can zoom in on the high-likelihood region with the following:

```
boxcox(male.earnings ~ 1, lambda = seq(0.3, 0.45, 1 / 100))
```

To find the MLE, run this R code:

```
bc = boxcox(male.earnings ~ 1, lambda = seq(0.3, 0.45, by = 1 / 100),
    interp = FALSE)
ind = (bc$y == max(bc$y))
ind2 = (bc$y > max(bc$y) - qchisq(0.95, df = 1) / 2)
bc$x[ind]
bc$x[ind2]
```

Problem 2 *(a) What are* ind *and* ind2 *and what purposes do they serve?*
(b) What is the effect of interp *on the output from* boxcox*?*
(c) What is the MLE of λ*?*
(d) What is a 95 % confidence interval for λ*?*
(e) Modify the code to find a 99 % confidence interval for λ*.*

Rather than trying to transform the variable male.earnings to a Gaussian distribution, we could fit a skewed Gaussian or skewed t-distribution. R code that fits a skewed t is listed below:

```
library("fGarch")
fit = sstdFit(male.earnings, hessian = TRUE)
```

Problem 3 *What are the estimates of the degrees-of-freedom parameter and of* ξ*?*

Problem 4 *Produce a plot of a kernel density estimate of the pdf of* male.earnings*. Overlay a plot of the skewed t-density with MLEs of the parameters. Make sure that the two curves are clearly labeled, say with a legend, so that it is obvious which curve is which. Include your plot with your work. Compare the parametric and nonparametric estimates of the pdf. Do they seem similar? Based on the plots, do you believe that the skewed t-model provides an adequate fit to* male.earnings*?*

Problem 5 *Fit a skewed GED model to* `male.earnings` *and repeat Problem 4 using the skewed GED model in place of the skewed t. Which parametric model fits the variable* `male.earnings` *best, skewed t or skewed GED?*

5.19.2 DAX Returns

This section uses log returns on the DAX index in the data set `EuStock-Markets`. Your first task is to fit the standardized *t*-distribution (std) to the log returns. This is accomplished with the following R code.

Here `loglik_std` is an R function that is defined in the code. This function returns minus the log-likelihood for the std model. The std density function is computed with the function `dstd` in the `fGarch` package. Minus the log-likelihood, which is called the objective function, is minimized by the function `optim`. The L-BFGS-B method is used because it allows us to place lower and upper bounds on the parameters. Doing this avoids the errors that would be produced if, for example, a variance parameter were negative. When `optim` is called, `start` is a vector of starting values. Use R's help to learn more about `optim`. In this example, `optim` returns an object `fit_std`. The component `fig_std$par` contains the MLEs and the component `fig_std$value` contains the minimum value of the objective function.

```
data(Garch, package = "Ecdat")
library("fGarch")
data(EuStockMarkets)
Y = diff(log(EuStockMarkets[ ,1]))  # DAX

##### std #####
loglik_std = function(x) {
    f = -sum(dstd(Y, x[1], x[2], x[3], log = TRUE))
    f}
start = c(mean(Y), sd(Y), 4)
fit_std = optim(start, loglik_std, method = "L-BFGS-B",
    lower = c(-0.1, 0.001, 2.1),
    upper = c(0.1, 1, 20), hessian = TRUE)
cat("MLE =", round(fit_std$par, digits = 5))
minus_logL_std = fit_std$value  # minus the log-likelihood
AIC_std = 2 * minus_logL_std + 2 * length(fit_std$par)
```

Problem 6 *What are the MLEs of the mean, standard deviation, and the degrees-of-freedom parameter? What is the value of AIC?*

Problem 7 *Modify the code so that the MLEs for the skewed t-distribution are found. Include your modified code with your work. What are the MLEs? Which distribution is selected by AIC, the t or the skewed t-distribution?*

Problem 8 *Compute and plot the TKDE of the density of the log returns using the methodology in Sects. 4.8 and 5.17. The transformation that you use should be $g(y) = \Phi^{-1}\{F(y)\}$, where F is the t-distribution with parameters estimated in Problem 6. Include your code and the plot with your work.*

Problem 9 *Plot the KDE, TKDE, and parametric estimator of the log-return density, all on the same graph. Zoom in on the right tail, specifically the region $0.035 < y < 0.06$. Compare the three densities for smoothness. Are the TKDE and parametric estimates similar? Include the plot with your work.*

Problem 10 *Fit the F-S skewed t-distribution to the returns on the FTSE index in* EuStockMarkets. *Find the MLE, the standard errors of the MLE, and AIC.*

5.19.3 McDonald's Returns

This section continues the analysis of McDonald's stock returns begun in Sect. 2.4.4 and continued in Sect. 4.10.2. Run the code below.

```
1  data = read.csv('MCD_PriceDaily.csv')
2  adjPrice = data[ ,7]
3  LogRet = diff(log(adjPrice))
4  library(MASS)
5  library(fGarch)
6  fit.T = fitdistr(LogRet, "t")
7  params.T = fit.T$estimate
8  mean.T = params.T[1]
9  sd.T = params.T[2] * sqrt(params.T[3] / (params.T[3] - 2))
10  nu.T = params.T[3]
11  x = seq(-0.04, 0.04, by = 0.0001)
12  hist(LogRet, 80, freq = FALSE)
13  lines(x, dstd(x, mean = mean.T, sd = sd.T, nu = nu.T),
14      lwd = 2, lty = 2, col = 'red')
```

Problem 11 *Referring to lines by number, describe in detail what the code does. Examine the plot and comment on the goodness of fit.*

Problem 12 *Is the mean significantly different than 0?*

Problem 13 *Discuss differences between the histogram and the parametric fit. Do you think that the parametric fit is adequate or should a nonparametric estimate be used instead?*

Problem 14 *How heavy is the tail of the parametric fit? Does it appear that the fitted t-distribution has a finite kurtosis? How confident are you that the kurtosis is finite?*

5.20 Exercises

1. Load the `CRSPday` data set in the `Ecdat` package and get the variable names with the commands

   ```
   library(Ecdat)
   data(CRSPday)
   dimnames(CRSPday)[[2]]
   ```

   Plot the IBM returns with the commands

   ```
   r = CRSPday[ ,5]
   plot(r)
   ```

   Learn the mode and class of the IBM returns with

   ```
   mode(r)
   class(r)
   ```

   You will see that the class of the variable `r` is "`ts`," which means "time series." Data of class `ts` are plotted differently than data not of this class. To appreciate this fact, use the following commands to convert the IBM returns to class `numeric` before plotting them:

   ```
   r2 = as.numeric(r)
   class(r2)
   plot(r2)
   ```

   The variable `r2` contains the same data as the variable `r`, but `r2` has class `numeric`.
   Find the covariance matrix, correlation matrix, and means of GE, IBM, and Mobil with the commands

   ```
   cov(CRSPday[ ,4:6])
   cor(CRSPday[ ,4:6])
   apply(CRSPday[ ,4:6], 2, mean)
   ```

   Use your R output to answer the following questions:
   (a) What is the mean of the Mobil returns?
   (b) What is the variance of the GE returns?
   (c) What is the covariance between the GE and Mobil returns?
   (d) What is the correlation between the GE and Mobil returns?

2. Suppose that $Y_1, \ldots, Y_n$ are i.i.d. $N(\mu, \sigma^2)$, where μ is *known*. Show that the MLE of σ^2 is

$$n^{-1} \sum_{i=1}^{n} (Y_i - \mu)^2.$$

3. Show that $f^*(y|\xi)$ given by Eq. (5.15) integrates to $(\xi + \xi^{-1})/2$.

4. Let X be a random variable with mean μ and standard deviation σ.
   (a) Show that the kurtosis of X is equal to 1 plus the variance of $\{(X - \mu)/\sigma\}^2$.
   (b) Show that the kurtosis of any random variable is at least 1.
   (c) Show that a random variable X has a kurtosis equal to 1 if and only if $P(X = a) = P(X = b) = 1/2$ for some $a \neq b$.

5. (a) What is the kurtosis of a normal mixture distribution that is 95 % $N(0, 1)$ and 5 % $N(0, 10)$?
   (b) Find a formula for the kurtosis of a normal mixture distribution that is $100p\%$ $N(0, 1)$ and $100(1 - p)\%$ $N(0, \sigma^2)$, where p and σ are parameters. Your formula should give the kurtosis as a function of p and σ.
   (c) Show that the kurtosis of the normal mixtures in part (b) can be made arbitrarily large by choosing p and σ appropriately. Find values of p and σ so that the kurtosis is 10,000 or larger.
   (d) Let $M > 0$ be arbitrarily large. Show that for any $p_0 < 1$, no matter how close to 1, there is a $p > p_0$ and a σ, such that the normal mixture with these values of p and σ has a kurtosis at least M. This shows that there is a normal mixture arbitrarily close to a normal distribution but with a kurtosis above any arbitrarily large value of M.

6. Fit the F-S skewed t-distribution to the gas flow data. The data set is in the file `GasFlowData.csv`, which can be found on the book's website.

7. Suppose that $X_1, \ldots, X_n$ are i.i.d. exponential(θ). Show that the MLE of θ is $\overline{X}$.

8. For any univariate parameter θ and estimator $\widehat{\theta}$, we define the bias to be $\mathrm{Bias}(\widehat{\theta}) = E(\widehat{\theta}) - \theta$ and the MSE (mean square error) to be $\mathrm{MSE}(\widehat{\theta}) = E(\widehat{\theta} - \theta)^2$. Show that

$$\mathrm{MSE}(\widehat{\theta}) = \{\mathrm{Bias}(\widehat{\theta})\}^2 + \mathrm{Var}(\widehat{\theta}).$$

9. Suppose that $X_1, \ldots, X_n \overset{iid}{\sim} Normal(\mu, \sigma^2)$, with $0 < \sigma^2 < \infty$, and define $\hat{\mu} = \frac{1}{n} \sum_{i=1}^{n} X_i$. What is $\mathrm{Bias}(\hat{\mu})$? What is $\mathrm{MSE}(\hat{\mu})$? What if the distribution of the X_i is not Normal, but Student's t distribution with the same mean μ and variance σ^2, and tail index (ν, df) of 5?

10. Assume that you have a sample from a t-distribution and the sample kurtosis is 9. Based on this information alone, what would you use as an estimate of ν, the tail-index parameter?

11. The number of small businesses in a certain region defaulting on loans was observed for each month over a 4-year period. In the R program below,

the variable y is the number of defaults in a month and x is the value for that month of an economic variable thought to affect the default rate. The function dpois computes the Poisson density.

```
start =c(1,1)
loglik = function(theta) {-sum(log(dpois(y,
    lambda = exp(theta[1] + theta[2] * x))))}
mle = optim(start, loglik, hessian = TRUE)
invFishInfo = solve(mle$hessian)
options(digits = 4)
mle$par
mle$value
mle$convergence
sqrt(diag(invFishInfo))
```

The output is

```
> mle$par
[1] 1.0773 0.4529
> mle$value
[1] 602.4
> mle$convergence
[1] 0
> sqrt(diag(invFishInfo))
[1] 0.08742 0.03912
```

(a) Describe the statistical model being used here.

(b) What are the parameter estimates?

(c) Find 95 % confidence intervals for the parameters in the model. Use a normal approximation.

12. In this problem you will fit a t-distribution by maximum likelihood to the daily log returns for BMW. The data are in the data set bmw that is part of the evir package. Run the following code:

```
library(evir)
library(fGarch)
data(bmw)
start_bmw = c(mean(bmw), sd(bmw), 4)
loglik_bmw = function(theta)
{
-sum(dstd(bmw, mean = theta[1], sd = theta[2],
    nu = theta[3], log = TRUE))
}
mle_bmw = optim(start_bmw, loglik_bmw, hessian = TRUE)
CovMLE_bmw = solve(mle_bmw$hessian)
```

Note: The R code defines a function loglik_bmw that is minus the log-likelihood. See Chap. 10 of *An Introduction to R* for more information about functions in R. Also, see page 59 of this manual for more about maximum likelihood estimation in R. optim minimizes this objective function and returns the MLE (which is mle_bmw$par) and other information, including the Hessian of the objective function evaluated at the MLE (because hessian=TRUE—the default is not to return the Hessian).

(a) What does the function dstd do, and what package is it in?
(b) What does the function solve do?
(c) What is the estimate of ν, the degrees-of-freedom parameter?
(d) What is the standard error of ν?

13. In this problem, you will fit a t-distribution to daily log returns of Siemens. You will estimate the degrees-of-freedom parameter graphically and then by maximum likelihood. Run the following code, which produces a 3 × 2 matrix of probability plots. If you wish, add reference lines as done in Sect. 4.10.1.

```
library(evir)
data(siemens)
n = length(siemens)
par(mfrow = c(3, 2))
qqplot(siemens, qt(((1 : n) - 0.5) / n, 2),
    ylab = "t(2) quantiles",
    xlab = "data quantiles")
qqplot(siemens,qt(((1:n)-.5)/n,3),ylab="t(3) quantiles",
    xlab="data quantiles")
qqplot(siemens,qt(((1:n)-.5)/n,4),ylab="t(4) quantiles",
    xlab="data quantiles")
qqplot(siemens,qt(((1:n)-.5)/n,5),ylab="t(5) quantiles",
    xlab="data quantiles")
qqplot(siemens,qt(((1:n)-.5)/n,8),ylab="t(8) quantiles",
    xlab="data quantiles")
qqplot(siemens,qt(((1:n)-.5)/n,12),ylab="t(12) quantiles",
    xlab="data quantiles")
```

R has excellent graphics capabilities—see Chap. 12 of *An Introduction to R* for more about R graphics and, in particular, pages 67 and 72 for more information about par and mfrow, respectively.

(a) Do the returns have lighter or heavier tails than a t-distribution with 2 degrees of freedom?
(b) Based on the QQ plots, what seems like a reasonable estimate of ν?
(c) What is the MLE of ν for the Siemens log returns?

References

Arellano-Valle, R. B., and Azzalini, A. (2013) The centred parameterization and related quantities of the skew-t distribution. *Journal of Multivariate Analysis*, 113, 73–90.

Azzalini, A. (2014) *The Skew-Normal and Related Families (Institute of Mathematical Statistics Monographs, Book 3)*, Cambridge University Press.

Azzalini, A., and Capitanio, A. (2003) Distributions generated by perturbation of symmetry with emphasis on a multivariate skew t distribution. *Journal of the Royal Statistics Society, Series B*, **65**, 367–389.

Boos, D. D., and Stefanski, L. A. (2013) *Essential Statistical Inference*, Springer.

Box, G. E. P., and Dox, D. R. (1964) An analysis of transformations. *Journal of the Royal Statistical Society, Series B*, **26** 211–246.

Buch-Larsen, T., Nielsen, J. P., Guillén, M., and Bolance, C. (2005), Kernel density estimation for heavy-tailed distributions using the champernowne transformation. *Statistics*, **39**, 503–518.

Burnham, K. P. and Anderson, D. R. (2002) *Model Selection and Multimodel Inference*, Springer, New York.

Casella, G. and Berger, R. L. (2002) *Statistical Inference*, 2nd ed., Duxbury/Thomson Learning, Pacific Grove, CA.

Efron, B., and Hinkley, D. V. (1978) Assessing the accuracy of the maximum likelihood estimator: Observed versus expected Fisher information. *Biometrika*, **65**, 457–487.

Fernandez, C., and Steel, M. F. J. (1998) On Bayesian Modelling of fat tails and skewness, *Journal of the American Statistical Association*, **93**, 359–371.

Hurvich, C. M., and Tsai, C-L. (1989) Regression and time series model selection in small samples. *Biometrika*, 76, 297–307.

Lehmann, E. L. (1999) *Elements of Large-Sample Theory*, Springer-Verlag, New York.

Ruppert, D., and Wand, M. P. (1992) Correction for kurtosis in density estimation. *Australian Journal of Statistics*, **34**, 19–29.

Self, S. G., and Liang, K. Y. (1987) Asymptotic properties of maximum likelihood estimators and likelihood ratio tests under non-standard conditions. *Journal of the American Statistical Association*, **82**, 605–610.

Serfling, R. J. (1980) *Approximation Theorems of Mathematical Statistics*, Wiley, New York.

van der Vaart, A. W. (1998) *Asymptotic Statistics*, Cambridge University Press, Cambridge.

Wasserman, L. (2004) *All of Statistics*, Springer, New York.

6

Resampling

6.1 Introduction

Finding a single set of estimates for the parameters in a statistical model is not enough. An assessment of the uncertainty in these estimates is also needed. Standard errors and confidence intervals are common methods for expressing uncertainty.[1] In the past, it was sometimes difficult, if not impossible, to assess uncertainty, especially for complex models. Fortunately, the speed of modern computers, and the innovations in statistical methodology inspired by this speed, have largely overcome this problem. In this chapter we apply a computer simulation technique called the "bootstrap" or "resampling" to find standard errors and confidence intervals. The bootstrap method is very widely applicable and will be used extensively in the remainder of this book. The bootstrap is one way that modern computing has revolutionized statistics. Markov chain Monte Carlo (MCMC) is another; see Chap. 20.

The term "bootstrap" was coined by Bradley Efron (1979) and comes from the phrase "pulling oneself up by one's bootstraps."

When statistics are computed from a randomly chosen sample, then these statistics are random variables. Students often do not appreciate this fact. After all, what could be random about $\overline{Y}$? We just averaged the data, so what is random? The point is that the sample is only one of many possible samples. Each possible sample gives a different value of $\overline{Y}$. Thus, although we only see one value of $\overline{Y}$, it was selected at random from the many possible values and therefore $\overline{Y}$ is a random variable.

Methods of statistical inference such as confidence intervals and hypothesis tests are predicated on the randomness of statistics. For example, the

---

[1] See Appendices A.16.2 and A.17 for introductions to standard errors and confidence intervals.

© Springer Science+Business Media New York 2015
D. Ruppert, D.S. Matteson, *Statistics and Data Analysis for Financial Engineering*, Springer Texts in Statistics,
DOI 10.1007/978-1-4939-2614-5_6

confidence coefficient of a confidence interval tells us the probability, before a random sample is taken, that an interval constructed from the sample will contain the parameter. Therefore, by the law of large numbers, the confidence coefficient is also the long-run frequency of intervals that cover their parameter. Confidence intervals are usually derived using probability theory. Often, however, the necessary probability calculations are intractable, and in such cases we can replace theoretical calculations by Monte Carlo simulation.

But how do we simulate sampling from an *unknown* population? The answer, of course, is that we cannot do this exactly. However, a sample is a good representative of the population, and we can simulate sampling from the population by sampling from the sample, which is called *resampling*.

Each resample has the same sample size n as the original sample. The reason for this is that we are trying to simulate the original sampling, so we want the resampling to be as similar as possible to the original sampling. By *bootstrap approximation*, we mean the approximation of the sampling process by resampling.

There are two basic resampling methods, model-free and model-based, which are also known, respectively, as nonparametric and parametric. In this chapter, we assume that we have an i.i.d. sample from some population. For dependent data, resampling requires different techniques, which will be discussed in Sect. 13.6.

In *model-free resampling*, the resamples are drawn *with replacement* from the original sample. Why with replacement? The reason is that only sampling with replacement gives independent observations, and we want the resamples to be i.i.d. just as the original sample. In fact, if the resamples were drawn without replacement, then every resample would be exactly the same as the original sample, so the resamples would show no random variation. This would not be very satisfactory, of course.

Model-based resampling does not take a sample from the original sample. Instead, one assumes that the original sample was drawn i.i.d. from a density in the parametric family, $\{f(\boldsymbol{y}|\boldsymbol{\theta}) : \boldsymbol{\theta} \in \boldsymbol{\Theta}\}$, so, for an unknown value of $\boldsymbol{\theta}$, $f(\boldsymbol{y}|\boldsymbol{\theta})$ is the population density. The resamples are drawn i.i.d. from the density $f(\boldsymbol{y}|\widehat{\boldsymbol{\theta}})$, where $\widehat{\boldsymbol{\theta}}$ is some estimate of the parameter vector $\boldsymbol{\theta}$.

The number of resamples taken should, in general, be large. Just how large depends on the context and is discussed more fully later. Sometimes thousands or even tens of thousands of resamples are used. We let B denote the number of resamples.

When reading the following section, keep in mind that with resampling, the original sample plays the role of the population, because the resamples are taken from the original sample. Therefore, estimates from the sample play the role of true population parameters.

6.2 Bootstrap Estimates of Bias, Standard Deviation, and MSE

Let θ be a one-dimensional parameter, let $\widehat{\theta}$ be its estimate from the sample, and let $\widehat{\theta}_1^*, \ldots, \widehat{\theta}_B^*$ be estimates from B resamples. Also, define $\overline{\widehat{\theta}^*}$ to be the mean of $\widehat{\theta}_1^*, \ldots, \widehat{\theta}_B^*$. An asterisk indicates a statistic calculated from a resample.

The bias of $\widehat{\theta}$ is defined as $\mathrm{BIAS}(\widehat{\theta}) = E(\widehat{\theta}) - \theta$. Since expectations, which are population averages, are estimated by averaging over resamples, the bootstrap estimate of bias is

$$\mathrm{BIAS}_{\mathrm{boot}}(\widehat{\theta}) = \overline{\widehat{\theta}^*} - \widehat{\theta}. \tag{6.1}$$

Notice that, as discussed in the last paragraph of the previous section, in the bootstrap estimate of bias, the unknown population parameter θ is replaced by the estimate $\widehat{\theta}$ from the sample. The bootstrap standard error for $\widehat{\theta}$ is the sample standard deviation of $\widehat{\theta}_1^*, \ldots, \widehat{\theta}_B^*$, that is,

$$s_{\mathrm{boot}}(\widehat{\theta}) = \sqrt{\frac{1}{B-1} \sum_{b=1}^{B} (\widehat{\theta}_b^* - \overline{\widehat{\theta}^*})^2}. \tag{6.2}$$

$s_{\mathrm{boot}}(\widehat{\theta})$ estimates the standard deviation of $\widehat{\theta}$.

The mean-squared error (MSE) of $\widehat{\theta}$ is $E(\widehat{\theta} - \theta)^2$ and is estimated by

$$\mathrm{MSE}_{\mathrm{boot}}(\widehat{\theta}) = \frac{1}{B} \sum_{b=1}^{B} (\widehat{\theta}_b^* - \widehat{\theta})^2.$$

As in the estimation of bias, when estimating MSE, the unknown θ is replaced by $\widehat{\theta}$. The MSE reflects both bias and variability and, in fact,

$$\mathrm{MSE}_{\mathrm{boot}}(\widehat{\theta}) \approx \mathrm{BIAS}_{\mathrm{boot}}^2(\widehat{\theta}) + s_{\mathrm{boot}}^2(\widehat{\theta}). \tag{6.3}$$

We would have equality in (6.3), rather than an approximation, if in the denominator of (6.1) we used B rather than $B - 1$. Since B is usually large, the error of the approximation is typically very small.

6.2.1 Bootstrapping the MLE of the t-Distribution

Functions that compute the MLE, such as, `fitdistr()` in R, usually compute standard errors for the MLE along with the estimates themselves. The standard errors are justified theoretically by an "asymptotic" or "large-sample" approximation, called the CLT (central limit theorem) for the maximum likelihood estimator.[2] This approximation becomes exact only as the sample size

---

[2] See Sect. 5.10.

increases to ∞. Since a sample size is always finite, one cannot be sure of the accuracy of the standard errors. Computing standard errors by the bootstrap can serve as a check on the accuracy of the large-sample approximation, as illustrated in the following example.

Example 6.1. Bootstrapping GE Daily Returns

This example uses the GE daily returns from January 3, 1969, to December 31, 1998, in the data set CRSPday in R's Ecdat package. The sample size is 2,528 and the number of resamples is $B = 1,000$. The t-distribution was fit using fitdistr() in R and the model-free bootstrap was used. The first and third lines in Table 6.1 are the estimates and standard errors returned by fitdistr(), which uses observed Fisher information to calculate standard errors. The second and fourth lines have the results from bootstrapping. The differences between "Estimate" and "Bootstrap mean" are the bootstrap estimates of bias. We can see that the biases are small relative to the standard errors in the row labeled "SE." Small, and even negligible, bias is common when the sample size is in the thousands, as in this example.

Table 6.1. *Estimates from fitting a t-distribution to the 2,528 GE daily returns. "Estimate" = MLE. "SE" is standard error from observed Fisher information returned by the R function* fitdistr()*. "Bootstrap mean" and "Bootstrap SE" are the sample mean and standard deviation of the maximum likelihood estimates from 1,000 bootstrap samples. ν is the degrees-of-freedom parameter. The model-free bootstrap was used.*

	μ	σ	ν
Estimate	0.000879	0.0113	6.34
Bootstrap mean	0.000874	0.0113	6.30
SE	0.000253	0.000264	0.73
Bootstrap SE	0.000252	0.000266	0.82

It is reassuring that "SE" and "Bootstrap SE" agree as closely as they do in Table 6.1. This is an indication that both are reliable estimates of the uncertainty in the parameter estimates. Such close agreement is more likely with samples as large as this one.   □

```
1 library(bootstrap)
2 library(MASS)
3 set.seed("3857")
4 data(CRSPday, package = "Ecdat")
5 ge = CRSPday[ ,4]
6 nboot = 1000
7 t_mle = function(x){as.vector(fitdistr(x, "t")$estimate)}
8 results = bootstrap(ge, nboot, t_mle)
```

```
 9 rowMeans(results$thetastar[ , ])
10 apply(results$thetastar[,], 1, sd)
11 fitdistr(ge, "t")
```

The code above computes the results reported in Table 6.1. The bootstrap was performed at line 8 by the function `bootstrap()` in the `bootstrap` package which is loaded at line 1. The function `bootstrap()` has three arguments, the data, the value of B, and the function that computes the statistic to be bootstrapped; in this example that function is `t_mle()` which is defined at line 7. The function `fitdistr()` is in the package `MASS` that is loaded at line 2. Lines 9, 10, and 11 compute, respectively, the bootstrap mean, the bootstrap SEs, and the MLE and its standard errors.

Example 6.2. Bootstrapping GE daily returns, continued

To illustrate the bootstrap for a smaller sample size, we now use only the first 250 daily GE returns, approximately the first year of data. The number of bootstrap samples is 1,000. The results are in Table 6.2. For μ and s, the results in Tables 6.1 and 6.2 are comparable though the standard errors in Table 6.2 are, of course, larger because of the smaller sample size. For the parameter ν, the results in Table 6.2 are different in two respects from those in Table 6.1. First, the estimate and the bootstrap mean differ by more than 1, a sign that there is some bias. Second, the bootstrap standard deviation is 2.99, considerably larger than the SE, which is only 1.97. This suggests that the SE, which is based on large-sample theory, specifically the CLT for the MLE, is not an accurate measure of uncertainty in the parameter ν, at least not for the smaller sample.

Table 6.2. *Estimates from fitting a t-distribution to the first 250 GE daily returns. Notation as in Table 6.1. The nonparametric bootstrap was used.*

	μ	σ	ν
Estimate	0.00142	0.01055	5.52
Bootstrap mean	0.00145	0.01064	6.77
SE	0.000764	0.000817	1.98
Bootstrap SE	0.000777	0.000849	2.99

To gain some insight about why the results of ν in these two tables disagree, kernel density estimates of the two bootstrap samples were plotted in Fig. 6.1. We see that with the smaller sample size in panel (a), the density is bimodal and has noticeable right skewness. The density with the full sample is unimodal and has much less skewness.

Tail-weight parameters such as ν are difficult to estimate unless the sample size is in the thousands. With smaller sample sizes, such as 250, there will not be enough extreme observations to obtain a precise estimate of the tail-weight parameters. This problem has been nicely illustrated by the bootstrap. The number of extreme observations will vary between bootstrap samples. The bootstrap samples with fewer extreme values will have larger estimates of ν, since larger values of ν correspond to thinner tails.

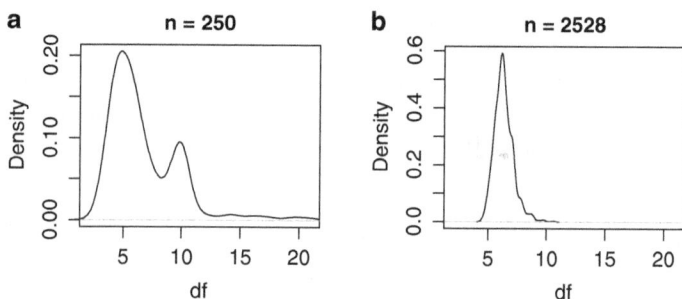

Fig. 6.1. *Kernel density estimates of* 1,000 *bootstrap estimates of df using (**a**) the first* 250 *daily GE returns and (**b**) all* 2,528 *GE returns. The default bandwidth was used in* R*'s* density *function to create the estimates.*

However, even with only 250 observations, ν can be estimated accurately enough to show, for example, that for the GE daily returns ν is very likely less than 13, the 98th percentile of the bootstrap distribution of ν. Therefore, the bootstrap provides strong evidence that the normal model corresponding to $\nu = \infty$ is not as satisfactory as a t-model.

By the CLT for the MLE, we know that the MLE is nearly normally distributed for large enough values of n. But this theorem does not tell us how large is large enough. To answer that question, we can use the bootstrap. We have seen here that $n = 250$ is not large enough for near normality of $\hat{\nu}$, and, though $n = 2,528$ is sufficiently large so that the bootstrap distribution is unimodal, there is still some right skewness when $n = 2,528$.   □

6.3 Bootstrap Confidence Intervals

Besides its use in estimating bias and finding standard errors, the bootstrap is widely used to construct confidence intervals. There are many bootstrap confidence intervals and some are quite sophisticated. We can only describe a few and the reader is pointed to the references in Sect. 6.4 for additional information.

Except in certain simple cases, confidence intervals are based on approximations such as the CLT for the MLE. The bootstrap is based on the approximation of the population's probability distribution using the sample. When a confidence interval uses an approximation, there are two coverage probabilities, the nominal one that is stated and the actual one that is unknown. Only for exact confidence intervals making no use of approximations will the two probabilities be equal. By the "accuracy" of a confidence interval, we mean the degree of agreement between the nominal and actual coverage probabilities. Even exact confidence intervals such as (A.44) for a normal mean and (A.45) for a normal variance are exact only when the data meet the assumptions exactly, e.g., are exactly normally distributed.

6.3.1 Normal Approximation Interval

Let $\widehat{\theta}$ be an estimate of θ and let $s_{\text{boot}}(\widehat{\theta})$ be the estimate of standard error given by (6.2). Then the normal theory confidence interval for θ is

$$\widehat{\theta} \pm s_{\text{boot}}(\widehat{\theta}) \, z_{\alpha/2}, \tag{6.4}$$

where $z_{\alpha/2}$ is the $\alpha/2$-upper quantile of the normal distribution. When $\widehat{\theta}$ is an MLE, this interval is essentially the same as (5.20) except that bootstrap, rather than the Fisher information, is used to find the standard error.

To avoid confusion, it should be emphasized that the normal approximation does not assume that the population is normally distributed but only that $\widehat{\theta}$ is normally distributed by a CLT.

6.3.2 Bootstrap-t Intervals

Often one has available a standard error for $\widehat{\theta}$, for example, from Fisher information. In this case, the bootstrap-t method can be used and, compared to normal approximation confidence intervals, offers the possibility of more accurate confidence intervals, that is, with nominal coverage probability closer to the actual coverage probability. We start by showing how the bootstrap-t method is related to the usual t-based confidence interval for a normal population mean, and then discuss the general theory.

Confidence Intervals for a Population Mean

Suppose we wish to construct a confidence interval for the population mean based on a random sample. One starts with the so-called "t-statistic,"[3] which is

$$t = \frac{\mu - \overline{Y}}{s/\sqrt{n}}. \tag{6.5}$$

The denominator of t, $s/\sqrt{n}$, is just the standard error of the mean, so that the denominator estimates the standard deviation of the numerator.

---

[3] Actually, t is not quite a statistic since it depends on the unknown μ, whereas a statistic, by definition, is something that depends only on the sample, not on unknown parameters. However, the term "t-statistic" is so widespread that we will use it here.

If we are sampling from a normally distributed population, then the probability distribution of t is known to be the t-distribution with $n-1$ degrees of freedom. Using the notation of Sect. 5.5.2, we denote by $t_{\alpha/2,n-1}$ the $\alpha/2$ upper t-value, that is, the $\alpha/2$-upper quantile of this distribution. Thus, t in (6.5) has probability $\alpha/2$ of exceeding $t_{\alpha/2,n-1}$. Because of the symmetry of the t-distribution, the probability is also $\alpha/2$ that t is less than $-t_{\alpha/2,n-1}$.

Therefore, for normally distributed data, the probability is $1-\alpha$ that

$$-t_{\alpha/2,n-1} \leq t \leq t_{\alpha/2,n-1}. \tag{6.6}$$

Substituting (6.5) into (6.6), after a bit of algebra we find that

$$1-\alpha = P\left\{ \overline{Y} - t_{\alpha/2,n-1}\frac{s}{\sqrt{n}} \leq \mu \leq \overline{Y} + t_{\alpha/2,n-1}\frac{s}{\sqrt{n}} \right\}, \tag{6.7}$$

which shows that

$$\overline{Y} \pm \frac{s}{\sqrt{n}} t_{\alpha/2,n-1}$$

is a $1-\alpha$ confidence interval for μ, assuming normally distributed data. This is the confidence interval given by Eq. (A.44). Note that in (6.7) the random variables are $\overline{Y}$ and s, and μ is fixed.

What if we are not sampling from a normal distribution? In that case, the distribution of t defined by (6.5) is *not* the t-distribution, but rather some other distribution that is not known to us. There are two problems. First, we do not know the distribution of the population. Second, even if the population distribution were known, it is a difficult, usually intractable, probability calculation to get the distribution of the t-statistic from the distribution of the population. This calculation has only been done for normal populations. Considering the difficulty of these two problems, can we still get a confidence interval? The answer is "yes, by resampling."

We start with a large number, say B, of resamples from the original sample. Let $\overline{Y}_{\text{boot},b}$ and $s_{\text{boot},b}$ be the sample mean and standard deviation of the bth resample, $b=1,\ldots,B$, and let $\overline{Y}$ be the mean of the original sample. Define

$$t_{\text{boot},b} = \frac{\overline{Y} - \overline{Y}_{\text{boot},b}}{s_{\text{boot},b}/\sqrt{n}}. \tag{6.8}$$

Notice that $t_{\text{boot},b}$ is defined in the same way as t except for two changes. First, $\overline{Y}$ and s in t are replaced by $\overline{Y}_{\text{boot},b}$ and $s_{\text{boot},b}$ in $t_{\text{boot},b}$. Second, μ in t is replaced by $\overline{Y}$ in $t_{\text{boot},b}$. The last point is a bit subtle, and uses the principle stated at the end of Sect. 6.1—a resample is taken using the original sample as the population. Thus, for the resample, the population mean is $\overline{Y}$!

Because the resamples are independent of each other, the collection $t_{\text{boot},1}$, $t_{\text{boot},2},\ldots$ can be treated as a random sample from the distribution of the t-statistic. After B values of $t_{\text{boot},b}$ have been calculated, one from each resample, we find the $\alpha/2$-lower and -upper quantiles of these $t_{\text{boot},b}$ values. Call these percentiles t_L and t_U.

If the original population is skewed, then there is no reason to suspect that the $\alpha/2$-lower quantile is minus the $\alpha/2$-upper quantile as happens for symmetric populations such as the t-distribution. In other words, we do not necessarily expect that $t_L = -t_U$, but this causes us no problem since the bootstrap allows us to estimate t_L and t_U without assuming any relationship between them. Now we replace $-t_{\alpha/2,n-1}$ and $t_{\alpha/2,n-1}$ in the confidence interval (6.7) by t_L and t_U, respectively. Finally, the bootstrap confidence interval for μ is

$$\left(\overline{Y} + t_L \frac{s}{\sqrt{n}}, \ \overline{Y} + t_U \frac{s}{\sqrt{n}}\right). \tag{6.9}$$

In (6.9), $\overline{Y}$ and s are the mean and standard deviation of the original sample, and only t_L and t_U are calculated from the B bootstrap resamples.

The bootstrap has solved both problems mentioned above. One does not need to know the population distribution since we can estimate it by the sample. A sample isn't a probability distribution. What is being done is creating a probability distribution, called the *empirical distribution*, from the sample by giving each observation in the sample probability $1/n$ where n is the sample size. Moreover, one doesn't need to calculate the distribution of the t-statistic using probability theory. Instead we can simulate from the empirical distribution.

Confidence Interval for a General Parameter

The method of constructing a t-confidence interval for μ can be generalized to other parameters. Let $\widehat{\theta}$ and $s(\widehat{\theta})$ be the estimate of θ and its standard error calculated from the sample. Let $\widehat{\theta}_b^*$ and $s_b(\widehat{\theta})$ be the same quantities from the bth bootstrap sample. Then the bth bootstrap t-statistic is

$$t_{\text{boot},b} = \frac{\widehat{\theta} - \widehat{\theta}_b^*}{s_b(\widehat{\theta})}. \tag{6.10}$$

As when estimating a population mean, let t_L and t_U be the $\alpha/2$-lower and $\alpha/2$-upper sample quantiles of these t-statistics. Then the confidence interval for θ is

$$\left(\widehat{\theta} + t_L s(\widehat{\theta}), \ \widehat{\theta} + t_U s(\widehat{\theta})\right)$$

since

$$1 - \alpha \approx P\left\{t_l \leq \frac{\widehat{\theta} - \widehat{\theta}_b^*}{s_b(\widehat{\theta})} \leq t_U\right\} \tag{6.11}$$

$$\approx P\left\{t_l \leq \frac{\theta - \widehat{\theta}}{s(\widehat{\theta})} \leq t_U\right\} \tag{6.12}$$

$$= P\left\{\widehat{\theta} + t_L s(\widehat{\theta}) \leq \theta \leq \widehat{\theta} + t_U s(\widehat{\theta})\right\}.$$

The approximation in (6.11) is due to Monte Carlo error and can be made small by choosing B large. The approximation in (6.12) is from the bootstrap approximation of the population's distribution by the empirical distribution. The error of the second approximation is independent of B and becomes small only as the sample size n becomes large. Though one generally has no control over the sample size, fortunately, sample sizes are often large in financial engineering.

6.3.3 Basic Bootstrap Interval

Let q_L and q_U be the $\alpha/2$-lower and -upper sample quantiles of $\widehat{\theta}_1^*, \ldots, \widehat{\theta}_B^*$. The fraction of bootstrap estimates that satisfy

$$q_L \leq \widehat{\theta}_b^* \leq q_U \tag{6.13}$$

is $1 - \alpha$. But (6.13) is algebraically equivalent to

$$\widehat{\theta} - q_U \leq \widehat{\theta} - \widehat{\theta}_b^* \leq \widehat{\theta} - q_L, \tag{6.14}$$

so that $\widehat{\theta} - q_U$ and $\widehat{\theta} - q_L$ are lower and upper quantiles for the distribution of $\widehat{\theta} - \widehat{\theta}_b^*$. The basic bootstrap interval uses them as lower and upper quantiles for the distribution of $\theta - \widehat{\theta}$. Using the bootstrap approximation, it is assumed that

$$\widehat{\theta} - q_U \leq \theta - \widehat{\theta} \leq \widehat{\theta} - q_L \tag{6.15}$$

will occur in a fraction $1 - \alpha$ of samples. Adding $\widehat{\theta}$ to each term in (6.15) gives $2\widehat{\theta} - q_U \leq \theta \leq 2\widehat{\theta} - q_L$, so that

$$(2\widehat{\theta} - q_U, 2\widehat{\theta} - q_L) \tag{6.16}$$

is a confidence interval for θ. Interval (6.16) is sometimes called the *basic bootstrap interval*.

6.3.4 Percentile Confidence Intervals

There are several bootstrap confidence intervals based on the so-called percentile method. Only one, the basic percentile interval, in discussed here in detail.

As in Sect. 6.3.3, let q_L and q_U be the $\alpha/2$-lower and -upper sample quantiles of $\widehat{\theta}_1^*, \ldots, \widehat{\theta}_B^*$. The basic percentile confidence interval is simply

$$(q_L, \ q_U). \tag{6.17}$$

By (6.13), the proportion of $\widehat{\theta}_b^*$-values in this interval is $1 - \alpha$. This interval can be justified by assuming that $\widehat{\theta}^*$ is distributed symmetrically about $\widehat{\theta}$. This assumption implies that for some $C > 0$, $q_L = \widehat{\theta} - C$ and $q_U = \widehat{\theta} + C$.

Then $2\widehat{\theta} - q_U = q_L$ and $2\widehat{\theta} - q_L = q_U$, so the basic bootstrap interval (6.16) coincides with the basic percentile interval (6.17).

What if $\widehat{\theta}^*$ is not distributed symmetrically about $\widehat{\theta}$? Fortunately, not all is lost. As discussed in Sect. 4.6, often random variables can be transformed to have a symmetric distribution. So, now assume only that for some monotonically increasing function g, $g(\widehat{\theta}^*)$ is symmetrically distributed about $g(\widehat{\theta})$. As we will now see, this weaker assumption is all that is needed to justify the basic percentile interval. Because g is monotonically strictly increasing and quantiles are transformation-respecting,[4] $g(q_L)$ and $g(q_U)$ are lower- and upper-$\alpha/2$ quantiles of $g(\widehat{\theta}_1^*), \ldots, g(\widehat{\theta}_B^*)$, and the basic percentile confidence interval for $g(\theta)$ is

$$\{g(q_L),\ g(q_U)\}. \tag{6.18}$$

Now, if (6.18) has coverage probability $(1 - \alpha)$ for $g(\theta)$, then, since g is monotonically increasing, (6.17) has coverage probability $(1 - \alpha)$ for θ. This justifies the percentile interval, at least if one is willing to assume the existence of a transformation to symmetry. Note that it is only assumed that such a g exists, not that it is known. No knowledge of g is necessary, since g is not used to construct the percentile interval.

The basic percentile method is simple, but it is not considered very accurate, except for large sample sizes. There are two problems with the percentile method. The first is an assumption of unbiasedness. The basic percentile interval assumes not only that $g(\widehat{\theta}^*)$ is distributed symmetrically, but also that it is symmetric about $g(\widehat{\theta})$ rather than $g(\widehat{\theta})$ plus some bias. Most estimators satisfy a CLT, e.g., the CLTs for sample quantiles and for the MLE in Sects. 4.3.1 and 5.10, respectively. Therefore, bias becomes negligible in large enough samples, but in practice the sample size might not be sufficiently large and bias can cause the nominal and actual coverage probabilities to differ.

The second problem is that $\widehat{\theta}$ may have a nonconstant variance, a problem called heteroskedasticity. If $\widehat{\theta}$ is the MLE, then the variance of $\widehat{\theta}$ is, at least approximately, the inverse of Fisher information and the Fisher information need not be constant—it often depends on θ.

More sophisticated percentile methods can correct for bias and heteroskedasticity. The BC_a and ABC (approximate bootstrap confidence) percentile intervals are improved percentile intervals in common use. In the name "BC_a," "BC" means "bias-corrected" and "a" means "accelerated," which refers to the rate at which the variance changes with θ. The BC_a method automatically estimates both the bias and the rate of change of the variance and then makes suitable adjustments. The theory behind the BCa and ABC intervals is beyond the scope of this book, but is discussed in references found in Sect. 6.4. Both the BC_a and ABC methods have been implemented in statistical software such as R. In R's bootstrap package, the functions bcanon(), abcpar(), and abcnon() implement the nonparametric BC_a, parametric ABC, and nonparametric ABC intervals, respectively.

---

[4] See Appendix A.2.2.

Example 6.3. Confidence interval for a quantile-based tail-weight parameter

It was mentioned in Sect. 5.8 that a quantile-based parameter quantifying tail weight can be defined as the ratio of two scale parameters:

$$\frac{s(p_1, 1 - p_1)}{s(p_2, 1 - p_2)}, \tag{6.19}$$

where

$$s(p_1, p_2) = \frac{F^{-1}(p_2) - F^{-1}(p_1)}{a},$$

a is a positive constant that does not affect the ratio (6.19) and so can be ignored, and $0 < p_1 < p_2 < 1/2$. We will call (6.19) quKurt. Finding a confidence interval for quKurt can be a daunting task without the bootstrap, but with the bootstrap it is simple. In this example, BC_a confidence intervals will be found for quKurt. The parameter is computed from a sample y by this R function, which has default values $p_1 = 0.025$ and $p_2 = 0.25$:

```
quKurt = function(y, p1 = 0.025, p2 = 0.25)
{
Q = quantile(y, c(p1, p2, 1 - p2, 1 - p1))
(Q[4] - Q[1]) / (Q[3] - Q[2])
}
```

The BC_a intervals are found with the bcanon() function in the bootstrap package using $B = 5,000$. The seed of the random number generator was fixed so that these results can be reproduced.

```
bmw = read.csv("bmw.csv")
library("bootstrap")
set.seed("5640")
bca_kurt = bcanon(bmwRet[ ,2], 5000, quKurt)
bca_kurt$confpoints
```

By default, the output gives four pairs of confidence limits.

```
> bca_kurt$confpoints
      alpha bca point
[1,] 0.025    4.07
[2,] 0.050    4.10
[3,] 0.100    4.14
[4,] 0.160    4.18
[5,] 0.840    4.41
[6,] 0.900    4.45
[7,] 0.950    4.50
[8,] 0.975    4.54
```

The results above show, for example, that the 90 % BC_a confidence interval is (4.10, 4.50). For reference, any normal distribution has quKurt equal 2.91, so these data have heavier than Gaussian tails, at least as measured by quKurt.
□

Example 6.4. Confidence interval for the ratio of two quantile-based tail-weight parameters

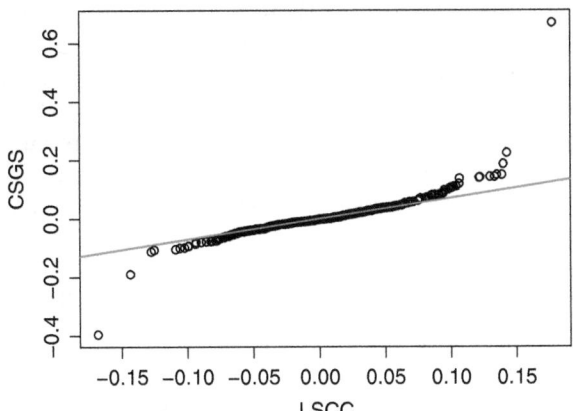

Fig. 6.2. *QQ plot of returns on two stocks in the* `midcapD.ts` *data set. The reference line goes through the first and third quartiles.*

This example uses the data set `midcapD.ts.csv` of returns on midcap stocks. Two of the stocks in this data set are LSCC and CSGS. From Fig. 6.2, which is a QQ plot comparing the returns from these two companies, it appears that LSCC returns have lighter tails than CSGS returns. The values of `quKurt` are 2.91 and 4.13 for LSCC and GSGS, respectively, and the ratio of the two values is 0.704. This is further evidence that LSCC returns have the lesser tail weight. A BC_a confidence interval for the ratio of `quKurt` for LSCC and CSGS is found with the following R program.

```
 1 midcapD.ts = read.csv("midcapD.ts.csv")
 2 attach(midcapD.ts)
 3 quKurt = function(y, p1 = 0.025, p2 = 0.25)
 4 {
 5    Q = quantile(y, c(p1, p2, 1 - p2, 1 - p1))
 6    as.numeric((Q[4] - Q[1]) / (Q[3] - Q[2]))
 7 }
 8 compareQuKurt = function(x, p1 = 0.025, p2 = 0.25, xdata)
 9 {
10    quKurt(xdata[x,1], p1, p2) / quKurt(xdata[x,2], p1, p2)
11 }
12 quKurt(LSCC)
13 quKurt(CSGS)
14 xdata = cbind(LSCC, CSGS)
15 compareQuKurt(1:n, xdata = xdata)
16 library("bootstrap")
```

```
17 set.seed("5640")
18 bca_kurt = bcanon((1:n), 5000, compareQuKurt, xdata = xdata)
19 bca_kurt$confpoints
```

The function compareQuKurt() (lines 8–11) computes a quKurt ratio. The function bcanon() is designed to bootstrap a vector, but this example has bivariate data in a matrix with two columns. To bootstrap multivariate data, there is a trick given in R's help for bcanon()—bootstrap the integers 1 to n where n is the sample size. This is done at line 18. The resamples of $1, \ldots, n$ allow one to resample the rows of the data vector. Thus, in this example bcanon() draws a random sample with replacement from $1, \ldots, n$ and selects the rows of xdata corresponding to these indices to create a resample.

The 95 % confidence interval for the quKurt ratio is 0.587 to 0.897, so with 95 % confidence it can be concluded that LSCC has a smaller value of quKurt.

```
> bca_kurt$confpoints
        alpha bca point
[1,] 0.025      0.587
[2,] 0.050      0.607
[3,] 0.100      0.634
[4,] 0.160      0.653
[5,] 0.840      0.811
[6,] 0.900      0.833
[7,] 0.950      0.864
[8,] 0.975      0.897
```

□

6.4 Bibliographic Notes

Efron (1979) introduced the name "bootstrap" and did much to popularize resampling methods. Efron and Tibshirani (1993), Davison and Hinkley (1997), Good (2005), and Chernick (2007) are introductions to the bootstrap that discuss many topics not treated here, including the theory behind the BC_a and ABC methods for confidence intervals. The R package bootstrap is described by its authors as "functions for Efron and Tibshirani (1993)" and the package contains the data sets used in that book. The R package boot is a more recent set of resampling functions and data sets to accompany Davison and Hinkley (1997).

6.5 R Lab

6.5.1 BMW Returns

This lab uses a data set containing 6146 daily returns on BMW stock from January 3, 1973 to July 23, 1996. Run the following code to fit a skewed t-distribution to the returns and check the fit with a QQ plot.

```
1  library("fGarch")
2  bmwRet = read.csv("bmwRet.csv")
3  n = dim(bmwRet)[1]
4
5  kurt = kurtosis(bmwRet[ ,2], method = "moment")
6  skew = skewness(bmwRet[,2], method = "moment")
7  fit_skewt = sstdFit(bmwRet[ ,2])
8
9  q.grid = (1:n) / (n+1)
10 qqplot(bmwRet[ ,2], qsstd(q.grid, fit_skewt$estimate[1],
11     fit_skewt$estimate[2],
12     fit_skewt$estimate[3], fit_skewt$estimate[4]),
13     ylab = "skewed-t quantiles" )
```

The function qsstd() is in the fGarch package loaded at line 1. The required package timeDate is also loaded and the function kurtosis() is in timeDate.

Problem 1 *What is the MLE of ν? Does the t-distribution with this value of ν have a finite skewness and kurtosis?*

Since the kurtosis coefficient based on the fourth central moment is infinite for some distributions, as in Sect. 6.4 we will define a quantile-based kurtosis:

$$\text{quKurt}(F) = \frac{F^{-1}(1 - p_1) - F^{-1}(p_1)}{F^{-1}(1 - p_2) - F^{-1}(p_2)},$$

where F is a CDF and $0 < p_1 < p_2 < 1/2$. Typically, p_1 is close to zero so that the numerator is sensitive to tail weight and p_2 is much larger and measures dispersion in the center of the distribution. Because the numerator and denominator of quKurt are each the difference between two quantiles, they are location-free and therefore scale parameters. Moreover, because quKurt is a ratio of two scale parameters, it is scale-free and therefore a shape parameter. A typical example would be $p_1 = 0.025$ and $p_2 = 0.25$. quKurt is estimated by replacing the population quantiles by sample quantiles.

Problem 2 *Write an R program to plot quKurt for the t-distribution as a function of ν. Use $p_1 = 0.025$ and $p_2 = 0.25$. Let ν take values from 1 to 10, incremented by 0.25. If you want to get fancy while labeling the axes, xlab=expression(nu) in the call to plot will put a "ν" on the x-axis.*

Run the following code, which defines a function to compute quKurt and bootstraps this function on the BMW returns. Note that p_1 and p_2 are given default values that are used in the bootstrap and that both model-free and model-based bootstrap samples are taken.

```
quKurt = function(y, p1 = 0.025, p2 = 0.25)
{
    Q = quantile(y, c(p1, p2, 1 - p2, 1 - p1))
    k = (Q[4] - Q[1]) / (Q[3] - Q[2])
    k
}
nboot = 5000
ModelFree_kurt  = rep(0, nboot)
ModelBased_kurt = rep(0, nboot)
set.seed("5640")
for (i in 1:nboot)
{
    samp_ModelFree = sample(bmwRet[,2], n, replace = TRUE)
    samp_ModelBased = rsstd(n, fit_skewt$estimate[1],
      fit_skewt$estimate[2],
      fit_skewt$estimate[3], fit_skewt$estimate[4])
    ModelFree_kurt[i] = quKurt(samp_ModelFree)
    ModelBased_kurt[i] = quKurt(samp_ModelBased)
}
```

Problem 3 *Plot KDEs of* ModelFree_kurt *and* ModelBased_kurt. *Also, plot side-by-side boxplots of the two samples. Describe any major differences between the model-based and model-free results. Include the plots with your work.*

Problem 4 *Find 90 % percentile method bootstrap confidence intervals for quKurt using the model-based and model-free bootstraps.*

Problem 5 BC_a *confidence intervals can be constructed using the function* bcanon() *in R's* bootstrap *package. Find a 90 % BC_a confidence interval for quKurt. Use 5,000 resamples. Compare the BC_a interval to the model-free percentile interval from Problem 4.*

6.5.2 Simulation Study: Bootstrapping the Kurtosis

The sample kurtosis is highly variable because it is based on the 4th moment. As a result, it is challenging to construct an accurate confidence interval for the kurtosis. In this section, five bootstrap confidence intervals for the kurtosis will be compared. The comparisons will be on widths of the intervals, where smaller is better, and actual coverage probabilities, where closer to nominal is better.

Run the following code. Warning: this simulation experiment takes a while to run, e.g., 5 to 10 minutes, and will have only moderate accuracy. To increase the accuracy, you might wish to increase `niter` and `nboot` and run the experiment over a longer period, even overnight.

```
library(bootstrap)
Kurtosis = function(x) mean((((x - mean(x)) / sd(x))^4)
set.seed(3751)
niter = 500
nboot = 400
n = 50
nu = 10
trueKurtosis =  3 + 6 / (nu - 4)
correct = matrix(nrow = niter, ncol = 5)
width   = matrix(nrow = niter, ncol = 5)
error = matrix(nrow = niter, ncol = 1)
t1 = proc.time()
for (i in 1:niter){
   y = rt(n,nu)
   int1 = boott(y, Kurtosis, nboott = nboot,
      nbootsd = 50)$confpoints[c(3, 9)]
   width[i,1] = int1[2] - int1[1]
   correct[i,1] = as.numeric((int1[1] < trueKurtosis) &
      (trueKurtosis < int1[2]))
   int2 = bcanon(y, nboot, Kurtosis)$confpoints[c(1, 8), 2]
   width[i,2] = int2[2] - int2[1]
   correct[i,2] = as.numeric((int2[1] < trueKurtosis) &
      (trueKurtosis < int2[2]))
   boot = bootstrap(y, nboot, Kurtosis)$thetastar
   int3 = Kurtosis(y) + 1.96 * c(-1, 1) * sd(boot)
   width[i,3] = int3[2] - int3[1]
   correct[i,3] = as.numeric((int3[1] < trueKurtosis) &
      (trueKurtosis < int3[2]))
   int4 = quantile(boot, c(0.025, 0.975))
   width[i,4] = int4[2] - int4[1]
   correct[i,4] = as.numeric((int4[1] < trueKurtosis) &
      (trueKurtosis < int4[2]))
   int5 = 2*Kurtosis(y) - quantile(boot, c(0.975, 0.025))
   width[i,5] = int5[2] - int5[1]
   correct[i,5] = as.numeric((int5[1] < trueKurtosis) &
      (trueKurtosis < int5[2]))
   error[i] = mean(boot) - Kurtosis(y)
}
t2 = proc.time()
(t2 - t1)/60
colMeans(width)
colMeans(correct)
options(digits = 3)
mean(error)
mean(error^2)
```

Problem 6 *Which five bootstrap intervals are being used here?*

Problem 7 *What is the value of B here?*

Problem 8 *How many simulations are used?*

Problem 9 *What are the estimates of bias?*

Problem 10 *What is the estimated MSE?*

Problem 11 *Estimate the actual coverage probability of the BC_a and bootstrap-t intervals. (Because this is a simulation experiment, it is subject to Monte Carlo errors, so the coverage probability is only estimated.)*

Problem 12 *Find a 95 % confidence interval for the actual coverage probability of the BC_a interval?*

Problem 13 *Which interval is most accurate? Would you consider any of the intervals as highly accurate?*

Problem 14 *How much clock time did the entire simulation take?*

As mentioned, kurtosis is difficult to estimate because it is based on the 4th moment and a quantile-based measure of tailweight might be a better alternative. The next problem investigates this conjecture.

Problem 15 *Repeat the simulation experiment with kurtosis replaced by* quKurt() *defined is Sect. 6.5.1 Which interval is most accurate now? Would you consider any of the intervals as highly accurate?*

6.6 Exercises

1. To estimate the risk of a stock, a sample of 50 log returns was taken and s was 0.31. To get a confidence interval for σ, 10,000 resamples were taken. Let $s_{b,\text{boot}}$ be the sample standard deviation of the bth resample. The 10,000 values of $s_{b,\text{boot}}/s$ were sorted and the table below contains selected values of $s_{b,\text{boot}}/s$ ranked from smallest to largest (so rank 1 is the smallest and so forth).

Rank	Value of $s_{b,\text{boot}}/s$
250	0.52
500	0.71
1,000	0.85
9,000	1.34
9,500	1.67
9,750	2.19

Find a 90 % confidence interval for σ.

2. In the following R program, resampling was used to estimate the bias and variance of the sample correlation between the variables in the vectors x and y.

```
samplecor = cor(x, y)
n = length(x)
nboot = 5000
resamplecor = rep(0, nboot)
for (b in (1:nboot))
{
   ind = sample(1:n, replace = TRUE)
   resamplecor[b] = cor(x[ind], y[ind])
}
samplecor
mean(resamplecor)
sd(resamplecor)
```

The output is

```
> n
[1] 20
> samplecor
[1] 0.69119
> mean(resamplecor)
[1] 0.68431
> sd(resamplecor)
[1] 0.11293
```

(a) Estimate the bias of the sample correlation coefficient.
(b) Estimate the standard deviation of the sample correlation coefficient.
(c) Estimate the MSE of the sample correlation coefficient.
(d) What fraction of the MSE is due to bias? How serious is the bias? Should something be done to reduce the bias? Explain your answer.

3. The following R code was used to bootstrap the sample standard deviation.

```
( code to read the variable x )
sampleSD = sd(x)
n = length(x)
nboot = 15000
resampleSD = rep(0, nboot)
```

```
for (b in (1:nboot))
{
    resampleSD[b] = sd(sample(x, replace = TRUE))
}
options(digits = 4)
sampleSD
mean(resampleSD)
sd(resampleSD)
```

The output is

```
> sampleSD
[1] 1.323
> mean(resampleSD)
[1] 1.283
> sd(resampleSD)
[1] 0.2386
```

(a) Estimate the bias of the sample standard deviation of x.
(b) Estimate the mean squared error of the sample standard deviation of x.

References

Chernick, M. R. (2007) *Bootstrap Methods: A Guide for Practitioners and Researchers,* 2nd ed., Wiley-Interscience, Hoboken, NJ.

Davison, A. C., and Hinkley, D. V. (1997) *Bootstrap Methods and Their Applications,* Cambridge University Press, Cambridge.

Efron, B. (1979) Bootstrap methods: Another look at the jackknife. *Annals of Statistics,* **7,** 1–26.

Efron, B., and Tibshirani, R. (1993) *An Introduction to the Bootstrap,* Chapman & Hall, New York.

Good, P. I. (2005) *Resampling Methods: A Practical Guide to Data Analysis,* 3rd ed., Birkhauser, Boston.

7

Multivariate Statistical Models

7.1 Introduction

Often we are not interested merely in a single random variable but rather in the joint behavior of several random variables, for example, returns on several assets and a market index. Multivariate distributions describe such joint behavior. This chapter is an introduction to the use of multivariate distributions for modeling financial markets data. Readers with little prior knowledge of multivariate distributions may benefit from reviewing Appendices A.12–A.14 before reading this chapter.

7.2 Covariance and Correlation Matrices

Let $\boldsymbol{Y} = (Y_1, \ldots, Y_d)^{\mathsf{T}}$ be a random vector. We define the expectation vector of $\boldsymbol{Y}$ to be

$$E(\boldsymbol{Y}) = \begin{pmatrix} E(Y_1) \\ \vdots \\ E(Y_d) \end{pmatrix}.$$

The *covariance matrix* of $\boldsymbol{Y}$ is the matrix whose (i,j)th entry is $\mathrm{Cov}(Y_i, Y_j)$ for $i, j = 1, \ldots, N$. Since $\mathrm{Cov}(Y_i, Y_i) = \mathrm{Var}(Y_i)$, the covariance matrix is

$$\mathrm{COV}(\boldsymbol{Y}) = \begin{pmatrix} \mathrm{Var}(Y_1) & \mathrm{Cov}(Y_1, Y_2) & \cdots & \mathrm{Cov}(Y_1, Y_d) \\ \mathrm{Cov}(Y_2, Y_1) & \mathrm{Var}(Y_2) & \cdots & \mathrm{Cov}(Y_2, Y_d) \\ \vdots & \vdots & \ddots & \vdots \\ \mathrm{Cov}(Y_d, Y_1) & \mathrm{Cov}(Y_d, Y_2) & \cdots & \mathrm{Var}(Y_d) \end{pmatrix}.$$

Similarly, the *correlation matrix* of $\boldsymbol{Y}$, denoted $\mathrm{CORR}(\boldsymbol{Y})$, has i, jth element $\rho_{Y_i Y_j}$. Because $\mathrm{Corr}(Y_i, Y_i) = 1$ for all i, the diagonal elements of a correlation

© Springer Science+Business Media New York 2015
D. Ruppert, D.S. Matteson, *Statistics and Data Analysis for Financial Engineering*, Springer Texts in Statistics,
DOI 10.1007/978-1-4939-2614-5_7

matrix are all equal to 1. Note the use of "COV" and "CORR" to denote matrices and "Cov" and "Corr" to denote scalars.

The covariance matrix can be written as

$$\text{COV}(\boldsymbol{Y}) = E\left[\{\boldsymbol{Y} - E(\boldsymbol{Y})\}\{\boldsymbol{Y} - E(\boldsymbol{Y})\}^{\mathsf{T}}\right]. \tag{7.1}$$

There are simple relationships between the covariance and correlation matrices. Let $\boldsymbol{S} = \text{diag}(\sigma_{Y_1}, \ldots, \sigma_{Y_d})$, where σ_{Y_i} is the standard deviation of Y_i. Then

$$\text{CORR}(\boldsymbol{Y}) = \boldsymbol{S}^{-1}\text{COV}(\boldsymbol{Y})\boldsymbol{S}^{-1} \tag{7.2}$$

and, equivalently,

$$\text{COV}(\boldsymbol{Y}) = \boldsymbol{S}\,\text{CORR}(\boldsymbol{Y})\,\boldsymbol{S}. \tag{7.3}$$

The *sample covariance* and *correlation matrices* replace $\text{Cov}(Y_i, Y_j)$ and $\rho_{Y_i Y_j}$ by their estimates given by (A.29) and (A.30).

A *standardized* variable is obtained by subtracting the variable's mean and dividing the difference by the variable's standard deviation. After standardization, a variable has a mean equal to 0 and a standard deviation equal to 1. The covariance matrix of standardized variables equals the correlation matrix of original variables, which is also the correlation matrix of the standardized variables.

Example 7.1. CRSPday covariances and correlations

This example uses the CRSPday data set in R's Ecdat package. There are four variables, daily returns from January 3, 1969, to December 31, 1998, on three stocks, GE, IBM, and Mobil, and on the CRSP value-weighted index, including dividends. CRSP is the Center for Research in Security Prices at the University of Chicago. The sample covariance matrix for these four series is

```
            ge       ibm      mobil     crsp
ge    1.88e-04 8.01e-05 5.27e-05 7.61e-05
ibm   8.01e-05 3.06e-04 3.59e-05 6.60e-05
mobil 5.27e-05 3.59e-05 1.67e-04 4.31e-05
crsp  7.61e-05 6.60e-05 4.31e-05 6.02e-05
```

It is difficult to get much information just by inspecting the covariance matrix. The covariance between two random variables depends on their variances as well as the strength of the linear relationship between them. Covariance matrices are extremely important as input to, for example, a portfolio analysis, but to understand the relationship between variables, it is much better to examine their sample correlation matrix. The sample correlation matrix in this example is

```
          ge    ibm mobil  crsp
ge     1.000 0.334 0.297 0.715
ibm    0.334 1.000 0.159 0.486
mobil  0.297 0.159 1.000 0.429
crsp   0.715 0.486 0.429 1.000
```

We can see that all sample correlations are positive and the largest correlations are between `crsp` and the individual stocks. GE is the stock most highly correlated with `crsp`. The correlations between individual stocks and a market index such as `crsp` are a key component of finance theory, especially the Capital Asset Pricing Model (CAPM) introduced in Chap. 17.   □

7.3 Linear Functions of Random Variables

Often we are interested in finding the expectation and variance of a linear combination (weighted average) of random variables. For example, consider returns on a set of assets. A *portfolio* is simply a weighted average of the assets with weights that sum to one. The weights specify what fractions of the total investment are allocated to the assets. For example, if a portfolio consists of 200 shares of Stock 1 selling at \$88/share and 150 shares of Stock 2 selling at \$67/share, then the weights are

$$w_1 = \frac{(200)(88)}{(200)(88) + (150)(67)} = 0.637 \quad \text{and} \quad w_2 = 1 - w_1 = 0.363. \quad (7.4)$$

Because the return on a portfolio is a linear combination of the returns on the individual assets in the portfolio, the material in this section is used extensively in the portfolio theory of Chaps. 16 and 17.

First, we look at a linear function of a single random variable. If Y is a random variable and a and b are constants, then

$$E(aY + b) = aE(Y) + b.$$

Also,
$$\text{Var}(aY + b) = a^2 \text{Var}(Y) \text{ and } \sigma_{aY+b} = |a|\sigma_Y.$$

Next, we consider linear combinations of two random variables. If X and Y are random variables and w_1 and w_2 are constants, then

$$E(w_1 X + w_2 Y) = w_1 E(X) + w_2 E(Y),$$

and

$$\text{Var}(w_1 X + w_2 Y) = w_1^2 \text{Var}(X) + 2w_1 w_2 \text{Cov}(X, Y) + w_2^2 \text{Var}(Y). \quad (7.5)$$

Check that (7.5) can be reexpressed as

$$\text{Var}(w_1 X + w_2 Y) = (\begin{matrix} w_1 & w_2 \end{matrix}) \begin{pmatrix} \text{Var}(X) & \text{Cov}(X, Y) \\ \text{Cov}(X, Y) & \text{Var}(Y) \end{pmatrix} \begin{pmatrix} w_1 \\ w_2 \end{pmatrix}. \quad (7.6)$$

Although formula (7.6) may seem unnecessarily complicated, we will show that this equation generalizes in an elegant way to more than two random variables; see (7.7) below. Notice that the matrix in (7.6) is the covariance matrix of the random vector $(\begin{matrix} X & Y \end{matrix})^\mathsf{T}$.

Let $\boldsymbol{w} = (w_1, \ldots, w_d)^\mathsf{T}$ be a vector of weights and let $\boldsymbol{Y} = (Y_1, \ldots, Y_d)$ be a random vector. Then

$$\boldsymbol{w}^\mathsf{T} \boldsymbol{Y} = \sum_{i=1}^N w_i Y_i$$

is a weighted average of the components of $\boldsymbol{Y}$. One can easily show that

$$E(\boldsymbol{w}^\mathsf{T} \boldsymbol{Y}) = \boldsymbol{w}^\mathsf{T} \{E(\boldsymbol{Y})\}$$

and

$$\text{Var}(\boldsymbol{w}^\mathsf{T} \boldsymbol{Y}) = \sum_{i=1}^N \sum_{j=1}^N w_i\, w_j\, \text{Cov}(Y_i, Y_j).$$

This last result can be expressed more succinctly using vector/matrix notation:

$$\text{Var}(\boldsymbol{w}^\mathsf{T} \boldsymbol{Y}) = \boldsymbol{w}^\mathsf{T} \text{COV}(\boldsymbol{Y}) \boldsymbol{w}. \quad (7.7)$$

Example 7.2. The variance of a linear combination of correlated random variables

Suppose that $\boldsymbol{Y} = (Y_1\ Y_2\ Y_3)^\mathsf{T}$, $\text{Var}(Y_1) = 2$, $\text{Var}(Y_2) = 3$, $\text{Var}(Y_3) = 5$, $\rho_{Y_1, Y_2} = 0.6$, and that Y_1 and Y_2 are independent of Y_3. Find $\text{Var}(Y_1 + Y_2 + 1/2\, Y_3)$.

Answer: The covariance between Y_1 and Y_3 is 0 by independence, and the same is true of Y_2 and Y_3. The covariance between Y_1 and Y_2 is $(0.6)\sqrt{(2)(3)} = 1.47$. Therefore,

$$\text{COV}(\boldsymbol{Y}) = \begin{pmatrix} 2 & 1.47 & 0 \\ 1.47 & 3 & 0 \\ 0 & 0 & 5 \end{pmatrix},$$

and by (7.7),

$$\text{Var}(Y_1 + Y_2 + Y_3/2) = (\begin{matrix} 1 & 1 & \frac{1}{2} \end{matrix}) \begin{pmatrix} 2 & 1.47 & 0 \\ 1.47 & 3 & 0 \\ 0 & 0 & 5 \end{pmatrix} \begin{pmatrix} 1 \\ 1 \\ \frac{1}{2} \end{pmatrix}$$

$$= (\begin{matrix} 1 & 1 & \frac{1}{2} \end{matrix}) \begin{pmatrix} 3.47 \\ 4.47 \\ 2.5 \end{pmatrix}$$

$$= 9.19.$$

□

An important property of covariance and correlation matrices is that they are symmetric and positive semidefinite. A matrix A is said to be positive semidefinite (definite) if $x^T A x \geq 0$ (> 0) for all vectors $x \neq 0$. By (7.7), any covariance matrix must be positive semidefinite, because otherwise there would exist a random variable with a negative variance, a contradiction. A nonsingular covariance matrix is positive definite. A covariance matrix must be symmetric because $\text{Cov}(Y_i, Y_j) = \text{Cov}(Y_j, Y_i)$ for every i and j. Since a correlation matrix is the covariance matrix of standardized variables, it is also symmetric and positive semidefinite.

7.3.1 Two or More Linear Combinations of Random Variables

More generally, suppose that $w_1^T Y$ and $w_2^T Y$ are two weighted averages of the components of Y, e.g., returns on two different portfolios. Then

$$\text{Cov}(w_1^T Y, w_2^T Y) = w_1^T \text{COV}(Y) w_2 = w_2^T \text{COV}(Y) w_1. \tag{7.8}$$

Example 7.3. (Example 7.2 continued)

Suppose that the random vector $Y = (Y_1, Y_2, Y_3)^T$ has the mean vector and covariance matrix used in the previous example and contains the returns on three assets. Find the covariance between a portfolio that allocates $1/3$ to each of the three assets and a second portfolio that allocates $1/2$ to each of the first two assets. That is, find the covariance between $(Y_1 + Y_2 + Y_3)/3$ and $(Y_1 + Y_2)/2$.

Answer: Let

$$w_1 = (\tfrac{1}{3} \quad \tfrac{1}{3} \quad \tfrac{1}{3})^T$$

and

$$w_2 = (\tfrac{1}{2} \quad \tfrac{1}{2} \quad 0)^T.$$

Then

$$\text{Cov}\left\{\frac{Y_1 + Y_2}{2}, \frac{Y_1 + Y_2 + Y_3}{3}\right\} = w_1^T \text{COV}(Y) w_2$$

$$= (1/3 \quad 1/3 \quad 1/3) \begin{pmatrix} 2 & 1.47 & 0 \\ 1.47 & 3 & 0 \\ 0 & 0 & 5 \end{pmatrix} \begin{pmatrix} 1/2 \\ 1/2 \\ 0 \end{pmatrix}$$

$$= (1.157 \quad 1.490 \quad 1.667) \begin{pmatrix} 1/2 \\ 1/2 \\ 0 \end{pmatrix}$$

$$= 1.323.$$

$\square$

Let $\boldsymbol{W}$ be a nonrandom $d \times q$ matrix so that $\boldsymbol{W}^\mathsf{T}\boldsymbol{Y}$ is a random vector of q linear combinations of $\boldsymbol{Y}$. Then (7.7) can be generalized to

$$\mathrm{COV}(\boldsymbol{W}^\mathsf{T}\boldsymbol{Y}) = \boldsymbol{W}^\mathsf{T}\mathrm{COV}(\boldsymbol{Y})\boldsymbol{W}. \tag{7.9}$$

Let $\boldsymbol{Y}_1$ and $\boldsymbol{Y}_2$ be two random vectors of dimensions n_1 and n_2, respectively. Then $\boldsymbol{\Sigma}_{Y_1,Y_2} = \mathrm{COV}(\boldsymbol{Y}_1, \boldsymbol{Y}_2)$ is defined as the $n_1 \times n_2$ matrix whose i,jth element is the covariance between the ith component of $\boldsymbol{Y}_1$ and the jth component of $\boldsymbol{Y}_2$, that is, $\boldsymbol{\Sigma}_{Y_1,Y_2}$ is the matrix of covariances between the random vectors $\boldsymbol{Y}_1$ and $\boldsymbol{Y}_2$.

It is not difficult to show that

$$\mathrm{Cov}(\boldsymbol{w}_1^\mathsf{T}\boldsymbol{Y}_1, \boldsymbol{w}_2^\mathsf{T}\boldsymbol{Y}_2) = \boldsymbol{w}_1^\mathsf{T}\mathrm{COV}(\boldsymbol{Y}_1, \boldsymbol{Y}_2)\boldsymbol{w}_2, \tag{7.10}$$

for constant vectors $\boldsymbol{w}_1$ and $\boldsymbol{w}_2$ of lengths n_1 and n_2.

7.3.2 Independence and Variances of Sums

If $Y_1, \ldots, Y_d$ are independent, or at least uncorrelated, then

$$\mathrm{Var}\left(\boldsymbol{w}^\mathsf{T}\boldsymbol{Y}\right) = \mathrm{Var}\left(\sum_{i=1}^{n} w_i Y_i\right) = \sum_{i=1}^{n} w_i^2 \mathrm{Var}(Y_i). \tag{7.11}$$

When $\boldsymbol{w}^\mathsf{T} = (1/n, \ldots, 1/n)$ so that $\boldsymbol{w}^\mathsf{T}\boldsymbol{Y} = \overline{Y}$, then we obtain that

$$\mathrm{Var}(\overline{Y}) = \frac{1}{n^2} \sum_{i=1}^{n} \mathrm{Var}(Y_i). \tag{7.12}$$

In particular, if $\mathrm{Var}(Y_i) = \sigma^2$ for all i, then we obtain the well-known result that if $Y_1, \ldots, Y_d$ are uncorrelated and have a constant variance σ^2, then

$$\mathrm{Var}(\overline{Y}) = \frac{\sigma^2}{n}. \tag{7.13}$$

Another useful fact that follows from (7.11) is that if Y_1 and Y_2 are uncorrelated, then

$$\mathrm{Var}(Y_1 - Y_2) = \mathrm{Var}(Y_1) + \mathrm{Var}(Y_2). \tag{7.14}$$

7.4 Scatterplot Matrices

A correlation coefficient is only a summary of the linear relationship between variables. Interesting features, such as nonlinearity or the joint behavior of extreme values, remain hidden when only correlations are examined. A solution to this problem is the so-called scatterplot matrix, which is a matrix of scatterplots, one for each pair of variables. A scatterplot matrix can be created easily with modern statistical software such as R. Figure 7.1 shows a scatterplot matrix for the CRSPday data set.

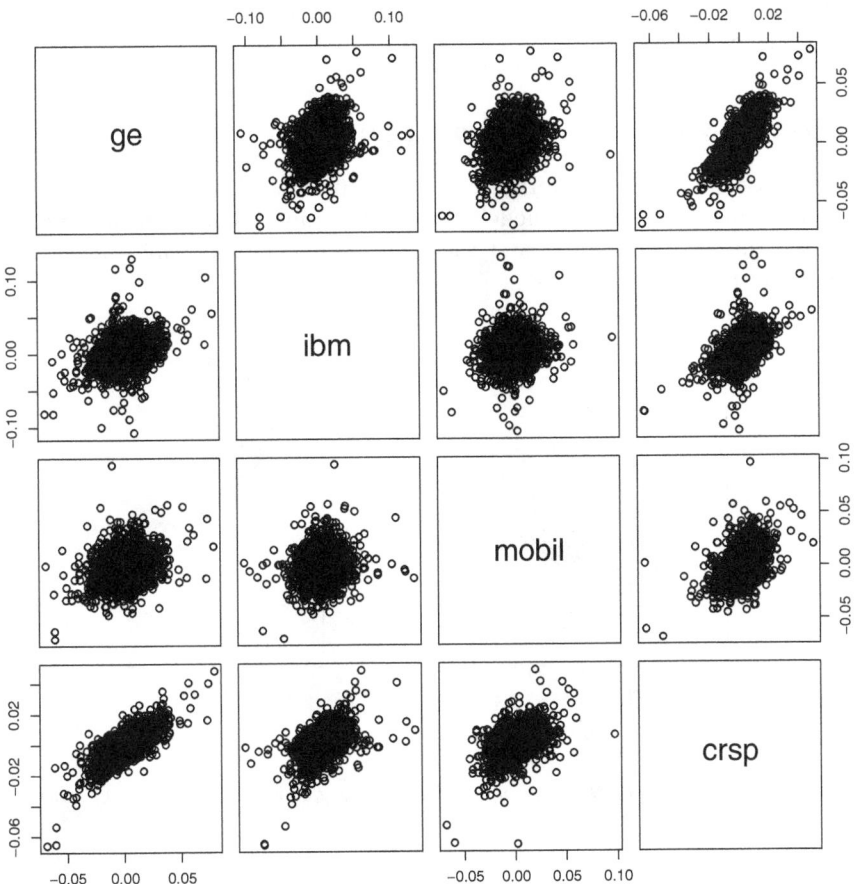

Fig. 7.1. *Scatterplot matrix for the* CRSPday *data set.*

One sees little evidence of nonlinear relationships in Fig. 7.1. This lack of nonlinearities is typical of returns on equities, but it should not be taken for granted—instead, one should always look at the scatterplot matrix. The strong linear association between GE and crsp, which was suggested before by their high correlation coefficient, can be seen also in their scatterplot.

A portfolio is riskier if large negative returns on its assets tend to occur together on the same days. To investigate whether extreme values tend to cluster in this way, one should look at the scatterplots. In the scatterplot for IBM and Mobil, extreme returns for one stock do not tend to occur on the same days as extreme returns on the other stock; this can be seen by noticing that the outliers tend to fall along the x- and y-axes. The extreme-value behavior is different with GE and `crsp`, where extreme values are more likely to occur together; note that the outliers have a tendency to occur together, that is, in the upper-right and lower-left corners, rather than being concentrated along the axes. The IBM and Mobil scatterplot is said to show *tail independence*. In contrast, the GE and `crsp` scatterplot is said to show *tail dependence*. Tail dependence is explored further in Chap. 8.

7.5 The Multivariate Normal Distribution

In Chap. 5 we saw the importance of having parametric families of univariate distributions as statistical models. Parametric families of multivariate distributions are equally useful, and the multivariate normal family is the best known of them.

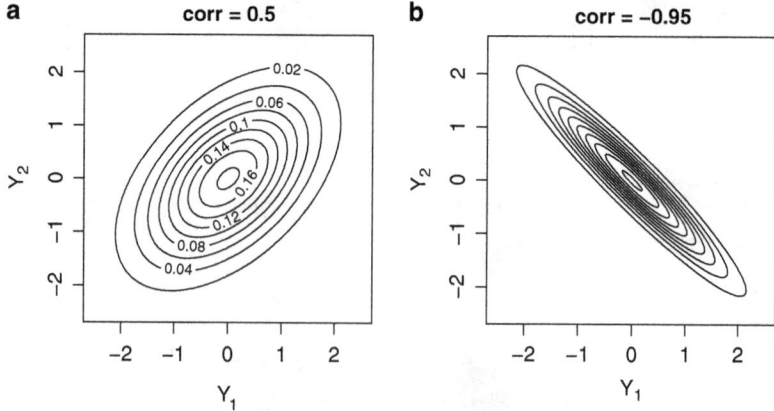

Fig. 7.2. *Contour plots of a bivariate normal densities with $N(0,1)$ marginal distributions and correlations of 0.5 or -0.95.*

The random vector $\boldsymbol{Y} = (Y_1, \ldots, Y_d)^{\mathsf{T}}$ has a d-dimensional *multivariate normal distribution* with mean vector $\boldsymbol{\mu} = (\mu_1, \ldots, \mu_d)^{\mathsf{T}}$ and covariance matrix $\boldsymbol{\Sigma}$ if its probability density function is

$$\phi_d(\boldsymbol{y}|\boldsymbol{\mu}, \boldsymbol{\Sigma}) = \left[\frac{1}{(2\pi)^{d/2}|\boldsymbol{\Sigma}|^{1/2}}\right] \exp\left\{-\frac{1}{2}(\boldsymbol{y} - \boldsymbol{\mu})^{\mathsf{T}}\boldsymbol{\Sigma}^{-1}(\boldsymbol{y} - \boldsymbol{\mu})\right\}, \quad (7.15)$$

where $|\boldsymbol{\Sigma}|$ is the determinant of $\boldsymbol{\Sigma}$. The quantity in square brackets is a constant that normalizes the density so that it integrates to 1. The density depends on $\boldsymbol{y}$ only through $(\boldsymbol{y}-\boldsymbol{\mu})^{\mathsf{T}}\boldsymbol{\Sigma}^{-1}(\boldsymbol{y}-\boldsymbol{\mu})$, and so the density is constant on each ellipse $\{\boldsymbol{y} : (\boldsymbol{y} - \boldsymbol{\mu})^{\mathsf{T}}\boldsymbol{\Sigma}^{-1}(\boldsymbol{y} - \boldsymbol{\mu}) = c\}$. Here $c > 0$ is a fixed constant that determines the size of the ellipse, with larger values of c giving larger ellipses, each centered at $\boldsymbol{\mu}$. Such densities are called *elliptically contoured*.

Figure 7.2 has contour plots of bivariate normal densities. Both Y_1 and Y_2 are $N(0,1)$ and the correlation between Y_1 and Y_2 is 0.5 in panel (a) or -0.95 in panel (b). Notice how the orientations of the contours depend on the sign and magnitude of the correlation. In panel (a) we can see that the height of the density is constant on ellipses and decreases with the distance from the mean, which is (0, 0). The same behavior occurs in panel (b), but, because of the high correlation, the contours are so close together that it was not possible to label them.

If $\boldsymbol{Y} = (Y_1, \ldots, Y_d)^\mathsf{T}$ has a multivariate normal distribution, then for *every* set of constants $\boldsymbol{c} = (c_1, \ldots, c_d)^\mathsf{T}$, the weighted average (linear combination) $\boldsymbol{c}^\mathsf{T}\boldsymbol{Y} = c_1 Y_1 + \cdots + c_d Y_d$ has a normal distribution with mean $\boldsymbol{c}^\mathsf{T}\boldsymbol{\mu}$ and variance $\boldsymbol{c}^\mathsf{T}\boldsymbol{\Sigma}\boldsymbol{c}$. In particular, the marginal distribution of Y_i is $N(\mu_i, \sigma_i^2)$, where σ_i^2 is the ith diagonal element of $\boldsymbol{\Sigma}$—to see this, take $c_i = 1$ and $c_j = 0$ for $j \neq i$.

The assumption of multivariate normality facilitates many useful probability calculations. If the returns on a set of assets have a multivariate normal distribution, then the return on any portfolio formed from these assets will be normally distributed. This is because the return on the portfolio is the weighted average of the returns on the assets. Therefore, the normal distribution could be used, for example, to find the probability of a loss of some size of interest, say, 10 % or more, on the portfolio. Such calculations have important applications in finding a value-at-risk; see Chap. 19.

Unfortunately, we saw in Chap. 5 that often individual returns are not normally distributed, which implies that a vector of returns will not have a multivariate normal distribution. In Sect. 7.6 we will look at an important class of heavy-tailed multivariate distributions.

7.6 The Multivariate t-Distribution

We have seen that the univariate t-distribution is a good model for the returns of individual assets. Therefore, it is desirable to have a model for vectors of returns such that the univariate marginals are t-distributed. The multivariate t-distribution has this property. The random vector $\boldsymbol{Y}$ has a multivariate $t_\nu(\boldsymbol{\mu}, \boldsymbol{\Lambda})$ distribution if

$$\boldsymbol{Y} = \boldsymbol{\mu} + \sqrt{\frac{\nu}{W}}\boldsymbol{Z}, \tag{7.16}$$

where W is chi-squared distributed with ν degrees of freedom, $\boldsymbol{Z}$ is $N_d(0, \boldsymbol{\Lambda})$ distributed, and W and $\boldsymbol{Z}$ are independent. Thus, the multivariate t-distribution is a continuous scale mixture of multivariate normal distributions. Extreme values of $\boldsymbol{Y}$ tend to occur when W is near zero. Since $W^{-1/2}$ multiplies all components of $\boldsymbol{Z}$, outliers in one component tend to occur with outliers in other components, that is, there is tail dependence.

For $\nu > 1$, $\boldsymbol{\mu}$ is the mean vector of $\boldsymbol{Y}$. For $0 < \nu \leq 1$, the expectation of $\boldsymbol{Y}$ does not exist, but $\boldsymbol{\mu}$ can still be regarded as the "center" of the distribution of Y because, for any value of ν, the vector $\boldsymbol{\mu}$ contains the medians of the components of Y and the contours of the density of Y are ellipses centered at $\boldsymbol{\mu}$. Also, $\boldsymbol{\mu}$ is the mode of the distribution, that is, the location where the density is maximized.

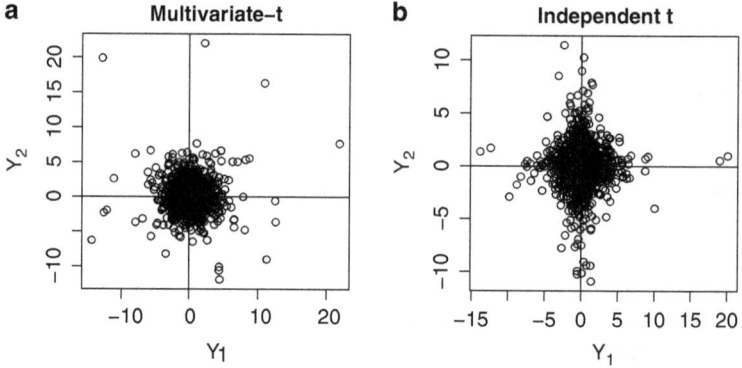

Fig. 7.3. *(a) Plot of a random sample from a bivariate t-distribution with $\nu = 3$, $\boldsymbol{\mu} = (0\ 0)^{\mathsf{T}}$ and identity covariate matrix. (b) Plot of a random sample of pairs of independent $t_3(0,1)$ random variables. Both sample sizes are 2,500.*

For $\nu > 2$, the covariance matrix of Y exists and is

$$\boldsymbol{\Sigma} = \frac{\nu}{\nu - 2}\boldsymbol{\Lambda}. \tag{7.17}$$

We will call $\boldsymbol{\Lambda}$ the *scale matrix*. The scale matrix exists for all values of ν. Since the covariance matrix $\boldsymbol{\Sigma}$ of Y is just a multiple of the covariance matrix $\boldsymbol{\Lambda}$ of Z, Y and Z have the same correlation matrices, assuming $\nu > 2$ so that the correlation matrix of Y exists. If $\Sigma_{i,j} = 0$, then Y_i and Y_j are uncorrelated, but they are dependent, nonetheless, because of the tail dependence. Tail dependence is illustrated in Fig. 7.3, where panel (a) is a plot of 2500 observations from an uncorrelated bivariate t-distribution with marginal distributions that are $t_3(0,1)$. For comparison, panel (b) is a plot of 2500 observations of pairs of independent $t_3(0,1)$ random variables—these pairs do not have a bivariate t-distribution. Notice that in (b), outliers in Y_1 are not associated with outliers in Y_2, since the outliers are concentrated near the x- and y-axes. In contrast, outliers in (a) are distributed uniformly in all directions. The univariate marginal distributions are the same in (a) and (b).

Tail dependence can be expected in equity returns. For example, on Black Monday, almost all equities had extremely large negative returns. Of course, Black Monday was an extreme, even among extreme events. We would not want to reach any general conclusions based upon Black Monday alone. However, in Fig. 7.1, we see little evidence that outliers are concentrated along the axes, with the possible exception of the scatterplot for IBM and Mobil. As another example of dependencies among stock returns, Fig. 7.4 contains a

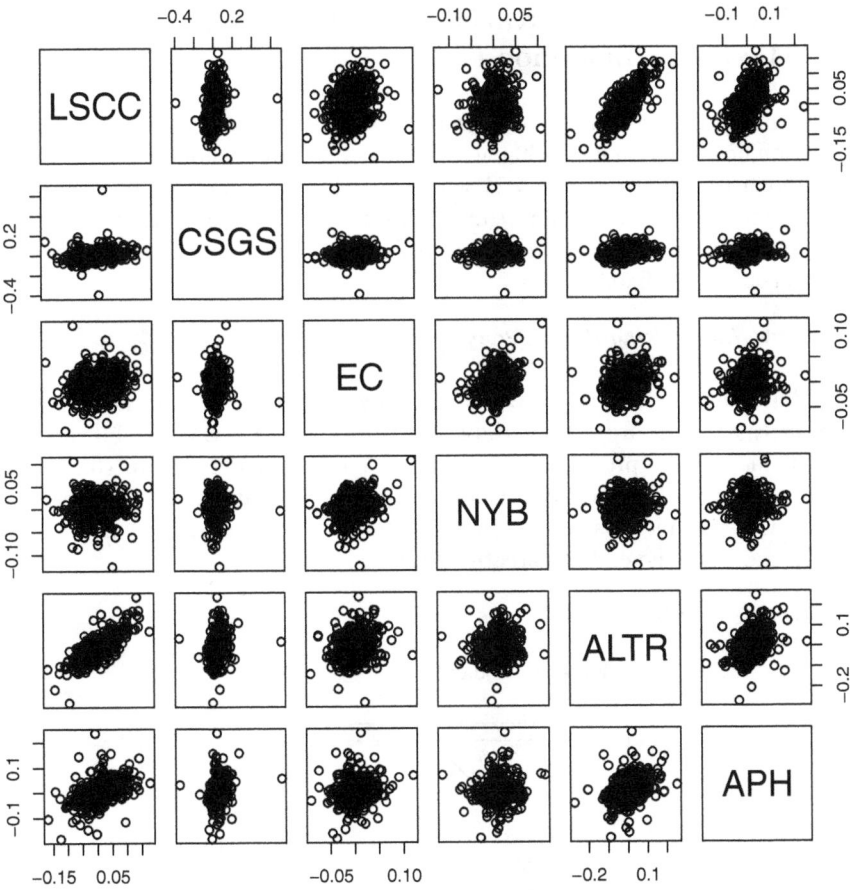

Fig. 7.4. *Scatterplot matrix of 500 daily returns on six midcap stocks in R's* `midcapD.ts` *data set.*

scatterplot matrix of returns on six midcap stocks in the `midcapD.ts` data set. Again, tail dependence can be seen. This suggests that tail dependence is common among equity returns and the multivariate t-distribution is a promising model for them.

7.6.1 Using the t-Distribution in Portfolio Analysis

If Y has a $t_\nu(\boldsymbol{\mu}, \boldsymbol{\Lambda})$ distribution, which we recall has covariance matrix $\boldsymbol{\Sigma} = \{\nu/(\nu - 2)\}\boldsymbol{\Lambda}$, and $\boldsymbol{w}$ is a vector of weights, then $\boldsymbol{w}^\mathsf{T}\boldsymbol{Y}$ has a univariate t-distribution with mean $\boldsymbol{w}^\mathsf{T}\boldsymbol{\mu}$ and variance $\{\nu/(\nu - 2)\}\boldsymbol{w}^\mathsf{T}\boldsymbol{\Lambda}\boldsymbol{w} = \boldsymbol{w}^\mathsf{T}\boldsymbol{\Sigma}\boldsymbol{w}$. This fact can be useful when computing risk measures for a portfolio. If the returns on the assets have a multivariate t-distribution, then the return on the portfolio will have a univariate t-distribution. We will make use of this result in Chap. 19.

7.7 Fitting the Multivariate t-Distribution by Maximum Likelihood

To estimate the parameters of a multivariate t-distribution, one can use the function cov.trob in R's MASS package. This function computes the maximum likelihood estimates of μ and Λ with ν fixed. To estimate ν, one computes the profile log-likelihood for ν and finds the value, $\hat{\nu}$ of ν that maximizes the profile log-likelihood. Then the MLEs of μ and Λ are the estimates from cov.trob with ν fixed at $\hat{\nu}$.

Example 7.4. Fitting the CRSPday *data*

This example uses the data set CRSPday analyzed earlier in Example 7.1. Recall that there are four variables, returns on GE, IBM, Mobil, and the CRSP index. The profile log-likelihood is plotted in Fig. 7.5. In that figure, one can see that the MLE of ν is 5.94, and there is relatively little uncertainty about this parameter's value—the 95 % profile likelihood confidence interval is (5.41, 6.55). The code to create this figure is below.

```
library(mnormt)
library(MASS)
data(CRSPday, package = "Ecdat")
```

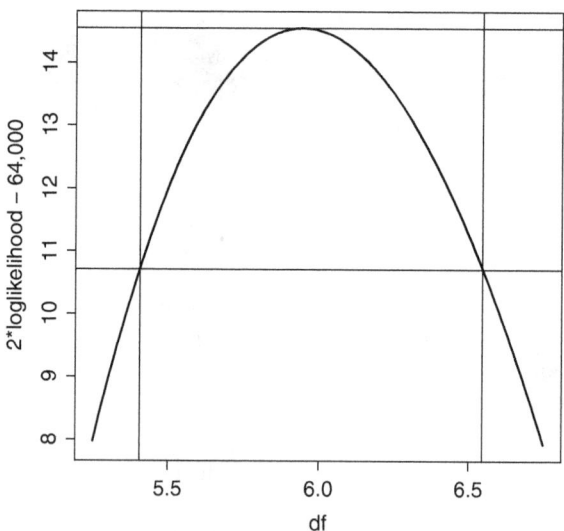

Fig. 7.5. CRSPday *data. A profile likelihood confidence interval for* ν. *The solid curve is* $2L_{\max}(\nu)$, *where* $L_{\max}(\nu)$ *is the profile likelihood minus 32,000. 32,000 was subtracted from the profile likelihood to simplify the labeling of the y-axis. The horizontal line intersects the y-axis at* $2L_{\max}(\hat{\nu}) - \chi^2_{\alpha,1}$, *where* $\hat{\nu}$ *is the MLE and* $\alpha = 0.05$. *All values of* ν *such that* $2L_{\max}(\nu)$ *is above the horizontal line are in the profile likelihood 95 % confidence interval. The two vertical lines intersect the x-axis at 5.41 and 6.55, the endpoints of the confidence interval.*

```
dat =CRSPday[ , 4:7]
df = seq(5.25, 6.75, 0.01)
n = length(df)
loglik = rep(0,n)
for(i in 1:n){
  fit = cov.trob(dat,nu=df)
  loglik[i] = sum(log(dmt(dat, mean=fit$center,
  S = fit$cov, df = df[i])))
}
aic_t = -max(2 * loglik) +  2 * (4 + 10 + 1) + 64000
z1 = (2 * loglik > 2 * max(loglik) - qchisq(0.95, 1))
plot(df, 2 * loglik - 64000, type = "l", cex.axis = 1.5,
    cex.lab = 1.5, ylab = "2 * loglikelihood - 64,000", lwd = 2)
abline(h = 2 * max(loglik) - qchisq(0.95, 1 ) - 64000)
abline(h = 2 * max(loglik) - 64000)
abline(v = (df[16] + df[17]) / 2)
abline(v = (df[130] + df[131]) / 2)
```

AIC for this model is 15.45 plus 64,000. Here AIC values are expressed as deviations from 64,000 to keep these values small. This is helpful when comparing two or more models via AIC. Subtracting the same constant from all AIC values, of course, has no effect on model comparisons.

The maximum likelihood estimates of the mean vector and the correlation matrix are called $center and $cor, respectively, in the following output:

```
$center
[1] 0.0009424 0.0004481 0.0006883 0.0007693

$cor
        [,1]   [,2]   [,3]   [,4]
[1,] 1.0000 0.3192 0.2845 0.6765
[2,] 0.3192 1.0000 0.1584 0.4698
[3,] 0.2845 0.1584 1.0000 0.4301
[4,] 0.6765 0.4698 0.4301 1.0000
```

These estimates were computed using cov.trob with ν fixed at 6.

When the data are *t*-distributed, the maximum likelihood estimates are superior to the sample mean and covariance matrix in several respects—the MLE is less variable and it is less sensitive to outliers. However, in this example, the maximum likelihood estimates are similar to the sample mean and correlation matrix. For example, the sample correlation matrix is

```
          ge    ibm  mobil   crsp
ge     1.0000 0.3336 0.2972 0.7148
ibm    0.3336 1.0000 0.1587 0.4864
mobil  0.2972 0.1587 1.0000 0.4294
crsp   0.7148 0.4864 0.4294 1.0000
```

□

7.8 Elliptically Contoured Densities

The multivariate normal and t-distributions have *elliptically contoured* densities, a property that will be discussed in this section. A d-variate multivariate density f is elliptically contoured if can be expressed as

$$f(\boldsymbol{y}) = |\boldsymbol{\Lambda}|^{-1/2} g \left\{ (\boldsymbol{y} - \boldsymbol{\mu})^{\mathsf{T}} \boldsymbol{\Lambda}^{-1} (\boldsymbol{y} - \boldsymbol{\mu}) \right\}, \tag{7.18}$$

where g is a nonnegative-valued function such that $1 = \int_{\Re^d} g\left(\|\boldsymbol{y}\|^2\right) d\boldsymbol{y}$, $\boldsymbol{\mu}$ is a $d \times 1$ vector, and $\boldsymbol{\Lambda}$ is a $d \times d$ symmetric, positive definite matrix. Usually, $g(x)$ is a decreasing function of $x \geq 0$, and we will assume this is true. We will also assume the finiteness of second moments, in which case $\boldsymbol{\mu}$ is the mean vector and the covariance matrix $\boldsymbol{\Sigma}$ is a scalar multiple of $\boldsymbol{\Lambda}$.

For each fixed $c > 0$,

$$\mathcal{E}(c) = \{\boldsymbol{y} : (\boldsymbol{y} - \boldsymbol{\mu})^{\mathsf{T}} \boldsymbol{\Sigma}^{-1} (\boldsymbol{y} - \boldsymbol{\mu}) = c\}$$

is an ellipse centered at $\boldsymbol{\mu}$, and if $c_1 > c_2$, then $\mathcal{E}(c_1)$ is inside $\mathcal{E}(c_2)$ because g is decreasing. The contours of f are concentric ellipses as can be seen in Fig. 7.6.

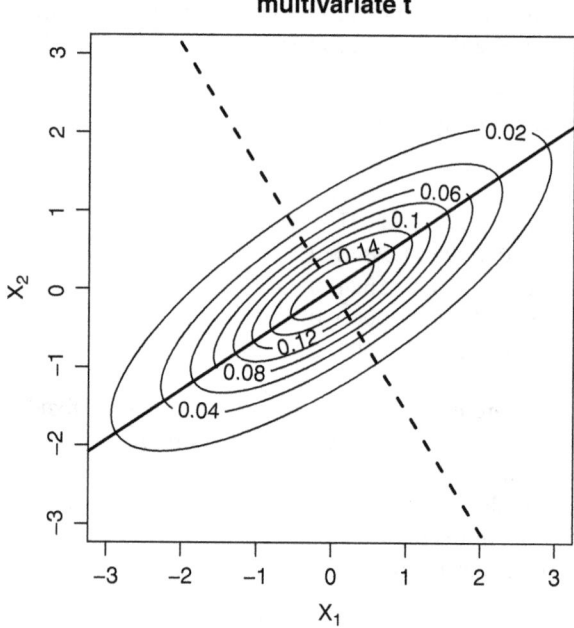

Fig. 7.6. *Contour plot of a multivariate t_4-density with $\mu = (0,0)^{\mathsf{T}}$, $\sigma_1^2 = 2$, $\sigma_2^2 = 1$, and $\sigma_{12} = 1.1$.*

That figure shows the contours of the bivariate t_4-density with $\boldsymbol{\mu} = (0,0)^\mathsf{T}$ and

$$\boldsymbol{\Sigma} = \begin{pmatrix} 2 & 1.1 \\ 1.1 & 1 \end{pmatrix}.$$

The major axis of the ellipses is a solid line and the minor axis is a dashed line.

How can the axes be found? From Appendix A.20, we know that $\boldsymbol{\Sigma}$ has an *eigenvalue-eigenvector decomposition*

$$\boldsymbol{\Sigma} = \boldsymbol{O}\,\mathrm{diag}(\lambda_i)\,\boldsymbol{O}^\mathsf{T},$$

where $\boldsymbol{O}$ is an orthogonal matrix whose columns are the eigenvectors of $\boldsymbol{\Sigma}$ and $\lambda_1, \ldots, \lambda_d$ are the eigenvalues of $\boldsymbol{\Sigma}$.

The columns of $\boldsymbol{O}$ determine the axes of the ellipse $\mathcal{E}(c)$. The decomposition can be found in R using the function `eigen()` and, for the matrix $\boldsymbol{\Sigma}$ in the example, the decomposition is

```
$values
[1] 2.708 0.292
```

which gives the eigenvalues, and

```
$vectors
        [,1]    [,2]
[1,] -0.841  0.541
[2,] -0.541 -0.841
```

which has the corresponding eigenvectors as columns; e.g., $(-0.841, -0.541)$ is an eigenvector with eigenvalue 2.708. The eigenvectors are normalized so have norm equal to 1. Nonetheless, the eigenvectors are only determined up to a sign change, so the first eigenvector could be taken as $(-0.841, -0.541)$, as in the R output, or $(0.841, 0.541)$.

If $\boldsymbol{o}_i$ is the ith column of $\boldsymbol{O}$, the ith axis of $\mathcal{E}(c)$ goes through the points $\boldsymbol{\mu}$ and $\boldsymbol{\mu} + \boldsymbol{o}_i$. Therefore, this axis is the line

$$\{\boldsymbol{\mu} + k\,\boldsymbol{o}_i : -\infty < k < \infty\}.$$

Because $\boldsymbol{O}$ is an orthogonal matrix, the axes are mutually perpendicular. The axes can be ordered according to the size of the corresponding eigenvalues. In the bivariate case the axis associated with the largest (smallest) eigenvalue is the major (minor) axis. We are assuming that there are no ties among the eigenvalues.

Since $\boldsymbol{\mu} = 0$, in our example the major axis is $k\,(0.841, 0.541)$, $-\infty < k < \infty$, and the minor axis is $k\,(0.541, -0.841)$, $-\infty < k < \infty$.

When there are ties among the eigenvalues, the eigenvectors are not unique and the analysis is somewhat more complicated and will not be discussed in detail. Instead two examples will be given. In the bivariate case if $\boldsymbol{\Sigma} = \boldsymbol{I}$, the contours are circles and there is no unique choice of the axes—any pair of perpendicular vectors will do. As a trivariate example, if $\boldsymbol{\Sigma} = \mathrm{diag}(1,1,3)$, then the first principle axis is $(0,0,1)$ with eigenvalue 3. The second and third principal axis can be any perpendicular pair of vectors with third coordinates equal to 0. The `eigen()` function in R returns (0,1,0) and (1,0,0) as the second and third axes.

7.9 The Multivariate Skewed t-Distributions

Azzalini and Capitanio (2003) have proposed a skewed extension of the multivariate t-distribution. The univariate special case was discussed in Sect. 5.7. In the multivariate case, in addition to the shape parameter ν determining tail weight, the skewed t-distribution has a vector $\boldsymbol{\alpha} = (\alpha_1, \ldots, \alpha_d)^{\mathsf{T}}$ of shape parameters determining the amounts of skewness in the components of the distribution. If $\boldsymbol{Y}$ has a skewed t-distribution, then Y_i is left-skewed, symmetric, or right-skewed depending on whether $\alpha_i < 0$, $\alpha_i = 0$, or $\alpha_i > 0$.

Figure 7.7 is a contour plot of a bivariate skewed t-distribution with $\boldsymbol{\alpha} = (-1, 0.25)^{\mathsf{T}}$ and df $= 4$. Notice that, because α_1 is reasonably large and negative, Y_1 has a considerable amount of left skewness, as can be seen in the contours, which are more widely spaced on the left side of the plot compared to the right. Also, Y_2 shows a lesser amount of right skewness, since the contours on top are slightly more widely spaced than on the bottom. This feature is to be expected since α_2 is positive but with a relatively small absolute value.

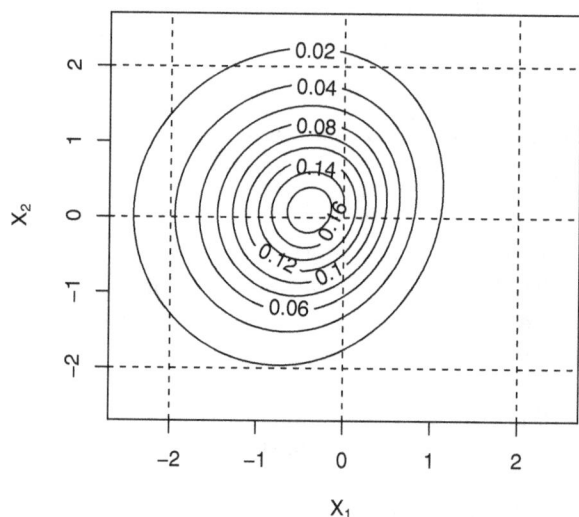

Fig. 7.7. *Contours of a bivariate skewed t-density. The contours are more widely spaced on the left compared to the right because X_1 is left-skewed. Similarly, the contours are more widely spaced on the top compared to the bottom because X_2 is right-skewed, but the skewness of X_2 is relatively small and less easy to see.*

Example 7.5. Fitting the skewed t-distribution to CRSPday

We now fit the skewed t-model to the CRSPday data set using the function mst.mple() in R's sn package. This function maximizes the likelihood over all parameters, so there is no need to use the profile likelihood as with cov.trob(). The code is below.

```
library(sn)
data(CRSPday, package = "Ecdat")
dat = CRSPday[ , 4:7]
fit = mst.mple(y = dat, penalty = NULL)
aic_skewt = -2 * fit$logL + 64000 + 2 * (4 + 10 + 4 + 1)
dp2cp(fit$dp,"st")
aic_skewt
```

The CP estimates are as follows.

```
> dp2cp(fit$dp,"st")
$beta
              ge          ibm         mobil         crsp
[1,] 0.0009459182 0.0004521179 0.0006917701 0.0007722816


$var.cov
                ge          ibm         mobil         crsp
ge      1.899520e-04 7.252242e-05 5.185778e-05 6.957078e-05
ibm     7.252242e-05 2.743354e-04 3.492763e-05 5.771567e-05
mobil   5.185778e-05 3.492763e-05 1.749708e-04 4.238468e-05
crsp    6.957078e-05 5.771567e-05 4.238468e-05 5.565159e-05


$gamma1
            ge          ibm         mobil         crsp
0.0010609438 0.0012389968 0.0007125122 0.0009920253


$gamma2M
[1] 25.24996
```

Here $beta is the estimate of the means, $var.cov is the estimate of covariance matrix, $gamma1 is the estimate of skewnesses, and $gamma2M estimates the common kurtosis of the four marginal distributions. The DP estimates are in fit$dp but are of less interest so are not included here.

AIC for the skewed t-model is 23.47885 (plus 64,000), larger than 15.45, the AIC found in Example 7.4 for the symmetric t-model. This result suggests that the symmetric t-model is adequate for this data set.

In summary, the CRSPday data are well fit by a symmetric t-distribution and no need was found for using a skewed t-distribution. Also, in the normal plots of the four variables in in Fig. 7.8, heavy tails are evident but there are no signs of serious skewness. Although this might be viewed as a negative result, since we have not found an improvement in fit by going to the more

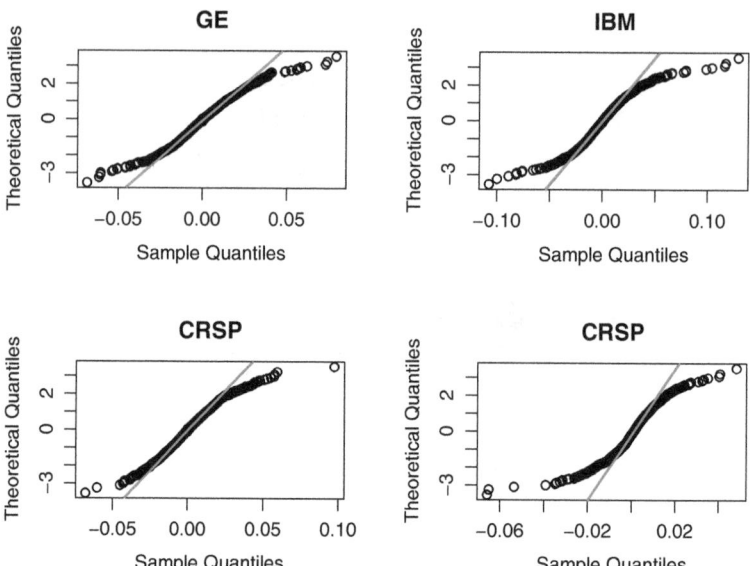

Fig. 7.8. *Normal plots of the four returns series in the* CRSPday *data set. The reference lines go through the first and third quartiles.*

flexible skewed t-distribution, the result does give us more confidence that the symmetric t-distribution is suitable for modeling this data set.   □

7.10 The Fisher Information Matrix

In the discussion of Fisher information in Sect. 5.10, θ was assumed to be one-dimensional. If θ is an m-dimensional parameter vector, then the Fisher information matrix is an $m \times m$ square matrix, $\mathcal{I}$, and is equal the matrix of expected second-order partial derivatives of $-\log\{L(\theta)\}$.[1] In other words, the i, jth entry of the Fisher information matrix is

$$\mathcal{I}_{ij}(\theta) = -E\left[\frac{\partial^2}{\partial\theta_i\,\partial\theta_j}\log\{L(\theta)\}\right]. \tag{7.19}$$

The standard errors are the square roots of the diagonal entries of the inverse of the Fisher information matrix. Thus, the standard error for θ_i is

$$s_{\widehat{\theta}_i} = \sqrt{\{\mathcal{I}(\widehat{\theta})^{-1}\}_{ii}}. \tag{7.20}$$

---

[1] The matrix of second partial derivatives of a function is called its *Hessian matrix*, so the Fisher information matrix is the expectation of the Hessian of the negative log-likelihood.

In the case of a single parameter, (7.20) reduces to (5.19). The central limit theorem for the MLE in Sect. 5.10 generalizes to the following multivariate version.

Result 7.6. *Under suitable assumptions, for large enough sample sizes, the maximum likelihood estimator is approximately normally distributed with mean equal to the true parameter vector and with covariance matrix equal to the inverse of the Fisher information matrix.*

Computation of the expectation in $\mathcal{I}(\boldsymbol{\theta})$ can be challenging. Programming the second derivatives can be difficult as well, especially for complex models. In practice, the observed Fisher information matrix, whose i, jth element is

$$\mathcal{I}_{ij}^{\mathrm{obs}}(\boldsymbol{\theta}) = -\frac{\partial^2}{\partial \theta_i \, \partial \theta_j} \, \log\{L(\boldsymbol{\theta})\} \qquad (7.21)$$

is often used. The observed Fisher information matrix is, of course, the multivariate analog of (5.21). Using observed information obviates the need to compute the expectation. Moreover, the Hessian matrix can be computed numerically by finite differences, for example, using R's fdHess() function in the nlme package. Also, as demonstrated in several examples in Chap. 5, if hessian=TRUE in the call to optim(), then the Hessian matrix is returned when the negative log-likelihood is minimized by that function.

Inverting the observed Fisher information matrix computed by finite differences is the most commonly used method for obtaining standard errors. The advantage of this approach is that only the computation of the log-likelihood is necessary, and of course this computation is necessary simply to compute the MLE.

The key point is that there is an explicit method of calculating standard errors for maximum likelihood estimators. The calculation of standard errors of maximum likelihood estimators by computing and then inverting the observed Fisher information matrix is routinely programmed into statistical software, e.g., by the R function fitdistr() used to fit univariate distributions.

7.11 Bootstrapping Multivariate Data

When resampling multivariate data, the dependencies within the observation vectors need to be preserved. Let the vectors $\boldsymbol{Y}_1, \ldots, \boldsymbol{Y}_n$ be an i.i.d. sample of multivariate data. In model-free resampling, the vectors $\boldsymbol{Y}_1, \ldots, \boldsymbol{Y}_n$ are sampled with replacement. There is no resampling of the components within a vector. Resampling within vectors would make their components mutually independent and would not mimic the actual data where the components are dependent. Stated differently, if the data are in a spreadsheet (or matrix) with rows corresponding to observations and columns to variables, then one samples entire rows.

Model-based resampling simulates vectors from the multivariate distribution of the $\boldsymbol{Y}_i$, for example, from a multivariate t-distribution with the mean vector, covariance matrix, and degrees of freedom equal to the MLEs.

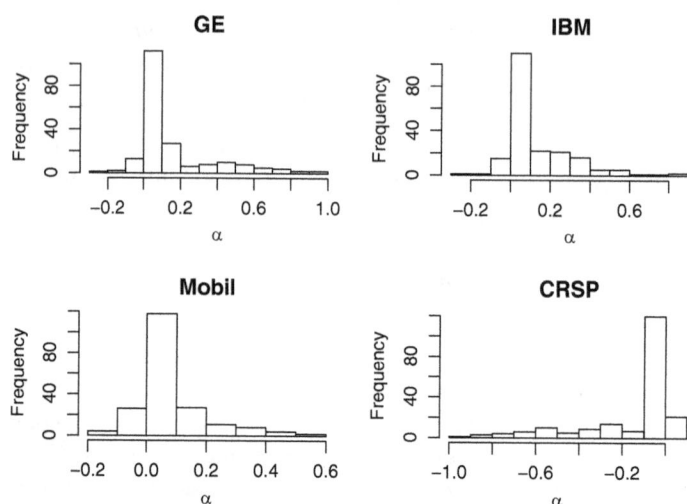

Fig. 7.9. *Histograms of* 200 *bootstrapped values of* $\widehat{\alpha}$ *for each of the returns series in the* CRSPday *data set.*

Example 7.7. Bootstrapping the skewed t fit to CRSPday

In Example 7.5 the skewed t-model was fit to the CRSPday data. This example continues that analysis by bootstrapping the estimator of α for each of the four returns series. Histograms of 200 bootstrap values of $\widehat{\alpha}$ are found in Fig. 7.9. Bootstrap percentile 95 % confidence intervals include 0 for all four stocks, so there is no strong evidence of skewness in any of the returns series.

Despite the large sample size of 2,528, the estimators of α do not appear to be normally distributed. We can see in Fig. 7.9 that they are right-skewed for the three stocks and left-skewed for the CRSP returns. The distribution of $\widehat{\alpha}$ also appears heavy-tailed. The excess kurtosis coefficient of the 200 bootstrap values of $\widehat{\alpha}$ is 2.38, 1.33, 3.18, and 2.38 for the four series.

The central limit theorem for the MLE guarantees that $\widehat{\alpha}$ is nearly normally distributed for sufficiently large samples, but this theorem does not tell us how large the sample size must be. Fortunately, the sample size needed for near normality is often small, but there are exceptions. We see in this example that in such cases the sample size must be very large indeed since 2,528 is not large enough. This is a major reason for preferring to construct confidence intervals using the bootstrap rather than a normal approximation.

A bootstrap sample of the returns was drawn with the following R code. The returns are in the matrix `dat` and `yboot` is a bootstrap sample chosen by taking a random sample of the rows of `dat`, with replacement of course.

```
yboot = dat[sample((1:n), n, replace = TRUE), ]
```

☐

7.12 Bibliographic Notes

The multivariate central limit theorem for the MLE is stated precisely and proved in textbooks on asymptotic theory such as Lehmann (1999) and van der Vaart (1998). The multivariate skewed t-distribution is in Azzalini and Capitanio (2003) and Azzalini (2014).

7.13 R Lab

7.13.1 Equity Returns

This section uses the data set `berndtInvest` on the book's web site and taken originally from R's `fEcofin` package. This data set contains monthly returns from January 1, 1987, to December 1, 1987, on 16 equities. There are 18 columns. The first column is the date and the last is the risk-free rate.

In the lab we will only use the first four equities. The following code computes the sample covariance and correlation matrices for these returns.

```
berndtInvest = read.csv("berndtInvest.csv")
Berndt = as.matrix(berndtInvest[, 2:5])
cov(Berndt)
cor(Berndt)
```

If you wish, you can also plot a scatterplot matrix with the following R code.

```
pairs(Berndt)
```

Problem 1 *Suppose the four variables being used are denoted by $X_1, \ldots, X_4$. Use the sample covariance matrix to estimate the variance of $0.5X_1 + 0.3X_2 + 0.2X_3$. (Useful R facts: "$t(a)$" is the transpose of a vector or matrix a and "a %*% b" is the matrix product of a and b.)*

Fit a multivariate-t model to the data using the function `cov.trob` in the MASS package. This function computes the MLE of the mean and covariance matrix with a fixed value of ν. To find the MLE of ν, the following code computes the profile log-likelihood for ν.

```
library(MASS)  # needed for cov.trob
library(mnormt)  # needed for dmt
df = seq(2.5, 8, 0.01)
n = length(df)
loglik_profile = rep(0, n)
for(i in 1:n)
{
  fit = cov.trob(Berndt, nu = df[i])
  mu = as.vector(fit$center)
  sigma = matrix(fit$cov, nrow = 4)
  loglik_profile[i] = sum(log(dmt(Berndt, mean = fit$center,
    S= f it$cov, df = df[i])))
}
```

Problem 2 *Using the results produced by the code above, find the MLE of
ν and a 90 % profile likelihood confidence interval for ν. Include your* R *code
with your work. Also, plot the profile log-likelihood and indicate the MLE and
the confidence interval on the plot.*

Section 7.13.3 demonstrates how the MLE for a multivariate t-model can
be fit directly with the `optim` function, rather than profile likelihood.

7.13.2 Simulating Multivariate t-Distributions

The following code generates and plots four bivariate samples. Each sample has univariate marginals that are standard t_3-distributions. However,
the dependencies are different.

```
library(MASS)  # need for mvrnorm
par(mfrow=c(1,4))
N =   2500
nu = 3

set.seed(5640)
cov=matrix(c(1, 0.8, 0.8, 1), nrow = 2)
x= mvrnorm(N, mu = c(0, 0), Sigma = cov)
w = sqrt(nu / rchisq(N, df = nu))
x = x * cbind(w, w)
plot(x, main = "(a)")

set.seed(5640)
cov=matrix(c(1, 0.8, 0.8, 1),nrow = 2)
x= mvrnorm(N, mu = c(0, 0), Sigma = cov)
w1 = sqrt(nu / rchisq(N, df = nu))
w2 = sqrt(nu / rchisq(N, df = nu))
x = x * cbind(w1, w2)
plot(x, main = "(b)")
```

```
set.seed(5640)
cov=matrix(c(1, 0, 0, 1), nrow = 2)
x= mvrnorm(N, mu = c(0, 0), Sigma = cov)
w1 = sqrt(nu / rchisq(N, df = nu))
w2 = sqrt(nu / rchisq(N, df = nu))
x = x * cbind(w1, w2)
plot(x, main = "(c)")

set.seed(5640)
cov=matrix(c(1, 0, 0, 1), nrow = 2)
x= mvrnorm(N, mu = c(0, 0), Sigma = cov)
w = sqrt(nu / rchisq(N, df = nu))
x = x * cbind(w, w)
plot(x, main = "(d)")
```

Note the use of these R commands: set.seed to set the seed of the random number generator, mvrnorm to generate multivariate normally distributed vectors, rchisq to generate χ^2-distributed random numbers, cbind to bind together vectors as the columns of a matrix, and matrix to create a matrix from a vector. In R, "a * b" is elementwise multiplication of same-size matrices a and b, and "a %*% b" is matrix multiplication of conforming matrices a and b.

Problem 3 *Which sample has independent variates? Explain your answer.*

Problem 4 *Which sample has variates that are correlated but do not have tail dependence? Explain your answer.*

Problem 5 *Which sample has variates that are uncorrelated but with tail dependence? Explain your answer.*

Problem 6 * *Suppose that* (X, Y) *are the returns on two assets and have a multivariate t-distribution with degrees of freedom, mean vector, and covariance matrix*

$$\nu = 5, \quad \mu = \begin{pmatrix} 0.001 \\ 0.002 \end{pmatrix}, \quad \Sigma = \begin{pmatrix} 0.10 & 0.03 \\ 0.03 & 0.15 \end{pmatrix}.$$

Then $R = (X + Y)/2$ *is the return on an equally weighted portfolio of the two assets.*

(a) What is the distribution of R?
(b) Write an R program to generate a random sample of size 10,000 from the distribution of R. Your program should also compute the 0.01 upper quantile of this sample and the sample average of all returns that exceed this quantile. This quantile and average will be useful later when we study risk analysis.

7.13.3 Fitting a Bivariate t-Distribution

When you run the R code that follows this paragraph, you will compute the MLE for a bivariate t-distribution fit to CRSP returns data. A challenge when fitting a multivariate distribution is enforcing the constraint that the scale matrix (or the covariance matrix) must be positive definite. One way to meet this challenge is to let the scale matrix be $A^T A$, where A is an upper triangular matrix. (It is easy to show that $A^T A$ is positive semidefinite if A is any square matrix. Because a scale or covariance matrix is symmetric, only the entries on and above the main diagonal are free parameters. In order for A to have the same number of free parameters as the covariance matrix, we restrict A to be upper triangular.)

```
library(mnormt)
data(CRSPday, package = "Ecdat")
Y = CRSPday[ , c(5, 7)]
loglik = function(par)
{
mu = par[1:2]
A = matrix(c(par[3], par[4], 0, par[5]), nrow = 2, byrow = T)
scale_matrix = t(A) %*% A
df = par[6]
-sum(log(dmt(Y, mean = mu, S = scale_matrix, df = df)))
}
A = chol(cov(Y))
start = as.vector(c(apply(Y, 2, mean),
   A[1, 1], A[1, 2], A[2, 2], 4))
fit_mvt = optim(start, loglik, method = "L-BFGS-B",
   lower = c(-0.02, -0.02, -0.1, -0.1, -0.1, 2),
   upper = c(0.02, 0.02, 0.1, 0.1, 0.1, 15), hessian = T)
```

Problem 7 * Let $\theta = (\mu_1, \mu_2, A_{1,1}, A_{1,2}, A_{2,2}, \nu)$, where μ_j is the mean of the jth variable, $A_{1,1}$, $A_{1,2}$, and $A_{2,2}$ are the nonzero elements of A, and ν is the degrees-of-freedom parameter.

(a) What does the code A = chol(cov(Y)) do?

(b) Find $\widehat{\theta}_{\mathrm{ML}}$, the MLE of θ.

(c) Find the Fisher information matrix for θ. (Hint: The Hessian is part of the object fit_mvt. Also, the R function solve will invert a matrix.)

(d) Find the standard errors of the components of $\widehat{\theta}_{\mathrm{ML}}$ using the Fisher information matrix.

(e) Find the MLE of the covariance matrix of the returns.

(f) Find the MLE of ρ, the correlation between the two returns (Y_1 and Y_2).

7.14 Exercises

1. Suppose that $E(X) = 1$, $E(Y) = 1.5$, $\text{Var}(X) = 2$, $\text{Var}(Y) = 2.7$, and $\text{Cov}(X, Y) = 0.8$.
   (a) What are $E(0.2X + 0.8Y)$ and $\text{Var}(0.2X + 0.8Y)$?
   (b) For what value of w is $\text{Var}\{wX + (1-w)Y\}$ minimized? Suppose that X is the return on one asset and Y is the return on a second asset. Why would it be useful to minimize $\text{Var}\{wX + (1-w)Y\}$?
2. Let X_1, X_2, Y_1, and Y_2 be random variables.
   (a) Show that $\text{Cov}(X_1 + X_2, Y_1 + Y_2) = \text{Cov}(X_1, Y_1) + \text{Cov}(X_1, Y_2) + \text{Cov}(X_2, Y_1) + \text{Cov}(X_2, Y_2)$.
   (b) Generalize part (a) to an arbitrary number of X_is and Y_is.
3. Verify formulas (A.24)–(A.27).
4. (a) Show that
$$E\{X - E(X)\} = 0$$
   for any random variable X.
   (b) Use the result in part (a) and Eq. (A.31) to show that if two random variables are independent then they are uncorrelated.
5. Show that if X is uniformly distributed on $[-a, a]$ for any $a > 0$ and if $Y = X^2$, then X and Y are uncorrelated but they are not independent.
6. Verify the following results that were stated in Sect. 7.3:
$$E(\boldsymbol{w}^\mathsf{T}\boldsymbol{X}) = \boldsymbol{w}^\mathsf{T}\{E(\boldsymbol{X})\}$$

and

$$\text{Var}(\boldsymbol{w}^\mathsf{T}\boldsymbol{X}) = \sum_{i=1}^{N}\sum_{j=1}^{N} w_i\, w_j\, \text{Cov}(X_i, X_j)$$
$$= \text{Var}(\boldsymbol{w}^\mathsf{T}\boldsymbol{X})\boldsymbol{w}^\mathsf{T}\text{COV}(\boldsymbol{X})\boldsymbol{w}.$$

7. Suppose $\mathbf{Y} = (Y_1, Y_2, Y_3)$ has covariance matrix

$$COV(\mathbf{Y}) = \begin{pmatrix} 1.0 & 0.9 & a \\ 0.9 & 1.0 & 0.9 \\ a & 0.9 & 1.0 \end{pmatrix}$$

for some unknown value a. Use Eq. (7.7) and the fact that the variance of a random variable is always ≥ 0 to show that a cannot equal 0.

References

Azzalini, A. (2014) *The Skew-Normal and Related Families (Institute of Mathematical Statistics Monographs, Book 3)*, Cambridge University Press.

Azzalini, A., and Capitanio, A. (2003) Distributions generated by perturbation of symmetry with emphasis on a multivariate skew t distribution. *Journal of the Royal Statistics Society, Series B*, **65**, 367–389.

Lehmann, E. L. (1999) *Elements of Large-Sample Theory*, Springer-Verlag, New York.

van der Vaart, A. W. (1998) *Asymptotic Statistics*, Cambridge University Press, Cambridge.

8

Copulas

8.1 Introduction

Copulas are a popular framework for both defining multivariate distributions and modeling multivariate data. A copula characterizes the dependence—and only the dependence—between the components of a multivariate distribution; they can be combined with any set of univariate marginal distributions to form a full joint distribution. Consequently, the use of copulas allows us to take advantage of the wide variety of univariate models that are available.

The primary financial application of copula models is risk assessment and management of portfolios that contain assets which exhibit co-movements in extreme behavior. For example, a pair of assets may have weakly correlated returns, but their largest losses may tend to occur in the same periods. They are commonly applied to portfolios of loans, bonds, and collateralized debt obligations (CDOs). Their misapplication in finance is also well-documented, as referenced in Sect. 8.8.

A *copula* is a multivariate CDF whose univariate marginal distributions are all Uniform(0,1). Suppose that $\boldsymbol{Y} = (Y_1, \ldots, Y_d)$ has a multivariate CDF F_Y with continuous marginal univariate CDFs $F_{Y_1}, \ldots, F_{Y_d}$. Then, by Eq. (A.9) in Appendix A.9.2, each of $F_{Y_1}(Y_1), \ldots, F_{Y_d}(Y_d)$ is distributed Uniform(0,1). Therefore, the CDF of $\{F_{Y_1}(Y_1), \ldots, F_{Y_d}(Y_d)\}$ is a copula. This CDF is called the copula of $\boldsymbol{Y}$ and denoted by C_Y. C_Y contains all information about dependencies among the components of $\boldsymbol{Y}$ but has no information about the marginal CDFs of $\boldsymbol{Y}$.

It is easy to find a formula for C_Y. To avoid technical issues, in this section we will assume that all random variables have continuous, strictly increasing CDFs. More precisely, the CDFs are assumed to be increasing on their support. For example, the standard exponential CDF

© Springer Science+Business Media New York 2015
D. Ruppert, D.S. Matteson, *Statistics and Data Analysis for Financial Engineering*, Springer Texts in Statistics,
DOI 10.1007/978-1-4939-2614-5_8

$$F(y) = \begin{cases} 1 - e^{-y}, & y \geq 0, \\ 0, & y < 0, \end{cases}$$

has support $[0, \infty)$ and is strictly increasing on that set. The assumption that the CDF is continuous and strictly increasing is reasonable in many financial applications, but it is avoided in more mathematically advanced texts; see Sect. 8.8.

Since C_Y is the CDF of $\{F_{Y_1}(Y_1), \ldots, F_{Y_d}(Y_d)\}$, by the definition of a CDF we have

$$\begin{aligned} C_Y(u_1, \ldots, u_d) &= P\{F_{Y_1}(Y_1) \leq u_1, \ldots, F_{Y_d}(Y_d) \leq u_d\} \\ &= P\{Y_1 \leq F_{Y_1}^{-1}(u_1), \ldots, Y_d \leq F_{Y_d}^{-1}(u_d)\} \\ &= F_Y\{F_{Y_1}^{-1}(u_1), \ldots, F_{Y_d}^{-1}(u_d)\}. \end{aligned} \tag{8.1}$$

Next, letting $u_j = F_{Y_j}(y_j)$, for $j = 1, \ldots, d$, in (8.1) we see that

$$F_Y(y_1, \ldots, y_d) = C_Y\{F_{Y_1}(y_1), \ldots, F_{Y_d}(y_d)\}. \tag{8.2}$$

Equation (8.2) is part of a famous theorem due to Sklar which states that the joint CDF F_Y can be decomposed into the copula C_Y, which contains all information about the dependencies among $(Y_1, \ldots, Y_d)$, and the univariate marginal CDFs $F_{Y_1}, \ldots, F_{Y_d}$, which contain all information about the univariate marginal distributions.

Let

$$c_Y(u_1, \ldots, u_d) = \frac{\partial^d}{\partial u_1 \cdots \partial u_d} C_Y(u_1, \ldots, u_d) \tag{8.3}$$

be the density associated with C_Y. By differentiating (8.2), we find that the density of $\boldsymbol{Y}$ is equal to

$$f_Y(y_1, \ldots, y_d) = c_Y\{F_{Y_1}(y_1), \ldots, F_{Y_d}(y_d)\} f_{Y_1}(y_1) \cdots f_{Y_d}(y_d), \tag{8.4}$$

in which $f_{Y_1}, \ldots, f_{Y_d}$ are the univariate marginal densities of $Y_1, \ldots, Y_d$, respectively.

One important property of copulas is that they are invariant to strictly increasing transformations of the component variables. More precisely, suppose that g_j is strictly increasing and $X_j = g_j(Y_j)$ for $j = 1, \ldots, d$. Then $\boldsymbol{X} = (X_1, \ldots, X_d)$ and $\boldsymbol{Y}$ have the same copulas. To see this, first note that the CDF of $\boldsymbol{X}$ is

$$\begin{aligned} F_X(x_1, \ldots, x_d) &= P\{g_1(Y_1) \leq x_1, \ldots, g_d(Y_d) \leq x_d\} \\ &= P\{Y_1 \leq g_1^{-1}(x_1), \ldots, Y_d \leq g_d^{-1}(x_d)\} \\ &= F_Y\{g_1^{-1}(x_1), \ldots, g_d^{-1}(x_d)\}, \end{aligned} \tag{8.5}$$

and therefore the CDF of X_j is

$$F_{X_j}(x_j) = F_{Y_j}\{g_j^{-1}(x_j)\}.$$

Consequently,

$$F_{X_j}^{-1}(u_j) = g_j\left\{F_{Y_j}^{-1}(u_j)\right\}$$

and

$$g_j^{-1}\left\{F_{X_j}^{-1}(u_j)\right\} = F_{Y_j}^{-1}(u_j), \tag{8.6}$$

and by applying (8.1) to X, followed by (8.5), (8.6), and then applying (8.1) to Y, we conclude that the copula of X is

$$
\begin{aligned}
C_X(u_1,\ldots,u_d) &= F_X\left\{F_{X_1}^{-1}(u_1),\ldots,F_{X_d}^{-1}(u_d)\right\} \\
&= F_Y\left[g_1^{-1}\left\{F_{X_1}^{-1}(u_1)\right\},\ldots,g_d^{-1}\left\{F_{X_d}^{-1}(u_d)\right\}\right] \\
&= F_Y\left\{F_{Y_1}^{-1}(u_1),\ldots,F_{Y_d}^{-1}(u_d)\right\} \\
&= C_Y(u_1,\ldots,u_d).
\end{aligned}
$$

8.2 Special Copulas

All d-dimensional copula functions C have domain $[0,1]^d$ and range $[0,1]$. There are three copulas of special interest because they represent independence and two extremes of dependence.

The d-dimensional *independence copula* C_0 is the CDF of d mutually independent Uniform(0,1) random variables. It equals

$$C_0(u_1,\ldots,u_d) = u_1\cdots u_d, \tag{8.7}$$

and the associated density is uniform on $[0,1]^d$; that is, $c_0(u_1,\ldots,u_d) = 1$ on $[0,1]^d$, and zero elsewhere.

The d-dimensional *co-monotonicity copula* C_+ characterizes perfect positive dependence. Let U be Uniform(0,1). Then, the co-monotonicity copula is the CDF of $\boldsymbol{U} = (U,\ldots,U)$; that is, $\boldsymbol{U}$ contains d copies of U so that all of the components of $\boldsymbol{U}$ are equal. Thus,

$$
\begin{aligned}
C_+(u_1,\ldots,u_d) &= P(U \le u_1,\ldots,U \le u_d) \\
&= P\{U \le \min(u_1,\ldots,u_d)\} = \min(u_1,\ldots,u_d).
\end{aligned}
$$

The co-monotonicity copula is also an upper bound for all copula functions: $C(u_1,\ldots,u_d) \le C_+(u_1,\ldots,u_d)$, for all $(u_1,\ldots,u_d) \in [0,1]^d$.

The two-dimensional *counter-monotonicity copula* C_- is defined as the CDF of $(U, 1-U)$, which has perfect negative dependence. Therefore,

$$
\begin{aligned}
C_-(u_1,u_2) &= P(U \le u_1,\ 1 - U \le u_2) \\
&= P(1 - u_2 \le U \le u_1) = \max(u_1 + u_2 - 1, 0). \tag{8.8}
\end{aligned}
$$

It is easy to derive the last equality in (8.8). If $1 - u_2 > u_1$, then the event $\{1 - u_2 \le U \le u_1\}$ is impossible, so the probability is 0. Otherwise, the probability is the length of the interval $(1 - u_2, u_1)$, which is $u_1 + u_2 - 1$. All two-dimensional copula functions are bounded below by (8.8). It is not possible to have a counter-monotonicity copula with $d > 2$. If, for example, U_1 is counter-monotonic to U_2 and U_2 is counter-monotonic to U_3, then U_1 and U_3 will be co-monotonic, not counter-monotonic. However, a lower bound for all copula functions is: $\max(u_1 + \cdots + u_d + 1 - d, 0) \leq C(u_1, \ldots, u_d)$, for all $(u_1, \ldots, u_d) \in [0, 1]^d$. This lower bound is obtainable only point-wise, but it is not itself a copula function for $d > 2$.

To use copulas to model multivariate dependencies, we next consider parametric families of copulas.

8.3 Gaussian and t-Copulas

Multivariate normal and multivariate t-distributions offer a convenient way to generate families of copulas. Let $\boldsymbol{Y} = (Y_1, \ldots, Y_d)$ have a multivariate normal distribution. Since C_Y depends only on the dependencies within $\boldsymbol{Y}$, not the univariate marginal distributions, C_Y depends only on the $d \times d$ correlation matrix of $\boldsymbol{Y}$, which will be denoted by $\boldsymbol{\Omega}$. Therefore, there is a one-to-one correspondence between correlation matrices and Gaussian copulas. The Gaussian copula[1] with correlation matrix $\boldsymbol{\Omega}$ will be denoted $C_{\text{Gauss}}(u_1 \ldots, u_d | \boldsymbol{\Omega})$.

If a random vector $\boldsymbol{Y}$ has a Gaussian copula, then $\boldsymbol{Y}$ is said to have a *meta-Gaussian distribution*. This does not, of course, mean that $\boldsymbol{Y}$ has a multivariate Gaussian distribution, since the univariate marginal distributions of $\boldsymbol{Y}$ could be any distributions at all. A d-dimensional Gaussian copula whose correlation matrix is the identity matrix, so that all correlations are zero, is the d-dimensional independence copula. A Gaussian copula will converge to the co-monotonicity copula C_+ if all correlations in $\boldsymbol{\Omega}$ converge to 1. In the bivariate case, as the pair-wise correlation converges to -1, the copula converges to the counter-monotonicity copula C_-.

Similarly, let $C_t(u_1 \ldots, u_d | \boldsymbol{\Omega}, \nu)$ denote the copula of a random vector that has a multivariate t-distribution with tail index[2] ν and correlation matrix $\boldsymbol{\Omega}$.[3] For multivariate t random vectors the tail index ν affects both the univariate marginal distributions and the tail dependence between components, so ν is a parameter of the t-copula C_t. We will see in Sect. 8.6 that ν similarly determines the amount of tail dependence of random vectors that have a t-copula. Such a vector is said to have a *meta-t-distribution*.

---

[1] Gaussian and normal distributions are synonymous and the Gaussian copula may also be referred to as the normal copula, especially in R functions.

[2] The tail index parameter for the t-distribution is also commonly referred to as the degrees-of-freedom parameter by its association with the theory of linear regression, and some R functions use the abbreviations df or nu.

[3] There is a minor technical issue here if $\nu \leq 2$. In this case, the t-distribution does not have covariance and correlation matrices. However, it still has a scale matrix and we will assume that the scale matrix is equal to some correlation matrix $\boldsymbol{\Omega}$.

8.4 Archimedean Copulas

An *Archimedean copula* with a strict generator has the form

$$C(u_1, \ldots, u_d) = \varphi^{-1}\{\varphi(u_1) + \cdots + \varphi(u_d)\}, \tag{8.9}$$

where the generator function φ satisfies the following conditions

1. φ is a continuous, strictly decreasing, and convex function mapping $[0,1]$ onto $[0, \infty]$,
2. $\varphi(0) = \infty$, and
3. $\varphi(1) = 0$.

A plot of a generator function is shown in Fig. 8.1 to illustrate these properties. It was generated using the iPsi() function from R's copula package with the following commands.

```
1 library(copula)
2 u = seq(0.000001, 1, length=500)
3 frank = iPsi(copula=archmCopula(family="frank", param=1), u)
4 plot(u, frank, type="l", lwd=3, ylab=expression(phi(u)))
5 abline(h=0) ; abline(v=0)
```

It is possible to relax assumption 2, but then the generator is not called strict and construction of the copula is more complex. The generator function φ is not unique; for example, $a\varphi$, in which a is any positive constant, generates the same copula as φ. The independence copula C_0 is an Archimedean copula with generator function $\varphi(u) = -\log(u)$. There are many families of Archimedean copulas, but we will only look at four, the Frank, Clayton, Gumbel, and Joe copulas.

Notice that in (8.9), the value of $C(u_1, \ldots, u_d)$ is unchanged if we permute $u_1, \ldots, u_d$. A distribution with this property is called *exchangeable*. One consequence of exchangeability is that both Kendall's and Spearman's rank correlation introduced later in Sect. 8.5 are the same for all pairs of variables. Archimedean copulas are most useful in the bivariate case or in applications where we expect all pairs to have similar dependencies.

8.4.1 Frank Copula

The Frank copula has generator

$$\varphi_{\text{Fr}}(u|\theta) = -\log\left(\frac{e^{-\theta u} - 1}{e^{-\theta} - 1}\right), \quad -\infty < \theta < \infty.$$

The inverse generator is

$$\varphi_{\text{Fr}}^{-1}(y|\theta) = -\frac{1}{\theta}\log\{e^{-y}(e^{-\theta} - 1) + 1\}.$$

Therefore, by (8.9), the bivariate Frank copula is

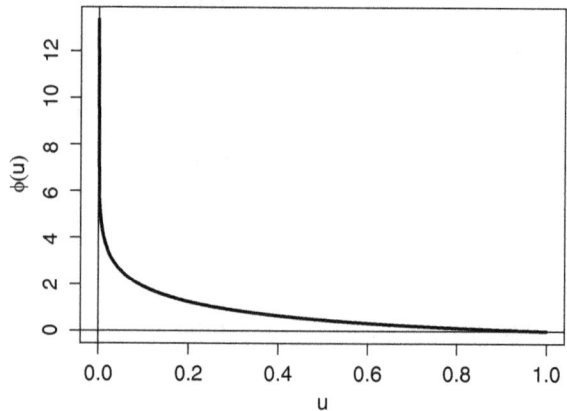

Fig. 8.1. *Generator function for the Frank copula with $\theta = 1$.*

$$C_{\mathrm{Fr}}(u_1, u_2|\theta) = -\frac{1}{\theta} \log\left\{1 + \frac{(e^{-\theta u_1} - 1)(e^{-\theta u_2} - 1)}{e^{-\theta} - 1}\right\}. \qquad (8.10)$$

The case $\theta = 0$ requires some care, since plugging this value into (8.10) gives $0/0$. Instead, one must evaluate the limit of (8.10) as $\theta \to 0$. Using the approximations $e^x - 1 \approx x$ and $\log(1+x) \approx x$ as $x \to 0$, one can show that as $\theta \to 0$, $C_{\mathrm{Fr}}(u_1, u_2|\theta) \to u_1 u_2$, the bivariate independence copula C_0. Therefore, for $\theta = 0$ we define the Frank copula to be the independence copula.

It is interesting to study the limits of $C_{\mathrm{Fr}}(u_1, u_2|\theta)$ as $\theta \to \pm\infty$. As $\theta \to -\infty$, the bivariate Frank copula converges to the counter-monotonicity copula C_-. To see this, first note that as $\theta \to -\infty$,

$$C_{\mathrm{Fr}}(u_1, u_2|\theta) \sim -\frac{1}{\theta} \log\left\{1 + e^{-\theta(u_1 + u_2 - 1)}\right\}. \qquad (8.11)$$

If $u_1 + u_2 - 1 > 0$, then as $\theta \to -\infty$, the exponent $-\theta(u_1 + u_2 - 1)$ in (8.11) converges to ∞ and

$$\log\left\{1 + e^{-\theta(u_1 + u_2 - 1)}\right\} \sim -\theta(u_1 + u_2 - 1),$$

so that $C_{\mathrm{Fr}}(u_1, u_2|\theta) \to u_1 + u_2 - 1$. Similarly, if $u_1 + u_2 - 1 < 0$, then $-\theta(u_1 + u_2 - 1) \to -\infty$, and $C_{\mathrm{Fr}}(u_1, u_2|\theta) \to 0$. Putting these results together, we see that $C_{\mathrm{Fr}}(u_1, u_2|\theta)$ converges to $\max(0, u_1 + u_2 - 1)$, the counter-monotonicity copula C_-, as $\theta \to -\infty$.

As $\theta \to \infty$, $C_{\mathrm{Fr}}(u_1, u_2|\theta) \to \min(u_1, u_2)$, the co-monotonicity copula C_+. Verification of this is left as an exercise for the reader.

Figure 8.2 contains scatterplots of nine bivariate random samples from various Frank copulas, all with a sample size of 200 and with values of θ that give dependencies ranging from strongly negative to strongly positive. Pseudo-random samples may be generated from the copula distributions discussed in this chapter using the rCopula() function from R's copula package. The convergence to the counter-monotonicity (co-monotonicity) copula as $\theta \to -\infty$ $(+\infty)$ can be seen in the scatterplots.

```
6  set.seed(5640)
7  theta = c(-100, -50, -10, -1, 0, 5, 20, 50, 500)
8  par(mfrow=c(3,3), cex.axis=1.2, cex.lab=1.2, cex.main=1.2)
9  for(i in 1:9){
10    U = rCopula(n=200,
11                copula=archmCopula(family="frank", param=theta[i]))
12    plot(U, xlab=expression(u[1]), ylab=expression(u[2]),
13         main=eval(substitute(expression(paste(theta," = ",j)),
14         list(j = as.character(theta[i])))))
15  }
```

8.4.2 Clayton Copula

The *Clayton copula*, with generator function $\varphi_{\mathrm{Cl}}(u|\theta) = \frac{1}{\theta}(u^{-\theta} - 1)$, $\theta > 0$, is

$$C_{\mathrm{Cl}}(u_1,\ldots,u_d|\theta) = (u_1^{-\theta} + \cdots + u_d^{-\theta} + 1 - d)^{-1/\theta}.$$

We define the Clayton copula for $\theta = 0$ as

$$\lim_{\theta \downarrow 0} C_{\mathrm{Cl}}(u_1,\ldots,u_d|\theta) = u_1 \cdots u_d$$

which is the independence copula C_0. There is another way to derive this result. As $\theta \downarrow 0$, l'Hôpital's rule shows that the generator $\frac{1}{\theta}(u^{-\theta}-1)$ converges to $\varphi_{\mathrm{Cl}}(u|\theta \downarrow 0) = -\log(u)$ with inverse $\varphi_{\mathrm{Cl}}^{-1}(y|\theta \downarrow 0) = \exp(-y)$. Therefore,

$$\lim_{\theta \downarrow 0} C_{\mathrm{Cl}}(u_1,\ldots,u_d|\theta) = \varphi_{\mathrm{Cl}}^{-1}\{\varphi_{\mathrm{Cl}}(u_1|\theta \downarrow 0) + \cdots + \varphi_{\mathrm{Cl}}(u_d|\theta \downarrow 0)|\theta \downarrow 0\}$$

$$= \exp\{-(-\log u_1 - \cdots - \log u_d)\} = u_1 \cdots u_d.$$

It is possible to extend the range of θ to include $-1 \le \theta < 0$, but then the generator $(u^{-\theta} - 1)/\theta$ is finite at $u = 0$ in violation of assumption 2, of strict generators. Thus, the generator is not strict if $\theta < 0$. As a result, it is necessary to define $C_{\mathrm{Cl}}(u_1,\ldots,u_d|\theta)$ to equal 0 for small values of u_i in this case. To appreciate this, consider the bivariate Clayton copula. If $-1 \le \theta < 0$, then $u_1^{-\theta} + u_2^{-\theta} - 1 < 0$ occurs when u_1 and u_2 are both small. In these cases, $C_{\mathrm{Cl}}(u_1,u_2|\theta)$ is set equal to 0. Therefore, there is no probability in the region $u_1^{-\theta} + u_2^{-\theta} - 1 < 0$. In the limit, as $\theta \to -1$, there is no probability in the region $u_1 + u_2 < 1$.

As $\theta \to -1$, the bivariate Clayton copula converges to the counter-monotonicity copula C_-, and as $\theta \to \infty$, the Clayton copula converges to the co-monotonicity copula C_+.

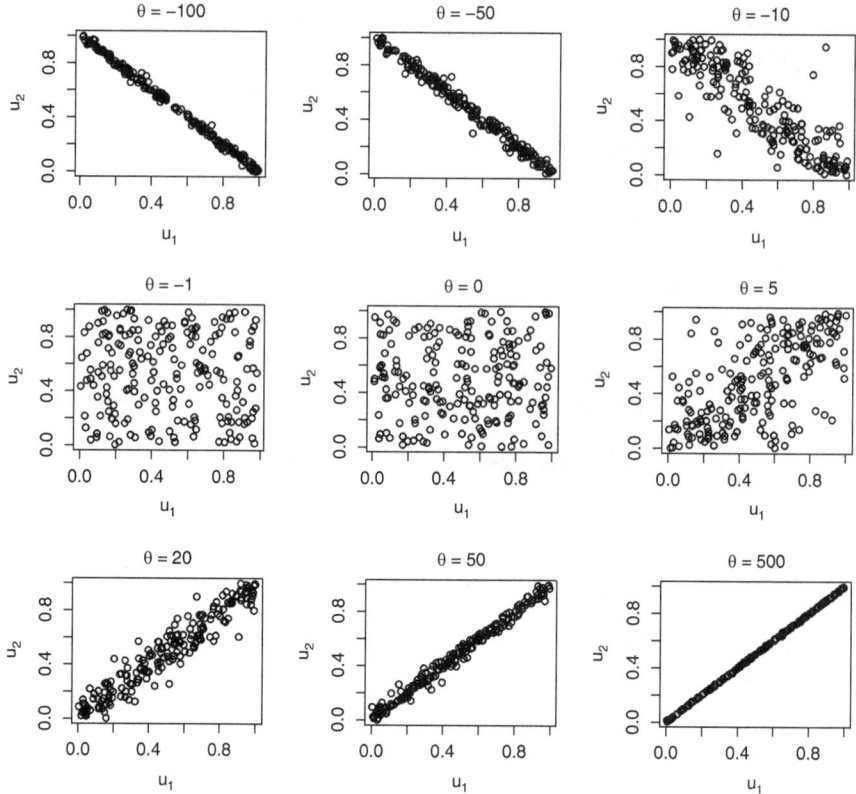

Fig. 8.2. *Bivariate random samples of size 200 from various Frank copulas.*

Figure 8.3 contains scatterplots of nine bivariate random samples from various Clayton copulas, all with a sample size of 200 and with values of θ that give dependencies ranging from counter-monotonicity to co-monotonicity.

```
16  set.seed(5640)
17  theta = c(-0.98, -0.7, -0.3, -0.1, 0.1, 1, 5, 15, 100)
18  par(mfrow=c(3,3), cex.axis=1.2, cex.lab=1.2, cex.main=1.2)
19  for(i in 1:9){
20    U = rCopula(n=200,
21              copula=archmCopula(family="clayton", param=theta[i]))
22    plot(U, xlab=expression(u[1]), ylab=expression(u[2]),
23        main=eval(substitute(expression(paste(theta," = ",j)),
24        list(j = as.character(theta[i]))))))
25  }
```

Comparing Figs. 8.2 and 8.3, we see that the Frank and Clayton copulas are rather different when the amount of dependence is somewhere between these two extremes. In particular, the Clayton copula's exclusion of the region $u_1^{-\theta} + u_2^{-\theta} - 1 < 0$ when $\theta < 0$ is evident, especially in the example with

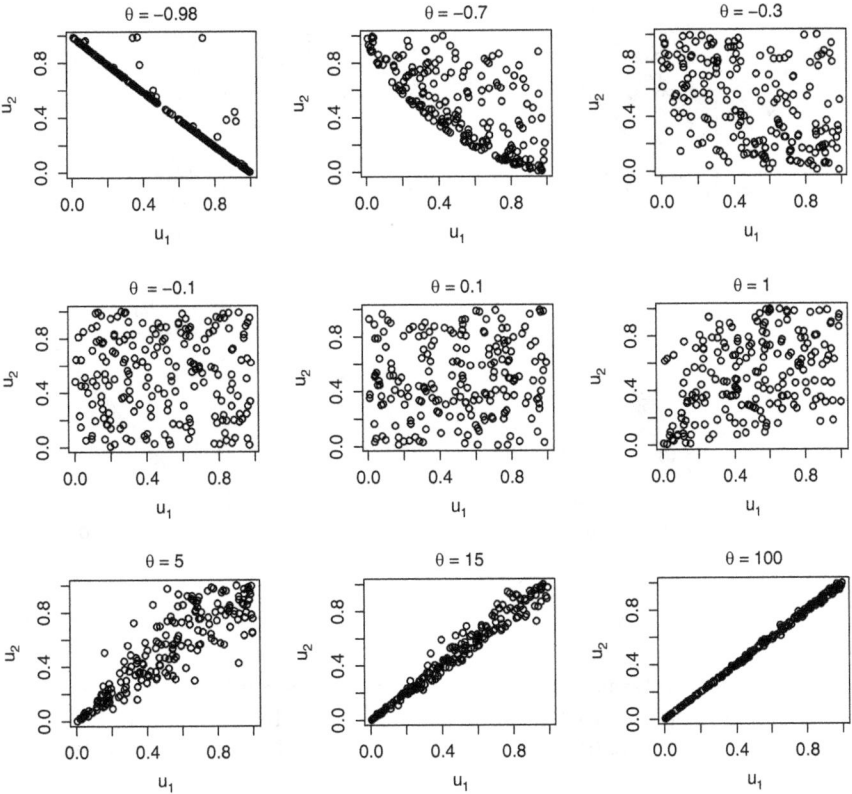

Fig. 8.3. *Bivariate random samples of size 200 from various Clayton copulas.*

$\theta = -0.7$. In contrast, the Frank copula has positive probability on the entire unit square. The Frank copula is symmetric about the diagonal from $(0,1)$ to $(1,0)$, but the Clayton copula does not have this symmetry.

8.4.3 Gumbel Copula

The Gumbel copula has the generator $\varphi_{\mathrm{Gu}}(u|\theta) = (-\log u)^{\theta}$, $\theta \geq 1$, and consequently is equal to

$$C_{\mathrm{Gu}}(u_1,\ldots,u_d|\theta) = \exp\left[-\left\{(-\log u_1)^{\theta} + \cdots + (-\log u_d)^{\theta}\right\}^{1/\theta}\right].$$

The Gumbel copula is the independence copula C_0 when $\theta = 1$, and converges to the co-monotonicity copula C_+ as $\theta \to \infty$, but the Gumbel copula cannot have negative dependence.

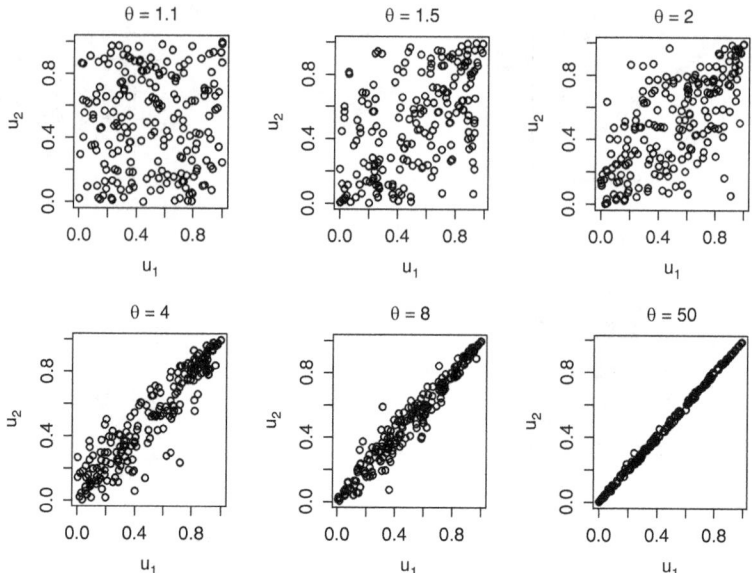

Fig. 8.4. *Bivariate random samples of size 200 from various Gumbel copulas.*

Figure 8.4 contains scatterplots of six bivariate random samples from various Gumbel copulas, with a sample size of 200 and with values of θ that give dependencies ranging from near independence to strong positive dependence.

```
26  set.seed(5640)
27  theta = c(1.1, 1.5, 2, 4, 8, 50)
28  par(mfrow=c(2,3), cex.axis=1.2, cex.lab=1.2, cex.main=1.2)
29  for(i in 1:6){
30    U = rCopula(n=200,
31                copula=archmCopula(family="gumbel", param=theta[i]))
32    plot(U, xlab=expression(u[1]), ylab=expression(u[2]),
33         main=eval(substitute(expression(paste(theta," = ",j)),
34         list(j = as.character(theta[i]))))))
35  }
```

8.4.4 Joe Copula

The Joe copula is similar to the Gumbel copula; it cannot have negative dependence, but it allows even stronger upper tail dependence and is closer to being the reverse of the Clayton copula in the positive dependence case. The Joe copula has the generator $\varphi_{\text{Joe}}(u|\theta) = -\log\{1 - (1 - u)^\theta\}$, $\theta \geq 1$. In the bivariate case, the Joe copula is equal to

$$C_{\text{Joe}}(u_1, u_2|\theta) = 1 - \left[(1 - u_1)^\theta + (1 - u_2)^\theta - (1 - u_1)^\theta(1 - u_2)^\theta\right]^{1/\theta}.$$

The Joe copula is the independence copula C_0 when $\theta = 1$, and converges to the co-monotonicity copula C_+ as $\theta \to \infty$.

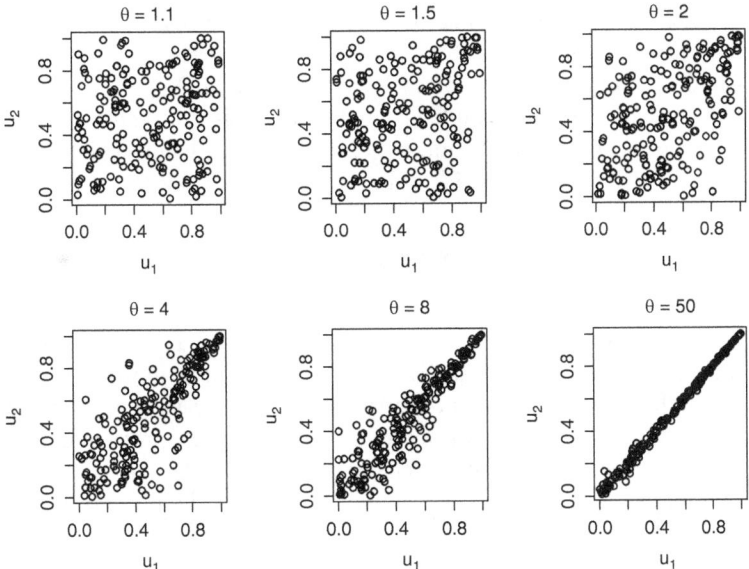

Fig. 8.5. *Bivariate random samples of size 200 from various Joe copulas.*

Figure 8.5 contains scatterplots of six bivariate random samples from various Joe copulas, with a sample size of 200 and with values of θ that give dependencies ranging from near independence to strong positive dependence.

```
36  set.seed(5640)
37  theta = c(1.1, 1.5, 2, 4, 8, 50)
38  par(mfrow=c(2,3), cex.axis=1.2, cex.lab=1.2, cex.main=1.2)
39  for(i in 1:6){
40    U = rCopula(n=200,
41              copula=archmCopula(family="joe", param=theta[i]))
42    plot(U, xlab=expression(u[1]), ylab=expression(u[2]),
43        main=eval(substitute(expression(paste(theta," = ",j)),
44        list(j = as.character(theta[i])))))
45  }
```

In applications, it is useful that the different copula families have different properties, since this increases our ability to find a copula that fits the data adequately.

8.5 Rank Correlation

The Pearson correlation coefficient defined by (4.4) is not convenient for fitting copulas to data, since it depends on the univariate marginal distributions as well as the copula. Rank correlation coefficients remedy this problem, since they depend only on the copula.

For each variable, the ranks of that variable are determined by ordering the observations from smallest to largest and giving the smallest rank 1, the next-smallest rank 2, and so forth. In other words, if $Y_1, \ldots, Y_n$ is a sample, then the *rank* of Y_i in the sample is equal to 1 if Y_i is the smallest observation, 2 if Y_i is the second smallest, and so forth. More mathematically, the rank of Y_i can also be defined by the formula

$$\mathrm{rank}(Y_i) = \sum_{j=1}^{n} I(Y_j \le Y_i), \tag{8.12}$$

which counts the number of observations (including Y_i itself) that are less than or equal to Y_i. A *rank statistic* is a statistic that depends on the data only through the ranks. A key property of ranks is that they are unchanged by strictly monotonic transformations of the variables. In particular, the ranks are unchanged by transforming each variable by its CDF, so the distribution of any rank statistic depends only on the copula of the observations, not on the univariate marginal distributions.

We will be concerned with rank statistics that measure statistical association between pairs of variables. These statistics are called *rank correlations*. There are two rank correlation coefficients in widespread usage, Kendall's tau and Spearman's rho.

8.5.1 Kendall's Tau

Let (Y_1, Y_2) be a bivariate random vector and let (Y_1^*, Y_2^*) be an independent copy of (Y_1, Y_2). Then (Y_1, Y_2) and (Y_1^*, Y_2^*) are called a *concordant pair* if the ranking of Y_1 relative to Y_1^* is the same as the ranking of Y_2 relative to Y_2^*, that is, either $Y_1 > Y_1^*$ and $Y_2 > Y_2^*$ or $Y_1 < Y_1^*$ and $Y_2 < Y_2^*$. In either case, $(Y_1 - Y_1^*)(Y_2 - Y_2^*) > 0$. Similarly, (Y_1, Y_2) and (Y_1^*, Y_2^*) are called a *discordant pair* if $(Y_1 - Y_1^*)(Y_2 - Y_2^*) < 0$. *Kendall's tau* is the probability of a concordant pair minus the probability of a discordant pair. Therefore, Kendall's tau for (Y_1, Y_2) is

$$\begin{aligned} \rho_\tau(Y_1, Y_2) &= P\{(Y_1 - Y_1^*)(Y_2 - Y_2^*) > 0\} - P\{(Y_1 - Y_1^*)(Y_2 - Y_2^*) < 0\} \\ &= E\left[\mathrm{sign}\{(Y_1 - Y_1^*)(Y_2 - Y_2^*)\}\right], \end{aligned} \tag{8.13}$$

where the *sign function* is

$$\mathrm{sign}(x) = \begin{cases} 1, & x > 0, \\ -1, & x < 0, \\ 0, & x = 0. \end{cases}$$

It is clear from (8.13) that ρ_τ is symmetric in its arguments and takes values in $[-1, 1]$. It is easy to check that if g and h are increasing functions, then

$$\rho_\tau\{g(Y_1), h(Y_2)\} = \rho_\tau(Y_1, Y_2). \tag{8.14}$$

Stated differently, Kendall's tau is invariant to monotonically increasing transformations. If g and h are the marginal CDFs of Y_1 and Y_2, then the left-hand side of (8.14) is the value of Kendall's tau for a pair of random variables distributed according to the copula of (Y_1, Y_2). This shows that Kendall's tau depends only on the copula of a bivariate random vector. For a random vector $\boldsymbol{Y}$, we define the *Kendall's tau correlation matrix* $\boldsymbol{\Omega}_\tau$ to be the matrix whose (j, k) entry is Kendall's tau for the jth and kth components of $\boldsymbol{Y}$, that is $[\boldsymbol{\Omega}_\tau(\boldsymbol{Y})]_{jk} = \rho_\tau(Y_j, Y_k)$.

If we have a bivariate sample $\boldsymbol{Y}_{1:n} = \{(Y_{i,1}, Y_{i,2}) : i = 1, \ldots, n\}$, then the sample Kendall's tau is

$$\widehat{\rho}_\tau(\boldsymbol{Y}_{1:n}) = \binom{n}{2}^{-1} \sum_{1 \le i < j \le n} \text{sign}\left\{(Y_{i,1} - Y_{j,1})(Y_{i,2} - Y_{j,2})\right\}. \qquad (8.15)$$

Note that $\binom{n}{2}$ is the number of summands in (8.15), so $\widehat{\rho}_\tau$ is the average of $\text{sign}\{(Y_{i,1} - Y_{j,1})(Y_{i,2} - Y_{j,2})\}$ across all distinct pairs of observations and is a sample version of (8.13). The sample Kendall's tau correlation matrix is defined analogously to $\boldsymbol{\Omega}_\tau$.

8.5.2 Spearman's Rank Correlation Coefficient

For a sample, Spearman's correlation coefficient is simply the usual Pearson correlation calculated from the marginal ranks of the data. For a distribution (that is, an infinite population rather than a finite sample), both variables are transformed by their univariate marginal CDFs and then the Pearson correlation is computed for the transformed variables. Transforming a random variable by its CDF is analogous to computing the ranks of a variable in a finite sample.

Stated differently, Spearman's rank correlation coefficient, also called *Spearman's rho*, for a bivariate random vector (Y_1, Y_2) will be denoted as $\rho_S(Y_1, Y_2)$ and is defined to be the Pearson correlation coefficient of $\{F_{Y_1}(Y_1), F_{Y_2}(Y_2)\}$:

$$\rho_S(Y_1, Y_2) = \text{Corr}\{F_{Y_1}(Y_1), F_{Y_2}(Y_2)\}.$$

Since the joint CDF of $\{F_{Y_1}(Y_1), F_{Y_2}(Y_2)\}$ is the copula of (Y_1, Y_2), Spearman's rho, like Kendall's tau, depends only on the copula function.

The sample version of Spearman's correlation coefficient can be computed from the ranks of the data and for a bivariate sample $\boldsymbol{Y}_{1:n} = \{(Y_{i,1}, Y_{i,2}) : i = 1, \ldots, n\}$, is

$$\widehat{\rho}_S(\boldsymbol{Y}_{1:n}) = \frac{12}{n(n^2 - 1)} \sum_{i=1}^n \left\{\text{rank}(Y_{i,1}) - \frac{n+1}{2}\right\}\left\{\text{rank}(Y_{i,2}) - \frac{n+1}{2}\right\}.$$
$$(8.16)$$

The set of ranks for any variable is, of course, the integers 1 to n, and hence $(n+1)/2$ is the mean of its ranks. It can be shown that $\widehat{\rho}_S(\boldsymbol{Y}_{1:n})$ is the sample Pearson correlation between the ranks of $\{Y_{i,1}\}$ and the ranks of $\{Y_{i,2}\}$.[4]

If $\boldsymbol{Y} = (Y_1, \ldots, Y_d)$ is a random vector, then the *Spearman's correlation matrix* $\boldsymbol{\Omega}_S$ of $\boldsymbol{Y}$ is the correlation matrix of $\{F_{Y_1}(Y_1), \ldots, F_{Y_d}(Y_d)\}$ and contains the Spearman's correlation coefficients for all pairs of components of $\boldsymbol{Y}$, such that $[\boldsymbol{\Omega}_S(\boldsymbol{Y})]_{jk} = \rho_S(Y_j, Y_k)$, for all $j, k = 1, \ldots, d$. The sample Spearman's correlation matrix is defined analogously.

8.6 Tail Dependence

Tail dependence measures association between the extreme values of two random variables and depends only on their copula. We will start with lower tail dependence, which uses extremes in the lower tail. Suppose that $\boldsymbol{Y} = (Y_1, Y_2)$ is a bivariate random vector with copula C_Y. Then the *coefficient of lower tail dependence* is denoted by λ_ℓ and defined as

$$\lambda_\ell := \lim_{q \downarrow 0} P\{Y_2 \leq F_{Y_2}^{-1}(q) \mid Y_1 \leq F_{Y_1}^{-1}(q)\} \tag{8.17}$$

$$= \lim_{q \downarrow 0} \frac{P\{Y_1 \leq F_{Y_1}^{-1}(q), Y_2 \leq F_{Y_2}^{-1}(q)\}}{P\{Y_1 \leq F_{Y_1}^{-1}(q)\}} \tag{8.18}$$

$$= \lim_{q \downarrow 0} \frac{P\{F_{Y_1}(Y_1) \leq q, F_{Y_2}(Y_2) \leq q\}}{P\{F_{Y_1}(Y_1) \leq q\}} \tag{8.19}$$

$$= \lim_{q \downarrow 0} \frac{C_Y(q, q)}{q}. \tag{8.20}$$

It is helpful to look at these equations individually. As elsewhere in this chapter, for simplicity we are assuming that F_{Y_1} and F_{Y_2} are strictly increasing on their supports and therefore have inverses.

First, (8.17) defines λ_ℓ as the limit as $q \downarrow 0$ of the conditional probability that Y_2 is less than or equal to its qth quantile, given that Y_1 is less than or equal to its qth quantile. Since we are taking a limit as $q \downarrow 0$, we are looking at the extreme left tail. What happens if Y_1 and Y_2 are independent? Then $P(Y_2 \leq y_2 \mid Y_1 \leq y_1) = P(Y_2 \leq y_2)$ for all y_1 and y_2. Therefore, the conditional probability in (8.17) equals the unconditional probability $P(Y_2 \leq F_{Y_2}^{-1}(q))$ and this probability converges to 0 as $q \downarrow 0$. Therefore, $\lambda_\ell = 0$ implies that in the extreme left tail, Y_1 and Y_2 behave as if they were independent.

Equation (8.18) is just the definition of conditional probability. Equation (8.19) is simply (8.18) after applying the probability transformation to each variable. The numerator in (8.19) is the copula by definition, and the

---

[4] If there are ties, then ranks are averaged among tied observations. For example, if there are two observations tied for smallest, then they each get a rank of 1.5. When there are ties, these results must be modified.

denominator in (8.20) is the result of $F_{Y_1}(Y_1)$ being distributed Uniform(0,1); see (A.9).

Deriving formulas for λ_ℓ for Gaussian and t-copulas is a topic best left for more advanced books. Here we give only the results; see Sect. 8.8 for further reading. For any bivariate Gaussian copula C_{Gauss} with $\rho \neq 1$, $\lambda_\ell = 0$, that is, Gaussian copulas do not have tail dependence except in the extreme case of perfect positive correlation. For a bivariate t-copula C_t with tail index ν and correlation ρ,

$$\lambda_\ell = 2F_{t,\nu+1}\left\{-\sqrt{\frac{(\nu+1)(1-\rho)}{1+\rho}}\right\}, \tag{8.21}$$

where $F_{t,\nu+1}$ is the CDF of the t-distribution with tail index $\nu + 1$.

Since $F_{t,\nu+1}(-\infty) = 0$, we see that $\lambda_\ell \to 0$ as $\nu \to \infty$, which makes sense since the t-copula converges to a Gaussian copula as $\nu \to \infty$. Also, $\lambda_\ell \to 0$ as $\rho \to -1$, which is also not too surprising, since $\rho = -1$ is perfect *negative* dependence and λ_ℓ measures *positive* tail dependence.

The *coefficient of upper tail dependence* λ_u is

$$\lambda_u := \lim_{q \uparrow 1} P\{Y_2 \geq F_{Y_2}^{-1}(q) \,|\, Y_1 \geq F_{Y_1}^{-1}(q)\} \tag{8.22}$$

$$= 2 - \lim_{q \uparrow 1} \frac{1 - C_Y(q,q)}{1-q}. \tag{8.23}$$

We see that λ_u is defined analogously to λ_ℓ; λ_u is the limit as $q \uparrow 1$ of the conditional probability that Y_2 is greater than or equal to its qth quantile, given that Y_1 is greater than or equal to its qth quantile. Deriving (8.23) is left as an exercise for the interested reader.

For Gaussian and t-copula, $\lambda_u = \lambda_\ell$, so that $\lambda_u = 0$ for any Gaussian copula and for a t-copula, λ_ℓ is given by the right-hand side of (8.21). Coefficients of tail dependence for t-copulas are plotted in Fig. 8.6. One can see $\lambda_\ell = \lambda_u$ depends strongly on both ρ and ν. For the independence copula C_0, λ_ℓ and λ_u are both equal to 0, and for the co-monotonicity copula C_+, both are equal to 1.

```
46 rho = seq(-1,1, by=0.01)
47 df = c(1, 4, 25, 240)
48 x1 = -sqrt((df[1]+1)*(1-rho)/(1+rho))
49 lambda1 = 2*pt(x1,df[1]+1)
50 x4 = -sqrt((df[2]+1)*(1-rho)/(1+rho))
51 lambda4 = 2*pt(x4,df[2]+1)
52 x25 = -sqrt((df[3]+1)*(1-rho)/(1+rho))
53 lambda25 = 2*pt(x25,df[3]+1)
54 x250 = -sqrt((df[4]+1)*(1-rho)/(1+rho))
55 lambda250 = 2*pt(x250,df[4]+1)
56 par(mfrow=c(1,1), lwd=2, cex.axis=1.2, cex.lab=1.2)
57 plot(rho, lambda1, type="l", lty=1, xlab=expression(rho),
```

```
58        ylab=expression(lambda[l]==lambda[u]))
59 lines(rho, lambda4, lty=2)
60 lines(rho, lambda25, lty=3)
61 lines(rho, lambda250, lty=4)
62 legend("topleft", c(expression(nu==1), expression(nu==4),
63        expression(nu==25), expression(nu==250)), lty=1:4)
```

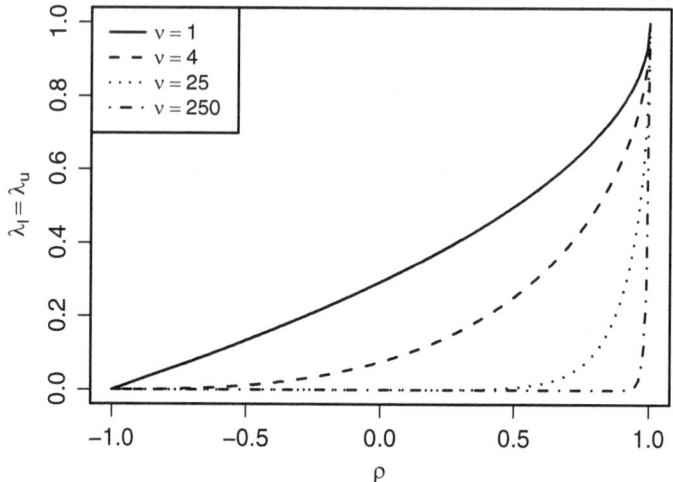

Fig. 8.6. *Coefficients of tail dependence for bivariate t-copulas as functions of ρ for $\nu = 1, 4, 25,$ and 250.*

Knowing whether or not there is tail dependence is important for risk management. If there are no tail dependencies among the returns on the assets in a portfolio, then there is little risk of simultaneous very negative returns, and the risk of an extreme negative return on the portfolio is low. Conversely, if there are tail dependencies, then the likelihood of extreme negative returns occurring simultaneously on several assets in the portfolio can be high. As such, tail dependencies should be considered when assessing the diversification and risk of any portfolio.

8.7 Calibrating Copulas

Assume that we have an i.i.d. sample $\boldsymbol{Y}_{1:n} = \{(Y_{i,1},\ldots,Y_{i,d}) : i = 1,\ldots,n\}$, and we wish to estimate the copula of $\boldsymbol{Y}$ and perhaps its univariate marginal distributions as well.

An important task is choosing a copula model. The various copula models differ notably from each other. For example, some have tail dependence and others do not. The Gumbel copula and Joe copula allow only positive dependence or independence. The Clayton copula with negative dependence excludes the region where both u_1 and u_2 are small. As will be seen in this section, an appropriate copula model can be selected via AIC, and by using graphical techniques.

8.7.1 Maximum Likelihood

Suppose we have parametric models $F_{Y_1}(\cdot \mid \boldsymbol{\theta}_1), \ldots, F_{Y_d}(\cdot \mid \boldsymbol{\theta}_d)$ for the marginal CDFs as well as a parametric model $c_Y(\cdot \mid \boldsymbol{\theta}_C)$ for the copula density. By taking logs of (8.4), we find that the log-likelihood is

$$
\log\{L(\boldsymbol{\theta}_1, \ldots, \boldsymbol{\theta}_d, \boldsymbol{\theta}_C)\} = \sum_{i=1}^{n} \Big(\log[c_Y\{F_{Y_1}(Y_{i,1}|\boldsymbol{\theta}_1), \ldots, F_{Y_d}(Y_{i,d}|\boldsymbol{\theta}_d)|\boldsymbol{\theta}_C\}]
$$
$$
+ \log\{f_{Y_1}(Y_{i,1}|\boldsymbol{\theta}_1)\} + \cdots + \log\{f_{Y_d}(Y_{i,d}|\boldsymbol{\theta}_d)\} \Big). \quad (8.24)
$$

Maximum likelihood estimation finds the maximum of $\log\{L(\boldsymbol{\theta}_1, \ldots, \boldsymbol{\theta}_d, \boldsymbol{\theta}_C)\}$ over the entire set of parameters $(\boldsymbol{\theta}_1, \ldots, \boldsymbol{\theta}_d, \boldsymbol{\theta}_C)$.

There are two potential problems with maximum likelihood estimation. First, because of the large number of parameters, especially for large values of d, numerically maximizing $\log\{L(\boldsymbol{\theta}_1, \ldots, \boldsymbol{\theta}_d, \boldsymbol{\theta}_C)\}$ can be challenging. This difficulty can be ameliorated by the use of starting values that are close to the MLEs. The pseudo-maximum likelihood estimates discussed in the next section are easier to compute than the MLE and can be used either as an alternative to the MLE or as starting values for the MLE.

Second, maximum likelihood estimation requires parametric models for both the copula and the univariate marginal distributions. If any of the univariate marginal distributions are not well fit by a convenient parametric family, this may cause biases in the estimated parameters of both the univariate marginal distributions and the copula. The semiparametric approach to pseudo-maximum likelihood estimation, where the univariate marginal distributions are estimated nonparametrically, provides a remedy to this problem.

8.7.2 Pseudo-Maximum Likelihood

Pseudo-maximum likelihood estimation is a two-step procedure. In the first step, each of the d univariate marginal distribution functions is estimated, one at a time. Let $\widehat{F}_{Y_j}$ be the estimate of the jth univariate marginal CDF, $j = 1, \ldots, d$. In the second step,

$$
\sum_{i=1}^{n} \log\Big[c_Y\Big\{\widehat{F}_{Y_1}(Y_{i,1}), \ldots, \widehat{F}_{Y_d}(Y_{i,d})|\boldsymbol{\theta}_C\Big\}\Big] \quad (8.25)
$$

is maximized over $\boldsymbol{\theta}_C$. Note that (8.25) is obtained from (8.24) by deleting terms that do not depend on $\boldsymbol{\theta}_C$ and replacing the univariate marginal CDFs by estimates. By estimating parameters in the univariate marginal distributions and in the copula separately, the pseudo-maximum likelihood approach avoids a high-dimensional optimization.

There are two approaches to the first step, parametric and nonparametric. In the parametric approach, parametric models $F_{Y_1}(\cdot \,|\, \boldsymbol{\theta}_1), \ldots, F_{Y_d}(\cdot \,|\, \boldsymbol{\theta}_d)$ for the univariate marginal CDFs are assumed as in maximum likelihood estimation. The data $Y_{1,j}, \ldots, Y_{n,j}$ for the jth variate are used to estimate $\boldsymbol{\theta}_j$, usually by maximum likelihood as discussed in Chap. 5. Then, $\widehat{F}_{Y_j}(\cdot) = F_{Y_j}(\cdot \,|\, \widehat{\boldsymbol{\theta}}_j)$. In the nonparametric approach, $\widehat{F}_{Y_j}$ is estimated by the empirical CDF of $Y_{1,j}, \ldots, Y_{n,j}$, except that the divisor n in (4.2) is replaced by $n+1$ so that

$$\widehat{F}_{Y_j}(y) = \frac{\sum_{i=1}^{n} I\{Y_{i,j} \leq y\}}{n+1}. \qquad (8.26)$$

With this modified divisor, the maximum value of $\widehat{F}_{Y_j}(Y_{i,j})$ is $n/(n+1)$ rather than 1. Avoiding a value of 1 is essential when, as is often the case, $c_Y(u_1, \ldots, u_d \,|\, \boldsymbol{\theta}_C) = \infty$ if some of $u_1, \ldots, u_d$ are equal to 1.

When both steps are parametric, the estimation method is called *parametric pseudo-maximum likelihood*. The combination of a nonparametric first step and a parametric second step is called *semiparametric pseudo-maximum likelihood*.

In the second step of pseudo-maximum likelihood, the maximization can be difficult when $\boldsymbol{\theta}_C$ is high-dimensional. For example, if one uses a Gaussian or t-copula, then there are $d(d-1)/2$ correlation parameters. One way to solve this problem is to assume some structure among the correlations. An extreme case of this is the *equi-correlation model* where all non-diagonal elements of the correlation matrix have a common value, call it ρ. If one is reluctant to assume some type of structured correlation matrix, then it is essential to have good starting values for the correlation matrix when maximizing (8.25). For Gaussian and t-copulas, starting values can be obtained via rank correlations as discussed in the next section.

The values $\widehat{F}_{Y_j}(Y_{i,j})$, $i = 1, \ldots, n$ and $j = 1, \ldots, d$, will be called the *uniform-transformed variables*, since they should be distributed approximately Uniform(0,1). The multivariate empirical CDF [see Eq. (A.38)] of the uniform-transformed variables is called the *empirical copula* and is a nonparametric estimate of the copula function. The empirical copula is useful for checking the goodness of fit of parametric copula models; see Example 8.1.

8.7.3 Calibrating Meta-Gaussian and Meta-t-Distributions

Gaussian Copulas

Rank correlation can be useful for estimating the parameters of a copula. Suppose $\boldsymbol{Y}_{1:n} = \{(Y_{i,1}, \ldots, Y_{i,d}) : i = 1, \ldots, n\}$, is an i.i.d. sample from a meta-Gaussian distribution. Then its copula is $C_{\text{Gauss}}(\cdot \,|\, \boldsymbol{\Omega})$ for some correlation matrix $\boldsymbol{\Omega}$. To estimate the distribution of $\boldsymbol{Y}$, we need to estimate the univariate marginal distributions and $\boldsymbol{\Omega}$. The marginal distribution can be estimated by the methods discussed in Chap. 5. Result (8.28) in the following theorem shows that $\boldsymbol{\Omega}$ can be estimated by the sample Spearman's correlation matrix.

Result 8.1 *Let* $\boldsymbol{Y} = (Y_1, \ldots, Y_d)$ *have a meta-Gaussian distribution with continuous univariate marginal distributions and copula* $C_{\text{Gauss}}(\cdot | \boldsymbol{\Omega})$, *and let* $\Omega_{ij} = [\boldsymbol{\Omega}]_{ij}$. *Then*

$$\rho_\tau(Y_i, Y_j) = \frac{2}{\pi} \arcsin(\Omega_{ij}), \text{ and} \tag{8.27}$$

$$\rho_S(Y_i, Y_j) = \frac{6}{\pi} \arcsin(\Omega_{ij}/2) \approx \Omega_{ij}. \tag{8.28}$$

Suppose, instead, that $\boldsymbol{Y}$ *has a meta-t-distribution with continuous univariate marginal distributions and copula* $C_t(\cdot | \boldsymbol{\Omega}, \nu)$. *Then (8.27) still holds, but (8.28) does not hold.*

The approximation in (8.28) uses the result that

$$\frac{6}{\pi} \arcsin(x/2) \approx x \text{ for } |x| \leq 1. \tag{8.29}$$

The left- and right-hand sides of (8.29) are equal when $x = -1, 0, 1$, and their maximum difference over the range $-1 \leq x \leq 1$ is 0.018. However, the relative error $\left\{ \frac{6}{\pi} \arcsin(x/2) - x \right\} / \frac{6}{\pi} \arcsin(x/2)$ can be larger, as much as 0.047, and is largest near $x = 0$.

By (8.28), the sample Spearman's rank correlation matrix $\widehat{\boldsymbol{\Omega}}(\boldsymbol{Y}_{1:n})$ can be used as an estimate of the correlation matrix $\boldsymbol{\Omega}$ associated with $C_{\text{Gauss}}(\cdot | \boldsymbol{\Omega})$. This estimate could be the final one or could be used as a starting value for numeric maximum likelihood or pseudo-maximum likelihood estimation.

t-Copulas

If $\boldsymbol{Y}_{1:n} = \{ (Y_{i,1}, \ldots, Y_{i,d}) : i = 1, \ldots, n \}$ is a sample from a distribution with a t-copula $C_t(\cdot | \boldsymbol{\Omega}, \nu)$ then we can use (8.27) and the sample Kendall's tau correlations to estimate $\boldsymbol{\Omega}$. Let $\widehat{\Omega}_{\tau, jk}$ be the sample Kendall's tau correlation of $\{ Y_{1,j}, \ldots, Y_{n,j} \}$ and $\{ Y_{1,k}, \ldots, Y_{n,k} \}$, the jth and kth components, and let $\widetilde{\boldsymbol{\Omega}}^{**}$ be defined such that $[\widetilde{\boldsymbol{\Omega}}^{**}]_{jk} = \sin\{ \frac{\pi}{2} \widehat{\Omega}_{\tau, jk} \}$ Then $\widetilde{\boldsymbol{\Omega}}^{**}$ will have two of the three properties of a correlation matrix; it will be symmetric, with all diagonal entries equal to 1. However, it may not be positive definite, or even semidefinite, because some of its eigenvalues may be negative.

If all of the eigenvalues of $\widetilde{\boldsymbol{\Omega}}^{**}$ are positive, then we will use $\widetilde{\boldsymbol{\Omega}}^{**}$ to estimate $\boldsymbol{\Omega}$. Otherwise, we alter $\widetilde{\boldsymbol{\Omega}}^{**}$ slightly to make it positive definite. By (A.50),

$$\widetilde{\boldsymbol{\Omega}}^{**} = \boldsymbol{O}\operatorname{diag}(\lambda_i)\,\boldsymbol{O}^{\mathsf{T}},$$

where $\boldsymbol{O}$ is an orthogonal matrix whose columns are the eigenvectors of $\widetilde{\boldsymbol{\Omega}}^{**}$ and $\lambda_1,\ldots,\lambda_d$ are the corresponding eigenvalues. We then define

$$\widetilde{\boldsymbol{\Omega}}^{*} = \boldsymbol{O}\operatorname{diag}\{\max(\epsilon,\lambda_i)\}\,\boldsymbol{O}^{\mathsf{T}},$$

where ϵ is some small positive quantity, for example, $\epsilon = 0.001$. Now, $\widetilde{\boldsymbol{\Omega}}^{*}$ is symmetric and positive definite, but its diagonal elements, $\widetilde{\Omega}^{*}_{ii}$, $i = 1,\ldots,d$, may not be equal to 1. This problem is easily fixed; multiply the ith row and the ith column of $\widetilde{\boldsymbol{\Omega}}^{*}$ by $\left(\widetilde{\Omega}^{*}_{ii}\right)^{-1/2}$, for $i = 1,\ldots,d$. The final result, which we denote as $\widetilde{\boldsymbol{\Omega}}$, is a bona fide correlation matrix; that is, it is symmetric, positive definite, and it has all diagonal entries equal to 1.

After $\boldsymbol{\Omega}$ has been estimated by $\widetilde{\boldsymbol{\Omega}}$, an estimate of the tail index ν is still needed. One can be obtained by plugging $\widetilde{\boldsymbol{\Omega}}$ into the log-likelihood (8.25) and then maximizing over ν.

Example 8.1. Flows in pipelines

In this example, we will continue the analysis of the pipeline flows data introduced in Example 4.2. Only the flows in the first two pipelines will be used.

In a fully parametric pseudo-likelihood analysis, the univariate skewed t-model will be used for flows 1 and 2. Let $\widehat{U}_{1,j},\ldots,\widehat{U}_{n,j}$ be the flows in pipeline j, $j = 1, 2$, transformed by their estimated skewed-t CDFs. We will call the $\widehat{U}_{i,j}$ "uniform-transformed flows." Define $\widehat{Z}_{i,j} = \Phi^{-1}(\widehat{U}_{i,j})$, where Φ^{-1} is the standard normal quantile function. The $\widehat{Z}_{i,j}$ should each be approximately $N(0,1)$-distributed, and we will call them "normal-transformed flows."

```
64 library(copula)
65 library(sn)
66 dat = read.csv("FlowData.csv")
67 dat = dat/10000
68 n = nrow(dat)
69 x1 = dat$Flow1
70 fit1 = st.mple(matrix(1,n,1), y=x1, dp=c(mean(x1), sd(x1), 0, 10))
71 est1 = fit1$dp
72 u1 = pst(x1, dp=est1)
73 x2 = dat$Flow2
74 fit2 = st.mple(matrix(1,n,1), y=x2, dp=c(mean(x2), sd(x2), 0, 10))
75 est2 = fit2$dp
76 u2 = pst(x2, dp=est2)
77 U.hat = cbind(u1, u2)
78 z1 = qnorm(u1)
79 z2 = qnorm(u2)
80 Z.hat = cbind(z1, z2)
```

Both sets of uniform-transformed flows should be approximately Uniform(0,1). Figure 8.7 shows density histograms of both samples of uniform-transformed flows as well as their scatterplot and two-dimensional KDE density contours. The histograms show some deviations from uniform distributions, which suggests that the skewed-t model may not provide adequate fits and that a semiparametric pseudo-maximum likelihood approach might be tried—this is considered below. However, the deviations may be due to random variation.

```
81  library(ks)
82  fhatU = kde(x=U.hat, H=Hscv(x=U.hat))
83  par(mfrow=c(2,2), cex.axis=1.2, cex.lab=1.2, cex.main=1.2)
84  hist(u1, main="(a)", xlab=expression(hat(U)[1]), freq = FALSE)
85  hist(u2, main="(b)", xlab=expression(hat(U)[2]), freq = FALSE)
86  plot(u1, u2, main="(c)", xlab = expression(hat(U)[1]),
87      ylab = expression(hat(U)[2]), mgp = c(2.5, 1, 0))
88  plot(fhatU, drawpoints=FALSE, drawlabels=FALSE,
89      cont=seq(10, 80, 10), main="(d)", xlab=expression(hat(U)[1]),
90      ylab=expression(hat(U)[2]), mgp = c(2.5, 1, 0))
```

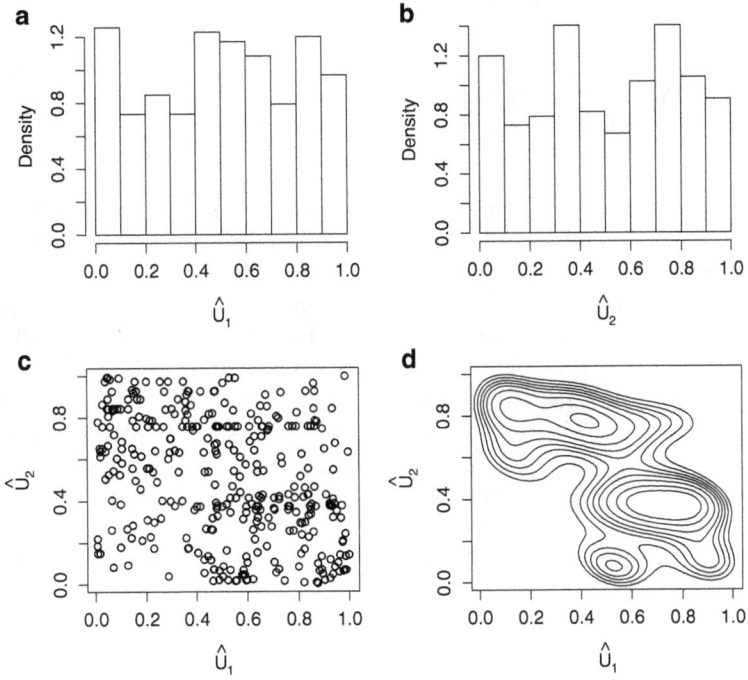

Fig. 8.7. *Pipeline data. Density histograms (a) and (b) and a scatterplot (c) of the uniform-transformed flows. The empirical copula $\widehat{C}$ is the empirical CDF of the data in (c). Contours (d) from an estimated copula density $\hat{c}$ via a two-dimensional KDE of (c).*

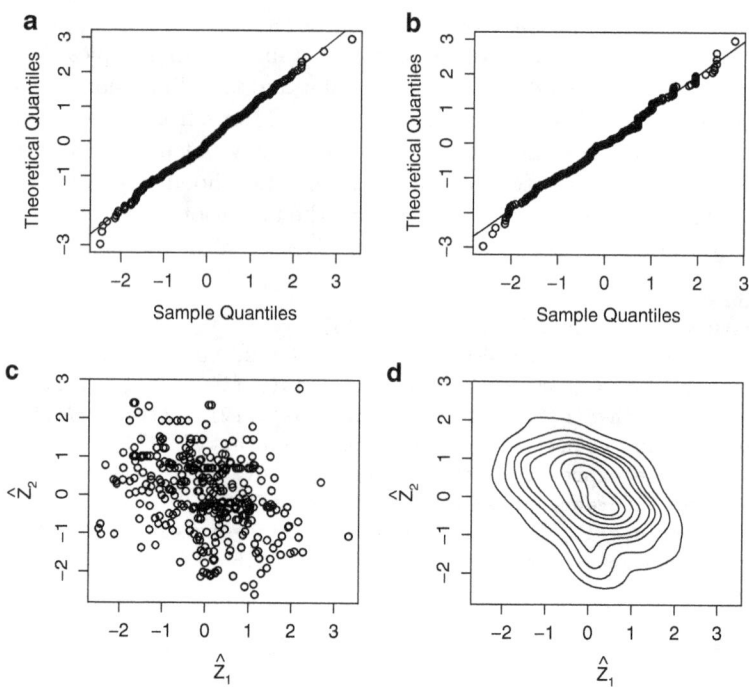

Fig. 8.8. *Pipeline data. Normal quantile plots (**a**) and (**b**), a scatterplot (**c**) and KDE density contours for the normal-transformed flows.*

The scatterplot in Fig. 8.7 shows some negative association as the data are somewhat concentrated along the diagonal from top left to bottom right. Thus, the Gumbel copula and Joe copula, which cannot have negative dependence, are not appropriate. Also, the Clayton copula may not fit well either, since the scatterplot shows data in the region where both $\widehat{U}_1$ and $\widehat{U}_2$ have small values, but this region is excluded by a Clayton copula with negative dependence. We will soon see that AIC agrees with these conclusions from a graphical analysis, since the Clayton model has higher (worse) AIC values compared to the Gaussian, t, and Frank copula models.

Figure 8.8 shows that the normal-transformed flows have approximately linear normal quantile plots, which would be expected if the estimated univariate marginal CDFs were adequate fits. Their scatterplot and KDE density contours again show negative association.

```
91  fhatZ = kde(x=Z.hat, H=Hscv(x=Z.hat))
92  par(mfrow=c(2,2), cex.axis=1.2, cex.lab=1.2, cex.main=1.2)
93  qqnorm(z1, datax=T, main="(a)") ; qqline(z1)
94  qqnorm(z2, datax=T, main="(b)") ; qqline(z2)
95  plot(z1, z2, main="(c)", xlab = expression(hat(Z)[1]),
96       ylab = expression(hat(Z)[2]), mgp = c(2.5, 1, 0))
97  plot(fhatZ, drawpoints=FALSE, drawlabels=FALSE,
```

```
98      cont=seq(10, 90, 10), main="(d)", xlab=expression(hat(Z)[1]),
99      ylab=expression(hat(Z)[2]), mgp = c(2.5, 1, 0))
```

We will assume for now that the two flows have a meta-Gaussian distribution. There are three ways to estimate the correlation in their Gaussian copula. The first, Spearman's rank correlation, is estimated -0.357. The second, which uses (8.27) is $\sin(\widehat{\rho}_\tau \pi/2)$, where $\widehat{\rho}_\tau$ is the sample Kendall's tau rank correlation; its value is -0.371. The third, Pearson correlation of the normal-transformed flows, is -0.335. There is reasonably close agreement among the three values, especially relative to their uncertainties; for example, the approximate $95\,\%$ confidence interval for the Pearson correlation of the normal-transformed flows is $(-0.426, -0.238)$, and the other two estimate are well within this interval.

```
100  cor.test(u1, u2, method="spearman")
101  cor.test(u1, u2, method="kendall")
102  sin(-0.242*pi/2)
103  cor.test(u1, u2, method="pearson")
104  cor.test(z1, z2, method="pearson")

     Pearson's product-moment correlation

     data:  z1 and z2
     t = -6.56, df = 340, p-value = 2.003e-10
     alternative hypothesis: true correlation is not equal to 0
     95 percent confidence interval:
      -0.426 -0.238
     sample estimates:
        cor
     -0.335
```

Four parametric copulas were fit to the uniform-transformed flows: t, Gaussian, Frank and Clayton. Estimation of the copula distributions discussed in this chapter may be performed using the `fitCopula()` function from R's `copula` package. The Gumbel and Joe copulas are not considered since they only allow positive dependence and these data show negative dependence; attempting to fit these models results in numerical failures. Since we used parametric estimates to transform the flows, we are fitting the copulas by parametric pseudo-maximum likelihood.

```
105  omega = -0.371
106  Ct = fitCopula(copula=tCopula(dim = 2), data=U.hat,
107                    method="ml", start=c(omega, 10))
108  Ct@estimate
109  loglikCopula(param=Ct@estimate, x=U.hat, copula=tCopula(dim = 2))
110  -2*.Last.value + 2*length(Ct@estimate)
111  #
112  Cgauss = fitCopula(copula=normalCopula(dim = 2), data=U.hat,
113                    method="ml", start=c(omega))
114  Cgauss@estimate
```

```
115 loglikCopula(param=Cgauss@estimate, x=U.hat,
116               copula=normalCopula(dim = 2))
117 -2*.Last.value + 2*length(Cgauss@estimate)
118 #
119 Cfr = fitCopula(copula=frankCopula(1, dim=2), data=U.hat,
120               method="ml")
121 Cfr@estimate
122 loglikCopula(param=Cfr@estimate, x=U.hat,
123               copula=frankCopula(dim = 2))
124 -2*.Last.value + 2*length(Cfr@estimate)
125 #
126 Ccl = fitCopula(copula=claytonCopula(1, dim=2), data=U.hat,
127               method="ml")
128 Ccl@estimate
129 loglikCopula(param=Ccl@estimate, x=U.hat,
130               copula=claytonCopula(dim = 2))
131 -2*.Last.value + 2*length(Ccl@estimate)
```

The results are summarized in Table 8.1. Looking at the maximized log-likelihood values, we see that the Frank copula fits best since it minimizes AIC, but the t and Gaussian fit reasonably well. Figure 8.9 shows the uniform-transformed flows scatterplot and contours of the distribution functions of five copulas: the independence copula and the four estimated parametric copulas; the empirical copula contours have been overlaid for comparison. The t-copula is similar to the Gaussian since $\widehat{\nu} = 22.247$ is large. The Frank copula fits best in the sense that its contours are closest to those of the empirical copula. This is in agreement with the AIC values.

Table 8.1. *Estimates of copula parameters, maximized log-likelihood, and AIC using the uniform-transformed pipeline flow data.*

Copula family	Estimates	Maximized log-likelihood	AIC
t	$\widehat{\rho} = -0.340$ $\widehat{\nu} = 22.247$	20.98	-37.96
Gaussian	$\widehat{\rho} = -0.331$	20.36	-38.71
Frank	$\widehat{\theta} = -2.249$	23.07	-44.13
Clayton	$\widehat{\theta} = -0.166$	9.86	-17.72

```
132 par(mfrow=c(2,3), mgp = c(2.5, 1, 0))
133 plot(u1, u2, main="Uniform-Transformed Data",
134      xlab = expression(hat(U)[1]), ylab = expression(hat(U)[2]))
135 Udex = (1:n)/(n+1)
136 Cn = C.n(u = cbind(rep(Udex, n), rep(Udex, each=n)) , U = U.hat,
137          offset=0, method="C")
138 EmpCop = expression(contour(Udex,Udex,matrix(Cn,n,n), col=2, add=T))
139 #
```

```
140  contour(normalCopula(param=0,dim=2), pCopula, main=expression(C[0]),
141          xlab = expression(hat(U)[1]), ylab = expression(hat(U)[2]))
142  eval(EmpCop)
143  #
144  contour(tCopula(param=Ct@estimate[1], dim=2,
145                  df=round(Ct@estimate[2])),
146          pCopula, main = expression(hat(C)[t]),
147          xlab = expression(hat(U)[1]), ylab = expression(hat(U)[2]))
148  eval(EmpCop)
149  #
150  contour(normalCopula(param=Cgauss@estimate[1], dim = 2),
151          pCopula, main = expression(hat(C)[Gauss]),
152          xlab = expression(hat(U)[1]), ylab = expression(hat(U)[2]))
153  eval(EmpCop)
154  #
155  contour(frankCopula(param=Cfr@estimate[1], dim = 2),
156          pCopula, main = expression(hat(C)[Fr]),
157          xlab = expression(hat(U)[1]), ylab = expression(hat(U)[2]))
158  eval(EmpCop)
159  #
160  contour(claytonCopula(param=Ccl@estimate[1], dim = 2),
161          pCopula, main = expression(hat(C)[Cl]),
162          xlab = expression(hat(U)[1]), ylab = expression(hat(U)[2]))
163  eval(EmpCop)
```

The analysis in the previous paragraph was repeated with the flows transformed by their empirical CDFs. Doing this yielded the semiparametric pseudo-maximum likelihood estimates. Since the results were very similar to those for parametric pseudo-maximum likelihood estimates, they are not presented here. □

8.8 Bibliographic Notes

For discussion of Archimedean copula with non-strict generators, see McNeil, Frey, and Embrechts (2005). These authors discuss a number of other topics in more detail than is done here. They discuss methods defining nonexchangeable Archimedean copulas. The coefficients of tail dependence for Gaussian and t-copulas are derived in their Sect. 5.2. The theorem and calibration methods in Sect. 8.7.3 are discussed in their Sect. 5.5.

Cherubini et al. (2004) treat the application of copulas to finance. Joe (1997) and Nelsen (2007) are standard references on copulas. Chapter 4 of Mari and Kotz (2001) discusses additional copula families.

Li (2000) developed a well-known but controversial model for credit risk using exponentially distributed default times with a Gaussian copula. An article in *Wired* magazine states that Li's Gaussian copula model was "a quick—and

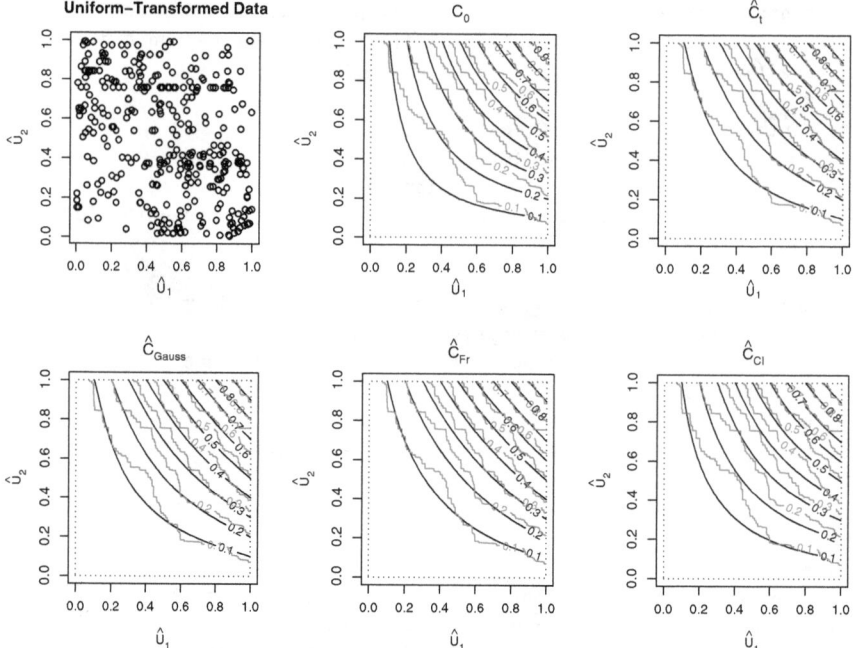

Fig. 8.9. *Uniform-transformed flows for pipeline data. Scatterplot; independence copula contours and four fitted copula contours via parametric models, versus the empirical copula contours.*

fatally flawed—way to assess risk" (Salmon 2009); in particular, the model does not include tail dependence. Duffie and Singleton's (2003, Section 10.4) also discusses copula-based methods for modeling dependent default times.

8.9 R Lab

8.9.1 Simulating from Copula Models

Run the R code that appears below to generate data from a copula. Line 1 loads the copula package. Lines 2–3 defines a copula object. At this point, nothing is done with the copula object—it is simply defined. However, the copula object is used in line 5 to generate a random sample from the specified copula model. The remaining lines create a scatterplot matrix of the sample and print its sample Pearson correlation matrix.

```
1 library(copula)
2 cop_t_dim3 = tCopula(dim = 3, param = c(-0.6,0.75,0),
3                      dispstr = "un", df = 1)
4 set.seed(5640)
```

```
5 rand_t_cop = rCopula(n = 500, copula = cop_t_dim3)
6 pairs(rand_t_cop)
7 cor(rand_t_cop)
```

You can use R's help to learn more about the functions tCopula() and rCopula().

Problem 1 *Consider the R code above.*

(a) *What type of copula model has been sampled? Give the copula family, the correlation matrix, and any other parameters that specify the copula.*
(b) *What is the sample size?*

Problem 2 *Examine the scatterplot matrix (generated by line 6) and answer the questions below. Include the scatterplot matrix with your answer.*

(a) *Components 2 and 3 are uncorrelated. Do they appear independent? Why or why not?*
(b) *Do you see signs of tail dependence? If so, where?*
(c) *What are the effects of dependence upon the plots?*
(d) *The nonzero correlations in the copula do not have the same values as the corresponding sample correlations. Do you think this is just due to random variation or is something else going on? If there is another cause besides random variation, what might that be? To help answer this question, you can get confidence intervals for the Pearson correlation: For example,*

```
8 cor.test(rand_t_cop[,1],rand_t_cop[,3])
```

will give a confidence interval (95 percent by default) for the correlation (Pearson by default) between components 1 and 3. Does this confidence interval include 0.75?

Lines 9–10 in the R code below defines a normal (Gaussian) copula. Lines 11–13 define a multivariate distribution by specifying its copula and its marginal distributions—the copula is the one just defined. Line 15 generates a random sample of size 1,000 from this distribution, which has three components. The remaining lines create a scatterplot matrix and kernel estimates of the marginal densities for each component.

```
9  cop_normal_dim3 = normalCopula(dim = 3, param = c(-0.6,0.75,0),
10                                 dispstr = "un")
11 mvdc_normal = mvdc(copula = cop_normal_dim3, margins = rep("exp",3),
12                paramMargins = list(list(rate=2), list(rate=3),
13                                     list(rate=4)))
14 set.seed(5640)
15 rand_mvdc = rMvdc(n = 1000, mvdc = mvdc_normal)
16 pairs(rand_mvdc)
17 par(mfrow = c(2,2))
18 for(i in 1:3) plot(density(rand_mvdc[,i]))
```

Problem 3 *Run the* R *code above to generate a random sample.*

(a) What are the marginal distributions of the three components in rand_mvdc*?*
     What are their expected values?
(b) Are the second and third components independent? Why or why not?

8.9.2 Fitting Copula Models to Bivariate Return Data

In this section, you will fit copula models to a bivariate data set of daily returns on IBM stock and the S&P 500 index.

First, you will fit a model with univariate marginal t-distributions and a t-copula. The model has three degrees-of-freedom (tail index) parameters, one for each of the two univariate models and a third for the copula. This means that the univariate distributions can have different tail indices and that their tail indices are independent of the tail dependence from the copula.

Run the following R code to load the data and necessary libraries, fit univariate t-distributions to the two components, and convert estimated scale parameters to estimated standard deviations:

```
1  library(MASS)      #  for fitdistr() and kde2d() functions
2  library(copula)    #  for copula functions
3  library(fGarch)    #  for standardized t density
4  netRtns = read.csv("IBM_SP500_04_14_daily_netRtns.csv", header = T)
5  ibm = netRtns[,2]
6  sp500 = netRtns[,3]
7  est.ibm = as.numeric( fitdistr(ibm,"t")$estimate )
8  est.sp500 = as.numeric( fitdistr(sp500,"t")$estimate )
9  est.ibm[2] = est.ibm[2] * sqrt( est.ibm[3] / (est.ibm[3]-2) )
10 est.sp500[2] = est.sp500[2] * sqrt(est.sp500[3] / (est.sp500[3]-2) )
```

The univariate estimates will be used as starting values when the meta-t-distribution is fit by maximum likelihood. You also need an estimate of the correlation coefficient in the t-copula. This can be obtained using Kendall's tau. Run the following code and complete line 12 so that omega is the estimate of the Pearson correlation based on Kendall's tau.

```
11 cor_tau = cor(ibm, sp500, method = "kendall")
12 omega =
```

Problem 4 *How did you complete line 12 of the code? What was the computed value of* omega*?*

Next, define the t-copula using omega as the correlation parameter and 4 as the degrees-of-freedom (tail index) parameter.

```
13  cop_t_dim2 = tCopula(omega, dim = 2, dispstr = "un", df = 4)
```

Now fit copulas to the uniform-transformed data.

```
14  data1 = cbind(pstd(ibm, est.ibm[1], est.ibm[2], est.ibm[3]),
15                pstd(sp500, est.sp500[1], est.sp500[2], est.sp500[3]))
16  n = nrow(netRtns) ; n
17  data2 = cbind(rank(ibm)/(n+1), rank(sp500)/(n+1))
18  ft1 = fitCopula(cop_t_dim2, data1, method="ml", start=c(omega,4) )
19  ft2 = fitCopula(cop_t_dim2, data2, method="ml", start=c(omega,4) )
```

Problem 5

(a) Explain the difference between methods used to obtain the two estimates
ft1 and ft2.
(b) Do the two estimates seem significantly different (in a practical sense)?

The next step defines a meta-t-distribution by specifying its t-copula and its univariate marginal distributions. Values for the parameters in the univariate margins are also specified. The values of the copula parameter were already defined in the previous step.

```
20  mvdc_t_t = mvdc( cop_t_dim2, c("std","std"), list(
21              list(mean=est.ibm[1],sd=est.ibm[2],nu=est.ibm[3]),
22              list(mean=est.sp500[1],sd=est.sp500[2],nu=est.sp500[3])))
```

Now fit the meta t-distribution. Be patient. This takes awhile; for instance, it took one minute on my laptop. The elapsed time in minutes will be printed.

```
23  start = c(est.ibm, est.sp500, ft1@estimate)
24  objFn = function(param) -loglikMvdc(param,cbind(ibm,sp500),mvdc_t_t)
25  tic = proc.time()
26  ft = optim(start, objFn, method="L-BFGS-B",
27            lower = c(-.1,0.001,2.2, -0.1,0.001,2.2,  0.2,2.5),
28            upper = c( .1,   10, 15,  0.1,   10, 15,  0.9, 15) )
29  toc = proc.time()
30  total_time = toc - tic ; total_time[3]/60
```

Lower and upper bounds are used to constrain the algorithm to stay inside a region where the log-likelihood is defined and finite. The function fitMvdc() in the copula package does not allow setting lower and upper bounds and did not converge on this problem.

Problem 6

(a) What are the estimates of the copula parameters in fit_cop?
(b) What are the estimates of the parameters in the univariate marginal
distributions?

(c) Was the estimation method maximum likelihood, semiparametric pseudo-maximum likelihood, or parametric pseudo-maximum likelihood?

(d) Estimate the coefficient of lower tail dependence for this copula.

Now fit normal (Gaussian), Frank, Clayton, Gumbel and Joe copulas to the data.

```
31 fnorm = fitCopula(copula=normalCopula(dim=2),data=data1,method="ml")
32 ffrank = fitCopula(copula = frankCopula(3, dim = 2),
33                         data = data1, method = "ml")
34 fclayton = fitCopula(copula = claytonCopula(1, dim=2),
35                         data = data1, method = "ml")
36 fgumbel = fitCopula(copula = gumbelCopula(3, dim=2),
37                         data = data1, method = "ml")
38 fjoe = fitCopula(copula=joeCopula(2,dim=2),data=data1,method="ml")
```

The estimated copulas (CDFs) will be compared with the empirical copula.

```
39 Udex = (1:n)/(n+1)
40 Cn = C.n(u=cbind(rep(Udex,n),rep(Udex,each=n)), U=data1, method="C")
41 EmpCop = expression(contour(Udex, Udex, matrix(Cn, n, n),
42                         col = 2, add = TRUE))
43 par(mfrow=c(2,3),  mgp = c(2.5,1,0))
44 contour(tCopula(param=ft$par[7],dim=2,df=round(ft$par[8])),
45         pCopula, main = expression(hat(C)[t]),
46         xlab = expression(hat(U)[1]), ylab = expression(hat(U)[2]) )
47 eval(EmpCop)
48 contour(normalCopula(param=fnorm@estimate[1], dim = 2),
49         pCopula, main = expression(hat(C)[Gauss]),
50         xlab = expression(hat(U)[1]), ylab = expression(hat(U)[2]) )
51 eval(EmpCop)
52 contour(frankCopula(param=ffrank@estimate[1], dim = 2),
53         pCopula, main = expression(hat(C)[Fr]),
54         xlab = expression(hat(U)[1]), ylab = expression(hat(U)[2]) )
55 eval(EmpCop)
56 contour(claytonCopula(param=fclayton@estimate[1], dim = 2),
57         pCopula, main = expression(hat(C)[Cl]),
58         xlab = expression(hat(U)[1]), ylab = expression(hat(U)[2]) )
59 eval(EmpCop)
60 contour(gumbelCopula(param=fgumbel@estimate[1], dim = 2),
61         pCopula, main = expression(hat(C)[Gu]),
62         xlab = expression(hat(U)[1]), ylab = expression(hat(U)[2]) )
63 eval(EmpCop)
64 contour(joeCopula(param=fjoe@estimate[1], dim = 2),
65         pCopula, main = expression(hat(C)[Joe]),
66         xlab = expression(hat(U)[1]), ylab = expression(hat(U)[2]) )
67 eval(EmpCop)
```

Problem 7 *Do you see any difference between the parametric estimates of the copula? If so, which seem closest to the empirical copula? Include the plot with your work.*

A two-dimensional KDE of the copula's density will be compared with the parametric density estimates (PDFs).

```
68  par(mfrow=c(2,3),  mgp = c(2.5,1,0))
69  contour(tCopula(param=ft$par[7],dim=2,df=round(ft$par[8])),
70          dCopula, main = expression(hat(c)[t]),
71      nlevels=25, xlab=expression(hat(U)[1]),ylab=expression(hat(U)[2]))
72  contour(kde2d(data1[,1],data1[,2]), col = 2, add = TRUE)
73  contour(normalCopula(param=fnorm@estimate[1], dim = 2),
74          dCopula, main = expression(hat(c)[Gauss]),
75      nlevels=25, xlab=expression(hat(U)[1]),ylab=expression(hat(U)[2]))
76  contour(kde2d(data1[,1],data1[,2]), col = 2, add = TRUE)
77  contour(frankCopula(param=ffrank@estimate[1], dim = 2),
78          dCopula, main = expression(hat(c)[Fr]),
79      nlevels=25, xlab=expression(hat(U)[1]),ylab=expression(hat(U)[2]))
80  contour(kde2d(data1[,1],data1[,2]), col = 2, add = TRUE)
81  contour(claytonCopula(param=fclayton@estimate[1], dim = 2),
82          dCopula, main = expression(hat(c)[Cl]),
83      nlevels=25, xlab=expression(hat(U)[1]),ylab=expression(hat(U)[2]))
84  contour(kde2d(data1[,1],data1[,2]), col = 2, add = TRUE)
85  contour(gumbelCopula(param=fgumbel@estimate[1], dim = 2),
86          dCopula, main = expression(hat(c)[Gu]),
87      nlevels=25, xlab=expression(hat(U)[1]),ylab=expression(hat(U)[2]))
88  contour(kde2d(data1[,1],data1[,2]), col = 2, add = TRUE)
89  contour(joeCopula(param=fjoe@estimate[1], dim = 2),
90          dCopula, main = expression(hat(c)[Joe]),
91      nlevels=25, xlab=expression(hat(U)[1]),ylab=expression(hat(U)[2]))
92  contour(kde2d(data1[,1],data1[,2]), col = 2, add = TRUE)
```

Problem 8 *Do you see any difference between the parametric estimates of the copula density? If so, which seem closest to the KDE? Include the plot with your work.*

Problem 9 *Find AIC for the t, (Gaussian), Frank, Clayton, Gumbel and Joe copulas. Which copula model fits best by AIC? (Hint: The* `fitCopula()` *function returns the log-likelihood.)*

8.10 Exercises

1. Kendall's tau rank correlation between X and Y is 0.55. Both X and Y are positive. What is Kendall's tau between X and $1/Y$? What is Kendall's tau between $1/X$ and $1/Y$?

2. Suppose that X is Uniform(0,1) and $Y = X^2$. Then the Spearman rank correlation and the Kendall's tau between X and Y will both equal 1, but the Pearson correlation between X and Y will be less than 1. Explain why.

3. Show that an Archimedean copula with generator function $\varphi(u) = -\log(u)$ is equal to the independence copula C_0. Does the same hold when the natural logarithm is replaced by the common logarithm, i.e., $\varphi(u) = -\log_{10}(u)$?

4. The co-monotonicity copula C_+ is not an Archimedean copula; however, in the two-dimensional case, the counter-monotonicity copula $C_-(u_1, u_2) = \max(u_1 + u_2 - 1, 0)$ is. What is its generator function?

5. Show that the generator of a Frank copula

$$\varphi_{\mathrm{Fr}}(u|\theta) = -\log\left\{\frac{e^{-\theta u} - 1}{e^{-\theta} - 1}\right\}, \quad \theta \in \{(-\infty, 0) \cup (0, \infty)\},$$

satisfies assumptions 1–3 of a strict generator.

6. Show that as $\theta \to \infty$, $C_{\mathrm{Fr}}(u_1, u_2|\theta) \to \min(u_1, u_2)$, the co-monotonicity copula C_+.

7. Suppose that $\varphi_1, \ldots, \varphi_k$ are k strict generator functions and define a new generator φ as a convex combination of these k generators, that is

$$\varphi(u) = a_1\varphi_1(u) + \cdots + a_k\varphi_k(u),$$

in which $a_1, \ldots, a_k$ are any non-negative constants which sum to 1. Show that $\varphi(u)$ is a strict generator function. For the case in which $k = 2$, what is the corresponding copula function for $\varphi(u)$?

8. Let $\varphi(u|\theta) = (1 - u)^\theta$, for some $\theta \geq 1$, and show that for the two-dimensional case this generates the copula

$$C(u_1, u_2|\theta) = \max[0, 1 - \{(1 - u_1)^\theta + (1 - u_2)^\theta\}^{1/\theta}].$$

Further, show that as $\theta \to \infty$, $C(u_1, u_2|\theta) \to \min(u_1, u_2)$, the co-monotonicity copula C_+, and that as $\theta \to 1$, $C(u_1, u_2|\theta) \to \max(u_1 + u_2 - 1, 0)$, the counter-monotonicity copula C_-.

9. A convex combination of k joint CDFs is itself a joint CDF (finite mixture), but is a convex combination of k copula functions a copula function itself?

10. Suppose $\mathbf{Y} = (Y_1, \ldots, Y_d)$ has a meta-Gaussian distribution with continuous marginal distributions and copula $C^{Gauss}(\cdot|\mathbf{\Omega})$. Show that if $\rho_\tau(Y_i, Y_j) = 0$ then Y_i and Y_j are independent.

References

Cherubini, U., Luciano, E., and Vecchiato, W. (2004) *Copula Methods in Finance*, John Wiley, New York.

Duffie, D. and Singleton, K. J. (2003) *Credit Risk*, Princeton University Press, Princeton and Oxford.

Joe, H. (1997) *Multivariate Models and Dependence Concepts*, Chapman & Hall, London.

Li, D (2000) On default correlation: A copula function approach, *Journal of Fixed Income*, **9**, 43–54.

Mari, D. D. and Kotz, S. (2001) *Correlation and Dependence*, World Scientific, London.

McNeil, A., Frey, R., and Embrechts, P. (2005) *Quantitative Risk Management*, Princeton University Press, Princeton and Oxford.

Nelsen, R. B. (2007) *An Introduction to Copulas,* 2nd ed., Springer, New York.

Salmon, F. (2009) Recipe for Disaster: The Formula That Killed Wall Street, *Wired* http://www.wired.com/techbiz/it/magazine/17-03/wp_quant?currentPage=all

9

Regression: Basics

9.1 Introduction

Regression is one of the most widely used of all statistical methods. For univariate regression, the available data are one response variable and p predictor variables, all measured on each of n observations. We let Y denote the response variable and $X_1, \ldots, X_p$ be the predictor or explanatory variables. Also, Y_i and $X_{i,1}, \ldots, X_{i,p}$ are the values of these variables for the ith observation. The goals of regression modeling include the investigation of how Y is related to $X_1, \ldots, X_p$, estimation of the conditional expectation of Y given $X_1, \ldots, X_p$, and prediction of future Y values when the corresponding values of $X_1, \ldots, X_p$ are already available. These goals are closely connected.

The *multiple linear regression* model relating Y to the predictor or regressor variables is

$$Y_i = \beta_0 + \beta_1 X_{i,1} + \cdots + \beta_p X_{i,p} + \epsilon_i, \tag{9.1}$$

where ϵ_i is called the noise, disturbances, or errors. The adjective "multiple" refers to the predictor variables. Multivariate regression, which has more than one response variable, is covered in Chap. 18. The ϵ_i are often called "errors" because they are the prediction errors when Y_i is predicted by $\beta_0 + \beta_1 X_{i,1} + \cdots + \beta_p X_{i,p}$. It is assumed that

$$E(\epsilon_i | X_{i,1}, \ldots, X_{i,p}) = 0, \tag{9.2}$$

which, with (9.1), implies that

$$E(Y_i | X_{i,1}, \ldots, X_{i,p}) = \beta_0 + \beta_1 X_{i,1} + \cdots + \beta_p X_{i,p}.$$

The parameter β_0 is the intercept. The regression coefficients $\beta_1, \ldots, \beta_p$ are the slopes. More precisely, β_j is the partial derivative of the expected response with respect to the jth predictor:

© Springer Science+Business Media New York 2015
D. Ruppert, D.S. Matteson, *Statistics and Data Analysis for Financial Engineering*, Springer Texts in Statistics,
DOI 10.1007/978-1-4939-2614-5_9

$$\beta_j = \frac{\partial\, E(Y_i | X_{i,1}, \ldots, X_{i,p})}{\partial\, X_{i,j}}.$$

Therefore, β_j is the change in the expected value of Y_i when $X_{i,j}$ changes one unit. It is assumed that the noise is i.i.d. white so that

$$\epsilon_1, \ldots, \epsilon_n \quad \text{are i.i.d. with mean 0 and variance } \sigma_\epsilon^2. \tag{9.3}$$

Often the ϵ_is are assumed to be normally distributed, which with (9.3) implies Gaussian white noise.

For the reader's convenience, the assumptions of the linear regression model are summarized:

1. linearity of the conditional expectation: $E(Y_i | X_{i,1}, \ldots, X_{i,p}) = \beta_0 + \beta_1 X_{i,1} + \cdots + \beta_p X_{i,p}$;
2. independent noise: $\epsilon_1, \ldots, \epsilon_n$ are independent;
3. constant variance: $\mathrm{Var}(\epsilon_i) = \sigma_\epsilon^2$ for all i;
4. Gaussian noise: ϵ_i is normally distributed for all i.

This chapter and, especially, the next two chapters discuss methods for checking these assumptions, the consequences of their violations, and possible remedies when they do not hold.

9.2 Straight-Line Regression

Straight-line regression is linear regression with only one predictor variable. The model is

$$Y_i = \beta_0 + \beta_1 X_i + \epsilon_i, \tag{9.4}$$

where β_0 and β_1 are the unknown intercept and slope of the line and ϵ_i is called the noise or error.

9.2.1 Least-Squares Estimation

The regression coefficients can be estimated by the *method of least squares*. The least-squares estimates are the values of $\widehat{\beta}_0$ and $\widehat{\beta}_1$ that minimize

$$\sum_{i=1}^{n} \left\{ Y_i - (\widehat{\beta}_0 + \widehat{\beta}_1 X_i) \right\}^2. \tag{9.5}$$

Geometrically, we are minimizing the sum of the squared lengths of the vertical lines in Fig. 9.1. The data points are shown as asterisks. The vertical lines connect the data points and the predictions using the linear equation. The predictions themselves are called the *fitted values* or "y-hats" and shown as open circles. The differences between the Y-values and the fitted values are called the *residuals*. Using calculus to minimize (9.5), one can show that

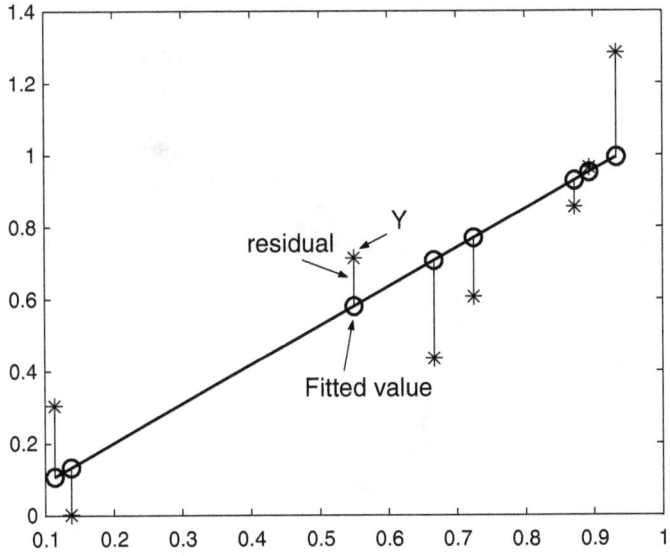

Fig. 9.1. *Least-squares estimation. The vertical lines connect the data (*) and the fitted values (o) represent the residuals. The least-squares line is defined as the line making the sum of the squared residuals as small as possible.*

$$\widehat{\beta}_1 = \frac{\sum_{i=1}^{n}(Y_i - \overline{Y})(X_i - \overline{X})}{\sum_{i=1}^{n}(X_i - \overline{X})^2} = \frac{\sum_{i=1}^{n} Y_i(X_i - \overline{X})}{\sum_{i=1}^{n}(X_i - \overline{X})^2}. \tag{9.6}$$

and

$$\widehat{\beta}_0 = \overline{Y} - \widehat{\beta}_1 \overline{X}. \tag{9.7}$$

The *least-squares line* is

$$\widehat{Y} = \widehat{\beta}_0 + \widehat{\beta}_1 X = \overline{Y} + \widehat{\beta}_1(X - \overline{X})$$

$$= \overline{Y} + \left\{ \frac{\sum_{i=1}^{n}(Y_i - \overline{Y})(X_i - \overline{X})}{\sum_{i=1}^{n}(X_i - \overline{X})^2} \right\} (X - \overline{X})$$

$$= \overline{Y} + \frac{s_{XY}}{s_X^2}(X - \overline{X}),$$

where $s_{XY} = (n-1)^{-1} \sum_{i=1}^{n}(Y_i - \overline{Y})(X_i - \overline{X})$ is the sample covariance between X and Y and s_X^2 is the sample variance of X.

Example 9.1. Weekly interest rates — least-squares estimates

Weekly interest rates from February 16, 1977, to December 31, 1993, were obtained from the Federal Reserve Bank of Chicago. Figure 9.2 is a plot of

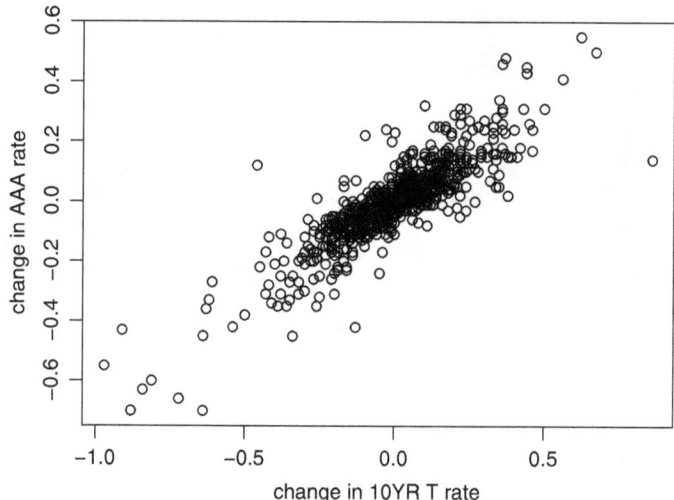

Fig. 9.2. *Changes in Moody's seasoned corporate AAA bond yields plotted against changes in 10-year Treasury constant maturity rate. Data from Federal Reserve Statistical Release H.15 and were taken from the Chicago Federal Bank's website.*

changes in the 10-year Treasury constant maturity rate and changes in the Moody's seasoned corporate AAA bond yield. The plot looks linear, so we try linear regression using R's lm() function. The code is:

```
options(digits = 3)
summary(lm(aaa_dif ~ cm10_dif))
```

The code aaa_dif ~ cm10_dif is an example of a formula in R with the outcome variable to the left of "~" and the explanatory variables to the right of "~." In this example, there is only one explanatory variable. In cases where there are multiple explanatory variables, they are separated by "+". Here is the output.

```
Call:
lm(formula = aaa_dif ~ cm10_dif)

Coefficients:
             Estimate Std. Error t value Pr(>|t|)
(Intercept) -0.000109   0.002221   -0.05     0.96
cm10_dif     0.615762   0.012117   50.82   <2e-16 ***
---
Signif. codes:  0 *** 0.001 ** 0.01 * 0.05 . 0.1  1

Residual standard error: 0.066 on 878 degrees of freedom
Multiple R-Squared: 0.746,      Adjusted R-squared: 0.746
F-statistic: 2.58e+03 on 1 and 878 DF, p-value: <2e-16
```

From the output we see that the least-squares estimates of the intercept and slope are -0.000109 and 0.616. The Residual standard error is 0.066; this is what we call $\widehat{\sigma}_\epsilon$ or s, the estimate of σ_ϵ; see Sect. 9.3. The remaining items of the output are explained shortly.   □

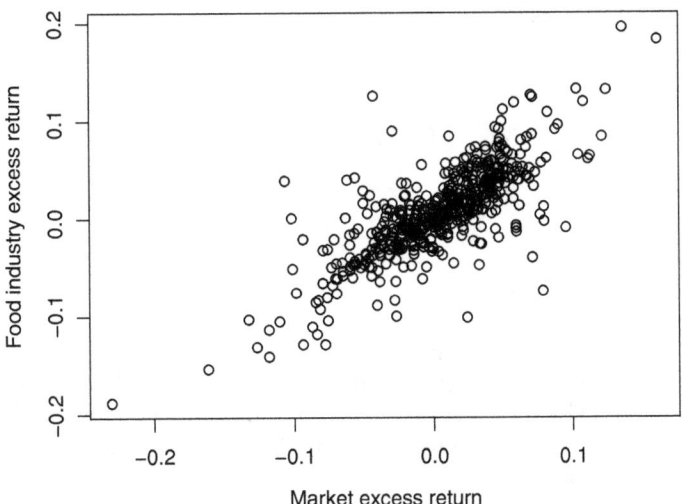

Fig. 9.3. *Plot of excess returns on the food industry versus excess returns on the market. Data from the data set* Capm *in R's* Ecdat *package.*

Example 9.2. Excess returns on the food sector and the market portfolio

The excess return on a security or market index is the return minus the risk-free interest rate. An important application of linear regression in finance is the regression of the excess return of an asset or market sector on the excess return of the entire market. This type of application will be discussed much more fully in Chap. 17. In this example, we will regress the excess monthly return of the food sector (rfood) on the excess monthly return of the market portfolio (rmrf). The data are in R's Capm data set in the Ecdat package and are plotted in Fig. 9.3. The returns are expressed as percentages in the data set but have been converted to fractions in this example. The output from lm is

```
Call:
lm(formula = rfood ~ rmrf)

Coefficients:
              Estimate Std. Error t value Pr(>|t|)
```

```
(Intercept)  0.00339    0.00128    2.66   0.0081 **
rmrf         0.78342    0.02835   27.63   <2e-16 ***
---
Signif. codes:  0 *** 0.001 ** 0.01 * 0.05 . 0.1 1

Residual standard error: 0.0289 on 514 degrees of freedom
Multiple R-Squared: 0.598,        Adjusted R-squared: 0.597
F-statistic:   763 on 1 and 514 DF,  p-value: <2e-16
```

Thus, the fitted regression equation is

$$\texttt{rfood} = 0.00339 + 0.78342 \, \texttt{rmrf} + \epsilon,$$

and $\widehat{\sigma}_\epsilon = 0.0289$. □

9.2.2 Variance of $\widehat{\beta}_1$

It is useful to have a formula for the variance of an estimator to show how the estimator's precision depends on various aspects of the data such as the sample size and the values of the predictor variables. Fortunately, it is easy to derive a formula for the variance of $\widehat{\beta}_1$. By (9.6), we can write $\widehat{\beta}_1$ as a weighted average of the responses

$$\widehat{\beta}_1 = \sum_{i=1}^{n} w_i Y_i,$$

where w_i is the weight given by

$$w_i = \frac{X_i - \overline{X}}{\sum_{i=1}^{n}(X_i - \overline{X})^2}.$$

We consider $X_1, \ldots, X_n$ as fixed, so if they are random we are conditioning upon their values. From the assumptions of the regression model, it follows that $\text{Var}(Y_i | X_1, \ldots, X_n) = \sigma_\epsilon^2$ and $Y_1, \ldots, Y_n$ are conditionally uncorrelated. Therefore,

$$\text{Var}(\widehat{\beta}_1 | X_1, \ldots, X_n) = \sigma_\epsilon^2 \sum_{i=1}^{n} w_i^2 = \frac{\sigma_\epsilon^2}{\sum_{i=1}^{n}(X_i - \overline{X})^2} = \frac{\sigma_\epsilon^2}{(n-1)s_X^2}. \quad (9.8)$$

It is worth taking some time to examine this formula. First, the numerator σ_ϵ^2 is simply the variance of the ϵ_i. This is not surprising. More variability in the noise means more variable estimators. The denominator shows us that the variance of $\widehat{\beta}_1$ is inversely proportional to $(n-1)$ and to s_X^2. So the precision of $\widehat{\beta}_1$ increases as σ_ϵ^2 is reduced, n is increased, or s_X^2 is increased. Why does increasing s_X^2 decrease $\text{Var}(\widehat{\beta}_1 | X_1, \ldots, X_n)$? The reason is that increasing s_X^2 means that the X_i are spread farther apart, which makes the slope of the line easier to estimate.

Example 9.3. Optimal sampling frequencies for regression

Here is an important application of (9.8). Suppose that we have two stationary time series, X_t and Y_t, and we wish to regress Y_t on X_t. We have just seen examples of this. A significant practical question is whether one should use daily or weekly data, or perhaps even monthly or quarterly data. Does it matter which sampling frequency we use? The answer is "yes" and the highest possible sampling frequency gives the most precise estimate of the slope. To understand why this is so, we compare daily and weekly data. Assume that the X_t and Y_t are white noise sequences. Since a weekly log return is simply the sum of the five daily log returns within a week, σ_ϵ^2 and s_X^2 will each increase by a factor of five if we change from daily to weekly log returns, so the ratio σ_ϵ^2/s_X^2 will not change. However, by changing from daily to weekly log returns, $(n-1)$ is reduced by approximately a factor of five. The result is that $\mathrm{Var}(\widehat{\beta}_1|X_1,\ldots,X_n)$ is approximately five times smaller using daily rather than weekly log returns. Similarly, $\mathrm{Var}(\widehat{\beta}_1|X_1,\ldots,X_n)$ is about four times larger using monthly rather than weekly returns.

The obvious conclusion is that one should use the highest sampling frequency available, which is often daily returns. We have assumed that the X_t and Y_t are white noise in order to simplify the calculations, but this conclusion still holds if they are stationary but autocorrelated. (Autocorrelation is discussed in Chap. 12.) However, the noise series, that is ϵ_i, $i = 1,\ldots,$ in Eq. (9.4) needs to be uncorrelated. If the noise is autocorrelated and becomes more highly correlated as the sampling frequency increases, then this conclusion need not hold. There may be a point of diminishing returns where more frequent sampling does not improve estimation accuracy. $\qquad\square$

9.3 Multiple Linear Regression

The multiple linear regression model is

$$Y_i = \beta_0 + \beta_1 X_{i,1} + \cdots + \beta_p X_{i,p} + \epsilon_i.$$

The least-squares estimates are the values $\widehat{\beta}_0, \widehat{\beta}_1, \ldots, \widehat{\beta}_p$ that minimize

$$\sum_{i=1}^{n}\left\{Y_i - (\widehat{\beta}_0 + \widehat{\beta}_1 X_{i,1} + \cdots + \widehat{\beta}_p X_{i,p})\right\}^2. \tag{9.9}$$

Calculation of the least-squares estimates is discussed in Sect. 11.1. For applications, the technical details are not important, since software for least-squares estimation is readily available.

The *i*th *fitted value* is

$$\widehat{Y}_i = \widehat{\beta}_0 + \widehat{\beta}_1 X_{i,1} + \cdots + \widehat{\beta}_p X_{i,p} \tag{9.10}$$

and estimates $E(Y_i|X_{i,1}, \ldots, X_{i,p})$. The ith residual is

$$\widehat{\epsilon}_i = Y_i - \widehat{Y}_i = Y_i - (\widehat{\beta}_0 + \widehat{\beta}_1 X_{i,1} + \cdots + \widehat{\beta}_p X_{i,p}) \tag{9.11}$$

and estimates ϵ_i. It is worth noting that (9.11) can be re-expressed as

$$Y_i = \widehat{Y}_i + \widehat{\epsilon}_i. \tag{9.12}$$

An unbiased estimate of σ_ϵ^2 is

$$\widehat{\sigma}_\epsilon^2 = \frac{\sum_{i=1}^n \widehat{\epsilon}_i^2}{n - 1 - p}. \tag{9.13}$$

The denominator in (9.13) is the sample size minus the number of regression coefficients that are estimated.

Example 9.4. Multiple linear regression with interest rates

As an example, we continue the analysis of the weekly interest-rate data but now with changes in the 30-year Treasury rate (cm30_dif) and changes in the Federal funds rate (ff_dif) as additional predictors. Thus $p = 3$. Figure 9.4 is a scatterplot matrix of the four time series. There is a strong linear relationship between all pairs of aaa_dif, cm10_dif, and cm30_dif, but ff_dif is not strongly related to the other series. The code is

```
summary(lm(aaa_dif ~ cm10_dif + cm30_dif + ff_dif))
```

The lm() output for this regression is

```
Call:
lm(formula = aaa_dif ~ cm10_dif + cm30_dif + ff_dif)

Coefficients:
              Estimate Std. Error t value Pr(>|t|)
(Intercept) -9.07e-05   2.18e-03   -0.04    0.97
cm10_dif     3.55e-01   4.51e-02    7.86  1.1e-14 ***
cm30_dif     3.00e-01   5.00e-02    6.00  2.9e-09 ***
ff_dif       4.12e-03   5.28e-03    0.78    0.44
---

Residual standard error: 0.0646 on 876 degrees of freedom
Multiple R-Squared: 0.756,      Adjusted R-squared: 0.755
F-statistic:  906 on 3 and 876 DF,  p-value: <2e-16
```

We see that $\widehat{\beta}_0 = -9.07 \times 10^{-05}$, $\widehat{\beta}_1 = 0.355$, $\widehat{\beta}_2 = 0.300$, and $\widehat{\beta}_3 = 0.00412$. $\square$

A commonly used special case of multiple regression is the polynomial regression model which uses powers of the predictors as well as the predictors

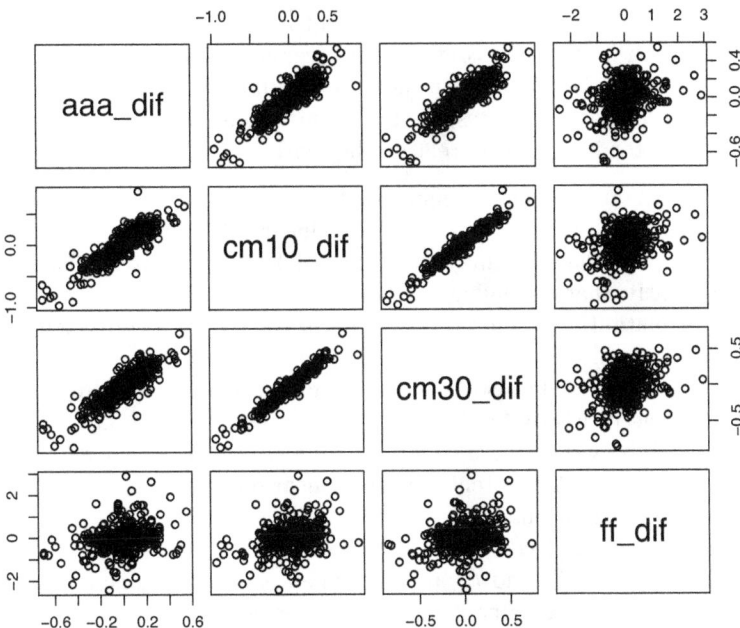

Fig. 9.4. *Scatterplot matrix of the changes in four weekly interest rates. The variable* aaa_dif *is the response in Example* 9.4.

themselves. For example, when there is one X-variable, the p-degree polynomial regression model is

$$Y_i = \beta_0 + \beta_1 X_i + \cdots + \beta_p X_i^p + \epsilon_i.$$

As another example, the quadratic regression model with two predictors is

$$Y_i = \beta_0 + \beta_1 X_{i,1} + \beta_2 X_{i,1}^2 + \beta_3 X_{i,1} X_{i,2} + \beta_4 X_{i,2} + \beta_5 X_{i,2}^2 + \epsilon_i.$$

9.3.1 Standard Errors, t-Values, and p-Values

In this section we explain the use of several statistics included in regression output. We use the output in Example 9.4 as an illustration.

As noted before, the estimated coefficients are $\widehat{\beta}_0 = -9.07 \times 10^{-05}$, $\widehat{\beta}_1 = 0.355$, $\widehat{\beta}_2 = 0.300$, and $\widehat{\beta}_3 = 0.00412$. Each of these coefficients has three other statistics associated with it.

- The standard error (SE), which is the estimated standard deviation of the least-squares estimator, tells us the precision of the estimator.
- The t-value, is the t-statistic for testing that the coefficient is 0. The t-value is the ratio of the estimate to its standard error. For example, for cm10_dif, the t-value is $7.86 = 0.355/0.0451$.

- The p-value (`Pr > |t|` in the `lm()` output), associated with testing the null hypothesis that the coefficient is 0 versus the alternative that it is not 0. If a p-value for a slope parameter is small, as it is here for β_1, then this is evidence that the corresponding coefficient is *not* 0, which means that the predictor has a *linear* relationship with the response.

It is important to keep in mind that the p-value only tells us if there is a linear relationship. The existence of a linear relationship between Y_i and $X_{i,j}$ means only that the linear predictor of Y_i has a nonzero slope on $X_{i,j}$, or, equivalently, that partial correlation between $X_{i,j}$ and Y_i is not zero. (The partial correlation between two variables is their correlation when all other variables are held fixed.) When the p-value is small (so a linear relationship exists), there could also be a strong nonlinear deviation from the linear relationship as in Fig. A.4g. Moreover, when the p-value is large (so no linear relationship exists), there could still be a strong nonlinear relationship in Fig. A.4f. Because of the potential for nonlinear relationships to go undetected in a linear regression analysis, graphical analysis of the data (e.g., Fig. 9.4) and residual analysis (see Chap. 10) are essential.

The p-values for β_1 and β_2 are *very* small, so we can conclude that these slopes are *not* 0. The p-value is large (0.97) for β_0, so we would not reject the hypothesis that the intercept is 0.

Similarly, we would not reject the null hypothesis that β_3 is zero. Stated differently, we can accept the null hypothesis that, conditional on `cm10_dif` and `cm30_dif`, `aaa_dif` and `ff_dif` are not linearly related. This result should *not* be interpreted as stating that `aaa_dif` and `ff_dif` are unrelated, but only that `ff_dif` is not useful for predicting `aaa_dif` when `cm10_dif` and `cm30_dif` are included in the regression model. (In fact, `aaa_dif` and `ff_dif` have a correlation of 0.25 (this is the full, not partial, correlation) and the linear regression of `aaa_dif` on `ff_dif` alone is highly significant; the p-value for testing that the slope is zero is 5.158×10^{-14}.)

Since the Federal Funds rate is a short-term (overnight) rate, it is not surprising that `ff_dif` is less useful than changes in the 10- and 30-year Treasury rates for predicting `aaa_dif`.

For regression with one predictor variable, by (9.8) the standard error of $\widehat{\beta}_1$ is $\widehat{\sigma}_\epsilon / \sqrt{\sum_{i=1}^n (X_i - \overline{X})^2}$. When there are more than two predictor variables, formulas of standard errors are more complex and are facilitated by the use of matrix notation. Because standard errors can be computed with standard software such as `lm`, the formulas are not needed for applications and so are postponed to Sect. 11.1.

9.4 Analysis of Variance, Sums of Squares, and R^2

9.4.1 ANOVA Table

Certain results of a regression fit are often displayed in an *analysis of variance table*, also called the ANOVA or AOV table. The idea behind the ANOVA table is to describe how much of the variation in Y is predictable if one knows $X_1, \ldots, X_p$. Here is the ANOVA table for the model in Example 9.4.

```
> anova(lm(aaa_dif ~ cm10_dif + cm30_dif + ff_dif))
Analysis of Variance Table

Response: aaa_dif
           Df Sum Sq Mean Sq F value   Pr(>F)
cm10_dif    1  11.21   11.21 2682.61 < 2e-16 ***
cm30_dif    1   0.15    0.15   35.46 3.8e-09 ***
ff_dif      1 0.0025  0.0025    0.61    0.44
Residuals 876   3.66  0.0042
---
```

The total variation in Y can be partitioned into two parts: the variation that can be predicted by $X_1, \ldots, X_p$ and the variation that cannot be predicted. The variation that can be predicted is measured by the regression sum of squares, which is

$$\text{regression SS} = \sum_{i=1}^{n} (\widehat{Y}_i - \overline{Y})^2.$$

The regression sum of squares for the model that uses only cm10_dif is in the first row of the ANOVA table and is 11.21. The entry, 0.15, in the second row is the increase in the regression sum of squares when cm30_dif is added to the model. Similarly, 0.0025 is the increase in the regression sum of squares when ff_dif is added. Thus, rounding to two decimal places, $11.36 = 11.21 + 0.15 + 0.00$ is the regression sum of squares with all three predictors in the model.

The amount of variation in Y that cannot be predicted by a linear function of $X_1, \ldots, X_p$ is measured by the residual error sum of squares, which is the sum of the squared residuals; i.e.,

$$\text{residual error SS} = \sum_{i=1}^{n} (Y_i - \widehat{Y}_i)^2.$$

In the ANOVA table, the residual error sum of squares is in the last row and is 3.66. The total variation is measured by the total sum of squares (total SS), which is the sum of the squared deviations of Y from its mean; that is,

$$\text{total SS} = \sum_{i=1}^{n} (Y_i - \overline{Y})^2. \tag{9.14}$$

It can be shown algebraically that

$$\text{total SS} = \text{regression SS} + \text{residual error SS}. \tag{9.15}$$

Therefore, in Example 9.4, the total SS is $11.36 + 3.66 = 15.02$.
R-squared, denoted by R^2, is

$$R^2 = \frac{\text{regression SS}}{\text{total SS}} = 1 - \frac{\text{residual error SS}}{\text{total SS}}$$

and measures the proportion of the total variation in Y that can be linearly predicted by X. In the example, R^2 is $0.746 = 11.21/15.02$ if only `cm10_dif` is the model and is $11.36/15.02 = 0.756$ if all three predictors are in the model. This value can be found in the output displayed in Example 9.4.

When there is only a single X variable, then $R^2 = r^2_{XY} = r^2_{\hat{Y}Y}$, where r_{XY} and $r_{\hat{Y}Y}$ are the sample correlations between Y and X and between Y and the predicted values, respectively. Put differently, R^2 is the squared correlation between Y and X and also between Y and $\hat{Y}$. When there are multiple predictors, then we still have $R^2 = r^2_{\hat{Y}Y}$. Since $\hat{Y}$ is a linear combination of the X variables, R can be viewed as the "multiple" correlation between Y and many Xs. The residual error sum of squares is also called the error sum of squares or sum of squared errors and is denoted by SSE.

It is important to understand that sums of squares in an ANOVA table depend upon the order of the predictor variables in the regression, because the sum of squares for any variable is the increase in the regression sum of squares when that variable is added to the predictors already in the model.

The table below has the same variables as before, but the order of the predictor variables is reversed. Now that `ff_dif` is the first predictor, its sum of squares is much larger than before and its p-value is highly significant; before it was nonsignificant, only 0.44. The sum of squares for `cm30_dif` is now much larger than that of `cm10_dif`, the reverse of what we saw earlier, since `cm10_dif` and `cm30_dif` are highly correlated and the first of them in the list of predictors will have the larger sum of squares.

```
> anova(lm(aaa_dif ~ ff_dif + cm30_dif + cm10_dif))
Analysis of Variance Table

Response: aaa_dif
            Df  Sum Sq  Mean Sq  F value    Pr(>F)
ff_dif       1    0.94     0.94    224.8  < 2e-16 ***
cm30_dif     1   10.16    10.16   2432.1  < 2e-16 ***
cm10_dif     1    0.26     0.26     61.8  1.1e-14 ***
Residuals  876    3.66   0.0042
```

The lesson here is that an ANOVA table is most useful for assessing the effects of adding predictors in some natural order. Since AAA bonds have maturities closer to 10 than to 30 years, and since the Federal Funds rate is an overnight rate, it made sense to order the predictors as `cm10_dif`, `cm30_dif`, and `ff_dif` as done initially.

9.4.2 Degrees of Freedom (DF)

There are degrees of freedom (DF) associated with each of these sources of variation. The degrees of freedom for regression is p, which is the number of predictor variables. The total degrees of freedom is $n - 1$. The residual error degrees of freedom is $n - p - 1$. Here is a way to think of degrees of freedom. Initially, there are n degrees of freedom, one for each observation. Then one degree of freedom is allocated to estimation of the intercept. This leaves a total of $n - 1$ degrees of freedom for estimating the effects of the X variables and σ_ϵ^2. Each regression parameter uses one degree of freedom for estimation. Thus, there are $(n - 1) - p$ degrees of freedom remaining for estimation of σ_ϵ^2 using the residuals. There is an elegant geometrical theory of regression where the responses are viewed as lying in an n-dimensional vector space and degrees of freedom are the dimensions of various subspaces. However, there is not sufficient space to pursue this subject here.

9.4.3 Mean Sums of Squares (MS) and F-Tests

As just discussed, every sum of squares in an ANOVA table has an associated degrees of freedom. The ratio of the sum of squares to the degrees of freedom is the mean sum of squares:

$$\text{mean sum of squares} = \frac{\text{sum of squares}}{\text{degrees of freedom}}.$$

The residual mean sum of squares is the unbiased estimate σ_ϵ^2 given by (9.13); that is,

$$\widehat{\sigma}_\epsilon^2 = \frac{\sum_{i=1}^n (Y_i - \widehat{Y}_i)^2}{n - 1 - p} \tag{9.16}$$
$$= \text{residual mean sum of squares}$$
$$= \frac{\text{residual error SS}}{\text{residual degrees of freedom}}.$$

Other mean sums of squares are used in testing. Suppose we have two models, I and II, and the predictor variables in model I are a subset of those in model II, so that model I is a submodel of II. A common null hypothesis is that the data are generated by model I. Equivalently, in model II the slopes are zero for variables not also in model I. To test this hypothesis, we use the excess regression sum of squares of model II relative to model I:

$$\text{SS(II}\,|\,\text{I)} = \text{regression SS for model II} - \text{regression SS for model I}$$
$$= \text{residual SS for model I} - \text{residual SS for model II}. \tag{9.17}$$

Equality (9.17) holds because (9.15) is true for all models and, in particular, for both model I and model II. The degrees of freedom for SS(II | I) is the number of extra predictor variables in model II compared to model I. The mean square is denoted as MS(II | I). Stated differently, if p_I and p_{II} are the number of parameters in models I and II, respectively, then $df_{II|I} = p_{II} - p_I$ and MS(II | I) = SS(II | I)/$df_{II|I}$. The F-statistic for testing the null hypothesis is

$$F = \frac{\text{MS(II|I)}}{\widehat{\sigma}_\epsilon^2},$$

where $\widehat{\sigma}_\epsilon^2$ is the mean residual sum of squares for model II. Under the null hypothesis, the F-statistic has an F-distribution with $df_{II|I}$ and $n - p_{II} - 1$ degrees of freedom and the null hypothesis is rejected if the F-statistic exceeds the α-upper quantile of this F-distribution.

Example 9.5. Weekly interest rates—Testing the one-predictor versus three-predictor model

In this example, the null hypothesis is that, in the three-predictor model, the slopes for cm30_dif and ff_dif are zero. The F-test can be computed using R's anova function. The output is

```
Analysis of Variance Table

Model 1: aaa_dif ~ cm10_dif
Model 2: aaa_dif ~ cm10_dif + cm30_dif + ff_dif
  Res.Df  RSS  Df Sum of Sq    F  Pr(>F)
1    878 3.81
2    876 3.66   2     0.15  18.0 2.1e-08 ***
---
Signif. codes:  0 *** 0.001 ** 0.01 * 0.05 . 0.1  1
```

In the last row, the entry 2 in the "Df" column is the difference between the two models in the number of parameters and 0.15 in the "Sum of Sq" column is the difference between the residual sum of squares (RSS) for the two models.

The very small p-value (2.1×10^{-8}) leads us to reject the null hypothesis and say that the result is "highly significant." It is important to be aware that this phrase refers to statistical significance. When the sample size is as large as it is here, it is common to reject the null hypothesis. The reason for this is that the null hypothesis is rarely true exactly, and with a large sample size it is highly likely that even a small deviation from the null hypothesis will be detected. Statistically significance must be distinguished from practical significance. The adjusted R^2 values for the two- and three-variable models are very similar, 0.746 and 0.755, respectively. Therefore, the rejection of the two-variable model may not be of practical importance.   □

Example 9.6. Weekly interest rates—Testing a two-predictor versus three-predictor model

In this example, the null hypothesis is that, in the three predictor model, the slope ff_dif is zero. The F-test is again computed using R's anova function with output:

```
Analysis of Variance Table

Model 1: aaa_dif ~ cm10_dif + cm30_dif
Model 2: aaa_dif ~ cm10_dif + cm30_dif + ff_dif
  Res.Df  RSS  Df  Sum of Sq    F Pr(>F)
1    877 3.66
2    876 3.66   1     0.0025 0.61    0.44
```

The large p-value (0.44) leads us to accept the null hypothesis. Notice that this is the same as the p-value for ff_dif in the ANOVA table in Sect. 9.4.1. This is not a coincidence. Both p-values are the same because they are testing the same hypothesis. □

9.4.4 Adjusted R^2

R^2 is biased in favor of large models, because R^2 is always increased by adding more predictors to the model, even if they are independent of the response. Recall that

$$R^2 = 1 - \frac{\text{residual error SS}}{\text{total SS}} = 1 - \frac{n^{-1}\text{residual error SS}}{n^{-1}\text{total SS}}.$$

The bias in R^2 can be reduced by using the following "adjustment," which replaces both occurrences of n by the appropriate degrees of freedom:

$$\text{adjusted } R^2 = 1 - \frac{(n-p-1)^{-1}\text{residual error SS}}{(n-1)^{-1}\text{total SS}} = 1 - \frac{\text{residual error MS}}{\text{total MS}}.$$

The presence of p in the adjusted R^2 penalizes the criterion for the number of predictor variables, so adjusted R^2 can either increase or decrease when predictor variables are added to the model. Adjusted R^2 increases if the added variables decrease the residual sum of squares enough to compensate for the increase in p.

9.5 Model Selection

When there are many potential predictor variables, often we wish to find a subset of them that provide a parsimonious regression model. F-tests are not very suitable for model selection. One problem is that there are many possible F-tests and the joint statistical behavior of all of them is not known. For model selection, it is more appropriate to use a model selection criterion such as AIC or BIC. For linear regression models, AIC is

$$\text{AIC} = n\log(\widehat{\sigma}^2) + 2(1+p),$$

where $1+p$ is the number of parameters in a model with p predictor variables; the intercept gives us the final parameter. BIC replaces $2(1+p)$ in AIC by $\log(n)(1+p)$. The first term, $n\log(\widehat{\sigma}^2)$, is equal to, up to an additive constant that does not affect model comparisons, -2 times the log-likelihood evaluated at the MLE, assuming that the noise is Gaussian.

In addition to AIC and BIC, there are two model selection criteria specialized for regression. One is adjusted R^2, which we have seen before. Another is C_p. C_p is related to AIC and usually C_p and AIC are minimized by the same model. The primary reason for using C_p instead of AIC is that some regression software computes only C_p, not AIC—this is true of the `regsubsets()` function in R's `leaps` package which will be used in the following example.

To define C_p, suppose there are M predictor variables. Let $\widehat{\sigma}^2_{\epsilon,M}$ be the estimate of σ^2_{ϵ} using all of them, and let $\text{SSE}(p)$ be the sum of squares for residual error for a model with some subset of only $p \leq M$ of the predictors. As usual, n is the sample size. Then C_p is

$$C_p = \frac{SSE(p)}{\widehat{\sigma}^2_{\epsilon,M}} - n + 2(p+1). \tag{9.18}$$

Of course, C_p will depend on which particular model is used among all of those with p predictors, so the notation "C_p" may not be ideal.

With C_p, AIC, and BIC, smaller values are better, but for adjusted R^2, larger values are better.

One should not use model selection criteria blindly. Model choice should be guided by economic theory and practical considerations, as well as by model selection criteria. It is important that the final model makes sense to the user. Subject-matter expertise might lead to adoption of a model not optimal according to the criterion being used but, instead, to a model slightly below optimal but more parsimonious or with a better economic rationale.

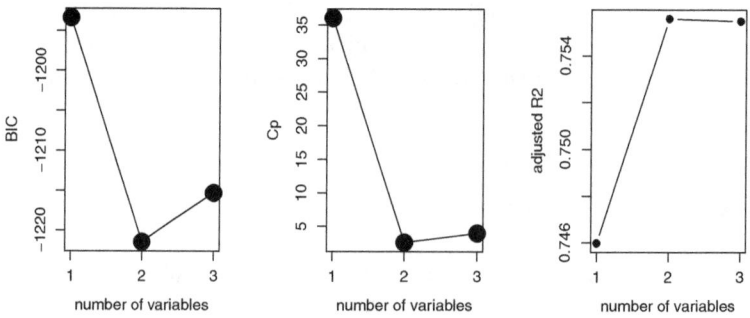

Fig. 9.5. *Changes in weekly interest rates. Plots for model selection.*

Example 9.7. Weekly interest rates—Model selection by AIC and BIC

Figure 9.5 contains plots of the number of predictors in the model versus the optimized value of a selection criterion. By "optimized value," we mean the best value among all models with the given number of predictor variables. "Best" means smallest for BIC and C_p and largest for adjusted R^2. There are three plots, one for each of BIC, C_p, and adjusted R^2. All three criteria are optimized by two predictor variables.

There are three models with two of the three predictors. The one that optimized the criteria[1] is the model with cm10_dif and cm30_dif, as can be seen in the following output from regsubsets. Here "*" indicates a variable in the model and " " indicates a variable not in the model, so the three rows of the table indicate that the best one-variable model is cm10_dif and the best two-variable model is cm10_dif and cm30_dif—the third row does not contain any real information since, with only three variables, there is only one possible three-variable model.

```
Selection Algorithm: exhaustive
         cm10_dif cm30_dif ff_dif
1  ( 1 )  "*"      " "      " "
2  ( 1 )  "*"      "*"      " "
3  ( 1 )  "*"      "*"      "*"
```

□

9.6 Collinearity and Variance Inflation

If two or more predictor variables are highly correlated with one another, then it is difficult to estimate their separate effects on the response. For example, cm10_dif and cm30_dif have a correlation of 0.96 and the scatterplot in Fig. 9.4 shows that they are highly related to each other. If we regress aaa_dif on cm10_dif, then the adjusted R^2 is 0.7460, but adjusted R^2 only increases to 0.7556 if we add cm30_dif as a second predictor. This suggests that cm30_dif might not be related to aaa_dif, but this is not the case. In fact, the adjusted R^2 is 0.7376 when cm30_dif is the only predictor, which indicates that cm30_dif is a good predictor of aaa_dif, nearly as good as cm10_dif.

Another effect of the high correlation between the predictor variables is that the regression coefficient for each variable is very sensitive to whether the other variable is in the model. For example, the coefficient of cm10_dif is 0.616 when cm10_dif is the sole predictor variable but only 0.360 if cm30_dif is also included.

---

[1] When comparing models with the same number of parameters, all three criteria are optimized by the same model.

The problem here is that cm10_dif and cm30_dif provide redundant information because of their high correlation. This problem is called *collinearity* or, in the case of more than two predictors, *multicollinearity*. Collinearity increases standard errors. The standard error of the β of cm10_dif is 0.01212 when only cm10_dif is in the model, but increases to 0.0451, a 372 % increase, if cm30_dif is added to the model.

The *variance inflation factor* (*VIF*) of a variable tells us how much the squared standard error, i.e., the variance of $\widehat{\beta}$, of that variable is increased by having the other predictor variables in the model. For example, if a variable has a VIF of 4, then the variance of its $\widehat{\beta}$ is four times larger than it would be if the other predictors were either deleted or were not correlated with it. The standard error is increased by a factor of 2.

Suppose we have predictor variables $X_1, \ldots, X_p$. Then the VIF of X_j is found by regressing X_j on the $p-1$ other predictors. Let R_j^2 be the R^2-value of this regression, so that R_j^2 measures how well X_j can be predicted from the other Xs. Then the VIF of X_j is

$$\text{VIF}_j = \frac{1}{1 - R_j^2}.$$

A value of R_j^2 close to 1 implies a large VIF. In other words, the more accurately that X_j can be predicted from the other Xs, the more redundant it is and the higher its VIF. The minimum value of VIF_j is 1 and occurs when R_j^2 is 0. There is, unfortunately, no upper bound to VIF_j. Variance inflation becomes infinite as R_j^2 approaches 1.

When interpreting VIFs, it is important to keep in mind that VIF_j tells us nothing about the relationship between the response and jth predictor. Rather, it tells us only how correlated the jth predictor is with the other predictors. In fact, the VIFs can be computed without knowing the values of the response variable.

The usual remedy to collinearity is to reduce the number of predictor variables by using one of the model selection criteria discussed in Sect. 9.5.

Example 9.8. Variance inflation factors for the weekly interest-rate example.

The function vif() in R's faraway library returned the following VIF values for the changes in weekly interest rates:

```
> library(faraway)
> options(digits = 2)
> vif(lm(aaa_dif ~ cm10_dif + cm30_dif + ff_dif))
cm10_dif cm30_dif   ff_dif
    14.4     14.1      1.1
```

cm10_dif and cm30_dif have large VIFs due to their high correlation with each other. The predictor ff_dif is not highly correlated with cm10_dif and cm30_dif and has a lower VIF.

VIF values give us information about linear relationships between the predictor variables, but not about their relationships with the response. In this example, ff_dif has a small VIF value but is not an important predictor because of its low correlation with the response. Despite their high VIF values, cm10_dif and cm30_dif are important predictors. The high VIF values tell us only that the regression coefficients for cm10_dif and cm30_dif are impossible to estimate with high precision.

The question is whether VIF values of 14.4 and 14.1 are so large that the number of predictor variables should be reduced to 1, that is, whether we should use only cm10_dif. The answer is "perhaps not" because the model with both cm10_dif and cm30_dif minimizes BIC. BIC generally selects a parsimonious model because of the high penalty BIC places on the number of predictor variables. Therefore, a model that minimizes BIC is unlikely to need further deletion of predictor variables simply to reduce VIF values. However, we saw earlier that adding cm30_dif to the model with cm10_dif offers only a minor increase in adjusted R^2, so the issue of whether or not to include cm30_dif is not clear. □

Example 9.9. Nelson–Plosser macroeconomic variables

To illustrate model selection, we now turn to an example with more predictors. We will start with six predictors but will find that a model with only two predictors fits rather well.

This example uses a subset of the well-known Nelson–Plosser data set of U.S. yearly macroeconomic time series. These data are available in the file nelsonplosser.csv. The variables we will use are:

1. sp-Stock Prices, [Index; 1941-43 = 100], [1871–1970].
2. gnp.r-Real GNP, [Billions of 1958 Dollars], [1909–1970],
3. gnp.pc-Real Per Capita GNP, [1958 Dollars], [1909–1970],
4. ip-Industrial Production Index, [1967 = 100], [1860–1970],
5. cpi-Consumer Price Index, [1967 = 100], [1860–1970],
6. emp-Total Employment, [Thousands], [1890–1970],
7. bnd-Basic Yields 30-year Corporate Bonds, [% pa], [1900–1970].

Since two of the time series start in 1909, we use only the data from 1909 until the end of the series in 1970, a total of 62 years. The response will be the differences of log(sp), the log returns on the stock prices. The regressors will be the differences of variables 2 through 7, with variables 4 and 5 log-transformed before differencing. A differenced log-series contains the approximate relative changes in the original variable, in the same way that a log return approximates a return that is the relative change in price.

How does one decide whether to difference the original series, the log-transformed series, or some other function of the series? Usually the aim is to stabilize the fluctuations in the differenced series. The top row of Fig. 9.6 has time series plots of changes in `gnp.r`, `log(gnp.r)`, and `sqrt(gnp.r)` and the bottom row has similar plots for `ip`. For `ip` the fluctuations in the differenced series increase steadily over time, but this is less true if one uses the square roots or logs of the series. This is the reason why `diff(log(ip))` is used here as a regressor. For `gnp.r`, the fluctuations in changes are more stable and we used `diff(gnp.r)` rather than `diff(log(gnp.r))` as a regressor. In this analysis, we did not consider using square-root transformations, since changes in the square roots are less interpretable than changes in the original variable or its logarithm. However, the changes in the square roots of both series are reasonably stable, so square-root transformations might be considered. Another possibility would be to use the transformation that gives the best-fitting model. One could, for example, put all three variables, `diff(ip)`, `diff(log(ip))`, and `diff(sqrt(ip))`, into the model and use model selection to decide which gives the best fit. The same could be done with `gnp.r` and the other regressors.

Notice that the variables are transformed first and then differenced. Differencing first and then taking logarithms or square roots would result in complex-valued variables, which would be difficult to interpret, to say the least.

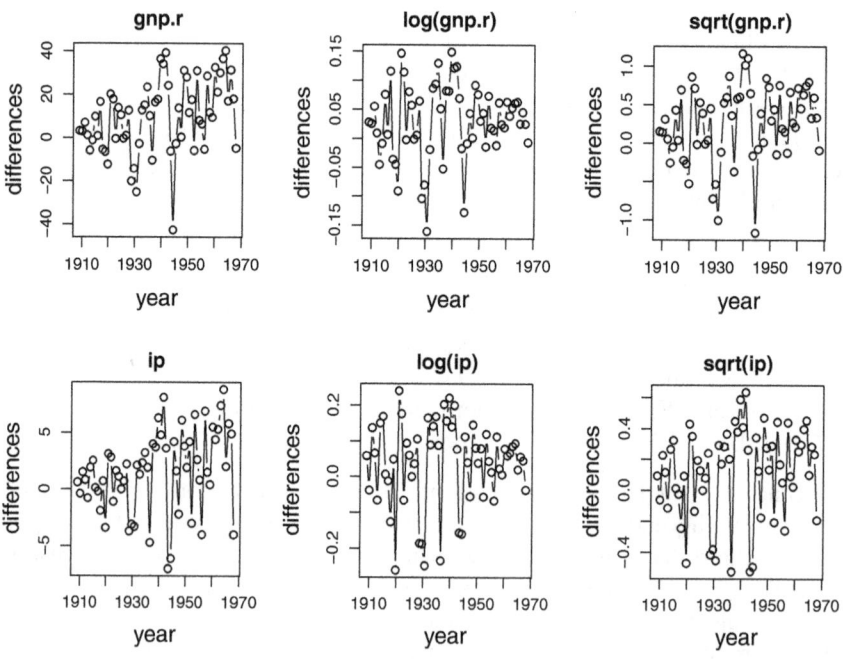

Fig. 9.6. *Differences in* `gnp.r` *and* `ip` *with and without transformations.*

There are additional variables in this data set that could be tried in the model. The analysis presented here is only an illustration and much more exploration is certainly possible with this rich data set.

Time series and normal plots of all eight differenced series did not reveal any outliers. The normal plots were only used to check for outliers, not to check for normal distributions. There is no assumption in a regression analysis that the regressors are normally distributed or that the response has a marginal normal distribution. It is only the conditional distribution of the response given the regressors that is assumed to be normal, and even that assumption can be weakened.

A linear regression with all of the regressors shows that only two, diff (log(ip)) and diff(bnd), are statistically significant at the 0.05 level and some have very large p-values:

```
Call:
lm(formula = diff(log(sp)) ~ diff(gnp.r) + diff(gnp.pc)
    + diff(log(ip)) + diff(log(cpi))
    + diff(emp) + diff(bnd), data = new_np)

Coefficients:
                Estimate Std. Error t value Pr(>|t|)
(Intercept)   -2.766e-02  3.135e-02  -0.882   0.3815
diff(gnp.r)    8.384e-03  4.605e-03   1.821   0.0742
diff(gnp.pc)  -9.752e-04  9.490e-04  -1.028   0.3087
diff(log(ip))  6.245e-01  2.996e-01   2.085   0.0418
diff(log(cpi)) 4.935e-01  4.017e-01   1.229   0.2246
diff(emp)     -9.591e-06  3.347e-05  -0.287   0.7756
diff(bnd)     -2.030e-01  7.394e-02  -2.745   0.0082
```

A likely problem here is multicollinearity, so variance inflation factors were computed:

```
diff(gnp.r)   diff(gnp.pc)  diff(log(ip)) diff(log(cpi))
      16.0           31.8            3.3            1.3
diff(emp)      diff(bnd)
     10.9            1.5
```

We see that diff(gnp.r) and diff(gnp.pc) have high VIF values, which is not surprising since they are expected to be highly correlated. In fact, their correlation is 0.96.

Next, we search for a more parsimonious model using stepAIC(), a variable selection procedure in R that starts with a user-specified model and adds or deletes variables sequentially. At each step it either makes the addition or deletion that most improves AIC. It this example, stepAIC() will start with all six predictors.

Here is the first step:

```
Start:  AIC=-224.92
diff(log(sp)) ~ diff(gnp.r) + diff(gnp.pc) + diff(log(ip)) +
    diff(log(cpi)) + diff(emp) + diff(bnd)
```

	Df	Sum of Sq	RSS	AIC
- diff(emp)	1	0.002	1.216	-226.826
- diff(gnp.pc)	1	0.024	1.238	-225.737
- diff(log(cpi))	1	0.034	1.248	-225.237
<none>			1.214	-224.918
- diff(gnp.r)	1	0.075	1.289	-223.284
- diff(log(ip))	1	0.098	1.312	-222.196
- diff(bnd)	1	0.169	1.384	-218.949

The listed models have either zero or one variable removed from the starting model with all regressors. The models are listed in order of their AIC values. The first model, which has diff(emp) removed (the minus sign indicates a variable that has been removed), has the best (smallest) AIC. Therefore, in the first step, diff(emp) is removed. Notice that the fourth-best model has no variables removed.

The second step starts with the model without diff(emp) and examines the effect on AIC of removing additional variables. The removal of diff(log(cpi)) leads to the largest improvement in AIC, so in the second step this variable is removed:

```
Step:  AIC=-226.83
diff(log(sp)) ~ diff(gnp.r) + diff(gnp.pc) + diff(log(ip)) +
    diff(log(cpi)) + diff(bnd)
```

	Df	Sum of Sq	RSS	AIC
- diff(log(cpi))	1	0.032	1.248	-227.236
<none>			1.216	-226.826
- diff(gnp.pc)	1	0.057	1.273	-226.025
- diff(gnp.r)	1	0.084	1.301	-224.730
- diff(log(ip))	1	0.096	1.312	-224.179
- diff(bnd)	1	0.189	1.405	-220.032

On the third step no variables are removed and the process stops:

```
Step:  AIC=-227.24
diff(log(sp)) ~ diff(gnp.r) + diff(gnp.pc) + diff(log(ip)) +
    diff(bnd)
```

	Df	Sum of Sq	RSS	AIC
<none>			1.248	-227.236
- diff(gnp.pc)	1	0.047	1.295	-227.001
- diff(gnp.r)	1	0.069	1.318	-225.942
- diff(log(ip))	1	0.122	1.371	-223.534
- diff(bnd)	1	0.157	1.405	-222.001

Notice that the removal of `diff(gnp.pc)` would cause only a very small increase in AIC. We should investigate whether this variable might be removed. The new model was refit to the data.

```
Coefficients:
              Estimate Std. Error t value Pr(>|t|)
(Intercept)  -0.018664   0.028723   -0.65    0.518
diff(gnp.r)   0.007743   0.004393    1.76    0.083
diff(gnp.pc) -0.001029   0.000712   -1.45    0.154
diff(log(ip)) 0.672924   0.287276    2.34    0.023
diff(bnd)    -0.177490   0.066840   -2.66    0.010

Residual standard error: 0.15 on 56 degrees of freedom
Multiple R-squared: 0.347,      Adjusted R-squared:  0.3
F-statistic: 7.44 on 4 and 56 DF,  p-value: 7.06e-05
```

Now three of the four variables are statistically significant at 0.1, though `diff(gnp.pc)` has a rather large p-value, and it seems to be worth exploring other possible models.

The R function `leaps()` in the `leaps` package will compute C_p for all possible models. To reduce the amount of output, only the **nbest** models with k regressors [for each $k = 1, \ldots, \dim(\boldsymbol{\beta})$] are printed. The value of **nbest** is selected by the user and in this analysis **nbest** was set at 1, so only the best model is given for each value of k. The following table gives the value of C_p (last column) for the best k-variable models, for $k = 1, \ldots, 6$ (k is in the first column). The remaining columns indicate with a "1" which variables are in the models. All predictors have been differenced, but to save space "`diff`" has been omitted from the variable names heading the columns.

	gnp.r	gnp.pc	log(ip)	log(cpi)	emp	bnd	Cp
1	0	0	1	0	0	0	6.3
2	0	0	1	0	0	1	3.8
3	1	0	1	0	0	1	4.6
4	1	1	1	0	0	1	4.5
5	1	1	1	1	0	1	5.1
6	1	1	1	1	1	1	7.0

We see that `stepAIC` stopping at the four-variable model was perhaps premature. The model selection process was stopped at the four-variable model because the three-variable model had a slightly larger C_p-value. However, if one continues to the best two-variable model, the minimum of C_p is obtained. Here is the fit to the best two-variable model:

```
Call:
lm(formula = diff(log(sp)) ~ +diff(log(ip)) + diff(bnd),
          data = new_np)

Residuals:
```

```
        Min        1Q    Median        3Q        Max
    -0.44254  -0.09786   0.00377   0.10525    0.28136
```

```
Coefficients:
                Estimate Std. Error t value Pr(>|t|)
(Intercept)       0.0166     0.0210    0.79  0.43332
diff(log(ip))     0.6975     0.1683    4.14  0.00011
diff(bnd)        -0.1322     0.0623   -2.12  0.03792
```

```
Residual standard error: 0.15 on 58 degrees of freedom
Multiple R-squared: 0.309,       Adjusted R-squared: 0.285
F-statistic: 12.9 on 2 and 58 DF,  p-value: 2.24e-05
```

Both variables are significant at 0.05. However, it is not crucial that all regressors be significant at 0.05 or at any other predetermined level. Other models could be used, especially if there were good economic reasons for doing so. One cannot say that the two-variable model is best, except in the narrow sense of minimizing C_p, and choosing instead the best three- or four-predictor model would not increase C_p by much. Also, which model is best depends on the criterion used. The best four-predictor model has a better adjusted R^2 than the best two-predictor model.   □

9.7 Partial Residual Plots

A partial residual plot is used to visualize the effect of a predictor on the response while removing the effects of the other predictors. The partial residual for the jth predictor variable is

$$Y_i - \left(\widehat{\beta}_0 + \sum_{j' \neq j} X_{i,j'} \widehat{\beta}_{j'} \right) = \widehat{Y}_i + \widehat{\epsilon}_i - \left(\widehat{\beta}_0 + \sum_{j' \neq j} X_{i,j'} \widehat{\beta}_{j'} \right) = X_{i,j} \widehat{\beta}_j + \widehat{\epsilon}_i,$$

(9.19)

where the first equality uses (9.12) and the second uses (9.10). Notice that the left-hand side of (9.19) shows that the partial residual is the response with the effects of all predictors but the jth subtracted off. The right-hand side of (9.19) shows that the partial residual is also equal to the residual with the effect of the jth variable added back. The partial residual plot is simply the plot of the responses against these partial residuals.

Example 9.10. Partial residual plots for the weekly interest-rate example

Partial residual plots for the weekly interest-rate example are shown in Fig. 9.7a, b. For comparison, scatterplots of cm10_dif and cm30_dif versus aaa_dif with the corresponding one-variable fitted lines are shown in panels (c) and (d). The main conclusion from examining the plots is that the slopes in (a) and (b) are shallower than the slopes in (c) and (d). What does this tell us? It says that, due to collinearity, the effect of cm10_dif on aaa_dif when cm30_dif is in the model [panel (a)] is less than when cm30_dif is not in the model [panel (c)], and similarly when the roles of cm10_dif and cm30_dif are reversed.

The same conclusion can be reached by looking at the estimated regression coefficients. From Examples 9.1 and 9.4, we can see that the coefficient of cm10_dif is 0.615 when cm10_dif is the only variable in the model, but the coefficient drops to 0.355 when cm30_dif is also in the model. There is a similar decrease in the coefficient for cm30_dif when cm10_dif is added to the model.   □

Example 9.11.   Nelson–Plosser macroeconomic variables—Partial residual Plots

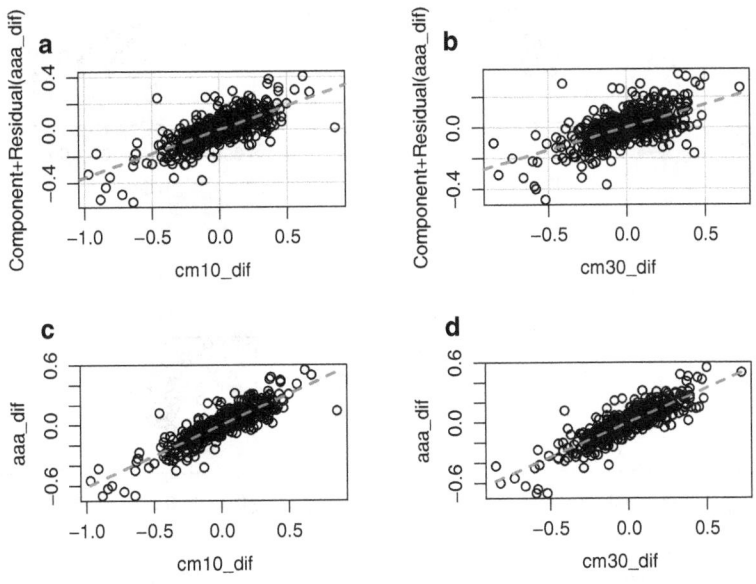

Fig. 9.7. *Partial residual plots for the weekly interest rates [panels (a) and (b)] and scatterplots of the predictors and the response [panels (c) and (d)].*

This example continues the analysis of the Nelson–Plosser macroeconomic variables. Partial residual plots for the four-variable model selected by stepAIC in Example 9.9 are shown in Fig. 9.8. One can see that all four variables have explanatory power, since the partial residuals have linear trends in the variables.

One puzzling aspect of this model is that the slope for gnp.pc is negative. However, the p-value for this regressor is large and the minimum C_p model does not contain either `gnp.r` or `gnp.pc`. Often, a regressor that is highly correlated with other regressors has an estimated slope that is counterintuitive. If used alone, both `gnp.r` and `gnp.pc` have positive slopes. The slope of `gnp.pc` is negative only when `gnp.r` is in the model.   □

9.8 Centering the Predictors

Centering or, more precisely, *mean-centering* a variable means expressing it as a deviation from its mean. Thus, if $X_{1,k}, \ldots, X_{n,k}$ are the values of the kth predictor and $\overline{X}_k$ is their mean, then $(X_{1,k} - \overline{X}_k), \ldots, (X_{n,k} - \overline{X}_k)$ are values of the centered predictor.

Centering is useful for two reasons:

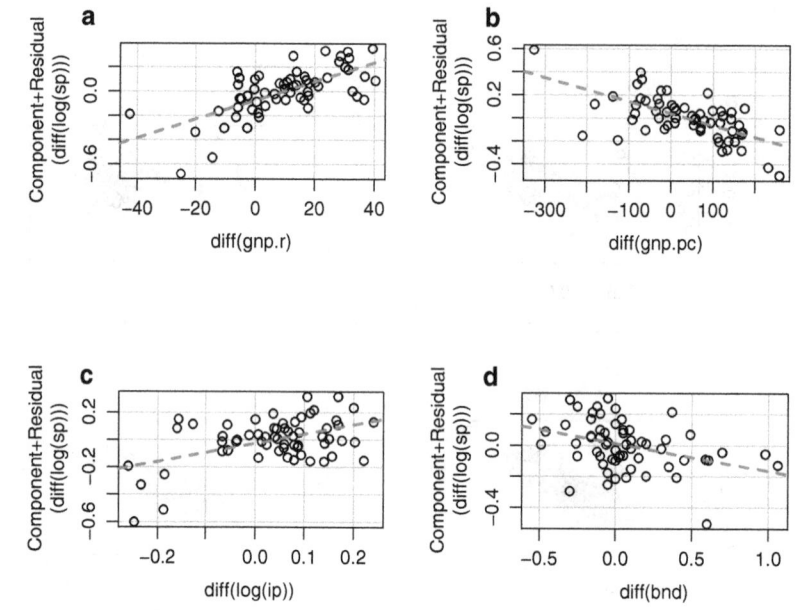

Fig. 9.8. *Partial residual plots for the Nelson–Plosser U.S. economic time series. (a) Change in* `gnp.r`. *(b) Change in* `gnp.pc`. *(c) Change in* `log(ip)`. *(d) Change in* `bnd`.

- centering can reduce collinearity in polynomial regression;
- if all predictors are centered, then β_0 is the expected value of Y when each of the predictors is equal to its mean. This gives β_0 an interpretable meaning. In contrast, if the variables are not centered, then β_0 is the expected value of Y when all of the predictors are equal to 0. Frequently, 0 is outside the range of some predictors, making the interpretation of β_0 of little real interest unless the variables are centered.

9.9 Orthogonal Polynomials

As just mentioned, centering can reduce collinearity in polynomial regression because, for example, if X is positive, then X and X^2 will be highly correlated but $X - \overline{X}$ and $(X - \overline{X})^2$ will be less correlated.

Orthogonal polynomials can eliminate correlation entirely, since they are defined in a way so that they are uncorrelated. This is done using the Gram–Schmidt orthogonalization procedure discussed in textbooks on linear algebra. Orthogonal polynomials can be created easily in most software packages, for instance, by using the `poly()` function in R. Orthogonal polynomials are particularly useful for polynomial regression of degree higher than 2 where centering is less successful at reducing collinearity. However, the use of polynomial models of degree 4 and higher is discouraged and nonparametric regression (see Chap. 21) is recommended instead. Even cubic regression can be problematic because cubic polynomials have only a limited range of shapes.

9.10 Bibliographic Notes

Harrell (2001) , Ryan (1997), Neter et al. (1996) and Draper and Smith (1998) are four of the many good introductions to regression. Faraway (2005) is an excellent modern treatment of linear regression with R. See Nelson and Plosser (1982) for information about their data set.

9.11 R Lab

9.11.1 U.S. Macroeconomic Variables

This section uses the data set `USMacroG` in R's `AER` package. This data set contains quarterly times series on 12 U.S. macroeconomic variables for the period 1950–2000. We will use the variables `consumption` = real consumption expenditures, `dpi` = real disposable personal income, `government` = real government expenditures, and `unemp` = unemployment rate. Our goal is to predict changes in `consumption` from changes in the other variables.

Run the following R code to load the data, difference the data (since we wish to work with changes in these variables), and create a scatterplot matrix.

```
library(AER)
data("USMacroG")
MacroDiff = as.data.frame(apply(USMacroG, 2, diff))
attach(MacroDiff)
pairs(cbind(consumption, dpi, cpi, government, unemp))
```

Problem 1 *Describe any interesting features, such as outliers, seen in the scatterplot matrix. Keep in mind that the goal is to predict changes in* consumption. *Which variables seem best suited for that purpose? Do you think there will be collinearity problems?*

Next, run the code below to fit a multiple linear regression model to consumption using the other four variables as predictors.

```
fitLm1 = lm(consumption ~ dpi + cpi + government + unemp)
summary(fitLm1)
confint(fitLm1)
```

Problem 2 *From the summary, which variables seem useful for predicting changes in* consumption?

Next, print an ANOVA table.

```
anova(fitLm1)
```

Problem 3 *For the purpose of variable selection, does the ANOVA table provide any useful information not already in the summary?*

Upon examination of the p-values, we might be tempted to drop several variables from the regression model, but we will not do that since variables should be removed from a model one at a time. The reason is that, due to correlation between the predictors, when one is removed the significance of the others changes. To remove variables sequentially, we will use the function stepAIC() in the MASS package.

```
library(MASS)
fitLm2 = stepAIC(fitLm1)
summary(fitLm2)
```

Problem 4 *Which variables are removed from the model, and in what order?*

Now compare the initial and final models by AIC.

```
AIC(fitLm1)
AIC(fitLm2)
AIC(fitLm1) - AIC(fitLm2)
```

Problem 5 *How much of an improvement in AIC was achieved by removing variables? Was the improvement large? Is so, can you suggest why? If not, why not?*

The function vif() in the car package will compute variance inflation factors. A similar function with the same name is in the faraway package. Run

```
library(car)
vif(fitLm1)
vif(fitLm2)
```

Problem 6 *Was there much collinearity in the original four-variable model? Was the collinearity reduced much by dropping two variables?*

Partial residual plots, which are also called *component plus residual* or *cr* plots, can be constructed using the function crPlot() in the car package. Run

```
par(mfrow = c(2, 2))
sp = 0.8
crPlot(fitLm1, dpi, span = sp, col = "black")
crPlot(fitLm1, cpi, span = sp, col = "black")
crPlot(fitLm1, government, span = sp, col = "black")
crPlot(fitLm1, unemp, span = sp, col = "black")
```

Besides dashed least-squares lines, the partial residual plots have solid lowess smooths through them unless this feature is turned off by specifying smooth=F, as was done in Fig. 9.8. Lowess is an earlier version of loess. The smoothness of the lowess curves is determined by the parameter span, with larger values of span giving smoother plots. The default is span $= 0.5$. In the code above, span is 0.8 but can be changed for all four plots by changing the variable sp. Lowess, loess, and span are described in Sect. 21.2.1. A substantial deviation of the lowess curve from the least-squares line is an indication that the effect of the predictor is nonlinear. The default color of the crPlot figure is red, but this can be changed as in the code above.

Problem 7 *What conclusions can you draw from the partial residual plots?*

9.12 Exercises

1. Suppose that $Y_i = \beta_0 + \beta_1 X_i + \epsilon_i$, where ϵ_i is $N(0, 0.3)$, $\beta_0 = 1.4$, and $\beta_1 = 1.7$.
   (a) What are the conditional mean and standard deviation of Y_i given that $X_i = 1$? What is $P(Y_i \le 3 | X_i = 1)$?
   (b) A regression model is a model for the conditional distribution of Y_i given X_i. However, if we also have a model for the marginal distribution of X_i, then we can find the marginal distribution of Y_i. Assume that X_i is $N(1, 0.7)$. What is the marginal distribution of Y_i? What is $P(Y_i \le 3)$?

2. Show that if $\epsilon_1, \ldots, \epsilon_n$ are i.i.d. $N(0, \sigma_\epsilon^2)$, then in straight-line regression the least-squares estimates of β_0 and β_1 are also the maximum likelihood estimates.

   Hint: This problem is similar to the example in Sect. 5.9. The only difference is that in that section, $Y_1, \ldots, Y_n$ are independent $N(\mu, \sigma^2)$, while in this exercise $Y_1, \ldots, Y_n$ are independent $N(\beta_0 + \beta_1 X_i, \sigma_\epsilon^2)$.

3. Use (7.11), (9.3), and (9.2) to show that (9.8) holds.

4. It was stated in Sect. 9.8 that centering reduces collinearity. As an illustration, consider the example of quadratic polynomial regression where X takes 30 equally spaced values between 1 and 15.

   (a) What is the correlation between X and X^2? What are the VIFs of X and X^2?

   (b) Now suppose that we center X before squaring. What is the correlation between $(X - \overline{X})$ and $(X - \overline{X})^2$? What are the VIFs of $(X - \overline{X})$ and $(X - \overline{X})^2$?

5. A linear regression model with three predictor variables was fit to a data set with 40 observations. The correlation between Y and $\widehat{Y}$ was 0.65. The total sum of squares was 100.

   (a) What is the value of R^2?

   (b) What is the value of the residual error SS?

   (c) What is the value of the regression SS?

   (d) What is the value of s^2?

6. A data set has 66 observations and five predictor variables. Three models are being considered. One has all five predictors and the others are smaller. Below is residual error SS for all three models. The total SS was 48. Compute C_p and R^2 for all three models. Which model should be used based on this information?

Number of predictors	Residual error SS
3	12.2
4	10.1
5	10.0

7. The quadratic polynomial regression model

   $$Y_i = \beta_0 + \beta_1 X_i + \beta_2 X_i^2 + \epsilon_i$$

   was fit to data. The p-value for β_1 was 0.67 and for β_2 was 0.84. Can we accept the hypothesis that β_1 and β_2 are both 0? Discuss.

8. Sometimes it is believed that β_0 is 0 because we think that $E(Y|X = 0) = 0$. Then the appropriate model is

   $$y_i = \beta_1 X_i + \epsilon_i.$$

   This model is usually called "regression through the origin" since the regression line is forced through the origin. The least-squares estimator of β_1 minimizes

$$\sum_{i=1}^{n}\{Y_i - \beta_1 X_i\}^2.$$

Find a formula that gives $\hat{\beta}_1$ as a function of the Y_is and the X_is.

9. Complete the following ANOVA table for the model $Y_i = \beta_0 + \beta_1 X_{i,1} + \beta_2 X_{i,2} + \epsilon_i$:

```
Source        df      SS      MS      F       P
Regression    ?       ?       ?       ?       0.04
Error         ?       5.66    ?
Total         15      ?

        R-sq = ?
```

10. Pairs of random variables (X_i, Y_i) were observed. They were assumed to follow a linear regression with $E(Y_i|X_i) = \theta_1 + \theta_2 X_i$ but with t-distributed noise, rather than the usual normally distributed noise. More specifically, the assumed model was that conditionally, given X_i, Y_i is t-distributed with mean $\theta_1 + \theta_2 X_i$, standard deviation θ_3, and degrees of freedom θ_4. Also, the pairs $(X_1, Y_1), \ldots, (X_n, Y_n)$ are mutually independent. The model could also be expressed as

$$Y_i = \theta_1 + \theta_2 X_i + \epsilon_i$$

where $\epsilon_1, \ldots, \epsilon_n$ are i.i.d. t with mean 0 and standard deviation θ_3 and degrees of freedom θ_4. The model was fit by maximum likelihood. The R code and output are

```
#(Code to input x and y not shown)
library(fGarch)
start = c(lmfit$coef, sd(lmfit$resid), 4)
loglik = function(theta)
{
-sum(log(dstd(y, mean = theta[1] + theta[2] * x, sd = theta[3],
    nu = theta[4])))
}
mle = optim(start, loglik, hessian = TRUE)
InvFishInfo = solve(mle$hessian)
mle$par
mle$value
mle$convergence
sqrt(diag(InvFishInfo))
qnorm(0.975)

> mle$par
[1] 0.511 1.042 0.152 4.133
> mle$value
[1] -188
```

```
> mle$convergence
[1] 0
> sqrt(diag(InvFishInfo))
[1] 0.00697 0.11522 0.01209 0.93492
>
> qnorm(.975)
[1] 1.96
>
```

(a) What is the MLE of the slope of Y_i on X_i?
(b) What is the standard error of the MLE of the degrees-of-freedom parameter?
(c) Find a 95 % confidence interval for the standard deviation of the noise.
(d) Did optim converge? Why or why not?

References

Draper, N. R. and Smith, H. (1998) *Applied Regression Analysis*, 3rd ed., Wiley, New York.

Faraway, J. J. (2005) *Linear Models with R*, Chapman & Hall, Boca Raton, FL.

Harrell, F. E., Jr. (2001) *Regression Modeling Strategies*, Springer-Verlag, New York.

Nelson C.R., and Plosser C.I. (1982) Trends and random walks in macroeconomic time series. *Journal of Monetary Economics*, **10**, 139–162.

Neter, J., Kutner, M. H., Nachtsheim, C. J., and Wasserman, W. (1996) *Applied Linear Statistical Models*, 4th ed., Irwin, Chicago.

Ryan, T. P. (1997) *Modern Regression Methods*, Wiley, New York.

10

Regression: Troubleshooting

10.1 Regression Diagnostics

Many things can, and often do, go wrong when data are analyzed. There may be data that were entered incorrectly, one might not be analyzing the data set one thinks, the variables may have been mislabeled, and so forth. In Example 10.5, presented shortly, one of the weekly time series of interest rates began with 371 weeks of zeros, indicating missing data. However, I was unaware of this problem when I first analyzed the data. The lesson here is that I should have plotted each of the data series first before starting to analyze them, but I hadn't. Fortunately, the diagnostics presented in this section showed quickly that there was some type of serious problem, and then after plotting each of the time series I easily discovered the nature of the problem.

Besides problems with the data, the assumed model may not be a good approximation to reality. The usual estimation methods, such as least squares in regression, are highly nonrobust, which means that they are particularly sensitive to problems with the data or the model.

Experienced data analysts know that they should always look at the raw data. Graphical analysis often reveals any problems that exist, especially the types of gross errors that can seriously degrade the analysis. However, some problems are only revealed by fitting a regression model and examining residuals.

Example 10.1. High-leverage points and residual outliers—Simulated data example

© Springer Science+Business Media New York 2015
D. Ruppert, D.S. Matteson, *Statistics and Data Analysis for Financial Engineering*, Springer Texts in Statistics,
DOI 10.1007/978-1-4939-2614-5_10

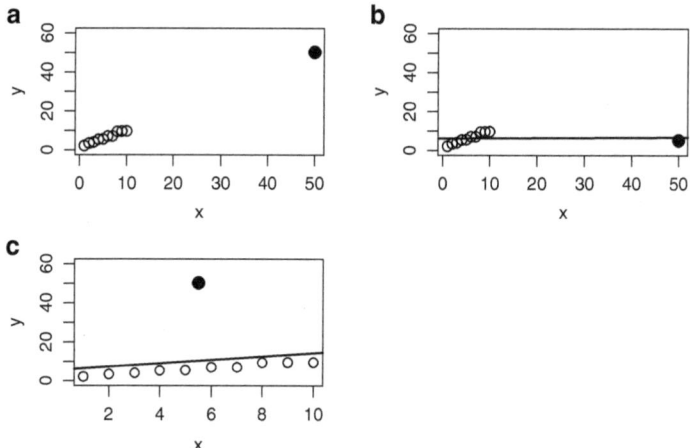

Fig. 10.1. *(a) Linear regression with a high-leverage point that is not a residual outlier (solid circle). (b) Linear regression with a high-leverage point that is a residual outlier (solid circle). (c) Linear regression with a low-leverage point that is a residual outlier (solid circle). Least-squares fits are shown as solid lines.*

Figure 10.1 uses data simulated to illustrate some of the problems that can arise in regression. There are 11 observations. The predictor variable takes on values 1, ..., 10 and 50, and $Y = 1 + X + \epsilon$, where $\epsilon \sim N(0, 1)$. The last observation is clearly an extreme value in X. Such a point is said to have *high leverage*. However, a high-leverage point is not necessarily a problem, only a potential problem. In panel (a), the data have been recorded correctly so that Y is linearly related to X and the extreme X-value is, in fact, helpful as it increases the precision of the estimated slope. In panel (b), the value of Y for the high-leverage point has been misrecorded as 5.254 rather than 50.254. This data point is called a *residual outlier*. As can be seen by comparing the least-squares lines in (a) and (b), the high-leverage point has an extreme influence on the estimated slope. In panel (c), X has been misrecorded for the high-leverage point as 5.5 instead of 50. Thus, this point is no longer high-leverage, but now it is a residual outlier. Its effect now is to bias the estimated intercept.

One should also look at the residuals after the model has been fit, because the residuals may indicate problems not visible in plots of the raw data. However, there are several types of residuals and, as explained soon, one type, called the *externally studentized residual* or *rstudent*, is best for diagnosing problems. Ordinary (or raw) residuals are not necessarily useful for diagnosing problems. For example, in Fig. 10.1b, none of the raw residuals is large, not even the one associated with the residual outlier. The problem is that the raw residuals are too sensitive to the outliers, particularly at high-leverage points, and problems can remain hidden when raw residuals are plotted.   □

Three important tools will be discussed for diagnosing problems with the model or the data:

- leverages;
- externally studentized residuals; and
- Cook's Distance (Cook's D), which quantifies the overall influence of each observation on the fitted values.

10.1.1 Leverages

The *leverage* of the ith observation, denoted by H_{ii}, measures how much influence Y_i has on its own fitted value $\widehat{Y}_i$. We will not go into the algebraic details until Sect. 11.1. An important result in that section is that there are weights H_{ij} depending on the values of the predictor variables but *not* on $Y_1, \ldots, Y_n$ such that

$$\widehat{Y}_i = \sum_{j=1}^{n} H_{ij} Y_j. \tag{10.1}$$

In particular, H_{ii} is the weight of Y_i in the determination of $\widehat{Y}_i$. It is a potential problem if H_{ii} is large since then $\widehat{Y}_i$ is determined too much by Y_i itself and not enough by the other data. The result is that the residual $\widehat{\epsilon}_i = Y_i - \widehat{Y}_i$ will be small and not a good estimate of ϵ_i. Also, the standard error of $\widehat{Y}_i$ is $\sigma_\epsilon \sqrt{H_{ii}}$, so a high value of H_{ii} means a fitted value with low accuracy.

The leverage value H_{ii} is large when the predictor variables for the ith case are atypical of those values in the data, for example, because one of the predictor variables for that case is extremely outlying. It can be shown by some elegant algebra that the average of $H_{11}, \ldots, H_{nn}$ is $(p+1)/n$, where $p+1$ is the number of parameters (one intercept and p slopes) and that therefore $0 < H_{ii} < 1$. A value of H_{ii} exceeding $2(p+1)/n$, that is, over twice the average value, is generally considered to be too large and therefore a cause for concern Belsley et al. (1980).

The square matrix with i, jth element equal to H_{ij} is called the hat matrix since by (10.1) it converts Y_j, $j = 1, \ldots, n$, to $\widehat{Y}_i$. The H_{ii} are sometimes called the *hat diagonals*.

Example 10.2. Leverages in Example 10.1

Figure 10.2 plots the leverages for the three cases in Fig. 10.1. Because the leverages depend only on the X-values, the leverages are the same in panels (a) and (b). In both panels, the high-leverage point has a leverage equal to 0.960. In these examples, the rule-of-thumb cutoff point for high leverage is only $2(p+1)/n = 2*2/11 = 0.364$, so 0.960 is a huge leverage and close to the maximum possible value of 1. In panel (c), none of the leverages is greater than 0.364.

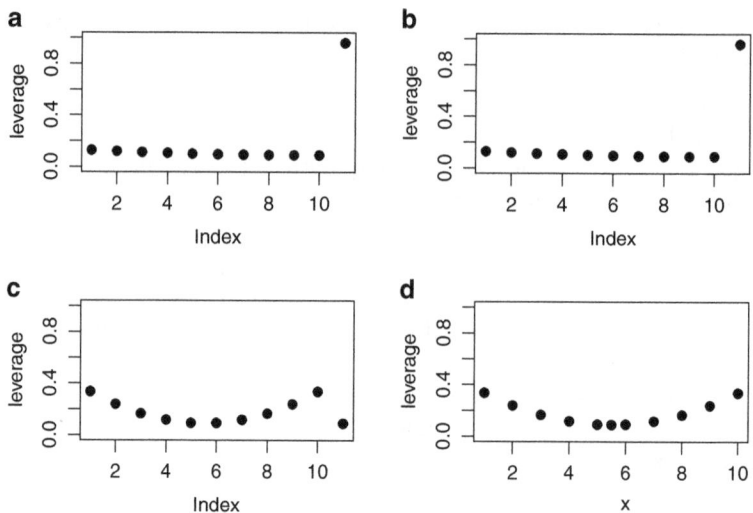

Fig. 10.2. *(a)–(c) Leverages plotted again case number (index) for the data sets in Fig. 10.1. Panels (a) and (b) are identical because leverages do not depend on the response values. Panel (d) plots the leverages in (c) against X_i.*

In the special case $p = 1$, there is a simple formula for the leverages:

$$H_{ii} = \frac{1}{n} + \frac{(X_i - \overline{X})^2}{\sum_{i=1}^{n}(X_i - \overline{X})^2},\tag{10.2}$$

It is easy to check that in this case, $H_{11} + \cdots + H_{nn} = p + 1 = 2$, so the average of the hat diagonals is, indeed, $(p + 1)/n$. Formula (10.2) shows that $H_{ii} \geq 1/n$, H_{ii} is equal $1/n$ if and only if $X_i = \overline{X}$, and H_{ii} increases quadratically with the distance between X_i and $\overline{X}$. This behavior can be seen in Fig. 10.2d. □

10.1.2 Residuals

The *raw residual* is $\widehat{\epsilon}_i = Y_i - \widehat{Y}_i$. Under ideal circumstances such as a reasonably large sample and no outliers or high-leverage points, the raw residuals are approximately $N(0, \sigma_\epsilon^2)$, so absolute values greater than $2\widehat{\sigma}_\epsilon^2$ are outlying and greater than $3\widehat{\sigma}_\epsilon^2$ are extremely outlying. However, circumstances are often not ideal. When residual outliers occur at high-leverage points, they can so distort the least-squares fit that they are not seen to be outlying. The problem in these cases is that $\widehat{\epsilon}_i$ is not close to ϵ_i because of the bias in the least-squares fit. The bias is due to residual outliers themselves. This problem can be seen in Fig. 10.1b.

The standard error of $\widehat{\epsilon}_i$ is $\widehat{\sigma}_\epsilon \sqrt{1 - H_{ii}}$, so the raw residuals do not have a constant variance, and those raw residuals with large leverages close to 1 are much less variable than the others. To fix the problem of nonconstant variance, one can use the *standardized residual*, sometimes called the *internally studentized residual*,[1] which is $\widehat{\epsilon}_i$ divided by its standard error, that is, $\widehat{\epsilon}_i/(\widehat{\sigma}_\epsilon \sqrt{1 - H_{ii}})$.

There is still another problem with standardized residuals. An extreme residual outlier can inflate $\widehat{\sigma}_\epsilon$, causing the standardized residual for the outlying point to appear too small. The solution is to redefine the ith studentized residual with an estimate of σ_ϵ that does not use the ith data point. Thus, the *externally studentized residual*, often called *rstudent*, is defined to be $\widehat{\epsilon}_i/\{\widehat{\sigma}_{\epsilon,(-i)} \sqrt{1 - H_{ii}}\}$, where $\widehat{\sigma}_{\epsilon,(-i)}$ is the estimate of σ_ϵ computed by fitting the model to the data with the ith observation deleted.[2] For diagnostics, rstudent is considered the best type of residual to plot and is the type of residual used in this book.

Warning: The terms "standardized residual" and "studentized residual" do not have the same definitions in all textbooks and software packages. The definitions used here agree with R's `influence.measures()` function. Other software, such as, SAS uses different definitions.

Example 10.3. Externally studentized and raw residuals in Example 10.1

The top row of Fig. 10.3 shows the externally studentized residuals in each of the three cases of simulated data in Fig. 10.1. Case #11 is correctly identified as a residual outlier in data sets (b) and (c) and also correctly identified in data set (a) as not being a residual outlier. The bottom row of Fig. 10.3 shows the raw residuals, rather than the externally studentized residuals. It is not apparent from the raw residuals that in data set (b), case #11 is a residual outlier. This shows the inappropriateness of raw residuals for the detection of outliers, especially when there are high-leverage points.   □

10.1.3 Cook's Distance

A high-leverage value or a large absolute externally studentized residual indicates only a *potential* problem with a data point. Neither tells how much influence the data point actually has on the estimates. For that information, we can use *Cook's distance*, often called *Cook's D*, which measures how much the fitted values change if the ith observation is deleted. We say that Cook's D measures influence, and any case with a large Cook's D is called a high-influence case. Leverage and rstudent alone do not measure influence.

Let $\widehat{Y}_j(-i)$ be the jth fitted value using estimates of the βs obtained with the ith observation deleted. Then Cook's D for the ith observation is

---

[1] *Studentization* means dividing a statistic by its standard error.
[2] The notation $(-i)$ signifies the deletion of the ith observation.

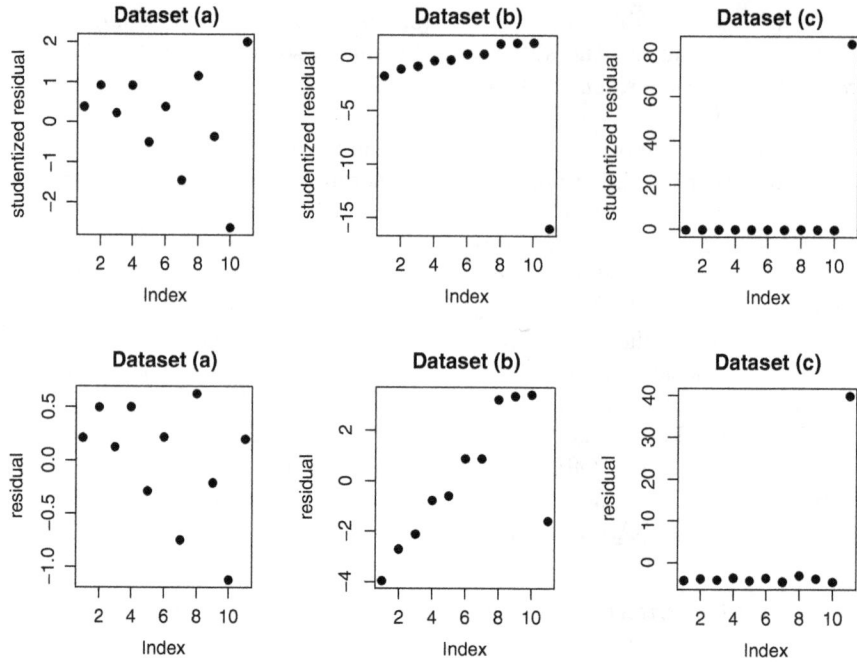

Fig. 10.3. Top row: *Externally studentized residuals for the data sets in Fig. 10.1; data set* **(a)** *is the data set in panel* **(a)** *of Fig. 10.1, and so forth. Case #11 is an outlier in data sets* **(b)** *and* **(c)** *but not in data set* **(a).** **Bottom row:** *Raw residuals for the same three data sets as in the top row. For data set* **(b),** *the raw residual does not reveal that case #11 is outlying.*

$$\frac{\sum_{j=1}^{n}\{\widehat{Y}_j - \widehat{Y}_j(-i)\}^2}{(p+1)s^2}.\tag{10.3}$$

The numerator in (10.3) is the sum of squared changes in the fitted values when the *i*th observation is deleted. The denominator standardizes this sum by dividing by the number of estimated parameters and an estimate of σ_ϵ^2.

One way to use Cook's D is to plot the values of Cook's D against case number and look for unusually large values. However, it can be difficult to decide which, if any, values of Cook's D are outlying. Of course, some Cook's D values will be larger than others, but are any so large as to be worrisome? To answer this question, a half-normal plot of values of Cook's D, or perhaps of their square roots, can be useful. Neither Cook's D nor its square root is normally distributed, so one does not check for linearity. Instead, one looks for values that are "detached" from the rest.

Example 10.4. Cook's D for simulated data in Example 10.1

The three columns of Fig. 10.4 show the values of square roots of Cook's D for the three simulated data examples in Fig. 10.1. In the top row, the square roots of Cook's D values are plotted versus case number (index). The bottom row contains half-normal plots of the square roots of the Cook's D values. In all panels, case #11 has the largest Cook's D, indicating that one should examine this case to see if there is a problem. In data set (a), case #11 is a high-leverage point and has high influence despite not being a residual outlier. In data set (b), where case #11 is both a high-leverage point and a residual outlier, the value of Cook's D for this case is very large, larger than in data set (a). In data set(c), where case #11 has low leverage, all 11 Cook's D values are reasonably small, at least in comparison with data sets (a) and (b), but case #11 is still somewhat outlying.   □

Example 10.5. *Weekly interest data with missing values recorded as zeros*

It was mentioned earlier that there were missing values of cm30 at the beginning of the data set that were coded as zeros. In fact, there were 371 weeks of missing data for cm30. I started to analyze the data without realizing this problem. This created a huge outlying value of cm30_dif (the first differences) at observation number 372 when cm30 jumps from 0 to the first nonmissing value. Fortunately, plots of rstudent, leverages, and Cook's D all reveal a serious problem somewhere between the 300th and 400th observations, and by zooming into this range of case numbers the problem was located in case #372; see Fig. 10.5. The nature of the problem is not evident from these plots, only its existence, so I plotted each of the series aaa, cm10, and cm30. After seeing the initial zero values of the latter series, the problem was obvious. Please remember this lesson: *ALWAYS look at the data.* Another lesson is that it is best to use nonnumeric values for missing values. For example, R uses "NA" for "not available."   □

10.2 Checking Model Assumptions

Because the ith residual $\widehat{\epsilon}_i$ estimates the "noise" ϵ_i, the residuals can be used to check the assumptions behind regression. Residual analysis generally consists of various plots of the residuals, each plot being designed to check one or more of the regression assumptions. Regression software will output the several types of residuals discussed in Sect. 10.1.2. Externally studentized residuals (rstudent) are recommended, for reasons given in that section.

Problems to look for include

1. nonnormality of the errors,
2. nonconstant variance of the errors,

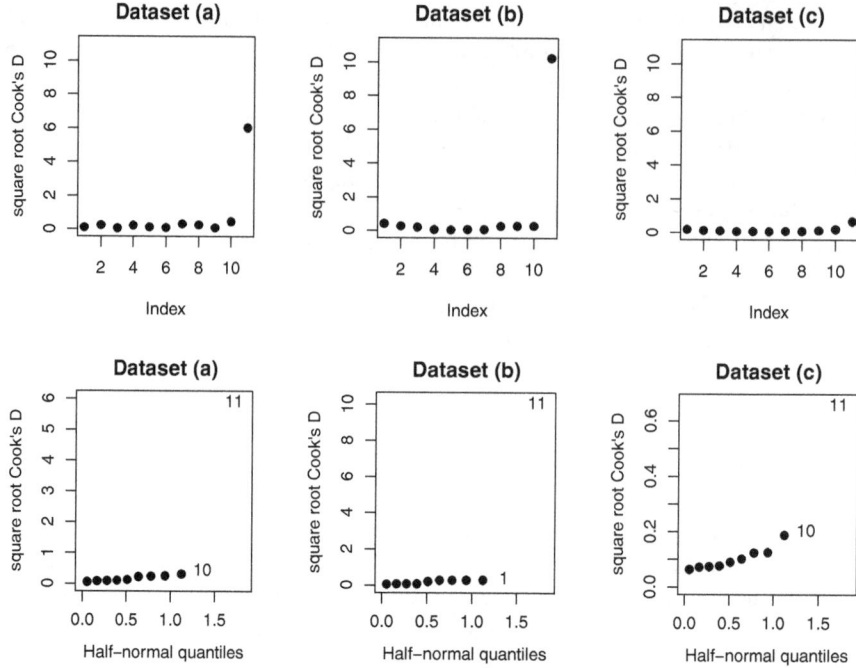

Fig. 10.4. Top row: *Square roots of Cook's D for the simulated data plotted against case number.* **Bottom row:** *Half-normal plots of square roots of Cook's D.* **Data set (a)** *Case #11 has high leverage. It is not a residual outlier but has high influence nonetheless.* **Data set (b)** *Case #11 has high leverage and is a residual outlier. It has higher influence (as measured by Cook's D) than in data set* **(a)**. **Data set (c)** *Case #11 has low leverage but is a residual outlier. It has much lower influence than in data sets* **(a)** *and* **(b)**. **Note:** *In the top row, the vertical scale is kept constant to emphasize differences among the three cases.*

3. nonlinearity of the effects of the predictor variables on the response, and
4. correlation of the errors.

The first three problems are discussed below; correlation of the errors is discussed later in Sect. 13.3.

10.2.1 Nonnormality

Nonnormality of the errors (noise) can be detected by a normal probability plot, boxplot, and histogram of the residuals. Not all three are needed, but looking at a normal plot is highly recommended. Moreover, inexperienced data analysts have trouble with the interpretation of normal plots. Looking at side-by-side normal plots and histograms (or KDEs) is helpful when learning to interpret normal probability plots.

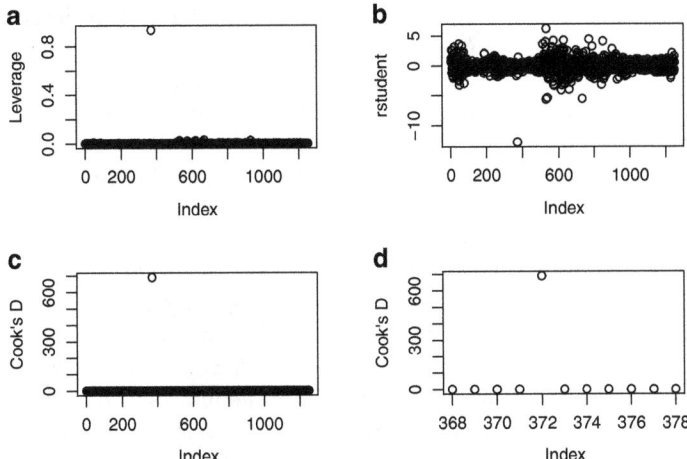

Fig. 10.5. *Weekly interest data. Regression of* aaa_dif *on* cm10_dif *and* cm30_dif. *Full data set including the first* 371 *weeks of data where* cm30 *was missing and assigned a value of* 0. *This caused severe problems at case number* 372, *which are detected by the leverages in* (**a**), *rstudent in* (**b**), *and Cook's D in* (**c**). *Panel* (**d**) *zooms in on the outlier case to identify the case number as* 372.

The residuals often appear nonnormal because there is an excess of outliers relative to the normal distribution. We have defined a value of rstudent to be outlying if its absolute value exceeds 2 and extremely outlying if it exceeds 3. Of course, these cutoffs of 2 and 3 are arbitrary and only intended to give rough guidelines.

It is the presence of outliers, particularly extreme outliers, that is a concern when we have nonnormality. A deficiency of outliers relative to the normal distribution is less of a problem, if it is a problem at all. Sometimes outliers are due to errors, such as mistakes in the entry of the data or, as in Example 10.5, misinterpreting a zero as a true data value rather than the indicator of a missing value. If possible, outliers due to mistakes should be corrected, of course. However, in financial time series, outliers are often "good observations" due, *inter alia*, to excess volatility in the markets on certain days.

Another possible reason for an excess of both positive and negative outlying residuals is nonconstant residual variance, a problem that is explained shortly. Normal probability plots assume that all observations come from the same distribution, in particular, that they have the same variance. The purpose of that plot is to determine if the common distribution is normal or not. If there is no common distribution, for example, because of nonconstant variance, then the normal plot is not readily interpretable. Therefore, one should check for a constant variance before making an extended effort to interpret a normal plot.

Outliers can be a problem because they have an unduly large influence on the estimation results. As discussed in Sect. 4.6, a common solution to the problem of outliers is transformation of the response. Data transformation can be very effective at handling outliers, but it does not work in all situations. Moreover, transformations can induce outliers. For example, if a log transformation is applied to positive data, values very close to 0 could be transformed to outlying negative values since $\log(x) \to -\infty$ as $x \downarrow 0$.

It is always wise to check whether outliers are due to erroneous data, for example, typing errors or other mistakes in data collection and entry. Of course, erroneous data should be corrected if possible and otherwise removed. Removal of outliers that are not known to be erroneous is dangerous and not recommended as routine statistical practice. However, reanalyzing the data with outliers removed is a sound practice. If the analysis changes drastically when the outliers are deleted, then one knows there is something about which to worry. On the other hand, if deletion of the outliers does not change the conclusions of the analysis, then there is less reason to be concerned with whether the outliers were erroneous data.

A certain amount of nonnormality of the errors is not necessarily a problem. Least-squares estimators are unbiased even without normality. Standard errors for regression coefficients are also correct and confidence intervals are nearly correct because the least-squares estimators obey a central limit theorem—they are nearly normally distributed even if the errors are not normally distributed. Nonetheless, outliers caused by highly skewed or heavy-tailed error distributions can cause the least-squares estimator to be highly variable and therefore inaccurate. Transformations of Y are commonly used when the errors have skewed distributions, especially when they also have a nonconstant variance. A common solution to heavy-tailed error distributions is robust regression; see Sect. 11.8.

10.2.2 Nonconstant Variance

Nonconstant residual variance means that the conditional variance of the response given the predictor variables is not constant as assumed by standard regression models. Nonconstant variance is also called *heteroskedasticity*. Nonconstant variance can be detected by an absolute residual plot, that is, by plotting the absolute residuals against the predicted values ($\widehat{Y}_i$) and, perhaps, also against the predictor variables. If the absolute residuals show a systematic trend, then this is an indication of nonconstant variance. Economic data often have the property that larger responses are more variable. A more technical way of stating this is that the conditional variance of the response (given the predictor variables) is an increasing function of the conditional mean of the response. This type of behavior can be detected by plotting the absolute residuals versus the predicted values and looking for an increasing trend.

Often, trends are difficult to detect just by looking at the plotted points and adding a so-called scatterplot smoother is very helpful. A *scatterplot smoother* fits a smooth curve to a scatterplot. Nonparametric regression estimators such as loess and smoothing splines are commonly used scatterplot smoothers available in statistical software packages. These are discussed more fully in Chap. 21.

A potentially serious problem caused by nonconstant variance is inefficiency, that is, too-variable estimates, if ordinary (that is, unweighted) least squares is used. Weighted least squares estimates β efficiently by minimizing

$$\sum_{i=1}^{n} w_i \{Y_i - f(\boldsymbol{X}_i; \widehat{\boldsymbol{\beta}})\}^2. \tag{10.4}$$

Here w_i an estimate of the inverse (that is, reciprocal) conditional variance of Y_i given $\boldsymbol{X}_i$, so that the more variable observations are given less weight. Estimation of the conditional variance function to determine the w_is is discussed in the more advanced textbooks mentioned in Sect. 10.3. Weighted least-squares for regression with GARCH errors is discussed in Sect. 14.12.

Another serious problem caused by heteroskedasticity is that standard errors and confidence intervals assume a constant variance and can be seriously wrong if there is substantial nonconstant variance.

Transformation of the response is a common solution to the problem of nonconstant variance; see Sect. 11.4. If the response can be transformed to constant variance, then unweighted least-squares will be efficient and standard errors and confidence intervals will be valid.

10.2.3 Nonlinearity

If a plot of the residuals versus a predictor variable shows a systematic nonlinear trend, then this is an indication that the effect of that predictor on the response is nonlinear. Nonlinearity causes biased estimates and a model that may predict poorly. Confidence intervals, which assume unbiasedness, can be seriously in error if there is nonlinearity. The value $100(1 - \alpha)\%$ is called the *nominal value* of the coverage probability of a confidence interval and is guaranteed to be the actual coverage probability only if all modeling assumptions are met.

Response transformation, polynomial regression, and nonparametric regression (e.g., splines and loess—see Chap. 21) are common solutions to the problem of nonlinearity.

Example 10.6. Detecting nonlinearity: A simulated data example

Data were simulated to illustrate some of the techniques for diagnosing problems. In the example there are two predictor variables, X_1 and X_2. The assumed model is multiple linear regression, $Y_i = \beta_0 + \beta_1 X_{i,1} + \beta_2 X_{i,2} + \epsilon_i$.

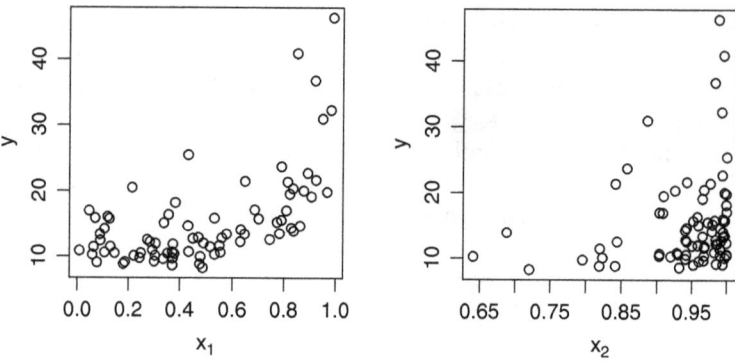

Fig. 10.6. *Simulated data. Responses plotted against the two predictor variables.*

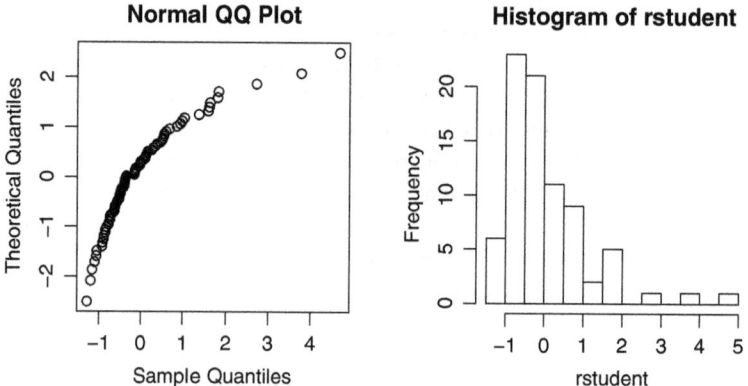

Fig. 10.7. *Simulated data. Normal plot and histogram of the studentized residuals. Right skewness is evident and perhaps a square root or log transformation of Y would be helpful.*

Figure 10.6, which shows the responses plotted against each of the predictors, suggests that the errors are heteroskedastic because there is more vertical scatter on the right sides of the plots. Otherwise, it is not clear whether there are other problems with the data or the model. The point here is that plots of the raw data often fail to reveal all problems. Rather, it is plots of the residuals that can more reliably detect heteroskedasticity, nonnormality, and other difficulties.

Figure 10.7 contains a normal plot and a histogram of the residuals—the externally standardized residuals (rstudents) are used in all examples of this chapter. Notice the right skewness which suggests that a response transformation to remove right skewness, such as, a square-root or log transformation, should be investigated.

Figure 10.8a is a plot of the residuals versus X_1. The residuals appear to have a nonlinear trend. This is better revealed by adding a loess curve to the residuals. The curvature of the loess fit is evident and indicates that Y is not linear in X_1. A possible remedy is to add X_1^2 as a third predictor. Figure 10.8a, a plot of the residuals against X_2, shows somewhat random scatter, indicating that Y appears to be linear in X_2. The concentration of the X_2-values near the right side is not a problem. This pattern only shows that the distribution of X_2 is left-skewed, but the regression model makes no assumptions about the distributions of the predictors.

Before doing any more plotting, the model was augmented by adding X_1^2 as a predictor, so the model is now

$$Y_i = \beta_0 + \beta_1 X_{i,1} + \beta_2 X_{i,2}^2 + \beta_3 X_{i,2} + \epsilon_i. \tag{10.5}$$

Figure 10.8c is a plot of the absolute residuals versus the predicted values for model (10.5). Note that the absolute residuals are largest where the fitted values are also largest, which is a clear sign of heteroskedasticity. A loess smooth has been added to make the heteroskedasticity clearer.

To remedy the problem of heteroskedasticity, Y_i was transformed to $\log(Y_i)$, so the model is now

$$\log(Y_i) = \beta_0 + \beta_1 X_{i,1} + \beta_2 X_{i,2}^2 + \beta_3 X_{i,2} + \epsilon_i. \tag{10.6}$$

Figure 10.9 shows residual plots for model (10.6). The plots in panels (a) and (b) of residuals versus X_1 and X_2 show no patterns, indicating that the

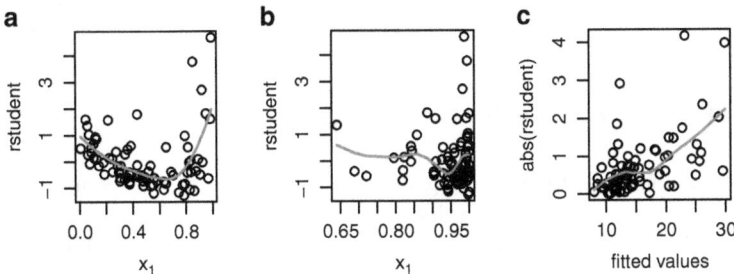

Fig. 10.8. *Simulated data.* **(a)** *Plot of externally studentized residuals versus X_1. This plot suggests that Y is not linearly related to X_1 and perhaps a model quadratic in X_1 is needed.* **(b)** *Plot of the residuals versus X_2 with a loess smooth. This plot suggests that Y is linearly related to X_2 so that the component of the model relating Y to X_2 is satisfactory.* **(c)** *Plot of the absolute residuals versus the predicted values using a model that is quadratic in X_1. This plot reveals heteroskedasticity. A loess smooth has been added to each plot.*

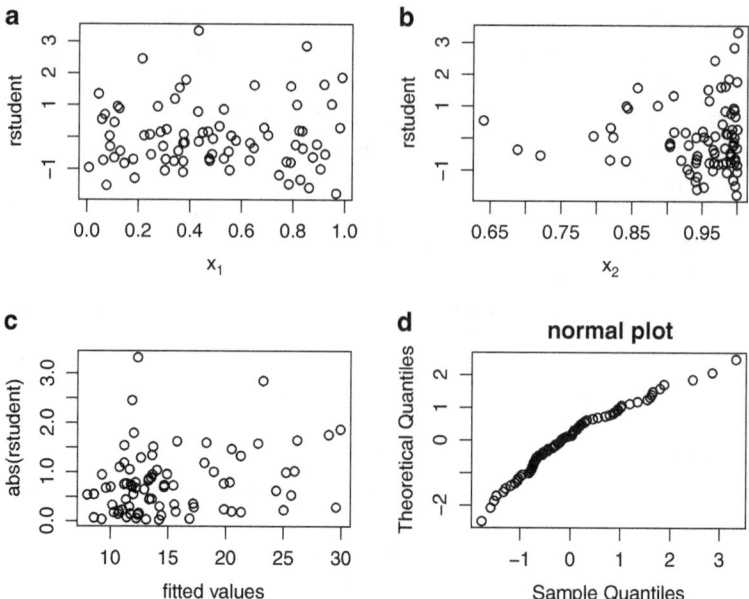

Fig. 10.9. *Simulated data. Residual plots for fit of* $\log(Y)$ *to* X_1, X_1^2, *and* X_2. *(a) Residuals versus* X_1. *(b) Residuals versus* X_2. *(c) Residuals versus* $\widehat{Y}$.

model that is quadratic in X_1 fits well. The plot in panel (c) of absolute residuals versus fitted values shows less heteroskedasticity than before, which shows the benefit of the log transformation. The normal plot of the residuals shown in panel (d) shows much less skewness than earlier, which is another benefit of the log transformation.   □

10.3 Bibliographic Notes

Graphical methods for detecting nonconstant variance, transform-both-sides regression, and weighting are discussed in Carroll and Ruppert (1988). The idea of using half-normal plots to detect usual values of Cook's D was borrowed from Faraway (2005).

Comprehensive treatments of regression diagnostics can be found in Belsley et al. (1980) and in Cook and Weisberg (1982). Although variance inflation factors detect collinearity, they do not indicate what correlations are causing the problem. For this purpose, one should use collinearity diagnostics. These are also discussed in Belsley et al. (1980).

10.4 R Lab

10.4.1 Current Population Survey Data

This section uses the CPS1988 data set from the March 1988 Current Population Survey by the U.S. Census Bureau and available in the AER package. These are cross-sectional data, meaning that the U.S. population was surveyed at a single time point. Cross-sectional data should be distinguished from longitudinal data where individuals are followed over time. Data collected and analyzed along two dimensions, that is, cross-sectionally and longitudinally, are called *panel data* by econometricians.

In this section, we will investigate how the variable wage (in dollars/week) depends on education (in years), experience (years of potential work experience), and ethnicity (Caucasian = "caus" or African-American = "afam"). Potential experience was (age − education − 6), the number of years of potential work experience assuming that education begins at age 6. Potential experience was used as a proxy for actual work experience, which was not available. The variable ethnicity is coded 0–1 for "cauc" and "afam," so its regression coefficient is the difference in the expected values of wage between an African-American and a Caucasian with the same values of education and experience. Run the code below to load the data and run a multiple linear regression.

```
library(AER)
data(CPS1988)
attach(CPS1988)
fitLm1 = lm(wage ~ education + experience + ethnicity)
```

Next, create residual plots with the following code. In some of these plots, the y-axis limits are set so as to eliminate outliers. This was done to focus attention on the bulk of the data. This is a very large data set with 28,155 observations, so scatterplots are very dense with data and almost solid black in places. Therefore, lowess smooths were added as thick, red lines so that they can be seen clearly. Also, thick blue reference lines were added as appropriate.

```
par(mfrow = c(3, 2))
resid1 = rstudent(fitLm1)
plot(fitLm1$fit, resid1,
   ylim = c(-1500, 1500), main = "(a)")
lines(lowess(fitLm1$fit, resid1, f = 0.2), lwd = 5, col = "red")
abline(h = 0, col = "blue", lwd = 5)

plot(fitLm1$fit, abs(resid1),
   ylim = c(0, 1500), main = "(b)")
lines(lowess(fitLm1$fit, abs(resid1), f = 0.2),
   lwd = 5, col = "red")
abline(h = mean(abs(resid1)), col = "blue", lwd = 5)
```

```
qqnorm(resid1, datax = FALSE, main = "(c)")
qqline(resid1, datax = FALSE, lwd = 5, col = "blue")

plot(education, resid1, ylim = c(-1000, 1500), main = "(d)")
lines(lowess(education, resid1), lwd = 5, col = "red")
abline(h = 0, col = "blue", lwd = 5)

plot(experience, resid1, ylim = c(-1000, 1500), main = "(e)")
lines(lowess(experience, resid1), lwd = 5, col = "red")
abline(h = 0, col = "blue", lwd = 5)
```

Problem 1 *For each of the panels (a)–(e) in the figure you have just created, describe what is being plotted and any conclusions that should be drawn from the plot. Describe any problems and discuss how they might be remedied.*

Problem 2 *Now fit a new model where the log of* wage *is regressed on* education *and* experience. *Create residual plots as done above for the first model. Describe differences between the residual plots for the two models. What do you suggest should be tried next?*

Problem 3 *Implement whatever you suggested to try next in Problem 2. Describe how well it worked. Are you satisfied with your model? If not, try further enhancements of the model until arriving at a model that you feel is satisfactory. What is your final model?*

Problem 4 *Use your final model to describe the effects of* education, exper- *ience, and* ethnicity *on the* wage. *Use graphs where appropriate.*

Check the data and your final model for possible problems or unusual features by examining the hat diagonals and Cook's D with the following code. Replace fitLm4 by the name of the lm object for your final model.

```
library(faraway)  #  required for halfnorm
par(mfrow=c(1, 3))
plot(hatvalues(fitLm4))
plot(sqrt(cooks.distance(fitLm4)))
halfnorm(sqrt(cooks.distance(fitLm4)))
```

Problem 5 *Do you see any high-leverage points or points with very high values of Cook's D? If you do, what is unusual about them?*

10.5 Exercises

1. Residual plots and other diagnostics are shown in Fig. 10.10 for a regression of Y on X. Describe any problems that you see and possible remedies.

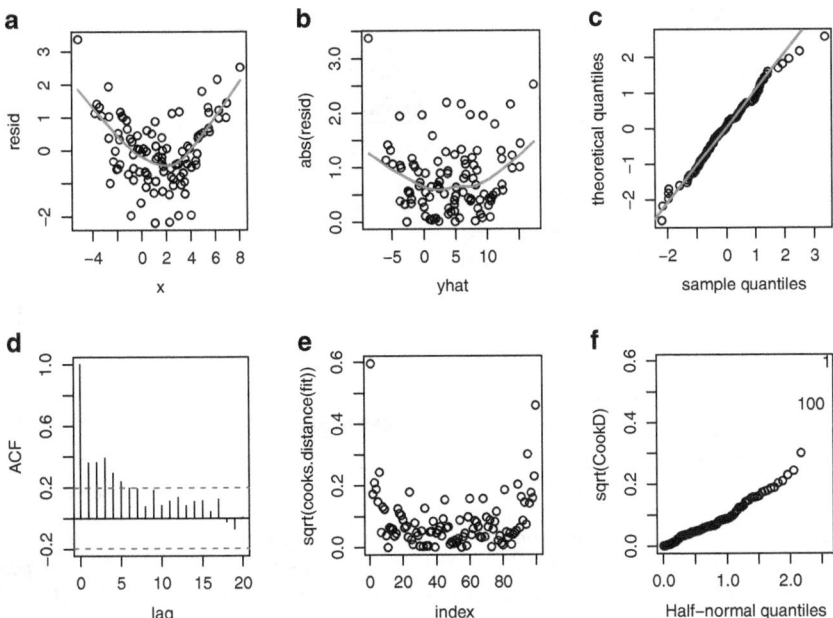

Fig. 10.10. *Residual plots and diagnostics for regression of Y on X in Problem 1. The residuals are rstudent values.* **(a)** *Plot of residuals versus x.* **(b)** *Plot of absolute residuals versus fitted values.* **(c)** *Normal QQ plot of residuals.* **(d)** *ACF plot of residuals.* **(e)** *Plot of the square root of Cook's D versus index (= observation number).* **(f)** *Half-normal plot of square root of Cook's D.*

2. Residual plots and other diagnostics are shown in Fig. 10.11 for a regression of Y on X. Describe any problems that you see and possible remedies.

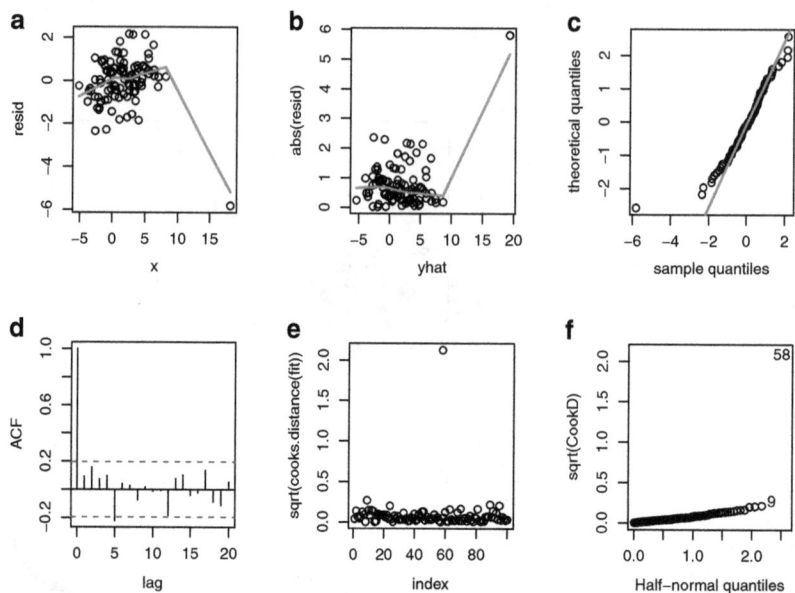

Fig. 10.11. *Residual plots and diagnostics for regression of Y on X in Problem 2. The residuals are rstudent values. (a) Plot of residual versus x. (b) Plot of absolute residuals versus fitted values. (c) Normal QQ plot of residuals. (d) ACF plot of residuals. (e) Plot of the square root of Cook's D versus index (= observation number). (f) Half-normal plot of square root of Cook's D.*

3. Residual plots and other diagnostics are shown in Fig. 10.12 for a regression of Y on X. Describe any problems that you see and possible remedies.

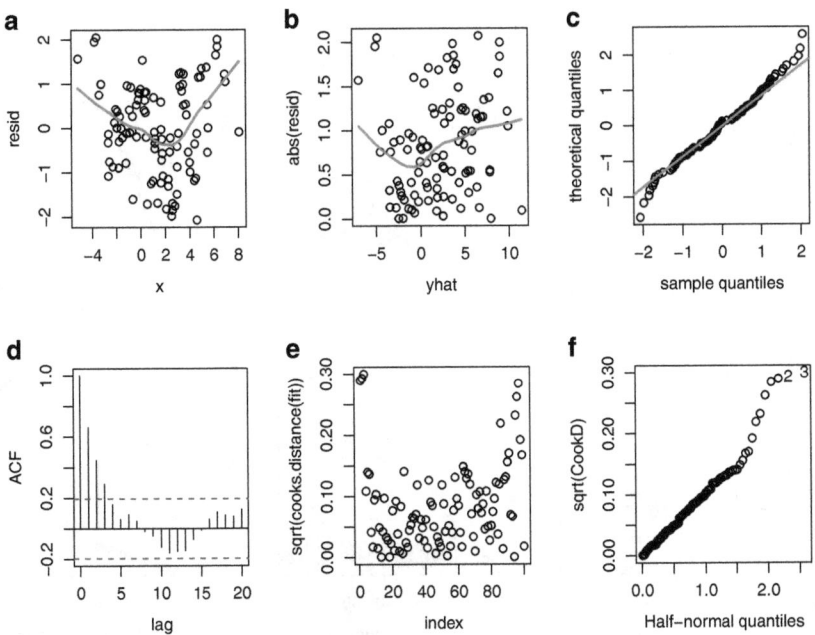

Fig. 10.12. *Residual plots and diagnostics for regression of Y on X in Problem 3. The residuals are rstudent values.* (**a**) *Plot of residual versus x.* (**b**) *Plot of absolute residuals versus fitted values.* (**c**) *Normal QQ plot of residuals.* (**d**) *ACF plot of residuals.* (**e**) *Plot of the square root of Cook's D versus index (= observation number).* (**f**) *Half-normal plot of square root of Cook's D.*

4. Residual plots and other diagnostics are shown in Fig. 10.13 for a regression of Y on X. Describe any problems that you see and possible remedies.

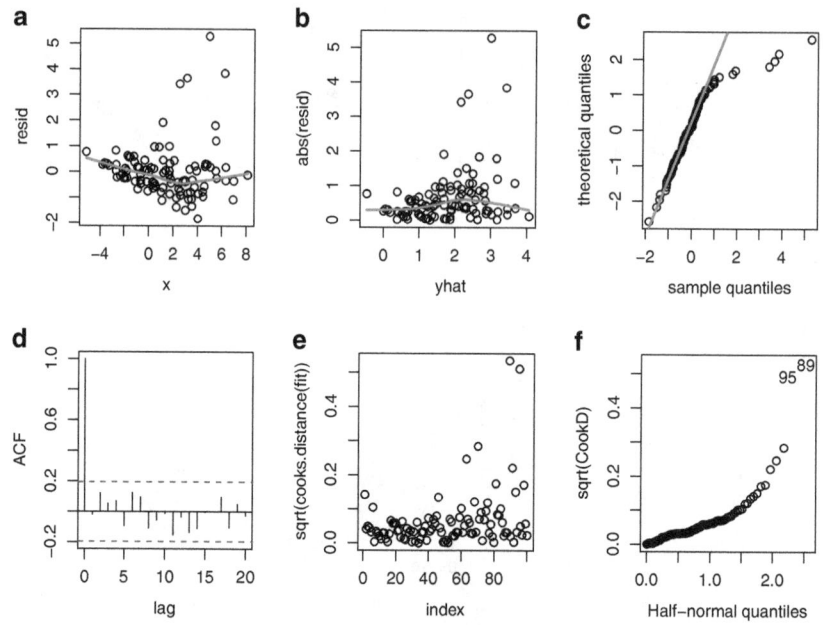

Fig. 10.13. *Residual plots and diagnostics for regression of Y on X in Problem 4. The residuals are rstudent values.* (**a**) *Plot of residual versus x.* (**b**) *Plot of absolute residuals versus fitted values.* (**c**) *Normal QQ plot of residuals.* (**d**) *ACF plot of residuals.* (**e**) *Plot of the square root of Cook's D versus index (= observation number).* (**f**) *Half-normal plot of square root of Cook's D.*

5. It was noticed that a certain observation had a large leverage (hat diagonal) but a small Cook's D. How could this happen?

References

Belsley, D. A., Kuh, E., and Welsch, R. E. (1980) *Regression Diagnostics*, Wiley, New York.

Carroll, R. J., and Ruppert, D. (1988) *Transformation and Weighting in Regression*, Chapman & Hall, New York.

Cook, R. D., and Weisberg, S. (1982) *Residuals and Influence in Regression*, Chapman & Hall, New York.

Faraway, J. J. (2005) *Linear Models with R*, Chapman & Hall, Boca Raton, FL.

Regression: Advanced Topics

11.1 The Theory Behind Linear Regression

This section provides some theoretical results about linear least-squares estimation. The study of linear regression is facilitated by the use of matrices.
Equation (9.1) can be written more succinctly as

$$Y_i = \boldsymbol{x}_i^{\mathsf{T}} \boldsymbol{\beta} + \epsilon_i, \quad i = 1, \ldots, n \tag{11.1}$$

where $\boldsymbol{x}_i = (1 \ X_{i,1} \ \cdots \ X_{i,p})^{\mathsf{T}}$ and $\boldsymbol{\beta} = (\beta_0 \ \beta_1 \ \cdots \ \beta_p)^{\mathsf{T}}$. Let

$$\boldsymbol{Y} = \begin{pmatrix} Y_1 \\ \vdots \\ Y_n \end{pmatrix}, \ \boldsymbol{X} = \begin{pmatrix} \boldsymbol{x}_1 \\ \vdots \\ \boldsymbol{x}_n \end{pmatrix}, \text{ and } \boldsymbol{\epsilon} = \begin{pmatrix} \epsilon_1 \\ \vdots \\ \epsilon_n \end{pmatrix}.$$

Then, the n equations in (11.1) can be expressed as

$$\underbrace{\boldsymbol{Y}}_{n \times 1} = \underbrace{\boldsymbol{X}}_{n \times (p+1)} \underbrace{\boldsymbol{\beta}}_{(p+1) \times 1} + \underbrace{\boldsymbol{\epsilon}}_{n \times 1}, \tag{11.2}$$

with the matrix dimensions indicated by underbraces.

The least-squares estimate of $\boldsymbol{\beta}$ minimizes

$$\|\boldsymbol{Y} - \boldsymbol{X}\boldsymbol{\beta}\|^2 = (\boldsymbol{Y} - \boldsymbol{X}\boldsymbol{\beta})^{\mathsf{T}}(\boldsymbol{Y} - \boldsymbol{X}\boldsymbol{\beta}) = \boldsymbol{Y}^{\mathsf{T}}\boldsymbol{Y} - 2\boldsymbol{\beta}^{\mathsf{T}}\boldsymbol{X}^{\mathsf{T}}\boldsymbol{Y} + \boldsymbol{\beta}^{\mathsf{T}}\boldsymbol{X}^{\mathsf{T}}\boldsymbol{X}\boldsymbol{\beta}. \tag{11.3}$$

By setting the derivatives of (11.3) with respect to $\beta_0, \ldots, \beta_p$ equal to 0
and simplifying the resulting equations, one finds that the least-squares
estimator is

$$\widehat{\boldsymbol{\beta}} = (\boldsymbol{X}^{\mathsf{T}}\boldsymbol{X})^{-1}\boldsymbol{X}^{\mathsf{T}}\boldsymbol{Y}. \tag{11.4}$$

© Springer Science+Business Media New York 2015
D. Ruppert, D.S. Matteson, *Statistics and Data Analysis for Financial
Engineering*, Springer Texts in Statistics,
DOI 10.1007/978-1-4939-2614-5_11

Using (7.9), one can find the covariance matrix of $\widehat{\boldsymbol{\beta}}$:

$$\begin{aligned}
\mathrm{COV}(\widehat{\boldsymbol{\beta}}|\boldsymbol{x}_1,\ldots,\boldsymbol{x}_n) &= (\boldsymbol{X}^{\mathsf{T}}\boldsymbol{X})^{-1}\boldsymbol{X}^{\mathsf{T}}\mathrm{COV}(\boldsymbol{Y}|\boldsymbol{x}_1,\ldots,\boldsymbol{x}_n)\boldsymbol{X}(\boldsymbol{X}^{\mathsf{T}}\boldsymbol{X})^{-1} \\
&= (\boldsymbol{X}^{\mathsf{T}}\boldsymbol{X})^{-1}\boldsymbol{X}^{\mathsf{T}}(\sigma_\epsilon^2\boldsymbol{I})\boldsymbol{X}(\boldsymbol{X}^{\mathsf{T}}\boldsymbol{X})^{-1} \\
&= \sigma_\epsilon^2(\boldsymbol{X}^{\mathsf{T}}\boldsymbol{X})^{-1},
\end{aligned}$$

since $\mathrm{COV}(\boldsymbol{Y}|\boldsymbol{x}_1,\ldots,\boldsymbol{x}_n) = \mathrm{COV}(\boldsymbol{\epsilon}) = \sigma_\epsilon^2\boldsymbol{I}$, where $\boldsymbol{I}$ is the $n \times n$ identity matrix. Therefore, the standard error of $\widehat{\beta}_j$ is the square root of the jth diagonal element of $\sigma_\epsilon^2(\boldsymbol{X}^{\mathsf{T}}\boldsymbol{X})^{-1}$.

The vector of fitted values is

$$\widehat{\boldsymbol{Y}} = \boldsymbol{X}\widehat{\boldsymbol{\beta}} = \{\boldsymbol{X}(\boldsymbol{X}^{\mathsf{T}}\boldsymbol{X})^{-1}\boldsymbol{X}^{\mathsf{T}}\}\boldsymbol{Y} = \boldsymbol{H}\boldsymbol{Y},$$

where $\boldsymbol{H} = \boldsymbol{X}(\boldsymbol{X}^{\mathsf{T}}\boldsymbol{X})^{-1}\boldsymbol{X}^{\mathsf{T}}$ is the *hat matrix*. The leverage of the ith observation is H_{ii}, the ith diagonal element of $\boldsymbol{H}$.

11.1.1 Maximum Likelihood Estimation for Regression

In this section, we assume a linear regression model with noise that may not be normally distributed and independent.

For example, consider the special case of i.i.d. errors. It is useful to put the scale parameter explicitly into the regression model, so we assume that

$$Y_i = \boldsymbol{x}_i^{\mathsf{T}}\beta + \sigma\epsilon_i,$$

where $\{\epsilon_i\}$ are i.i.d. with a known density f that has variance equal to 1 and σ is the unknown noise standard deviation. For example, f could be a standardized t-density. Then the likelihood of $Y_1,\ldots,Y_n$ is

$$\prod_{i=1}^n \frac{1}{\sigma} f\left\{\frac{Y_i - \boldsymbol{x}_i^{\mathsf{T}}\boldsymbol{\beta}}{\sigma}\right\}.$$

The maximum likelihood estimator maximizes the log-likelihood

$$L(\boldsymbol{\beta},\sigma) = -n\log(\sigma) + \sum_{i=1}^n \log\left[f\left\{\frac{Y_i - \boldsymbol{x}_i^{\mathsf{T}}\boldsymbol{\beta}}{\sigma}\right\}\right].$$

For normally distributed errors, $\log\{f(x)\} = -\frac{1}{2}x^2 - \frac{1}{2}\log(2\pi)$, and for the purpose of maximization, the constant $-\frac{1}{2}\log(2\pi)$ can be ignored. Therefore, the log-likelihood is

$$L^{\mathrm{GAUSS}}(\boldsymbol{\beta},\sigma) = -n\log(\sigma) - \frac{1}{2}\sum_{i=1}^n \left(\frac{Y_i - \boldsymbol{x}_i^{\mathsf{T}}\boldsymbol{\beta}}{\sigma}\right)^2.$$

It should be obvious that the least-squares estimator is the MLE of $\boldsymbol{\beta}$. Also, maximizing $L^{\mathrm{GAUSS}}(\widehat{\boldsymbol{\beta}},\sigma)$ in σ, where $\boldsymbol{\beta}$ has been replaced by the least-squares estimate, is a standard calculus exercise and the result is

$$\widehat{\sigma}^2_{\mathrm{MLE}} = n^{-1} \sum_{i=1}^{n} (Y_i - \boldsymbol{x}_i^\mathsf{T} \widehat{\boldsymbol{\beta}})^2.$$

In can be shown that $\widehat{\sigma}^2_{\mathrm{MLE}}$ is biased but that the bias is eliminated if n^{-1} is replaced by $\{n - (p+1)\}^{-1}$ where $p+1$ is the dimension of $\boldsymbol{\beta}$. This give us the estimator (9.16).

Now assume that $\boldsymbol{\epsilon}$ has a covariance matrix $\boldsymbol{\Sigma}$ and, for some function f, density

$$|\boldsymbol{\Sigma}|^{-1/2} f\{(\boldsymbol{Y} - \boldsymbol{X}\boldsymbol{\beta})^\mathsf{T} \boldsymbol{\Sigma}^{-1} (\boldsymbol{Y} - \boldsymbol{X}\boldsymbol{\beta})\}.$$

Then the log-likelihood is

$$-\frac{1}{2} \log |\boldsymbol{\Sigma}| + \log \left[f\{(\boldsymbol{Y} - \boldsymbol{X}\boldsymbol{\beta})^\mathsf{T} \boldsymbol{\Sigma}^{-1} (\boldsymbol{Y} - \boldsymbol{X}\boldsymbol{\beta})\} \right].$$

In the important special case where $\boldsymbol{\epsilon}$ has a mean-zero multivariate normal distribution, the density of $\boldsymbol{\epsilon}$ is

$$\left[\frac{1}{|\boldsymbol{\Sigma}|^{1/2} (2\pi)^{p/2}} \right] \exp \left\{ -\frac{1}{2} \boldsymbol{\epsilon}^\mathsf{T} \boldsymbol{\Sigma}^{-1} \boldsymbol{\epsilon} \right\}, \tag{11.5}$$

If $\boldsymbol{\Sigma}$ is known, then the MLE of $\boldsymbol{\beta}$ minimizes

$$(\boldsymbol{Y} - \boldsymbol{X}\boldsymbol{\beta})^\mathsf{T} \boldsymbol{\Sigma}^{-1} (\boldsymbol{Y} - \boldsymbol{X}\boldsymbol{\beta})$$

and is called the *generalized least-squares estimator* (GLS estimator). If $\epsilon_1, \ldots, \epsilon_n$ are uncorrelated but with possibly different variances, then $\boldsymbol{\Sigma}$ is the diagonal matrix of these variances and the generalized least-squares estimator is the weighted least-squares estimator (10.4).

The GLS estimator is

$$\widehat{\boldsymbol{\beta}}_{\mathrm{GLS}} = (\boldsymbol{X}^\mathsf{T} \boldsymbol{\Sigma}^{-1} \boldsymbol{X})^{-1} \boldsymbol{X}^\mathsf{T} \boldsymbol{\Sigma}^{-1} \boldsymbol{Y}. \tag{11.6}$$

Typically, $\boldsymbol{\Sigma}$ is unknown and must be replaced by an estimate, for example, from an ARMA model for the errors.

11.2 Nonlinear Regression

Often we can derive a theoretical model relating predictor variables and a response, but the model we derive is not linear. In particular, models derived from economic theory are commonly used in finance and many are not linear.

The nonlinear regression model is

$$Y_i = f(\boldsymbol{X}_i; \boldsymbol{\beta}) + \epsilon_i, \tag{11.7}$$

where Y_i is the response measured on the ith observation, $\boldsymbol{X}_i$ is a vector of observed predictor variables for the ith observation, $f(\cdot \, ; \, \cdot)$ is a *known* function, β is an unknown parameter vector, and $\epsilon_1, \ldots, \epsilon_n$ are i.i.d. with mean 0 and variance σ_ϵ^2. The least-squares estimate $\widehat{\beta}$ minimizes

$$\sum_{i=1}^{n} \{Y_i - f(\boldsymbol{X}_i; \beta)\}^2 .$$

The predicted values are $\widehat{Y}_i = f(\boldsymbol{X}_i; \widehat{\beta})$ and the residuals are $\widehat{\epsilon}_i = Y_i - \widehat{Y}_i$.

Since the model is nonlinear, finding the least-squares estimate requires nonlinear optimization. Because of the importance of nonlinear regression, almost every statistical software package will have routines for nonlinear least-squares estimation. This means that most of the difficult programming has already been done for us. However, we do need to write an equation that specifies the model we are using.[1] In contrast, when using linear regression only the predictor variables need to be specified.

Example 11.1. Simulated bond prices

Consider prices of par \$1000 zero-coupon bonds issued by a particular borrower, perhaps the Federal government or a corporation. Suppose that there are several times to maturity, the ith being denoted by T_i. Suppose also that the yield to maturity is a constant, say r. The assumption that $Y_T = r$ for all T is not realistic and is used only to keep this example simple. In Sect. 11.3 more realistic models will be used.

The rate r is determined by the market and can be estimated from prices. Under the assumption of a constant value of r, the present price of a bond with maturity T_i is

$$P_i = 1000 \, \exp(-rT_i). \tag{11.8}$$

There is some random variation in the observed prices. One reason is that the price of a bond can only be determined by the sale of the bond, so the observed prices have not been determined simultaneously. Prices that may no longer reflect current market values are called *stale*. Each bond's price was determined at the time of the last trade of a bond of that maturity, and r may have had a somewhat different value then. It is only as a function of time to maturity that r is assumed constant, so r may vary with calendar time. Thus, we augment model (11.8) by including a noise term to obtain the regression model

$$P_i = 1000 \, \exp(-rT_i) + \epsilon_i. \tag{11.9}$$

An estimate of r can be determined by least squares, that is, by minimizing over r the sum of squares:

$$\sum_{i=1}^{n} \left\{ P_i - 1{,}000 \, \exp(-rT_i) \right\}^2 .$$

---

[1] Even this work can sometimes be avoided, since some nonlinear regression software has many standard models already programmed.

The least-squares estimator is denoted by $\hat{r}$.

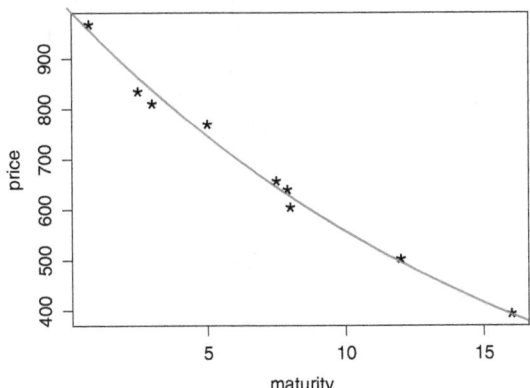

Fig. 11.1. *Plot of bond prices against maturities with the predicted price from the nonlinear least-squares fit.*

Since it is unlikely that market data will have a constant r, this example uses simulated data. The data were generated with r fixed at 0.06 and plotted in Fig. 11.1. The nonlinear least-squares estimate of r was found using R's nls() function. Nonlinear optimization requires starting values for the parameters, and a starting value of 0.04 was used for r.

```
bondprices = read.table("bondprices.txt", header = TRUE)
attach(bondprices)
fit = nls(price ~ 1000 * exp(-r * maturity), start = list(r = 0.04))
summary(fit)
detach(bondprices)
```

The output is:

```
Formula: price ~ 1000 * exp(-r * maturity)

Parameters:
  Estimate Std. Error t value Pr(>|t|)
r  0.05850    0.00149    39.3  1.9e-10 ***
---


Residual standard error: 20 on 8 degrees of freedom

Number of iterations to convergence: 4
Achieved convergence tolerance: 5.53e-08
```

Notice that $\widehat{r} = 0.0585$ and the standard error of this estimate is 0.00149. The predicted price curve using nonlinear regression is shown in Fig. 11.1.   □

As mentioned, in *nonlinear regression* the form of the regression function is nonlinear but *known* up to a few unknown parameters. For example, the regression function has an exponential form in model (11.9). For this reason, nonlinear regression would best be called *nonlinear parametric regression* to distinguish it from nonparametric regression, where the regression function is also nonlinear but not of a known parametric form. Nonparametric regression is discussed in Chap. 21.

Polynomial regression may appear to be nonlinear since polynomials are nonlinear functions. For example, the quadratic regression model

$$Y_i = \beta_0 + \beta_1 X_i + \beta_2 X_i^2 + \epsilon_i \qquad (11.10)$$

is nonlinear in X_i. However, by defining X_i^2 as a second predictor variable, this model is linear in (X_i, X_i^2) and therefore is an example of multiple *linear* regression. What makes model (11.10) linear is that the right-hand side is a linear function of the parameters β_0, β_1, and β_2, and therefore can be interpreted as a linear regression with the appropriate definition of the variables. In contrast, the exponential model

$$Y_i = \beta_0 e^{\beta_1 X_i} + \epsilon_i$$

is nonlinear in the parameter β_1, so it cannot be made into a linear model by redefining the predictor variable.

Example 11.2. Estimating default probabilities

This example illustrates both nonlinear regression and the detection of heteroskedasticity by residual plotting.

Credit risk is the risk to a lender that a borrower will default on contractual obligations, for example, that a loan will not be repaid in full. A key parameter in the determination of credit risk is the probability of default. Bluhm, Overbeck, and Wagner (2003) illustrate how one can calibrate Moody's credit rating to estimate default probabilities. These authors use observed default frequencies for bonds in each of 16 Moody's ratings from Aaa (best credit rating) to B3 (worse rating). They convert the credit ratings to a 1 to 16 scale (Aaa $= 1, \ldots$, B3 $= 16$). Figure 11.2a shows default frequencies (as fractions, not percentages) plotted against the ratings. The data are from Bluhm, Overbeck, and Wagner (2003). The relationship is clearly nonlinear. Not surprisingly, Bluhm, Overbeck, and Wagner used a nonlinear model, specifically

$$Pr\{\text{default}|\text{rating}\} = \exp\{\beta_0 + \beta_1 \text{rating}\}. \qquad (11.11)$$

To use this model they fit a linear function to the logarithms of the default frequencies. One difficulty with doing this is that six of the default frequencies are zero giving a log transformation of $-\infty$.

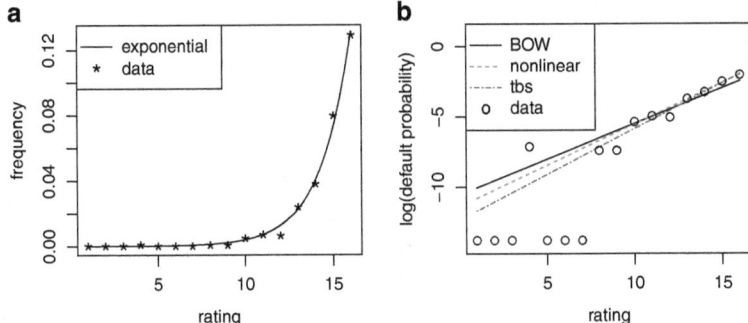

Fig. 11.2. *(a) Default frequencies with an exponential fit. "Rating" is a conversion of the Moody's rating to a 1 to 16-point scale as follows:* $1 = Aaa$, $2 = Aa1$, $3 = Aa3$, $4 = A1$, ..., $16 = B3$. *(b) Estimation of default probabilities by Bluhm, Overbeck, and Wagner's (2003) linear regression with ratings removed that have no observed defaults (BOW) and by nonlinear regression with all data (nonlinear). Because some default frequencies are zero, when plotting the data on a semilog plot,* 10^{-6} *was added to the default frequencies. This constant was not added when estimating default frequencies, only for plotting the raw data. The six observations along the bottom of the plot are the ones removed by Bluhm, Overbeck, and Wagner. "TBS" is the transform-both-sides estimate, which will be discussed soon.*

Bluhm, Overbeck, and Wagner (2003) address this issue by labeling default frequencies equal to zero as "unobserved" and not using them in the estimation process. The problem with their technique is that they have deleted the data with the lowest observed default frequencies. This biases their estimates of default probabilities in an upward direction. Bluhm, Overbeck, and Wagner argue that an observed default frequency of zero does not imply that the true default probability is zero. This is certainly true. However, the default frequencies, even when they are zero, are unbiased estimates of the true default probabilities. There is no intent here to be critical of their book, which is well-written and useful. However, one can avoid the bias of their method by using nonlinear regression with model (11.11). The advantage of fitting (11.11) by nonlinear regression is that it avoids the use of a logarithm transformation thus allowing the use of all the data, even data with a default frequency of zero. The fits by the Bluhm, Overbeck, and Wagner method and by nonlinear regression using model (11.11) are shown in Fig. 11.2b with a log scale on the vertical axis so that the fitted functions are linear. Notice that at good credit ratings the estimated default probabilities are lower using nonlinear regression compared to Bluhm, Overbeck, and Wagner's biased method. The differences between the two sets of estimated default probabilities can be substantial. Bluhm,

Overbeck, and Wagner estimate the default probability of an Aaa bond as
0.005 %. In contrast, the unbiased estimate by nonlinear regression is only 40 %
of that figure, specifically, 0.0020 %. Thus, the bias in the Bluhm, Overbeck,
and Wagner estimate leads to a substantial overestimate of the credit risk of
Aaa bonds and similar overestimation at other good credit ratings.

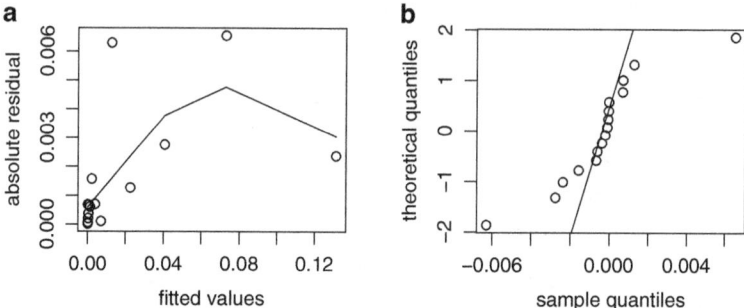

Fig. 11.3. *(a) Residuals for estimation of default probabilities by nonlinear regression. Absolute studentized residuals plotted against fitted values with a loess smooth.
Substantial heteroskedasticity is indicated because the data on the left side are less
scattered than elsewhere. (b) Normal probability plot of the residuals. Notice the
outliers caused by the nonconstant variance.*

A plot of the absolute residuals versus the fitted values in Fig. 11.3a gives
a clear indication of heteroskedasticity. Heteroskedasticity does not cause bias
but it does cause inefficient estimates. In Sect. 11.4, this problem is fixed by a
variance-stabilizing transformation. Figure 11.3b is a normal probability plot
of the residuals. Outliers with both negative and positive values can be seen.
These are due to the nonconstant variance and are not necessarily a sign of
nonnormality. This plot illustrates the danger of attempting to interpret a
normal plot when the data have a nonconstant variance. One should apply a
variance-stabilizing transformation first before checking for normality.   □

11.3 Estimating Forward Rates from Zero-Coupon Bond Prices

In practice, the forward-rate function $r(t)$ is unknown. Only bond prices are
known. If the prices $P(T_i)$ of zero-coupon bonds are available on a relatively
fine grid of values of $T_1 < T_2 < \cdots < T_n$, then using (3.24) we can estimate
the forward-rate curve at T_i with

$$-\frac{\Delta \log\{P(T_i)\}}{\Delta T_i} = -\frac{\log\{P(T_i)\} - \log\{P(T_{i-1})\}}{T_i - T_{i-1}}. \tag{11.12}$$

We will call these the *empirical forward-rate estimates.* Figure 11.4 shows prices and empirical forward-rate estimates from data to be described soon in Example 11.3. As can be seen in the plot, the empirical forward-rate estimates can be rather noisy when the denominators in (11.12) are small because the maturities are spaced closely together. If the maturities were more widely spaced, then bias rather than variance would be the major problem. Despite these difficulties, the empirical forward-rate estimates give a general impression of the forward-rate curve and are useful for comparing with estimates from parametric models, which are discussed next.

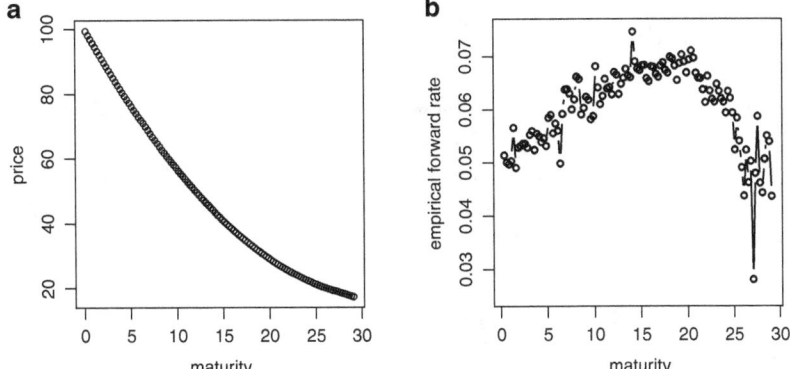

Fig. 11.4. *(a) U.S. STRIPS prices. (b) Empirical forward-rate estimates from the prices.*

We can estimate $r(t)$ from the bond prices using nonlinear regression. An example of estimating $r(t)$ was given in Sect. 11.2 assuming that $r(t)$ was constant and using as data the prices of zero-coupon bonds of different maturities. In this section, we estimate $r(t)$ without assuming it is constant.

Parametric estimation of the forward-rate curves starts with a parametric family $r(t; \boldsymbol{\theta})$ of forward rates and the correspond yield curves

$$y_T(\boldsymbol{\theta}) = T^{-1} \int_0^T r(t; \boldsymbol{\theta}) \, dt$$

and model for the price of a par-\$1 bond:

$$P_T(\boldsymbol{\theta}) = \exp\{-Ty_T(\boldsymbol{\theta})\} = \exp\left(-\int_0^T r(t; \boldsymbol{\theta}) \, dt\right).$$

For example, suppose that $r(t; \boldsymbol{\theta})$ is a pth-degree polynomial, so that

$$r(t; \boldsymbol{\theta}) = \theta_0 + \theta_1 t + \cdots + \theta_p t^p$$

for some unknown parameters $\theta_0, \ldots, \theta_p$. Then

$$\int_0^T r(t; \boldsymbol{\theta})\, dt = \theta_0 T + \theta_1 \frac{T^2}{2} + \cdots + \theta_p \frac{T^{p+1}}{p},$$

and the yield curve is

$$y_T = T^{-1} \int_0^T r(t; \boldsymbol{\theta})\, dt = \theta_0 + \theta_1 \frac{T}{2} + \cdots + \theta_p \frac{T^p}{p}.$$

A popular model is the Nelson–Siegel family with forward-rate and yield curves

$$r(t; \boldsymbol{\theta}) = \theta_0 + (\theta_1 + \theta_2 t) \exp(-\theta_3 t),$$

$$y_t(\boldsymbol{\theta}) = \theta_0 + \left(\theta_1 + \frac{\theta_2}{\theta_3} \right) \frac{1 - \exp(-\theta_3 t)}{\theta_3 t} - \frac{\theta_2}{\theta_3} \exp(-\theta_3 t).$$

The six-parameter Svensson model extends the Nelson–Siegel model by adding the term $\theta_4 t \exp(-\theta_5 t)$ to the forward rate.

The nonlinear regression model for estimating the forward-rate curve states that the price of the ith bond in the sample with maturity T_i expressed as a fraction of par value is

$$P_i = D(T_i) + \epsilon_i = \exp \left(- \int_0^{T_i} r(t; \boldsymbol{\theta})\, dt \right) + \epsilon_i, \qquad (11.13)$$

where D is the discount function and ϵ_i is an "error" due to problems such as prices being somewhat stale and the bid–ask spread.[2]

Example 11.3. Estimating forward rates from STRIPS prices

We now look at an example using data on U.S. STRIPS, a type of zero-coupon bond. STRIPS is an acronym for "Separate Trading of Registered Interest and Principal of Securities." The interest and principal on Treasury bills, notes, and bonds are traded separately through the Federal Reserve's book-entry system, in effect creating zero-coupon bonds by repackaging coupon bonds.[3]

The data are from December 31, 1995. The prices are given as a percentage of par value. Price is plotted against maturity in years in Fig. 11.4a. There are 117 prices and the maturities are nearly equally spaced from 0 to 30 years. We can see that the price drops smoothly with maturity and that there is not much noise in the price data. The empirical forward-rate estimates in Fig. 11.4b are much noisier than the prices.

---

[2] A bond dealer buys bonds at the bid price and sells them at the ask price, which is slightly higher than the bid price. The difference is called the bid–ask spread and covers the trader's administrative costs and profit.

[3] Jarrow(2002, p. 15).

Three models for the forward curve were fit: quadratic polynomial, cubic polynomial, and quadratic polynomial spline with a knot at $T = 15$. The latter splices two quadratic functions together at $T = 15$ so that the resulting curve is continuous and with a continuous first derivative. The spline's second derivative jumps at $T = 15$. One way to write the spline is

$$r(t) = \beta_0 + \beta_1 t + \beta_2 t^2 + \beta_3 (t - 15)_+^2, \tag{11.14}$$

where the positive-part function is $x_+ = x$ if $x \geq 0$ and $x_+ = 0$ if $x < 0$. Also, x_+^2 means $(x_+)^2$, that is, take the positive part first. See Chap. 21 for further information about splines. From (11.14), one obtains

$$\int_0^T r(t)\, dt = \beta_0 T + \beta_1 \frac{T^2}{2} + \beta_2 \frac{T^3}{3} + \beta_3 \frac{(T - 15)_+^3}{3}, \tag{11.15}$$

and therefore the yield curve is

$$y_T = \beta_0 + \beta_1 \frac{T}{2} + \beta_2 \frac{T^2}{3} + \beta_3 \frac{(T - 15)_+^3}{3T}. \tag{11.16}$$

From (11.15), the model for a bond price (as a percentage of par) is

$$100 \exp\left\{ -\left(\beta_0 T + \beta_1 \frac{T^2}{2} + \beta_2 \frac{T^3}{3} + \beta_3 \frac{(T - 15)_+^3}{3} \right) \right\}. \tag{11.17}$$

R code to fit the quadratic spline and plot its forward-rate estimate is

```
fitSpline = nls(price ~ 100 * exp(-beta0 * T
    - (beta1 * T^2)/2 - (beta2 * T^3) / 3
    - (T > 15) * (beta3 * (T - 15)^3) / 3), data = dat,
    start = list(beta0 = 0.03, beta1 = 0, beta2 = 0, beta3 = 0))
coefSpline = summary(fitSpline)$coef[ , 1]
forwardSpline = coefSpline[1] + (coefSpline[2] * t) +
    (coefSpline[3] * t^2)  + (t > 15) * (coefSpline[4] * (t - 15)^2)
plot(t, forwardSpline, lty = 2, lwd = 2)
```

Only slight changes in the code are needed to fit the quadratic or cubic polynomial models.

Figure 11.5 contains all three estimates of the forward rate and the empirical forward rates. The cubic polynomial and quadratic spline models follow the empirical forward rates much more closely than the quadratic polynomial model. The cubic polynomial and quadratic spline fits both use four parameters and are similar to each other, though the spline has a slightly smaller residual sum of squares. The summary of the spline model's fit is

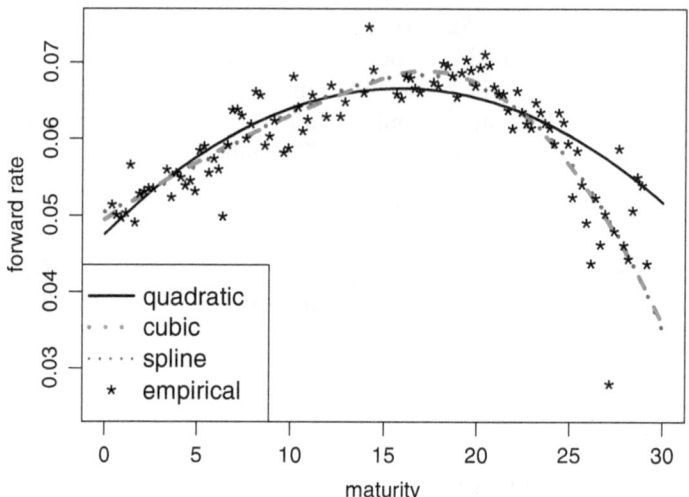

Fig. 11.5. *Polynomial and spline estimates of forward rates of U.S. Treasury bonds. The empirical forward rates are also shown.*

```
> summary(fitSpline)

Formula: price ~ 100 * exp(-beta0 * T - (beta1 * T^2)/2
   - (beta2 * T^3)/3 - (T > 15) * (beta3 * (T - 15)^3)/3)

Parameters:
          Estimate Std. Error t value Pr(>|t|)
beta0  4.947e-02  9.221e-05  536.52   <2e-16 ***
beta1  1.605e-03  3.116e-05   51.51   <2e-16 ***
beta2 -2.478e-05  1.820e-06  -13.62   <2e-16 ***
beta3 -1.763e-04  5.755e-06  -30.64   <2e-16 ***
---

Residual standard error: 0.0667 on 113 degrees of freedom

Number of iterations to convergence: 5
Achieved convergence tolerance: 1.181e-07
```

Notice that all coefficients have very small p-values. The small p-value of beta3 is further evidence that the spline model fits better than the quadratic polynomial model, since the two models differ only in that beta3 is 0 for the quadratic model.

R's nls function could not find the least-squares estimator for the Nelson–Siegel model, but the least-squares estimator was found using the optim non-linear optimization function with the sum of squares as the objective function. The fit of the Nelson–Siegel model was noticeably inferior to that of the cubic polynomial and quadratic spline models. In fact, the Nelson–Siegel model did not fit even as well as the quadratic polynomial model.

The Svensson model is likely to fit better than the Nelson–Siegel model, but the four-parameter cubic polynomial and quadratic spline models fit sufficiently well that it did not seem worthwhile to try the six-parameter Svensson model.   □

11.4 Transform-Both-Sides Regression

Suppose we have a theoretical model that states that in the absence of any noise,

$$Y_i = f(\boldsymbol{X}_i; \boldsymbol{\beta}). \tag{11.18}$$

Model (11.18) is identical to the model

$$h\{Y_i\} = h\{f(\boldsymbol{X}_i; \boldsymbol{\beta})\}, \tag{11.19}$$

where h is *any* one-to-one function, such as, a strictly increasing function. In the absence of noise, one choice of h is as good as any other and one might as well stick with model (11.18), but when noise exists, this is no longer true.

When we have noisy data, Eq. (11.19) can be converted to the nonlinear regression model

$$h\{Y_i\} = h\{f(\boldsymbol{X}_i; \boldsymbol{\beta})\} + \epsilon_i. \tag{11.20}$$

Model (11.20) is called *the transform-both-sides (TBS) regression model* because both sides of Eq. (11.19) have been transformed by the same function h. Typically, h will be one of the Box–Cox transformations and h is chosen to stabilize the variation and to induce nearly normally distributed errors. To estimate $\boldsymbol{\beta}$ for a fixed h, one minimizes

$$\sum_{i=1}^{n} \left[h\{Y_i\} - h\left\{ f(\boldsymbol{X}_i; \widehat{\boldsymbol{\beta}}) \right\} \right]^2. \tag{11.21}$$

Various choices of h can be compared by residual plots. The h that gives approximately normally distributed residuals with a constant variance is used for the final analysis.

Example 11.4. TBS regression for the default frequency data

TBS regression was applied to the default frequency data. The Box–Cox transformation $h(y) = y^{(\alpha)}$ was tried with various positive values of α. It was found that $\alpha = 1/2$ gave residuals that appeared normally distributed with a constant variance, so the square-root transformation was used for estimation; see Fig. 11.6. With this transformation, $\boldsymbol{\beta}$ is estimated by minimizing

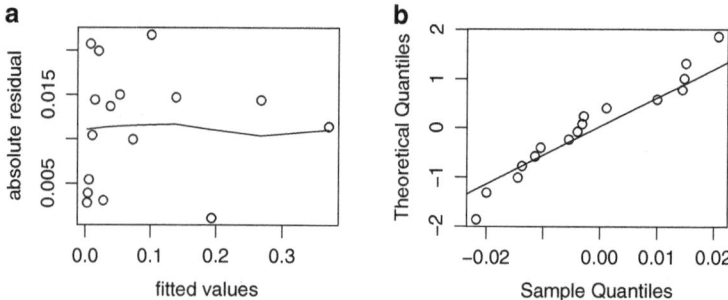

Fig. 11.6. (a) Transform-both-sides regression (TBS) with $h(y) = \sqrt{y}$. Absolute studentized residuals plotted against fitted values with a loess smooth. (b) Normal plot of the studentized residuals.

$$\sum_{i=1}^{n} \left[\sqrt{Y_i} - \exp\{\beta_0/2 + (\beta_1/2)X_i\} \right]^2, \tag{11.22}$$

where Y_i is the ith default frequency and X_i is the ith rating. The square-root transformation of the model is accomplished by dividing β_0 and β_1 by 2.

The R code to fit the TBS model and create Fig. 11.6 is below. The fitted values fit_tbs are computed by subtracting the residuals from the responses; this is done because the function summary() does not return the fitted values.

```
DefaultData = read.table("DefaultData.txt", header = TRUE)
attach(DefaultData)
freq2 = freq / 100
fit_tbs = nls(sqrt(freq2) ~ exp(b1 / 2 + b2 * rating / 2),
    start = list(b1 = -6, b2 = 0.5))
sum_tbs = summary(fit_tbs)
par(mfrow = c(1, 2))
fitted_tbs = sqrt(freq2) - sum_tbs$resid
plot(fitted_tbs,abs(sum_tbs$resid), xlab = "fitted values",
    ylab = "absolute residual")
fit_loess_tbs  = loess( abs(sum_tbs$resid) ~ fitted_tbs,
    span = 1, deg = 1)
ord_tbs = order(fitted_tbs)
lines(fitted_tbs[ord_tbs], fit_loess_tbs$fit[ord_tbs])
qqnorm(sum_tbs$resid, datax = TRUE, main = "")
qqline(sum_tbs$resid, datax = TRUE)
detact(DefaultData)
```

Using TBS regression, the estimated default probability of Aaa bonds is 0.0008 %, only 16 % of the estimate given by Bluhm, Overbeck, and Wagner (2003) and only 40 % of the estimate given by nonlinear regression without a transformation. Of course, a reduction in estimated risk by 84 % is a huge change. This shows how proper statistical modeling—e.g., using all the data and an appropriate transformation—can have a major impact on financial risk analysis. TBS allows one to use all the data (for unbiasedness) and, as described next, to effectively weight the data by the reciprocals of their variances for high efficiency.

□

11.4.1 How TBS Works

TBS in effect weights the data. To appreciate this, we use a Taylor series linearization[4] to obtain

$$\sum_{i=1}^{n} \left[h(Y_i) - h\left\{ f(\boldsymbol{X}_i; \widehat{\boldsymbol{\beta}}) \right\} \right]^2 = \sum_{i=1}^{n} \left[h^{(1)}\left\{ f(\boldsymbol{X}_i; \widehat{\boldsymbol{\beta}}) \right\} \right]^2 \left\{ Y_i - f(\boldsymbol{X}_i; \widehat{\boldsymbol{\beta}}) \right\}^2.$$

The weight of the ith observation is $\left[h^{(1)}\{f(\boldsymbol{X}_i; \widehat{\boldsymbol{\beta}})\} \right]^2$. Since the best weights are inverse variances, the most appropriate transformation h solves

$$\text{Var}(Y_i|\boldsymbol{X}_i) \propto \left[h^{(1)}\{f(\boldsymbol{X}_i; \widehat{\boldsymbol{\beta}})\} \right]^{-2}. \tag{11.23}$$

For example, if $h(y) = \log(y)$, then $h^{(1)}(y) = 1/y$ and (11.23) becomes

$$\text{Var}(Y_i|\boldsymbol{X}_i) \propto \{f(\boldsymbol{X}_i; \widehat{\boldsymbol{\beta}})\}^2, \tag{11.24}$$

so that the conditional standard deviation of the response is proportional to its conditional mean. This occurs frequently. For example, if the response is exponentially distributed then (11.24) must hold. Equation (11.24) holds also if the response is lognormally distributed and the log-variance is constant. In this case, it is not surprising that the log transformation is best since the log transforms to i.i.d. normal noise.

The *coefficient of variation* of a random variable is the ratio of its standard deviation to its expected value. When (11.24) holds, the response has a constant coefficient of variation.

A transformation that causes that conditional variance to be constant is called the *variance-stabilizing transformation*. We have just shown that when the coefficient of variation is constant, then the variance-stabilizing transformation is the logarithm.

Example 11.5. Poisson responses

Assume $Y_i|\boldsymbol{X}_i$ is Poisson distributed with mean $f(\boldsymbol{X}_i; \boldsymbol{\beta})$, as might, for example, happen if Y_i were of the number of companies declaring bankruptcy

---

[4] A Taylor series linearization of the function h about the point x is $h(y) \approx h(x) + h^{(1)}(x)(y - x)$, where $h^{(1)}$ is the first derivative of h. See any calculus textbook for further discussion of Taylor series.

in a year, with $f(\boldsymbol{X}_i; \boldsymbol{\beta})$ modeling how that expected number depends on macroeconomic variables in $\boldsymbol{X}_i$. The variance equals the mean for the Poisson distribution, so

$$\text{Var}(Y_i | \boldsymbol{X}_i) = f(\boldsymbol{X}_i; \boldsymbol{\beta}).$$

Using the same type of reasoning as in the previous example, it follows that one should use $\alpha = 1/2$; the square-root transformation is the variance-stabilizing transformation for Poisson-distributed responses.   □

11.5 Transforming Only the Response

The so-called Box–Cox transformation model is

$$Y_i^{(\alpha)} = \beta_0 + X_{i,1}\beta_1 + \cdots + X_{i,p}\beta_p + \epsilon_i, \qquad (11.25)$$

where $\epsilon_1, \ldots, \epsilon_n$ are i.i.d. $N(0, \sigma_\epsilon^2)$ for some σ_ϵ. In contrast to the TBS model, only the response is transformed. The goal of transforming the response is to achieve three objectives:

1. a simple model: $Y_i^{(\alpha)}$ is linear in predictors $X_{i,1}, \ldots, X_{i,p}$ and in the parameters $\beta_1, \ldots, \beta_p$;
2. constant residual variance; and
3. Gaussian noise.

In contrast, 2 and 3 but *not* 1 are the goals of the TBS model.

   Model (11.25) was introduced by Box and Cox (1964) who suggested estimation of α by maximum likelihood. The function boxcox() in R's MASS package will compute the profile log-likelihood for α along with a confidence interval. Usually, $\widehat{\alpha}$ is taken to be some round number, e.g., $-1, -1/2, 0, 1/2$, or 1, in the confidence interval. The reason for selecting one of these numbers is that then the transformation is readily interpretable, that is, it is the square root, log, inverse, or some other familiar function. Of course, one can use the value of α that maximizes the profile log-likelihood if one is not concerned with having a familiar transformation. After $\widehat{\alpha}$ has been selected in this way, $\beta_0, \ldots, \beta_p$ and σ_ϵ^2 can be estimated by regressing $Y_i^{(\widehat{\alpha})}$ on $X_{i,1}, \ldots, X_{i,p}$.

Example 11.6. Simulated data—Box Cox transformation

   This example uses the simulated data introduced in Example 10.6. The model is

$$Y_i^{(\alpha)} = \beta_0 + \beta_1 X_{i,1} + \beta_2 X_{i,1}^2 + \beta_3 X_{i,2} + \epsilon_i. \qquad (11.26)$$

The profile likelihood for α was produced by the boxcox() function in R and is plotted in Fig. 11.7. The code to produce the figure is:

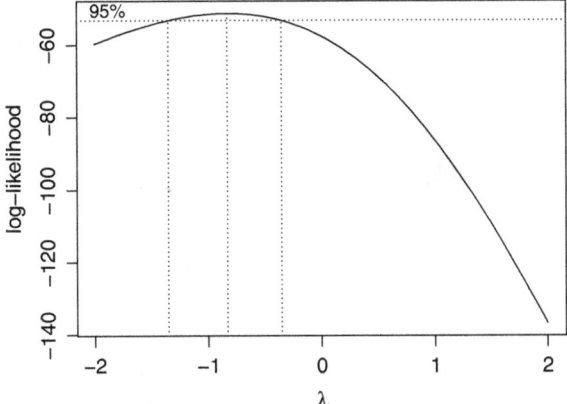

Fig. 11.7. *Profile likelihood for the Box–Cox model applied to the simulated data.*

```
boxcox(y ~ poly(x1,2) + x2, ylab = "log-likelihood")
```

We see that the MLE is near -1 and -1 is well within the confidence interval; these results suggest that we use $-1/Y_i$ as the response.

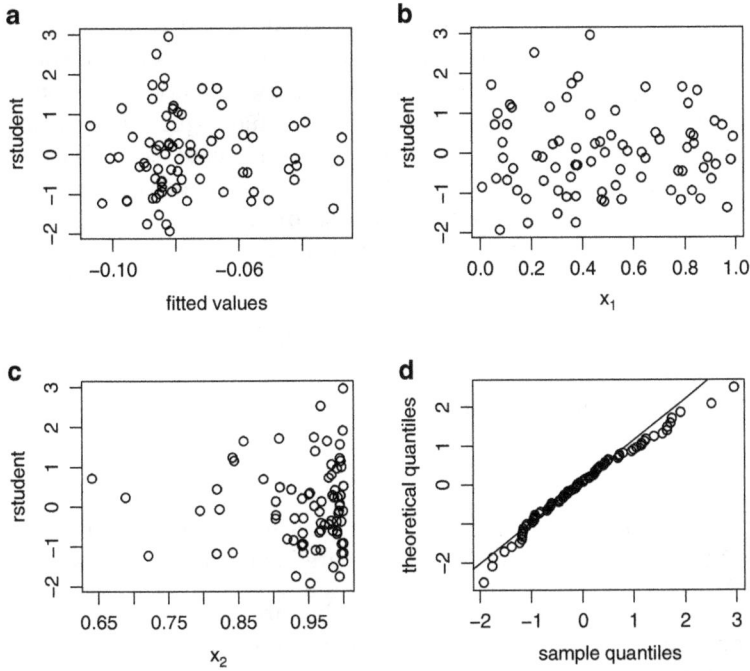

Fig. 11.8. *Residuals for the Box–Cox model applied to the simulated data.*

Residual plots with response $-1/Y_i$ are shown in Fig. 11.8. We see in panel (a) that there is no sign of heteroskedasticity, since the vertical scatter of the residuals does not change from left to right. In panels (b) and (c) we see uniform vertical scatter which shows that the model that is quadratic in X_1 and linear in X_2 fits $-1/Y_i$ well. Finally, in panel (d), we see that the residuals appear normally distributed.   □

11.6 Binary Regression

A binary response Y can take only two values, 0 or 1, which code two possible outcomes, for example, that a company goes into default on its loans or that it does not default. Binary regression models the conditional probability that a binary response is 1, given the values of the predictors $X_{i,1}, \ldots, X_{i,p}$. Since a probability is constrained to lie between 0 and 1, a linear model is not appropriate for a binary response. However, linear models are so convenient that one would like a model that has many of the features of a linear model. This has motivated the development of *generalized linear models*, often called GLMs.

Generalized linear models for binary responses are of the form

$$P(Y_i = 1 | X_{i,1}, \ldots, X_{i,p}) = H(\beta_0 + \beta_1 X_{i,1} + \cdots + \beta_p X_{i,p}) = H(\boldsymbol{x}_i^\mathsf{T} \boldsymbol{\beta}),$$

where $H(x)$ is a function that increases from 0 to 1 as x increases from $-\infty$ to ∞, so that $H(x)$ is a CDF, and the last expression uses the vector notation of (11.1). The most common GLMs for a binary responses are probit regression, where $H(x) = \Phi(x)$, the $N(0,1)$ CDF; and logistic regression, where $H(x)$ is the logistic CDF, which is $H(x) = 1/\{1 + \exp(-x)\}$. The parameter vector $\boldsymbol{\beta}$ can be estimated by maximum likelihood. Assume that conditional on $\boldsymbol{x}_1, \ldots, \boldsymbol{x}_n$ the binary responses $Y_1, \ldots, Y_n$ are mutually independent. Then, using (A.8), the likelihood (conditional on $\boldsymbol{x}_1, \ldots, \boldsymbol{x}_n$) is

$$\prod_{i=1}^{n} H\left(\boldsymbol{x}_i^\mathsf{T}\boldsymbol{\beta}\right)^{Y_i} \left\{1 - H(\boldsymbol{x}_i^\mathsf{T}\boldsymbol{\beta})\right\}^{1-Y_i}. \tag{11.27}$$

The MLEs can be found by standard software, e.g., the function `glm()` in R.

Example 11.7. Who gets a credit card?

In this example, we will analyze the data in the `CreditCard` data set in R's `AER` package. The following variables are included in the data set:

1. `card` = Was the application for a credit card accepted?
2. `reports` = Number of major derogatory reports
3. `income` = Yearly income (in USD 10,000)
4. `age` = Age in years plus 12ths of a year

5. `owner` = Does the individual own his or her home?
6. `dependents` = Number of dependents
7. `months` = Months living at current address
8. `share` = Ratio of monthly credit card expenditure to yearly income
9. `selfemp` = Is the individual self-employed?
10. `majorcards` = Number of major credit cards held
11. `active` = Number of active credit accounts
12. `expenditure` = Average monthly credit card expenditure

The first variable, `card`, is binary and will be the response. Variables 2–8 will be used as predictors. The goal of the analysis is to discover which of the predictors influences the probability that an application is accepted. R's documentation mentions that there are some values of the variable `age` under one year. These cases must be in error and they were deleted from the analysis. Figure 11.9 contains histograms of the predictors. The variable `share` is highly right-skewed, so `log(share)` will be used in the analysis. The variable `reports`

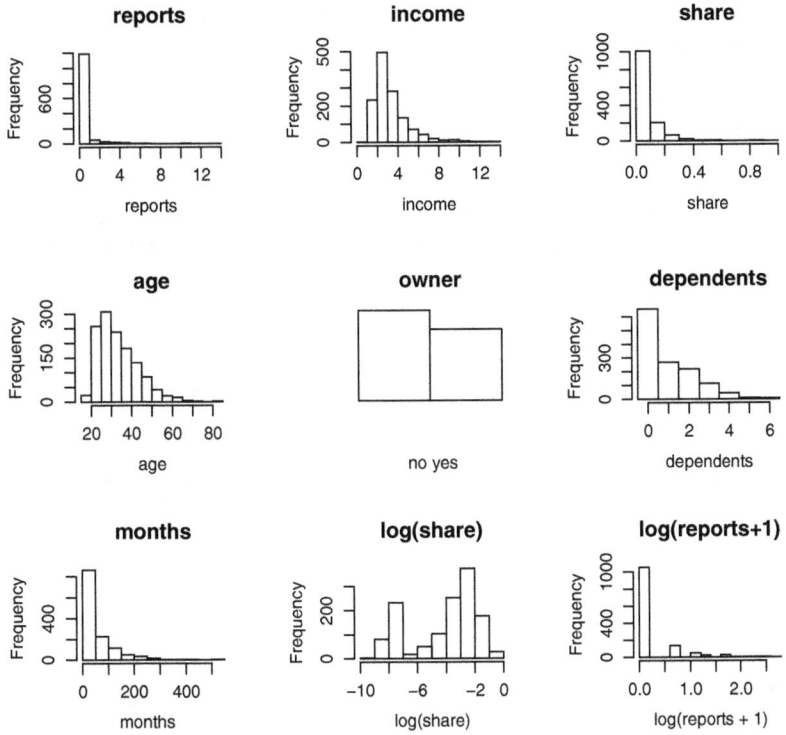

Fig. 11.9. *Histograms of variables for potential use in a model to predict whether a credit card application will be accepted.*

is also extremely right-skewed; most values of reports are 0 or 1 but the maximum value is 14. To reduce the skewness, log(reports+1) will be used instead of reports. The "1" is added to avoid taking the logarithm of 0. There are no assumptions in regression about the distributions of the predictors, so skewed predictor variables can, in principle, be used. However, highly skewed predictors have high-leverage points and are less likely to be linearly related to the response. It is a good idea at least to consider transformation of highly skewed predictors. In fact, the logistic model was also fit with reports and share untransformed, but this increased AIC by more than 3 compared to using the transformed predictors.

First, a logistic regression model is fit with all seven predictors using the glm() function. The R code is:

```
library("AER")
library("faraway")
data("CreditCard")
CreditCard_clean = CreditCard[CreditCard$age > 18, ]
names(CreditCard)
fit1 = glm(card ~ log(reports + 1) + income + log(share) + age
    + owner + dependents + months,
    family = "binomial", data = CreditCard_clean)
summary(fit1)
stepAIC(fit1)

Call:
glm(formula = card ~ log(reports + 1) + income + log(share) +
    age + owner + dependents + months, family = "binomial",
    data = CreditCard_clean)

Coefficients:
                    Estimate Std. Error z value Pr(>|z|)
(Intercept)        21.473930   3.674325   5.844 5.09e-09 ***
log(reports + 1)   -2.908644   1.097604  -2.650  0.00805 **
income              0.903315   0.189754   4.760 1.93e-06 ***
log(share)          3.422980   0.530499   6.452 1.10e-10 ***
age                 0.022682   0.021895   1.036  0.30024
owneryes            0.705171   0.533070   1.323  0.18589
dependents         -0.664933   0.267404  -2.487  0.01290 *
months             -0.005723   0.003988  -1.435  0.15130
---

(Dispersion parameter for binomial family taken to be 1)

    Null deviance: 1398.53  on 1311  degrees of freedom
Residual deviance:  139.79  on 1304  degrees of freedom
AIC: 155.79

Number of Fisher Scoring iterations: 11
```

Several of the regressors have large p-values, so stepAIC() was used to find a more parsimonious model. The final step where no more variables were deleted is

```
Step:  AIC=154.22
card ~ log(reports + 1) + income + log(share) + dependents

                    Df Deviance     AIC
<none>                   144.22  154.22
- dependents         1   150.28  158.28
- log(reports + 1)   1   164.18  172.18
- income             1   173.62  181.62
- log(share)         1  1079.61 1087.61
```

Below is the fit using the model selected by stepAIC(). For convenience later, each of the regressors was mean-centered; "_c" appended to a variable name indicates centering.

```
glm(formula = card ~ log_reports_c + income_c + log_share_c +
    dependents_c, family = "binomial", data = CreditCard_clean)

Coefficients:
              Estimate Std. Error z value Pr(>|z|)
(Intercept)     9.5238     1.7213   5.533 3.15e-08 ***
log_reports_c  -2.8953     1.0866  -2.664  0.00771 **
income_c        0.8717     0.1724   5.056 4.28e-07 ***
log_share_c     3.3102     0.4942   6.698 2.11e-11 ***
dependents_c   -0.5506     0.2505  -2.198  0.02793 *
---

(Dispersion parameter for binomial family taken to be 1)

    Null deviance: 1398.53  on 1311  degrees of freedom
Residual deviance:  144.22  on 1307  degrees of freedom
AIC: 154.22

Number of Fisher Scoring iterations: 11
```

It is important to understand what the logistic regression model is telling us about the probability of an application being accepted. Qualitatively, we see that the probability of having an application accepted increases with income and share and decreases with reports and dependents. To understand these effects quantitatively, first consider the intercept. Since the predictors have been mean-centered, the probability of an application being accepted when all variables are at their mean is simply $H(9.5238) = 0.999927$. Since reports and dependents are integer-valued and cannot exactly equal their means, this probability only provides an idea of what the intercept 9.5238 signifies. Figure 11.10 plots the probability that a credit card application is accepted as functions of `reports`, `income`, `log(share)`, and `dependents`. In each plot, the other variables are fixed at their means. Clearly, the variable with the largest effect is `share`, the ratio of monthly credit card expenditure to yearly income. We see that applicants who spend little of their income through credit cards are unlikely to have their applications accepted.

In Fig. 11.11, panel (a) is a plot of `card`, which takes value 0 if an application is rejected and 1 if it is accepted, versus `log(share)`. It should be emphasized that panel (a) is a plot of the data, not a fit from the model. We see that an application is always accepted if `log(share)` exceeds -6, which translates into `share` exceeding 0.0025. Thus, in this data set, among the group of applicants whose average monthly credit card expenses exceeded 0.25 % of yearly income, all credit card applications were accepted. How do

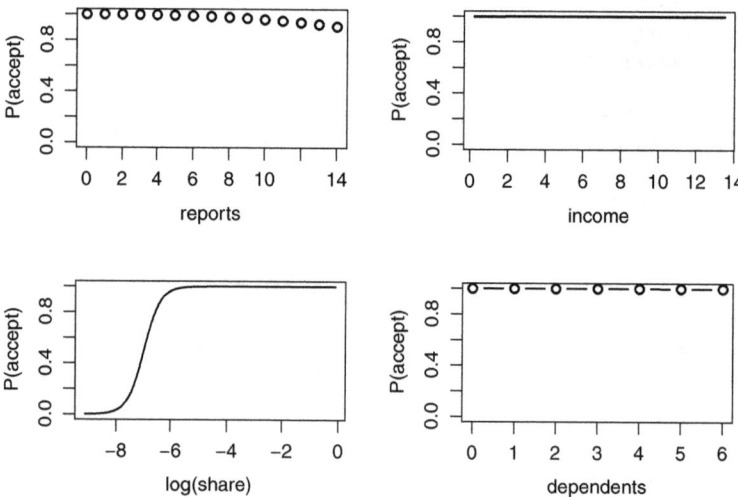

Fig. 11.10. *Plots of probabilities of a credit card application being accepted as functions of single predictors with other predictors fixed at their means. The variables vary over their ranges in the data.*

these applicants look on the other variables? Panels (b)–(d) plot `reports`, `income`, and `majorcards` versus `log(share)`. The variable `majorcards` was not used in the logistic regression analysis but is included here.

An odd feature in Fig. 11.11c is a group of points following a smooth curve. This is a group of 316 applications who had the product of `share` times `income` exactly equal to 0.0012, the minimum value of this product. Oddly, `share` is never 0. Perhaps because of some coding artifact, these 316 had 0 credit card expenditures rather than the reported values. Another interesting feature of the data is that among these 316 applications, only 21 were accepted. Among the remaining 996 applications, all were accepted.

Besides illustrating logistic regression, this example demonstrates that real-world data often contain errors, or perhaps we should call them idiosyncracies, and that a thorough graphical analysis of the data is always a good thing.   □

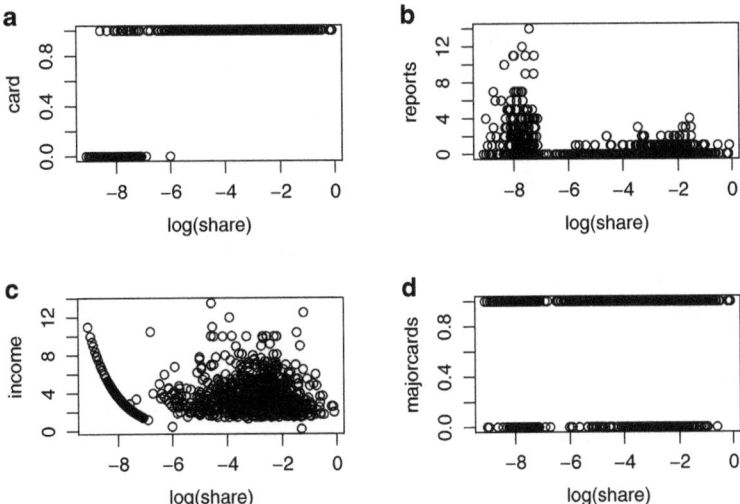

Fig. 11.11. *Plots of* `log(share)` *versus other variables.*

11.7 Linearizing a Nonlinear Model

Sometimes a nonlinear model can be linearized by applying a transformation to both the model and the response. In such cases, should one use a linearizing transformation or, instead, apply nonlinear regression to the original model? The answer is that linearization can sometimes be a good thing, but not always. Fortunately, residual analysis can help us decide whether a linearizing transformation should be used.

For example, consider the model

$$Y_i = \beta_1 \exp(\beta_2 X_i). \tag{11.28}$$

This model is "equivalent" to the linear model

$$\log(Y_i) = \alpha + \beta_2 X_i, \tag{11.29}$$

where $\alpha = \log(\beta_1)$. "Equivalent" is in quotes, because the two models are no longer equivalent when noise is present.

Suppose (11.28) has i.i.d. additive noise, so that

$$Y_i = \beta_1 \exp(\beta_2 X_i) + \epsilon_i, \tag{11.30}$$

where $\epsilon_1, \ldots, \epsilon_n$ are i.i.d. Then applying the log transformation to (11.29) gives us the model

$$\log(Y_i) = \log\{\beta_1 \exp(\beta_2 X_i) + \epsilon_i\} \tag{11.31}$$

with nonadditive noise. Because the noise is not additive, the variation of $\log(Y_i)$ about the model $\log\{\beta_1 \exp(\beta_2 X_i)\}$ will have nonconstant variation and skewness, even if $\epsilon_1, \ldots, \epsilon_n$ are i.i.d. Gaussian.

Example 11.8. Linearizing transformation—Simulated data

Figure 11.12a shows a simulated sample from model (11.28) with $\beta_1 = 1$, $\beta_2 = -1$, and $\sigma_\epsilon = 0.02$. The X_i are equally spaced from -1 to 2.5 by increments of 0.025. Panel (b) shows $\log(Y_i)$ plotted against X_i. One can see that the transformation has linearized the relationship between the variables but has introduced nonconstant residual variation. Panels (c) and (d) show residual plots using the linearized model. Notice the nonlinear normal plot and the severe nonconstant variance. □

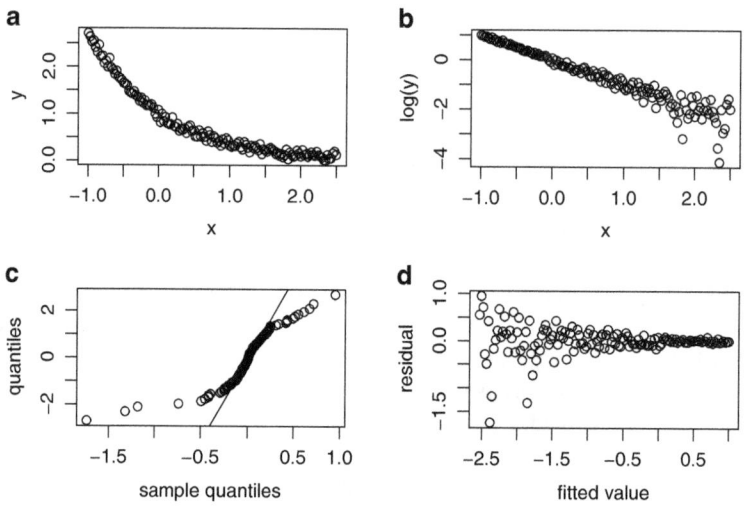

Fig. 11.12. *Example where the log transformation linearizes a model but induces substantial heteroskedasticity and skewness. (a) Raw data. (b) Data after log transformation of the response. (c) Normal plot of residuals after linearization. (d) Absolute residual plot after linearization.*

Linearizing is not always a bad thing. Suppose the noise is multiplicative and lognormal so that (11.28) becomes

$$Y_i = \beta_1 \exp(\beta_2 X_i) \exp(\epsilon_i) = \beta_1 \exp(\beta_2 X_i + \epsilon_i), \qquad (11.32)$$

where $\epsilon_1, \ldots, \epsilon_n$ are i.i.d. Gaussian. Then the log transformation converts (11.32) to

$$\log(Y_i) = \alpha + \beta_2 X_i + \epsilon_i, \qquad (11.33)$$

which is a linear model satisfying all of the usual assumptions.

In summary, a linearizing transformation may or may not cause the data to better follow the assumptions of regression analysis. Residual analysis can help one decide whether a transformation is appropriate.

11.8 Robust Regression

A robust regression estimator should be relatively immune to two types of outliers. The first are *bad data*, meaning *contaminants* that are not part of the population, for example, due to undetected recording errors. The second are outliers due to the noise distribution having heavy tails. There are a large number of robust regression estimators, and their sheer number has been an impediment to their use. Many data analysts are confused as to which robust estimator is best and consequently are reluctant to use any. Rather than describe many of these estimators, which might contribute to this problem, we mention just one, the *least-trimmed sum of squares estimator*, often called the *LTS*.

Recall the trimmed mean, a robust estimator of location for a univariate sample. The trimmed mean is simply the mean of the sample after a certain percentage of the largest observations and the same percentage of the smallest observations have been removed. This trimming removes some non-outliers, which, under the ideal conditions of no outliers, causes some loss of precision, but not an unacceptable amount. The trimming also removes outliers, and this causes the estimator to be robust. Trimming is easy for a univariate sample because we know which observations to trim, the very largest and the very smallest. This is not the case in regression. Consider the data in Fig. 11.13. There are 26 observations that fall closely along a line plus two *residual outliers* that are far from this line. Notice that the residual outliers have neither extreme X-values nor extreme Y-values. They are outlying only relative to the linear regression fit to the other data.

The residual outliers are obvious in Fig. 11.13 because there is only a single predictor. When there are many predictors, outliers can only be identified when we have a model *and* good estimates of the parameters in that model. The difficulty, then, is that estimation of the parameters requires the identification of the outliers, and vice versa. One can see from the figure that the least-squares line is changed by including the residual outliers in the data used for estimation. In some cases, e.g., Fig. 10.1b, the effect of a residual outlier can be so severe that it totally changes the least-squares estimates. This is likely to happen if the residual outlier occurs at a high-leverage point.

The LTS estimator simultaneously identifies residual outliers and estimates robustly the parameters of a model. Let $0 < \alpha \leq 1/2$ be the trimming proportion and let k equal $n\alpha$ rounded to an integer. The trimmed sum of squares about a set of values of the regression parameters is defined as follows: Form the residuals from the model evaluated at these parameters, square the residuals, then order the squared residuals and remove the k largest, and finally sum the remaining squared residuals. The LTS estimates are the set of parameter values that minimize the trimmed sum of squares. The LTS estimator can be computed using the ltsReg() function in R's robust package.

If the noise distribution is heavy-tailed, then an alternative to a robust regression analysis is to use a heavy-tailed distribution as a model for the noise and then to estimate the parameters by maximum likelihood. For example, one could assume that the noise has a double-exponential or t-distribution. In the latter case, one could either estimate the degrees of freedom or simply fix the degrees of freedom at a low value, which implies heavier tails; see Lange, Little, and Taylor (1989). This strategy is called *robust modeling* rather than robust estimation. The distinction is that in robust estimation one assumes a fairly restrictive model such as a normal noise distribution, but finds a robust alternative to maximum likelihood. In robust modeling, one uses a more flexible model so that maximum likelihood estimation is itself robust. When there is a single gross residual outlier, particularly at a high-leverage point, robust regression is a better alternative than the MLE with a heavy-tailed noise distribution; see the next example.

Another possibility is that residual outliers are due to nonconstant standard deviations, with the outliers mainly in the data with a higher noise standard deviation. The remedy to this problem is to apply a variance stabilization transformation or to model the nonconstant standard deviation, say by one of the GARCH models discussed in Chap. 14.

Example 11.9. Simulated data in Example 10.1—Robust regression

Figure 11.14 compares least-squares fit, the LTS fit, and the MLE assuming t-distributed noise for the simulated data in Example 10.1. In panel (a) with no residuals outliers, the three fits coincide. In panels (b) and (c), the LTS fits are not affected by the residual outliers and fit the nonoutlying data very well. In these panels, the LS and MLE fits are highly affected by the outlier and nearly identical. For these examples, the MLE assuming t-distributed noise is not robust.   □

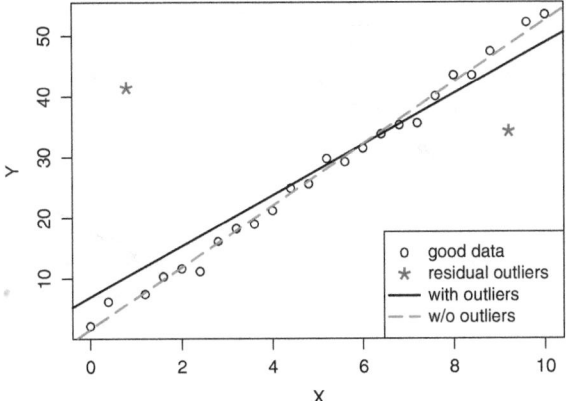

Fig. 11.13. *Straight-line regression with two residual outliers showing least-squares fits with and without the outliers.*

11.9 Regression and Best Linear Prediction

11.9.1 Best Linear Prediction

Often we observe a random variable X and we want to predict an unobserved random variable Y that is related to X. For example, Y could be the future price of an asset and X might be the most recent change in that asset's price. Prediction has many practical uses, and it is also important in theoretical studies.

The predictor of Y that minimizes the expected squared prediction error is $E(Y|X)$ (see Appendix A.19), but $E(Y|X)$ is often a nonlinear function of X and difficult to compute. A common solution to this difficulty it to consider only linear functions of X as possible predictors. This is called *linear prediction*. In this section, we will show that linear prediction is closely related to linear regression.

A linear predictor of Y based on X is a function $\beta_0 + \beta_1 X$ where β_0 and β_1 are parameters that we can choose. *Best linear prediction* means finding β_0 and β_1 so that expected squared prediction error, which is given by

$$E\{Y - (\beta_0 + \beta_1 X)\}^2, \tag{11.34}$$

is minimized. Doing this makes the predictor as close as possible, on average, to Y. The expected squared prediction error can be rewritten as

$$E\{Y - (\beta_0 + \beta_1 X)\}^2$$
$$= E(Y^2) - 2\beta_0 E(Y) - 2\beta_1 E(XY) + \beta_0^2 + 2\beta_0 \beta_1 E(X) + \beta_1^2 E(X^2).$$

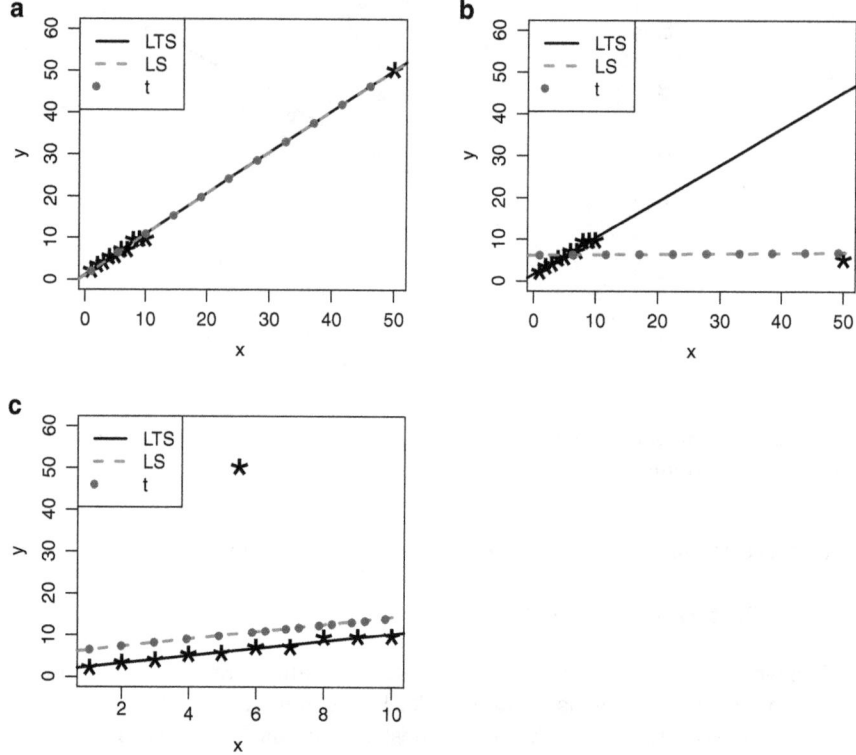

Fig. 11.14. *Simulated data in Example 10.1 with LS fits (dashed red) and LTS fits (solid black) and MLEs assuming t-distributed noise (dotted blue). In (a) the three fits are too close together to distinguish between them. In (b) and (c) the LS and t fits are nearly identical and difficult to distinguish.*

To find the minimizers, we set the partial derivatives of this expression to zero to obtain

$$0 = -E(Y) + \beta_0 + \beta_1 E(X), \tag{11.35}$$
$$0 = -E(XY) + \beta_0 E(X) + \beta_1 E(X^2). \tag{11.36}$$

After some algebra we find that

$$\beta_1 = \sigma_{XY}/\sigma_X^2 \tag{11.37}$$

and

$$\beta_0 = E(Y) - \beta_1 E(X) = E(Y) - \sigma_{XY}/\sigma_X^2 \, E(X). \tag{11.38}$$

One can check that the matrix of second derivatives of (11.34) is positive definite so that the solution (β_0, β_1) to (11.35) and (11.36) minimizes (11.34). Thus, the best linear predictor of Y is

$$\widehat{Y}^{\mathrm{Lin}}(X) = \beta_0 + \beta_1 X = E(Y) + \frac{\sigma_{XY}}{\sigma_X^2}\{X - E(X)\}. \qquad (11.39)$$

In practice, (11.39) cannot be used directly unless $E(X)$, $E(Y)$, σ_{XY}, and σ_X^2 are known, which is often not the case. Linear regression analysis is essentially the use of (11.39) with these unknown parameters replaced by least-squares estimates—see Sect. 11.9.3.

11.9.2 Prediction Error in Best Linear Prediction

In this section, assume that $\widehat{Y}$ is the best linear predictor of Y. The *prediction error* is $Y - \widehat{Y}$. It is easy to show that $E\{Y - \widehat{Y}\} = 0$ so that the prediction is unbiased. With a little algebra we can show that the expected squared prediction error is

$$E\{Y - \widehat{Y}\}^2 = \sigma_Y^2 - \frac{\sigma_{XY}^2}{\sigma_X^2} = \sigma_Y^2(1 - \rho_{XY}^2). \qquad (11.40)$$

How much does X help us predict Y? To answer this question, notice first that if we do not observe X, then we must predict Y using a constant, which we denote by c. It is easy to show that the best predictor has c equal to $E(Y)$. Notice first that the expected squared prediction error is $E(Y - c)^2$. Some algebra shows that

$$E(Y - c)^2 = \mathrm{Var}(Y) + \{c - E(Y)\}^2, \qquad (11.41)$$

which, since $\mathrm{Var}(Y)$ does not depend on c, shows that the expected squared prediction error is minimized by $c = E(Y)$. Thus, when X is unobserved, the best predictor of Y is $E(Y)$ and the expected squared prediction error is σ_Y^2, but when X is observed, then the expected squared prediction error is smaller, $\sigma_Y^2(1 - \rho_{XY}^2)$. Therefore, ρ_{XY}^2 is the fraction by which the prediction error is reduced when X is known. This is an important fact that we will see again.

Result 11.1 Prediction when Y is independent of all available information:
    If Y is independent of all presently available information, that is, Y is independent of all random variables that have been observed, then the best predictor of Y is $E(Y)$ and the expected value of the squared prediction error is σ_Y^2. We say that Y "cannot be predicted" when there exists no predictor better than its expected value.

11.9.3 Regression Is Empirical Best Linear Prediction

For the case of a single predictor, note the similarity between the best linear predictor,

$$\widehat{Y} = E(Y) + \frac{\sigma_{XY}}{\sigma_X^2}\{X - E(X)\},$$

and the least-squares line,

$$\widehat{Y} = \overline{Y} + \frac{s_{XY}}{s_X^2}(X - \overline{X}).$$

The least-squares line is a sample version of the best linear predictor. Also, ρ_{XY}^2, the squared correlation between X and Y, is the fraction of variation in Y that can be predicted using the linear predictor, and the sample version of ρ_{XY}^2 is $R^2 = r_{XY}^2 = r_{\widehat{Y}Y}^2$.

11.9.4 Multivariate Linear Prediction

So far we have assumed that there is only a single random variable, X, available to predict Y. More commonly, Y is predicted using a set of observed random variables, $X_1, \ldots, X_n$.

Let $\boldsymbol{Y}$ and $\boldsymbol{X}$ by $p \times 1$ and $q \times 1$ random vectors. As before in Sect. 7.3.1, define

$$\boldsymbol{\Sigma}_{Y,X} = E\{\boldsymbol{Y} - E(\boldsymbol{Y})\}\{\boldsymbol{X} - E(\boldsymbol{X})\}^\mathsf{T},$$

so that the i,jth element of $\boldsymbol{\Sigma}_{Y,X}$ is the covariance between Y_i and X_j. Then the best linear predictor of $\boldsymbol{Y}$ given $\boldsymbol{X}$ is

$$\widehat{\boldsymbol{Y}} = E(\boldsymbol{Y}) + \boldsymbol{\Sigma}_{Y,X}\boldsymbol{\Sigma}_X^{-1}\{\boldsymbol{X} - E(\boldsymbol{X})\}. \tag{11.42}$$

Note the similarity between (11.39) and (11.42), the best linear predictors in the univariate and multivariate cases.

The sample analog of multivariate linear prediction is multiple regression.

11.10 Regression Hedging

An interesting application of regression is determining the optimal hedge of a bond position. Market makers buy securities at a *bid price* and make a profit by selling them at a higher *ask price*. Suppose a market maker has just purchased a bond from a pension fund. Ideally, the market maker would sell the bond immediately after purchasing it. However, many bonds are illiquid, so it may take some time before the bond can be sold. During the period that a market maker is holding a bond, the market maker is at risk that the bond price could drop due to a change in interest rates. The change could wipe out the profit due to the small bid–ask spread. The market maker would prefer to hedge this risk by assuming another risk which is likely to be in the opposite direction. To hedge the interest-rate risk of the bond being held, the market maker can sell other, more liquid, bonds short. Suppose that the market maker decides to sell short a 30-year Treasury bond, which is more liquid.

Regression hedging determines the optimal amount of the 30-year Treasury bonds to sell short to hedge the risk of the bond just purchased. The goal is that the price of the portfolio long in the first bond and short in the Treasury bond changes as little as possible as yields change. Suppose the first bond has a maturity of 25 years. One can determine the sensitivity of price to yield changes using results from Sect. 3.8. Let y_{30} be the yield on 30-year bonds, let P_{30} be the price of \$1 in face amount of 30-year bonds, and let DUR_{30} be the duration. The change in price, ΔP_{30}, and the change in yield, Δy_{30}, are related by

$$\Delta P_{30} \approx -P_{30}\,\mathrm{DUR}_{30}\,\Delta y_{30}$$

for small values of Δy_{30}. A similar result holds for 25-year bonds.

Consider a portfolio that holds face amount F_{25} in 25-year bonds and is short face amount F_{30} in 30-year bonds. The value of the portfolio is

$$F_{25}P_{25} - F_{30}P_{30}.$$

If Δy_{25} and Δy_{30} are the changes in the yields, then the change in value of the portfolio is approximately

$$\{F_{30}P_{30}\,\mathrm{DUR}_{30}\,\Delta y_{30} - F_{25}P_{25}\,\mathrm{DUR}_{25}\,\Delta y_{25}\}. \tag{11.43}$$

Suppose that the regression of Δy_{30} on Δy_{25} is

$$\Delta y_{30} = \widehat{\beta}_0 + \widehat{\beta}_1 \Delta y_{25} \tag{11.44}$$

and $\widehat{\beta}_0 \approx 0$, as is usually the case for regression of changes in interest rates, as in Example 9.1. Substituting (11.44) into (11.43), the change in price of the portfolio is approximately

$$\{F_{30}P_{30}\,\mathrm{DUR}_{30}\widehat{\beta}_1 - F_{25}P_{25}\,\mathrm{DUR}_{25}\}\Delta y_{25}. \tag{11.45}$$

This change is approximately zero for all values of Δy_{25} if

$$F_{30} = F_{25}\frac{P_{25}\,\mathrm{DUR}_{25}}{P_{30}\,\mathrm{DUR}_{30}\widehat{\beta}_1}. \tag{11.46}$$

Equation (11.46) tells us how much face value of the 30-year bond to sell short in order to hedge F_{25} face value of the 25-year bond. All quantities on the right-hand side of (11.46) are known or readily calculated: F_{25} is the current position in the 25-year bond, P_{25} and P_{30} are known bond prices, calculation of DUR_{25} and DUR_{30} is discussed in Chap. 3, and $\widehat{\beta}_1$ is the slope of the regression of Δy_{30} on Δy_{25}.

The higher the R^2 of the regression, the better the hedge works. Hedging with two or more liquid bonds, say a 30-year and a 10-year, can be done by multiple regression and might produce a better hedge.

11.11 Bibliographic Notes

Atkinson (1985) has nice coverage of transformations and residual plotting and many good examples. For more information on nonlinear regression, see Bates and Watts (1988) and Seber and Wild (1989). Graphical methods for detecting a nonconstant variance, transform-both-sides regression, and weighting are discussed in Carroll and Ruppert (1988). Hosmer and Lemeshow (2000) is an in-depth treatment of logistic regression. Faraway (2006) covers generalized linear models including logistic regression. See Tuckman (2002) for more discussion of regression hedging.

The Nelson–Siegel and Svensson models are from Nelson and Siegel (1985) and Svensson (1994).

11.12 R Lab

11.12.1 Nonlinear Regression

In this section, you will be fitting short-rate models. Let r_t be the short rate (the risk-free rate for short-term borrowing) at time t. It is assumed that the short rate satisfies the stochastic differential equation

$$dr_t = \mu(t, r_t)\,dt + \sigma(t, r_t)\,dW_t, \tag{11.47}$$

where $\mu(t, r_t)$ is a drift function, $\sigma(t, r_t)$ is a volatility function, and W_t is a standard Brownian motion. We will use a discrete approximation to (11.47):

$$(r_t - r_{t-1}) = \mu(t - 1, r_{t-1}) + \sigma(t - 1, r_{t-1})\,\epsilon_{t-1} \tag{11.48}$$

where $\epsilon_1, \ldots, \epsilon_{n-1}$ are i.i.d. $N(0, 1)$.

We will start with the Chan, Karolyi, Longstaff, and Sanders (1992) (CKLS) model, which assumes that

$$\mu(t, r) = \mu(r) = a\,(\theta - r) \tag{11.49}$$

for some unknown parameters a and θ, and

$$\sigma(t, r) = \sigma r^{\gamma} \tag{11.50}$$

for some σ and γ. Be careful to distinguish between the volatility function $\sigma(t, r)$ and the constant volatility parameter σ.

We will use the Irates data set in the Ecdat package. This data set has interests rates for maturities from 1 to 120 months. We will use the first column, which has the one-month maturity rates, since we want the short rate.

Run the following code to input the data, compute the lagged and differenced short-rate series, and construct some basic plots.

```
library(Ecdat)
data(Irates)
r1 = Irates[,1]
n = length(r1)
lag_r1 = lag(r1)[-n]
delta_r1 = diff(r1)
n = length(lag_r1)
par(mfrow = c(3, 2))
plot(r1, main = "(a)")
plot(delta_r1, main = "(b)")
plot(delta_r1^2, main = "(c)")
plot(lag_r1, delta_r1, main = "(d)")
plot(lag_r1, delta_r1^2, main = "(e)")
```

Problem 1 *What is the maturity of the interest rates in the first column? What is the sampling frequency of this data set—daily, weekly, monthly, or quarterly? What country are the data from? Are the rates expressed as percentages or fractions (decimals)?*

In the plot you have just created, panels (a), (b), and (c) show how the short rate, changes in the short rate, and squared changes in the short rate depend on time. The plots of changes in the short rate are useful for choosing the drift $\mu(t-1, r_{t-1})$ while squared changes in the short rate are helpful for selecting the volatility $\sigma(t-1, r_{t-1})$.

Problem 2 *Model (11.49) states that $\mu(t, r) = \mu(r)$, that is, that the drift does not depend on t. Use your plots to discuss whether this assumption seems valid. Assuming for the moment that this assumption is valid, any trend in the plot in panel (d) would give us information about the form of $\mu(r)$. Do you see any trend?*

Now run the following code to fit model (11.49) and fill in the first two panels of a figure. This figure will be continued next.

```
#  CKLS (Chan, Karolyi, Longstaff, Sanders)

nlmod_CKLS = nls(delta_r1 ~ a * (theta-lag_r1),
    start=list(theta = 5,    a = 0.01),
    control = list(maxiter = 200))
param = summary(nlmod_CKLS)$parameters[ , 1]
par(mfrow = c(2, 2))
t = seq(from = 1946, to = 1991 + 2 / 12, length = n)
plot(lag_r1, ylim = c(0, 16), ylab = "rate and theta",
    main = "(a)", type = "l")
abline(h = param[1], lwd = 2, col = "red")
```

Problem 3 *What are the estimates of a and θ and their 95 % confidence intervals?*

Note that the nonlinear regression analysis estimates $\sigma^2(r)$, not $\sigma(r)$, since the response variable is the squared residual. Here $A = \sigma^2$ and $B = 2\gamma$.

```
res_sq = residuals(nlmod_CKLS)^2
nlmod_CKLS_res <- nls(res_sq ~ A*lag_r1^B,
    start = list(A = 0.2, B = 1/2))
param2 = summary(nlmod_CKLS_res)$parameters[ , 1]
plot(lag_r1, sqrt(res_sq), pch = 5, ylim = c(0, 6),
    main = "(b)")
attach(as.list(param2))
curve(sqrt(A * x^B), add = T, col = "red", lwd = 3)
```

Problem 4 *What are the estimates of σ and γ and their 95 % confidence intervals?*

Finally, refit model (11.49) using weighted least squares.

```
nlmod_CKLS_wt = nls(delta_r1 ~ a * (theta-lag_r1),
    start = list(theta = 5,  a = 0.01),
    control = list(maxiter = 200),
    weights = 1 / fitted(nlmod_CKLS_res))

plot(lag_r1, ylim = c(0, 16), ylab = "rate and theta",
    main = "(c)", type = "l")
param3 = summary(nlmod_CKLS_wt)$parameters[ , 1]
abline(h = param3[1], lwd = 2, col = "red")
```

Problem 5 *How do the unweighted estimate of θ shown in panel (a) and the weighted estimate plotted in panel (d) differ? Why do they differ in this manner?*

11.12.2 Response Transformations

This section uses the `HousePrices` data set in the `AER` package. This is a cross-sectional data set on house prices and other features, e.g., the number of bedrooms of houses in Windsor, Ontario. The data were gathered during the summer of 1987. Accurate modeling of house prices is important for the mortgage industry. Run the code below to read the data and regress `price` on the other variables; the period on the right-hand side of the formula "`price~.`" specifies that the predictors should include all variables except, of course, the response.

```
library(AER)
data(HousePrices)
fit1 = lm(price ~ ., data = HousePrices)
summary(fit1)
```

Next construct a profile log-likelihood plot for the transformation parameter α in model (11.25)

```
library(MASS)
fit2 = boxcox(fit1, xlab = expression(alpha))
```

Problem 6 *What is the MLE of α? (Hint: Type* ?boxcox *to learn what is returned by this function.)*

Next, fit a linear model with price transformed by $\hat{\alpha}$ (the MLE). Here the function bcPower() in the AER package computes a Box–Cox transformation for a given value of α and must be distinguished from boxcox(), which computes the profile log-likelihood for α. In the following code, replace $1/2$ by the MLE of α.

```
library(car)
alphahat = 1/2
fit3 = lm(bcPower(price, alphahat) ~ ., data = HousePrices)
summary(fit3)
AIC(fit1)
AIC(fit3)
```

Problem 7 *Does the Box–Cox transformation offer a substantial improvement in fit compared to the regression with no transformation of* price*?*

Problem 8 *Would it be worthwhile to check the residuals for correlation?*

11.12.3 Binary Regression: Who Owns an Air Conditioner?

This section uses the HousePrices data set used in Sect. 11.12.2. The goal here is to investigate how the presence or absence of air conditioning is related to the other variables. The code below fits a logistic regression model to all potential predictor variables and then uses stepAIC() to find a parsimonious model.

```
library(AER)
data(HousePrices)
fit1 = glm(aircon ~ ., family = "binomial",
   data = HousePrices)
summary(fit1)
library(MASS)
fit2 = stepAIC(fit1)
summary(fit2)
```

Problem 9 *Which variables are most useful for predicting whether a home has air conditioning? Describe qualitatively the relationships between these variables and the variable* `aircon`*. Are there any variables in the model selected by* `stepAIC()` *that you think might be dropped?*

Problem 10 *Estimate the probability that a house will have air conditioning if it has the following characteristics:*

price	lotsize	bedrooms	bathrooms	stories	driveway	recreation
42000	5850	3	1	2	yes	no

fullbase	gasheat	garage	prefer
yes	no	1	no

(Hint: The R *function* `plogis()` *computes the logistic function.)*

11.13 Exercises

1. When we were finding the best linear predictor of Y given X, we derived the equations

$$0 = -E(Y) + \beta_0 + \beta_1 E(X)$$
$$0 = -E(XY) + \beta_0 E(X) + \beta_1 E(X^2).$$

Show that their solution is

$$\beta_1 = \frac{\sigma_{XY}}{\sigma_X^2}$$

and

$$\beta_0 = E(Y) - \beta_1 E(X) = E(Y) - \frac{\sigma_{XY}}{\sigma_X^2} E(X).$$

2. Suppose one has a long position of F_{20} face value in 20-year Treasury bonds and wants to hedge this with short positions in both 10- and 30-year Treasury bonds. The prices and durations of 10-, 20-, and 30-year Treasury bonds are P_{10}, DUR_{10}, P_{20}, DUR_{20}, P_{30}, and DUR_{30} and are assumed to be known. A regression of changes in the 20-year yield on changes in the 10- and 30-year yields is $\Delta y_{20} = \widehat{\beta}_0 + \widehat{\beta}_1 \Delta y_{10} + \widehat{\beta}_2 \Delta y_{30}$. The p-value of $\widehat{\beta}_0$ is large and it is assumed that β_0 is close enough to zero to be ignored. What face amounts F_{10} and F_{30} of 10- and 30-year Treasury bonds should be shorted to hedge the long position in 20-year Treasury bonds? (Express F_{10} and F_{30} in terms of the known quantities P_{10}, P_{20}, P_{30}, DUR_{10}, DUR_{20}, DUR_{30}, $\widehat{\beta}_1$, $\widehat{\beta}_2$, and F_{20}.)

3. The maturities (T) in years and prices in dollars of zero-coupon bonds are in file `ZeroPrices.txt` on the book's website. The prices are expressed

as percentages of par. A popular model is the Nelson–Siegel family with forward rate

$$r(T; \theta_1, \theta_2, \theta_3, \theta_4) = \theta_1 + (\theta_2 + \theta_3 T) \exp(-\theta_4 T).$$

Fit this forward rate to the prices by nonlinear regression using R's `optim()` function.

(a) What are your estimates of θ_1, θ_2, θ_3, and θ_4?

(b) Plot the estimated forward rate and estimated yield curve on the same figure. Include the figure with your work.

4. Least-squares estimators are unbiased in linear models, but in nonlinear models they can be biased. Simulation studies (including bootstrap resampling) can be used to estimate the amount of bias. In Example 11.1, the data were simulated with $r = 0.06$ and $\hat{r} = 0.0585$. Do you think this is a sign of bias or simply due to random variability? Justify your answer.

References

Atkinson, A. C. (1985) *Plots, Transformations and Regression*, Clarendon, Oxford.

Bates, D. M., and Watts, D. G. (1988) *Nonlinear Regression Analysis and Its Applications*, Wiley, New York.

Bluhm, C., Overbeck, L., and Wagner, C. (2003) *An Introduction to Credit Risk Modelling*, Chapman & Hall/CRC, Boca Raton, FL.

Box, G. E. P., and Dox, D. R. (1964) An analysis of transformations. *Journal of the Royal Statistical Society, Series B*, **26** 211–246.

Carroll, R. J., and Ruppert, D. (1988) *Transformation and Weighting in Regression*, Chapman & Hall, New York.

Chan, K. C., Karolyi, G. A., Longstaff, F. A., and Sanders, A. B. (1992) An empirical comparison of alternative models of the short-term interest rate. *Journal of Finance*, **47**, 1209–1227.

Faraway, J. J. (2006) *Extending the Linear Model with R*, Chapman & Hall, Boca Raton, FL.

Hosmer, D., and Lemeshow, S. (2000) *Applied Logistic Regression*, 2nd ed., Wiley, New York.

Jarrow, R. (2002) *Modeling Fixed-Income Securities and Interest Rate Options, 2nd Ed.*, Stanford University Press, Stanford, CA.

Lange, K. L., Little, R. J. A., and Taylor, J. M. G. (1989) Robust statistical modeling using the *t*-distribution. *Journal of the American Statistical Association*, **84**, 881–896.

Nelson, C. R., and Siegel, A. F. (1985) Parsimonious modelling of yield curves. *Journal of Business*, **60**, 473–489.

Seber, G. A. F., and Wild, C. J. (1989) *Nonlinear Regression*, Wiley, New York.

Svensson, L. E. (1994) Estimating and interpreting forward interest rates: Sweden 1992–94, Working paper. International Monetary Fund, 114.

Tuckman, B. (2002) *Fixed Income Securities*, 2nd ed., Wiley, Hoboken, NJ.

12

Time Series Models: Basics

12.1 Time Series Data

A *time series* is a sequence of observations in chronological order, for example, daily log returns on a stock or monthly values of the Consumer Price Index (CPI). A common simplifying assumption is that the data are equally spaced with a discrete-time observation index; however, this may only hold approximately. For example, daily log returns on a stock may only be available for weekdays, with additional gaps on holidays, and monthly values of the CPI are equally spaced by month, but unequally spaced by days. In either case, the consecutive observations are commonly regarded as equally spaced, for simplicity. In this chapter, we study statistical models for time series. These models are widely used in econometrics, business forecasting, and many scientific applications.

A *stochastic process* is a sequence of random variables and can be viewed as the "theoretical" or "population" analog of a time series—conversely, a time series can be considered a sample from a stochastic process. "Stochastic" is a synonym for random. One of the most useful methods for obtaining parsimony in a time series model is to assume some form of distributional invariance over time, or *stationarity*, a property discussed next.

12.2 Stationary Processes

When we observe a time series, the fluctuations appear random, but often with the same type of stochastic behavior from one time period to the next. For example, returns on stocks or changes in interest rates can be very different from the previous year, but the mean, standard deviation, and other statistical

D. Ruppert, D.S. Matteson, *Statistics and Data Analysis for Financial Engineering*, Springer Texts in Statistics, DOI 10.1007/978-1-4939-2614-5_12

properties often are similar from one year to the next.[1] Similarly, the demand for many consumer products, such as sunscreen, winter coats, and electricity, has random as well as seasonal variation, but each summer is similar to past summers, each winter to past winters, at least over shorter time periods. *Stationary stochastic processes* are probability models for time series with time-invariant behavior.

A process is said to be *strictly stationary* if all aspects of its behavior are unchanged by shifts in time. Mathematically, stationarity is defined as the requirement that for every m and n, the distributions of $(Y_1, \ldots, Y_n)$ and $(Y_{1+m}, \ldots, Y_{n+m})$ are the same; that is, the probability distribution of a sequence of n observations does not depend on their time origin. Strict stationarity is a very strong assumption, because it requires that "all aspects" of stochastic behavior be constant in time. Often, it will suffice to assume less, namely, weak stationarity. A process is *weakly stationary* if its mean, variance, and covariance are unchanged by time shifts. More precisely, $Y_1, Y_2, \ldots$ is a *weakly stationary process* if

- $E(Y_t) = \mu$ (a finite constant) for all t;
- $\mathrm{Var}(Y_t) = \sigma^2$ (a positive finite constant) for all t; and
- $\mathrm{Cov}(Y_t, Y_s) = \gamma(|t - s|)$ for all t and s for some function $\gamma(h)$.

Thus, the mean and variance do not change with time and the covariance between two observations depends only on the *lag*, the time distance $|t - s|$ between them, not the indices t or s directly. For example, if the process is weakly stationary, then the covariance between Y_2 and Y_5 is the same as the covariance between Y_7 and Y_{10}, since each pair is separated by three units of time. The adjective "weakly" in "weakly stationary" refers to the fact that we are only assuming that means, variance, and covariances, not other distributional characteristics such as quantiles, skewness, and kurtosis, are stationary. Weakly stationary is also sometimes referred to as *covariance stationary*. The term *stationary* will sometimes be used as a shorthand for strictly stationary.

The function γ is called the *autocovariance function* of the process. Note that $\gamma(h) = \gamma(-h)$. Why? Assuming weak stationarity, the correlation between Y_t and Y_{t+h} is denoted by $\rho(h)$. The function *rho* is called the *autocorrelation function*. Note that $\gamma(0) = \sigma^2$ and that $\gamma(h) = \sigma^2 \rho(h)$. Also, $\rho(h) = \gamma(h)/\sigma^2 = \gamma(h)/\gamma(0)$.

As mentioned, many financial time series do not exhibit stationarity, but often the *changes* in them, perhaps after applying a log transformation, are approximately stationary. For this reason, stationary time series models have broad applicability and wide ranging applications. From the viewpoint of

---

[1] It is the returns, not the stock prices, that have time-invariant behavior. Stock prices themselves tend to increase over time, so this year's stock prices tend to be higher and more variable than those a decade or two ago.

statistical modeling, it is not important whether it is the time series itself or changes in the time series that are stationary, because either way we get a parsimonious model.

The beauty of a stationary process is that it can be modeled with relatively few parameters. For example, we do not need a different expectation for each Y_t; rather they all have a common expectation, μ. This implies that μ can be estimated accurately by $\overline{Y}$. If instead we did not assume stationarity and each Y_t had its own unique expectation, μ_t, then it would not be possible to estimate μ_t accurately—μ_t could only be estimated by the single observation Y_t itself.

When a time series is observed, a natural question is whether it appears to be stationary. This is not an easy question to address, and we can never be absolutely certain of the answer. However, visual inspection of the time series and changes in the time series can be helpful. A *time series plot* is a plot of the series in chronological order. This very basic plot is useful for assessing stationary behavior, though it can be supplemented with other plots, such as the plot of the sample autocorrelation function that will be introduced later. In addition, there are statistical tests of stationarity—these are discussed in Sect. 12.10.

A time series plot of a stationary series should show random oscillation around some fixed level, a phenomenon called *mean-reversion*. If the series wanders without returning repeatedly to some fixed level, then the series should not be modeled as a stationary process.

Example 12.1. Inflation rates and changes in inflation rates—time series plots

The one-month inflation rate (annual rate, in percent) is shown in Fig. 12.1a. The data come from the Mishkin data set in R's Ecdat package. The series may be wandering without reverting to a fixed mean, or it may be slowly reverting to a mean of approximately 4 %, as would be expected with a stationary time series. In panel (b), the first differences, that is, the changes from one month to the next, are shown. In contrast to the original series, the differenced series certainly oscillate around a fixed mean that is 0 %, or nearly so. The differenced series appears stationary, but whether or not the original series is stationary needs further investigation. We will return to this question later.   □

Example 12.2. Air passengers

Figure 12.2 is a plot of monthly total international airline passengers for the years 1949 to 1960. The data come from the AirPassengers data set in R's Datasets package. There are three types of nonstationarity seen in the plot. First is the obvious upward trend, second is the seasonal variation with local peaks in summer and troughs in winter months, and third is the increase over time in the size of the seasonal oscillations.   □

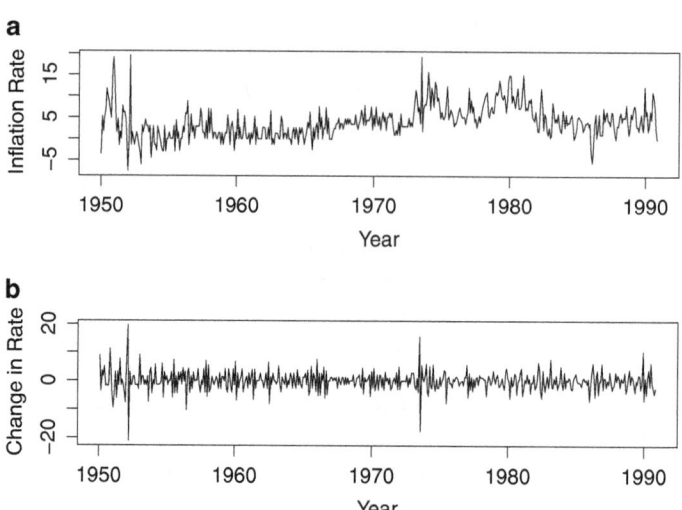

Fig. 12.1. *Time series plots of (a) one-month inflation rate (annual rate, in percent) and (b) first differences (changes) in the one-month inflation rate. It is unclear if the series in (a) is stationary, but the differenced series in (b) seems suitable for modeling as stationary.*

12.2.1 White Noise

White noise is the simplest example of a stationary process. We will define several types of white noise with increasingly restrictive assumptions.

The sequence $Y_1, Y_2, \ldots$ is a *weak white noise process* with mean μ and variance σ^2, which will be shortened to "weak WN(μ, σ^2)," if

- $E(Y_t) = \mu$ (a finite constant) for all t;
- $\mathrm{Var}(Y_t) = \sigma^2$ (a positive finite constant) for all t; and
- $\mathrm{Cov}(Y_t, Y_s) = 0$ for all $t \neq s$.

If the mean is not specified, then it is assumed that $\mu = 0$. A weak white noise process is weakly stationary with

$$\gamma(0) = \sigma^2,$$
$$\gamma(h) = 0 \quad \text{if} \quad h \neq 0,$$

so that

$$\rho(0) = 1,$$
$$\rho(h) = 0 \ \text{if} \ h \neq 0.$$

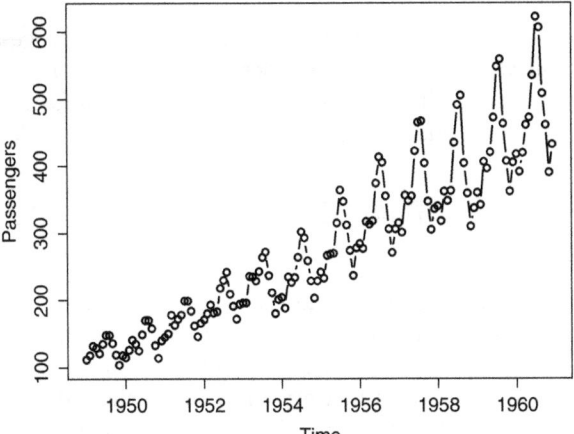

Fig. 12.2. *Time series plot of monthly totals of air passengers (in thousands).*

If $Y_1, Y_2, \ldots$ is an i.i.d. process, then we call it an *i.i.d. white noise process* or simply *i.i.d.* $\text{WN}(\mu, \sigma^2)$. Weak white noise is weakly stationary, while i.i.d. white noise is strictly stationary. An i.i.d. white noise process with σ^2 finite is also a weak white noise process, but not vice versa.

If, in addition, $Y_1, Y_2 \ldots$ is an i.i.d. process with a specific marginal distribution, then this might be noted. For example, if $Y_1, Y_2 \ldots$ are i.i.d. normal random variables, then the process is called a *Gaussian white noise process*. Similarly, if $Y_1, Y_2 \ldots$ are i.i.d. t random variables with ν degrees of freedom, then it is called a t_ν white noise process.

12.2.2 Predicting White Noise

Because of the lack of dependence, past values of a white noise process contain no information that can be used to predict future values. More precisely, suppose that $Y_1, Y_2, \ldots$ is an i.i.d. $\text{WN}(\mu, \sigma^2)$ process. Then

$$E(Y_{t+h}|Y_1, \ldots, Y_t) = \mu \ \text{for all} \ h \geq 1. \tag{12.1}$$

What this equation is saying is that one cannot predict the future deviations of a white noise process from its mean, because its future is independent of its past and present. Therefore, the best predictor of any future value of the process is simply the mean μ, what you would use even if $Y_1, \ldots, Y_t$ had not been observed. For weak white noise, (12.1) need not be true, but it is still true that the best linear predictor[2] of Y_{t+h} given $Y_1, \ldots, Y_t$ is μ.

12.3 Estimating Parameters of a Stationary Process

Suppose we observe $Y_1, \ldots, Y_n$ from a weakly stationary process. To estimate the mean μ and variance σ^2 of the process, we can use the sample mean $\overline{Y}$ and sample variance s^2. To estimate the autocovariance function, we use the *sample autocovariance function*

$$\widehat{\gamma}(h) = n^{-1} \sum_{t=1}^{n-h} (Y_{t+h} - \overline{Y})(Y_t - \overline{Y}) = n^{-1} \sum_{t=h+1}^{n} (Y_t - \overline{Y})(Y_{t-h} - \overline{Y}). \quad (12.2)$$

Equation (12.2) is an example of the usefulness of parsimony induced by the stationarity assumption. Because the covariance between Y_t and Y_{t+h} does not depend on t, *all* $n - h$ pairs of data points that are separated by a lag of h time units can be used to estimate $\gamma(h)$. Some authors define $\widehat{\gamma}(h)$ with the factor n^{-1} in (12.2) replaced by $(n-h)^{-1}$, but this change has little effect if n is reasonably large and h is small relative to n, as is typically the case.

To estimate $\rho(\cdot)$, we use the *sample autocorrelation function (sample ACF)* defined as

$$\widehat{\rho}(h) = \frac{\widehat{\gamma}(h)}{\widehat{\gamma}(0)}.$$

12.3.1 ACF Plots and the Ljung–Box Test

Most statistical software will plot a sample ACF with *test bounds*. These bounds are used to test the null hypothesis that an autocorrelation coefficient is 0. The null hypothesis is rejected if the sample autocorrelation is outside the bounds. The usual level of the test is 0.05, so one can expect to see about 1 out of 20 sample autocorrelations outside the test bounds simply by chance.

An alternative to using the bounds to test the autocorrelations one at a time is to use a simultaneous test. A *simultaneous test* is one that tests whether a group of null hypotheses are all true versus the alternative that at least one of them is false. The null hypothesis of the Ljung–Box test is $H_0 : \rho(1) = \rho(2) = \cdots = \rho(K) = 0$ for some K, say $K = 5$ or 10. If the Ljung–Box test rejects, then we conclude that one or more of $\rho(1), \rho(2), \cdots, \rho(K)$ is nonzero.

---

[2] Best linear prediction is discussed in Sect. 11.9.1.

If, in fact, the autocorrelations 1 to K are all zero, then there is only a 1 in 20 chance of falsely concluding that they are not all zero, assuming a level 0.05 test. In contrast, if the autocorrelations are tested one at time, then there is a much higher chance of concluding that one or more is nonzero.

The Ljung–Box test is sometimes called simply the Box test, though the former name is preferable since the test is based on a joint paper of Ljung and Box.

Example 12.3. Inflation rates and changes in the inflation rate—sample ACF plots and the Ljung–Box test

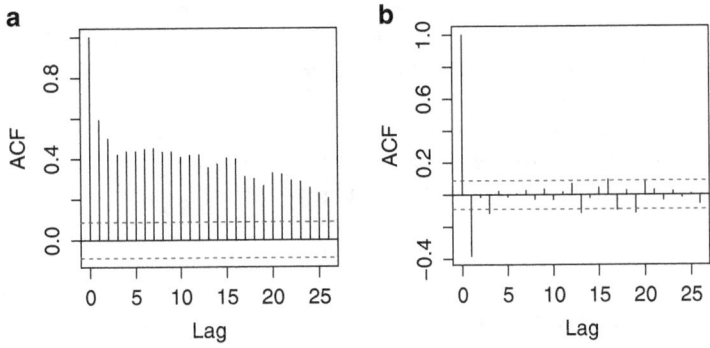

Fig. 12.3. *Sample ACF plots of the one-month inflation rate (a) and changes in the inflation rate (b).*

We return to the inflation rate data used in Example 12.1. Figure 12.3 contains plots of (a) the sample ACF of the one-month inflation rate and (b) the sample ACF of changes in the inflation rate.

```
1 data(Mishkin, package = "Ecdat")
2 y = as.vector(Mishkin[,1])
3 par(mfrow=c(1,2))
4 acf(y)
5 acf(diff(y))
```

In (a) we see that the sample ACF decays to zero slowly. This is a sign of either nonstationarity or possibly of stationarity with long-memory dependence, which is discussed in Sect. 13.5. In contrast, the sample ACF in (b) decays to zero quickly, indicating clearly that the differenced series is stationary. Thus, the sample ACF plots agree with the conclusions reached by examining the time series plots in Fig. 12.1, specifically that the differenced series is stationary and the original series might not be. In Sect. 12.10 we will use hypothesis testing to further address the question of whether or not the original series is stationary.

Several of the autocorrelations of the rate change series fall outside the test bounds, which suggests that the series is not white noise. To check, the Ljung–Box test was implemented using R's `Box.test()` function. K is called `lag` when `Box.test()` is called and `df` in the output, and we specify `type="Ljung-Box"` for the Ljung–Box test.

```
6 Box.test(diff(y), lag = 10, type = "Ljung-Box")
```

The Ljung–Box test statistic with $K = 10$ is 79.92, which has an extremely small p-value, 5.217e−13, so the null hypothesis of white noise is strongly rejected. Other choices of K give similar results.   □

Although a stationary process is somewhat parsimonious with parameters, at least relative to a general nonstationary process, a stationary process is still not sufficiently parsimonious for most purposes. The problem is that there are still an infinite number of parameters, $\rho(1), \rho(2), \ldots$. What we need is a class of stationary time series models with only a finite, preferably small, number of parameters. The ARIMA models of this chapter are precisely such a class. The simplest ARIMA models are autoregressive (AR) models, and we turn to these first.

12.4 AR(1) Processes

Time series models with correlation can be constructed from white noise. The simplest correlated stationary processes are *autoregressive processes*, where Y_t is modeled as a weighted average of past observations plus a white noise "error," which is also called the "noise" or "disturbance." We start with AR(1) processes, the simplest autoregressive processes.

Let $\epsilon_1, \epsilon_2, \ldots$ be weak $\text{WN}(0, \sigma_\epsilon^2)$. We say that $Y_1, Y_2, \ldots$ is an *AR(1) process* if for some constant parameters μ and ϕ,

$$Y_t - \mu = \phi(Y_{t-1} - \mu) + \epsilon_t \tag{12.3}$$

for all t. The parameter μ is the mean of the process, hence $(Y_t - \mu)$ has mean zero for all t. We may interpret the term $\phi(Y_{t-1} - \mu)$ as representing "memory" or "feedback" of the past into the present value of the process. The process $\{Y_t\}_{t=-\infty}^{+\infty}$ is correlated because the deviation of Y_{t-1} from its mean is fed back into Y_t. The parameter ϕ determines the amount of feedback, with a larger absolute value of ϕ resulting in more feedback and $\phi = 0$ implying that $Y_t = \mu + \epsilon_t$, so that Y_t is weak $\text{WN}(\mu, \sigma_\epsilon^2)$. In applications in finance, one can think of ϵ_t as representing the effect of "new information." For example, if Y_t is the log return on an asset at time t, then ϵ_t represents the effect on the asset's price of business and economic information that is revealed at time t. Information that is truly new cannot be anticipated, so the effects of today's new information should be independent of the effects of yesterday's news. This is why we model new information as white noise.

If $Y_1, Y_2, \ldots$ is a weakly stationary process, then $|\phi| < 1$. To see this, note that stationarity implies that the variances of $(Y_t - \mu)$ and $(Y_{t-1} - \mu)$ in (12.3) are equal, say, to σ_Y^2. Therefore, $\sigma_Y^2 = \phi^2 \sigma_Y^2 + \sigma_\epsilon^2$, which requires that $|\phi| < 1$. The mean of this process is μ. Simple algebra shows that (12.3) can be rewritten as

$$Y_t = (1 - \phi)\mu + \phi Y_{t-1} + \epsilon_t. \tag{12.4}$$

Recall the linear regression model $Y_t = \beta_0 + \beta_1 X_t + \epsilon_t$ from your statistics courses or see Chap. 9 for an introduction to regression analysis. Equation (12.4) is just a linear regression model with intercept $\beta_0 = (1 - \phi)\mu$ and slope $\beta_1 = \phi$. The term *autoregression* refers to the regression of the process on its own past values.

If $|\phi| < 1$, then repeated substitution of (12.3) shows that

$$Y_t = \mu + \epsilon_t + \phi \epsilon_{t-1} + \phi^2 \epsilon_{t-2} + \cdots = \mu + \sum_{h=0}^{\infty} \phi^h \epsilon_{t-h}, \tag{12.5}$$

assuming that time index t of Y_t and ϵ_t can be extended to negative values so that the white noise process is $\{\epsilon_t\}_{t=-\infty}^{+\infty}$ and (12.3) is true for all integers t. Equation (12.5) is called *the infinite moving average* [MA(∞)] representation of the process. This equation shows that Y_t is a weighted average of *all* past values of the white noise process. This representation should be compared to the AR(1) representation that shows Y_t as depending only on Y_{t-1} and ϵ_t. Since $|\phi| < 1$, $\phi^h \to 0$ as the lag $h \to \infty$. Thus, the weights given to the distant past are small. In fact, they are quite small. For example, if $\phi = 0.5$, then $\phi^{10} = 0.00098$, so ϵ_{t-10} has virtually no effect on Y_t. For this reason, the sum in (12.5) could be truncated at a finite number of terms, with no practical need to assume that the processes existed in the infinite past.

12.4.1 Properties of a Stationary AR(1) Process

When an AR(1) process is weakly stationary, which implies that $|\phi| < 1$, then

$$E(Y_t) = \mu \quad \forall t, \tag{12.6}$$

$$\operatorname{Var}(Y_t) = \gamma(0) = \sigma_Y^2 = \frac{\sigma_\epsilon^2}{1 - \phi^2} \quad \forall t, \tag{12.7}$$

$$\operatorname{Cov}(Y_t, Y_{t+h}) = \gamma(h) = \phi^{|h|} \frac{\sigma_\epsilon^2}{1 - \phi^2} \quad \forall t \text{ and } \forall h, \text{ and} \tag{12.8}$$

$$\operatorname{Corr}(Y_t, Y_{t+h}) = \rho(h) = \phi^{|h|} \quad \forall t \text{ and } \forall h. \tag{12.9}$$

It is important to remember that formulas (12.6) to (12.9) hold only if $|\phi| < 1$ and only for AR(1) processes. Moreover, for Y_t to be stationary, Y_0 must start in the stationary distribution so that $E(Y_0) = \mu$ and $\operatorname{Var}(Y_0) = \sigma_\epsilon^2/(1 - \phi^2)$. Otherwise, Y_t is not stationary though it eventually converges to stationarity.

These formulas can be proved using (12.5). For example, using (7.11) in Sect. 7.3.2,

$$\text{Var}(Y_t) = \text{Var}\left(\sum_{h=0}^{\infty} \phi^h \epsilon_{t-h}\right) = \sigma_\epsilon^2 \sum_{h=0}^{\infty} \phi^{2h} = \frac{\sigma_\epsilon^2}{1 - \phi^2}, \tag{12.10}$$

which proves (12.7). In (12.10) the formula for summation of a geometric series was used. This formula is

$$\sum_{i=0}^{\infty} r^i = \frac{1}{1 - r} \quad \text{if } |r| < 1. \tag{12.11}$$

Also, for $h > 0$,

$$\text{Cov}\left(\sum_{i=0}^{\infty} \epsilon_{t-i}\phi^i, \sum_{j=0}^{\infty} \epsilon_{t+h-j}\phi^j\right) = \phi^{|h|} \frac{\sigma_\epsilon^2}{1 - \phi^2}, \tag{12.12}$$

thus verifying (12.8). Then (12.9) follows by dividing (12.8) by (12.7).

Be sure to distinguish between σ_ϵ^2, which is the variance of the white noise process $\epsilon_1, \epsilon_2, \ldots$, and $\gamma(0)$, which is the variance, σ_Y^2, of the stationary AR(1) process $Y_1, Y_2, \ldots$. We can see from (12.7) that $\gamma(0)$ is larger than σ_ϵ^2 unless $\phi = 0$, in which case $Y_t = \mu + \epsilon_t$, such that Y_t and ϵ_t have the same variance.

The ACF (autocorrelation function) of an AR(1) process depends upon only one parameter, ϕ. This is a remarkable amount of parsimony, but it comes at a price. The ACF of an AR(1) process has only a limited range of shapes, as can be seen in Fig. 12.4. The magnitude of its ACF decays geometrically to zero, either slowly as when $\phi = 0.95$, moderately slowly as when $\phi = 0.75$, or rapidly as when $\phi = 0.2$. If $\phi < 0$, then the sign of the ACF alternates as its magnitude decays geometrically. If the sample ACF of the data does not behave in one of these ways, then an AR(1) model is unsuitable. The remedy is to use more AR parameters, or to switch to another class of models such as the moving average (MA) or autoregressive moving average (ARMA) models. We investigate these alternatives in this chapter.

12.4.2 Convergence to the Stationary Distribution

Suppose that Y_0 is an arbitrary starting value not chosen from the stationary distribution and that (12.3) holds for $t = 1, 2, \ldots$. Then the process is not stationary, but converges to the stationary distribution satisfying (12.6) to (12.9) as $t \to \infty$.[3] For example, since $Y_t - \mu = \phi(Y_{t-1} - \mu) + \epsilon_t$, we have

---

[3] However, there is a technical issue here. It must be assumed that Y_0 has a finite mean and variance, since otherwise Y_t will not have a finite mean and variance for any $t > 0$.

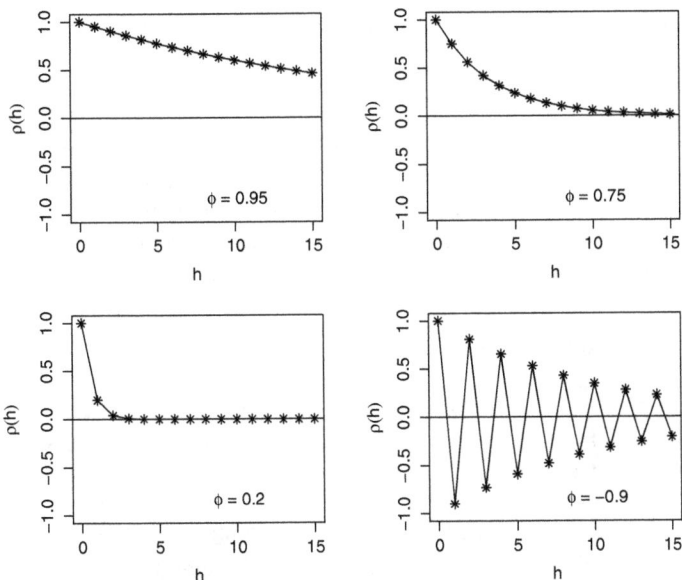

Fig. 12.4. *Autocorrelation functions of AR(1) processes with ϕ equal to 0.95, 0.75, 0.2, and −0.9.*

$E(Y_1) - \mu = \phi\{E(Y_0) - \mu\}$, and $E(Y_2) - \mu = \phi^2\{E(Y_0) - \mu\}$, and so forth, so that

$$E(Y_t) = \mu + \phi^t\{E(Y_0) - \mu\} \text{ for all } t > 0. \tag{12.13}$$

Since $|\phi| < 1$, $\phi^t \to 0$ and $E(Y_t) \to \mu$ as $t \to \infty$. The convergence of $\mathrm{Var}(Y_t)$ to $\sigma_\epsilon^2/(1 - \phi^2)$ can be proved in a somewhat similar manner. The convergence to the stationary distribution can be very rapid when $|\phi|$ is not too close to 1. For example, if $\phi = 0.5$, then $\phi^{10} = 0.00097$, so by (12.13), $E(Y_{10})$ is very close to μ unless $E(Y_0)$ was extremely far from μ.

12.4.3 Nonstationary AR(1) Processes

If $|\phi| \geq 1$, then the AR(1) process is nonstationary, and the mean, variance, covariances and and correlations are not constant.

Random Walk ($\phi = 1$)

If $\phi = 1$, then

$$Y_t = Y_{t-1} + \epsilon_t$$

and the process is *not* stationary. This is the random walk process we saw in Chap. 2.

Suppose we start the process at an arbitrary point Y_0. It is easy to see that

$$Y_t = Y_0 + \epsilon_1 + \cdots + \epsilon_t.$$

Then $E(Y_t|Y_0) = Y_0$ for all t, which is constant but depends entirely on the arbitrary starting point. Moreover, $\text{Var}(Y_t|Y_0) = t\sigma_\epsilon^2$, which is not stationary but rather increases linearly with time. The increasing variance makes the random walk "wander" in that Y_t takes increasingly longer excursions away from its conditional mean of Y_0 and therefore is not mean-reverting.

AR(1) Processes When $|\phi| > 1$

When $|\phi| > 1$, an AR(1) process has explosive behavior. This can be seen in Fig. 12.5. This figure shows simulations of 200 observations from AR(1) processes with various values of ϕ. The explosive case where $\phi = 1.01$ clearly is different from the other cases where $|\phi| \leq 1$. However, the case where $\phi = 1$ is not that much different from $\phi = 0.98$ even though the former is nonstationary while the latter is stationary. Longer time series would help distinguish between $\phi = 0.98$ and $\phi = 1$.

12.5 Estimation of AR(1) Processes

R has the function `arima()` for fitting AR and other time series models. The function `arima()` and similar functions in other software packages have two primary estimation methods, conditional least-squares and maximum likelihood. The two methods are explained in Sect. 12.5.2. They are similar and generally give nearly the same estimates. In this book, we use the default method in R's `arima()`, which is the MLE with the conditional least-squares estimate used as the starting value for computing the MLE by iterative nonlinear optimization.

12.5.1 Residuals and Model Checking

Once μ and ϕ have been estimated, one can estimate the white noise process $\epsilon_1, \ldots, \epsilon_n$. Rearranging Eq. (12.3), we have

$$\epsilon_t = (Y_t - \mu) - \phi(Y_{t-1} - \mu). \tag{12.14}$$

In analogy with (12.14), the residuals, $\widehat{\epsilon}_2, \widehat{\epsilon}_3, \ldots, \widehat{\epsilon}_n$, are defined as

$$\widehat{\epsilon}_t = (Y_t - \widehat{\mu}) - \widehat{\phi}(Y_{t-1} - \widehat{\mu}), \quad t \geq 2, \tag{12.15}$$

and estimate $\epsilon_2, \ldots, \epsilon_n$. The first noise, ϵ_1, cannot be estimated directly since it is assumed that the observations start at Y_1 so that Y_0 is not available.

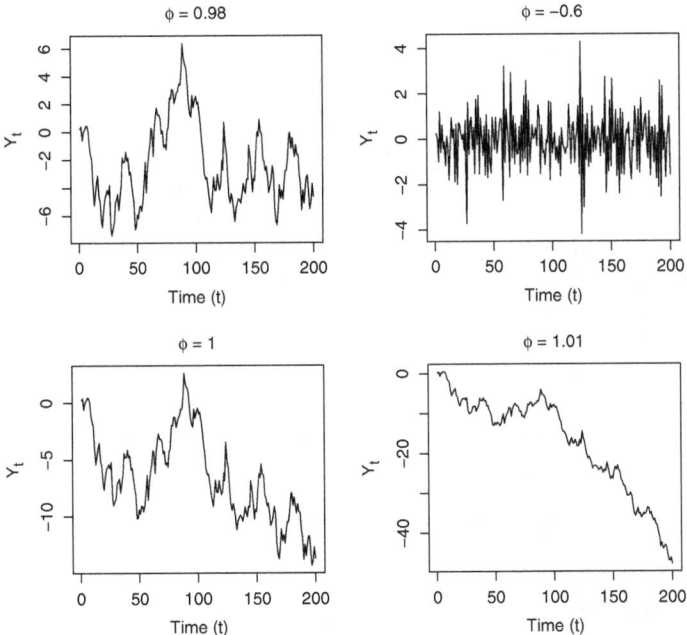

Fig. 12.5. *Simulations of* 200 *observations from AR*(1) *processes with various values of ϕ and $\mu = 0$. The white noise process $\epsilon_1, \epsilon_2, \ldots, \epsilon_{200}$ is the same for all four AR*(1) *processes.*

The residuals can be used to check the assumption that $Y_1, Y_2, \ldots, Y_n$ is an AR(1) process; any autocorrelation in the residuals is evidence against the assumption of an AR(1) process.

To appreciate why residual autocorrelation indicates a possible problem with the model, suppose that we are fitting an AR(1) model, $Y_t = \mu + \phi$ $(Y_{t-1} - \mu) + \epsilon_t$, but the true model is an AR(2) process[4] given by

$$Y_t - \mu = \phi_1(Y_{t-1} - \mu) + \phi_2(Y_{t-2} - \mu) + \epsilon_t.$$

Since we are fitting the incorrect AR(1) model, there is no hope of estimating ϕ_2 since it is not in the model. Moreover, $\widehat{\phi}$ does not necessarily estimate ϕ_1 because of bias caused by model misspecification. Let ϕ^* be the expected value of $\widehat{\phi}$. For the purpose of illustration, assume that $\widehat{\mu} \approx \mu$ and $\widehat{\phi} \approx \phi^*$. This is a sensible approximation if the sample size n is large enough. Then

$$
\begin{aligned}
\widehat{\epsilon}_t &\approx (Y_t - \mu) - \phi^*(Y_{t-1} - \mu) \\
&= \phi_1(Y_{t-1} - \mu) + \phi_2(Y_{t-2} - \mu) + \epsilon_t - \phi^*(Y_{t-1} - \mu) \\
&= (\phi_1 - \phi^*)(Y_{t-1} - \mu) + \phi_2(Y_{t-2} - \mu) + \epsilon_t.
\end{aligned}
$$

---

[4] We discuss higher-order AR models in more detail soon.

Thus, the residuals do not estimate the white noise process as they would if the correct AR(2) model were used. Even if there is no bias in the estimation of ϕ_1 by $\widehat{\phi}$ so that $\phi_1 = \phi^*$ and the term $(\phi_1 - \phi^*)(Y_{t-1} - \mu)$ drops out, the presence of $\phi_2(Y_{t-2} - \mu)$ in the residuals causes them to be autocorrelated.

To check for residual autocorrelation, one can use the *test bounds* of ACF plots. Any residual ACF value outside the test bounds is significantly different from 0 at the 0.05 level. As discussed earlier, the danger here is that some sample ACF values will be significant merely by chance, and to guard against this danger, one can use the Ljung–Box test that *simultaneously* tests that all autocorrelations up to a specified lag are zero. When the Ljung–Box test is applied to residuals, a correction is needed to account for the use of $\widehat{\phi}$ in place of the unknown ϕ. Some software makes this correction automatically. In R the correction is not automatic but is done by setting the `fitdf` parameter in `Box.test()` to the number of autoregressive coefficient parameters that were estimated, so for an AR(1) model `fitdf` should be 1.

Example 12.4. Daily log returns for BMW stock—ACF plots and AR fit

The daily log returns for BMW stock between January 1973 and July 1996 from the `bmw` data set in R's `evir` package are shown in Fig. 12.6a. Their sample ACF and quantiles are shown in Fig. 12.6b and c, respectively. The estimated autocorrelation coefficient at lag 1 is well outside the test bounds, so the series has some dependence. Also, the Ljung–Box test that the first `lag` autocorrelations are 0 was performed using R's `Box.test()` function.

```
7 data(bmw, package = "evir")
8 Box.test(bmw, lag = 5, type = "Ljung-Box")
```

The parameter `lag`, which specifies the number of autocorrelation coefficients to test, was set equal to 5, though other choices give similar results. The output was

```
        Box-Ljung test

data:   bmw
X-squared = 44.987, df = 5, p-value = 1.460e-08
```

The p-value is very small, indicating that at least one of the first five autocorrelations is nonzero. Whether the amount of dependence is of any practical importance is debatable, but an AR(1) model to account for the small amount of autocorrelation might be appropriate.

Next, an AR(1) model was fit using the `arima()` command in R.

```
9  fitAR1 = arima(bmw, order = c(1,0,0))
10 print(fitAR1)
```

The `order` parameter will be explained later, but for an AR(1) process it should be `c(1,0,0)`. A summary of the output is below.

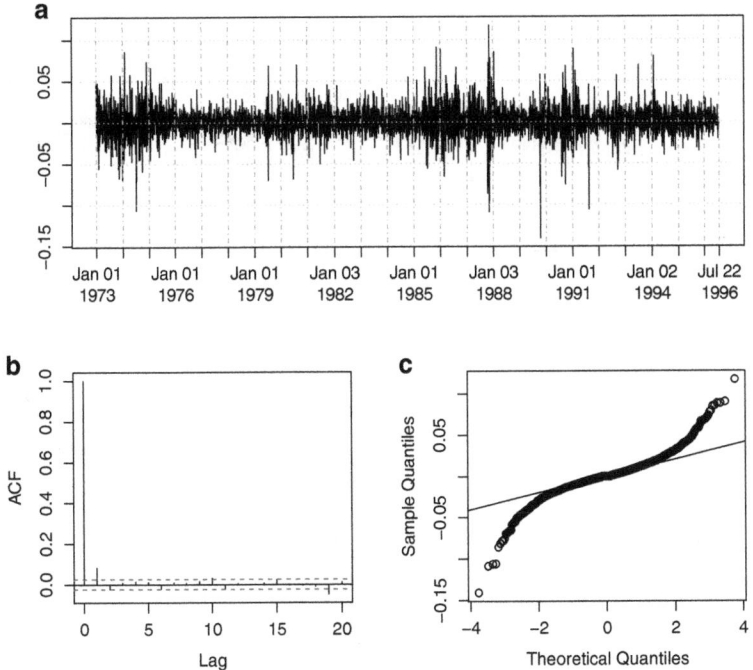

Fig. 12.6. *(a) Daily log returns for BMW stock from January 1973 until July 1996, and their (b) sample ACF and (c) sample quantiles relative to the normal distribution.*

```
Call:
arima(x = bmw, order = c(1, 0, 0))

Coefficients:
              ar1    intercept
         0.081116    0.000340
s.e.     0.012722    0.000205

sigma^2 estimated as 0.000216260:  log-likelihood = 17212.34,
aic = -34418.68
```

We see that $\widehat{\phi} = 0.081$ and $\widehat{\sigma}^2 = 0.00022$. Although $\widehat{\phi}$ is small, it is statistically significant since it is 6.4 times its standard error 0.013, so its p-value is near zero. As just mentioned, whether this small, but nonzero, value of $\widehat{\phi}$ is of practical significance is another matter. A non-zero value of ϕ means that there is some information in today's return that could be used for prediction of tomorrow's return, but a small value of ϕ means that the prediction will not be very accurate. The potential for profit might be negated by trading costs.

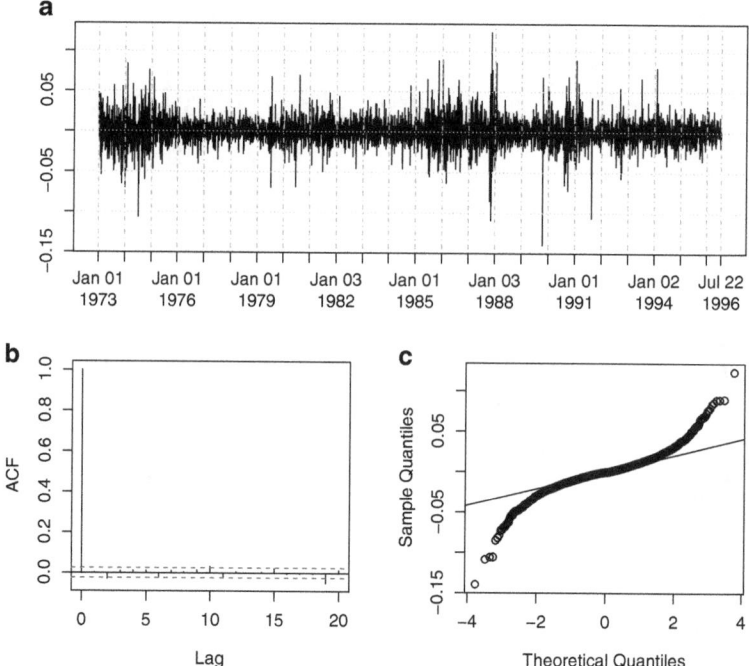

Fig. 12.7. *A (a) time series plot, (b) sample ACF and (c) normal quantile plot of residuals from an AR(1) fit to the daily log returns for BMW stock.*

The sample ACF of the residuals is plotted in Fig. 12.7b. None of the autocorrelations at low lags are outside the test bounds. A few at higher lags are outside the bounds, but this type of behavior is expected to occur by chance or because, with a large sample size, very small but nonzero true correlations can be detected. The Ljung–Box test was applied, with `lag` equal to 5 and `fitdf=1`.

```
11 Box.test(residuals(fitAR1), lag = 5, type = "Ljung-Box", fitdf = 1)

   Box-Ljung test

   data:  residuals(fitAR1)
   X-squared = 6.8669, df = 4, p-value = 0.1431
```

The large p-value indicates that we should accept the null hypothesis that the residuals are uncorrelated, at least at small lags. This is a sign that the AR(1) model provides an adequate fit. However, the Ljung–Box test was repeated with `lag` equal to 10, 15, and 20 and the p-values were 0.041, 0.045, and 0.004, respectively. These values are "statistically significant" using the conventional cutoff of 0.05. The sample size is 6146, so it is not surprising that even a small amount of autocorrelation can be statistically significant. The practical significance of this autocorrelation is very doubtful.

We conclude that the AR(1) model is adequate for the BMW daily returns, but at longer lags some slight amount of autocorrelation appears to remain. However, the time series plot and normal quantile plot of the AR(1) residuals in Fig. 12.7a and c show volatility clustering and heavy tails. These are common features of economic data and will be modeled in subsequent chapters.   □

Example 12.5. Inflation rate—AR(1) fit and checking residuals

This example uses the inflation rate time series used earlier in Example 12.1. Although there is some doubt as to whether this series is stationary, we will fit an AR(1) model. The ACF of the residuals are shown in Fig. 12.8 and there is considerable residual autocorrelation, which indicates that the AR(1) model is not adequate. A Ljung–Box test confirms this result.

```
12  data(Mishkin, package = "Ecdat")
13  y = as.vector(Mishkin[,1])
14  fit = arima(y, order = c(1,0,0))
15  Box.test(fit$resid, type = "Ljung", lag = 24, fitdf = 1)

    Box-Ljung test

    data:  fit$resid
    X-squared = 138.5776, df = 23, p-value < 2.2e-16
```

One might try fitting an AR(1) to the changes in the inflation rate, since this series is clearly stationary. However, the AR(1) model also does not fit the changes in the inflation rate. We will return to this example when we have a larger collection of models in our statistics toolbox.   □

12.5.2 Maximum Likelihood and Conditional Least-Squares

Estimators for AR processes can be computed automatically by most statistical software packages, and the user need not know what is "under the hood" of the software. Nonetheless, for readers interested in the estimation methodology, this section has been provided.

To find the joint density for $Y_1, \ldots, Y_n$, we use (A.41) and the fact that

$$f_{Y_t|Y_1,\ldots,Y_{t-1}}(y_t|y_1,\ldots,y_{t-1}) = f_{Y_t|Y_{t-1}}(y_t|y_{t-1}) \qquad (12.16)$$

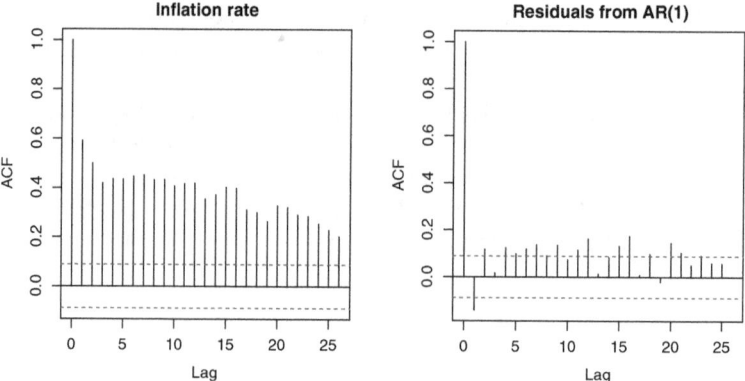

Fig. 12.8. *Sample ACF for the inflation rate time series and residual series from an AR(1) fit.*

for $t = 2, 3, \ldots, n$. A stochastic process with property (12.16) is called a *Markov process*. By (A.41) and (12.16), we have

$$f_{Y_1,\ldots,Y_n}(y_1, \ldots, y_n) = f_{Y_1}(y_1) \prod_{t=2}^{n} f_{Y_t|Y_{t-1}}(y_t|y_{t-1}). \tag{12.17}$$

If we assume the errors are from a Gaussian white noise process, then by (12.6) and (12.7), we know that Y_1 is $N\{\mu, \sigma_\epsilon^2/(1 - \phi^2)\}$ for a stationary AR(1) process. Given Y_{t-1}, the only random component of Y_t is ϵ_t, so that Y_t given Y_{t-1} is $N\{\mu + \phi(Y_{t-1} - \mu), \sigma_\epsilon^2\}$. It then follows that the joint density for $Y_1, \ldots, Y_n$ is

$$\left(\frac{1}{\sqrt{2\pi}\sigma_\epsilon}\right)^n \sqrt{1 - \phi^2} \exp\left\{-\frac{(Y_1 - \mu)^2}{2\sigma_\epsilon^2/(1 - \phi^2)}\right\} \prod_{i=2}^{n} \exp\left(-\frac{\left[Y_i - \{\mu + \phi(Y_{i-1} - \mu)\}\right]^2}{2\sigma_\epsilon^2}\right).$$
$$\tag{12.18}$$

The maximum likelihood estimator maximizes the logarithm of (12.18) over $(\mu, \phi, \sigma_\epsilon)$. A somewhat simpler estimator deletes the marginal density of Y_1 from the likelihood and maximizes the logarithm of

$$\left(\frac{1}{\sqrt{2\pi}\sigma_\epsilon}\right)^{n-1} \prod_{t=2}^{n} \exp\left(-\frac{\left[Y_t - \{\mu + \phi(Y_{t-1} - \mu)\}\right]^2}{2\sigma_\epsilon^2}\right). \tag{12.19}$$

This estimator is called the conditional least-squares estimator. It is "conditional" because it uses the conditional density of $Y_2, \ldots, Y_n$ given Y_1. It is a least-squares estimator because the estimates of μ and ϕ minimize

$$\sum_{t=2}^{n} \left[Y_t - \left\{ \mu + \phi(Y_{t-1} - \mu) \right\} \right]^2. \qquad (12.20)$$

The default method for the function `arima()` in R is to use the conditional least-squares estimates as starting values for maximum likelihood. The MLE is returned, along with approximate standard errors. The default option is used in the examples in this book.

12.6 AR(p) Models

We have seen that the ACF of an AR(1) process decays geometrically to zero if $|\phi| < 1$ and also alternates in sign if $\phi < 0$. This is a limited range of behavior and many time series do not behave in this way. To get a more flexible class of models, but one that is still parsimonious, we can use a model that regresses the current value of the process on several of the recent past values, not just the most recent. Thus, we let the last p values of the process, $Y_{t-1}, \ldots, Y_{t-p}$, feed back into the current value Y_t.

Here is a formal definition. The stochastic process Y_t is an *AR(p) process* if

$$Y_t - \mu = \phi_1(Y_{t-1} - \mu) + \phi_2(Y_{t-2} - \mu) + \cdots + \phi_p(Y_{t-p} - \mu) + \epsilon_t,$$

where $\epsilon_1, \epsilon_2, \ldots$ is weak WN$(0, \sigma_\epsilon^2)$.

This is a multiple linear regression[5] model with lagged values of the time series as the "x-variables." The model can also be expressed as

$$Y_t = \beta_0 + \phi_1 Y_{t-1} + \cdots + \phi_p Y_{t-p} + \epsilon_t,$$

where $\beta_0 = \{1 - (\phi_1 + \cdots + \phi_p)\}\mu$. The parameter β_0 is called the "constant" or "intercept" as in an AR(1) model. It can be shown that $\{1 - (\phi_1 + \cdots + \phi_p)\} > 0$ for a stationary process, so $\mu = 0$ if and only if β_0 is zero.

Formulas for the ACFs of AR(p) processes with $p > 1$ are more complicated than for an AR(1) process and can be found in the time series textbooks listed in the "References" section. However, software is available for computing and plotting the ACF of any AR processes, as well as for the MA and ARMA processes to be introduced soon. Figure 12.9 is a plot of the ACFs of three AR(2) process. The ACFs were computed using R's `ARMAacf()` function. Notice the wide variety of ACFs that are possible with two AR parameters.

Most of the concepts we have discussed for AR(1) models generalize easily to AR(p) models. The conditional least squares or maximum likelihood estimators can be calculated using software such as R's `arima()` function. The residuals are defined by

$$\widehat{\epsilon}_t = Y_t - \{\widehat{\beta}_0 + \widehat{\phi}_1 Y_{t-1} + \cdots + \widehat{\phi}_{t-p} Y_{t-p}\}, \quad t \geq p+1.$$

---

[5] See Chap. 9 for an introduction to multiple regression.

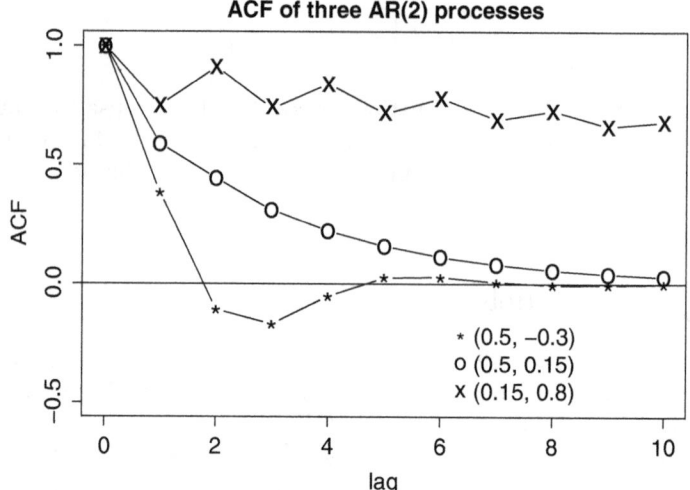

Fig. 12.9. *ACF of three AR(2) processes; the legend gives the values of ϕ_1 and ϕ_2.*

If the AR(p) model fits the time series well, then the residuals should look like white noise. Residual autocorrelation can be detected by examining the sample ACF of the residuals and using the Ljung–Box test. Any significant residual autocorrelation is a sign that the AR(p) model does not fit well.

One problem with AR models is that they often need a rather large value of p to fit a data set. The problem is illustrated by the following two examples.

Example 12.6. Changes in the inflation rate—AR(p) models

Figure 12.10 is a plot of AIC and BIC versus p for AR(p) fits to the changes in the inflation rate. Both criteria suggest that p should be large. AIC decreases steadily as p increases from 1 to 19, though there is a local minimum at 8. Even the conservative BIC criterion indicates that p should be as large as 6. Thus, AR models are not parsimonious for this example. The remedy is to use a MA or ARMA model, which are the topics of the next sections.

Many statistical software packages have functions to automate the search for the AR model that optimizes AIC or other criteria. The `auto.arima` function in R's `forecast` package found that $p = 8$ is the first local minimum of AIC.

```
16 library(forecast)
17 auto.arima(diff(y), max.p = 20, max.q = 0, d = 0, ic = "aic")

   Series: diff(y)
   ARIMA(8,0,0) with zero mean
```

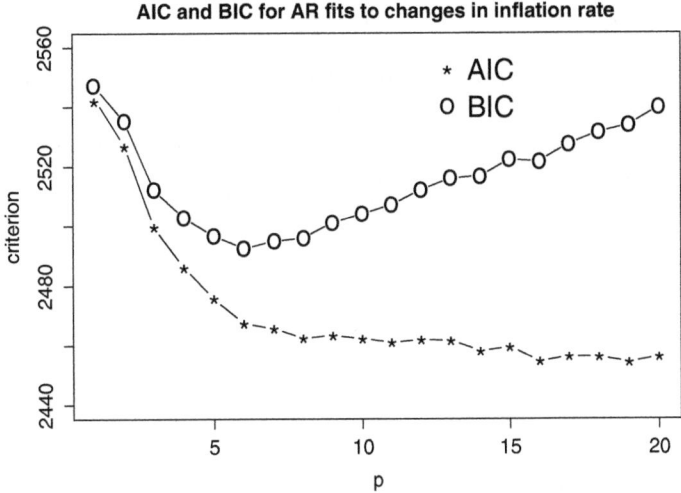

Fig. 12.10. *Fitting AR(p) models to changes in the one-month inflation rate; AIC and BIC plotted against p.*

```
Coefficients:
             ar1       ar2       ar3       ar4       ar5       ar6
         -0.6274   -0.4977   -0.5158   -0.4155   -0.3443   -0.2560
s.e.      0.0456    0.0536    0.0576    0.0606    0.0610    0.0581
             ar7       ar8
         -0.1557   -0.1051
s.e.      0.0543    0.0459

sigma^2 estimated as 8.539:  log likelihood=-1221.2
AIC=2460.4    AICc=2460.78   BIC=2498.15
```

The first local minimum of BIC is at $p = 6$.

```
18 auto.arima(diff(y), max.p = 20, max.q = 0, d = 0, ic = "bic")

Series: diff(y)
ARIMA(6,0,0) with zero mean

Coefficients:
             ar1       ar2       ar3       ar4       ar5       ar6
         -0.6057   -0.4554   -0.4558   -0.3345   -0.2496   -0.1481
s.e.      0.0454    0.0522    0.0544    0.0546    0.0526    0.0457

sigma^2 estimated as 8.699:  log likelihood=-1225.67
AIC=2465.33   AICc=2465.56   BIC=2494.69
```

We will see later that a more parsimonious fit can be obtained by going beyond AR models. □

Example 12.7. Inflation rates—AR(p) models

Since it is uncertain whether or not the inflation rates are stationary, one might fit an AR model to the inflation rates themselves, rather than their differences. An AR(p) models was fit to the inflation rates with p selected via an information criterion by `auto.arima()`. The BIC method chose $p = 2$ and AIC selected $p = 7$. The results for $p = 7$ are below.

```
Series: y
ARIMA(7,0,0) with non-zero mean

Coefficients:
         ar1    ar2     ar3    ar4    ar5    ar6    ar7  intercept
      0.366  0.129  -0.020  0.099  0.065  0.080  0.119      3.987
s.e.  0.045  0.048   0.048  0.048  0.049  0.048  0.046      0.784

sigma^2 estimated as 8.47:  log likelihood=-1221.8
AIC=2461.6   AICc=2461.9   BIC=2499.3
```

The inflation rate and its residual series from an AR(7) fit and their sample ACFs are shown in Fig. 12.11.   □

12.7 Moving Average (MA) Processes

As we saw in Example 12.6, there is a potential need for large values of p when fitting AR processes. A remedy for this problem is to add a moving average component to an AR(p) process. The result is an *autoregressive-moving average process*, often called an *ARMA process*. Before introducing ARMA processes, we start with pure moving average (MA) processes.

12.7.1 MA(1) Processes

The idea behind AR processes is to feed past data back into the current value of the process. This induces correlation between the past and present. The effect is to have at least some correlation at *all* lags. Sometimes data show correlation at only short lags, for example, only at lag 1 or only at lags 1 and 2. See, for example, Fig. 12.3b where the sample ACF of changes in the inflation rate is approximately -0.4 at lag 1, but then is approximately 0.1 or less in magnitude after one lag. AR processes do not behave this way and, as already seen in Example 12.6, do not provide a parsimonious fit. In such situations, a useful alternative to an AR model is a moving average (MA) model. A process Y_t is a *moving average process* if Y_t can be expressed as a weighted average (moving average) of the past values of the white noise process $\{\epsilon_t\}$.

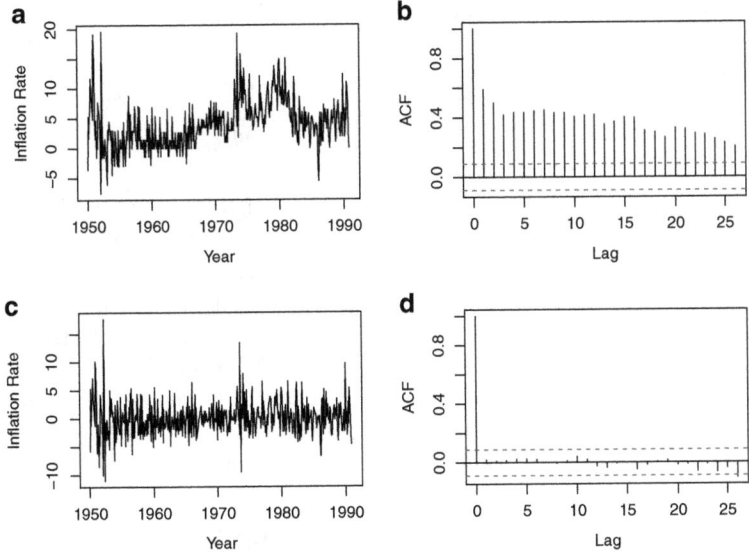

Fig. 12.11. *(a) The inflation rate series and (b) its sample ACF; (c) the residuals series from an AR(7) fit to the inflation rates (d) ACF of residuals.*

The **MA(1)** (moving average of order 1) process is

$$Y_t - \mu = \epsilon_t + \theta\epsilon_{t-1}, \tag{12.21}$$

where as before the ϵ_t are weak $\mathrm{WN}(0, \sigma_\epsilon^2)$.[6]

One can show that

$$
\begin{aligned}
E(Y_t) &= \mu, \\
\mathrm{Var}(Y_t) &= \sigma_\epsilon^2(1 + \theta^2), \\
\gamma(1) &= \theta\sigma_\epsilon^2, \\
\gamma(h) &= 0 \text{ if } |h| > 1, \\
\rho(1) &= \frac{\theta}{1 + \theta^2}, \tag{12.22} \\
\rho(h) &= 0 \text{ if } |h| > 1. \tag{12.23}
\end{aligned}
$$

Notice the implication of (12.22) and (12.23)—an MA(1) model has zero correlation at all lags except lag 1 (and of course lag 0). It is relatively easy to derive these formulas and this is left as an exercise for the reader.

---

[6] Some textbooks and some software write MA models with the signs reversed so that model (12.21) is written as $Y_t - \mu = \epsilon_t - \theta\epsilon_{t-1}$. We have adopted the same form of MA models as R's `arima()` function. These remarks apply as well to the general MA and ARMA models given by Eqs. (12.24) and (12.25).

12.7.2 General MA Processes

The **MA**(q) process is

$$Y_t = \mu + \epsilon_t + \theta_1\epsilon_{t-1} + \cdots + \theta_q\epsilon_{t-q}. \tag{12.24}$$

One can show that $\gamma(h) = 0$ and $\rho(h) = 0$ if $|h| > q$. Formulas for $\gamma(h)$ and $\rho(h)$ when $|h| \leq q$ are given in time series textbooks and these functions can be computed in R by the function `ARMAacf()`.

Unlike AR(p) models where the "constant" in the model is not the same as the mean, in an MA(q) model μ, the mean of the process, is the same as β_0, the "constant" in the model. This fact can be appreciated by examining the right-hand side of Eq. (12.24), where μ is the "intercept" or "constant" in the model and is also the mean of Y_t because $\epsilon_t, \ldots, \epsilon_{t-q}$ have mean zero. MA(q) models can be fit easily using, for example, the `arima()` function in R.

Example 12.8. Changes in the inflation rate—MA models

MA(q) models were fit to the changes in the inflation rate. Figure 12.12 shows plots of AIC and BIC versus q. BIC suggests that an MA(2) model is adequate, while AIC suggests an MA(3) model. We fit the MA(3) model. The Ljung–Box test was applied to the residuals with `fitdf = 3` and `lag` equal to 5, 10, and 15 and gave p-values of 0.65, 0.76, and 0.32, respectively. The MA(2) also provided an adequate fit with the p-values from the Ljung–Box test all above 0.08. The output for the MA(3) model is below.

```
19 fitMA3 = arima(diff(y), order = c(0,0,3))
20 fitMA3

   Series: diff(y)
   ARIMA(0,0,3) with non-zero mean
     -

   Coefficients:
            ma1      ma2      ma3   intercept
          -0.633   -0.103   -0.108      0.000
   s.e.    0.046    0.051    0.047      0.021

   sigma^2 estimated as 8.5:  log likelihood=-1220.3
   AIC=2450.5   AICc=2450.7   BIC=2471.5
```

Thus, if an MA model is used, then only two or three MA parameters are needed. This is a strong contrast with AR models, which require far more parameters, perhaps as many as six. □

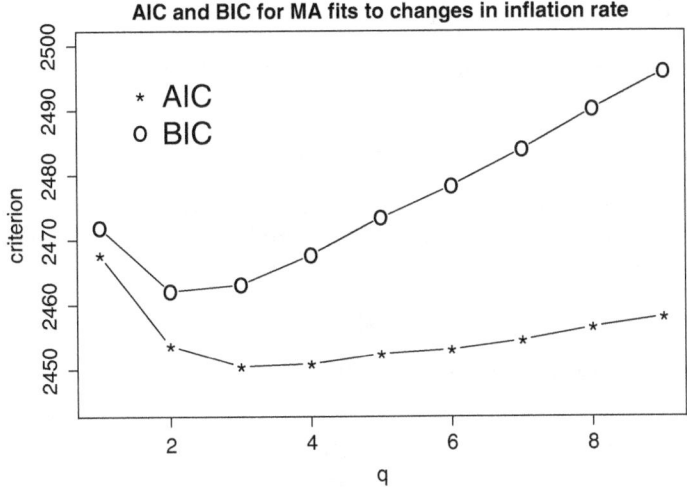

Fig. 12.12. *Fitting MA(q) models to changes in the one-month inflation rate; AIC and BIC plotted against q.*

12.8 ARMA Processes

Stationary time series with complex autocorrelation behavior often are more parsimoniously modeled by mixed autoregressive and moving average (ARMA) processes than by either a pure AR or pure MA process. For example, it is sometimes the case that a model with one AR and one MA parameter, called an ARMA(1, 1) model, will provide a more parsimonious fit than a pure AR or pure MA model. This section introduces ARMA processes.

12.8.1 The Backwards Operator

The *backwards operator* B is a simple notation with a fancy name. It is useful for describing ARMA (and ARIMA) models. The backwards operator is defined by

$$B\,Y_t = Y_{t-1}$$

and, more generally,

$$B^h\,Y_t = Y_{t-h}.$$

Thus, B backs up time one unit while B^h does this repeatedly so that time is backed up h time units. Note that $B\,c = c$ for any constant c, since a constant does not change with time. The backwards operator is sometimes called the *lag operator*.

12.8.2 The ARMA Model

An $ARMA(p, q)$ model combines both AR and MA terms and is defined by the equation

$$(Y_t - \mu) = \phi_1(Y_{t-1} - \mu) + \cdots + \phi_p(Y_{t-p} - \mu) + \epsilon_t + \theta_1 \epsilon_{t-1} + \cdots + \theta_q \epsilon_{t-q}, \quad (12.25)$$

which shows how Y_t depends on lagged values of itself and lagged values of the white noise process. Equation (12.25) can be written more succinctly with the backwards operator as

$$(1 - \phi_1 B - \cdots - \phi_p B^p)(Y_t - \mu) = (1 + \theta_1 B + \cdots + \theta_q B^q)\epsilon_t. \quad (12.26)$$

A white noise process is ARMA(0,0) since if $p = q = 0$, then (12.26) reduces to

$$(Y_t - \mu) = \epsilon_t.$$

12.8.3 ARMA(1,1) Processes

The ARMA(1,1) model is commonly used in practice and is simple enough to study theoretically. In this section, formulas for its variance and ACF will be derived. Without loss of generality, one can assume that $\mu = 0$ when computing the variance and ACF. Multiplying the model

$$Y_t = \phi Y_{t-1} + \theta \epsilon_{t-1} + \epsilon_t \quad (12.27)$$

by ϵ_t and taking expectations, one has

$$\mathrm{Cov}(Y_t, \epsilon_t) = E(Y_t \epsilon_t) = \sigma_\epsilon^2, \quad (12.28)$$

since ϵ_t is independent of ϵ_{t-1} and Y_{t-1}. From (12.27) and (12.28),

$$\gamma(0) = \phi^2 \gamma(0) + (1 + \theta^2)\sigma_\epsilon^2 + 2\phi\theta\sigma_\epsilon^2, \quad (12.29)$$

and then solving (12.29) for $\gamma(0)$ gives us the formula

$$\gamma(0) = \frac{(1 + \theta^2 + 2\phi\theta)\sigma_\epsilon^2}{1 - \phi^2}. \quad (12.30)$$

By similar calculations, multiplying (12.27) by Y_{t-1} and taking expectations yields a formula for $\gamma(1)$. Dividing this formula by the right-hand side of (12.29) gives us

$$\rho(1) = \frac{(1 + \phi\theta)(\phi + \theta)}{1 + \theta^2 + 2\phi\theta}. \quad (12.31)$$

For $h \geq 2$, multiplying (12.27) by Y_{t-h} and taking expectations results in the formula

$$\rho(h) = \phi\rho(h - 1), \quad h \geq 2. \quad (12.32)$$

By (12.32), after one lag, the ACF of an ARMA(1,1) process decays in the same way as the ACF of an AR(1) process with the same ϕ.

12.8.4 Estimation of ARMA Parameters

The parameters of ARMA models can be estimated by maximum likelihood or conditional least-squares. These methods were introduced for AR(1) processes is Sect. 12.5. The estimation methods for AR(p) models are very similar to those for AR(1) models. For MA and ARMA, because the noise terms $\epsilon_1, \ldots, \epsilon_n$ are unobserved, there are complications that are best left for advanced time series texts.

Example 12.9. Changes in risk-free returns–ARMA models

This example uses the monthly changes in the risk-free returns shown in Fig. 4.3. In Table 12.1, AIC and BIC are shown for ARMA models with $p, q = 0, 1, 2$. We see that AIC and BIC are both minimized by the ARMA(1,1) model, though the MA(2) model is a very close second. The ARMA(1,1) and MA(2) fit nearly equally well, and it is difficult to decide between them.

Sample ACF, normal, and time series plots of the residuals from the ARMA(1,1) model are shown in Fig. 12.13. The ACF plot shows no short-term autocorrelation, which is another sign that the ARMA(1,1) model is satisfactory. However, the normal plot shows heavy tails and the residual time series plot shows volatility clustering. These problems will be addressed in later chapters. □

Table 12.1. *AIC and BIC for ARMA models fit to the monthly changes in the risk-free interest returns. The minimum values of both criteria are shown in boldface. To improve the appearance of the table, 1290 was added to all AIC and BIC values.*

p	q	AIC	BIC
0	0	29.45	37.8
0	1	9.21	21.8
0	2	3.00	19.8
1	0	14.86	27.5
1	1	**2.67**	**19.5**
1	2	4.67	25.7
2	0	5.61	22.4
2	1	6.98	28.0
2	2	4.89	30.1

12.8.5 The Differencing Operator

The *differencing operator* is another useful notation and is defined as $\Delta = 1 - B$, where B is the backwards operator, so that

$$\Delta Y_t = Y_t - B Y_t = Y_t - Y_{t-1}.$$

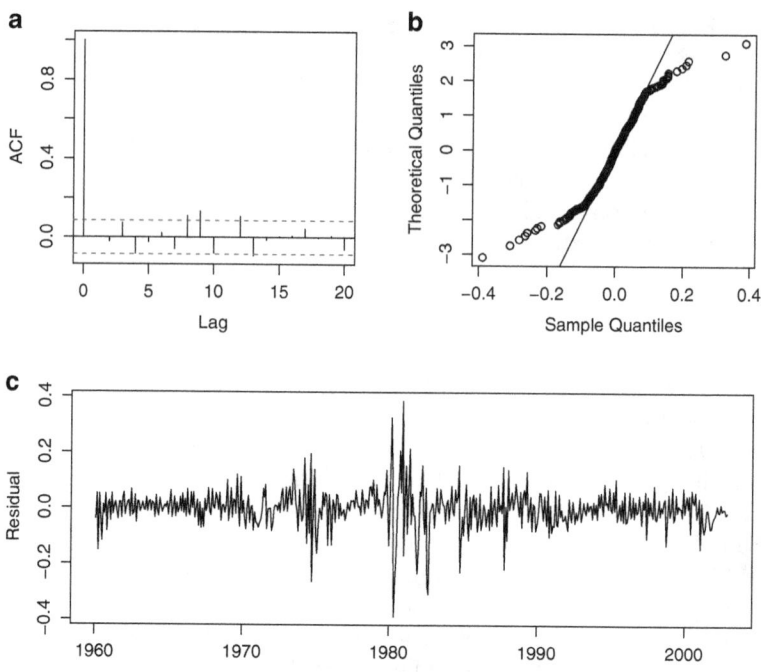

Fig. 12.13. *Residual sample ACF, normal quantile, and time series plots for the ARMA(1, 1) fit to the monthly changes in the risk-free returns.*

For example, if $p_t = \log(P_t)$ is the log price, then the log return is

$$r_t = \Delta p_t.$$

Differencing can be iterated. For example,

$$\Delta^2 Y_t = \Delta(\Delta Y_t) = \Delta(Y_t - Y_{t-1}) = (Y_t - Y_{t-1}) - (Y_{t-1} - Y_{t-2})$$
$$= Y_t - 2Y_{t-1} + Y_{t-2}.$$

Δ^k is called the kth-order differencing operator. A general formula for Δ^k can be derived from a binomial expansion:

$$\Delta^k Y_t = (1 - B)^k Y_t = \sum_{\ell=0}^{k} \binom{k}{\ell} (-1)^\ell Y_{t-\ell}. \tag{12.33}$$

12.9 ARIMA Processes

Often the first or perhaps second differences of nonstationary time series are stationary. For example, the first differences of a random walk (nonstationary) are white noise (stationary). In this section, *autoregressive integrated moving average* (ARIMA) processes are introduced. They include stationary as well as nonstationary processes.

A time series Y_t is said to be an *ARIMA(p, d, q)* process if $\Delta^d Y_t$ is ARMA(p, q). For example, if log returns on an asset are ARMA(p, q), then the log prices are ARIMA$(p, 1, q)$. An ARIMA(p, d, q) is stationary only if $d = 0$. Otherwise, only its differences of order d or above are stationary.

Notice that an ARIMA$(p, 0, q)$ model is the same as an ARMA(p, q) model. ARIMA$(p, 0, 0)$, ARMA$(p, 0)$, and AR(p) models are the same. Similarly, ARIMA$(0, 0, q)$, ARMA$(0, q)$, and MA(q) models are the same. A random walk is an ARIMA$(0, 1, 0)$ model, and white noise is an ARIMA$(0, 0, 0)$ model.

The inverse of differencing is "integrating." The integral of a process Y_t is the process w_t, where

$$w_t = w_{t_0} + Y_{t_0+1} + \cdots + Y_t. \tag{12.34}$$

Here t_0 is an arbitrary starting time point and w_{t_0} is the starting value of the w_t process. It is easy to check that

$$\Delta w_t = Y_t, \tag{12.35}$$

so integrating and differencing are inverse processes.[7]

We will say that a process is I(d) if it is stationary after being differenced d times. For example, a stationary process is I(0). An ARIMA(p, d, q) process is I(d). An I(d) process is said to be "integrated to order d."

Figure 12.14 shows an AR(1) process, its integral, and its second integral, meaning the integral of its integral. These three processes are I(0), I(1), and I(2), respectively. The three processes behave in entirely different ways. The AR(1) process is stationary and varies randomly about its mean, which is 0; one says that the process *reverts* to its mean. The integral of this process behaves much like a random walk in having no fixed level to which it reverts. The second integral has *momentum*. Once the process starts moving upward or downward, it tends to continue in that direction. If data show momentum like this, then the momentum is an indication that $d = 2$. The AR(1) process was generated by the R function `arima.sim()`. This process was integrated twice with R's `cumsum()` function.

```
21 set.seed(4631)
22 y1 = arima.sim(n = 500, list(ar = c(0.4)))
23 y2 = cumsum(y1)
24 y3 = cumsum(y2)
```

Example 12.10. Fitting an ARIMA model to CPI data

---

[7] An analog is, of course, differentiation and integration in calculus, which are inverses of each other.

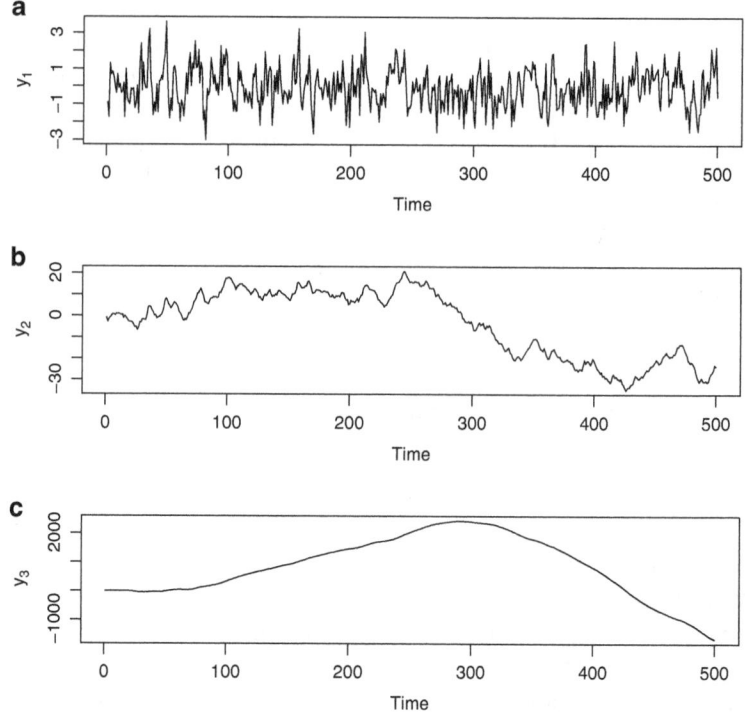

Fig. 12.14. *The (a) top plot is of an AR(1) process with $\mu = 0$ and $\phi = 0.4$. The (b) middle and (c) bottom plots are, respectively, the integral and second integral of this AR(1) process. Thus, from top to bottom, the series are I(0), I(1), and I(2), respectively.*

This example uses the `CPI.dat.csv` data set. CPI is a seasonally adjusted U.S. Consumer Price Index. The data are monthly. Only data from January 1977 to December 1987 are used in this example. Figure 12.15 shows time series plots of log(CPI) and the first and second differences of this series. The original series shows the type of momentum that is characteristic of an I(2) series. The first differences show no momentum, but they do not appear to be mean-reverting and so they may be I(1). The second differences appear to be mean-reverting and therefore seem to be I(0). ACF plots in Fig. 12.16a,b, and c provide additional evidence that the log(CPI) is I(2).

Notice that the ACF of $\Delta^2 \log(\text{CPI})$ has large correlations at the first two lags and then small autocorrelations after that. This suggests using an MA(2) for $\Delta^2 \log(\text{CPI})$ or, equivalently, an ARIMA(0,2,2) model for log(CPI). The ACF of the residuals from this fit is shown in Fig. 12.16d. The residual ACF has small correlations at short lags, which is an indication that the ARIMA(0,2,2) model fits well. Also, the residuals pass Ljung–Box tests for various choices of `lag`, for example, with a p-value of 0.08 at `lag = 20`, with `fitdf = 2`.   □

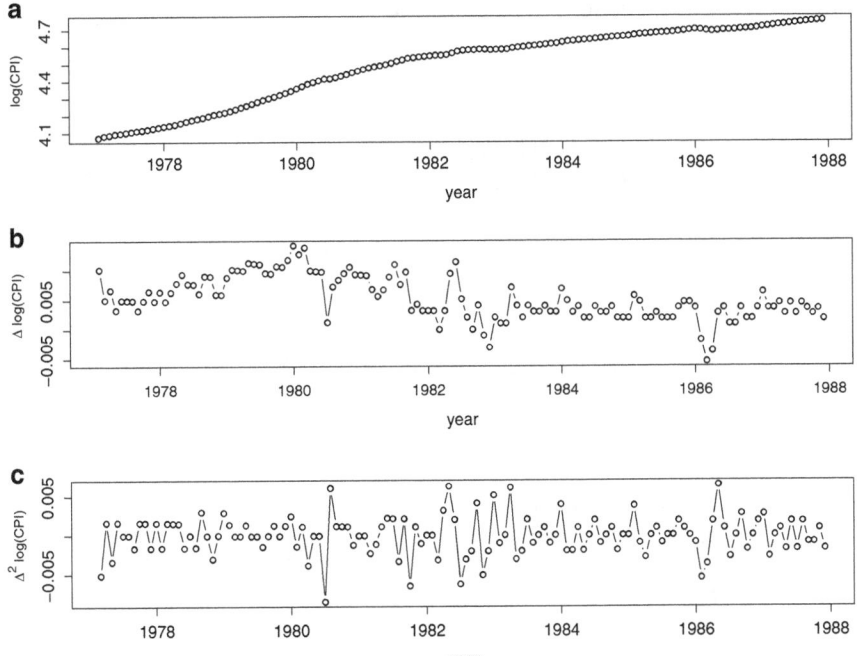

Fig. 12.15. *(a) log(CPI), (b) first differences of log(CPI), and (c) second differences of log(CPI).*

Example 12.11. Fitting an ARIMA model to industrial production (IP) data

This example uses the `IP.dat` data set. The variable, IP, is a seasonally adjusted U.S. industrial production index. Figure 12.17 panels (a) and (b) show time series plots of log(IP) and Δlog(IP) and panel (c) has the sample ACF of Δlog(IP). The log(IP) series appears to be I(1), implying that we should fit an ARMA model to Δlog(IP). AR(1), AR(2), and ARMA(1,1) each fit Δlog(IP) reasonably well and the ARMA(1,0) model is selected using the BIC criterion with R's `auto.arima()` function. The ACF of the residuals in Fig. 12.17d indicates a satisfactory fit to the ARMA(1,0) model since it shows virtually no short-term autocorrelation. In summary, log(IP) is well fit by an ARIMA(1,1,0) model. □

12.9.1 Drifts in ARIMA Processes

If a nonstationary process has a constant mean, then the first differences of this process have mean zero. For this reason, it is often assumed that a differenced process has mean zero. The `arima()` function in R makes this assumption.

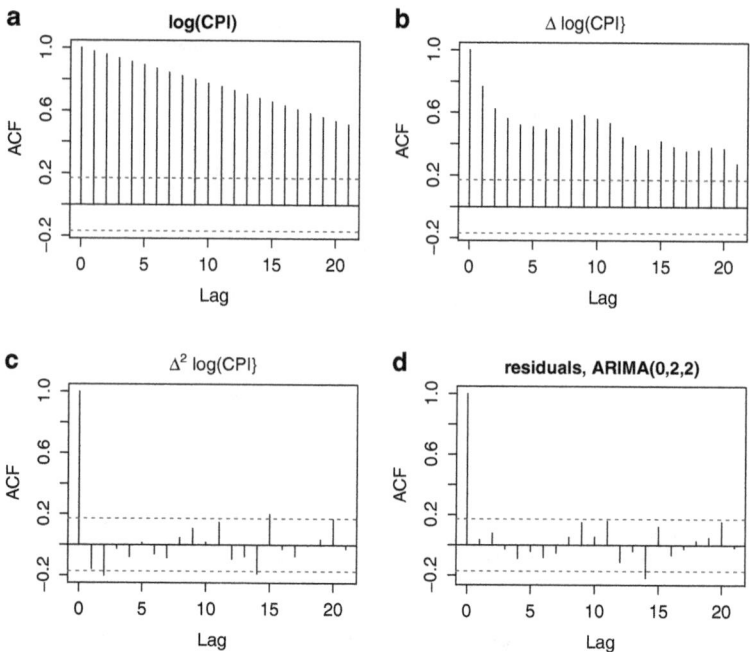

Fig. 12.16. *Sample ACF of (a) log(CPI), (b) first differences of log(CPI), (c) second differences of log(CPI), and (d) residuals from an ARIMA(0,2,2) model fit to log(CPI).*

Instead of a constant mean, sometimes a nonstationary process has a mean with a deterministic linear trend, e.g., $E(Y_t) = \beta_0 + \beta_1 t$. Then, β_1 is called the *drift* of Y_t. Note that $E(\Delta Y_t) = \beta_1$, so if Y_t has a nonzero drift then ΔY_t has a nonzero mean. The R function `auto.arima()` discussed in Sect. 12.11 allows a differenced process to have a nonzero mean, which is called the `drift` in the output.

These ideas can be extended to higher-degree polynomial trends and higher-order differencing. If $E(Y_t)$ has an mth-degree polynomial trend, then the mean of $E(\Delta^d Y_t)$ has an $(m-d)$th-degree trend for $d \leq m$. For $d > m$, $E(\Delta^d Y_t) = 0$.

12.10 Unit Root Tests

We have seen that it can be difficult to tell whether a time series is best modeled as stationary or nonstationary. To help decide between these two possibilities, it can be helpful to use hypothesis testing.

What is meant by a unit root? Recall that an ARMA(p, q) process can be written as

$$(Y_t - \mu) = \phi_1(Y_{t-1} - \mu) + \cdots + \phi_p(Y_{t-p} - \mu) + \epsilon_t + \theta_1 \epsilon_{t-1} + \cdots + \theta_q \epsilon_{t-q}. \quad (12.36)$$

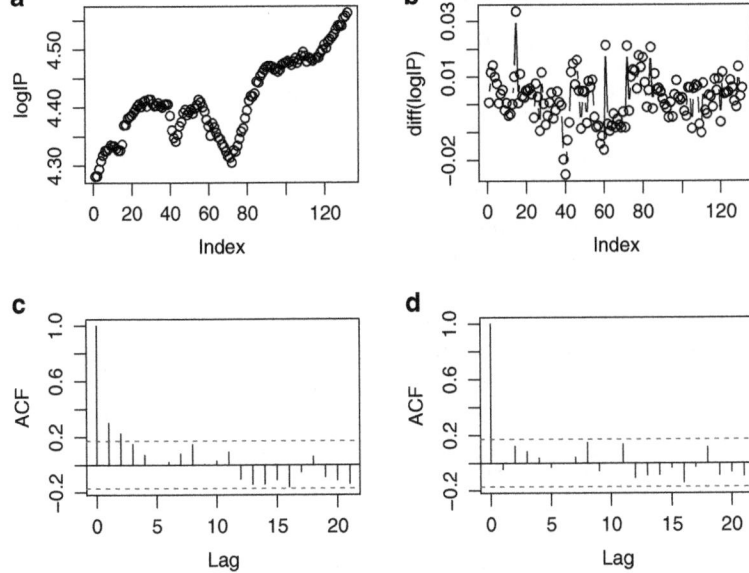

Fig. 12.17. *Time series plots of (a) log(IP) and (b) $\Delta \log(IP)$, and sample ACF plots of (c) $\Delta \log(IP)$ and (d) residual from ARMA(1,0) fit to $\Delta \log(IP)$.*

The condition for $\{Y_t\}$ to be stationary is that all roots of the polynomial

$$1 - \phi_1 x - \cdots - \phi_p x^p \tag{12.37}$$

have absolute values greater than one. (See Appendix A.21 for information about complex roots of polynomials and the absolute value of a complex number.) For example, when $p = 1$, then (12.37) is

$$1 - \phi x$$

and has one root, $1/\phi$. We know that the process is stationary if $|\phi| < 1$, which, of course, is equivalent to $|1/\phi| > 1$.

If there is a unit root, that is, a root with an absolute value equal to 1, then the ARMA process is nonstationary and behaves much like a random walk. Not surprisingly, this is called the unit root case. The explosive case is when a root has an absolute value less than 1.

Example 12.12. Inflation rates

Recall from Examples 12.1 and 12.3 that we have had difficulty deciding whether the inflation rates are stationary or not. If we fit stationary ARMA models to the inflation rates, then `auto.arima()` selects an ARMA(2,1) model and the AR coefficients are $\widehat{\phi}_1 = 1.229$ and $\widehat{\phi}_2 = -0.233$. The roots of

$$1 - \widehat{\phi}_1 x - \widehat{\phi}_2 x^2$$

can be found easily using R's `polyroot()` function and are 1.0053 and 4.2694.

```
25 polyroot(c(1, -1.229, +0.233))
```

Both roots have absolute values greater than 1, indicating possible stationarity; however, the first is very close to 1, and since the roots are estimated with error, there is reason to suspect that this series may be nonstationary. □

Unit root tests are used to decide if an AR model has an absolute root equal to 1. One popular unit root test is the augmented Dickey–Fuller test, often called the ADF test. The null hypothesis is that there is a unit root. The usual alternative is that the process is stationary, but one can instead use the alternative that the process is explosive.

Another unit root test is the Phillips–Perron test. It is similar to the Dickey–Fuller test, but differs in some details.

A third test is the KPSS test. The null hypothesis for the KPSS test is stationarity and the alternative is a unit root, just the opposite of the hypotheses for the Dickey–Fuller and Phillips–Perron tests.

Example 12.13. Inflation rates—unit root tests

Recall that we were undecided as to whether or not the inflation rate time series was stationary. The unit root tests might help resolve this issue, but unfortunately they do not provide unequivocal evidence in favor of stationarity. Both the augmented Dickey–Fuller and Phillips–Perron tests, which were implemented in R with the functions `adf.test()` and `pp.test()`, respectively, have small p-values, 0.016 for the former and less than 0.01 for the latter; see the output below. The functions `adf.test()`, `pp.test()`, and `kpss.test()` (used below) are in R's `tseries` package. Therefore, at level 0.05 the null hypothesis of a unit root is rejected by both tests in favor of the alternative of stationarity, the default alternative hypothesis for both `adf.test()` and `pp.test()`.

```
26 library(tseries)
27 adf.test(y)

   Augmented Dickey-Fuller Test
   data:  y
   Dickey-Fuller = -3.8651, Lag order = 7, p-value = 0.01576
   alternative hypothesis: stationary

28 pp.test(y)

   Phillips-Perron Unit Root Test
   data:  y
   Dickey-Fuller Z(alpha) = -248.75, Truncation lag parameter = 5,
      p-value = 0.01
   alternative hypothesis: stationary
```

Although the augmented Dickey–Fuller and Phillips–Perron tests suggest that the inflation rate series is stationary since the null hypothesis of a unit root is rejected, the KPSS test leads one to the opposite conclusion. The null hypothesis for the KPSS is stationarity and it is rejected with a p-value smaller than 0.01. Here is the R output.

```
29 kpss.test(y)

   KPSS Test for Level Stationarity
   data:  y
   KPSS Level = 2.51, Truncation lag parameter = 5, p-value = 0.01
```

Thus, the unit root tests are somewhat contradictory. Perhaps the inflation rates are stationary with long-term memory. Long-memory processes will be introduced in Sect. 13.5.   □

12.10.1 How Do Unit Root Tests Work?

A full discussion of the theory behind unit root tests is beyond the scope of this book. Here, only the basic idea will be mentioned. See Sect. 12.14 for more information. The Dickey–Fuller test is based on the AR(1) model

$$Y_t = \phi Y_{t-1} + \epsilon_t. \tag{12.38}$$

The null hypothesis (H_0) is that there is a unit root, that is, $\phi = 1$, and the alternative (H_1) is stationarity, which is equivalent to $\phi < 1$, assuming, as seems reasonable, that $\phi > -1$. The AR(1) model (12.38) is equivalent to $\Delta Y_t = (\phi - 1)Y_{t-1} + \epsilon_t$, or

$$\Delta Y_t = \pi Y_{t-1} + \epsilon_t, \tag{12.39}$$

where $\pi = \phi - 1$. Stated in terms of π, H_0 is $\pi = 0$ and H_1 is $\pi < 0$. The Dickey–Fuller test regresses ΔY_t on Y_{t-1} and tests H_0. Because Y_{t-1} is nonstationary under H_0, the t-statistic for π has a nonstandard distribution so special tables need to be developed in order to compute p-values.

The augmented Dickey–Fuller test expands model (12.39) by adding a time trend and lagged values of ΔY_t. Typically, the time trend is linear so that the expanded model is

$$\Delta Y_t = \beta_0 + \beta_1 t + \pi Y_{t-1} + \sum_{j=1}^{p} \gamma_j \Delta Y_{t-j} + \epsilon_t. \tag{12.40}$$

The hypotheses are still H_0: $\pi = 0$ and H_1: $\pi < 0$. There are several methods for selecting p. The adf.test() function has a default value of p equal to trunc((length(y)-1)^(1/3)), where y is the input series (Y_t in our notation).

12.11 Automatic Selection of an ARIMA Model

It is useful to have an automatic method for selecting an ARIMA model. As always, an automatically selected model should not be accepted blindly, but it makes sense to start model selection with something chosen quickly and by an objective criterion.

The R function `auto.arima()` can select all three parameters, p, d, and q, for an ARIMA model. The differencing parameter d is selected using the KPSS test. If the null hypothesis of stationarity is accepted when the KPSS is applied to the original time series, then $d = 0$. Otherwise, the series is differenced until the KPSS accepts the null hypothesis. After that, p and q are selected using either AIC or BIC.

Example 12.14. Inflation rates—automatic selection of an ARIMA model

In this example, `auto.arima()` is applied to the inflation rates. The ARIMA (1,1,1) model is selected by `auto.arima()` using either AIC or BIC to select p and q after $d = 1$ is selected by the KPSS test.

```
30 auto.arima(y, max.p = 5, max.q = 5, ic = "bic", trace = FALSE)

    Series: y
    ARIMA(1,1,1)

    Coefficients:
              ar1      ma1
            0.238   -0.877
    s.e.    0.055    0.027

    sigma^2 estimated as 8.55:   log likelihood=-1221.6
    AIC=2449.2    AICc=2449.3    BIC=2461.8
```

This is a very parsimonious model and residual diagnostics (not shown) show that it fits well.

`AICc` in the output from `auto.arima()` is the value of the corrected AIC criterion defined by (5.34). The sample size is 491 so, not surprisingly, AICc is equal to AIC, at least after rounding to the nearest integer.   □

12.12 Forecasting

Forecasting means predicting future values of a time series using the current *information set*, which is the set of present and past values of the time series. In some contexts, the information set could include other variables related to the time series, but in this section the information set contains only the past and present values of the time series that is being predicted.

ARIMA models are often used for forecasting. Consider forecasting using an AR(1) process. Suppose that we have data $Y_1, \ldots, Y_n$ and estimates $\widehat{\mu}$ and $\widehat{\phi}$. We know that

$$Y_{n+1} = \mu + \phi(Y_n - \mu) + \epsilon_{n+1}. \tag{12.41}$$

Since ϵ_{n+1} is independent of the past and present, by Result 11.1 in Sect. 11.9.2 the best predictor of ϵ_{n+1} is its expected value, which is 0. We know, of course, that ϵ_{n+1} is not 0, but 0 is our best guess at its value. On the other hand, we know or have estimates of all other quantities in (12.41). Therefore, we predict Y_{n+1} by

$$\widehat{Y}_{n+1} = \widehat{\mu} + \widehat{\phi}(Y_n - \widehat{\mu}).$$

By the same reasoning we forecast Y_{n+2} by

$$\widehat{Y}_{n+2} = \widehat{\mu} + \widehat{\phi}(\widehat{Y}_{n+1} - \widehat{\mu}) = \widehat{\mu} + \widehat{\phi}\{\widehat{\phi}(Y_n - \widehat{\mu})\}, \tag{12.42}$$

and so forth. Notice that in (12.42) we do not use Y_{n+1}, which is unknown at time n, but rather the forecast $\widehat{Y}_{n+1}$. Continuing in this way, we find the general formula for the k-step-ahead forecast:

$$\widehat{Y}_{n+k} = \widehat{\mu} + \widehat{\phi}^k(Y_n - \widehat{\mu}). \tag{12.43}$$

If $|\widehat{\phi}| < 1$, as is true for a stationary series, then as k increases, the forecasts will converge geometrically fast to $\widehat{\mu}$.

Formula (12.43) is valid only for AR(1) processes, but forecasting other AR(p) processes is similar. For an AR(2) process,

$$\widehat{Y}_{n+1} = \widehat{\mu} + \widehat{\phi}_1(Y_n - \widehat{\mu}) + \widehat{\phi}_2(Y_{n-1} - \widehat{\mu})$$

and

$$\widehat{Y}_{n+2} = \widehat{\mu} + \widehat{\phi}_1(\widehat{Y}_{n+1} - \widehat{\mu}) + \widehat{\phi}_2(Y_n - \widehat{\mu}),$$

and so on.

Forecasting ARMA and ARIMA processes is similar to forecasting AR processes. Consider the MA(1) process, $Y_t - \mu = \epsilon_t + \theta\epsilon_{t-1}$. Then the next observation will be

$$Y_{n+1} = \mu + \epsilon_{n+1} + \theta\epsilon_n. \tag{12.44}$$

In the right-hand side of (12.44) we replace μ and θ by estimates and ϵ_n by the residual $\widehat{\epsilon}_n$. Also, since ϵ_{n+1} is independent of the observed data, it is replaced by its mean 0. Then the forecast is

$$\widehat{Y}_{n+1} = \widehat{\mu} + \widehat{\theta}\,\widehat{\epsilon}_n.$$

The two-step-ahead forecast of $Y_{n+2} = \mu + \epsilon_{n+2} + \theta\epsilon_{n+1}$ is simply $\widehat{Y}_{n+2} = \widehat{\mu}$, since ϵ_{n+1} and ϵ_{n+2} are independent of the observed data. Similarly, $\widehat{Y}_{n+k} = \widehat{\mu}$ for all $k > 2$.

To forecast the ARMA(1,1) process

$$Y_t - \mu = \phi(Y_{t-1} - \mu) + \epsilon_t + \theta\epsilon_{t-1},$$

we use

$$\widehat{Y}_{n+1} = \widehat{\mu} + \widehat{\phi}(Y_n - \widehat{\mu}) + \widehat{\theta}\widehat{\epsilon}_n$$

as the one-step-ahead forecast and

$$\widehat{Y}_{n+k} = \widehat{\mu} + \widehat{\phi}(\widehat{Y}_{n+k-1} - \widehat{\mu}), \ k \geq 2$$

for forecasting two or more steps ahead.

As a final example, suppose that Y_t is ARIMA(1,1,0), so that ΔY_t is AR(1). To forecast Y_{n+k}, $k \geq 1$, one first fits an AR(1) model to the ΔY_t process and forecasts ΔY_{n+k}, $k \geq 1$. Let the forecasts be denoted by $\widehat{\Delta Y}_{n+k}$, $k \geq 1$. Then, since

$$Y_{n+1} = Y_n + \Delta Y_{n+1},$$

the forecast of Y_{n+1} is

$$\widehat{Y}_{n+1} = Y_n + \widehat{\Delta Y}_{n+1},$$

and similarly

$$\widehat{Y}_{n+2} = \widehat{Y}_{n+1} + \widehat{\Delta Y}_{n+2} = Y_n + \widehat{\Delta Y}_{n+1} + \widehat{\Delta Y}_{n+2},$$

and so on.

Most time series software packages offer functions to automate forecasting. R's `predict()` function forecasts using an "object" returned by the `arima()` fitting function.

12.12.1 Forecast Errors and Prediction Intervals

When making forecasts, one would of course like to know the uncertainty of the predictions. To this end, one first computes the variance of the forecast error. Then a $(1 - \alpha)100\,\%$ prediction interval is the forecast itself plus or minus the forecast error's standard deviation times $z_{\alpha/2}$ (the normal upper quantile). The use of $z_{\alpha/2}$ assumes that $\epsilon_1, \epsilon_2, \ldots$ is Gaussian white noise. If the residuals are heavy-tailed, then we might be reluctant to make the Gaussian assumption. One way to avoid this assumption is discussed in Sect. 12.12.2.

Computation of the forecast error variance and the prediction interval is automated by modern statistical software, so we need not present general formulas for the forecast error variance. However, to gain some understanding of general principles, we will look at two special cases, one stationary and the other nonstationary.

Stationary AR(1) Forecast Errors

We will first consider the errors made when forecasting a stationary AR(1) process. The error in the first prediction is

$$Y_{n+1} - \widehat{Y}_{n+1} = \{\mu + \phi(Y_n - \mu) + \epsilon_{n+1}\} - \{\widehat{\mu} + \widehat{\phi}(Y_n - \widehat{\mu})\}$$
$$= (\mu - \widehat{\mu}) + (\phi - \widehat{\phi})Y_n - (\phi\mu - \widehat{\phi}\widehat{\mu}) + \epsilon_{n+1} \qquad (12.45)$$
$$\approx \epsilon_{n+1}. \qquad (12.46)$$

Here (12.45) is the exact error and (12.46) is a "large-sample" approximation. The basis for (12.46) is that as the sample size increases $\widehat{\mu} \to \mu$ and $\widehat{\phi} \to \phi$, so the first three terms in (12.45) converge to 0, but the last term remains unchanged. The large-sample approximation simplifies formulas and helps us focus on the main components of the forecast error. Using the large-sample approximation again, so $\widehat{\mu}$ is replaced by μ and $\widehat{\phi}$ by ϕ, the error in the two-steps-ahead forecast is

$$Y_{n+2} - \widehat{Y}_{n+2} = \{\mu + \phi(Y_{n+1} - \mu) + \epsilon_{n+2}\} - \{\mu + \phi(\widehat{Y}_{n+1} - \mu)\}$$
$$= \phi(Y_{n+1} - \widehat{Y}_{n+1}) + \epsilon_{n+1}$$
$$= \phi\epsilon_{n+1} + \epsilon_{n+2}. \qquad (12.47)$$

Continuing in this manner, we find that the k-step-ahead forecasting error is

$$Y_{n+k} - \widehat{Y}_{n+k} \approx \{\mu + \phi(Y_{n+k-1} - \mu) + \epsilon_{n+k}\} - \{\mu + \phi(\widehat{Y}_{n+k-1} - \mu)\}$$
$$= \phi^{k-1}\epsilon_{n+1} + \phi^{k-2}\epsilon_{n+2} + \cdots + \phi\epsilon_{n+k-1} + \epsilon_{n+k}. \qquad (12.48)$$

By the formula for the sum of a finite geometric series, the variance of the right-hand side of (12.48) is

$$\left\{\phi^{2(k-1)} + \phi^{2(k-2)} + \cdots + \phi^2 + 1\right\}\sigma_\epsilon^2 = \left(\frac{1 - \phi^{2k}}{1 - \phi^2}\right)\sigma_\epsilon^2$$
$$\to \frac{\sigma_\epsilon^2}{1 - \phi^2} \text{ as } k \to \infty. \quad (12.49)$$

An important point here is that the variance of the forecast error does not diverge as $k \to \infty$, but rather the variance converges to $\gamma(0)$, the marginal covariance of the AR(1) process given by (12.7). This is an example of the general principle that for any stationary ARMA process, the variance of the forecast error converges to the marginal variance.

Forecasting a Random Walk

For the random walk process, $Y_{n+1} = \mu + Y_n + \epsilon_{n+1}$, many of the formulas just derived for the AR(1) process still hold, but with $\phi = 1$. An exception is that the last result in (12.49) does not hold because the summation formula for a geometric series does not apply when $\phi = 1$. One result that does still hold is

$$Y_{n+k} - \widehat{Y}_{n+k} = \epsilon_{n+1} + \epsilon_{n+2} + \cdots + \epsilon_{n+k-1} + \epsilon_{n+k}$$

so the variance of the k-step-ahead forecast error is $k\sigma_\epsilon^2$ and, unlike for the stationary AR(1) case, the forecast error variance diverges to ∞ as $k \to \infty$.

Forecasting ARIMA Processes

As mentioned before, in practice we do not need general formulas for the forecast error variance of ARIMA processes, since statistical software can compute the variance. However, it is worth repeating a general principle: For stationary ARMA processes, the variance of the k-step-ahead forecast error variance converges to a finite value as $k \to \infty$, but for a nonstationary ARIMA process this variance converges to ∞. The result of this principle is that for a nonstationary process, the forecast limits diverge away from each other as $k \to \infty$, but for a stationary process the forecast limits converge to parallel horizontal lines.

Example 12.15.   Forecasting the one-month inflation rate

We saw in Example 12.8 that an MA(3) model provided a parsimonious fit to the changes in the one-month inflation rate. This implies that an ARIMA(0,1,3) model will be a good fit to the inflation rates themselves. The two models are, of course, equivalent, but they forecast different series. The first model gives forecasts and confidence limits for the changes in the inflation rate, while the second model provides forecasts and confidence limits for the inflation rate itself.

Figures 12.18 and 12.19 plot forecasts and forecast limits from the two models out to 100 steps ahead. One can see that the forecast limits diverge for the second model and converge to parallel horizontal lines for the first model. □

12.12.2 Computing Forecast Limits by Simulation

Simulation can be used to compute forecasts limits. This is done by simulating random forecasts and finding their $\alpha/2$-upper and -lower sample quantiles. A set of random forecasts up to m time units ahead is generated for an ARMA process by recursion:

$$\begin{aligned}
\widehat{Y}_{n+t} = {}& \widehat{\mu} + \widehat{\phi}_1(\widehat{Y}_{n+t-1} - \widehat{\mu}) + \cdots + \widehat{\phi}_p(\widehat{Y}_{n+t-p} - \widehat{\mu}) \\
& + \widehat{\epsilon}_{n+t} + \widehat{\theta}_1\widehat{\epsilon}_{n+t-1} + \cdots + \widehat{\theta}_q\widehat{\epsilon}_{n+t-q}, \quad t = 1, \ldots, m, \quad (12.50)
\end{aligned}$$

in which

- $\widehat{\epsilon}_k$ is the kth residual if $k \leq n$,
- $\{\widehat{\epsilon}_k : k = n+1, \ldots, n+m\}$ is a resample from the residuals.

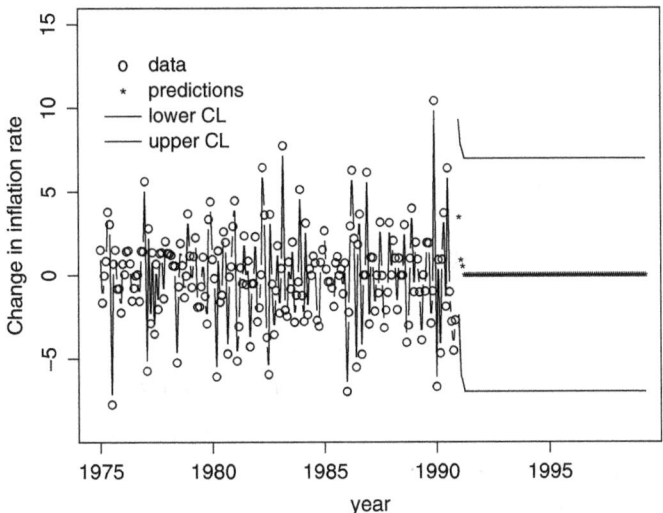

Fig. 12.18. *Forecasts of changes in inflation rate.*

Thus, $\widehat{Y}_{n+1}$ is generated from $Y_{n+1-p}, \ldots, Y_n, \widehat{\epsilon}_{n+1-q}, \ldots, \widehat{\epsilon}_{n+1}$, then $\widehat{Y}_{n+2}$ is generated from $Y_{n+2-p}, \ldots, Y_n, \widehat{Y}_{n+1}, \widehat{\epsilon}_{n+2-q}, \ldots, \widehat{\epsilon}_{n+2}$, then $\widehat{Y}_{n+3}$ is generated from $Y_{n+3-p}, \ldots, Y_n, \widehat{Y}_{n+1}, \widehat{Y}_{n+2}, \widehat{\epsilon}_{n+3-q}, \ldots, \widehat{\epsilon}_{n+3}$, and so forth.

A large number, call it B, of sets of random forecasts are generated in this way. They differ because their sets of future errors generated in step two are mutually independent. For each $t = 1, \ldots, m$, the $\alpha/2$-upper and -lower sample quantiles of the B random values of $\widehat{Y}_{n+h}$ are the forecast limits for Y_{n+h}.

To obtain forecasts, rather than forecast limits, one uses $\widehat{\epsilon}_k = 0$, for $k = n+1, \ldots, n+m$, in step two. The forecasts are nonrandom, conditional given the data, and therefore need to be computed only once.

If $Y_t = \Delta W_t$ for some nonstationary series $\{W_1, \ldots, W_n\}$, then random forecasts of $\{W_{n+1}, \ldots\}$ can be obtained as partial sums of $\{W_n, \widehat{Y}_{n+1}, \ldots\}$. For example,

$$\widehat{W}_{n+1} = W_n + \widehat{Y}_{n+1},$$
$$\widehat{W}_{n+2} = \widehat{W}_{n+1} + \widehat{Y}_{n+2} = W_n + \widehat{Y}_{n+1} + \widehat{Y}_{n+2},$$
$$\widehat{W}_{n+3} = \widehat{W}_{n+2} + \widehat{Y}_{n+3} = W_n + \widehat{Y}_{n+1} + \widehat{Y}_{n+2} + \widehat{Y}_{n+3},$$

and so forth. Then, upper and lower quantiles of the randomly generated $\widehat{W}_{n+h}$ can be used as forecast limits for W_{n+h}.

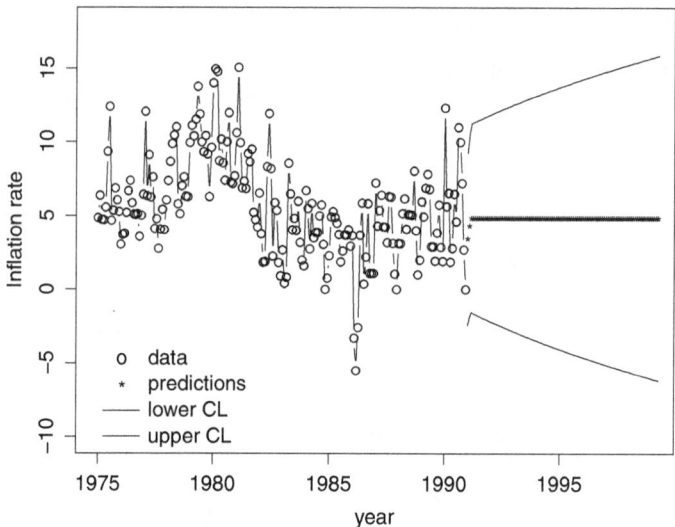

Fig. 12.19. *Forecasts of inflation rate.*

Example 12.16. Forecasting the one-month inflation rate and changes in the inflation rate by simulation

To illustrate the amount of random variation in the forecasts, a small number (five) of sets of random forecasts of the changes in the inflation rate were generated out to 30 months ahead. These are plotted in Fig. 12.20. Notice the substantial random variation between the random forecasts. Because of this large variation, to calculate forecasts limits a much larger number of random forecasts should be used. In this example, $B = 50,000$ sets of random forecasts are generated. Figure 12.21 shows the forecast limits, which are the 2.5 % upper and lower sample quantiles. For comparison, the forecast limits generated by R's function `arima()` are also shown. The two sets of forecast limits are very similar even though the `arima()` limits assume Gaussian noise, but the residuals are heavy-tailed. Thus, the presence of heavy tails does not invalidate the Gaussian limits in this example with 95 % forecast limits. If a larger confidence coefficient were used, that is, one very close to 1, then the forecast intervals based on sampling heavy-tailed residuals would be wider than those based on a Gaussian assumption.

As described above, forecasts for future inflation rates were obtained by taking partial sums of random forecasts of changes in the inflation rate and the forecast limits (upper and lower quantiles) are shown in Fig. 12.22. As expected for a nonstationary process, the forecast limits diverge.   □

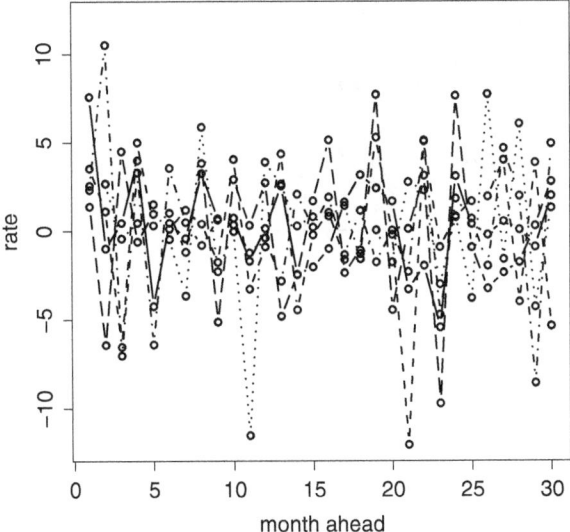

Fig. 12.20. *Five random sets of forecasts of changes in the inflation rate computed by simulation.*

There are two important advantages to using simulation for forecasting:

1. simulation can be used in situations where standard software does not compute forecast limits, and
2. simulation does not require that the noise series be Gaussian.

The first advantage will be important in some future examples, such as, multivariate AR processes fit by R's `ar()` function. The second advantage is less important if one is generating 90 % or 95 % forecast limits, but if one wishes more extreme quantiles, say 99 % forecast limits, then the second advantage could be more important since in most applications the noise series has heavier than Gaussian tails.

12.13 Partial Autocorrelation Coefficients

The partial autocorrelation function (PACF) can be useful for identifying the order of an AR process. The kth partial autocorrelation, denoted by $\phi_{k,k}$, for a stationary process Y_t is the correlation between Y_t and Y_{t+k}, conditional on $Y_{t+1}, \ldots, Y_{t+k-1}$. For $k = 1$, $Y_{t+1}, \ldots, Y_{t+k-1}$ is an empty set, so the partial autocorrelation coefficient is simply equal to the autocorrelation coefficient, that is, $\phi_{1,1} = \rho(1)$. Let $\widehat{\phi}_{k,k}$ denote the estimate of $\phi_{k,k}$. $\widehat{\phi}_{k,k}$ can be calculated by fitting the regression model

$$Y_t = \phi_{0,k} + \phi_{1,k} Y_{t-1} + \cdots + \phi_{k,k} Y_{t-k} + \epsilon_{k,t}.$$

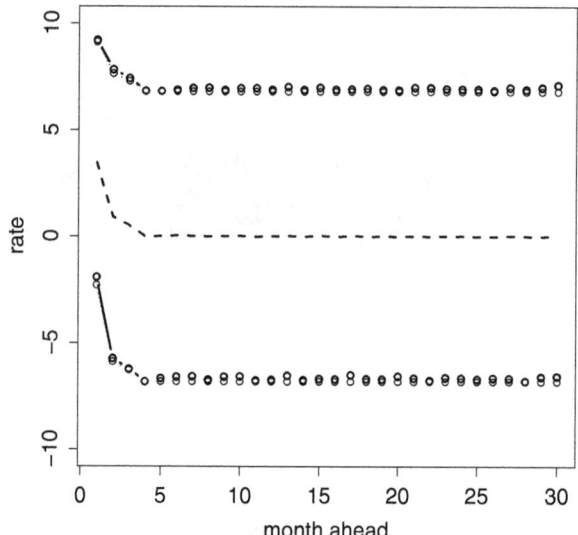

Fig. 12.21. *Forecast limits of changes in the inflation rate computed by simulation (solid), computed by* `arima()` *(dotted), and the mean of the forecast (dashed). Notice that the two sets of future limits are very similar and nearly overprint each other, so they are difficult to distinguish visually.*

If Y_t is an AR(p) process, then $\phi_{k,k} = 0$ for $k > p$. Therefore, a sign that a time series can be fit by an AR(p) model is that the sample PACF will be nonzero up to p and then will be nearly zero for larger lags.

Example 12.17. PACF for BMW log returns

Figure 12.23 is the sample PACF for the BMW log returns computed using the R function `pacf()`. The large value of $\widehat{\phi}_{1,1}$ and the smaller values of $\widehat{\phi}_{k,k}$ for $k = 2, \ldots, 9$ are a sign that this time series can be fit by an AR(1) model, in agreement with the results in Example 12.4. Note that $\widehat{\phi}_{k,k}$ is outside the test bounds for some values of $k > 9$, particularly for $k = 19$. This is likely due to random variation. □

When computing resources were expensive, the standard practice was to identify a tentative ARMA model using the sample ACF and PACF, fit this model, and then check the ACF and PACF of the residuals. If the residual ACF and PACF revealed some lack of fit, then the model could be enlarged. As computing has become much cheaper and faster and the use of information-based model selection criteria has become popular, this practice has changed. Now many data analysts prefer to start with a relatively large set of models and compare them with selection criteria such as AIC and BIC. This can be done automatically by `auto.arima()` in R or similar functions in other software packages.

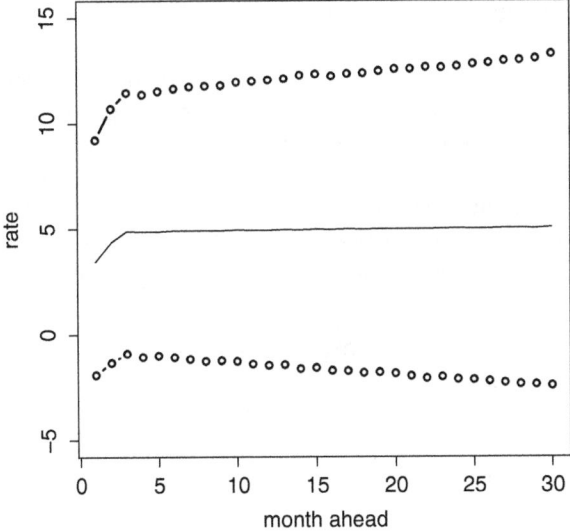

Fig. 12.22. *Forecast limits for the inflation rate computed by simulation.*

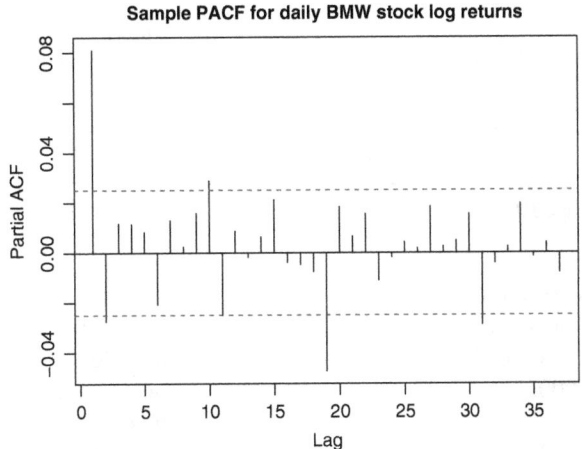

Fig. 12.23. *Sample PACF for the daily BMW stock log returns.*

Example 12.18. PACF for changes in the inflation rate

Figure 12.24 is the sample PACF for the changes in the inflation rate. The sample PACF decays slowly to zero, rather than dropping abruptly to zero as for an AR process. This is an indication that this time series should not be fit by a pure AR process. An MA or ARMA process would be preferable. In fact, we saw previously that an MA(2) or MA(3) model provides a parsimonious fit. □

12.14 Bibliographic Notes

There are many books on time series analysis and only a few will be mentioned. Box, Jenkins, and Reinsel (2008) did so much to popularize ARIMA models that these are often called "Box–Jenkins models." Hamilton (1994) is a comprehensive treatment of time series. Brockwell and Davis (1991) is particularly recommended for those with a strong mathematical preparation wishing to understand the theory of time series analysis. Brockwell and Davis (2003) is a gentler introduction to time series and is suited for those wishing

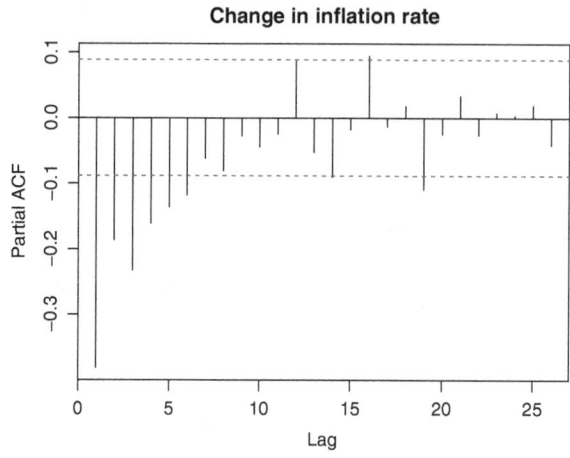

Fig. 12.24. *Sample PACF for changes in the inflation rate.*

to concentrate on applications. Enders (2004) and Tsay (2005) are time series textbooks concentrating on economic and financial applications; Tsay (2005) is written at a somewhat more advanced level than Enders (2004). Gourieroux and Jasiak (2001) has a chapter on the applications of univariate time series in financial econometrics, and Alexander (2001) has a chapter on time series models. Pfaff (2006) covers both the theory and application of unit root tests.

12.15 R Lab

12.15.1 T-bill Rates

Run the following code to input the `TbGdpPi.csv` data set and plot the three quarterly time series, as well as their auto- and cross-correlation functions. The last three lines of code run augmented Dickey–Fuller tests on the three series.

```
1  TbGdpPi = read.csv("TbGdpPi.csv", header=TRUE)
2  #   r = the 91-day treasury bill rate
3  #   y = the log of real GDP
4  #   pi = the inflation rate
5  TbGdpPi = ts(TbGdpPi, start = 1955, freq = 4)
6  library(tseries)
7  plot(TbGdpPi)
8  acf(TbGdpPi)
9  adf.test(TbGdpPi[,1])
10 adf.test(TbGdpPi[,2])
11 adf.test(TbGdpPi[,3])
```

Problem 1

(a) Describe the signs of nonstationarity seen in the time series and ACF plots.

(b) Use the augmented Dickey–Fuller tests to decide which of the series are nonstationary. Do the tests corroborate the conclusions of the time series and ACF plots?

Next run the augmented Dickey–Fuller test on the differenced series and plot the differenced series using the code below. Notice that the **pairs()** function creates a scatterplot matrix, but the **plot()** function applied to time series creates time series plots. [The **plot()** function would create a scatterplot matrix if the data were in a data.frame rather than having "class" time series (ts). Check the class of **diff_rate** with **attr(diff_rate,"class")**.] Both types of plots are useful. The former shows cross-sectional associations, while the time series plots are helpful when deciding whether differencing once is enough to induce stationarity. You should see that the first-differenced data look stationary.

```
12 diff_rate = diff(TbGdpPi)
13 adf.test(diff_rate[,1])
14 adf.test(diff_rate[,2])
15 adf.test(diff_rate[,3])
16 pairs(diff_rate)          #  scatterplot matrix
17 plot(diff_rate)           #  time series plots
```

Next look at the autocorrelation functions of the differenced series. These will be on the diagonal of a 3 × 3 matrix of plots. The off-diagonal plots are cross-correlation functions, which will be discussed in Chap. 13 and can be ignored for now.

```
18 acf(diff_rate)            # auto- and cross-correlations
```

Problem 2

1. *Do the differenced series appear stationary according to the augmented Dickey–Fuller tests?*
2. *Do you see evidence of autocorrelations in the differenced series? If so, describe these correlations.*

For the remainder of this lab, we will focus on the analysis of the 91-day T-bill rate. Since the time series are quarterly, it is good to see whether the mean depends on the quarter. One way to check for such effects is to compare boxplots of the four quarters. The following code does this. Note the use of the `cycle()` function to obtain the quarterly period of each observation; this information is embedded in the data and `cycle()` simply extracts it.

```
19  par(mfrow=c(1,1))
20  boxplot(diff_rate[,1] ~ cycle(diff_rate))
```

Problem 3 *Do you see any seasonal differences in the boxplots? If so, describe them.*

Regardless of whether seasonal variation is present, for now we will look at nonseasonal models. Seasonal models are introduced in Sect. 13.1. Next, use the `auto.arima()` function in the `forecast` package to find a "best-fitting" nonseasonal ARIMA model for the T-bill rates. The specifications `max.P=0` and `max.Q=0` force the model to be nonseasonal, since `max.P` and `max.Q` are the number of seasonal AR and MA components.

```
21  library(forecast)
22  auto.arima(TbGdpPi[,1], max.P=0, max.Q=0, ic="aic")
```

Problem 4

1. *What order of differencing is chosen? Does this result agree with your previous conclusions?*
2. *What model was chosen by AIC?*
3. *Which goodness-of-fit criterion is being used here?*
4. *Change the criterion to BIC. Does the best-fitting model then change?*
5. *Carefully express the fitted model chosen by the BIC criterion in mathematical notation.*

Finally, refit the best-fitting model with the following code, and check for any residual autocorrelation. You will need to replace the three question marks by the appropriate numerical values for the best-fitting model.

```
23  fit1 = arima(TbGdpPi[,1], order=c(?,?,?))
24  acf(residuals(fit1))
25  Box.test(residuals(fit1), lag = 12, type="Ljung", fitdf=?)
```

Problem 5 *Do you think that there is residual autocorrelation? If so, describe this autocorrelation and suggest a more appropriate model for the T-bill series.*

GARCH effects, that is, volatility clustering, can be detected by looking for auto-correlation in the mean-centered squared residuals. Another possibility is that some quarters are more variable than others. This can be detected for quarterly data by autocorrelation in the squared residuals at time lags that are a multiple of 4. Run the following code to look at autocorrelation in the mean-centered squared residuals.

```
26 resid2 = (residuals(fit1) - mean(residuals(fit1)))^2
27 acf(resid2)
28 Box.test(resid2, lag = 12, type="Ljung")
```

Problem 6 *Do you see evidence of GARCH effects?*

12.15.2 Forecasting

This example shows how to forecast a time series using R. Run the following code to fit a nonseasonal ARIMA model to the quarterly inflation rate. The code also uses the predict() function to forecast 36 quarters ahead. The standard errors of the forecasts are also returned by predict() and can be used to create prediction intervals. Note the use of col to specify colors. Replace c(?,?,?) by the specification of the ARIMA model that minimizes BIC.

```
1  TbGdpPi = read.csv("TbGdpPi.csv", header=TRUE)
2  attach(TbGdpPi)
3  #  r = the 91-day treasury bill rate
4  #  y = the log of real GDP
5  #  pi = the inflation rate
6  #  fit the non-seasonal ARIMA model found by auto.arima()
7  #  quarterly observations from 1955-1 to 2013-4
8  year = seq(1955,2013.75, by=0.25)
9  library(forecast)
10 auto.arima(pi, max.P=0, max.Q=0, ic="bic")
11 fit = arima(pi, order=c(?,?,?))
12 forecasts = predict(fit, 36)
13 plot(year,pi,xlim=c(1980,2023), ylim=c(-7,12), type="b")
14 lines(seq(from=2014, by=.25, length=36), forecasts$pred, col="red")
15 lines(seq(from=2014, by=.25, length=36),
16        forecasts$pred + 1.96*forecasts$se, col="blue")
17 lines(seq(from=2014, by=.25, length=36),
18        forecasts$pred - 1.96*forecasts$se, col="blue")
```

Problem 7 *Include the plot with your work.*

(a) *Why do the prediction intervals (blue curves) widen as one moves farther into the future?*

(b) *Why are the predictions (red) constant throughout?*

12.16 Exercises

1. This problem and the next use CRSP daily returns. First, get the data and plot the ACF in two ways:

```
1 library(Ecdat)
2 data(CRSPday)
3 crsp=CRSPday[,7]
4 acf(crsp)
5 acf(as.numeric(crsp))
```

   (a) Explain what "lag" means in the two ACF plots. Why does lag differ between the plots?

   (b) At what values of lag are there significant autocorrelations in the CRSP returns? For which of these values do you think the statistical significance might be due to chance?

2. Next, fit AR(1) and AR(2) models to the CRSP returns:

```
6 arima(crsp,order=c(1,0,0))
7 arima(crsp,order=c(2,0,0))
```

   (a) Would you prefer an AR(1) or an AR(2) model for this time series? Explain your answer.

   (b) Find a 95 % confidence interval for ϕ for the AR(1) model.

3. Consider the AR(1) model

$$Y_t = 5 - 0.55Y_{t-1} + \epsilon_t$$

   and assume that $\sigma_\epsilon^2 = 1.2$.

   (a) Is this process stationary? Why or why not?

   (b) What is the mean of this process?

   (c) What is the variance of this process?

   (d) What is the covariance function of this process?

4. Suppose that $Y_1, Y_2, \ldots$ is an AR(1) process with $\mu = 0.5$, $\phi = 0.4$, and $\sigma_\epsilon^2 = 1.2$.

   (a) What is the variance of Y_1?

   (b) What are the covariances between Y_1 and Y_2 and between Y_1 and Y_3?

   (c) What is the variance of $(Y_1 + Y_2 + Y_3)/2$?

5. An AR(3) model has been fit to a time series. The estimates are $\hat{\mu} = 104$, $\hat{\phi}_1 = 0.4$, $\hat{\phi}_2 = 0.25$, and $\hat{\phi}_3 = 0.1$. The last four observations were $Y_{n-3} = 105$, $Y_{n-2} = 102$, $Y_{n-1} = 103$, and $Y_n = 99$. Forecast Y_{n+1} and Y_{n+2} using these data and estimates.

6. Let Y_t be an MA(2) process,

$$Y_t = \mu + \epsilon_t + \theta_1 \epsilon_{t-1} + \theta_2 \epsilon_{t-2}.$$

Find formulas for the autocovariance and autocorrelation functions of Y_t.

7. Let Y_t be a stationary AR(2) process,

$$(Y_t - \mu) = \phi_1(Y_{t-1} - \mu) + \phi_2(Y_{t-2} - \mu) + \epsilon_t.$$

(a) Show that the ACF of Y_t satisfies the equation

$$\rho(k) = \phi_1 \rho(k-1) + \phi_2 \rho(k-2)$$

for all values of $k > 0$. (These are a special case of the Yule–Walker equations.)

[*Hint:* $\gamma(k) = \mathrm{Cov}(Y_t, Y_{t-k}) = \mathrm{Cov}\{\phi_1(Y_{t-1} - \mu) + \phi_2(Y_{t-2} - \mu) + \epsilon_t, Y_{t-k}\}$ and ϵ_t and Y_{t-k} are independent if $k > 0$.]

(b) Use part (a) to show that (ϕ_1, ϕ_2) solves the following system of equations:

$$\begin{pmatrix} \rho(1) \\ \rho(2) \end{pmatrix} = \begin{pmatrix} 1 & \rho(1) \\ \rho(1) & 1 \end{pmatrix} \begin{pmatrix} \phi_1 \\ \phi_2 \end{pmatrix}.$$

(c) Suppose that $\rho(1) = 0.4$ and $\rho(2) = 0.2$. Find ϕ_1, ϕ_2, and $\rho(3)$.

8. Use (12.11) to verify Eq. (12.12).

9. Show that if w_t is defined by (12.34) then (12.35) is true.

10. For a univariate, discrete time process, what is the difference between a strictly stationary process and a weakly stationary process?

11. The time series in the middle and bottom panels of Fig. 12.14 are both nonstationary, but they clearly behave in different manners. The time series in the bottom panel exhibits "momentum" in the sense that once it starts moving upward or downward, it often moves consistently in that direction for a large number of steps. In contrast, the series in the middle panel does not have this type of momentum and a step in one direction is quite likely to be followed by a step in the opposite direction. Do you think the time series model with momentum would be a good model for the price of a stock? Why or why not?

12. The MA(2) model $Y_t = \mu + \epsilon_t + \theta_1 \epsilon_{t-1} + \theta_2 \epsilon_{t-2}$ was fit to data and the estimates are

Parameter	Estimate
μ	45
θ_1	0.3
θ_2	−0.15

The last two values of the observed time series and residuals are

t	Y_t	$\widehat{\epsilon}_t$
$n-1$	39.8	-4.3
n	42.7	1.5

Find the forecasts of Y_{n+1} and Y_{n+2}.

13. The ARMA(1,2) model $Y_t = \mu + \phi_1 Y_{t-1} + \epsilon_t + \theta_1 \epsilon_{t-1} + \theta_2 \epsilon_{t-2}$ was fit to data and the estimates are

Parameter	Estimate
μ	103
ϕ_1	0.2
θ_1	0.4
θ_2	-0.25

The last two values of the observed time series and residuals are

t	Y_t	$\widehat{\epsilon}_t$
$n-1$	120.1	-2.3
n	118.3	2.6

Find the forecasts of Y_{n+1} and Y_{n+2}.

14. To decide the value of d for an ARIMA(p, d, q) model for a time series y, plots were created using the R program:

```
 8  par(mfrow=c(3,2))
 9  plot(y,type="l")
10  acf(y)
11  plot(diff(y),type="l")
12  acf(diff(y))
13  plot(diff(y,d=2),type="l")
14  acf(diff(y,d=2))
```

The output was the following figure:

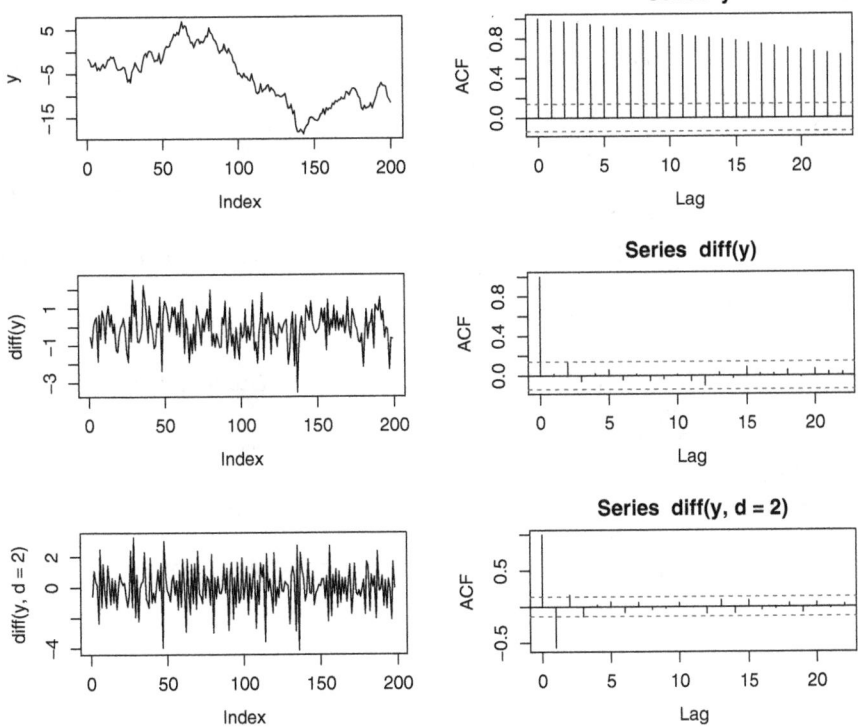

What value of d do you recommend? Why?

15. This problem fits an ARIMA model to the logarithms monthly one-month T-bill rates in the data set Mishkin in the Ecdat package. Run the following code to get the variable:

```
15  library(Ecdat)
16  data(Mishkin)
17  tb1 = log(Mishkin[,3])
```

   (a) Use time series and ACF plots to determine the amount of differencing needed to obtain a stationary series.
   (b) Next use auto.arima to determine the best-fitting nonseasonal ARIMA models. Use both AIC and BIC and compare the results.
   (c) Examine the ACF of the residuals for the model you selected. Do you see any problems?

16. Suppose you fit an AR(2) model to a time series Y_t, $t = 1, \ldots, n$, and the estimates were $\hat{\mu} = 100.1$, $\hat{\phi}_1 = 0.5$, and $\hat{\phi}_2 = 0.1$. The last three observations were $Y_{n-2} = 101.0$, $Y_{n-1} = 99.5$, and $Y_n = 102.3$. What are the forecasts of Y_{n+1}, Y_{n+2}, and Y_{n+3}?

17. In Sect. 12.9.1, it was stated that "if $E(Y_t)$ has an mth-degree polynomial trend, then the mean of $E(\Delta^d Y_t)$ has an $(m-d)$th-degree trend for $d \leq m$. For $d > m$, $E(\Delta^d Y_t) = 0$." Prove these assertions.

References

Alexander, C. (2001) *Market Models: A Guide to Financial Data Analysis*, Wiley, Chichester.

Box, G. E. P., Jenkins, G. M., and Reinsel, G. C. (2008) *Times Series Analysis: Forecasting and Control*, 4th ed., Wiley, Hoboken, NJ.

Brockwell, P. J. and Davis, R. A. (1991) *Time Series: Theory and Methods*, 2nd ed., Springer, New York.

Brockwell, P. J. and Davis, R. A. (2003) *Introduction to Time Series and Forecasting*, 2nd ed., Springer, New York.

Enders, W. (2004) *Applied Econometric Time Series, 2nd Ed.*, Wiley, New York.

Gourieroux, C., and Jasiak, J. (2001) *Financial Econometrics*, Princeton University Press, Princeton, NJ.

Hamilton, J. D. (1994) *Time Series Analysis*, Princeton University Press, Princeton, NJ.

Pfaff, B (2006) *Analysis of Integrated and Cointegrated Time Series with R*, Springer, New York.

Tsay, R. S. (2005) *Analysis of Financial Time Series*, 2nd ed., Wiley, New York.

13

Time Series Models: Further Topics

13.1 Seasonal ARIMA Models

Economic time series often exhibit strong seasonal variation. For example, an investor in mortgage-backed securities might be interested in predicting future housing starts, and these are usually much lower in the winter months compared to the rest of the year. Figure 13.1a is a time series plot of the logarithms of quarterly urban housing starts in Canada from the first quarter of 1960 to final quarter of 2001. The data are in the data set `Hstarts` in R's `Ecdat` package.

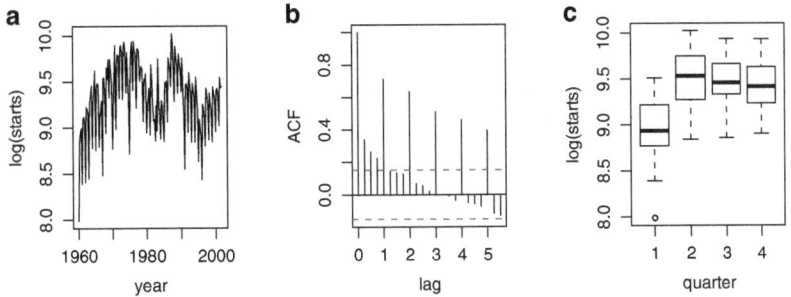

Fig. 13.1. *Logarithms of quarterly urban housing starts in Canada: (a) time series plot; (b) sample ACF; (c) boxplots by quarter.*

Figure 13.1 shows one and perhaps two types of nonstationarity: (1) There is strong seasonality, and (2) it is unclear whether the seasonal sub-series revert to a fixed mean and, if not, then this is a second type of nonstationarity because the process is integrated. These effects can also be seen in the ACF plot in Fig. 13.1b. At lags that are a multiples of four, the autocorrelations

© Springer Science+Business Media New York 2015
D. Ruppert, D.S. Matteson, *Statistics and Data Analysis for Financial Engineering*, Springer Texts in Statistics,
DOI 10.1007/978-1-4939-2614-5_13

are large, and decay slowly to zero. At other lags, the autocorrelations are smaller but also decay somewhat slowly. The boxplots in Fig. 13.1c give us a better picture of the seasonal effects. Housing starts are much lower in the first quarter than other quarters, jump to a peak in the second quarter, and then drop off slightly in the last two quarters.

Other time series might have only seasonal nonstationarity. For example, monthly average temperatures in a city with a temperate climate will show a strong seasonal effect, but if we plot temperatures for any single month of the year, say July, we will see mean-reversion.

13.1.1 Seasonal and Nonseasonal Differencing

Nonseasonal differencing is the type of differencing that we have been using so far. The series Y_t is replaced by $\Delta Y_t = Y_t - Y_{t-1}$ if the differencing is first-order, and so forth for higher-order differencing. Nonseasonal differencing does not remove seasonal nonstationarity and does not alone create a stationary series; see the top row of Fig. 13.2.

To remove seasonal nonstationarity, one uses seasonal differencing. Let s be the period. For example, $s = 4$ for quarterly data and $s = 12$ for monthly data. Define $\Delta_s = 1 - B^s$ so that $\Delta_s Y_t = Y_t - Y_{t-s}$.

Be careful to distinguish between $\Delta_s = 1 - B^s$ and $\Delta^s = (1-B)^s$. Note that $\Delta_s = 1 - B^s$ is the first-order seasonal differencing operator and $\Delta^s = (1-B)^s$ is the sth-order nonseasonal differencing operator. For example, $\Delta_2 Y_t = Y_t - Y_{t-2}$ but $\Delta^2 Y_t = \Delta(\Delta Y_t) = Y_t - 2Y_{t-1} + Y_{t-2}$.

The series $\Delta_s Y_t$ is called the seasonally differenced series. See the middle row of Fig. 13.2 for the seasonally differenced logarithm of housing starts and its Sample ACF.

One can combine seasonal and nonseasonal differencing by using, for example, for first-order differences

$$\Delta(\Delta_s Y_t) = \Delta(Y_t - Y_{t-s}) = (Y_t - Y_{t-s}) - (Y_{t-1} - Y_{t-s-1}).$$

The order in which the seasonal and nonseasonal difference operators are applied does not matter, since one can show that

$$\Delta(\Delta_s Y_t) = \Delta_s(\Delta Y_t).$$

For a seasonal time series, seasonal differencing is necessary, but whether also to use nonseasonal differencing will depend on the particular time series. For the housing starts data, the seasonally differenced series appears stationary so only seasonal differencing is absolutely needed, but combining seasonal and nonseasonal differencing might be preferred since it results in a simpler model.

13.1.2 Multiplicative ARIMA Models

One of the simplest seasonal models is the ARIMA$\{(1, 1, 0) \times (1, 1, 0)_s\}$ model, which puts together the nonseasonal ARIMA$(1,1,0)$ model

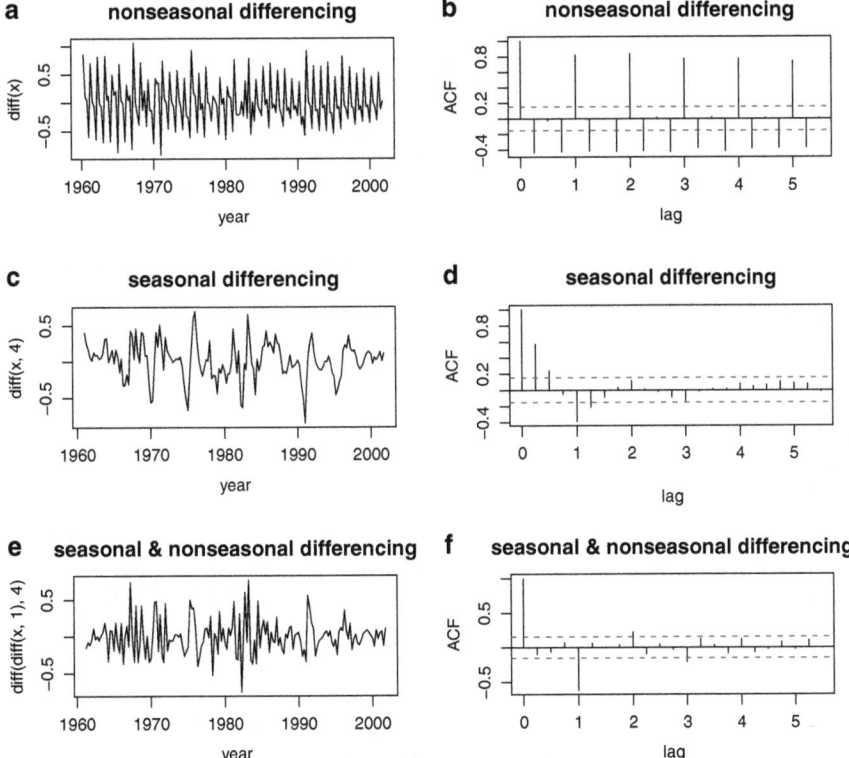

Fig. 13.2. *Time series (left column) and sample ACF plots (right column) of the logarithms of quarterly urban housing starts with nonseasonal differencing (top row), seasonal differencing (middle row), and both seasonal and nonseasonal differencing (bottom row). Note: in the sample ACF plots, lag = 1 means a lag of one year, which is four observations for quarterly data.*

$$(1 - \phi B)(\Delta Y_t - \mu) = \epsilon_t \tag{13.1}$$

and a purely seasonal $\mathrm{ARIMA}(1,1,0)_s$ model

$$(1 - \phi^* B^s)(\Delta_s Y_t - \mu) = \epsilon_t \tag{13.2}$$

to obtain the multiplicative model

$$(1 - \phi B)\,(1 - \phi^* B^s)\,\{\Delta(\Delta_s Y_t) - \mu\} = \epsilon_t. \tag{13.3}$$

Model (13.2) is called "purely seasonal" and has a subscript "s" since it uses only B^s and Δ_s; it is obtained from the $\mathrm{ARIMA}(1,1,0)$ by replacing B and Δ by B^s and Δ_s. For a monthly time series ($s = 12$), model (13.2) gives 12 independent processes, one for Januaries, a second for Februaries, and so forth. Model (13.3) uses the components from (13.1) to tie these 12 series together.

The ARIMA$\{(p, d, q) \times (p_s, d_s, q_s)_s\}$ process is

$$(1 - \phi_1 B - \cdots - \phi_p B^p) \{1 - \phi_1^* B^s - \cdots - \phi_{p_s}^* (B^s)^{p_s}\} \{\Delta^d(\Delta_s^{d_s} Y_t) - \mu\}$$
$$= (1 + \theta_1 B + \ldots + \theta_q B^q) \{1 + \theta_1^* B^s + \ldots + \theta_{q_s}^* (B^s)^{q_s}\} \epsilon_t. \tag{13.4}$$

This process multiplies together the AR components, the MA components, and the differencing components of two processes: the nonseasonal ARIMA (p, d, q) process

$$(1 - \phi_1 B - \cdots - \phi_p B^p) \{(\Delta^d Y_t) - \mu\} = (1 + \theta_1 B + \ldots + \theta_q B^q) \epsilon_t$$

and the seasonal ARIMA$(p_s, d_s, q_s)_s$ process

$$\{1 - \phi_1^* B^s - \cdots - \phi_{p_s}^* (B^s)^{p_s}\} \{(\Delta_s^{d_s} Y_t) - \mu\} = \{1 + \theta_1^* B^s + \ldots + \theta_{q_s}^* (B^s)^{q_s}\} \epsilon_t.$$

Example 13.1. ARIMA$\{(1, 1, 1) \times (0, 1, 1)_4\}$ model for housing starts

We return to the housing starts data. The first question is whether to difference only seasonally, or both seasonally and nonseasonally. The seasonally differenced quarterly series in the middle row of Fig. 13.2 is possibly stationary, so perhaps seasonal differencing is sufficient. However, the ACF of the seasonally and nonseasonally differenced series in the bottom row has a simpler ACF than the data that are only seasonally differenced. By differencing both ways, we should be able find a more parsimonious ARMA model.

Two models with seasonal and nonseasonal differencing were tried, ARIMA $\{(1, 1, 1) \times (1, 1, 1)_4\}$ and ARIMA$\{(1, 1, 1) \times (0, 1, 1)_4\}$. Both provided good fits and had residuals that passed the Ljung–Box test. The second of the two models was selected, because it has one fewer parameter than the first, though the other model would have been a reasonable choice. The results from fitting the chosen model are below.

```
1 data(Hstarts, package="Ecdat")
2 x = ts(Hstarts[,1], start=1960, frequency=4)
3 fit2 = arima(x, c(1,1,1), seasonal = list(order = c(0,1,1),
4   period = 4))
5 fit2

Call:
arima(x = hst, order = c(1, 1, 1), seasonal
= list(order = c(0, 1, 1), period = 4))

Coefficients:
          ar1      ma1     sma1
       0.675   -0.890   -0.822
s.e.   0.142    0.105    0.051

sigma^2 estimated as 0.0261: log-likelihood = 62.9,
    aic = -118
```

Thus, the fitted model is

$$(1 - 0.675\,B)Y_t^* = (1 - 0.890\,B)(1 - 0.822\,B_4)\,\epsilon_t$$

where $Y_t^* = \Delta(\Delta_4 Y_t)$ and ϵ_t is white noise with mean zero and variance 0.0261. Figure 13.3 shows forecasts from this model for the four years following the end of the time series. □

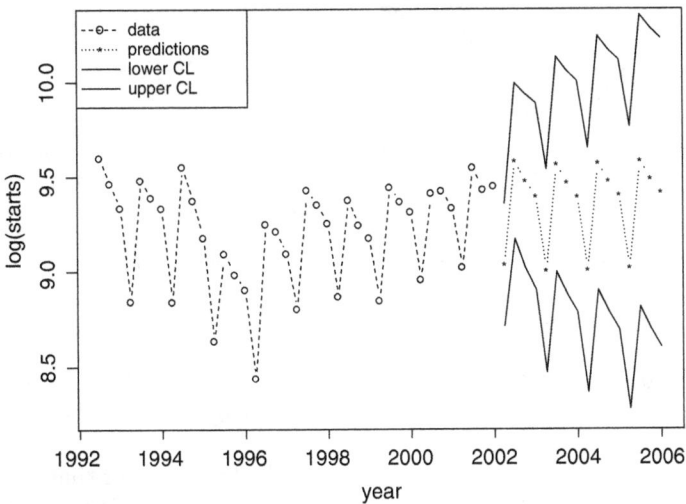

Fig. 13.3. *Forecasting logarithms of quarterly urban housing starts using the ARIMA{$(1,1,1) \times (0,1,1)_4$} model. The dashed line connects the data, the dotted line connects the forecasts, and the solid lines are the forecast limits.*

When the size of the seasonal oscillations increases, as with the air passenger data in Fig. 12.2, some type of preprocessing is needed before differencing. Often, taking logarithms stabilizes the size of the oscillations. This can be seen in Fig. 13.4. Box, Jenkins, and Reinsel (2008) obtain a parsimonious fit to the log passengers with an $ARIMA(0,1,1) \times (0,1,1)_{12}$ model.

For the housing starts series, the data come as logarithms in the `Ecdat` package. If they had come untransformed, then we would have needed to apply some type of transformation.

13.2 Box–Cox Transformation for Time Series

As just discussed, it is often desirable to transform a time series to stabilize the size of the variability, both seasonal and random. Although a transformation can be selected by trial-and-error, another possibility is automatic selection by maximum likelihood estimation using the model

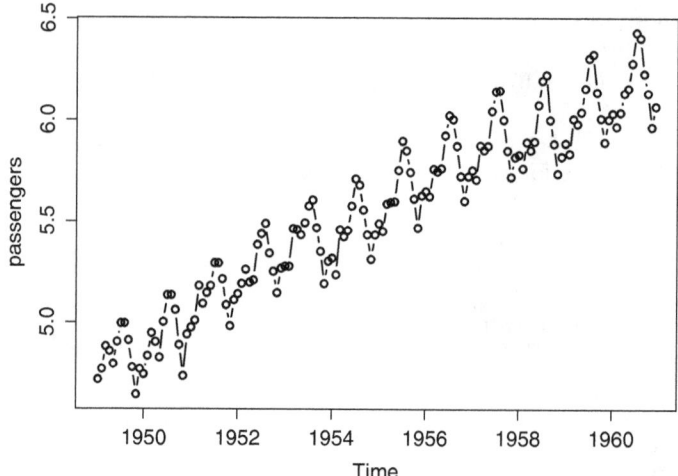

Fig. 13.4. *Time series plot of the logarithms of the monthly total international airline passengers (in thousands).*

$$(\Delta^d Y_t^{(\alpha)} - \mu) = \phi_1(\Delta^d Y_{t-1}^{(\alpha)} - \mu) + \cdots + \phi_p(\Delta^d Y_{t-p}^{(\alpha)} - \mu)$$
$$+ \epsilon_t + \theta_1 \epsilon_{t-1} + \cdots + \theta_q \epsilon_{t-q}, \tag{13.5}$$

where $\epsilon_1, \epsilon_2, \ldots$ is Gaussian white noise. Model (13.5) states that after a Box–Cox transformation, Y_t follows an ARIMA model with Gaussian noise that has a constant variance. The transformation parameter α is considered unknown and is estimated by maximum likelihood along with the AR and MA parameters and the noise variance. For notational simplicity, (13.5) uses a nonseasonal model, but a seasonal ARIMA model could just as easily have been used.

Example 13.2. Selecting a transformation for the housing starts

Figure 13.5 show the profile likelihood for α for the housing starts series (not the logarithms). The ARIMA model was ARIMA$\{(1,1,1) \times (1,1,1)_4\}$. The figure was created by the `BoxCox.Arima()` function in R's `FitAR` package. This function denotes the transformation parameter by λ. The MLE of α is 0.34 and the 95 % confidence interval is roughly from 0.15 to 0.55. Thus, the log transformation ($\alpha = 0$) is somewhat outside the confidence interval, but the square-root transformation is in the interval. Nonetheless, the log transformation worked satisfactorily in Example 13.1 and might be retained.

Without further analysis, it is not clear why $\alpha = 0.34$ achieves a better fit than the log transformation. Better fit could mean that the ARIMA model fits better, that the noise variability is more nearly constant, that the noise is closer to being Gaussian, or some combination of these effects.

It would be interesting to compare forecasts using the log and square-root transformations to see in what ways, if any, the square-root transformation outperforms the log transformation for forecasting. The forecasts would need to be back-transformed to the original scale in order for them to be comparable. One might use the final year as test data to see how well housing starts in that year are forecast. ☐

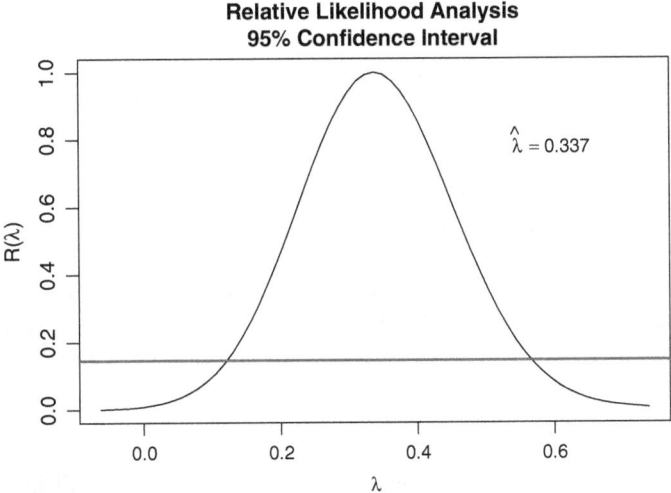

Fig. 13.5. *Profile likelihood for* α *(called* λ *in the legend) in the housing start example. Values of* λ *with* R(λ) *(the profile likelihood) above the horizontal line are in the 95 % confidence limit.*

Data transformations can stabilize some types of variation in time series, but not all types. For example, in Fig. 12.2 the seasonal oscillations in the numbers of air passengers increase as the series itself increases, and we can see in Fig. 13.4 that a log transformation stabilizes these oscillations. In contrast, the S&P 500 returns in Fig. 4.1 exhibit periods of low and high volatility even though the returns maintain a mean near 0. Transformations cannot remove this type of volatility clustering. Instead, these changes of volatility could be modeled by a GARCH process; this topic is pursued in Chap. 14.

13.3 Time Series and Regression

In a multiple linear regression model (9.1) the errors ϵ_i are assumed to be mutually independent. However, if the data $\{(\boldsymbol{X}_i, Y_i),\ i = 1, \ldots, n\}$ are time series, then it is likely that the errors are correlated, a problem we will call *residual correlation*; see Sect. 13.3.1.

Residual correlation causes standard errors and confidence intervals (which incorrectly assume uncorrelated noise) to be incorrect. In particular, the coverage probability of confidence intervals can be much lower than the nominal value. A solution to this problem is to adjust or correct the estimated covariance matrix of the coefficient estimates; see Sect. 13.3.2. An alternative solution is to model the noise as an ARMA process, assuming that the residuals are stationary; see Sect. 13.3.3.

13.3.1 Residual Correlation and Spurious Regressions

In the extreme case where the residuals are an integrated process, the least-squares estimator is inconsistent, meaning that it will not converge to the true parameter as the sample size converges to ∞. If an $I(1)$ process is regressed on another $I(1)$ process and the two processes are independent (so that the regression coefficient is 0), it is quite possible to obtain a highly significant result, that is, to strongly reject the true null hypothesis that the regression coefficient is 0. This is called a *spurious regression*. The problem, of course, is that the test is based on the incorrect assumption of independent error. The residuals from the The problem of correlated noise can be detected by looking at the sample ACF of the residuals. Sometimes the presence of residual correlation is obvious. In other cases, it is not so clear and a statistical test is desirable. The Durbin–Watson test can be used to test the null hypothesis of no residual autocorrelation. More precisely, the null hypothesis of the Durbin–Watson test is that the first p autocorrelation coefficients are all 0, where p can be selected by the user. The p-value for a Durbin–Watson test is not trivial to compute, and different implementations use different computational methods. In the R function durbinWatsonTest() in the car package, p is called max.lag and has a default value of 1. The p-value is computed by durbinWatsonTest() using bootstrapping. The lmtest package of R has another function, dwtest(), that computes the Durbin–Watson test, but only with $p = 1$. The function dwtest() uses either a normal approximation (default) or an exact algorithm to calculate the p-value.

Example 13.3. Residual plots for weekly interest changes

Using the interest rate data from Chap. 9, Fig. 13.6 contains residual plots for the regression of aaa_dif on cm10_dif and cm30_dif. The normal plot in panel (a) shows heavy tails. A t-distribution was fit to the residuals, and the estimated degrees of freedom was 2.99, again indicating heavy tails. Panel (b) shows a QQ plot of the residuals and the quantiles of the fitted t-distribution with a 45° reference line. There is excellent agreement between the data and the t-distribution.

Panel (c) is a plot of the ACF of the residuals. There is some evidence of autocorrelation. The Durbin–Watson test was performed three times with

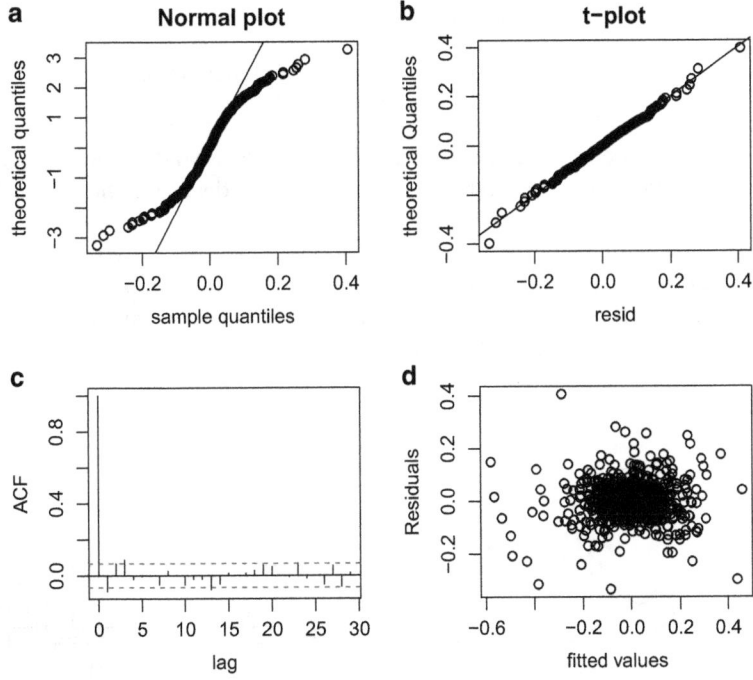

Fig. 13.6. *Residual plots for the regression of* `aaa_dif` *on* `cm10_dif` *and* `cm30_dif`.

R's `durbinWatsonTest()` using `max.lag` =1 and gave p-values of 0.006, 0.004, and 0.012. This shows the substantial random variation due to bootstrapping with the default of $B = 1000$ resamples. Using a larger number of resamples will compute the p-value with more accuracy. For example, when the number of resamples was increased to 10,000, three p-values were 0.0112, 0.0096, and 0.0106. Using `dwtest()`, the approximate p-value was 0.01089 and the exact p-value could not be computed. Despite some uncertainty about the p-value, it is clear that the p-value is small, so there is at least some residual autocorrelation.

To further investigate autocorrelation, ARMA models were fit to the residuals using the `auto.arima()` function in R to automatically select the order. Using BIC, the selected model is ARIMA(0,0,0), that is, white noise. Using AIC, the selected model is ARIMA(0,0,3) with estimates:

```
6 auto.arima(resid, ic="aic")

 Series: resid
 ARIMA(0,0,3) with zero mean

 Coefficients:
          ma1     ma2     ma3
      -0.0857  0.0770  0.0888
```

```
s.e.    0.0336   0.0338   0.0342

sigma^2 estimated as 0.004075:  log likelihood=1172.54
AIC=-2337.09    AICc=-2337.04    BIC=-2317.97
```

Several of the coefficients are large relative to their standard errors. There is evidence of some autocorrelation, but not a great deal and the BIC-selected model does not have any autocorrelation. The sample size is 880, so there are enough data to detect small autocorrelations. The autocorrelation that was found seems of little practical significance and could perhaps be ignored; see Sect. 13.3.2 for further investigation. The plot of residuals versus fitted values in panel (d) shows no sign of heteroskedasticity.   □

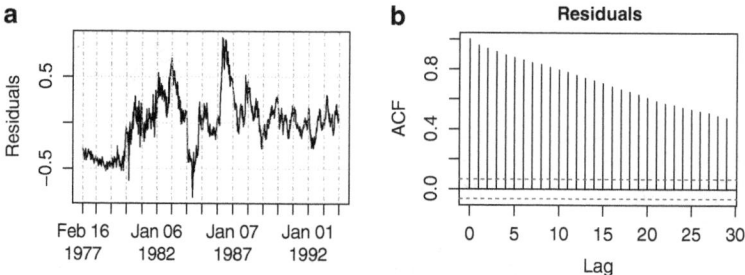

Fig. 13.7. *Time series plot and ACF plot of residuals when* **aaa** *is regressed on* **cm10** *and* **cm30**. *The plots indicate that the residuals are nonstationary.*

Example 13.4. Residual plots for weekly interest rates without differencing

The reader may have noticed that differenced time series have been used in the examples. There is a good reason for this. Many, if not most, financial time series are nonstationary or, at least, have very high and long-term autocorrelation. When one nonstationary series is regressed upon another, it happens frequently that the residuals are nonstationary. This is a substantial violation of the assumption of uncorrelated noise and can lead to serious problems. An estimator is said to be consistent if it converges to the true value of the parameter as the sample size increases to ∞. The least-squares estimator is not consistent when the errors are an integrated process.

As an example, we regressed **aaa** on **cm10** and **cm30**. These are the weekly time series of AAA, 10-year Treasury, and 30-year Treasury interest rates, which, when differenced, gave us **aaa_dif**, **cm10_dif**, and **cm30_dif** used in previous examples. Figure 13.7 contains time series and ACF plots of the residuals. The residuals are very highly correlated and perhaps are nonstationary. Unit root tests provide more evidence that the residuals are nonstationary. The *p*-values of augmented Dickey–Fuller tests are on one side of 0.05 or the other, depending on the order. With the default lag order in the adf.test() function from the tseries package in R, the p-value is 0.12, so one would not reject the null hypothesis of nonstationarity at level 0.05 or even level 0.1. The kpss.test() function does reject the null hypothesis of stationarity.

Let us compare the estimates from regression using the original series with the estimates from the differenced series. First, what should we expect when we make this comparison? Suppose that X_t and Y_t are time series following the regression model

$$Y_t = \alpha + \beta_0 t + \beta_1 X_i + \epsilon_t. \tag{13.6}$$

Note the linear time trend $\beta_0 t$. Then, upon differencing, we have

$$\Delta Y_t = \beta_0 + \beta_1 \Delta X_i + \Delta \epsilon_t, \tag{13.7}$$

so the original intercept α is removed, and the time trend's slope β_0 in (13.6) becomes an intercept in (13.7). The time trend could be omitted in (13.6) if the intercept in (13.7) is not significant, as happens in this example. The slope β_1 in (13.6) remains unchanged in (13.7). However, if ϵ_t is $I(1)$, then the regression of Y_t on X_t will not provide a consistent estimate of β_1, but the regression of ΔY_t on ΔX_i will consistently estimate β_1, so the estimates from the two regressions could be very different. This is what happens with this example.

The results from regression with the original series without the time trend are

```
Call:
lm(formula = aaa ~ cm10 + cm30)

Coefficients:
            Estimate Std. Error t value Pr(>|t|)
(Intercept)   0.9803     0.0700   14.00  < 2e-16 ***
cm10          0.3183     0.0445    7.15  1.9e-12 ***
cm30          0.6504     0.0498   13.05  < 2e-16 ***
```

The results with the differenced series are

```
Call:
lm(formula = aaa_dif ~ cm10_dif + cm30_dif)

Coefficients:
              Estimate Std. Error t value Pr(>|t|)
(Intercept) -9.38e-05   2.18e-03   -0.04     0.97
cm10_dif     3.60e-01   4.45e-02    8.09  2.0e-15 ***
cm30_dif     2.97e-01   4.98e-02    5.96  3.7e-09 ***
```

The estimated slopes for cm10 and cm10_dif, 0.3183 and 0.360, are somewhat similar. However, the estimated slopes for cm30 and cm30_dif, 0.650 and 0.297, are quite dissimilar relative to their standard errors. This is to be expected if the estimators using the undifferenced series are not consistent; also, their standard errors are not valid because they are based on the assumption of uncorrelated noise. In the analysis with the differenced data, the p-value for the intercept is 0.97, so we can accept the null hypothesis that the intercept is zero; this justifies the omission of the time trend when using the undifferenced series.   □

Example 13.5. Simulated independent AR processes

To illustrate further the problems caused by regressing nonstationary series, or even stationary series with high correlation, we simulated two independent AR process, both of length 200 with $\phi = 0.99$.

```
7  set.seed(997711)
8  n = 200
9  x = arima.sim(list(order=c(1,0,0),ar=.99),n=n)
10 y = arima.sim(list(order=c(1,0,0),ar=.99),n=n)
11 fit1 = lm(y~x)
12 fit5 = lm(diff(y)~diff(x))
```

These processes are stationary but near the borderline of being nonstationary. After simulating these processes, one process was regressed on the other. We repeated this three more times. Since the processes are independent, the true slope is 0. In each case, the estimated slope was far from the true value of 0 and was statistically significant according to the (incorrect) p-value. The results are below.

	Estimate	Std. Error	t value	Pr(>|t|)
(Intercept)	-8.40	0.269	-31	1.9e-78
x	0.48	0.036	13	1.6e-29

	Estimate	Std. Error	t value	Pr(>|t|)
(Intercept)	5.96	0.328	18.2	4.9e-44
x	-0.43	0.088	-4.8	2.6e-06

	Estimate	Std. Error	t value	Pr(>|t|)
(Intercept)	-5.154	0.213	-24.2	4.5e-61
x	0.095	0.031	3.1	2.3e-03

	Estimate	Std. Error	t value	Pr(>|t|)
(Intercept)	-0.51	0.312	-1.6	1.1e-01
x	-0.53	0.079	-6.7	2.3e-10

Notice how the estimated intercepts and slope randomly vary between the four simulations. The standard errors and p-values are based on the invalid assumption of independent errors and are erroneous and very misleading, a problem that is called *spurious regression*. Fortunately, the violation of the independence assumption would be easy to detect by plotting the residuals.

We also regressed the differenced series and obtained completely different results:

	Estimate	Std. Error	t value	Pr(>|t|)
(Intercept)	0.082	0.069	1.18	0.24
diff(x)	-0.023	0.068	-0.34	0.73

	Estimate	Std. Error	t value	Pr(>|t|)
(Intercept)	-0.027	0.064	-0.41	0.68
diff(x)	-0.021	0.063	-0.33	0.74

	Estimate	Std. Error	t value	Pr(>|t|)
(Intercept)	-0.015	0.071	-0.21	0.83
diff(x)	-0.022	0.076	-0.29	0.77

	Estimate	Std. Error	t value	Pr(>|t|)
(Intercept)	-0.025	0.077	-0.32	0.75
diff(x)	0.022	0.078	0.28	0.78

Notice that now the estimated slopes are all near the true value of 0. All the p-values are large and lead one to the correct conclusion that the true slope is 0.

When the noise process is stationary, an alternative to differencing is to use an ARMA model for the noise process; see Sect. 13.3.3. □

13.3.2 Heteroscedasticity and Autocorrelation Consistent (HAC) Standard Errors

We now consider the effect of correlated noise and heteroskedasticity on standard errors and confidence intervals in multiple linear regression models. If $\mathrm{COV}(\epsilon) \neq \sigma_\epsilon^2 I$ but rather $\mathrm{COV}(\epsilon) = \Sigma_\epsilon$ for some matrix Σ_ϵ, then

$$\begin{aligned}
\mathrm{COV}(\widehat{\beta}|x_1,\ldots,x_n) &= (X^\mathsf{T}X)^{-1}X^\mathsf{T}\mathrm{COV}(Y|x_1,\ldots,x_n)X(X^\mathsf{T}X)^{-1} \\
&= (X^\mathsf{T}X)^{-1}X^\mathsf{T}\Sigma_\epsilon X(X^\mathsf{T}X)^{-1}.
\end{aligned} \tag{13.8}$$

This result lets us see the effect of correlation or nonconstant variance among $\epsilon_1,\ldots,\epsilon_n$.

Example 13.6. Regression with AR(1) errors

Suppose that $\epsilon_1,\ldots,\epsilon_n$ is a stationary AR(1) process so that $\epsilon_t = \phi\epsilon_{t-1} + u_t$, where $|\phi| < 1$ and $u_1, u_2, \ldots$ is weak $\mathrm{WN}(0,\sigma_u^2)$. Then

$$\Sigma_\epsilon = \sigma_\epsilon^2 \begin{pmatrix}
1 & \phi & \phi^2 & \cdots & \phi^{n-1} \\
\phi & 1 & \phi & \cdots & \phi^{n-2} \\
\vdots & \vdots & \vdots & \ddots & \vdots \\
\phi^{n-1} & \phi^{n-2} & \phi^{n-3} & \cdots & 1
\end{pmatrix}. \tag{13.9}$$

As an example, suppose that $n = 21$, $X_1, \ldots, X_n$ are equally spaced between -10 and 10, and $\sigma_\epsilon^2 = 1$. Substituting (13.9) into (13.8) gives the covariance matrix of the estimator $(\widehat{\beta}_0, \widehat{\beta}_1)$, and taking the square roots of the diagonal elements gives the standard errors. This was done with $\phi = -0.75, -0.5, -0.25, 0, 0.25, 0.5, 0.75$.

Figure 13.8 plots the ratios of standard errors for the independent case ($\phi = 0$) to the standard errors for the true value of ϕ. These ratios are the factors by which the standard errors are miscalculated if we assume that $\phi = 0$, but it is not. Notice that negative values of ϕ result in a conservative (too large) standard error, but positive values of ϕ give a standard error that is too small. In the case of $\phi = 0.75$, assuming independence gives standard errors that are only about half as large as they should be. As discussed in Sect. 13.3.3, this problem can be fixed by assuming (correctly) that the noise process is AR(1).   □

As discussed in Sect. 13.3.1, if the errors in a regression model are an integrated process, such as a random walk, the least-squares estimator is inconsistent. However, when the dependence between the errors is not too strong, there are mild conditions under which the least-squares estimator is consistent, meaning that it will converge to the true parameter as the sample size converges to ∞. For the latter case there are methods available to estimate consistent standard errors for the coefficient estimates. Two simple and widely used approaches are the heteroskedasticity consistent (HC) and heteroskedasticity and autocorrelation consistent (HAC) estimators.

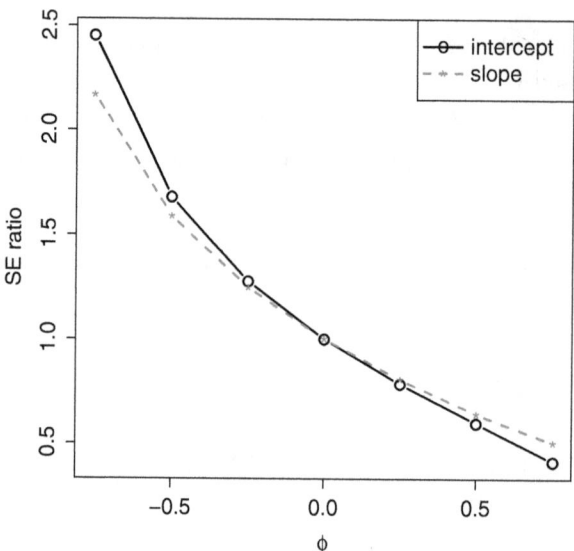

Fig. 13.8. *Factor by which the standard error is changed when ϕ deviates from 0 for intercept (solid) and slope (dashed).*

Let $\hat{\epsilon}_i$ denote the OLS residuals, let $\widehat{\boldsymbol{\Sigma}}_{\hat{\epsilon}} = \text{diag}\{\hat{\epsilon}_1^2, \ldots, \hat{\epsilon}_n^2\}$ denote a diagonal matrix of the squared residuals, and let $\widehat{\boldsymbol{C}}_{HC} = \boldsymbol{X}^\mathsf{T} \widehat{\boldsymbol{\Sigma}}_{\hat{\epsilon}} \boldsymbol{X}$. Then a heteroskedasticity consistent (HC) estimator (see White, 1980) of the covariance matrix for the coefficient estimates is

$$\widehat{\text{COV}}_{HC}(\hat{\boldsymbol{\beta}} | \boldsymbol{x}_1, \ldots, \boldsymbol{x}_n) = (\boldsymbol{X}^\mathsf{T} \boldsymbol{X})^{-1} \widehat{\boldsymbol{C}}_{HC} (\boldsymbol{X}^\mathsf{T} \boldsymbol{X})^{-1}. \qquad (13.10)$$

The corresponding HC standard errors are defined as the square roots of the diagonal entries of (13.10).

A heteroskedasticity and autocorrelation consistent (HAC) estimator (see Newey and West, 1987) of the covariance matrix for the coefficient estimates is similarly defined as

$$\widehat{\text{COV}}_{HAC}(\hat{\boldsymbol{\beta}} | \boldsymbol{x}_1, \ldots, \boldsymbol{x}_n) = (\boldsymbol{X}^\mathsf{T} \boldsymbol{X})^{-1} \widehat{\boldsymbol{C}}_{HAC} (\boldsymbol{X}^\mathsf{T} \boldsymbol{X})^{-1}, \qquad (13.11)$$

in which

$$\widehat{\boldsymbol{C}}_{HAC} = \widehat{\boldsymbol{C}}_{HC} + \sum_{\ell=1}^{L} w_\ell \sum_{i=\ell+1}^{n} \left(\boldsymbol{X}_i \hat{\epsilon}_i \hat{\epsilon}_{i-\ell} \boldsymbol{X}_{i-\ell}^\mathsf{T} + \boldsymbol{X}_{i-\ell} \hat{\epsilon}_{i-\ell} \hat{\epsilon}_i \boldsymbol{X}_i^\mathsf{T} \right),$$
$$(13.12)$$

where $w_\ell = 1 - \ell/(L+1)$ denotes the Bartlett weight function, although other weight functions w_ℓ can also be used. The corresponding HAC standard errors are defined as the square root of the diagonal entries of (13.11).

Example 13.7. HC and HAC estimates for regression of weekly interest changes

In Sect. 13.3.1 a regression of `aaa_dif` on `cm10_dif` and `cm30_dif` produced residuals that exhibited minor autocorrelation; AIC suggested an MA(3) model for the residuals while BIC selected ARIMA(0,0,0), i.e., white noise. We now consider whether ignoring the small autocorrelations has a practical impact on inference. The previous regression results are obtained from the following R commands.

```
13  dat = read.table(file="WeekInt.txt", header=T)
14  attach(dat)
15  cm10_dif = diff(cm10)
16  aaa_dif = diff(aaa)
17  cm30_dif = diff(cm30)
18  fit = lm(aaa_dif ~ cm10_dif + cm30_dif)
19  round(summary(fit)$coef, 4)
```

The HC and HAC covariance matrix estimates can be computed using the `NeweyWest()` function from the R package **sandwich**. The first argument is a fitted model object, in this case `fit`. In both cases we set `prewhite = F`. For the HAC estimate, the argument `lag` corresponds to the maximal lag L used in the Bartlett weight function above. If no value is specified, one is selected automatically via the `bwNeweyWest()` function (see the help file for more information). For the HC estimate we specify `lag = 0`. The HC estimate and HAC estimate with $L = 3$ are shown below.

```
20 library(sandwich)
21 options(digits=2)
22 NeweyWest(fit, lag = 0, prewhite = F)
```

```
                (Intercept) cm10_dif cm30_dif
   (Intercept)     4.7e-06  7.3e-06 -1.1e-05
   cm10_dif        7.3e-06  6.3e-03 -6.2e-03
   cm30_dif       -1.1e-05 -6.2e-03  6.7e-03
```

```
23 NeweyWest(fit, lag = 3, prewhite = F)
```

```
                (Intercept) cm10_dif cm30_dif
   (Intercept)     4.6e-06 -0.00003  2.6e-05
   cm10_dif       -3.0e-05  0.00666 -6.6e-03
   cm30_dif        2.6e-05 -0.00662  7.0e-03
```

The OLS regression results, as well as the HC and HAC estimated standard errors, and their corresponding t values are summarized in Table 13.1. Recall that the HC and HAC standard error estimates are computed as the square roots of the diagonal entries of the covariance matrix estimates.

```
24 sqrt(diag(NeweyWest(fit, lag = 0, prewhite = F)))
25 sqrt(diag(NeweyWest(fit, lag = 3, prewhite = F)))
```

The corresponding t values are the OLS coefficient estimates divided by their standard error estimates.

```
26 coef(fit)/sqrt(diag(NeweyWest(fit, lag = 0, prewhite = F)))
27 coef(fit)/sqrt(diag(NeweyWest(fit, lag = 3, prewhite = F)))
```

Table 13.1. *Regression estimates of* aaa_dif *on* cm10_dif *and* cm30_dif*: the OLS estimates, estimated standard errors, and t values are shown on the left; the estimated HC standard errors and corresponding t values are shown in the middle; and the estimated HAC standard errors with $L = 3$ and corresponding t values are shown on the right.*

Coefficient	OLS			HC		HAC$_{L=3}$	
	Estimate	Std. Err.	t value	Std. Err.	t value	Std. Err.	t value
(Intercept)	−0.0001	0.0022	−0.043	0.0022	−0.043	0.0021	−0.044
cm10_dif	0.3602	0.0445	8.091	0.0791	4.553	0.0816	4.415
cm30_dif	0.2968	0.0498	5.956	0.0816	3.637	0.0836	3.551

From Table 13.1 we see that the HC and HAC estimates produced similar results. The estimated standard error for the intercept are stable, while the estimated HC and HAC standard errors for the cm10_dif and cm30_dif coefficients are about twice as large as the OLS estimates of the standard errors, and as a result, the corresponding t values are about half as large in magnitude. In this case, however, the cm10_dif and cm30_dif coefficients remain statistically significant, with both estimates over three standard errors above zero. The minor serial correlation (and heteroskedasticity) in the OLS residuals does not appear to have a practical impact on inference in this example. □

13.3.3 Linear Regression with ARMA Errors

When residual analysis shows that the residuals are correlated, then one of the key assumptions of the linear model does not hold, and tests and confidence intervals based on this assumption are invalid and cannot be trusted. Fortunately, there is a solution to this problem: replace the assumption of independent noise by the weaker assumption that the noise process is stationary but possibly correlated. One could, for example, assume that the noise is an ARMA process. This is the strategy we will discuss in this section; this approach is referred to as an ARMAX model, in which the X indicates the inclusion of exogenous regression variables.

The linear regression model with ARMA errors combines the linear regression model (9.1) and the ARMA model (12.26) for the noise, so that

$$Y_t = \beta_0 + \beta_1 X_{t,1} + \cdots + \beta_p X_{t,p} + \epsilon_t, \tag{13.13}$$

where

$$(1 - \phi_1 B - \cdots - \phi_p B^p)\, \epsilon_t = (1 + \theta_1 B + \cdots + \theta_q B^q)\, u_t, \tag{13.14}$$

and $u_1, \ldots, u_n$ is white noise.

Example 13.8. Demand for ice cream

This example uses the data set `Icecream` in R's `Ecdat` package. The data are four-weekly observations from March 18, 1951, to July 11, 1953 on four variables, `cons` = U.S. consumption of ice cream per head in pints; `income` = average family income per week (in U.S. Dollars); `price` = price of ice cream (per pint); and `temp` = average temperature (in Fahrenheit). There is a total of 30 observations. Since there are 13 four-week periods per year, there are slightly over two years of data.

First, a linear model was fit with `cons` as the response and `income`, `price`, and `temp` as the predictor variables. One can see that `income` and `temp` are significant, especially `temp` (not surprisingly).

```
Call:
lm(formula = cons ~ income + price + temp, data = Icecream)

Residuals:
      Min       1Q    Median       3Q      Max
-0.06530 -0.01187  0.00274  0.01595  0.07899

Coefficients:
             Estimate Std. Error t value Pr(>|t|)
(Intercept)  0.197315   0.270216    0.73    0.472
income       0.003308   0.001171    2.82    0.009 **
price       -1.044414   0.834357   -1.25    0.222
```

```
temp           0.003458   0.000446    7.76  3.1e-08 ***
---
```

```
Residual standard error: 0.0368 on 26 degrees of freedom
Multiple R-squared: 0.719,      Adjusted R-squared: 0.687
F-statistic: 22.2 on 3 and 26 DF,  p-value: 2.45e-07
```

A Durbin–Watson test has a very small p-value, so we can reject the null hypothesis that the noise is uncorrelated.

```
28 options(digits=3)
29 library("car")
30 durbinWatsonTest(fit_ic_lm)
```

```
   lag Autocorrelation D-W Statistic p-value
    1            0.33          1.02        0
 Alternative hypothesis: rho != 0
```

Next, the linear regression model with AR(1) errors was fit and the AR(1) coefficient was over three times its standard error, indicating statistical significance. This was done using R's `arima()` function, which specifies the regression model with the `xreg` argument. It is interesting to note that the coefficient of `income` is now nearly equal to 0 and no longer significant. The effect of `temp` is similar to that of the linear model fit, though its standard error is now larger.

```
Series: cons
ARIMA(1,0,0) with non-zero mean

Coefficients:
          ar1  intercept  income   price   temp
        0.732      0.538   0.000  -1.086  0.003
s.e.    0.237      0.325   0.003   0.734  0.001

sigma^2 estimated as 0.00091:  log likelihood=62.1
AIC=-112    AICc=-109    BIC=-104
```

Finally, the linear regression model with MA(1) errors was fit and the MA(1) coefficient was also over three times its standard error, again indicating statistical significance. The model with AR(1) errors has a slightly better (smaller) AIC and BIC values than the model with MA(1), but there is not much of a difference between the models in terms of AIC or BIC. However, the two models imply rather different types of noise autocorrelation. The MA(1) model has no correlation beyond lag 1. The AR(1) model with coefficient 0.732 has autocorrelation persisting much longer. For example, the autocorrelation is $0.732^2 = 0.536$ at lag 2, $0.732^3 = 0.392$ at lag 3, and still $0.732^4 = 0.287$ at lag 4.

```
Series: cons
ARIMA(0,0,1) with non-zero mean
```

```
Coefficients:
          ma1   intercept   income   price   temp
        0.503       0.332    0.003  -1.398  0.003
  s.e.  0.160       0.270    0.001   0.798  0.001

sigma^2 estimated as 0.000957:  log likelihood=61.6
AIC=-111    AICc=-107    BIC=-103
```

Interestingly, the estimated effect of income is larger and significant, much like its effect as estimated by the linear model with independent errors but unlike the result for the linear model with AR(1) errors.

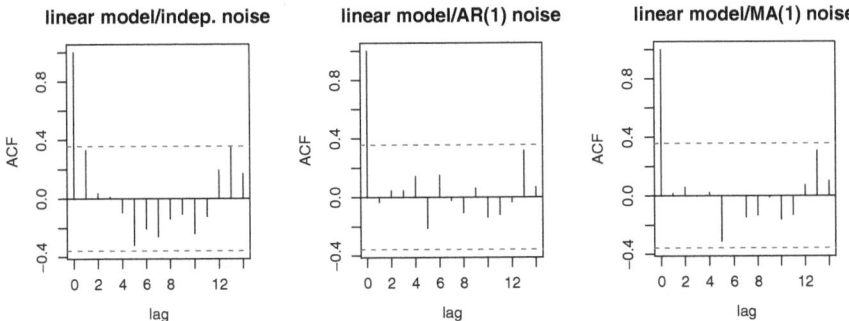

Fig. 13.9. *Ice cream consumption example. Residual ACF plots for the linear model with independent noise, the linear model with AR(1) noise, and the linear model with MA(1) noise.*

The ACFs of the residuals from the linear model and from the linear models with AR(1) and MA(1) errors are shown in Fig. 13.9. The residuals from the linear model estimate $\epsilon_1, \ldots, \epsilon_n$ in (13.13), and show some autocorrelation. The residuals from the linear models with either AR(1) or MA(1) errors estimate $u_1, \ldots, u_n$ in (13.14), and show little autocorrelation. One concludes that the linear model with either AR(1) or MA(1) errors fits well and either an AR(1) or MA(1) term is needed.

Why is the effect of income larger and significant if the noise is assumed to be either independent or MA(1) but smaller and insignificant if the noise is AR(1)? To attempt an answer, time series plots of the four variables were examined. The plots are shown in Fig. 13.10. The strong seasonal trend in temp is obvious and cons follows this trend. There is a slightly increasing trend in cons, which appears to have two possible explanations. The trend might be explained by the increasing trend in income. However, with the strong residual autocorrelation implied by the AR(1) model, the trend in cons could also be explained by noise autocorrelation. One problem here is that we have a small sample size, only 30 observations. With more data it might be possible to separate the effects on ice cream consumption of income and noise autocorrelation.

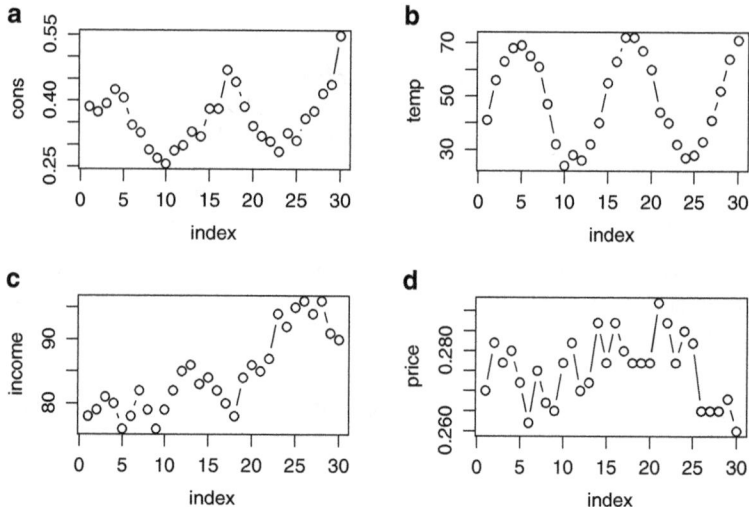

Fig. 13.10. *Time series plots for the ice cream consumption example and the variables used to predict consumption.*

In summary, there is a strong seasonal component to ice cream consumption, with consumption increasing, as would be expected, with warmer temperatures. Ice cream consumption does not depend much, if at all, on `price`, though it should be noted that `price` has not varied much in this study; see Fig. 13.10. Greater variation in `price` might cause `cons` to depend more on `price`. Finally, it is uncertain whether ice cream consumption increases with family income.   □

13.4 Multivariate Time Series

Suppose that for each t, $\boldsymbol{Y}_t = (Y_{1,t}, \ldots, Y_{d,t})'$ is a d-dimensional random vector representing quantities that were measured at time t, e.g., returns on d equities. Then $\boldsymbol{Y}_1, \boldsymbol{Y}_2 \ldots$ is called a d-dimensional *multivariate time series*.

The definition of stationarity for multivariate time series is the same as given before for univariate time series. A multivariate time series is said to be *stationary* if for every n and m, $\boldsymbol{Y}_1, \ldots, \boldsymbol{Y}_n$ and $\boldsymbol{Y}_{1+m}, \ldots, \boldsymbol{Y}_{n+m}$ have the same distributions.

13.4.1 The Cross-Correlation Function

Suppose that Y_j and Y_i are the two component series of a stationary multivariate time series. The *cross-correlation function* (CCF) between Y_j and Y_i is defined as

$$\rho_{Y_j, Y_i}(h) = \mathrm{Corr}\{Y_j(t), Y_i(t-h)\} \tag{13.15}$$

and is the correlation between Y_j at a time t and Y_i at h time units earlier. As with autocorrelation, h is called the *lag*. However, unlike the ACF, the CCF is not symmetric in the lag variable h, that is, $\rho_{Y_j,Y_i}(h) \neq \rho_{Y_j,Y_i}(-h)$. Instead, as a direct consequence of definition (13.15), we have that $\rho_{Y_j,Y_i}(h) = \rho_{Y_i,Y_j}(-h)$.

The CCF can be defined for multivariate time series that are not stationary, but only weakly stationary. A multivariate time series $\boldsymbol{Y}_1, \boldsymbol{Y}_2, \ldots$ is said to be weakly stationary if the mean and covariance matrix of $\boldsymbol{Y}_t$ are finite and do not depend on t, and if the right-hand side of (13.15) is independent of t for all j, i, and h.

Cross-correlations can suggest how the component series might be influencing each other or might be influenced by a common factor. Like all correlations, cross-correlations only show statistical association, not causation, but a causal relationship might be deduced from other knowledge.

Example 13.9. Cross-correlation between changes in CPI (consumer price index) and IP (industrial production)

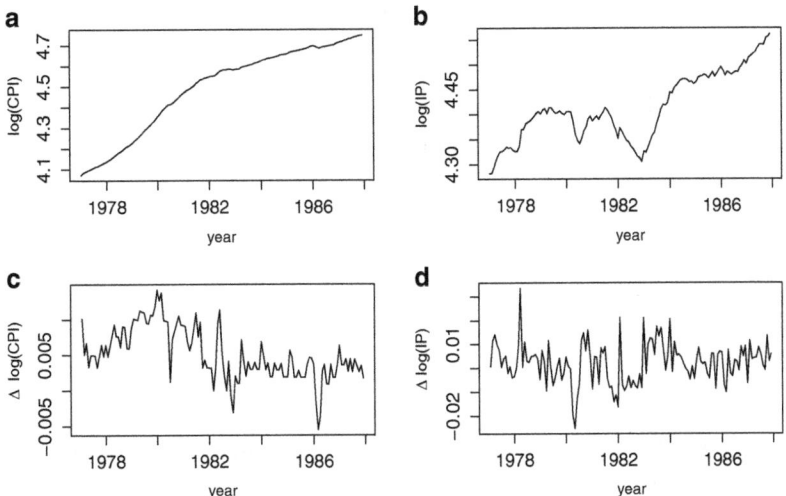

Fig. 13.11. *(a) Time series plot of log(CPI) (b) Time series plot of log(IP) (c) Time series plot of changes in log(CPI) (d) Time series plot of changes in log(IP).*

Time series plots for the logarithm of CPI (*cpi*), the logarithm of IP (*ip*), and changes in *cpi* and *ip*, are shown in Fig. 13.11 panels (a)–(d), respectively. The cross-correlation function between changes in the logarithm of CPI (Δcpi) and changes in the logarithm of IP (Δip) is shown in Fig. 13.12. It was created by the ccf() function in R.

```
31  CPI.dat = read.csv("CPI.dat.csv")
32  CPI_diff1 = diff(log(as.matrix(CPI.dat$CPI)[769:900,])) # 1977--1987
```

```
33 IP.dat = read.csv("IP.dat.csv")
34 IP_diff1 = diff(log(as.matrix(IP.dat$IP)[697:828,]))    # 1977--1987
35 ccf(CPI_diff1, IP_diff1)
```

The largest absolute cross-correlations are at negative lags and these correlations are negative. This means that an above-average (below-average) change in *cpi* predicts a future change in *ip* that is below (above) average. As just emphasized, correlation does not imply causation, so we cannot say that changes in *cpi* cause opposite changes in future *ip*, but the two series behave as if this were happening. Correlation does imply predictive ability. Therefore, if we observe an above-average change in *cpi*, then we should predict future changes in *ip* that will be below average. In practice, we should use the currently observed changes in both *cpi* and *ip*, not just *cpi*, to predict future changes in *ip*. We will discuss prediction using two or more related time series in Sect. 13.4.5.   □

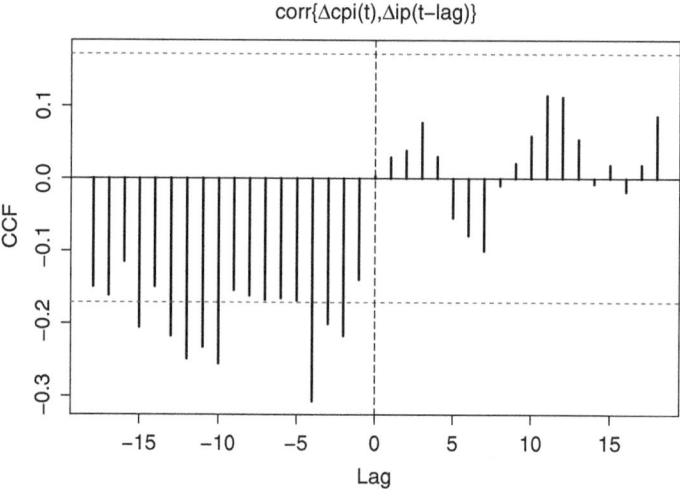

Fig. 13.12. *Sample CCF for Δcpi and Δip. Note the negative correlation at negative lags, that is, between the cpi and future values of ip.*

13.4.2 Multivariate White Noise

A d-dimensional multivariate time series $\boldsymbol{Y}_1, \boldsymbol{Y}_2, \ldots$ is a weak $\mathrm{WN}(\boldsymbol{\mu}, \boldsymbol{\Sigma})$ process if

1. $E(\boldsymbol{Y}_t) = \boldsymbol{\mu}$ (constant and finite) for all t;
2. $\mathrm{COV}(\boldsymbol{Y}_t) = \boldsymbol{\Sigma}$ (constant and finite) for all t; and
3. for all $t \neq s$, all components of $\boldsymbol{Y}_t$ are uncorrelated with all components of $\boldsymbol{Y}_s$.

Notice that if $\boldsymbol{\Sigma}$ is not diagonal, then there is cross-correlation between the components of $\boldsymbol{Y}_t$ because $\text{Corr}(Y_{j,t}, Y_{i,t}) = \Sigma_{j,i}$; in other words, there may be nonzero *contemporaneous* correlations. However, for all $1 \leq j, i \leq d$, $\text{Corr}(Y_{j,t}, Y_{i,s}) = 0$ if $t \neq s$.

Furthermore, $\boldsymbol{Y}_1, \boldsymbol{Y}_2, \ldots$ is an i.i.d. $\text{WN}(\boldsymbol{\mu}, \boldsymbol{\Sigma})$ process if, in addition to conditions 1–3, $\boldsymbol{Y}_1, \boldsymbol{Y}_2, \ldots$ are independent and identically distributed. If $\boldsymbol{Y}_1, \boldsymbol{Y}_2, \ldots$ are also multivariate normally distributed, then they are a Gaussian $\text{WN}(\boldsymbol{\mu}, \boldsymbol{\Sigma})$ process.

13.4.3 Multivariate ACF Plots and the Multivariate Ljung-Box Test

The ACF for multivariate time series includes the d marginal ACFs for each univariate series $\{\rho_{Y_i}(h) : i = 1, \ldots, d\}$, and the $d(d-1)/2$ CCFs for all unordered pairs of the univariate series $\{\rho_{Y_j, Y_i}(h) : 1 \leq j < i \leq d\}$. It is sufficient to only consider the unordered pairs because $\rho_{Y_j, Y_i}(h) = \rho_{Y_i, Y_j}(-h)$.

In R, if two (or more) univariate time series with matching time indices are stored as $n \times 1$ vectors, the cbind() function can by used to make an $n \times 2$ matrix consisting of the joint time series. The acf() function may be applied to such multivariate time series. The sample ACF for $(\Delta cpi, \Delta ip)'$ is shown in Fig. 13.13, and generated by the following commands in R.

```
36  CPI_IP = cbind(CPI_diff1, IP_diff1)
37  acf(CPI_IP)
```

The marginal sample ACF for Δcpi and Δip are shown in the first and second diagonal panels, respectively. Both show significant serial correlation, but there is much more persistence in the first. The sample CCF for Δcpi and Δip has been split between the top right and bottom left panels by positive and negative lags, respectively. Notice that combining the off-diagonal panels in Fig. 13.13 reproduce the CCF shown in Fig. 13.12.

Each of the panels in Fig. 13.13 include *test bounds* to test the null hypothesis that an individual autocorrelation or lagged cross-correlation coefficient is 0. As in the univariate case, the usual level of the test is 0.05, and one can expect to see about 1 out of 20 sample correlations outside the test bounds simply by chance. Also, as in the univariate case, a simultaneous test is available.

Let $\rho(h)$ denote the $d \times d$ lag-h cross-correlation matrix for a d-dimensional multivariate time series. The null hypothesis of the multivariate Ljung–Box test is $H_0 : \rho(1) = \rho(2) = \cdots = \rho(K) = \boldsymbol{0}$ for some K, say $K = 5$ or 10. If the multivariate Ljung–Box test rejects, then we conclude that one or more of $\rho(1), \cdots, \rho(K)$ is nonzero. If, in fact, the lagged cross-correlation 1 to K are all zero, then there is only a 1 in 20 chance of falsely concluding that they are not all zero, assuming a level 0.05 test. In contrast, if the lagged cross-correlation are tested one at time, then there is a much higher chance of concluding that one or more is nonzero.

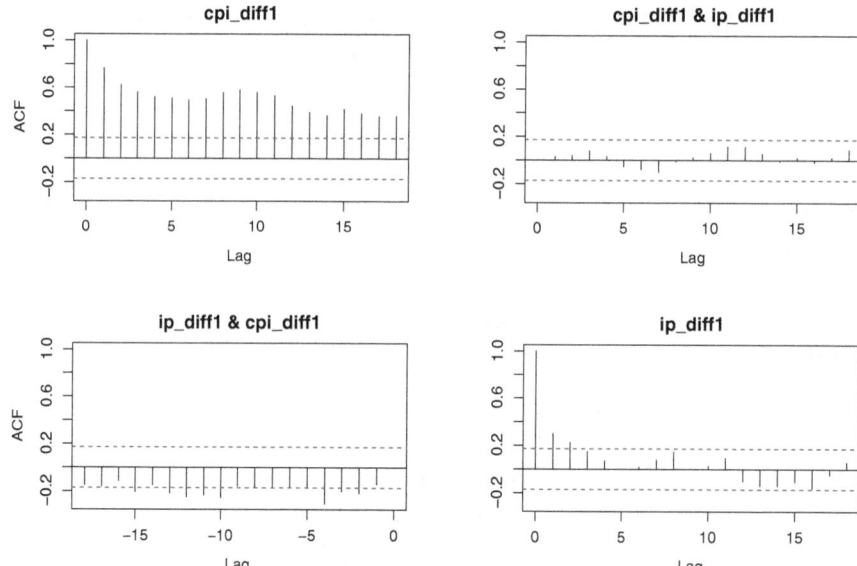

Fig. 13.13. *Sample ACF for* $(\Delta cpi, \Delta ip)'$. *The marginal sample ACF for* Δcpi *and* Δip *are shown in the first and second diagonal panels, respectively; the sample CCF for* Δcpi *and* Δip *has been split between the top right and bottom left panels by positive and negative lags, respectively.*

The following commands will conduct the multivariate Ljung–Box test in R for the bivariate series $(\Delta cpi, \Delta ip)'$.

```
38 source("SDAFE2.R")
39 mLjungBox(CPI_IP, lag = 10)

    K   Q(K) d.f. p-value
 1 10 532.48   40       0
```

The multivariate Ljung–Box test statistic was 532.48, and the approximate p-value was 0, confirming that there is significant serial correlation in the first $K = 10$ lags.

13.4.4 Multivariate ARMA Processes

A d-dimensional multivariate time series $\boldsymbol{Y}_1, \boldsymbol{Y}_2, \ldots$ is a multivariate ARMA (p, q) process with mean $\boldsymbol{\mu}$ if for $d \times d$ matrices $\boldsymbol{\Phi}_1, \ldots, \boldsymbol{\Phi}_p$ and $\boldsymbol{\Theta}_1, \ldots, \boldsymbol{\Theta}_q$,

$$\boldsymbol{Y}_t - \boldsymbol{\mu} = \boldsymbol{\Phi}_1(\boldsymbol{Y}_{t-1} - \boldsymbol{\mu}) + \cdots + \boldsymbol{\Phi}_p(\boldsymbol{Y}_{t-p} - \boldsymbol{\mu}) + \boldsymbol{\epsilon}_t + \boldsymbol{\Theta}_1 \boldsymbol{\epsilon}_{t-1} + \cdots + \boldsymbol{\Theta}_q \boldsymbol{\epsilon}_{t-q}, \tag{13.16}$$

where $\boldsymbol{\epsilon}_1, \ldots, \boldsymbol{\epsilon}_n$ is a multivariate weak $\mathrm{WN}(\boldsymbol{0}, \boldsymbol{\Sigma})$ process. Multivariate AR processes (the case $q = 0$) are also called vector AR or VAR processes and are widely used in practice.

As an example, a bivariate AR(1) process can be written as

$$\begin{pmatrix} Y_{1,t} - \mu_1 \\ Y_{2,t} - \mu_2 \end{pmatrix} = \begin{pmatrix} \phi_{1,1} & \phi_{1,2} \\ \phi_{2,1} & \phi_{2,2} \end{pmatrix} \begin{pmatrix} Y_{1,t-1} - \mu_1 \\ Y_{2,t-1} - \mu_2 \end{pmatrix} + \begin{pmatrix} \epsilon_{1,t} \\ \epsilon_{2,t} \end{pmatrix},$$

where

$$\boldsymbol{\Phi} = \boldsymbol{\Phi}_1 = \begin{pmatrix} \phi_{1,1} & \phi_{1,2} \\ \phi_{2,1} & \phi_{2,2} \end{pmatrix}.$$

Therefore,

$$Y_{1,t} = \mu_1 + \phi_{1,1}(Y_{1,t-1} - \mu_1) + \phi_{1,2}(Y_{2,t-1} - \mu_2) + \epsilon_{1,t}$$

and

$$Y_{2,t} = \mu_2 + \phi_{2,1}(Y_{1,t-1} - \mu_1) + \phi_{2,2}(Y_{2,t-1} - \mu_2) + \epsilon_{2,t},$$

so that $\phi_{i,j}$ is the amount of "influence" of $Y_{j,t-1}$ on $Y_{i,t}$. Similarly, for a bivariate AR(p) process, $\phi_{i,j}^k$ (the (i,j)th component of $\boldsymbol{\Phi}_k$) is the influence of $Y_{j,t-k}$ on $Y_{i,t}$, $k = 1, \ldots, p$.

For a d-dimensional AR(1), it follows from (13.16) with $p = 1$ and $\boldsymbol{\Phi} = \boldsymbol{\Phi}_1$ that

$$E(\boldsymbol{Y}_t | \boldsymbol{Y}_{t-1}, \ldots, \boldsymbol{Y}_1) = E(\boldsymbol{Y}_t | \boldsymbol{Y}_{t-1}) = \boldsymbol{\mu} + \boldsymbol{\Phi}(\boldsymbol{Y}_{t-1} - \boldsymbol{\mu}). \tag{13.17}$$

How does $E(\boldsymbol{Y}_t)$ depend on the more distant past, say on $\boldsymbol{Y}_{t-2}$? To answer this question, we can generalize (13.17). To keep notation simple, assume that the mean has been subtracted from $\boldsymbol{Y}_t$ so that $\boldsymbol{\mu} = \boldsymbol{0}$. Then

$$\boldsymbol{Y}_t = \boldsymbol{\Phi}\boldsymbol{Y}_{t-1} + \boldsymbol{\epsilon}_t = \boldsymbol{\Phi}\{\boldsymbol{\Phi}\boldsymbol{Y}_{t-2} + \boldsymbol{\epsilon}_{t-1}\} + \boldsymbol{\epsilon}_t$$

and, because $E(\boldsymbol{\epsilon}_{t-1} | \boldsymbol{Y}_{t-2}) = \boldsymbol{0}$ and $E(\boldsymbol{\epsilon}_t | \boldsymbol{Y}_{t-2}) = \boldsymbol{0}$,

$$E(\boldsymbol{Y}_t | \boldsymbol{Y}_{t-2}) = \boldsymbol{\Phi}^2 \boldsymbol{Y}_{t-2}.$$

By similar calculations,

$$E(\boldsymbol{Y}_t | \boldsymbol{Y}_{t-k}) = \boldsymbol{\Phi}^k \boldsymbol{Y}_{t-k}, \text{ for all } k > 0. \tag{13.18}$$

It can be shown using (13.18), that the mean will explode if any of the eigenvectors of $\boldsymbol{\Phi}$ are greater than 1 in magnitude. In fact, an AR(1) process is stationary if and only if all of the eigenvalues of $\boldsymbol{\Phi}$ are less than 1 in absolute value. The eigen() function in R can be used to find the eigenvalues.

Example 13.10. A bivariate AR model for $\Delta\,cpi$ and $\Delta\,ip$

This example uses the CPI and IP data sets discussed in earlier examples (*cpi* and *ip* denote the log transformed series). Bivariate AR processes were fit to $(\Delta cpi, \Delta ip)'$ using R's function ar(). AIC as a function of p is shown below. The two best-fitting models are AR(1) and AR(5), with the latter being slightly better by AIC. Although BIC is not part of $ar()$'s output, it can be calculated easily since BIC = AIC + $\{\log(n)-2\}p$. Because $\{\log(n)-2\} = 2.9$ in this example, it is clear that BIC is much smaller for the AR(1) model than for the AR(5) model. For this reason and because the AR(1) model is so much simpler to analyze, we will use the AR(1) model.

```
40  CPI_IP = cbind(CPI_diff1,IP_diff1)
41  arFit = ar(CPI_IP,order.max=10)
42  options(digits=2)
43  arFit$aic
```

p	0	1	2	3	4
AIC	127.99	0.17	1.29	5.05	3.40

	5	6	7	8	9	10
	0.00	6.87	9.33	10.83	13.19	14.11

The commands and results for fitting the bivariate AR(1) model are

```
44  arFit1 = ar(CPI_IP, order.max = 1) ; arFit1
```

with

$$\widehat{\Phi} = \begin{pmatrix} 0.767 & 0.0112 \\ -0.330 & 0.3014 \end{pmatrix}$$

and

$$\widehat{\Sigma} = \begin{pmatrix} 5.68e-06 & 3.33e-06 \\ 3.33e-06 & 6.73e-05 \end{pmatrix}. \tag{13.19}$$

The function $ar()$ does not estimate μ, but μ can be estimated by the sample mean, which is $(0.0052, 0.0021)'$.

```
45  colMeans(CPI_IP)
```

It is useful to look at the two off-diagonals of $\widehat{\Phi}$. Since $\Phi_{1,2} = 0.01 \approx 0$, $Y_{2,t-1}$ (lagged ip) has little influence on $Y_{1,t}$ (cpi), and since $\Phi_{2,1} = -0.330$, $Y_{1,t-1}$ (lagged cpi) has a substantial negative effect on $Y_{2,t}$ (ip), given the other variables in the model. It should be emphasized that "effect" means statistical association, not necessarily causation. This agrees with what we found when looking at the CCF for these series in Example 13.9.

How does ip depend on cpi further back in time? To answer this question we look at the $(1, 2)$ elements of the following powers of Φ:

```
46  bPhi = arFit1$ar[,,] ; bPhi
47  bPhi2 = bPhi %*% bPhi ; bPhi2
48  bPhi3 = bPhi2 %*% bPhi ; bPhi3
49  bPhi4 = bPhi3 %*% bPhi ; bPhi4
50  bPhi5 = bPhi4 %*% bPhi ; bPhi5
```

$$\widehat{\boldsymbol{\Phi}}^2 = \begin{pmatrix} 0.58 & 0.012 \\ -0.35 & 0.087 \end{pmatrix}, \quad \widehat{\boldsymbol{\Phi}}^3 = \begin{pmatrix} 0.44 & 0.010 \\ -0.30 & 0.022 \end{pmatrix},$$

$$\widehat{\boldsymbol{\Phi}}^4 = \begin{pmatrix} 0.34 & 0.0081 \\ -0.24 & 0.0034 \end{pmatrix}, \quad \text{and} \quad \widehat{\boldsymbol{\Phi}}^5 = \begin{pmatrix} 0.26 & 0.0062 \\ -0.18 & -0.0017 \end{pmatrix}.$$

What is interesting here is that the (1,2) elements, that is, -0.35, -0.30, -0.24, and -0.18, decay to zero slowly, much like the CCF. This helps explain why the AR(1) model fits the data well. This behavior where the cross-correlations are all negative and decay only slowly to zero is quite different from the behavior of the ACF of a univariate AR(1) process. For the latter, the correlations either are all positive or else alternate in sign, and in either case, unless the lag-1 correlation is nearly equal to 1, the correlations decay rapidly to 0.

In contrast to these negative correlations between Δcpi and future Δip, it follows from (13.19) that the white noise series has a positive, albeit small, correlation of $3.33/\sqrt{(5.68)(67.3)} = 0.17$. The white noise series represents unpredictable changes in the Δcpi and Δip series, so we see that the unpredictable changes have positive correlation. In contrast, the negative correlations between Δcpi and future Δip concern predictable changes.

Figure 13.14 shows the ACF of the Δcpi and Δip residuals and the CCF of these residuals. There is little auto- or cross-correlation in the residuals at nonzero lags, indicating that the AR(1) has a satisfactory fit. Figure 13.14 was produced by the `acf()` function in R. When applied to a multivariate time series, `acf()` creates a matrix of plots. The univariate ACFs are on the main diagonal, the CCFs at positive lags are above the main diagonal, and the CCFs at negative values of lag are below the main diagonal.   □

13.4.5 Prediction Using Multivariate AR Models

Forecasting with multivariate AR processes is much like forecasting with univariate AR processes. Given a multivariate AR(p) time series $\boldsymbol{Y}_1, \ldots, \boldsymbol{Y}_n$, the forecast of $\boldsymbol{Y}_{n+1}$ is

$$\widehat{\boldsymbol{Y}}_{n+1} = \widehat{\boldsymbol{\mu}} + \widehat{\boldsymbol{\Phi}}_1(\boldsymbol{Y}_n - \widehat{\boldsymbol{\mu}}) + \cdots + \widehat{\boldsymbol{\Phi}}_p(\boldsymbol{Y}_{n+1-p} - \widehat{\boldsymbol{\mu}}),$$

the forecast of $\boldsymbol{Y}_{n+2}$ is

$$\widehat{\boldsymbol{Y}}_{n+2} = \widehat{\boldsymbol{\mu}} + \widehat{\boldsymbol{\Phi}}_1(\widehat{\boldsymbol{Y}}_{n+1} - \widehat{\boldsymbol{\mu}}) + \cdots + \widehat{\boldsymbol{\Phi}}_p(\boldsymbol{Y}_{n+2-p} - \widehat{\boldsymbol{\mu}}),$$

and so forth, so that for all h,

$$\widehat{\boldsymbol{Y}}_{n+h} = \widehat{\boldsymbol{\mu}} + \widehat{\boldsymbol{\Phi}}_1(\widehat{\boldsymbol{Y}}_{n+h-1} - \widehat{\boldsymbol{\mu}}) + \cdots + \widehat{\boldsymbol{\Phi}}_p(\widehat{\boldsymbol{Y}}_{n+h-p} - \widehat{\boldsymbol{\mu}}), \qquad (13.20)$$

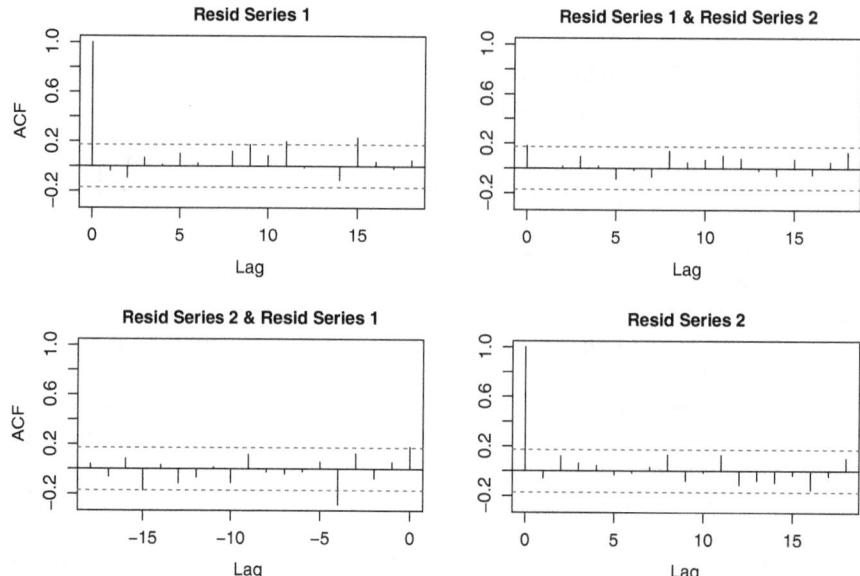

Fig. 13.14. *The ACF and CCF for the residuals when fitting a bivariate AR(1) model to $(\Delta cpi, \Delta ip)'$. Top left: The ACF of Δcpi residuals. Top right: The CCF of Δcpi and Δip residuals with positive values of lag. Bottom left: The CCF of Δcpi and Δip residuals with negative values of lag. Bottom right: The ACF of Δip residuals.*

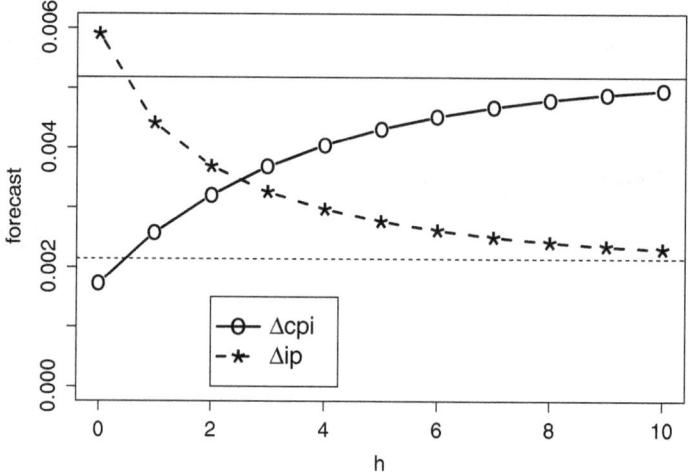

Fig. 13.15. *Forecasts of Δcpi (solid) and Δip (dashed) using a bivariate AR(1) model. The number of time units ahead is h. At $h = 0$, the last observed values of the time series are plotted. The two horizontal lines are at the means of the series, and the forecasts will asymptote to these lines as $h \to \infty$ since this model is stationary.*

where we use the convention that $\widehat{\boldsymbol{Y}}_t = \boldsymbol{Y}_t$ if $t \leq n$. For an AR(1) model, repeated application of (13.20) shows that

$$\widehat{\boldsymbol{Y}}_{n+h} = \widehat{\boldsymbol{\mu}} + \widehat{\boldsymbol{\Phi}}_1^h (\boldsymbol{Y}_n - \widehat{\boldsymbol{\mu}}).\qquad(13.21)$$

Example 13.11. Using a bivariate AR(1) model to predict CPI and IP

The ΔCPI and ΔIP series were forecast using (13.21) with estimates found in Example 13.10. Figure 13.15 shows forecasts up to 10 months ahead for both CPI and IP. Figure 13.16 shows forecast limits computed by simulation using the techniques described in Sect. 12.12.2 generalized to a multivariate time series.   □

13.5 Long-Memory Processes

13.5.1 The Need for Long-Memory Stationary Models

In Chap. 12, ARMA processes were used to model stationary time series. Stationary ARMA processes have only short memories in that their autocorrelation functions decay to zero exponentially fast. That is, there exist a $D > 0$ and $r < 1$ such that

$$\rho(k) < D|r|^k$$

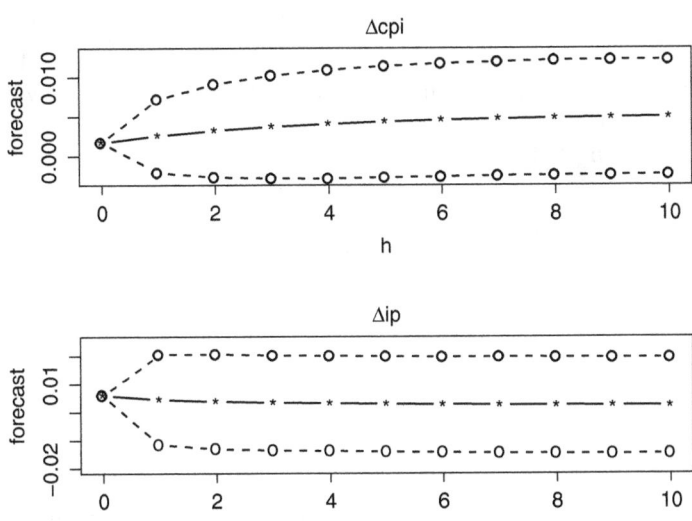

Fig. 13.16. *Forecast limits (dashed) for Δcpi and Δip computed by simulation, and forecasts (solid). At $h = 0$, the last observed changes are plotted so the widths of the forecast intervals are zero.*

for all k. In contrast, many financial time series appear to have long memory since their ACFs decay at a (slow) polynomial rate rather than a (fast) geometric rate, that is,

$$\rho(k) \sim Dk^{-\alpha}$$

for some D and $\alpha > 0$. A polynomial rate of decay is sometimes called a hyperbolic rate. In this section, we will introduce the fractional ARIMA models, which include stationary processes with long memory.

13.5.2 Fractional Differencing

The most widely used models for stationary, long-memory processes use fractional differencing. For integer values of d we have

$$\Delta^d = (1 - B)^d = \sum_{k=0}^{d} \binom{d}{k} (-B)^k. \tag{13.22}$$

In this subsection, the definition of Δ^d will be extended to noninteger values of d. The only restriction on d will be that $d > -1$.

We define

$$\binom{d}{k} = \frac{d(d-1)\cdots(d-k+1)}{k!} \tag{13.23}$$

for any d except negative integers and any integer $k \geq 0$, except if d is an integer and $k > d$, in which case $d - k$ is a negative integer and $(d-k)!$ is not defined. In the latter case, we define $\binom{d}{k}$ to be 0, so $\binom{d}{k}$ is defined for all d except negative integers and for all integer $k \geq 0$. Only values of d greater than -1 are needed for modeling long-memory processes, so we will restrict attention to this case.

The function $f(x) = (1 - x)^d$ has an infinite Taylor series expansion

$$(1 - x)^d = \sum_{k=0}^{\infty} \binom{d}{k} (-x)^k. \tag{13.24}$$

Since $\binom{d}{k} = 0$ if $k > d$ and $d > -1$ is an integer, when d is an integer we have

$$(1 - x)^d = \sum_{k=0}^{\infty} \binom{d}{k} (-x)^k = \sum_{k=0}^{d} \binom{d}{k} (-x)^k. \tag{13.25}$$

The right-hand side of (13.25) is the usual finite binomial expansion for d a nonnegative integer, so (13.24) extends the binomial expansion to all $d > -1$. Since $(1 - x)^d$ is defined for all $d > -1$, we can define $\Delta^d = (1 - B)^d$ for any $d > -1$. In summary, if $d > -1$, then

$$\Delta^d Y_t = \sum_{k=0}^{\infty} \binom{d}{k} (-1)^k Y_{t-k}. \qquad (13.26)$$

13.5.3 FARIMA Processes

A process Y_t is a fractional ARIMA(p, d, q) process, also called an ARFIMA or FARIMA(p, d, q) process, if $\Delta^d Y_t$ is an ARMA(p, q) process. We say that Y_t is a fractionally integrated process of order d or, simply, $I(d)$ process. This is, of course, the previous definition of an ARIMA process extended to noninteger values of d. Usually, $d \geq 0$, with $d = 0$ being the ordinary ARMA case, but d could be negative. If $-1/2 < d < 1/2$, then the process is stationary. If $0 < d < 1/2$, then it is a long-memory stationary process.

If $d > \frac{1}{2}$, then Y_t can be differenced an integer number of times to become a stationary process, though perhaps with long-memory. For example, if $\frac{1}{2} < d < 1\frac{1}{2}$, then ΔY_t is fractionally integrated of order $d - 1 \in (-\frac{1}{2}, \frac{1}{2})$ and ΔY_t. has long-memory if $1 < d < 1\frac{1}{2}$ so that $d - 1 \in (0, \frac{1}{2})$.

Figure 13.17 shows time series plots and sample ACFs for simulated FARIMA$(0, d, 0)$ processes with $n = 2{,}500$ and $d = -0.35$, 0.35, and 0.7. The last case is nonstationary. The R function simARMA0() in the longmemo package was used to simulate the stationary series. For the case $d = 0.7$, simARMA0() was used to simulate a FARIMA$(0, -0.3, 0)$ series and this was integrated to create a FARIMA$(0, d, 0)$ with $d = -0.3 + 1 = 0.7$. As explained in Sect. 12.9, integration is implemented by taking partial sums, and this was done with R's function cumsum().

The FARIMA$(0, 0.35, 0)$ process has a sample ACF which drops below 0.5 almost immediately but then persists well beyond 30 lags. This behavior is typical of stationary processes with long memory. A short-memory stationary process would not have autocorrelations persisting that long, and a nonstationary processes would not have a sample ACF that dropped below 0.5 so quickly.

Note that the case $d = -0.35$ in Fig. 13.17 has an ACF with a negative lag-1 autocorrelation and little additional autocorrelation. This type of ACF is often found when a time series is differenced once. After differencing, an MA term is needed to accommodate the negative lag-1 autocorrelated. A more parsimonious model can sometimes be used if the differencing is fractional. For example, consider the third series in Fig. 13.17. If it is differenced once, then a series with $d = -0.3$ is the result. However, if it is differenced with $d = 0.7$, then white noise is the result. This can be seen in the ACF plots in Fig. 13.18.

Example 13.12. Inflation rates—FARIMA modeling

This example uses the inflation rates that have been studied already in Chap. 12. From the analysis in that chapter it was unclear whether to model the series as $I(0)$ or $I(1)$. Perhaps it would be better to have a compromise

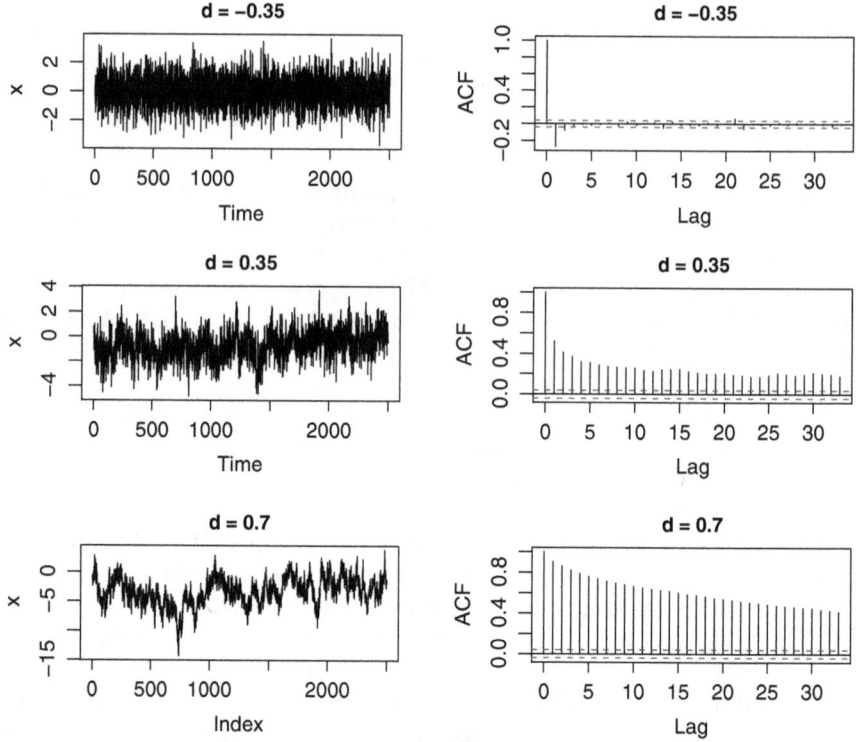

Fig. 13.17. *Time series plots (left) and sample ACFs (right) for simulated FARIMA$(0, d, 0)$: the top series is stationary with short-term memory; the middle series is stationary with long-term memory; the bottom series is nonstationary.*

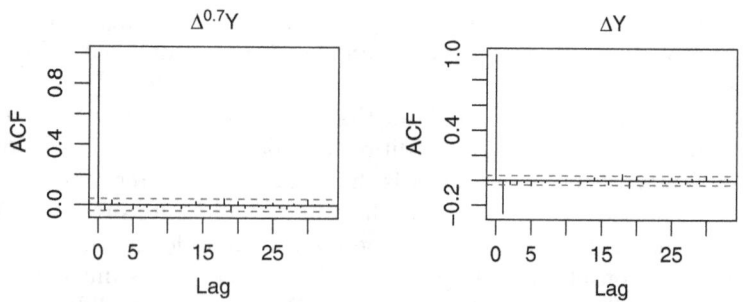

Fig. 13.18. *Sample ACF plots for the simulated FARIMA$(0, 0.7, 0)$ series in Figure 13.17 after differencing using $d = 0.7$ and 1.*

between these alternatives. Now, with the new tool of fractional integration, we can try differencing with d between 0 and 1. There is some reason to believe that fractional differencing is suitable for this example, since the ACF plot in Fig. 12.3 is similar to that of the $d = 0.35$ plot in Fig. 13.17.

The function `fracdiff()` in R's `fracdiff` package will fit a FARIMA (p, d, q) process. The values of p, d, and q must be input; we are not aware of any R function that will chose p, d, and q automatically in the way this can be done for an ARIMA process (that is, with d restricted to be an integer) using `auto.arima()`. First, a trial value of d was chosen by using `fracdiff()` with $p = q = 0$, the default values. The estimate was $\hat{d} = 0.378$. Then, the inflation rates were fractionally differenced using this value of d and `auto.arima()` was applied to the fractionally differenced series. The result was that BIC selected $p = q = d = 0$. The value $d = 0$ means that no further differencing is applied to the already fractionally differenced series. Fractional differencing was done with the `diffseries()` function in R's `fracdiff` package.

Figure 13.19 has sample ACF plots of the original series and the series differenced with $d = 0$, 0.4 (from rounding 0.378), and 1. The first series has a slowly decaying ACF typical of a long-memory process, the second series looks like white noise, and the third series has negative autocorrelation at lag-1 which indicates overdifferencing.

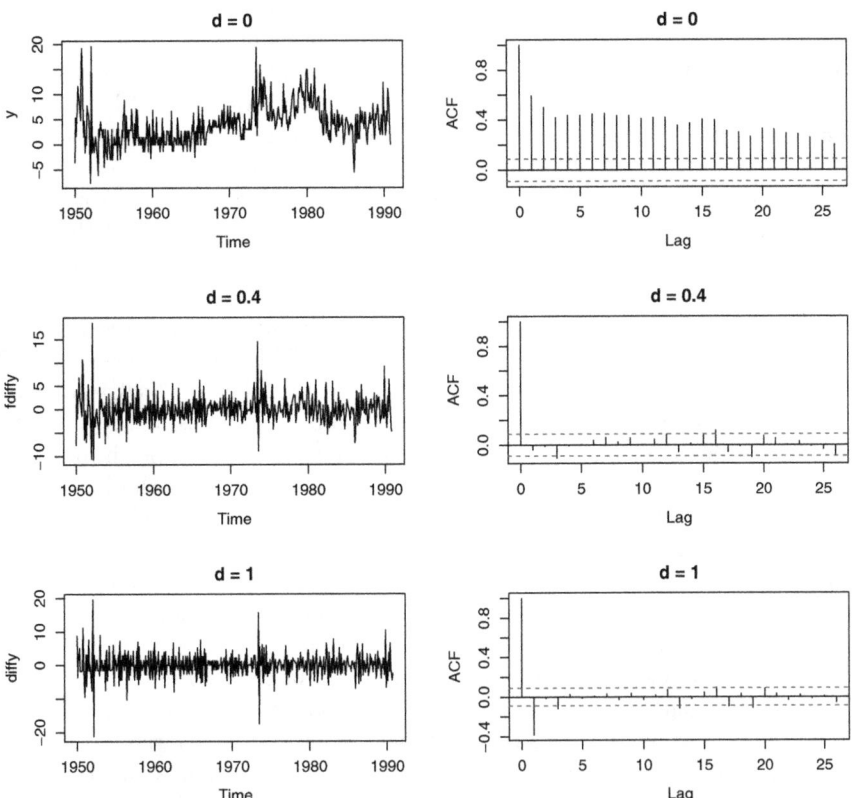

Fig. 13.19. *Time series plots (left) and sample ACF plots (right) for the inflation rates series with differencing using $d = 0$, 0.4, and 1.*

The conclusion is that a white noise process seems to be a suitable model for the fractionally differenced series and the original series can be model as FARIMA(0,0.378,0), or, perhaps, more simply as FARIMA(0,0.4,0).

Differencing a stationary process creates another stationary process, but the differenced process often has a more complex autocorrelation structure than the original process. Therefore, one should not *overdifference* a time series. However, if d is restricted to integer values, then often, as in this example, overdifferencing cannot be avoided.   □

13.6 Bootstrapping Time Series

The resampling methods introduced in Chap. 6 are designed for i.i.d. univariate data but are easily extended to multivariate data. As discussed in Sect. 7.11, if $\boldsymbol{Y}_1, \ldots, \boldsymbol{Y}_n$ is a sample of vectors, then one resamples the $\boldsymbol{Y}_i$ themselves, not their components, to maintain the covariance structure of the data in the resamples.

It is not immediately obvious whether one can resample a time series $Y_1, \ldots, Y_n$. A time series is essentially a sample of size 1 from a stochastic process. Resampling a sample of size 1 in the usual way is a futile exercise— each resample is the original sample, so one learns nothing by resampling. Therefore, resampling of a time series requires new ideas.

Model-based resampling is easily adapted to time series. The resamples are obtained by simulating the time series model. For example, if the model is ARIMA$(p, 1, q)$, then the resamples start with simulated samples of an ARMA(p, q) model with MLEs (from the differenced series) of the autoregressive and moving average coefficients and the noise variance. The resamples are the sequences of partial sums of the simulated ARMA(p, q) process.

Model-free resampling of a time series is accomplished by *block resampling*, also called the *block bootstrap*, which can be implemented using the `tsboot()` function in R's boot package. The idea is to break the time series into roughly equal-length blocks of consecutive observations, to resample the blocks with replacement, and then to paste the blocks together. For example, if the time series is of length 200 and one uses 10 blocks of length 20, then the blocks are the first 20 observations, the next 20, and so forth. A possible resample is the fourth block (observations 61 to 80), then the last block (observations 181 to 200), then the second block (observations 21 to 40), then the fourth block again, and so on until there are 10 blocks in the resample.

A major issue is how best to select the block length. The correlations in the original sample are preserved only within blocks, so a large block size is desirable. However, the number of possible resamples depends on the number of blocks, so a large number of blocks is also desirable. Obviously, there must be a tradeoff between the block size and the number of blocks. A full discussion of block bootstrapping is beyond the scope of this book, but see Sect. 13.7 for further reading.

13.7 Bibliographic Notes

Beran (1994) is a standard reference for long-memory processes, and Beran (1992) is a good introduction to this topic. Most of the time series text-books listed in the "References" section discuss seasonal ARIMA models. For more details on HC and HAC covariance matrix estimators and the R package sandwich see Zeileis (2004). Enders (2004) has a section on bootstrapping time series and a chapter on multivariate time series. Reinsel (2003) is an in-depth treatment of multivariate time series; see also Hamilton (1994) for this topic. Transfer function models are another method for analyzing multi-variate time series; see Box, Jenkins, and Reinsel (2008). Davison and Hinkley (1997) discuss both model-based and block resampling of time series and other types of dependent data. Lahiri (2003) provides an advanced and comprehensive account of block resampling. Bühlmann (2002) is a review article about bootstrapping time series.

13.8 R Lab

13.8.1 Seasonal ARIMA Models

This section uses seasonally non-adjusted quarterly data on income and consumption in the UK. Run the following code to load the data and plot the variable consumption.

```
1 library("Ecdat")
2 library("forecast")
3 data(IncomeUK)
4 consumption = IncomeUK[,2]
5 plot(consumption)
```

Problem 1 *Describe the behavior of* consumption. *What types of differencing, seasonal, nonseasonal, or both, would you recommend? Do you recommend fitting a seasonal ARIMA model to the data with or without a log transformation? Consider also using ACF plots to help answer these questions.*

Problem 2 *Regardless of your answers to Problem 1, find an ARIMA model that provides a good fit to* log(consumption). *What order model did you select? (Give the orders of the nonseasonal and seasonal components.)*

Problem 3 *Check the ACF of the residuals from the model you selected in Problem 2. Do you see any residual autocorrelation?*

Problem 4 *Apply* auto.arima() *to* log(consumption) *using BIC. Which model is selected?*

Problem 5 *Forecast* `log(consumption)` *for the next eight quarters using the models you found in Problems 2 and 4. Plot the two sets of forecasts in side-by-side plots with the same limits on the x- and y-axes. Describe any differences between the two sets of forecasts.*

Note: To predict an `arima` object (an object returned by the `arima()` function), use the `predict` function. To learn how the predict() function works on an `arima` object, use `?predict.Arima`. To forecast an object returned by `auto.arima()`, use the `forecast()` function in the `forecast` package. For example, the following code will forecast eight quarters ahead using the object returned by `auto.arima()` and then plot the forecasts.

```
6  logConsumption = log(consumption)
7  fitAutoArima = auto.arima(logConsumption, ic="bic")
8  foreAutoArima = forecast(fitAutoArima, h=8)
9  plot(foreAutoArima, xlim=c(1985.5,1987.5), ylim=c(10.7,11.2))
```

13.8.2 Regression with HAC Standard Errors

Run the following commands in R to compute the OLS estimates of the regression of the differenced one-month T-bill rates, `tb1_diff`, on the differenced three-month T-bill rates, `tb3_diff`.

```
1  data(Mishkin, package="Ecdat")
2  tb1_dif = diff(as.vector(Mishkin[,3]))
3  tb3_dif = diff(as.vector(Mishkin[,4]))
4  fit = lm(tb1_dif ~ tb3_dif )
5  round(summary(fit)$coef, 4)
6  acf(fit$resid)
```

Problem 6 *Is there evidence of significant autocorrelation among the residuals? Why?*

Now run the following commands to compute the HC standard error estimates and their associated t values.

```
7  library(sandwich)
8  sqrt(diag(NeweyWest(fit, lag = 0, prewhite = F)))
9  coef(fit)/sqrt(diag(NeweyWest(fit, lag = 0, prewhite = F)))
```

Problem 7 *How do these t values compare to the t values from the OLS fit? Does the HC adjustment change the conclusions of the hypothesis tests?*

Problem 8 *Run the commands again, but with* `lag` *equal to 1,2, and 3 to obtain the corresponding HAC t values. How do the t values vary with* `lag`*?*

13.8.3 Regression with ARMA Noise

This section uses the USMacroG data set used earlier in Sect. 9.11.1. In the earlier analysis, we did not investigate residual correlation, but now we will. The model will be the regression of changes in unemp = unemployment rate on changes in government = real government expenditures and changes in invest = real investment by the private sector. Run the following R code to read the data, compute differences, and then fit a linear regression model with AR(1) errors.

```
1 library(AER)
2 data("USMacroG")
3 MacroDiff = as.data.frame(apply(USMacroG, 2, diff))
4 attach(MacroDiff)
5 fit1 = arima(unemp, order=c(1,0,0), xreg=cbind(invest, government))
```

Problem 9 *Fit a linear regression model using* lm()*, which assumes uncorrelated errors. Compare the two models by AIC and residual ACF plots. Which model fits better?*

Problem 10 *What are the values of BIC for the model with uncorrelated errors and for the model with AR(1) errors? Does the conclusion in Problem 9 about which model fits better change if one uses BIC instead of AIC?*

Problem 11 *Does the model with AR(2) noise or the model with ARMA(1,1) noise offer a better fit than the model with AR(1) noise?*

13.8.4 VAR Models

This section uses data on the 91-day Treasury bill, the real GDP, and the inflation rate. Run the following R code to read the data, find the best-fitting multivariate AR to changes in the three series, and check the residual correlations.

```
1 TbGdpPi = read.csv("TbGdpPi.csv", header=TRUE)
2 #   r = the 91-day treasury bill rate
3 #   y = the log of real GDP
4 #   pi = the inflation rate
5 TbGdpPi = ts(TbGdpPi, start = 1955, freq = 4)
6 del_dat = diff(TbGdpPi)
7 var1 = ar(del_dat, order.max=4, aic=T)
8 var1
9 acf(na.omit(var1$resid))
```

Problem 12 *For this problem, use the notation of Eq. (13.16) with $q = 0$.*

(a) What is p and what are the estimates $\Phi_1, \ldots, \Phi_p$?
(b) What is the estimated covariance matrix of ϵ_t?
(c) If the model fits adequately, then there should be no residual auto- or cross-correlation. Do you believe that the model does fit adequately?

Problem 13 *The last three changes in* r*,* y*, and* pi *are given next. What are the predicted values of the next set of changes in these series?*

```
10  tail(TbGdpPi, n = 4)

             r    y     pi
    [233,] 0.07  9.7   1.38
    [234,] 0.04  9.7   0.31
    [235,] 0.02  9.7   0.28
    [236,] 0.07  9.7  -0.47
```

Now fit a VAR(1) using the following commands.

```
11  var1 = ar(del_dat, order.max=1)
```

Suppose we observe changes in r, y, and pi that are each 10 % above the mean changes:

```
12  yn = var1$x.mean * 1.1 ; yn
```

Problem 14 *Compute the h-step forecasts for $h = 1, 2$, and 5 using* yn *as the most recent observation. How do these forecasts compare to the mean* var1$x.mean*? For each h, compute ratios between the forecasts and the mean. How do these values compare to the starting value,* yn/var1$x.mean = 1.1*? Are they closer to or farther from* 1.0 = var1$x.mean/var1$x.mean*? What does this suggest?*

Using the fitted VAR(1) from above, examine the estimate of $\hat{\Phi}$:

```
13  Phi_hat = var1$ar[1,,] ; Phi_hat
```

Problem 15 *What do the elements of* Phi_hat *suggest about the relationships among the changes in* r*,* y*, and* pi*?*

A VAR(1) process is stationary provided that the eigenvalues of Φ are less than one in magnitude. Compute the eigenvalues of $\hat{\Phi}$:

```
14  eigen.values = eigen(Phi_hat)$values
15  abs(eigen.values)
```

Problem 16 *Is the estimated process stationary? How does this result relate to the forecast calculations in Problem 14 above?*

The dataset `MacroVars.csv` contains three US macroeconomic indicators from Quarter 1 of 1959 to Quarter 4 of 1997: Real Gross Domestic Product (a measure of economic activity), Consumer Price Index (a measure of inflation), and Federal Funds Rate (a proxy for monetary policy). Each series has been transformed to stationary based on the procedures suggested by Stock and Watson (2005).

```
16 MacroVars = read.csv("MacroVars.csv", head=TRUE)
```

Problem 17 *Fit a VAR(p) model using the* `ar()` *function in* R *using AIC (the default) to select lag order.*

Problem 18
By modifying the output of the `ar()` *function as discussed in Example 13.10, use BIC to select the lag order. Comment on any differences.*

13.8.5 Long-Memory Processes

This section uses changes in the square root of the Consumer Price Index. The following code creates this time series.

```
1 data(Mishkin, package="Ecdat")
2 cpi = as.vector(Mishkin[,5])
3 DiffSqrtCpi = diff(sqrt(cpi))
```

Problem 19 *Plot* `DiffSqrtCpi` *and its ACF. Do you see any signs of long memory? If so, describe them.*

Run the following code to estimate the amount of fractional differencing, fractionally difference `DiffSqrtCpi` appropriately, and check the ACF of the fractionally differenced series.

```
4 library("fracdiff")
5 fit.frac = fracdiff(DiffSqrtCpi,nar=0,nma=0)
6 fit.frac$d
7 fdiff = diffseries(DiffSqrtCpi,fit.frac$d)
8 acf(fdiff)
```

Problem 20 *Do you see any short- or long-term autocorrelation in the fractionally differenced series?*

Problem 21 *Fit an ARIMA model to the fractionally differenced series using* `auto.arima()`. *Compare the models selected using AIC and BIC.*

13.8.6 Model-Based Bootstrapping of an ARIMA Process

This exercise uses the price of frozen orange juice. Run the following code to fit an ARIMA model.

```
1  library(AER)
2  library(forecast)
3  data("FrozenJuice")
4  price = FrozenJuice[,1]
5  plot(price)
6  auto.arima(price, ic="bic")
```

The output from `auto.arima()`, which is needed for model-based bootstrapping, is

```
Series: price
ARIMA(2,1,0)

Coefficients:
          ar1      ar2
       0.2825   0.0570
s.e.   0.0407   0.0408

sigma^2 estimated as 9.989:  log likelihood = -1570.11
AIC = 3146.23   AICc = 3146.27   BIC = 3159.47
```

Next, we will use the model-based bootstrap to investigate how well BIC selects the "correct" model, which is ARIMA(2,1,0). Since we will be looking at the output of each fitted model, only a small number of resamples will be used. Despite the small number of resamples, we will get some sense of how well BIC works in this context. To simulate 10 model-based resamples from the ARIMA(2,1,0) model, run the following commands.

```
7   n = length(price)
8   sink("priceBootstrap.txt")
9   set.seed(1998852)
10  for (iter in 1:10){
11    eps = rnorm(n+20)
12    y = rep(0,n+20)
13    for (t in 3:(n+20)){
14      y[t] = 0.2825*y[t-1] + 0.0570*y[t-2] + eps[t]
15    }
16    y = y[101:n+20]
17    y = cumsum(y)
18    y = ts(y, frequency=12)
19    fit = auto.arima(y, d=1, D=0, ic="bic")
20    print(fit)
21  }
22  sink()
```

The results will be sent to the file `priceBootstrap.txt`. The first two values of y are independent and are used to initialize the process. A burn-in period of 20 is used to remove the effect of initialization. Note the use of `cumsum()` to integrate the simulated AR(2) process and the use of `ts()` to convert a vector to a monthly time series.

Problem 22 *How often is the "correct" AR(2) model selected?*

Now we will perform a bootstrap where the correct model AR(2) is known and study the accuracy of the estimators. Since the correct model is known, it can be fit by `arima()`. The estimates will be stored in a matrix called `estimates`. In contrast to earlier when model-selection was investigated by resampling, now a large number of bootstrap samples can be used, since `arima()` is fast and only the estimates are stored. Run the following:

```
23  set.seed(1998852)
24  niter = 1000
25  estimates=matrix(0, nrow=niter, ncol=2)
26  for (iter in 1:niter){
27    eps = rnorm(n+20)
28    y = rep(0, n+20)
29    for (t in 3:(n+20)){
30      y[t] = .2825 *y[t-1] + 0.0570*y[t-2] + eps[t]
31    }
32    y = y[101:n+20]
33    y = cumsum(y)
34    y = ts(y, frequency=12)
35    fit=arima(y, order=c(2,1,0))
36    estimates[iter,] = fit$coef
37  }
```

Problem 23 *Find the biases, standard deviations, and MSEs of the estimators of the two coefficients.*

13.9 Exercises

1. Figure 13.20 contains ACF plots of 40 years of quarterly data, with all possible combinations of first-order seasonal and nonseasonal differencing. Which combination do you recommend in order to achieve stationarity?
2. Figure 13.21 contains ACF plots of 40 years of quarterly data, with all possible combinations of first-order seasonal and nonseasonal differencing. Which combination do you recommend in order to achieve stationarity?

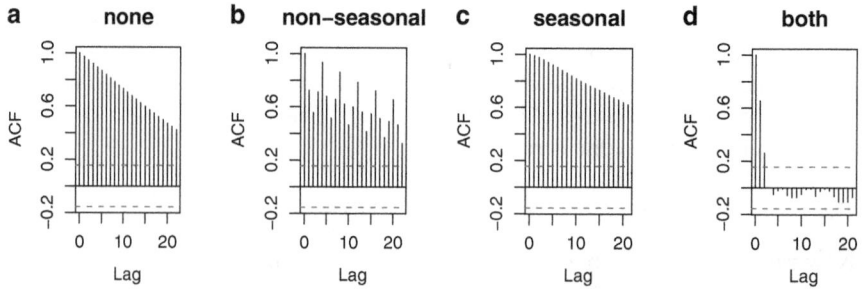

Fig. 13.20. *ACF plots of quarterly data with no differencing, nonseasonal differencing, seasonal differencing, and both seasonal and nonseasonal differencing.*

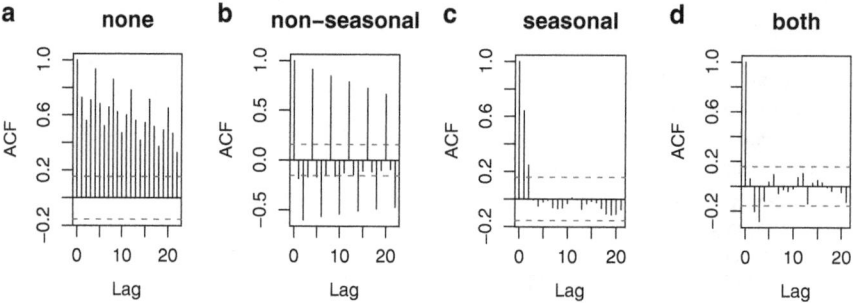

Fig. 13.21. *ACF plots of quarterly data with no differencing, nonseasonal differencing, seasonal differencing, and both seasonal and nonseasonal differencing.*

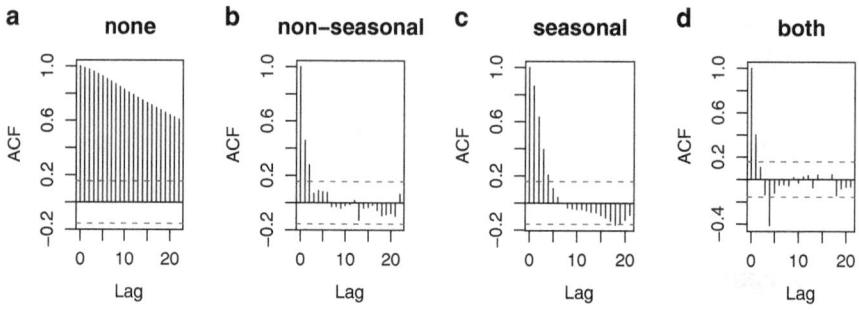

Fig. 13.22. *ACF plots of quarterly data with no differencing, nonseasonal differencing, seasonal differencing, and both seasonal and nonseasonal differencing.*

3. Figure 13.22 contains ACF plots of 40 years of quarterly data, with all possible combinations of first-order seasonal and nonseasonal differencing. Which combination do you recommend in order to achieve stationarity?

4. In Example 13.10, a bivariate AR(1) model was fit to $(\Delta cpi, \Delta ip)'$ and

$$\widehat{\boldsymbol{\Phi}} = \begin{pmatrix} 0.767 & 0.0112 \\ -0.330 & 0.3014 \end{pmatrix}.$$

The mean of $(\Delta cpi, \Delta ip)'$ is $(0.0052, 0.0021)'$ and the last observation of $(\Delta cpi, \Delta ip)'$ is $(0.0017, 0.0059)'$. Forecast the next two values of Δip. (The forecasts are shown in Fig. 13.15, but you should compute numerical values.)

5. Fit an ARIMA model to `income`, which is in the first column of the `IncomeUK` data set in the `Ecdat` package. Explain why you selected the model you did. Does your model exhibit any residual correlation?

6. (a) Find an ARIMA model that provides a good fit to the variable `unemp` in the `USMacroG` data set in the `AER` package.

   (b) Now perform a small model-based bootstrap to see how well `auto.arima()` can select the true model. To do this, simulate eight data sets from the ARIMA model selected in part (a) of this problem. Apply `auto.arima()` with BIC to each of these data sets. How often is the "correct" amount of differencing selected, that is, d and D are correctly selected? How often is the "correct" model selected? "Correct" means in agreement with the simulation model. "Correct model" means both the correct amount of differencing and the correct orders for all the seasonal and nonseasonal AR and MA components.

7. This exercise uses the `TbGdpPi.csv` data set. In Sect. 12.15.1, nonseasonal models were fit. Now use `auto.arima()` to find a seasonal model. Which seasonal model is selected by AIC and by BIC? Do you feel that a seasonal model is needed, or is a nonseasonal model sufficient?

References

Beran, J. (1992) Statistical methods for data with long-range dependence. *Statistical Science*, **7**, 404–427.

Beran, J. (1994) *Statistics for Long-Memory Processes*, Chapman & Hall, Boca Raton, FL.

Box, G. E. P., Jenkins, G. M., and Reinsel, G. C. (2008) *Times Series Analysis: Forecasting and Control*, 4th ed., Wiley, Hoboken, NJ.

Bühlmann, P. (2002) Bootstraps for time series. *Statistical Science*, **17**, 52–72.

Davison, A. C. and Hinkley, D. V. (1997) *Bootstrap Methods and Their Applications*, Cambridge University Press, Cambridge.

Enders, W. (2004) *Applied Econometric Time Series*, 2nd ed., Wiley, New York.

Hamilton, J. D. (1994) *Time Series Analysis*, Princeton University Press, Princeton, NJ.

Lahiri, S. N. (2003) *Resampling Methods for Dependent Data*, Springer, New York.

Newey, W. and West, K. (1987) A simple, positive semidefinite, heteroscedasticity and autocorrelation consistent covariance matrix. *Econometrica*, **55**, 703–708.

Reinsel, G. C. (2003) *Elements of Multivariate Time Series Analysis*, 2nd ed., Springer, New York.

Stock, J. H. and Watson, M. W. (2005). *An empirical comparison of methods for forecasting using many predictors*, manuscript `http://www4.ncsu.edu/~arhall/beb_4.pdf`

White, H. (1980) A heteroscedasticity consistent covariance matrix estimator and a direct test for heteroscedasticity. *Econometrica*, **48**, 827–838.

Zeileis, A. (2004) Econometric computing with HC and HAC covariance matrix estimators. *Journal of Statistical Software*, **11**(10), 1–17.

14

GARCH Models

14.1 Introduction

As seen in earlier chapters, financial market data often exhibits volatility clustering, where time series show periods of high volatility and periods of low volatility; see, for example, Fig. 14.1. In fact, with economic and financial data, time-varying volatility is more common than constant volatility, and accurate modeling of time-varying volatility is of great importance in financial engineering.

As we saw in Chap. 12, ARMA models are used to model the conditional expectation of a process given the past, but in an ARMA model the conditional variance given the past is constant. What does this mean for, say, modeling stock returns? Suppose we have noticed that recent daily returns have been unusually volatile. We might expect that tomorrow's return is also more variable than usual. However, an ARMA model cannot capture this type of behavior because its conditional variance is constant. So we need better time series models if we want to model the nonconstant volatility. In this chapter we look at GARCH time series models that are becoming widely used in econometrics and finance because they have randomly varying volatility.

ARCH is an acronym meaning Auto-Regressive Conditional Heteroskedasticity. In ARCH models the conditional variance has a structure very similar to the structure of the conditional expectation in an AR model. We first study the first order ARCH(1) model, which is the simplest GARCH model, and analogous to an AR(1) model. Then we look at ARCH(p) models, which are analogous to AR(p) models, and GARCH (Generalized ARCH) models, which model conditional variances much as the conditional expectation is modeled by an ARMA model. Finally, we consider several multivariate GARCH processes.

© Springer Science+Business Media New York 2015
D. Ruppert, D.S. Matteson, *Statistics and Data Analysis for Financial Engineering*, Springer Texts in Statistics,
DOI 10.1007/978-1-4939-2614-5_14

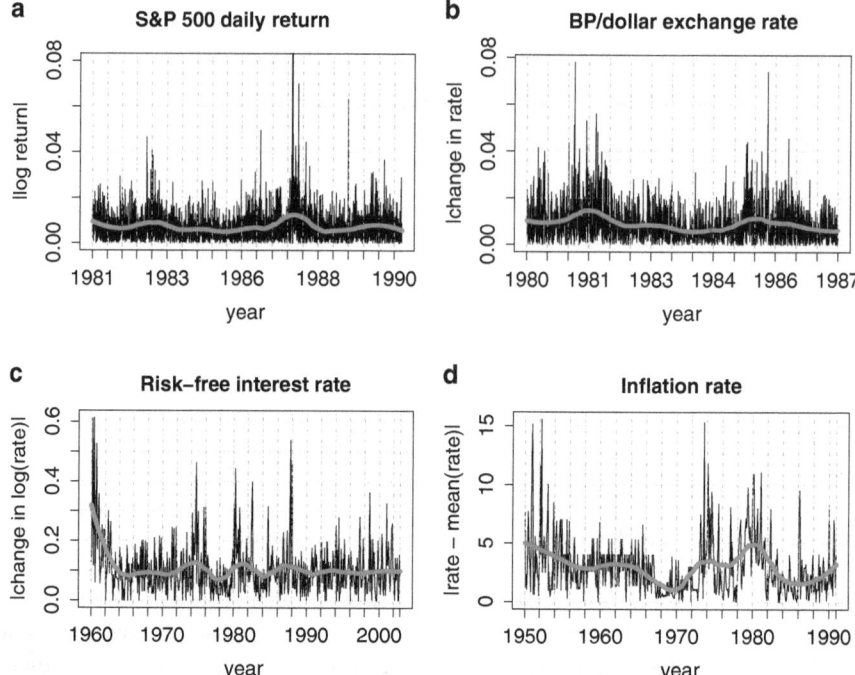

Fig. 14.1. *Examples of financial markets and economic data with time-varying volatility: (a) absolute values of S&P 500 log returns; (b) absolute values of changes in the BP/dollar exchange rate; (c) absolute values of changes in the log of the risk-free interest rate; (d) absolute deviations of the inflation rate from its mean. Loess (see Section 21.2) smooths have been added in red.*

14.2 Estimating Conditional Means and Variances

Before looking at GARCH models, we study some general principles about modeling nonconstant conditional variance. Consider regression modeling with a *constant* conditional variance, $\mathrm{Var}(Y_t|\,X_{1,t},\ldots,X_{p,t}) = \sigma^2$. Then the general form for the regression of Y_t on $X_{1,t},\ldots,X_{p,t}$ is

$$Y_t = f(X_{1,t},\ldots,X_{p,t}) + \epsilon_t, \qquad (14.1)$$

where ϵ_t is independent of $X_{1,t},\ldots,X_{p,t}$ and has expectation equal to 0 and a constant conditional variance σ_ϵ^2. The function $f(\cdot)$ is the conditional expectation of Y_t given $X_{1,t},\ldots,X_{p,t}$. Moreover, the conditional variance of Y_t is σ_ϵ^2.

Equation (14.1) can be modified to allow conditional heteroskedasticity. Let $\sigma^2(X_{1,t},\ldots,X_{p,t})$ be the conditional variance of Y_t given $X_{1,t},\ldots,X_{p,t}$. Then the model

$$Y_t = f(X_{1,t},\ldots,X_{p,t}) + \epsilon_t\,\sigma(X_{1,t},\ldots,X_{p,t}), \qquad (14.2)$$

where ϵ_t has conditional (given $X_{1,t}, \ldots, X_{p,t}$) mean equal to 0 and conditional variance equal to 1, gives the correct conditional mean and variance of Y_t.

The function $\sigma(X_{1,t}, \ldots, X_{p,t})$ should be nonnegative since it is a standard deviation. If the function $\sigma(\cdot)$ is linear, then its coefficients must be constrained to ensure nonnegativity. Such constraints are cumbersome to implement, so nonlinear nonnegative functions are usually used instead. Models for conditional variances are often called *variance function models*. The GARCH models of this chapter are an important class of variance function models.

14.3 ARCH(1) Processes

Suppose for now that $\epsilon_1, \epsilon_2, \ldots$ is Gaussian white noise with unit variance. Later we will allow the noise to be i.i.d. white noise with a possibly non-normal distribution, such as, a standardized t-distribution. Then

$$E(\epsilon_t | \epsilon_{t-1}, \ldots) = 0,$$

and

$$\mathrm{Var}(\epsilon_t | \epsilon_{t-1}, \ldots) = 1. \tag{14.3}$$

Property (14.3) is called *conditional homoskedasticity*.

The process a_t is an ARCH(1) process under the model

$$a_t = \epsilon_t \sqrt{\omega + \alpha a_{t-1}^2}, \tag{14.4}$$

which is a special case of (14.2) with f equal to 0 and σ equal to $\sqrt{\omega + \alpha a_{t-1}^2}$. We require that $\omega > 0$ and $\alpha \geq 0$ so that $\omega + \alpha a_{t-1}^2 > 0$ for all t. It is also required that $\alpha < 1$ in order for $\{a_t\}$ to be stationary with a finite variance. Equation (14.4) can be written as

$$a_t^2 = \epsilon_t^2 (\omega + \alpha a_{t-1}^2),$$

which is similar to an AR(1), but in a_t^2, not a_t, and with multiplicative noise with a mean of 1 rather than additive noise with a mean of 0. In fact, the ARCH(1) model induces an ACF for a_t^2 that is the same as an AR(1)'s ACF, as we will see from the calculations below.

Define

$$\sigma_t^2 = \mathrm{Var}(a_t | a_{t-1}, \ldots)$$

to be the conditional variance of a_t given past values. Since ϵ_t is independent of a_{t-1} and $E(\epsilon_t^2) = \mathrm{Var}(\epsilon_t) = 1$, we have

$$E(a_t | a_{t-1}, \ldots) = 0, \tag{14.5}$$

and

$$\sigma_t^2 = E\{(\omega + \alpha a_{t-1}^2)\,\epsilon_t^2|a_{t-1}, a_{t-2}, \dots\}$$
$$= (\omega + \alpha a_{t-1}^2)E\{\epsilon_t^2|a_{t-1}, a_{t-2}, \dots\}$$
$$= \omega + \alpha a_{t-1}^2. \tag{14.6}$$

Equation (14.6) is crucial to understanding how GARCH processes work. If a_{t-1} has an unusually large absolute value, then σ_t is larger than usual and so a_t is also expected to have an unusually large magnitude. This volatility propagates since when a_t has a large magnitude that makes σ_{t+1}^2 large, then a_{t+1} tends to be large in magnitude, and so on. Similarly, if a_{t-1}^2 is unusually small, then σ_t^2 is small, and a_t^2 is also expected to be small, and so forth. Because of this behavior, unusual volatility in a_t tends to persist, though not forever. The conditional variance tends to revert to the unconditional variance provided that $\alpha < 1$, so that the process is stationary with a finite variance.

The unconditional, that is, marginal, variance of a_t denoted by $\gamma_a(0)$ is obtained by taking expectations in (14.6), which gives us

$$\gamma_a(0) = \omega + \alpha\gamma_a(0)$$

for a stationary model. This equation has a positive solution if $\alpha < 1$:

$$\gamma_a(0) = \omega/(1-\alpha).$$

If $\alpha = 1$, then $\gamma_a(0)$ is infinite, but a_t is stationary nonetheless and is called an integrated GARCH (I-GARCH) model.

Straightforward calculations using (14.5) show that the ACF of a_t is

$$\rho_a(h) = 0 \text{ if } h \neq 0.$$

In fact, any process in which the conditional expectation of the present observation given the past is constant is an uncorrelated process.

In introductory statistics courses, it is often mentioned that independence implies zero correlation but not vice versa. A process, such as a GARCH process, in which the conditional mean is constant but the conditional variance is nonconstant is an example of an uncorrelated but dependent process. The dependence of the conditional variance on the past causes the process to be dependent. The independence of the conditional mean on the past is the reason that the process is uncorrelated.

Although a_t is an uncorrelated process, the process a_t^2 has a more interesting ACF. If $\alpha < 1$, then

$$\rho_{a^2}(h) = \alpha^{|h|}, \quad \forall h.$$

If $\alpha \geq 1$, then a_t^2 either is nonstationary or has an infinite variance, so it does not have an ACF. This geometric decay in the ACF of a_t^2 for an ARCH(1) process is analogous to the geometric decay in the ACF of an AR(1) process. To complete the analogy, define $\eta_t = a_t^2 - \sigma_t^2$, and note that $\{\eta_t\}$ is a mean zero weak white noise process, but not an i.i.d. white noise process. Adding η_t to both sides of (14.6) and simplifying we have

$$\sigma_t^2 + \eta_t = a_t^2 = \omega + \alpha a_{t-1}^2 + \eta_t, \tag{14.7}$$

which is a direct representation of $\{a_t^2\}$ as an AR(1) process.

14.4 The AR(1)+ARCH(1) Model

As we have seen, an AR(1) process has a nonconstant conditional mean but a constant conditional variance, while an ARCH(1) process is just the opposite. If both the conditional mean and variance of the data depend on the past, then we can combine the two models. In fact, we can combine any ARMA model with any of the GARCH models in Sect. 14.6. In this section we combine an AR(1) model with an ARCH(1) model.

Let a_t be an ARCH(1) process so that $a_t = \epsilon_t \sqrt{\omega + \alpha a_{t-1}^2}$, where ϵ_t is i.i.d. $N(0, 1)$, and suppose that

$$y_t - \mu = \phi(y_{t-1} - \mu) + a_t.$$

The process y_t is an AR(1) process, except that the noise term (a_t) is not i.i.d. white noise, but rather an ARCH(1) process which is only weak white noise.

Because a_t is an uncorrelated process, it has the same ACF as independent white noise, and therefore, y_t has the same ACF as an AR(1) process with independent white noise

$$\rho_y(h) = \phi^{|h|} \quad \forall h,$$

in the stationary case. Moreover, a_t^2 has the ARCH(1) ACF:

$$\rho_{a^2}(h) = \alpha^{|h|} \quad \forall h.$$

The ACF of y_t^2 also decays with $|h|$ at a geometric rate in the stationary case, provided some additional assumptions hold, however, the exact expressions are more complicated (see Palma and Zevallos, 2004). We need to assume that both $|\phi| < 1$ and $\alpha < 1$ in order for y_t to be stationary with a finite variance. Of course, $\omega > 0$ and $\alpha \geq 0$ are also assumed for positiveness of the conditional variance process σ_t^2. The process y_t is such that its conditional mean and variance, given the past, are both nonconstant, so a wide variety of time series can be modeled.

Example 14.1. A simulated ARCH(1) process and AR(1)+ARCH(1) process

A simulated ARCH(1) process is shown in Fig. 14.2. Panel (a) shows the i.i.d. white noise process ϵ_t, (b) shows $\sigma_t = \sqrt{1 + 0.55a_{t-1}^2}$, the conditional standard deviation process, and (c) shows $a_t = \sigma_t \epsilon_t$, the ARCH(1) process. As discussed in the previous section, an ARCH(1) process can be used as the noise term of an AR(1) process. This process is shown in panel (d). The AR(1) parameters are $\mu = 0.1$ and $\phi = 0.8$. The unconditional variance of a_t is $\gamma_a(0) = 1/(1 - 0.55) = 2.22$, so the unconditional standard deviation is $\sqrt{2.22} = 1.49$. Panels (e)–(h) are sample ACF plots of the ARCH and AR+ARCH processes and squared processes. Notice that for the ARCH series, the process is uncorrelated but the squared series has autocorrelation. Also notice that for the AR(1)+ARCH(1) series the ACFs of the process and the squared process, panels (g) and (h), both show autocorrelation. While the true ACFs have an exact geometric decay, this is only approximately true for the sample ACFs in panels (f)–(h); similarly, negative values are not present in the true ACFs, but the sample ACF has sampling error and may result in negative values. The processes were all started at 0 and simulated for 10,200 observations. The first 10,000 observations were treated as a burn-in period and discarded.   □

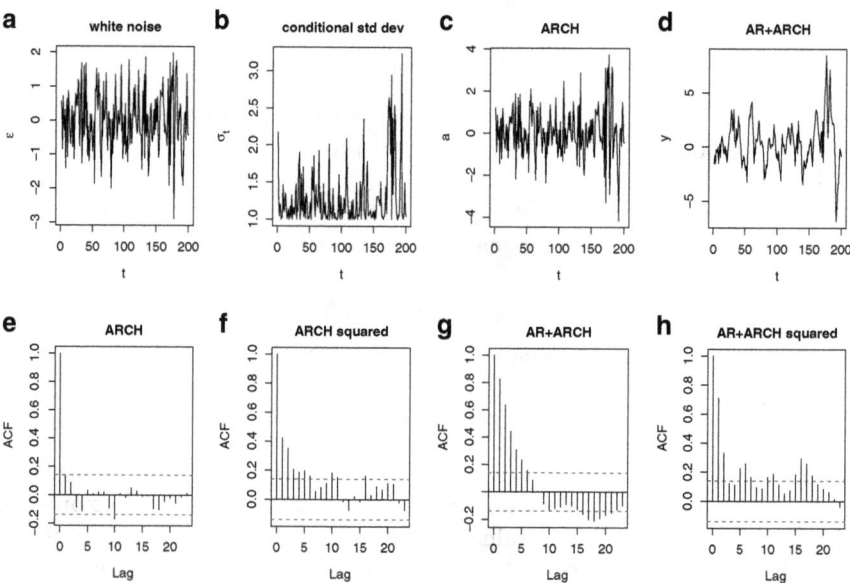

Fig. 14.2. *Simulation of 200 observations from an ARCH(1) process and an AR(1)+ARCH(1) process. The parameters are $\omega = 1$, $\alpha = 0.55$, $\mu = 0.1$, and $\phi = 0.8$. Sample ACF plots of the ARCH and AR+ARCH processes and squared processes are shown in the bottom row.*

14.5 ARCH(p) Models

As before, let ϵ_t be Gaussian white noise with unit variance. Then a_t is an ARCH(p) process if

$$a_t = \sigma_t \epsilon_t,$$

where

$$\sigma_t = \sqrt{\omega + \sum_{i=1}^{p} \alpha_i a_{t-i}^2}$$

is the conditional standard deviation of a_t given the past values $a_{t-1}, a_{t-2}, \ldots$ of this process. Like an ARCH(1) process, an ARCH(p) process is uncorrelated and has a constant mean (both conditional and unconditional) and a constant unconditional variance, but its conditional variance is nonconstant. In fact, the ACF of a_t^2 has the same structure as the ACF of an AR(p) process; see Sect. 14.9.

14.6 ARIMA(p_M, d, q_M)+GARCH(p_V, q_V) Models

A deficiency of ARCH(p) models is that the conditional standard deviation process has high-frequency oscillations with high volatility coming in short bursts. This behavior can be seen in Fig. 14.2b. GARCH models permit a wider range of behavior, in particular, more persistent volatility. The GARCH(p, q) model is

$$a_t = \sigma_t \epsilon_t,$$

in which

$$\sigma_t = \sqrt{\omega + \sum_{i=1}^{p} \alpha_i a_{t-i}^2 + \sum_{j=1}^{q} \beta_j \sigma_{t-j}^2}. \qquad (14.8)$$

Because past values of the σ_t process are fed back into the present value (with nonnegative coefficients β_j), the conditional standard deviation can exhibit more persistent periods of high or low volatility than seen in an ARCH process. In the stationary case, the process a_t is uncorrelated with a constant unconditional mean and variance and a_t^2 has an ACF like an ARMA process (see Sect. 14.9). GARCH models include ARCH models as a special case, and we use the term "GARCH" to refer to both ARCH and GARCH models.

A very general time series model lets a_t be GARCH(p_V, q_V) and uses a_t as the noise term in an ARIMA(p_M, d, q_M) model. The subscripts on p and q distinguish between the conditional variance (V) or GARCH parameters and the conditional mean (M) or ARIMA parameters. We will call such a process an ARIMA(p_M, d, q_M)+GARCH(p_V, q_V) model.

Figure 14.3 is a simulation of 500 observations from a GARCH(1,1) process and from a AR(1)+GARCH(1,1) process. The GARCH parameters are $\omega = 1$, $\alpha = 0.08$, and $\beta = 0.9$. The large value of β causes σ_t to be highly correlated with σ_{t-1} and gives the conditional standard deviation process a relatively long-term persistence, at least compared to its behavior under an ARCH model. In particular, notice that the conditional standard deviation is less "bursty" than for the ARCH(1) process in Fig. 14.2.

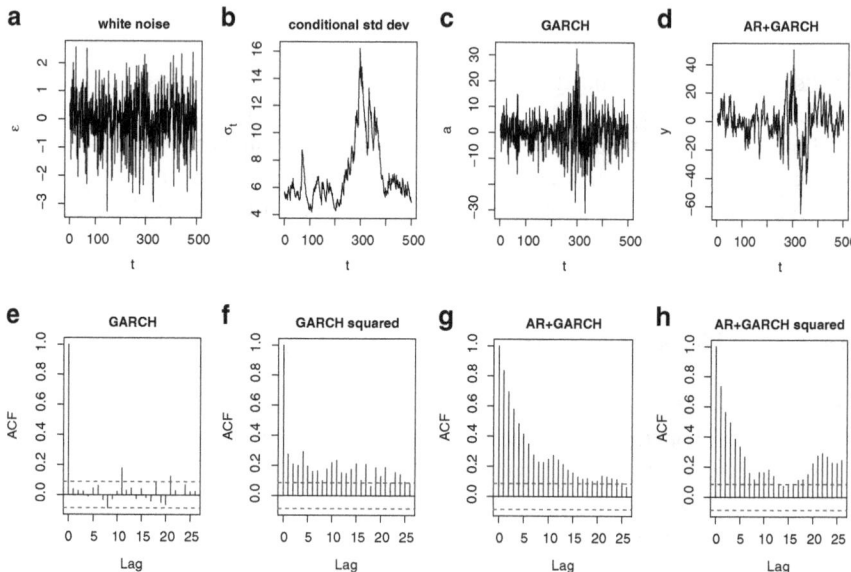

Fig. 14.3. *Simulation of GARCH(1,1) and AR(1)+GARCH(1,1) processes. The parameters are $\omega = 1$, $\alpha = 0.08$, $\beta = 0.9$, and $\phi = 0.8$.*

14.6.1 Residuals for ARIMA(p_M, d, q_M)+GARCH(p_V, q_V) Models

When one fits an ARIMA(p_M, d, q_M)+GARCH(p_V, q_V) model to a time series Y_t, there are two types of residuals. The ordinary residual, denoted $\widehat{a}_t$, is the difference between Y_t and its conditional expectation. As the notation implies, $\widehat{a}_t$ estimates a_t. A standardized residual, denoted $\widehat{\epsilon}_t$, is an ordinary residual $\widehat{a}_t$ divided by its estimated conditional standard deviation $\widehat{\sigma}_t$. A standardized residual estimates ϵ_t. The standardized residuals should be used for model checking. If the model fits well, then neither $\widehat{\epsilon}_t$ nor $\widehat{\epsilon}_t^2$ should exhibit serial correlation. Moreover, if ϵ_t has been assumed to have a normal distribution, then this assumption can be checked by a normal plot of the standardized residuals $\widehat{\epsilon}_t$. The $\widehat{a}_t$ are the residuals of the ARIMA process and are used when forecasting via the methods in Sect. 12.12.

14.7 GARCH Processes Have Heavy Tails

Researchers have long noticed that stock returns have "heavy-tailed" or "outlier-prone" probability distributions, and we have seen this ourselves in earlier chapters. One reason for outliers may be that the conditional variance is not constant, and the outliers occur when the variance is large, as in the normal mixture example of Sect. 5.5. In fact, GARCH processes exhibit heavy tails even if $\{\epsilon_t\}$ is Gaussian. Therefore, when we use GARCH models, we can model both the conditional heteroskedasticity and the heavy-tailed distributions of financial market data. Nonetheless, many financial time series have tails that are heavier than implied by a GARCH process with Gaussian $\{\epsilon_t\}$. To handle such data, one can assume that, instead of being Gaussian white noise, $\{\epsilon_t\}$ is an i.i.d. white noise process with a heavy-tailed distribution.

14.8 Fitting ARMA+GARCH Models

Example 14.2. AR(1)+GARCH(1,1) model fit to daily BMW stock log returns

This example uses the daily BMW stock log returns. The `ugarchfit()` function from R's `rugarch` package is used to fit an AR(1)+GARCH(1,1) model to this series. Although `ugarchfit()` allows the white noise to have a nonGaussian distribution, we begin this example using Gaussian white noise (the default). First the model is specified using the `ugarchspec()` function; for an AR(1)+GARCH(1,1) model we specify `armaOrder=c(1,0)` and `garchOrder=c(1,1)`. The commands and abbreviated output are below.

```
1 library(rugarch)
2 data(bmw, package="evir")
3 arma.garch.norm = ugarchspec(mean.model=list(armaOrder=c(1,0)),
4                              variance.model=list(garchOrder=c(1,1)))
5 bmw.garch.norm = ugarchfit(data=bmw, spec=arma.garch.norm)
6 show(bmw.garch.norm)

  GARCH Model  : sGARCH(1,1)
  Mean Model   : ARFIMA(1,0,0)
  Distribution : norm

  Optimal Parameters
  -------------------------------------
          Estimate  Std. Error  t value Pr(>|t|)
  mu      0.000453    0.000175   2.5938 0.009493
  ar1     0.098135    0.014261   6.8813 0.000000
  omega   0.000009    0.000000  23.0613 0.000000
  alpha1  0.099399    0.005593  17.7730 0.000000
  beta1   0.863672    0.006283 137.4591 0.000000
```

```
LogLikelihood : 17752

Information Criteria
-----------------------------------
Akaike         -5.7751
Bayes          -5.7696
Shibata        -5.7751
Hannan-Quinn  -5.7732
```

In the output, $\widehat{\phi}_1$ is denoted by ar1, the estimated mean $\widehat{\mu}$ is mean, and $\widehat{\omega}$ is called omega. Note that $\widehat{\phi}_1 = 0.0981$ and is statistically significant, implying that there is a small amount of positive autocorrelation. Both α_1 and β_1 are highly significant and $\widehat{\beta}_1 = 0.8636$, which implies rather persistent volatility clustering. There are two additional information criteria reported, Shibata's information criterion and Hannan–Quinn information criterion (HQIC). These are less widely used than AIC and BIC and will not be discussed here.

In the output from ugarchfit(), the AIC and BIC values have been normalized by dividing by n, so these values should be multiplied by $n = 6146$ to have their usual values. In particular, AIC and BIC will not be so close to each other after multiplication by 6146. The daily BMW stock log return series Y_t, with two estimated conditional standard deviations superimposed, and the estimated conditional standard deviation series $\widehat{\sigma}_t$ (vs. the absolute value of the log return series $|Y_t|$) are shown in the top row of Fig. 14.4.

The output also includes the following tests applied to the standardized and squared standardized residuals.

```
Weighted Ljung-Box Test on Standardized Residuals
-----------------------------------
                            statistic p-value
Lag[1]                         0.7786  0.3776
Lag[2*(p+q)+(p+q)-1][2]        0.9158  0.7892
Lag[4*(p+q)+(p+q)-1][5]        3.3270  0.3536
d.o.f=1
H0 : No serial correlation
Weighted Ljung-Box Test on Standardized Squared Residuals
-----------------------------------
                            statistic p-value
Lag[1]                         0.277   0.5987
Lag[2*(p+q)+(p+q)-1][5]        1.026   0.8537
Lag[4*(p+q)+(p+q)-1][9]        1.721   0.9356
d.o.f=2
Weighted ARCH LM Tests
-----------------------------------
             Statistic Shape Scale P-Value
ARCH Lag[3]     0.1922 0.500 2.000  0.6611
ARCH Lag[5]     1.1094 1.440 1.667  0.7008
ARCH Lag[7]     1.2290 2.315 1.543  0.8737
Adjusted Pearson Goodness-of-Fit Test:
```

```
----------------------------------
    group statistic p-value(g-1)
1     20     493.1    1.563e-92
2     30     513.4    5.068e-90
3     40     559.3    2.545e-93
4     50     585.6    5.446e-93
```

Weighted versions of the Ljung-Box (and ARCH-LM) test statistics[1] and their approximate p-values all indicate that the estimated model for the conditional mean and variance are adequate for removing serial correlation from the series and squared series, respectively. The sample ACF of the standardized residuals $\widehat{\epsilon}_t$, and the squared standardized residuals $\widehat{\epsilon}_t^2$ are shown in the middle row of Fig. 14.4. The Goodness-of-Fit tests[2] compare the empirical distribution of the standardized residuals with the theoretical ones from the specified density, which is Gaussian by default. The small p-values strongly reject the null hypothesis that the white noise standardized innovation process $\{\epsilon_t\}$ is Gaussian. Empirical density estimates and a normal quantile plot of the standardized residuals $\widehat{\epsilon}_t$ are shown in the bottom row of Fig. 14.4.

Figure 14.5 shows a t-plot with 4 df for the standardized residuals $\widehat{\epsilon}_t$. Unlike the normal quantile plot in the last panel of Fig. 14.4, this plot is nearly a straight line except for four outliers in the left tail. The sample size is 6146, so the outliers are a very small fraction of the data. Thus, it seems like a t-distribution would be suitable for the innovation process ϵ_t. A t-distribution was fit to the standardized residuals by maximum likelihood using the `fitdistr()` function from the Rpackage MASS.

```
7 library(MASS)
8 e = residuals(bmw.garch.norm, standardize=TRUE)
9 fitdistr(e,"t")
```

```
       m          s          df
   -0.0243     0.7269     4.1096
  ( 0.0109)  ( 0.0121)  ( 0.2359)
```

The MLE of the degrees-of-freedom parameter was 4.1. This confirms the good fit by this distribution seen in Fig. 14.5. The AR(1)+GARCH(1,1) model was refit assuming t-distributed errors, so `distribution.model = "std"` in `ugarchspec()`. The commands and abbreviated results are below.

```
10 arma.garch.t = ugarchspec(mean.model=list(armaOrder=c(1,0)),
11                           variance.model=list(garchOrder=c(1,1)),
12                           distribution.model = "std")
13 bmw.garch.t = ugarchfit(data=bmw,spec=arma.garch.t)
14 show(bmw.garch.t)
```

---

[1] Weighted Ljung-Box and ARCH-LM statistics of Fisher and Gallagher (2012) are provided by the `ugarchfit()` function to better account for the distribution of the statistics when applied to residuals from a fitted model; their use and interpretation remains unchanged.

[2] These Chi-squared tests are based on the tests of Palm (1996); group indicates the number of bins used in the implementation.

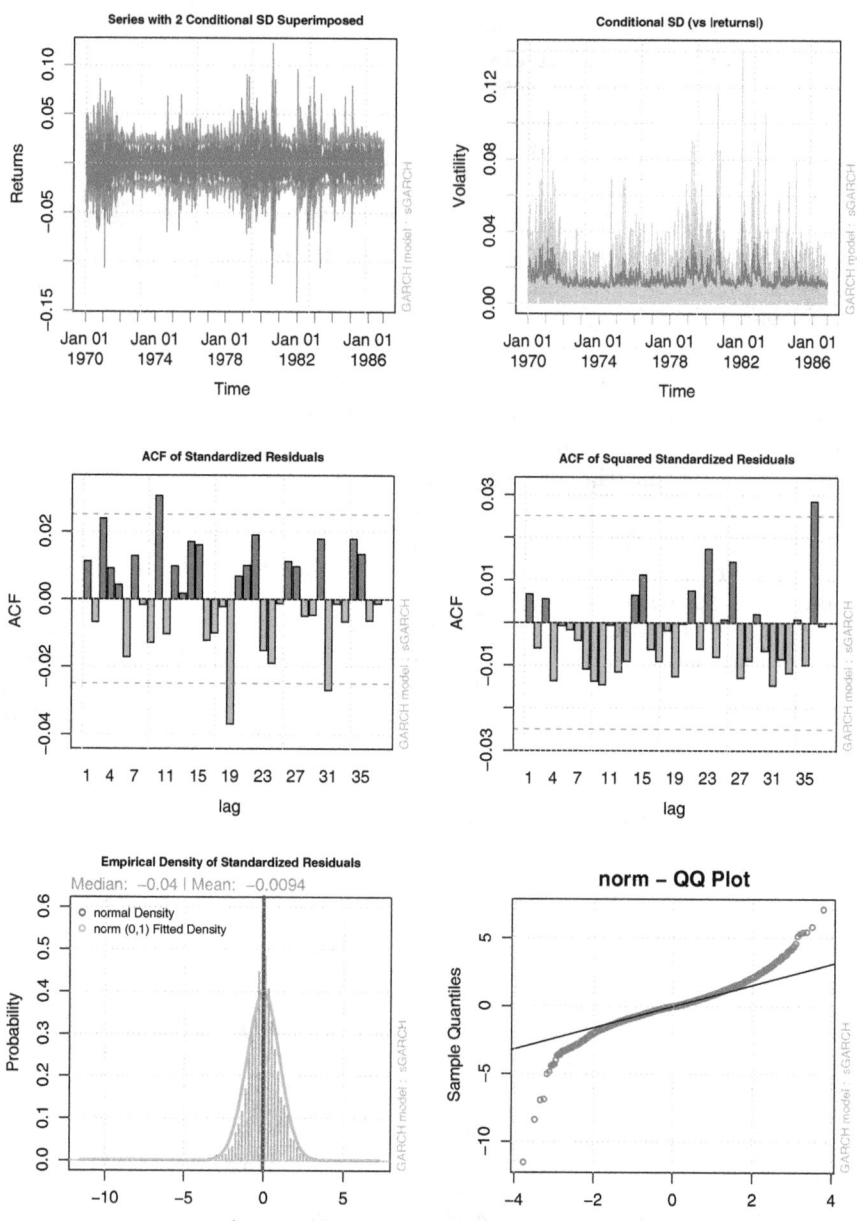

Fig. 14.4. *The daily BMW stock log return series* Y_t, *with two estimated conditional standard deviations superimposed; the estimated conditional standard deviation* $\widehat{\sigma}_t$ *series (vs. the absolute value of the log return series* $|Y_t|$*); the sample ACF of the standardized residuals* $\widehat{\epsilon}_t$ *and the squared standardized residuals* $\widehat{\epsilon}_t^2$*; empirical density estimates of the standardized residuals* $\widehat{\epsilon}_t$*; and a normal quantile plot of the standardized residuals* $\widehat{\epsilon}_t$.

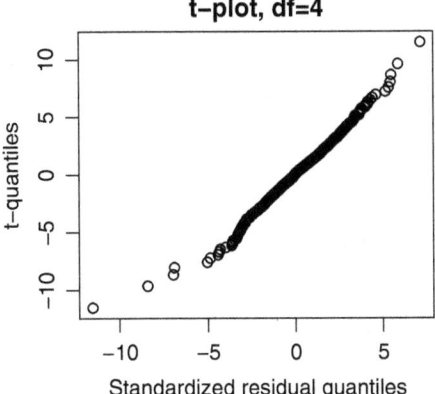

Fig. 14.5. *A t-plot with 4 df for the standardized residuals* $\hat{\epsilon}_t$ *from an AR(1)+GARCH(1,1) model fit to daily BMW stock log return; the reference lines go through the first and third quartiles.*

```
GARCH Model  : sGARCH(1,1)
Mean Model   : ARFIMA(1,0,0)
Distribution : std

Optimal Parameters
-------------------------------------
        Estimate  Std. Error  t value  Pr(>|t|)
mu      0.000135  0.000144    0.93978  0.347333
ar1     0.063911  0.012521    5.10436  0.000000
omega   0.000006  0.000003    1.69915  0.089291
alpha1  0.090592  0.012479    7.25936  0.000000
beta1   0.889887  0.014636   60.80228  0.000000
shape   4.070078  0.301306   13.50813  0.000000

LogLikelihood : 18152

Information Criteria
-------------------------------------
Akaike        -5.9048
Bayes         -5.8983
Shibata       -5.9048
Hannan-Quinn  -5.9026

Weighted Ljung-Box Test on Standardized Residuals
-------------------------------------
                          statistic   p-value
Lag[1]                        9.640   1.904e-03
Lag[2*(p+q)+(p+q)-1][2]       9.653   3.367e-09
Lag[4*(p+q)+(p+q)-1][5]      11.983   1.455e-04
```

```
d.o.f=1
HO : No serial correlation
```

```
Weighted Ljung-Box Test on Standardized Squared Residuals
------------------------------------
                            statistic p-value
Lag[1]                         0.5641  0.4526
Lag[2*(p+q)+(p+q)-1][5]        1.2964  0.7898
Lag[4*(p+q)+(p+q)-1][9]        2.0148  0.9032
d.o.f=2
```

```
Adjusted Pearson Goodness-of-Fit Test:
------------------------------------
   group statistic p-value(g-1)
1    20    229.0     5.460e-38
2    30    279.6     8.428e-43
3    40    313.8     1.230e-44
4    50    374.6     1.037e-51
```

The weighted Ljung–Box tests for the residuals have small p-values. These are due to small autocorrelations that should not be of practical importance. The sample size here is 6146 so, not surprisingly, small autocorrelations are statistically significant. The goodness-of-fit test statistics are much smaller but still significant; the large sample size again makes rejection likely even when the discrepancies are negligible from a practical standpoint. However, both AIC and BIC decreased substantially, and the refit model with a t conditional distribution offers an improvement over the original fit with a Gaussian conditional distribution. $\square$

14.9 GARCH Models as ARMA Models

The similarities seen in this chapter between GARCH and ARMA models are not a coincidence. If a_t is a GARCH process, then a_t^2 is an ARMA process, but with weak white noise, not i.i.d. white noise. To show this, we will start with the GARCH(1,1) model, where $a_t = \sigma_t \epsilon_t$. Here ϵ_t is i.i.d. white noise and

$$E(a_t^2|\mathcal{F}_{t-1}) = \sigma_t^2 = \omega + \alpha a_{t-1}^2 + \beta \sigma_{t-1}^2, \qquad (14.9)$$

where $\mathcal{F}_{t-1}$ is the information set at time $t - 1$. Define $\eta_t = a_t^2 - \sigma_t^2$. Since $E(\eta_t|\mathcal{F}_{t-1}) = E(a_t^2|\mathcal{F}_{t-1}) - \sigma_t^2 = 0$ by (A.33), η_t is an uncorrelated process, that is, a weak white noise process. The conditional heteroskedasticity of a_t is inherited by η_t, so η_t is not i.i.d. white noise.

Simple algebra shows that

$$\sigma_t^2 = \omega + (\alpha + \beta)a_{t-1}^2 - \beta \eta_{t-1} \qquad (14.10)$$

and therefore

$$a_t^2 = \sigma_t^2 + \eta_t = \omega + (\alpha + \beta)a_{t-1}^2 - \beta\eta_{t-1} + \eta_t. \tag{14.11}$$

Assume that $\alpha + \beta < 1$. If $\upsilon = \omega/\{1 - (\alpha + \beta)\}$, then

$$a_t^2 - \upsilon = (\alpha + \beta)(a_{t-1}^2 - \upsilon) + \beta\eta_{t-1} + \eta_t. \tag{14.12}$$

From (14.12) one sees that a_t^2 is an ARMA(1,1). Using the notation of (12.25), the mean is $\mu = \upsilon$, the AR(1) coefficient is $\phi = \alpha + \beta$ and the MA(1) coefficient is $\theta = -\beta$.

For the general case, assume that σ_t follows (14.8) such that

$$\sigma_t^2 = \omega + \sum_{i=1}^{p} \alpha_i a_{t-i}^2 + \sum_{j=1}^{q} \beta_j \sigma_{t-j}^2. \tag{14.13}$$

To simplify notation, if $q > p$, then define $\alpha_i = 0$ for $i = p + 1, \ldots, q$. Similarly, if $p > q$, then define $\beta_j = 0$ for $j = q + 1, \ldots, p$. Define $\upsilon = \omega/\{1 - \sum_{i=1}^{\max(p,q)}(\alpha_i + \beta_i)\}$. Straightforward algebra similar to the GARCH(1,1) case shows that

$$a_t^2 - \upsilon = \sum_{i=1}^{\max(p,q)} (\alpha_i + \beta_i)(a_{t-i}^2 - \upsilon) - \sum_{j=1}^{q} \beta_j \eta_{t-j} + \eta_t, \tag{14.14}$$

so that a_t^2 is an ARMA(max$(p, q), q$) process with mean $\mu = \upsilon$, AR coefficients $\phi_i = \alpha_i + \beta_i$ and MA coefficients $\theta_j = -\beta_j$. As a byproduct of these calculations, we obtain a necessary condition for a_t to be stationary:

$$\sum_{i=1}^{\max(p,q)} (\alpha_i + \beta_i) < 1. \tag{14.15}$$

14.10 GARCH(1,1) Processes

The GARCH(1,1) is the most widely used GARCH process, so it is worthwhile to study it in some detail. If a_t is GARCH(1,1), then as we have just seen, a_t^2 is ARMA(1,1). Therefore, the ACF of a_t^2 can be obtained from formulas (12.31) and (12.32). After some algebra, one finds that

$$\rho_{a^2}(1) = \frac{\alpha(1 - \alpha\beta - \beta^2)}{1 - 2\alpha\beta - \beta^2} \tag{14.16}$$

and

$$\rho_{a^2}(h) = (\alpha + \beta)^{h-1}\rho_{a^2}(1), \quad h \geq 2. \tag{14.17}$$

These formulas also hold in an AR(1)+GARCH(1,1) model, and the ACF of y_t^2 also decays with $h \geq 2$ at a geometric rate in the stationary case, provided some additional assumptions hold, however, the exact expressions are more complicated (see Palma and Zevallos, 2004).

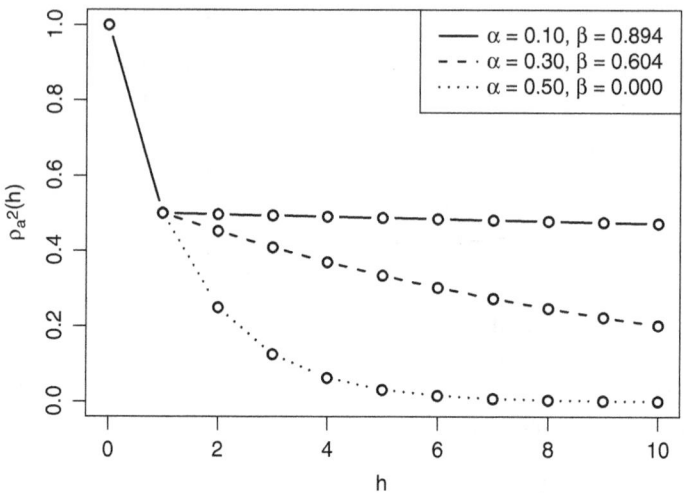

Fig. 14.6. *ACFs of three GARCH(1,1) processes with $\rho_{a^2}(1) = 0.5$.*

By (14.16), there are infinitely many values of (α, β) with the same value of $\rho_{a^2}(1)$. By (14.17), a higher value of $\alpha + \beta$ means a slower decay of $\rho_{a^2}(\cdot)$ after the first lag. This behavior is illustrated in Fig. 14.6, which contains the ACF of a_t^2 for three GARCH(1,1) processes with a lag-1 autocorrelation of 0.5. The solid curve has the highest value of $\alpha + \beta$ and the ACF decays very slowly. The dotted curve is a pure ARCH(1) process and has the most rapid decay.

In Example 14.2, an AR(1)+GARCH(1,1) model was fit to the BMW daily log returns. The GARCH parameters were estimated to be $\widehat{\alpha} = 0.10$ and $\widehat{\beta} = 0.86$. By (14.16) the $\widehat{\rho}_{a^2}(1) = 0.197$ for this process and the high value of $\widehat{\beta}$ suggests slow decay. The sample ACF of the squared residuals [from an AR(1) model] is plotted in Fig. 14.7. In that figure, we see the lag-1 autocorrelation is slightly below 0.2 and after one lag the ACF decays slowly, exactly as expected.

The capability of the GARCH(1,1) model to fit the lag-1 autocorrelation and the subsequent rate of decay separately is important in practice. It appears to be the main reason that the GARCH(1,1) model fits so many financial time series.

14.11 APARCH Models

In some financial time series, large negative returns appear to increase volatility more than do positive returns of the same magnitude. This is called the

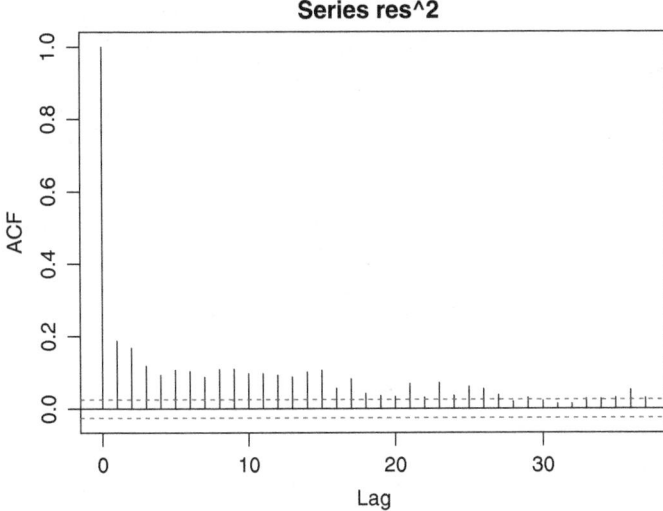

Fig. 14.7. *ACF of the squared residuals from an AR(1) fit to the BMW log returns.*

leverage effect. Standard GARCH models, that is, the models given by (14.8), cannot model the leverage effect because they model σ_t as a function of past values of a_t^2—whether the past values of a_t are positive or negative is not taken into account. The problem here is that the square function x^2 is symmetric in x. The solution is to replace the square function with a flexible class of nonnegative functions that include asymmetric functions. The APARCH (asymmetric power ARCH) models do this. They also offer more flexibility than GARCH models by modeling σ_t^δ, where $\delta > 0$ is another parameter.

The APARCH(p, q) model for the conditional standard deviation is

$$\sigma_t^\delta = \omega + \sum_{i=1}^{p} \alpha_i (|a_{t-i}| - \gamma_i a_{t-i})^\delta + \sum_{j=1}^{q} \beta_j \sigma_{t-j}^\delta, \tag{14.18}$$

where $\delta > 0$ and $-1 < \gamma_i < 1$, $i = 1, \ldots, p$. Note that $\delta = 2$ and $\gamma = \cdots = \gamma_p = 0$ give a standard GARCH model.

The effect of a_{t-i} upon σ_t is through the function g_{γ_i}, where $g_\gamma(x) = |x| - \gamma x$. Figure 14.8 shows $g_\gamma(x)$ for several values of γ. When $\gamma > 0$, $g_\gamma(-x) > g_\gamma(x)$ for any $x > 0$, so there is a leverage effect. If $\gamma < 0$, then there is a leverage effect in the opposite direction to what is expected—positive past values of a_t increase volatility more than negative past values of the same magnitude.

Example 14.3. AR(1)+APARCH(1,1) fit to daily BMW stock log returns

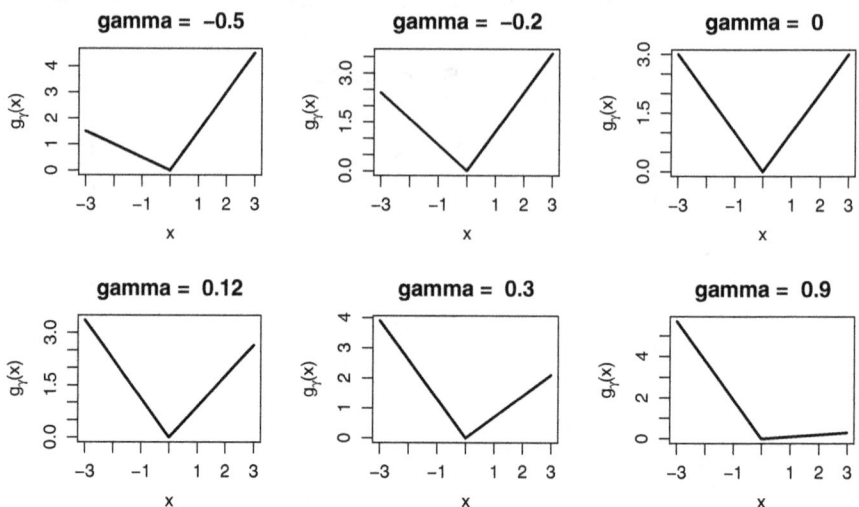

Fig. 14.8. *Plots of $g_\gamma(x)$ for various values of γ.*

In this example, an AR(1)+APARCH(1,1) model with t-distributed errors is fit to the BMW log returns. The commands and abbreviated output from ugarchfit() is below. The estimate of δ is 1.48 with a standard error of 0.14, so there is strong evidence that δ is not 2, the value under a standard GARCH model. Also, $\widehat{\gamma}_1$ is 0.12 with a standard error of 0.045, so there is a statistically significant leverage effect, since we reject the null hypothesis that $\gamma_1 = 0$. However, the leverage effect is small, as can be seen in the plot in Fig. 14.8 with $\gamma = 0.12$. The leverage might not be of practical importance.

```
15 arma.aparch.t = ugarchspec(mean.model=list(armaOrder=c(1,0)),
16                  variance.model=list(model="apARCH",
17                                 garchOrder=c(1,1)),
18                           distribution.model = "std")
19 bmw.aparch.t = ugarchfit(data=bmw, spec=arma.aparch.t)
20 show(bmw.aparch.t)

   GARCH Model : apARCH(1,1)
   Mean Model : ARFIMA(1,0,0)
   Distribution : std
   Optimal Parameters
   ------------------------------------
            Estimate  Std. Error  t value Pr(>|t|)
   mu       0.000048    0.000147   0.3255 0.744801
   ar1      0.063666    0.012352   5.1543 0.000000
```

```
omega    0.000050    0.000032    1.5541 0.120158
alpha1   0.098839    0.012741    7.7574 0.000000
beta1    0.899506    0.013565   66.3105 0.000000
gamma1   0.121947    0.044664    2.7303 0.006327
delta    1.476643    0.142442   10.3666 0.000000
shape    4.073809    0.234417   17.3784 0.000000
```

```
LogLikelihood : 18161
```

```
Information Criteria
-----------------------------------
Akaike         -5.9073
Bayes          -5.8985
Shibata        -5.9073
Hannan-Quinn -5.9042
```

```
Weighted Ljung-Box Test on Standardized Residuals
-----------------------------------
                        statistic   p-value
Lag[1]                      9.824 1.723e-03
Lag[2*(p+q)+(p+q)-1][2]     9.849 2.003e-09
Lag[4*(p+q)+(p+q)-1][5]    12.253 1.100e-04
d.o.f=1
H0 : No serial correlation
```

```
Weighted Ljung-Box Test on Standardized Squared Residuals
-----------------------------------
                        statistic p-value
Lag[1]                      1.456  0.2276
Lag[2*(p+q)+(p+q)-1][5]     2.363  0.5354
Lag[4*(p+q)+(p+q)-1][9]     3.258  0.7157
d.o.f=2
```

As mentioned earlier, in the output from ugarchfit(), the Information Criteria values have been normalized by dividing by n, though this is not noted in the output.

The normalized BIC for this model (-5.8985) is very nearly the same as the normalized BIC for the GARCH model with t-distributed errors (-5.8983), but after multiplying by $n = 6146$, the difference in the BIC values is 1.23. The difference between the two normalized AIC values, -5.9073 and -5.9048, is even larger, 15.4, after multiplication by n. Therefore, AIC and BIC support using the APARCH model instead of the GARCH model.

ACF plots (not shown) for the standardized residuals and their squares showed little correlation, so the AR(1) model for the conditional mean and the APARCH(1,1) model for the conditional variance fit well. Finally, shape is the estimated degrees of freedom of the t-distribution and is 4.07 with a small standard error, so there is very strong evidence that the conditional distribution is heavy-tailed. $\square$

14.12 Linear Regression with ARMA+GARCH Errors

When using time series regression, one often observes autocorrelated residuals. For this reason, linear regression with ARMA disturbances was introduced in Sect. 13.3.3. The model considered was

$$Y_t = \beta_0 + \beta_1 X_{t,1} + \cdots + \beta_p X_{t,p} + e_t, \tag{14.19}$$

where

$$(1 - \phi_1 B - \cdots - \phi_p B^p)(e_t - \mu) = (1 + \theta_1 B + \ldots + \theta_q B^q)a_t, \tag{14.20}$$

and $\{a_t\}$ is i.i.d. white noise. This model is sufficient for serially correlated errors, but it does not accommodate volatility clustering, which is often found in the residuals.

One solution is to model the noise as an ARMA+GARCH process. Therefore, we will now assume that, instead of being i.i.d. white noise, $\{a_t\}$ is a GARCH process so that

$$a_t = \sigma_t \epsilon_t, \tag{14.21}$$

where

$$\sigma_t = \sqrt{\omega + \sum_{i=1}^{p} \alpha_i a_{t-i}^2 + \sum_{j=1}^{q} \beta_j \sigma_{t-j}^2}, \tag{14.22}$$

and $\{\epsilon_t\}$ is i.i.d. white noise. The model given by (14.19)–(14.22) is a *linear regression model with ARMA+GARCH disturbances*.

Some software, including the `ugarchfit()` function from R's `rugarch` package, can fit the linear regression model with ARMA+GARCH disturbances in one step. Another solution is to adjust or correct the estimated covariance matrix of the regression coefficients, via the HAC estimator from Sect. 13.3.2, by using the `NeweyWest()` function from the R package `sandwich`. However, if such software is not available, then a three-step estimation method is the following:

1. estimate the parameters in (14.19) by ordinary least-squares;
2. fit model (14.20)–(14.22) to the ordinary least-squares residuals;
3. reestimate the parameters in (14.19) by weighted least-squares with weights equal to the reciprocals of the conditional variances from step 2.

Example 14.4. Regression analysis with ARMA+GARCH errors of the Nelson–Plosser data

In Example 9.9, we saw that a parsimonious model for the yearly log returns on the stock index `diff(log(sp))` used `diff(log(ip))` and `diff(bnd)` as predictors. Figure 14.9 contains ACF plots of the residuals [panel (a)] and squared residuals [panel (b)]. Externally studentized residuals were used, but the plots for the raw residuals are similar. There is some autocorrelation in both the residuals and squared residuals.

```
21  nelsonplosser = read.csv("nelsonplosser.csv", header = TRUE)
22  new_np = na.omit(nelsonplosser)
23  attach(new_np)
24  fit.lm1 = lm(diff(log(sp)) ~ diff(log(ip)) + diff(bnd))
25  summary(fit.lm1)

    Coefficients:
                   Estimate Std. Error t value Pr(>|t|)
    (Intercept)     0.01657    0.02100   0.789 0.433316
    diff(log(ip))   0.69748    0.16834   4.143 0.000113 ***
    diff(bnd)      -0.13224    0.06225  -2.124 0.037920 *
    ---
    Signif. codes:  0 *** 0.001 ** 0.01 * 0.05 . 0.1   1

    Residual standard error: 0.1509 on 58 degrees of freedom
    Multiple R-squared:  0.3087,Adjusted R-squared:  0.2848
    F-statistic: 12.95 on 2 and 58 DF,  p-value: 2.244e-05
```

The `auto.arima()` function from R's `forecast` package selected an MA(1) model [i.e., ARIMA(0,0,1)] for the residuals. Next an MA(1)+ARCH(1) model was fit to the regression model's raw residuals. Sample ACF plots of the standardized residuals from the MA(1)+ARCH(1) model are in Fig. 14.9c and d. One sees essentially no short-term autocorrelation in the ARMA+GARCH standardized or squared standardized residuals, which indicates that the ARMA+GARCH model accounts for the observed dependence in the regression residuals satisfactorily. A normal plot showed that the standardized residuals are close to normally distributed, which is not unexpected for yearly log returns.

Finally, the linear model was refit with the reciprocals of the conditional variances as weights. The estimated regression coefficients are given below along with their standard errors and p-values.

```
26  fit.lm3 = lm(diff(log(sp)) ~ diff(log(ip)) + diff(bnd),
27                weights = 1/sigma.arch^2)
28  summary(fit.lm3)

    Coefficients:
                   Estimate Std. Error t value Pr(>|t|)
    (Intercept)     0.03216    0.02052   1.567 0.12263
    diff(log(ip))   0.55464    0.16942   3.274 0.00181 **
    diff(bnd)      -0.12215    0.05827  -2.096 0.04051 *
    ---
    Signif. codes:  0 *** 0.001 ** 0.01 * 0.05 . 0.1   1

    Residual standard error: 1.071 on 57 degrees of freedom
    Multiple R-squared:  0.2416,Adjusted R-squared:  0.2149
    F-statistic: 9.077 on 2 and 57 DF,  p-value: 0.0003783
```

There are no striking differences between these results and the unweighted fit in Example 9.9. In this situation, the main reason for using the GARCH

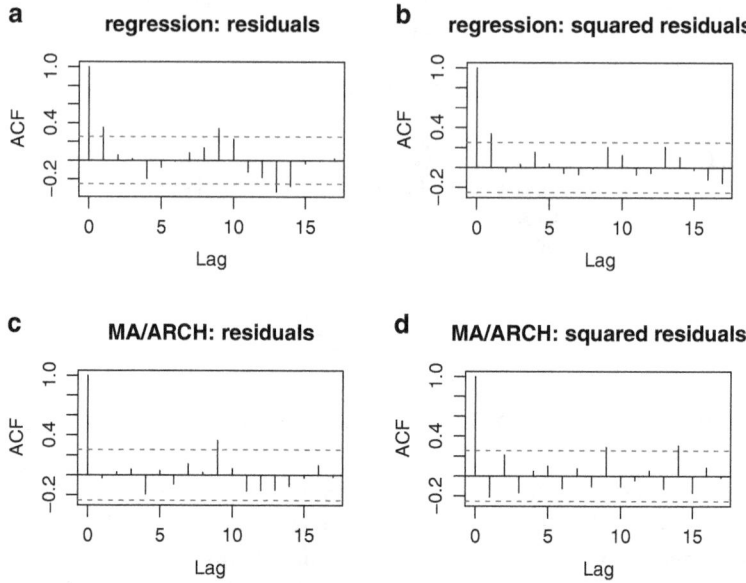

Fig. 14.9. *(a) Sample ACF of the externally studentized residuals and (b) their squared values, from a linear model; (c) Sample ACF of the standardized residuals and (d) their squared values, from an MA(1)+ARCH(1) fit to the regression residuals.*

model for the residuals would be in providing more accurate prediction intervals if the model were to be used for forecasting; see Sect. 14.13.   □

14.13 Forecasting ARMA+GARCH Processes

Forecasting ARMA+GARCH processes is in one way similar to forecasting ARMA processes—point estimates, e.g., forecasts of the conditional mean, are the same because a GARCH process is weak white noise. What differs between forecasting ARMA+GARCH and ARMA processes is the behavior of the prediction intervals. In times of high volatility, prediction intervals using an ARMA+GARCH model will widen to take into account the higher amount of uncertainty. Similarly, the prediction intervals will narrow in times of lower volatility. Prediction intervals using an ARMA model without conditional heteroskedasticity cannot adapt in this way.

To illustrate, we will compare the prediction of a Gaussian white noise process and the prediction of a GARCH(1,1) process with Gaussian innovations.

Both have an ARMA(0,0) model for the conditional mean so their forecasts are equal to the marginal mean, which will be called μ. For Gaussian white noise, the prediction limits are $\mu \pm z_{\alpha/2}\sigma$, where σ is the marginal standard deviation. For a GARCH(1,1) process $\{Y_t\}$, the prediction limits at time origin n for h-steps ahead forecasting are $\mu \pm z_{\alpha/2}\sigma_{n+h|n}$ where $\sigma_{n+h|n}$ is the conditional standard deviation of Y_{n+h} given the information available at time n. As h increases, $\sigma_{n+h|n}$ converges to σ, so for long lead times the prediction intervals for the two models are similar. For shorter lead times, however, the prediction limits can be quite different.

Example 14.5. Forecasting BMW log returns

In this example, we will return to the daily BMW stock log returns used in several earlier examples. We have seen in Example 14.2 that an AR(1)+GARCH(1,1) model fits the returns well. Also, the estimated AR(1) coefficient is small, less than 0.1. Therefore, it is reasonable to use a GARCH (1,1) model for forecasting.

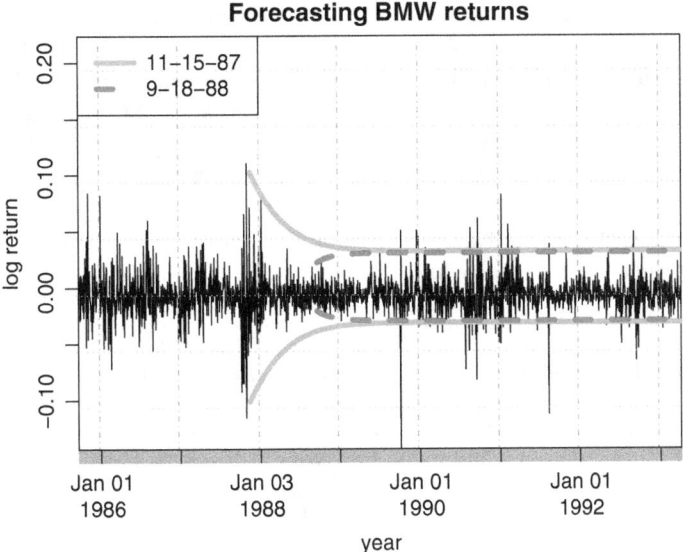

Fig. 14.10. *Prediction limits for forecasting daily BMW stock log returns from two different time origins.*

Figure 14.10 plots the returns from 1986 until 1992. Forecast limits are also shown for two time origins, November 15, 1987 and September 18, 1988. At the first time origin, which is soon after Black Monday, the markets were very volatile. The forecast limits are wide initially but narrow as the conditional standard deviation converges downward to the marginal standard deviation. At the second time origin, the markets were less volatile than usual and the prediction intervals are narrow initially but then widen. In theory, both sets of prediction limits should converge to the same values, $\mu \pm z_{\alpha/2}\sigma$ where σ is the marginal standard deviation for a stationary process. In this example, they do not quite converge to each other because the estimates of σ differ between the two time origins.   □

14.14 Multivariate GARCH Processes

Financial asset returns tend to move together over time, as do their respective volatilities, across both assets and markets. Modeling a time-varying conditional covariance matrix, or volatility matrix, is important in many financial applications, including asset pricing, hedging, portfolio selection, and risk management.

Multivariate volatility modeling has major challenges to overcome. First, the curse of dimensionality; there are $d(d+1)/2$ variances and covariances for a d-dimensional process, e.g., 45 for $d = 9$, all of which may vary over time. Further, unlike returns, all of these variances and covariances are unobserved, or latent. Many parameterizations for the evolution of the volatility matrix use such a large number of parameters that estimation becomes infeasible for $d > 10$. In addition to empirical adequacy (i.e., goodness of fit of the model to the data), ease and feasibility of estimation are important considerations.

Analogous to positivity constraints in univariate GARCH models, a well-defined multivariate volatility matrix process must be positive-definite at each time point, and model-based forecasts should as well. From a practical perspective, a well-defined inverse of a volatility matrix is frequently needed in applications. Additionally, a positive conditional variance estimate for a portfolio's return, which are a linear combination of asset returns, is essential; fortunately, this is guaranteed by positive definiteness.

14.14.1 Multivariate Conditional Heteroscedasticity

Figures 14.11a and b are time series plots of daily returns (in percentage) for IBM stock and the Center for Research in Security Prices (CRSP) value-weighted index, including dividends, from January 3, 1989 to December 31, 1998, respectively. The data are from the Ecdat package in R. Each series clearly exhibits volatility clustering. Let $\boldsymbol{Y}_t$ denote the vector time series of these returns.

```
29 data(CRSPday, package="Ecdat")
30 CRSPday = ts(CRSPday, start = c(1989, 1), frequency = 253)
31 ibm  = CRSPday[,5] * 100
32 crsp = CRSPday[,7] * 100
33 Y = cbind(ibm, crsp)
34 par(mfrow = c(2,1))
35 plot(Y[,1], type='l', xlab="year", ylab="return (%)", main="(a)")
36 plot(Y[,2], type='l', xlab="year", ylab="return (%)", main="(b)")
```

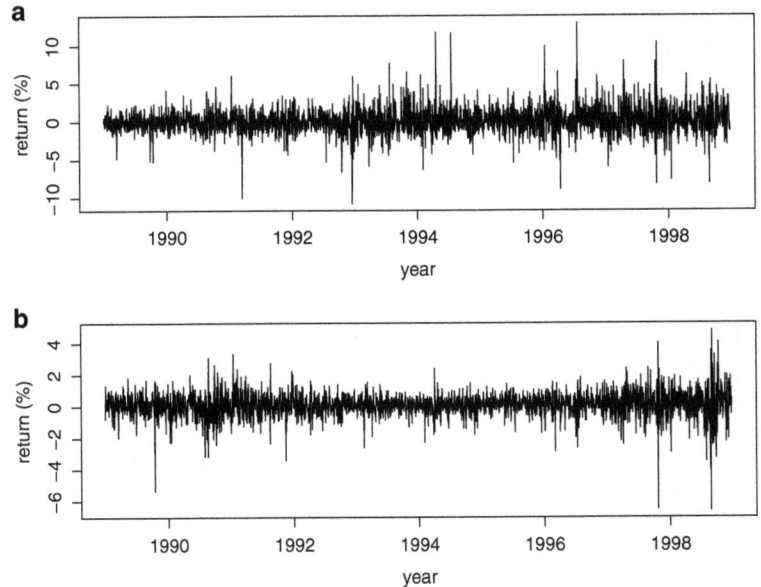

Fig. 14.11. Daily returns (in percentage) for (a) IBM stock and (b) the CRSP value-weighted index, including dividends.

Figures 14.12a and b are the sample ACF plots for the IBM stock and CRSP index returns, respectively. There is some evidence of minor serial correlation at low lags. Next, we consider the lead-lag linear relationship between pairs of returns. Figure 14.12c is the sample cross-correlation function (CCF) between IBM and CRSP. The lag zero estimate for contemporaneous correlation is approximately 0.49. There is also some evidence of minor cross-correlation at low lags.

```
37 layout(rbind(c(1,2), c(3,3)),widths=c(1,1,2),heights=c(1,1))
38 acf(as.numeric(Y[,1]), ylim=c(-0.1,0.1), main="(a)")
39 acf(as.numeric(Y[,2]), ylim=c(-0.1,0.1), main="(b)")
40 ccf(as.numeric(Y[,1]),as.numeric(Y[,2]),
41     type=c("correlation"), main="(c)", ylab="CCF", lag=20)
42 cor(ibm, crsp)
```

```
[1] 0.4863639
```

The multivariate Ljung-Box test (see Sect. 13.4.3) is applied to simultaneously test that the first K auto-correlations, as well as the lagged cross-correlations, are all zero. The multivariate Ljung-Box test statistic at lag five is 50.15. The associate p-value is very close to zero, which provides strong evidence to reject the null hypothesis and indicates there is significant serial correlation in the vector process.

```
43 source("SDAFE2.R")
44 mLjungBox(Y, 5)
```

```
   K  Q(K) d.f. p-value
 1 5 50.15   20      0
```

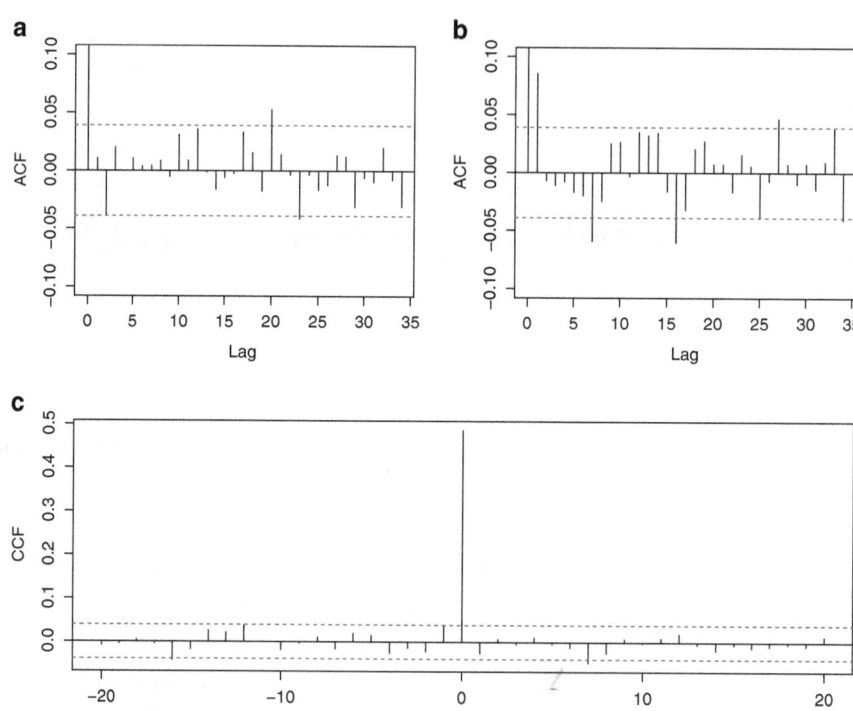

Fig. 14.12. ACFs for (a) the IBM stock and (b) CRSP index returns; (c) CCF between IBM and CRSP returns.

For simplicity, we use ordinary least squares to fit a VAR(1) model (see Sect. 13.4.4) to remove the minor serial correlation and focus on the conditional variance and covariance. Let $\widehat{a}_t$ denote the estimated residuals from the regression; $\widehat{a}_t$ is an estimate of the innovation process a_t, which is described more fully below. The multivariate Ljung-Box test statistic at lag five is now 16.21, which has an approximate p-value of 0.704, indicating there is no significant serial correlation in the vector residual process.

```
45 fit.AR1 = ar(Y, aic = FALSE, order.max=1)
46 A = fit.AR1$resid[-1,]
47 mLjungBox(A, 5)
```

```
   K  Q(K) d.f. p-value
 1 5 16.21   20   0.704
```

Although the residual series $\boldsymbol{a}_t$ is serially uncorrelated, Fig. 14.13 shows it is not an independent process. Figures 14.13a and b are sample ACF plots for the squared residual series $\widehat{a}_{it}^2$. They both show substantial positive autocorrelation because of the volatility clustering. Figure 14.13c is the sample CCF for the squared series; this figure shows there is a dynamic relationship between the squared series at low lags. Figure 14.13d is the sample ACF for the product series $\widehat{a}_{1t}\widehat{a}_{2t}$ and shows that there is also evidence of positive autocorrelation in the conditional covariance series. The multivariate volatility models described below attempt to account for these forms of dependence exhibited in the vector residual series.

14.14.2 Basic Setting

Let $\boldsymbol{Y}_t = (Y_{1,t}, \ldots, Y_{d,t})'$ denote a d-dimensional vector process and let $\mathcal{F}_t$ denote the information set at time index t, generated by $\boldsymbol{Y}_t, \boldsymbol{Y}_{t-1}, \ldots$. We may partition the process as

$$\boldsymbol{Y}_t = \boldsymbol{\mu}_t + \boldsymbol{a}_t, \tag{14.23}$$

in which $\boldsymbol{\mu}_t = \mathrm{E}(\boldsymbol{Y}_t|\mathcal{F}_{t-1})$ is the conditional mean vector at time index t, and $\{\boldsymbol{a}_t\}$ is the mean zero weak white noise innovation vector process with unconditional covariance matrix $\boldsymbol{\Sigma_a} = \mathrm{Cov}(\boldsymbol{a}_t)$. Let

$$\boldsymbol{\Sigma}_t = \mathrm{Cov}(\boldsymbol{a}_t|\mathcal{F}_{t-1}) = \mathrm{Cov}(\boldsymbol{Y}_t|\mathcal{F}_{t-1}) \tag{14.24}$$

denote the conditional covariance or volatility matrix at time index t. Multivariate time series modeling is primarily concerned with the time evolutions of $\boldsymbol{\mu}_t$ and $\boldsymbol{\Sigma}_t$, the conditional mean and conditional covariance matrix. For a stationary process, the unconditional mean and unconditional covariance matrix are constant, even though the conditional mean and conditional covariance matrix may be time-varying.

Throughout this section we assume that $\boldsymbol{\mu}_t$ follows a stationary VAR(p) model with $\boldsymbol{\mu}_t = \boldsymbol{\mu} + \sum_{\ell=1}^{p} \boldsymbol{\Phi}_\ell(\boldsymbol{Y}_{t-\ell} - \boldsymbol{\mu})$, where p is a non-negative integer, $\boldsymbol{\mu}$ is the $d \times 1$ unconditional mean vector, and the $\boldsymbol{\Phi}_\ell$ are $d \times d$ coefficient matrices, respectively. Recall, the residual series considered in Fig. 14.13 were from a VAR model with $p = 1$.

The relationship between the innovation process and the volatility process is defined by

$$\boldsymbol{a}_t = \boldsymbol{\Sigma}_t^{1/2}\boldsymbol{\epsilon}_t, \quad \boldsymbol{\epsilon}_t \overset{iid}{\sim} F(\boldsymbol{0}, \boldsymbol{I}_d), \tag{14.25}$$

in which $\boldsymbol{\Sigma}_t^{1/2}$ is a symmetric *matrix square-root* of $\boldsymbol{\Sigma}_t$, such that $\boldsymbol{\Sigma}_t^{1/2}\boldsymbol{\Sigma}_t^{1/2} = \boldsymbol{\Sigma}_t$. The iid white noise $\boldsymbol{\epsilon}_t$ are *standardized* innovations from a multivariate distribution F with mean zero and a covariance matrix equal to the identity. The models detailed below describe dynamic evolutions for the volatility matrix $\boldsymbol{\Sigma}_t$.

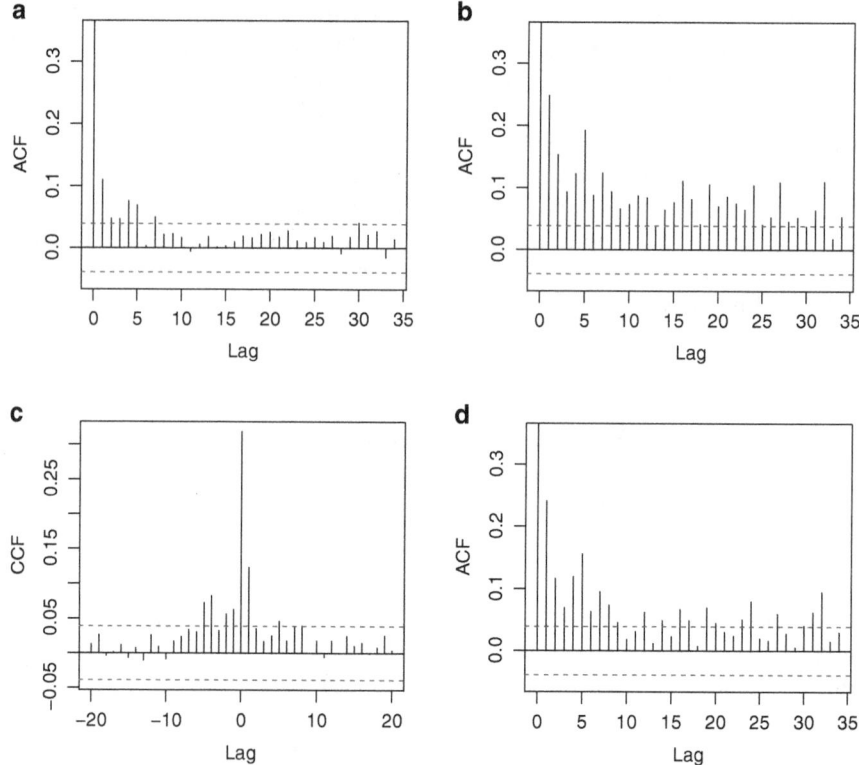

Fig. 14.13. ACFs of squared residuals from a VAR(1) model for (a) IBM and (b) CRSP; (c) CCF between the squared residuals; (d) ACF for the product of the residuals.

14.14.3 Exponentially Weighted Moving Average (EWMA) Model

The simplest matrix generalization of a univariate volatility model is the exponentially weighted moving average (EWMA) model. It is indexed by a single parameter $\lambda \in (0, 1)$, and is defined by the recursion

$$\boldsymbol{\Sigma}_t = (1 - \lambda)\boldsymbol{a}_{t-1}\boldsymbol{a}_{t-1}' + \lambda\boldsymbol{\Sigma}_{t-1} \tag{14.26}$$

$$= (1 - \lambda)\sum_{\ell=1}^{\infty} \lambda^{\ell-1}\boldsymbol{a}_{t-\ell}\boldsymbol{a}_{t-\ell}'.$$

When the recursion in (14.26) is initialized with a positive-definite (p.d.) matrix the sequence remains p.d. This single parameter model is simple to estimate regardless of the dimension, with large values of λ indicating high persistence in the volatility process. However, the dynamics can be too restrictive in practice, since the component-wise evolutions all have the same discounting factor (i.e., persistence parameter) λ.

Figure 14.14 shows the in-sample fitted EWMA model for $\widehat{a}_t$ assuming a multivariate standard normal distribution for ϵ_t and using conditional maximum likelihood estimation. The estimated conditional standard deviations are shown in (a) and (d), and the conditional covariances and implied conditional correlations are shown in (b) and (c), respectively. The persistence parameter λ was estimated as 0.985. Estimation and Fig. 14.14 were calculated using the following commands in R.

```
48  source("SDAFE2.R")
49  EWMA.param = est.ewma(lambda.0=0.95, innov=A)
50  EWMA.Sigma = sigma.ewma(lambda=EWMA.param$lambda.hat, innov=A)
51  par(mfrow = c(2,2))
52  plot(ts(EWMA.Sigma[1,1,]^.5, start = c(1989, 1), frequency = 253),
53      type = 'l', xlab = "year", ylab = NULL,
54      main = expression(paste("(a) ", hat(sigma)["1,t"])))
55  plot(ts(EWMA.Sigma[1,2,], start = c(1989, 1), frequency = 253),
56      type = 'l', xlab = "year", ylab = NULL,
57      main = expression(paste("(b) ", hat(sigma)["12,t"])))
58  plot(ts(EWMA.Sigma[1,2,]/(sqrt(EWMA.Sigma[1,1,]* EWMA.Sigma[2,2,])),
59          start = c(1989, 1), frequency = 253),
60      type = 'l', xlab = "year", ylab = NULL,
61      main = expression(paste("(c) ", hat(rho)["12,t"])))
62  points(ts(mvwindow.cor(A[,1],A[,2], win = 126)$correlation,
63          start = c(1989, 1), frequency = 253),
64      type = 'l', col = 2, lty = 2, lwd=2)
65  plot(ts(EWMA.Sigma[2,2,]^.5, start = c(1989, 1), frequency = 253),
66      type = 'l', xlab = "year", ylab = NULL,
67      main = expression(paste("(d) ", hat(sigma)["2,t"])))
68  EWMA.param$lambda.hat

    [1] 0.985046
```

14.14.4 Orthogonal GARCH Models

Several factor and orthogonal models have been proposed to reduce the number of parameters and parameter constraints by imposing a common dynamic structure on the elements of the volatility matrix. The orthogonal GARCH (O-GARCH) model of Alexander (2001) is among the most popular because of its simplicity. It is assumed that the innovations a_t can be decomposed into orthogonal components z_t via a linear transformation U. This is done in conjunction with principal component analysis (PCA, see Sect. 18.2) as follows. Let O be the matrix of eigenvectors and Λ the diagonal matrix of the corresponding eigenvalues of Σ_a. Then, take $U = \Lambda^{-1/2}O'$, and let

$$z_t = Ua_t.$$

The components are constructed such that $\mathrm{Cov}(z_t) = I_d$. The sample estimate of Σ_a is typically used to estimate U.

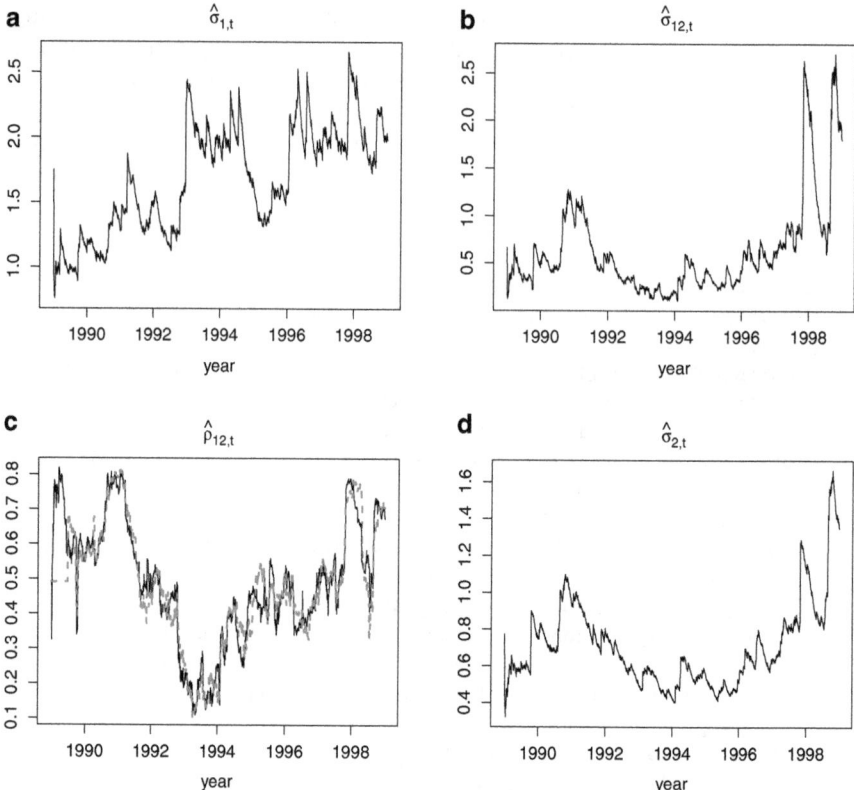

Fig. 14.14. A fitted EWMA model with $\lambda = 0.985$. The red line in (c) is the sample correlation estimate over the previous six months for comparison.

Next, univariate GARCH(1,1) models are individually fit to each orthogonal component to estimate the *conditional* covariance $\boldsymbol{V}_t = \text{Cov}(\boldsymbol{z}_t | \mathcal{F}_{t-1})$. Let

$$v_{it}^2 = \omega_i + \alpha_i z_{i,t-1}^2 + \beta_i v_{i,t-1}^2$$
$$\boldsymbol{V}_t = \text{diag}\{v_{1,t}^2, \ldots, v_{d,t}^2\}$$
$$\boldsymbol{\Sigma}_t = U^{-1} \boldsymbol{V}_t U^{-1'}.$$

In summary, a linear transformation U is estimated, using PCA, such that the components of $\boldsymbol{z}_t = U\boldsymbol{a}_t$ have unconditional correlation approximately equal to zero. It is then also *assumed* that the *conditional* correlations of $\boldsymbol{z}_t$ are also zero; however, this is not at all assured to be true. Under this additional stronger assumption, $\boldsymbol{V}_t$, the conditional covariance matrix for $\boldsymbol{z}_t$, is diagonal. For simplicity, univariate models are then fit to model the conditional variance v_{it}^2 for each component of $\boldsymbol{z}_t$.

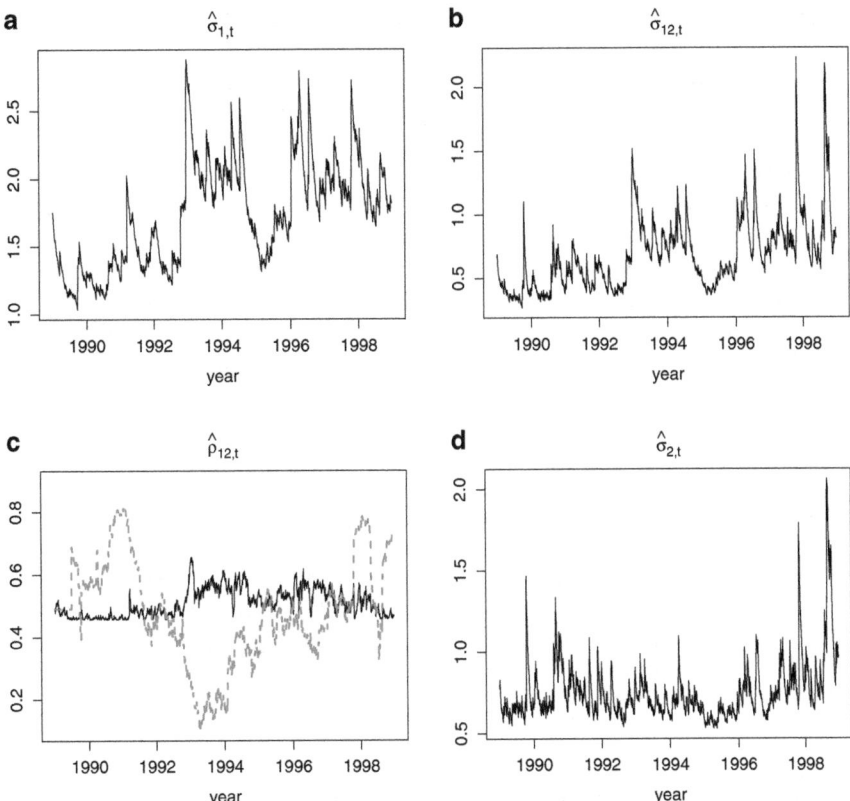

Fig. 14.15. A fitted first order orthogonal GARCH model with $(\omega_1, \alpha_1, \beta_1)' = (0.0038, 0.0212, 0.9758)'$, $(\omega_2, \alpha_2, \beta_2)' = (0.0375, 0.0711, 0.8913)'$, and $U^{-1} = ((1.7278, 0.2706)', (0.2706, 0.7241)')$. The red line in (c) is the sample correlation estimate over the previous six months for comparison.

The main drawback of this model is that the orthogonal components are uncorrelated unconditionally, but they may still be conditionally correlated. The O-GARCH model implicitly assumes the conditional correlations for z_t are zero. Figure 14.15 shows a fitted O-GARCH model for $\hat{a}_t$ using PCA followed by univariate conditional maximum likelihood estimation. The estimated conditional standard deviations are shown in (a) and (d), and the conditional covariances and conditional correlations are shown in (b) and (c), respectively. The implied conditional correlations do not appear adequate for this fitted model compared to the sample correlation estimate over the previous six months (used as a proxy for the conditional correlation process).

14.14.5 Dynamic Orthogonal Component (DOC) Models

To properly apply univariate modeling after estimating a linear transformation in the spirit of the O-GARCH model above, the resulting component processes must not only be orthogonal contemporaneously, the conditional correlations must also be zero. Additionally, the lagged cross-correlations for the squared components must also be zero. In Matteson and Tsay (2011), if the components of a time series s_t satisfy these conditions, then they are called dynamic orthogonal components (DOCs) in volatility.

Let $s_t = (s_{1,t}, \ldots, s_{d,t})'$ denote a vector time series of DOCs. Without loss of generality, s_t is assumed to be standardized such that $E(s_{i,t}) = 0$ and $Var(s_{i,t}) = 1$ for $i = 1, \ldots, d$. A Ljung-Box type statistic, defined below, is used to test for the existence of DOCs in volatility. Including lag zero in the test implies that the pairwise product processes among stationary DOCs $s_{i,t}s_{j,t}$ has zero serial correlation since the Cauchy-Schwarz inequality gives

$$|Cov(s_{i,t}s_{j,t}, s_{i,t-h}s_{j,t-h})| \leq Var(s_{i,t}s_{j,t}) = E(s_i^2 s_j^2), \tag{14.27}$$

and $E(s_{i,t}s_{j,t}) = E(s_{i,t-h}s_{j,t-h}) = 0$ by the assumption of a DOC model.

Let $\rho_{s_i^2, s_j^2}(h) = Corr(s_{i,t}^2, s_{j,t-h}^2)$. The joint lag-$K$ null and alternative hypotheses to test for the existence of DOCs in volatility are

$$H_0 : \rho_{s_i^2, s_j^2}(h) = 0 \text{ for all } i \neq j, \; h = 0, \ldots, K$$
$$H_A : \rho_{s_i^2, s_j^2}(h) \neq 0 \text{ for some } i \neq j, \; h = 0, \ldots, K.$$

The corresponding Ljung-Box type test statistic is

$$Q_d^0(s^2; K) = n \sum_{i<j} \rho_{s_i^2, s_j^2}(0)^2 + n(n+2) \sum_{h=1}^{K} \sum_{i \neq j} \rho_{s_i^2, s_j^2}(h)^2 / (n-h). \tag{14.28}$$

Under H_0, $Q_d^0(s^2; K)$ is asymptotically distributed as Chi-squared with $d(d-1)/2 + Kd(d-1)$ degrees of freedom. The null hypothesis is rejected for a large value of Q_d^0. When H_0 is rejected, one must seek an alternative modeling procedure.

As expected from Fig. 14.13, the DOCs in volatility hypothesis is rejected for the VAR(1) residuals. The test statistic is $Q_2^0(\widehat{a}^2, 5) = 356.926$ with a p-value near zero. DOCs in volatility is also rejected for the principal components used in the O-GARCH model, the test statistic is $Q_2^0(z^2, 5) = 135.492$ with a p-value near zero. Starting with the uncorrelated principal components z_t, Matteson and Tsay (2011) propose estimating an orthogonal matrix W such that the components $s_t = W z_t$ are as close to DOCs in volatility as possible. This is done by minimizing a reweighted version of the Ljung-Box type test statistic (14.28), with respect to the separating matrix W. The null hypothesis of DOCs in volatility is accepted for the estimated components s_t, with $Q_2^0(s^2, 5) = 7.845$ which has a p-value approximately equal to 0.727.

After DOCs are identified, a univariate volatility model is considered for each process $v_{i,t}^2 = \text{Var}(s_{i,t}|\mathcal{F}_{t-1})$. For example, the following model was fit

$$a_t = Ms_t = MV_t^{1/2}\epsilon_t,$$
$$V_t = \text{diag}\{v_{1,t}^2, \ldots, v_{d,t}^2\}, \quad \epsilon_{it} \overset{iid}{\sim} t_{\nu_i}(0,1)$$
$$v_{i,t}^2 = \omega_i + \alpha_i s_{i,t-1}^2 + \beta_i v_{i,t-1}^2$$
$$\Sigma_t = MV_tM',$$

in which $t_{\nu_i}(0,1)$ denotes the standardized Student-t distribution with tail-index ν_i. Each Σ_t is positive-definite if $v_{i,t}^2 > 0$ for all components. The fundamental motivation is that empirically the dynamics of a_t can often be well approximated by an invertible linear combination of DOCs $a_t = Ms_t$, in which $M = U^{-1}W'$ by definition.

In summary, U is estimated by PCA to uncorrelate a_t, W is estimated to minimize a reweighted version of (14.28) defined above (giving more weight to lower lags). The matrices U and W are combined to estimate DOCs s_t, of which univariate volatility modeling may then be appropriately applied. This approach allows modeling of a d-dimensional multivariate volatility process with d univariate volatility models, while greatly reducing both the number of parameters and the computational cost of estimation, and at the same time maintaining adequate empirical performance.

Figure 14.16 shows a fitted DOCs in volatility GARCH model for $\hat{a}_t$ using generalized decorrelation followed by univariate conditional maximum likelihood estimation. The estimated conditional standard deviations are shown in (a) and (d), and the conditional covariances and implied correlations are shown in (b) and (c), respectively. Unlike the O-GARCH fit, the implied conditional correlations appear adequate compared to the rolling estimator. Estimation of the O-GARCH and DOC models and Figs. 14.15 and 14.16 were calculated using the following commands in R.

```
69 source("SDAFE2.R")
70 DOC.fit = doc.garch(E = A, L = 4., c = 2.25, theta.ini = NULL)

71 par(mfrow = c(2,2)) # O-GARCH
72 plot(ts(DOC.fit$Sigma.pca[1,1,]^.5, start=c(1989,1), frequency=253),
73      type = 'l', xlab = "year", ylab = NULL,
74      main = expression(paste("(a) ", hat(sigma)["1,t"])))
75 plot(ts(DOC.fit$Sigma.pca[2,1,], start=c(1989,1), frequency=253),
76      type = 'l', xlab = "year", ylab = NULL,
77      main = expression(paste("(b) ", hat(sigma)["12,t"])))
78 plot(ts(DOC.fit$Sigma.pca[2,1,]/(sqrt(DOC.fit$Sigma.pca[1,1,]*
79                              DOC.fit$Sigma.pca[2,2,])),
80          start=c(1989,1), frequency=253),
81      type = 'l', xlab = "year", ylab = NULL, ylim = c(0.1,0.9),
82      main = expression(paste("(c) ", hat(rho)["12,t"])))
83 points(ts(mvwindow.cor(A[,1],A[,2], win = 126)$correlation,
84          start = c(1989, 1), frequency = 253),
```

```
85        type = 'l', col = 2, lty = 2, lwd = 2)
86 plot(ts(DOC.fit$Sigma.pca[2,2,]^.5, start=c(1989,1), frequency=253),
87        type = 'l', xlab = "year", ylab = NULL,
88        main = expression(paste("(d) ", hat(sigma)["2,t"])))
```

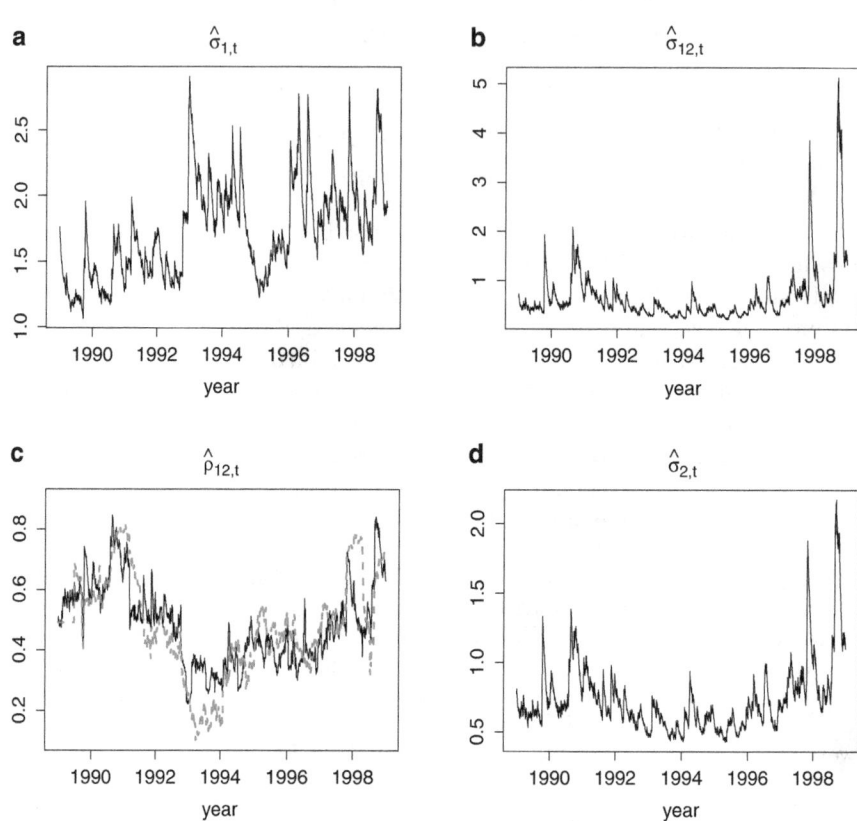

Fig. 14.16. A fitted first order DOCs in volatility GARCH model with $(\omega_1, \alpha_1, \beta_1, \nu_1)' = (0.0049, 0.0256, 0.9703, 4.3131)'$, $(\omega_2, \alpha_2, \beta_2, \nu_2)' = (0.0091, 0.0475, 0.9446, 5.0297)'$, and $\boldsymbol{M} = ((1.5350, 0.838)', (0.0103, 0.7730)')$. The red line in (c) is the sample correlation estimate over the previous six months for comparison.

```
89 par(mfrow = c(2,2)) # DOCs in volatility
90 plot(ts(DOC.fit$Sigma.doc[1,1,]^.5, start=c(1989,1), frequency=253),
91        type = 'l', xlab = "year", ylab = NULL,
92        main = expression(paste("(a) ", hat(sigma)["1,t"])))
93 plot(ts(DOC.fit$Sigma.doc[2,1,], start=c(1989,1), frequency=253),
94        type = 'l', xlab = "year", ylab = NULL,
95        main = expression(paste("(b) ", hat(sigma)["12,t"])))
96 plot(ts(DOC.fit$Sigma.doc[2,1,]/(sqrt(DOC.fit$Sigma.doc[1,1,]*
```

```
97                                        DOC.fit$Sigma.doc[2,2,])),
98          start=c(1989,1), frequency=253),
99        type = 'l', xlab = "year", ylab = NULL, ylim = c(0.1,0.9),
100       main = expression(paste("(c) ", hat(rho)["12,t"])))
101 points(ts(mvwindow.cor(A[,1],A[,2], win = 126)$correlation,
102            start=c(1989,1), frequency=253),
103        type = 'l', col = 2, lty = 2,lwd=2)
104 plot(ts(DOC.fit$Sigma.doc[2,2,]^.5, start=c(1989,1), frequency=253),
105       type = 'l', xlab = "year", ylab = NULL,
106       main = expression(paste("(d) ", hat(sigma)["2,t"])))
107 DOC.fit$coef.pca
```

```
             omega       alpha1      beta1
[1,] 0.003845283 0.02118369 0.9758129
[2,] 0.037473820 0.07101731 0.8913321
```

```
108 DOC.fit$coef.doc
```

```
             omega        alpha1      beta1     shape
[1,] 0.004874403 0.02560464 0.9702966 4.313164
[2,] 0.009092705 0.04740792 0.9446408 5.030019
```

```
109 DOC.fit$W.hat
```

```
           [,1]        [,2]
[1,] 0.9412834 -0.3376174
[2,] 0.3376174  0.9412834
```

```
110 DOC.fit$U.hat
```

```
            [,1]        [,2]
[1,]  0.6147515 -0.2297417
[2,] -0.2297417  1.4669516
```

```
111 DOC.fit$M.hat
```

```
            [,1]        [,2]
[1,] 1.53499088 0.8380397
[2,] 0.01024854 0.7729063
```

```
112 solve(DOC.fit$U.hat)
```

```
           [,1]       [,2]
[1,] 1.7277983 0.2705934
[2,] 0.2705934 0.7240638
```

14.14.6 Dynamic Conditional Correlation (DCC) Models

Nonlinear combinations of univariate volatility models have been proposed to allow for time-varying correlations, a feature that is prevalent in many financial applications. Both Tse and Tsui (2002) and Engle (2002) generalize the constant correlation model of Bollerslev (1990) to allow for such dynamic conditional correlations (DCC).

Analogously to the GARCH(1,1) model, the first order form of the DCC model in Engle (2002) may be represented by the following equations

$$\sigma_{i,t}^2 = \omega_i + \alpha_i a_{i,t-1}^2 + \beta_i \sigma_{i,t-1}^2,$$
$$D_t = \text{diag}\{\sigma_{1,t}, \dots, \sigma_{d,t}\},$$
$$\varepsilon_t = D_t^{-1} a_t,$$
$$Q_t = (1 - \lambda)\varepsilon_{t-1}\varepsilon_{t-1}' + \lambda Q_{t-1},$$
$$R_t = \text{diag}\{Q_t\}^{-\frac{1}{2}} Q_t \, \text{diag}\{Q_t\}^{-\frac{1}{2}},$$
$$\Sigma_t = D_t R_t D_t.$$

The main idea is to first model the conditional variance of each individual series σ_{it}^2 with a univariate volatility model, estimate the *scaled* innovations ε_t (not to be confused with *standardized* innovations ϵ_t) from these models, then focus on modeling the conditional correlation matrix R_t. These are then combined at each time point t to estimate the volatility matrix Σ_t. The recursion Q_t is an EWMA model applied to the scaled innovations ε_t. It is indexed by a single parameter $\lambda \in (0, 1)$. The matrix Q_t needs to be rescaled to form a proper correlation matrix R_t with the value 1 for all elements on the main diagonal.

The DCC model parameters can be estimated consistently in two stages using quasi-maximum likelihood. First, a univariate GARCH(1, 1) model is fit to each series to estimate σ_{it}^2. Then, given ε_t, λ is estimated by maximizing the components of the quasi-likelihood that only depend on the correlations. This is justified since the squared residuals do not depend on the correlation parameters.

In the form above, the variance components only condition on their own individual lagged returns and not the joint returns. Also, the dynamics for each of the conditional correlations are constrained to have equal persistence parameters, similar to the EWMA model. An explicit parameterization of the conditional correlation matrix R_t, with flexible dynamics, is just as difficult to estimate in high dimensions as Σ_t itself. Figure 14.17 shows a fitted DCC model for $\hat{a}_t$ using quasi-maximum likelihood estimation. The estimated conditional standard deviations are shown in (a) and (d), and the conditional covariances and conditional correlations are shown in (b) and (c), respectively. Estimation and Fig. 14.17 were calculated using the following commands in R.

```
113 source("SDAFE2.R")
114 DCCe.fit = fit.DCCe(theta.0=0.95, innov=A)
115 DCCe.fit$coef

          omega      alpha1      beta1
    [1,] 0.07435095 0.05528162 0.9231251
    [2,] 0.02064808 0.08341755 0.8822517

116 DCCe.fit$lambda
```

```
    [1] 0.9876297

117 par(mfrow = c(2,2))
118 plot(ts(DCCe.fit$Sigma.t[1,1,]^.5, start=c(1989, 1), frequency=253),
119     type = 'l', xlab = "year", ylab = NULL,
120     main = expression(paste("(a) ", hat(sigma)["1,t"])))
121 plot(ts(DCCe.fit$Sigma.t[2,1,], start=c(1989, 1), frequency=253),
122     type = 'l', xlab = "year", ylab = NULL,
123     main = expression(paste("(b) ", hat(sigma)["12,t"])))
124 plot(ts(DCCe.fit$R.t[2,1,], start=c(1989, 1), frequency=253),
125     type = 'l', xlab = "year", ylab = NULL,
126     main = expression(paste("(c) ", hat(rho)["12,t"])))
127 points(ts(mvwindow.cor(A[,1],A[,2], win = 126)$correlation,
128         start=c(1989, 1), frequency=253),
129     type = 'l', col = 2, lty = 2, lwd=2)
130 plot(ts(DCCe.fit$Sigma.t[2,2,]^.5, start=c(1989, 1), frequency=253),
131     type = 'l', xlab = "year", ylab = NULL,
132     main = expression(paste("(d) ", hat(sigma)["2,t"])))
```

14.14.7 Model Checking

For a fitted volatility sequence $\widehat{\Sigma}_t$, the *standardized* residuals are defined as

$$\widehat{\epsilon}_t = \widehat{\Sigma}_t^{-1/2} a_t, \tag{14.29}$$

in which $\widehat{\Sigma}_t^{-1/2}$ denotes the inverse of the matrix $\widehat{\Sigma}_t^{1/2}$. To verify the adequacy of a fitted volatility model, lagged cross-correlations of the squared standardized residuals should be zero. The product process $\widehat{\epsilon}_{it}\widehat{\epsilon}_{jt}$ should also have no serial correlation. Additional diagnostic checks for time series are considered in Li (2003). Since the standardized residuals are estimated and not observed, all p-values given in this section are only approximate.

To check the first condition we can apply a multivariate Ljung-Box test to the squared standardized residuals. For the EWMA model, $Q_2(\widehat{\epsilon}_t^2, 5) = 26.40$ with a p-value of 0.153, implying no significant serial correlation. For the O-GARCH model, $Q_2(\widehat{\epsilon}_t^2, 5) = 30.77$ with a p-value 0.058. In this case, there is some minor evidence of serial correlation. For the DOC in volatility model, $Q_2(\widehat{\epsilon}_t^2, 5) = 18.68$ with a p-value 0.543, implying no significant serial correlation. For the DCC model, $Q_2(\widehat{\epsilon}_t^2, 5) = 10.54$ with a p-value 0.957, implying no significant serial correlation.

```
133 n = dim(A)[1] ; d = dim(A)[2]
134 stdResid.EWMA = matrix(0,n,d)
135 stdResid.PCA = matrix(0,n,d)
136 stdResid.DOC = matrix(0,n,d)
137 stdResid.DCCe = matrix(0,n,d)
138 for(t in 1:n){
139  stdResid.EWMA[t,] = A[t,] %*% matrix.sqrt.inv(EWMA.Sigma[,,t])
140  stdResid.PCA[t,] = A[t,] %*% matrix.sqrt.inv(DOC.fit$Sigma.pca[,,t])
141  stdResid.DOC[t,] = A[t,] %*% matrix.sqrt.inv(DOC.fit$Sigma.doc[,,t])
```

```
142  stdResid.DCCe[t,] = A[t,] %*% matrix.sqrt.inv(DCCe.fit$Sigma.t[,,t])
143  }
144  mLjungBox(stdResid.EWMA^2, lag=5)
145  mLjungBox(stdResid.PCA^2, lag=5)
146  mLjungBox(stdResid.DOC^2, lag=5)
147  mLjungBox(stdResid.DCCe^2, lag=5)
```

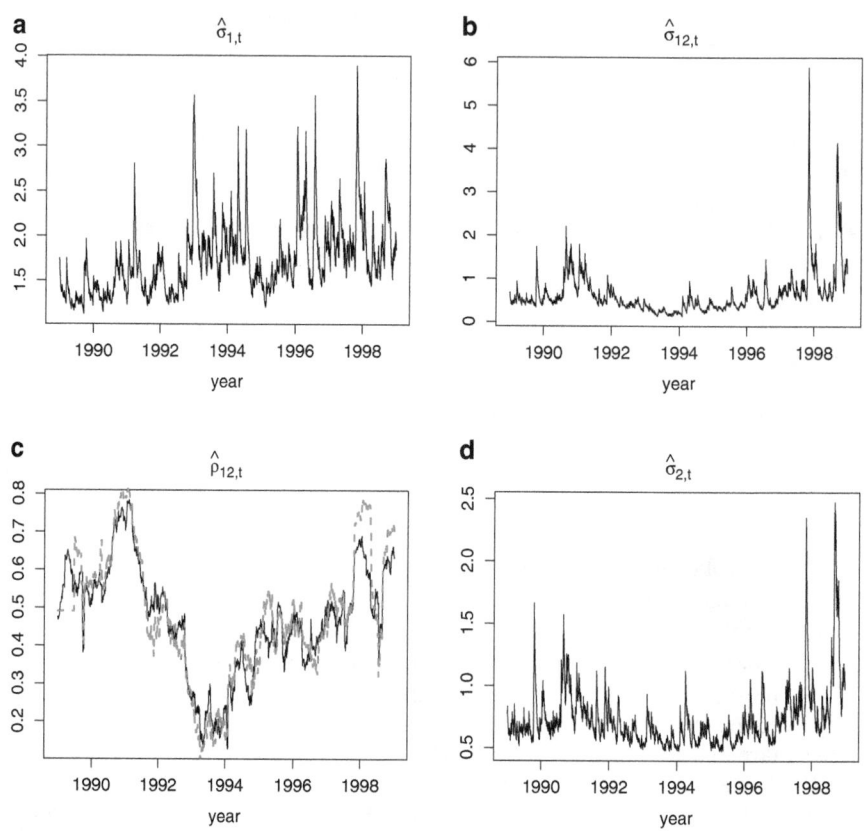

Fig. 14.17. A fitted first order DCC model with $(\omega_1, \alpha_1, \beta_1)' = (0.0741, 0.0552, 0.9233)'$, $(\omega_2, \alpha_2, \beta_2)' = (0.0206, 0.0834, 0.8823)'$, and $\lambda = 0.9876$. The red line in (c) is the sample correlation estimate over the previous six months for comparison.

The multivariate Ljung-Box test for the squared standardized residuals is not sensitive to misspecification of the conditional correlation structure. To check this condition, we apply univariate Ljung-Box tests to the product of each pair of standardized residuals. For the EWMA model, $Q(\widehat{\epsilon}_{1t}\widehat{\epsilon}_{2t}, 5) = 16.45$ with a p-value of 0.006. This model has not adequately accounted for the time-varying conditional correlation. For the O-GARCH model, $Q(\widehat{\epsilon}_{1t}\widehat{\epsilon}_{2t}, 5) = 63.09$ with a p-value near zero. This model also fails to account for the observed time-varying conditional correlation. For the DOC in volatility model, $Q(\widehat{\epsilon}_{1t}\widehat{\epsilon}_{2t}, 5) = 9.07$ with a p-value of 0.106, implying no significant serial correlation. For the DCC model, $Q(\widehat{\epsilon}_{1t}\widehat{\epsilon}_{2t}, 5) = 8.37$ with a p-value of 0.137, also implying no significant serial correlation.

```
148 mLjungBox(stdResid.EWMA[,1] * stdResid.EWMA[,2], lag=5)
149 mLjungBox(stdResid.PCA[,1] * stdResid.PCA[,2], lag=5)
150 mLjungBox(stdResid.DOC[,1] * stdResid.DOC[,2], lag=5)
151 mLjungBox(stdResid.DCCe[,1] * stdResid.DCCe[,2], lag=5)
```

14.15 Bibliographic Notes

Modeling nonconstant conditional variances in regression is treated in depth in the book by Carroll and Ruppert (1988).

There is a vast literature on GARCH processes beginning with Engle (1982), where ARCH models were introduced. Hamilton (1994), Enders (2004), Pindyck and Rubinfeld (1998), Gourieroux and Jasiak (2001), Alexander (2001), and Tsay (2005) have chapters on GARCH models. There are many review articles, including Bollerslev (1986), Bera and Higgins (1993), Bollerslev, Engle, and Nelson (1994), and Bollerslev, Chou, and Kroner (1992). Jarrow (1998) and Rossi (1996) contain a number of papers on volatility in financial markets. Duan (1995), Ritchken and Trevor (1999), Heston and Nandi (2000), Hsieh and Ritchken (2000), Duan and Simonato (2001), and many other authors study the effects of GARCH errors on options pricing, and Bollerslev, Engle, and Wooldridge (1988) use GARCH models in the CAPM.

For a thorough review of multivariate GARCH modeling see Bauwens, Laurent, and Rombouts (2006), and Silvennoinen and Teräsvirta (2009).

14.16 R Lab

14.16.1 Fitting GARCH Models

Run the following code to load the data set TbGdpPi.csv, which has three variables: the 91-day T-bill rate, the log of real GDP, and the inflation rate. In this lab you will use only the T-bill rate.

```
1 TbGdpPi = read.csv("TbGdpPi.csv", header=TRUE)
2 #  r = the 91-day treasury bill rate
3 #  y = the log of real GDP
4 #  pi = the inflation rate
5 TbGdpPi = ts(TbGdpPi, start = 1955, freq = 4)
6 Tbill = TbGdpPi[,1]
7 Tbill.diff = diff(Tbill)
```

Problem 1 *Plot both* `Tbill` *and* `Tbill.diff`*. Use both time series and ACF plots. Also, perform ADF and KPSS tests on both series. Which series do you think are stationary? Why? What types of heteroskedasticity can you see in the* `Tbill.diff` *series?*

In the following code, the variable `Tbill` can be used if you believe that series is stationary. Otherwise, replace `Tbill` by `Tbill.diff`. This code will fit an ARMA+
GARCH model to the series.

```
8  library(rugarch)
9  arma.garch.norm = ugarchspec(mean.model=list(armaOrder=c(1,0)),
10                               variance.model=list(garchOrder=c(1,1)))
11 Tbill.arma.garch.norm = ugarchfit(data=Tbill, spec=arma.garch.norm)
12 show(Tbill.arma.garch.norm)
```

Problem 2 *(a) Which ARMA+GARCH model is being fit? Write down the model using the same parameter names as in the* R *output.*
(b) What are the estimates of each of the parameters in the model?

Next, plot the residuals (ordinary or raw) and standardized residuals in various ways using the code below. The standardized residuals are best for checking the model, but the residuals are useful to see if there are GARCH effects in the series.

```
13 res = ts(residuals(Tbill.arma.garch.norm, standardize=FALSE),
14           start = 1955, freq = 4)
15 res.std = ts(residuals(Tbill.arma.garch.norm, standardize=TRUE),
16               start = 1955, freq = 4)
17 par(mfrow=c(2,3))
18 plot(res)
19 acf(res)
20 acf(res^2)
21 plot(res.std)
22 acf(res.std)
23 acf(res.std^2)
```

Problem 3 *(a) Describe what is plotted by* `acf(res)`*. What, if anything, does the plot tell you about the fit of the model?*
(b) Describe what is plotted by `acf(res^2)`*. What, if anything, does the plot tell you about the fit of the model?*
(c) Describe what is plotted by `acf(res_std^2)`*. What, if anything, does the plot tell you about the fit of the model?*
(d) Is there anything noteworthy in the figure produced by the command `plot(res.std)`*?*

Problem 4 *Now find an ARMA+GARCH model for the series* diff.log. Tbill, *which we will define as* diff(log(Tbill)). *Do you see any advantages of working with the differences of the logarithms of the T-bill rate, rather than with the difference of* Tbill *as was done earlier?*

14.16.2 The GARCH-in-Mean (GARCH-M) Model

A GARCH-in-Mean or *GARCH-M* model takes the form

$$Y_t = \mu + \delta\sigma_t + a_t$$
$$a_t = \sigma_t\epsilon_t$$
$$\sigma_t^2 = \omega + \alpha a_{t-1}^2 + \beta\sigma_{t-1}^2$$

in which $\epsilon_t \overset{iid}{\sim} (0,1)$. The GARCH-M model directly incorporates volatility as a regression variable. The parameter δ represents the *risk premium*, or reward for additional risk. Modern portfolio theory dictates that increased volatility leads to increased risk, requiring larger expected returns. The presence of volatility as a statistically significant predictor of returns is one of the primary contributors to serial correlation in historic return series. The data set GPRO.csv() contains the adjusted daily closing price of GoPro stock from June 26, 2014 to January 28, 2015.

Run the following R commands to fit a GARCH-M model to the GoPro stock returns.

```
1 library(rugarch)
2 GPRO = read.table("GPRO.csv")
3 garchm = ugarchspec(mean.model=list(armaOrder=c(0,0),
4                                   archm=T,archpow=1),
5                   variance.model=list(garchOrder=c(1,1)))
6 GPRO.garchm = ugarchfit(garchm, data=GPRO)
7 show(GPRO.garchm)
```

Problem 5 *Write out the fitted model. The parameter δ is equal to* archm *in the R output.*

Problem 6 *Test the one-sided hypothesis that $\delta > 0$ verses the alternative that $\delta = 0$. Is the risk premium significant?*

14.16.3 Fitting Multivariate GARCH Models

Run the following code to again load the data set TbGdpPi.csv, which has three variables: the 91-day T-bill rate, the log of real GDP, and the inflation rate. In this lab you will now use the first and third series after taking first differences.

```
1 TbGdpPi = read.csv("TbGdpPi.csv", header=TRUE)
2 TbPi.diff = ts(apply(TbGdpPi[,-2],2,diff), start=c(1955,2), freq=4)
3 plot(TbPi.diff)
4 acf(TbPi.diff^2)
5 source("SDAFE2.R")
6 mLjungBox(TbPi.diff^2, lag=8)
```

Problem 7 *Does the joint series exhibit conditional heteroskedasticity? Why?*

Now fit and plot a EWMA model with the following R commands.

```
7  EWMA.param = est.ewma(lambda.0=0.95, innov=TbPi.diff)
8  EWMA.param$lambda.hat
9  EWMA.Sigma=sigma.ewma(lambda=EWMA.param$lambda.hat,innov=TbPi.diff)
10 par(mfrow = c(2,2))
11 plot(ts(EWMA.Sigma[1,1,]^.5, start = c(1955, 2), frequency = 4),
12     type = 'l', xlab = "year", ylab = NULL,
13     main = expression(paste("(a) ", hat(sigma)["1,t"])))
14 plot(ts(EWMA.Sigma[1,2,], start = c(1955, 2), frequency = 4),
15     type = 'l', xlab = "year", ylab = NULL,
16     main = expression(paste("(b) ", hat(sigma)["12,t"])))
17 plot(ts(EWMA.Sigma[1,2,]/(sqrt(EWMA.Sigma[1,1,]* EWMA.Sigma[2,2,])),
18         start = c(1955, 2), frequency = 4),
19     type = 'l', xlab = "year", ylab = NULL,
20     main = expression(paste("(c) ", hat(rho)["12,t"])))
21 plot(ts(EWMA.Sigma[2,2,]^.5, start = c(1955, 2), frequency = 4),
22     type = 'l', xlab = "year", ylab = NULL,
23     main = expression(paste("(d) ", hat(sigma)["2,t"])))
```

Problem 8 *What is the estimated persistence parameter λ?*

Now estimate standardized residuals and check whether they exhibit any conditional heteroskedasticity

```
24 n = dim(TbPi.diff)[1]
25 d = dim(TbPi.diff)[2]
26 stdResid.EWMA = matrix(0,n,d)
27 for(t in 1:n){
28   stdResid.EWMA[t,] = TbPi.diff[t,] %*% matrix.sqrt.inv
29   (EWMA.Sigma[,,t])
30 }
31 mLjungBox(stdResid.EWMA^2, lag=8)
```

Problem 9 *Based on the output of the Ljung-Box test for the squared standardized residuals, is the EWMA model adequate?*

Run the following command in R to determine whether the joint series are DOCs in volatility.

```
32 DOC.test(TbPi.diff^2, 8)
```

Problem 10 *Is the null hypothesis rejected? Based on this conclusion, how should the conditional heteroskedasticity in the bivariate series be modeled, jointly or separately?*

14.17 Exercises

1. Let Z have an $N(0,1)$ distribution. Show that

$$E(|Z|) = \int_{-\infty}^{\infty} \frac{1}{\sqrt{2\pi}} |z| e^{-z^2/2} dz = 2 \int_{0}^{\infty} \frac{1}{\sqrt{2\pi}} z e^{-z^2/2} dz = \sqrt{\frac{2}{\pi}}.$$

Hint: $\frac{d}{dz} e^{-z^2/2} = -z e^{-z^2/2}$.

2. Suppose that $f_X(x) = 1/4$ if $|x| < 1$ and $f_X(x) = 1/(4x^2)$ if $|x| \geq 1$. Show that

$$\int_{-\infty}^{\infty} f_X(x) dx = 1,$$

so that f_X really is a density, but that

$$\int_{-\infty}^{0} x f_X(x) dx = -\infty$$

and

$$\int_{0}^{\infty} x f_X(x) dx = \infty,$$

so that a random variable with this density does not have an expected value.

3. Suppose that ϵ_t is an i.i.d. WN(0, 1) process, that

$$a_t = \epsilon_t \sqrt{1 + 0.35 a_{t-1}^2},$$

and that

$$y_t = 3 + 0.72 y_{t-1} + a_t.$$

   (a) Find the mean of y_t.
   (b) Find the variance of y_t.
   (c) Find the autocorrelation function of y_t.
   (d) Find the autocorrelation function of a_t^2.

4. Let y_t be the AR(1)+ARCH(1) model

$$a_t = \epsilon_t \sqrt{\omega + \alpha a_{t-1}^2},$$
$$(y_t - \mu) = \phi(y_{t-1} - \mu) + a_t,$$

where ϵ_t is i.i.d. WN(0,1). Suppose that $\mu = 0.4$, $\phi = 0.45$, $\omega = 1$, and $\alpha_1 = 0.3$.

(a) Find $E(y_2|y_1 = 1, y_0 = 0.2)$.

(b) Find $\mathrm{Var}(y_2|y_1 = 1, y_0 = 0.2)$.

5. Suppose that ϵ_t is white noise with mean 0 and variance 1, that $a_t = \epsilon_t \sqrt{7 + a_{t-1}^2/2}$, and that $Y_t = 2 + 0.67Y_{t-1} + a_t$.

(a) What is the mean of Y_t?

(b) What is the ACF of Y_t?

(c) What is the ACF of a_t?

(d) What is the ACF of a_t^2?

6. Let Y_t be a stock's return in time period t and let X_t be the inflation rate during this time period. Assume the model

$$Y_t = \beta_0 + \beta_1 X_t + \delta\sigma_t + a_t, \tag{14.30}$$

where

$$a_t = \epsilon_t\sqrt{1 + 0.5a_{t-1}^2}. \tag{14.31}$$

Here the ϵ_t are independent $N(0,1)$ random variables. Model (14.30)–(14.31) is called a *GARCH-in-mean* model or a GARCH-M model. Assume that $\beta_0 = 0.06$, $\beta_1 = 0.35$, and $\delta = 0.22$.

(a) What is $E(Y_t|X_t = 0.1$ and $a_{t-1} = 0.6)$?

(b) What is $\mathrm{Var}(Y_t|X_t = 0.1$ and $a_{t-1} = 0.6)$?

(c) Is the conditional distribution of Y_t given X_t and a_{t-1} normal? Why or why not?

(d) Is the marginal distribution of Y_t normal? Why or why not?

7. Suppose that $\epsilon_1, \epsilon_2, \ldots$ is a Gaussian white noise process with mean 0 and variance 1, and a_t and y_t are stationary processes such that

$$a_t = \sigma_t\epsilon_t \quad \text{where} \quad \sigma_t^2 = 2 + 0.3a_{t-1}^2,$$

and

$$y_t = 2 + 0.6y_{t-1} + a_t.$$

(a) What type of process is a_t?

(b) What type of process is y_t?

(c) Is a_t Gaussian? If not, does it have heavy or lighter tails than a Gaussian distribution?

(d) What is the ACF of a_t?

(e) What is the ACF of a_t^2?

(f) What is the ACF of y_t?

8. On Black Monday, the return on the S&P 500 was -22.8%. Ouch! This exercise attempts to answer the question, "what was the conditional probability of a return this small or smaller on Black Monday?" "Conditional" means given the information available the previous trading day. Run the following R code:

```
1  library(rugarch)
2  library(Ecdat)
3  data(SP500,package="Ecdat")
4  returnBlMon = SP500$r500[1805] ; returnBlMon
5  x = SP500$r500[(1804-2*253+1):1804]
6  ts.plot(c(x,returnBlMon))
7  spec = ugarchspec(mean.model=list(armaOrder=c(1,0)),
8                    variance.model=list(garchOrder=c(1,1)),
9                    distribution.model = "std")
10 fit = ugarchfit(data=x, spec=spec)
11 dfhat = coef(fit)[6]
12 forecast = ugarchforecast(fit, data=x, n.ahead=1)
```

The S&P 500 returns are in the data set SP500 in the Ecdat package. The returns are the variable r500 (this is the only variable in this data set). Black Monday is the 1805th return in this data set. This code fits an AR(1)+GARCH(1,1) model to the last two years of data before Black Monday, assuming 253 trading days/year. The conditional distribution of the white noise is the t-distribution (called "std" in ugarchspec()). The code also plots the returns during these two years and on Black Monday. From the plot you can see that Black Monday was highly unusual. The parameter estimates are in coef(fit) and the sixth parameter is the degrees of freedom of the t-distribution. The ugarchforecast() function is used to predict one-step ahead, that is, to predict the return on Black Monday; the input variable n.ahead specifies how many days ahead to forecast, so n.ahead=5 would forecast the next five days. The object forecast will contain fitted(forecast), which is the conditional expected return on Black Monday, and sigma(forecast), which is the conditional standard deviation of the return on Black Monday.

(a) Use the information above to calculate the conditional probability of a return less than or equal to -0.228 on Black Monday.
(b) Compute and plot the standardized residuals. Also plot the ACF of the standardized residuals and their squares. Include all three plots with your work. Do the standardized residuals indicate that the AR(1)+GARCH(1,1) model fits adequately?
(c) Would an AR(1)+ARCH(1) model provide an adequate fit?
(d) Does an AR(1) model with a Gaussian conditional distribution provide an adequate fit? Use the arima() function to fit the AR(1) model. This function only allows a Gaussian conditional distribution.

9. This problem uses monthly observations of the two-month yield, that is, Y_T with T equal to two months, in the data set Irates in the Ecdat package. The rates are log-transformed to stabilize the variance. To fit a GARCH model to the changes in the log rates, run the following R code.

```
13 library(rugarch)
14 library(Ecdat)
15 data(Irates)
```

```
16 r = as.numeric(log(Irates[,2]))
17 n = length(r)
18 lagr = r[1:(n-1)]
19 diffr = r[2:n] - lagr
20 spec = ugarchspec(mean.model=list(armaOrder=c(1,0)),
21                   variance.model=list(garchOrder=c(1,1)),
22                   distribution.model = "std")
23 fit = ugarchfit(data=diffr, spec=spec)
24 plot(fit, which="all")
```

(a) What model is being fit to the changes in r? Describe the model in detail.
(b) What are the estimates of the parameters of the model?
(c) What is the estimated ACF of Δr_t?
(d) What is the estimated ACF of a_t?
(e) What is the estimated ACF of a_t^2?

10. Consider the daily log returns on the S&P 500 index (GSPC). Begin by running the following commands in R, then answer the questions below for the series y.

```
25 library(rugarch)
26 library(quantmod)
27 getSymbols("^GSPC", from="2005-01-01", to="2014-12-31")
28 head(GSPC)
29 sp500 = xts( diff( log( GSPC[,6] ) )[-1] )
30 plot(sp500)
31 y = as.numeric(sp500)
```

(a) Is there any serial correlation in the log returns of S&P 500 index? Why?
(b) Is there any ARCH effect (evidence of conditional heteroskedasticity) in the log returns of S&P 500 index? Why?
(c) Specify and fit an ARCH model to the log returns of S&P 500 index. Write down the fitted model.
(d) Is your fitted ARCH model stationary? Why?
(e) Fit a GARCH(1,1) model for the log returns on the S&P 500 index using the Gaussian distribution for the innovations. Write down the fitted model.
(f) Perform model checking to ensure that the model is adequate using 20 lags in a Ljung-Box test of the standardized residuals and the squared standardized residuals.
(g) Is the fitted GARCH model stationary? Why?
(h) Make a Normal quantile plot for the standardized residuals. Use qqnorm() and qqline() in R. Is the Gaussian distribution appropriate for the standardized innovations?
(i) Plot the fitted conditional standard deviation process $\hat{\sigma}_t$ and comment.

(j) Calculate the 1–10 step ahead forecasts from the end of the series for both the process y_t and the conditional variance using the ugarchforecast() function.

References

Alexander, C. (2001) *Market Models: A Guide to Financial Data Analysis*, Wiley, Chichester.

Bauwens, L., Laurent, S., and Rombouts, J. V. (2006) Multivariate GARCH models: a survey. *Journal of Applied Econometrics*, **21**(1), 79–109.

Bera, A. K., and Higgins, M. L. (1993) A survey of Arch models. *Journal of Economic Surveys*, **7**, 305–366. [Reprinted in Jarrow (1998).]

Bollerslev, T. (1986) Generalized autoregressive conditional heteroskedasticity. *Journal of Econometrics*, **31**, 307–327.

Bollerslev, T. (1990) Modelling the coherence in short-run nominal exchange rates: a multivariate generalized ARCH model. *The Review of Economics and Statistics*, **72**(3), 498–505.

Bollerslev, T., Chou, R. Y., and Kroner, K. F. (1992) ARCH modelling in finance. *Journal of Econometrics*, **52**, 5–59. [Reprinted in Jarrow (1998)]

Bollerslev, T., Engle, R. F., and Nelson, D. B. (1994) ARCH models, In *Handbook of Econometrics, Vol IV*, Engle, R.F., and McFadden, D.L., Elsevier, Amsterdam.

Bollerslev, T., Engle, R. F., and Wooldridge, J. M. (1988) A capital asset pricing model with time-varying covariances. *Journal of Political Economy*, **96**, 116–131.

Carroll, R. J., and Ruppert, D. (1988) *Transformation and Weighting in Regression*, Chapman & Hall, New York.

Duan, J.-C. (1995) The GARCH option pricing model. *Mathematical Finance*, **5**, 13–32. [Reprinted in Jarrow (1998).]

Duan, J-C., and Simonato, J. G. (2001) American option pricing under GARCH by a Markov chain approximation. *Journal of Economic Dynamics and Control*, **25**, 1689–1718.

Enders, W. (2004) *Applied Econometric Time Series*, 2nd ed., Wiley, New York.

Engle, R. F. (1982) Autoregressive conditional heteroskedasticity with estimates of variance of U.K. inflation. *Econometrica*, **50**, 987–1008.

Engle, R. F. (2002) Dynamic conditional correlation: A simple class of multivariate generalized autoregressive conditional heteroskedasticity models. *Journal of Business & Economic Statistics*, **20**(3), 339–350.

Fisher, T.J., and Gallagher, C.M. (2012) New weighted portmanteau statistics for time series goodness of fit testing. *Journal of the American Statistical Association*, **107**(498), 777–787.

Gourieroux, C. and Jasiak, J. (2001) *Financial Econometrics*, Princeton University Press, Princeton, NJ.

Hamilton, J. D. (1994) *Time Series Analysis*, Princeton University Press, Princeton, NJ.

Heston, S. and Nandi, S. (2000) A closed form GARCH option pricing model. *The Review of Financial Studies*, **13**, 585–625.

Hsieh, K. C. and Ritchken, P. (2000) An empirical comparison of GARCH option pricing models. working paper.

Jarrow, R. (1998) *Volatility: New Estimation Techniques for Pricing Derivatives*, Risk Books, London. (This is a collection of articles, many on GARCH models or on stochastic volatility models, which are related to GARCH models.)

Li, W. K. (2003) *Diagnostic checks in time series*, CRC Press.

Matteson, D. S. and Tsay, R. S. (2011) Dynamic orthogonal components for multivariate time series. *Journal of the American Statistical Association*, **106**(496), 1450–1463.

Palm, F.C. (1996) GARCH models of volatility. *Handbook of Statistics*, **14**, 209–240.

Palma, W. and Zevallos, M. (2004). Analysis of the correlation structure of square time series. *Journal of Time Series Analysis*, **25**(4), 529–550.

Pindyck, R. S. and Rubinfeld, D. L. (1998) *Econometric Models and Economic Forecasts*, Irwin/McGraw Hill, Boston.

Ritchken, P. and Trevor, R. (1999) Pricing options under generalized GARCH and stochastic volatility processes. *Journal of Finance*, **54**, 377–402.

Rossi, P. E. (1996) *Modelling Stock Market Volatility*, Academic Press, San Diego.

Silvennoinen, A. and Teräsvirta, T (2009) Multivariate GARCH models. In *Handbook of Financial Time Series*, 201–229, Springer, Berlin.

Tsay, R. S. (2005) *Analysis of Financial Time Series*, 2nd ed., Wiley, New York.

Tse, Y. K. and Tsui, A. K. C. (2002) A multivariate generalized autoregressive conditional heteroscedasticity model with time-varying correlations. *Journal of Business & Economic Statistics*, **20**(3), 351–362.

15

Cointegration

15.1 Introduction

Cointegration analysis is a technique that is frequently applied in econometrics. In finance it can be used to find trading strategies based on mean-reversion.

Suppose one could find a stock whose price (or log-price) series was stationary and therefore mean-reverting. This would be a wonderful investment opportunity. Whenever the price was below the mean, one could buy the stock and realize a profit when the price returned to the mean. Similarly, one could realize profits by selling short whenever the price was above the mean. Alas, returns are stationary but not prices. We have seen that log-prices are integrated. However, not all is lost. Sometimes one can find two or more assets with prices so closely connected that a linear combination of their prices is stationary. Then, a portfolio with weights assigned by the *cointegrating vector*, which is the vector of coefficients of this linear combination, will have a stationary price. Cointegration analysis is a means for finding cointegration vectors.

Two time series, $Y_{1,t}$ and $Y_{2,t}$, are cointegrated if each is $I(1)$ but there exists a λ such that $Y_{1,t} - \lambda Y_{2,t}$ is stationary. For example, the common trends model is that

$$Y_{1,t} = \beta_1 W_t + \epsilon_{1,t},$$
$$Y_{2,t} = \beta_2 W_t + \epsilon_{2,t},$$

where β_1 and β_2 are nonzero, the trend W_t common to both series is $I(1)$, and the noise processes $\epsilon_{1,t}$ and $\epsilon_{2,t}$ are $I(0)$. Because of the common trend,

© Springer Science+Business Media New York 2015
D. Ruppert, D.S. Matteson, *Statistics and Data Analysis for Financial Engineering*, Springer Texts in Statistics,
DOI 10.1007/978-1-4939-2614-5_15

$Y_{1,t}$ and $Y_{2,t}$ are nonstationary but there is a linear combination of these two series that is free of the trend, so they are cointegrated. To see this, note that if $\lambda = \beta_1/\beta_2$, then

$$\beta_2(Y_{1,t} - \lambda Y_{2,t}) = \beta_2 Y_{1,t} - \beta_1 Y_{2,t} = \beta_2 \epsilon_{1,t} - \beta_1 \epsilon_{2,t} \qquad (15.1)$$

is free of the trend W_t, and therefore is $I(0)$.

The definition of cointegration extends to more than two time series. A d-dimensional multivariate time series is cointegrated of order r if the component series are $I(1)$ but r independent linear combinations of the components are $I(0)$ for some r, $0 < r \leq d$. Somewhat different definitions of cointegration exist, but this one is best for our purposes.

In Sect. 13.3.1 we saw the danger of spurious regression when the residuals are integrated. This problem should make one cautious about regression with nonstationary time series. However, if Y_t is regressed on X_t and the two series are cointegrated, then the residuals will be $I(0)$ so that the least-squares estimator will be consistent.

The Phillips–Ouliaris cointegration test regresses one integrated series on others and applies the Phillips–Perron unit root test to the residuals. The null hypothesis is that the residuals are unit root nonstationary, which implies that the series are *not* cointegrated. Therefore, a small p-value implies that the series *are* cointegrated and therefore suitable for regression analysis. The residuals will still be correlated and so they should be modeled as such; see Sect. 13.3.3.

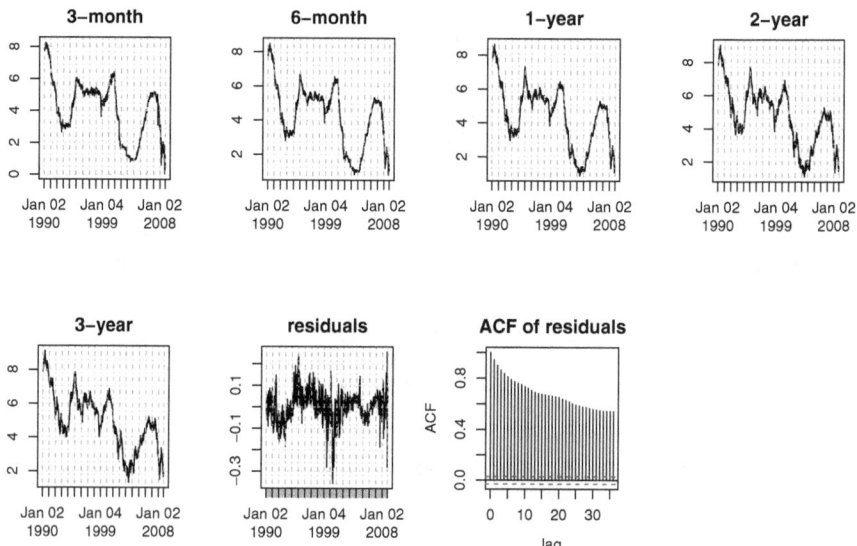

Fig. 15.1. *Time series plots of the five yields and the residuals from a regression of the 1-year yields on the other four yields. Also, a sample ACF plot of the residuals.*

Example 15.1. Phillips–Ouliaris test on bond yields

This example uses three-month, six-month, one-year, two-year, and three-year bond yields recorded daily from January 2, 1990 to October 31, 2008, for a total of 4,714 observations. The five yield series are plotted in Fig. 15.1, and one can see that they track each other somewhat closely. This suggests that the five series may be cointegrated. The one-year yields were regressed on the four others and the residuals and their ACF are also plotted in Fig. 15.1. The two residual plots are ambiguous about whether the residuals are stationary, so a test of cointegration would be helpful.

```
1  library(forecast)
2  library(tseries)
3  library(urca)
4  library(xts)
5  yieldDat = read.table("treasury_yields.txt", header=T)
6  date = as.Date(yieldDat[,1], format = "%m/%d/%y")
7  dat = as.xts(yieldDat[,3:7], date)
8  res = residuals(lm(dat[,3]~dat[,1]+dat[,2]+dat[,4]+dat[,5]))
```

Next, the Phillips–Ouliaris test was run using the R function po.test() in the tseries package.

```
9  po.test(dat[,c(3,1,2,4,5)])

            Phillips-Ouliaris Cointegration Test

data:  dat[, c(3, 1, 2, 4, 5)]
Phillips-Ouliaris demeaned = -323.546, Truncation lag
parameter = 47, p-value = 0.01

Warning message:
In po.test(dat[, c(3, 1, 2, 4, 5)]) : p-value smaller
than printed p-value
```

The p-value is computed by interpolation if it is within the range of a table in Phillips and Ouliaris (1990). In this example, the p-value is outside the range and we know only that it is below 0.01, the lower limit of the table. The small p-value leads to the conclusion that the residuals are stationary and so the five series are cointegrated.

Though stationary, the residuals have a large amount of autocorrelation and may have long-term memory. They take a long time to revert to their mean of zero. Devising a profitable trading strategy from these yields seems problematic.   □

15.2 Vector Error Correction Models

The regression approach to cointegration is somewhat unsatisfactory, since one series must be chosen as the dependent variable, and this choice must be somewhat arbitrary. In Example 15.1, the middle yield, ordered by maturity, was used but for no compelling reason. Moreover, regression will find only one cointegration vector, but there could be more than one.

An alternative approach to cointegration that treats the series symmetrically uses a *vector error correction model* (VECM). In these models, the deviation from the mean is called the "error" and whenever the stationary linear combination deviates from its mean, it is subsequently pushed back toward its mean (the error is "corrected").

The idea behind error correction is simplest when there are only two series, $Y_{1,t}$ and $Y_{2,t}$. In this case, the error correction model is

$$\Delta Y_{1,t} = \phi_1(Y_{1,t-1} - \lambda Y_{2,t-1}) + \epsilon_{1,t}, \tag{15.2}$$

$$\Delta Y_{2,t} = \phi_2(Y_{1,t-1} - \lambda Y_{2,t-1}) + \epsilon_{2,t}, \tag{15.3}$$

where $\epsilon_{1,t}$ and $\epsilon_{2,t}$ are white noise. Subtracting λ times (15.3) from (15.2) gives

$$\Delta(Y_{1,t} - \lambda Y_{2,t}) = (\phi_1 - \lambda\phi_2)(Y_{1,t-1} - \lambda Y_{2,t-1}) + (\epsilon_{1,t} - \lambda\epsilon_{2,t}). \tag{15.4}$$

Let $\mathcal{F}_t$ denote the information set at time t. If $(\phi_1 - \lambda\phi_2) < 0$, then $E\{\Delta(Y_{1,t} - \lambda Y_{2,t})|\mathcal{F}_{t-1}\}$ is opposite in sign to $Y_{1,t-1} - \lambda Y_{2,t-1}$. This causes error correction because whenever $Y_{1,t-1} - \lambda Y_{2,t-1}$ is positive, its expected change is negative and vice versa.

A rearrangement of (15.4) shows that $Y_{1,t-1} - \lambda Y_{2,t-1}$ is an AR(1) process with coefficient $1 + \phi_1 - \lambda\phi_2$. Therefore, the series $Y_{1,t} - \lambda Y_{2,t}$ is $I(0)$, unit-root nonstationary, or an explosive series in the cases where $|1 + \phi_1 - \lambda\phi_2|$ is less than 1, equal to 1, and greater than 1, respectively.

- If $\phi_1 - \lambda\phi_2 > 0$, then $1 + \phi_1 - \lambda\phi_2 > 1$ and $Y_{1,t} - \lambda Y_{2,t}$ is explosive.
- If $\phi_1 - \lambda\phi_2 = 0$, then $1 + \phi_1 - \lambda\phi_2 = 1$ and $Y_{1,t} - \lambda Y_{2,t}$ is a random walk.
- If $\phi_1 - \lambda\phi_2 < 0$, then $1 + \phi_1 - \lambda\phi_2 < 1$ and $Y_{1,t} - \lambda Y_{2,t}$ is stationary, unless $\phi_1 - \lambda\phi_2 \leq -2$, so that $1 + \phi_1 - \lambda\phi_2 \leq -1$.

The case $\phi_1 - \lambda\phi_2 \leq -2$ is "over-correction." The change in $Y_{1,t} - \lambda Y_{2,t}$ is in the correct direction but too large, so the series oscillates in sign but diverges to ∞ in magnitude.

Example 15.2. Simulation of an error correction model

Model (15.2)–(15.3) was simulated with $\phi_1 = 0.5$, $\phi_2 = 0.55$, and $\lambda = 1$. A total of 5,000 observations were simulated, but, for visual clarity, only every 10th observation is plotted in Fig. 15.2. Neither $Y_{1,t}$ nor $Y_{2,t}$ is stationary, but $Y_{1,t} - \lambda Y_{2,t}$ is stationary. Notice how closely $Y_{1,t}$ and $Y_{2,t}$ track one another.

```
10  n = 5000
11  set.seed(12345)
12  a1 = 0.5
```

```
13  a2 = 0.55
14  lambda  = 1
15  y1 = rep(0,n)
16  y2 = y1
17  e1 = rnorm(n)
18  e2 = rnorm(n)
19  for (i in 2:n){
20      y1[i] = y1[i-1] + a1 * (y1[i-1] - lambda*y2[i-1]) + e1[i]
21      y2[i] = y2[i-1] + a2 * (y1[i-1] - lambda*y2[i-1]) + e2[i]
22  }
```

$\square$

To see how to generalize error correction to more than two series, it is useful to rewrite Eqs. (15.2) and (15.3) in vector form. Let $\boldsymbol{Y}_t = (Y_{1,t}, Y_{2,t})'$ and $\boldsymbol{\epsilon}_t = (\epsilon_{1,t}, \epsilon_{2,t})'$. Then

$$\Delta \boldsymbol{Y}_t = \boldsymbol{\alpha}\boldsymbol{\beta}'\boldsymbol{Y}_{t-1} + \boldsymbol{\epsilon}_t, \tag{15.5}$$

where

$$\boldsymbol{\alpha} = \begin{pmatrix} \phi_1 \\ \phi_2 \end{pmatrix} \quad \text{and} \quad \boldsymbol{\beta} = \begin{pmatrix} 1 \\ -\lambda \end{pmatrix}, \tag{15.6}$$

so that $\boldsymbol{\beta}$ is the cointegration vector, and $\boldsymbol{\alpha}$ specifies the speed of mean-reversion and is called the *loading matrix* or *adjustment matrix*.

Model (15.5) also applies when there are d series such that $\boldsymbol{Y}_t$ and $\boldsymbol{\epsilon}_t$ are d-dimensional. In this case $\boldsymbol{\beta}$ and $\boldsymbol{\alpha}$ are each full-rank $d \times r$ matrices for

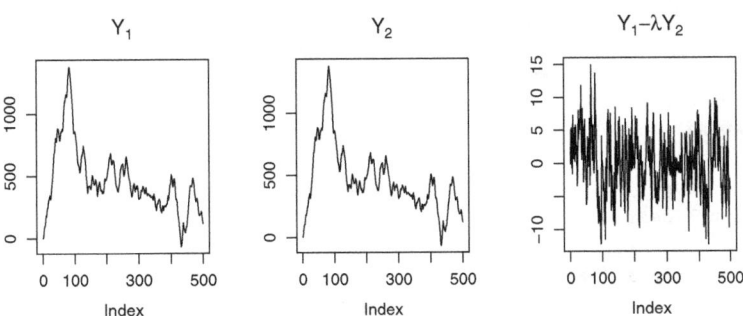

Fig. 15.2. *Simulation of an error correction model. 5,000 observations were simulated but only every 10th is plotted.*

some $r \leq d$ which is the number of linearly independent cointegration vectors. The columns of $\boldsymbol{\beta}$ are the cointegration vectors.

Model (15.5) is a vector AR(1) [that is, VAR(1)] model but, for added flexibility, can be extended to a VAR(p) model, and there are several ways to do this. We will use the notation and the second of two forms of the VECM from the function `ca.jo()` in R's `urca` package. This VECM is

$$\Delta Y_t = \Gamma_1 \Delta Y_{t-1} + \cdots + \Gamma_{p-1} \Delta Y_{t-p+1} + \Pi Y_{t-1} + \mu + \Phi D_t + \epsilon_t, \quad (15.7)$$

where μ is a mean vector, D_t is a vector of nonstochastic regressors, and

$$\Pi = \alpha \beta'. \tag{15.8}$$

As before, β and α are each full-rank $d \times r$ matrices and α is called the loading matrix.

It is easy to show that the columns of β are the cointegration vectors. Since Y_t is $I(1)$, ΔY_t on the left-hand side of (15.7) is $I(0)$ and therefore $\Pi Y_{t-1} = \alpha \beta' Y_{t-1}$ on the right-hand side of (15.7) is also $I(0)$. It follows that each of the r components of $\beta' Y_{t-1}$ is $I(0)$.

Example 15.3. VECM test on bond yields

A VECM was fit to the bond yields using R's `ca.jo()` function. The output is below. The eigenvalues are used to test null hypotheses of the form H_0: $r \leq r_0$. The values of the test statistics and critical values (for 1%, 5%, and 10% level tests) are listed below the eigenvalues. The null hypothesis is rejected when the test statistic exceeds the critical level. In this case, regardless of whether one uses a 1%, 5%, or 10% level test, one accepts that r is less than or equal to 3 but rejects that r is less than or equal to 2, so one concludes that $r = 3$. Although five cointegration vectors are printed, only the first three would be meaningful. The cointegration vectors are the columns of the matrix labeled "Eigenvectors, normalized to first column." The cointegration vectors are determined only up to multiplication by a nonzero scalar and so can be normalized so that their first element is 1.

```
#####################
# Johansen-Procedure #
#####################

Test type: maximal eigenvalue statistic (lambda max),
with linear trend

Eigenvalues (lambda):
[1] 0.03436 0.02377 0.01470 0.00140 0.00055

Values of test statistic and critical values of test:

          test 10pct  5pct 1pct
r <= 4 |   2.59   6.5  8.18 11.6
r <= 3 |   6.62  12.9 14.90 19.2
r <= 2 |  69.77  18.9 21.07 25.8
r <= 1 | 113.36  24.8 27.14 32.1
r = 0  | 164.75  30.8 33.32 38.8

Eigenvectors, normalised to first column:
```

(These are the cointegration relations)

```
           X3mo.12 X6mo.12 X1yr.12 X2yr.12 X3yr.12
X3mo.12    1.000    1.00    1.00   1.0000   1.000
X6mo.12   -1.951    2.46    1.07   0.0592   0.897
X1yr.12    1.056   14.25   -3.95  -2.5433  -1.585
X2yr.12    0.304  -46.53    3.51  -3.4774  -0.118
X3yr.12   -0.412   30.12   -1.71   5.2322   1.938
```

Weights W:
(This is the loading matrix)

```
          X3mo.12   X6mo.12    X1yr.12    X2yr.12    X3yr.12
X3mo.d  -0.03441  -0.002440  -0.011528  -0.000178  -0.000104
X6mo.d   0.01596  -0.002090  -0.007066   0.000267  -0.000170
X1yr.d  -0.00585  -0.001661  -0.001255   0.000358  -0.000289
X2yr.d   0.00585  -0.000579  -0.003673  -0.000072  -0.000412
X3yr.d   0.01208  -0.000985  -0.000217  -0.000431  -0.000407
```

□

15.3 Trading Strategies

As discussed previously, price series that are cointegrated can be used in *sta-tistical arbitrage*. Unlike pure arbitrage, statistical arbitrage means an opportunity where a profit is only likely, not guaranteed. Pairs trading uses pairs of cointegrated asset prices and has been a popular statistical arbitrage technique. Pairs trading requires the trader to find cointegrated pairs of assets, to select from these the pairs that can be traded profitably after accounting for transaction costs, and finally to design a trading strategy which includes the buy and sell signals. A full discussion of statistical arbitrage is outside the scope of this book, but see Sect. 15.4 for further reading.

Although many firms have been very successful using statistical arbitrage, one should be mindful of the risks. One is model risk; the error-correction model may be incorrect. Even if the model is correct, one must use estimates based on past data and the parameters might change, perhaps rapidly. If statistical arbitrage opportunities exist, then it is possible that other traders have discovered them and their trading activity is one reason to expect parameters to change. Another risk is that one can go bankrupt before a stationary process reverts to its mean. This risk is especially large because firms engaging in statistical arbitrage are likely to be heavily leveraged. High leverage will magnify a small loss caused when a process diverges even farther from its mean before reverting. See Sects. 2.4.2 and 15.5.4.

15.4 Bibliographic Notes

Alexander (2001), Enders (2004), and Hamilton (1994) contain useful discussions of cointegration. Pfaff (2006) is a good introduction to the analysis of cointegrated time series using R.

The MLEs and likelihood ratio tests of the parameters in (15.7) were developed by Johansen (1991), Johansen (1995) and Johansen and Juselius (1990).

The applications of cointegration theory in statistical arbitrage are discussed by Vidyamurthy (2004) and Alexander, Giblin, and Weddington (2001). Pole (2007) is a less technical introduction to statistical arbitrage.

15.5 R Lab

15.5.1 Cointegration Analysis of Midcap Prices

The data set `midcapD.ts.csv` has daily log returns on 20 midcap stocks in columns 2–21. Columns 1 and 22 contain the date and market returns, respectively. In this section, we will use returns on the first 10 stocks. To find the stock prices from the returns, we use the relationship

$$P_t = P_0 \exp(r_1 + \cdots + r_t),$$

where P_t and r_t are the price and log return at time t. The returns will be used as approximations to the log returns. The prices at time 0 are unknown, so we will use $P_0 = 1$ for each stock. This means that the price series we use will be off by multiplicative factors. This does not affect the number of cointegration vectors. If we find that there are cointegration relationships, then it would be necessary to get the price data to investigate trading strategies.

Johansen's cointegration analysis will be applied to the prices with the `ca.jo()` function in the `urca` package. Run

```
1  library(urca)
2  midcapD.ts = read.csv("midcapD.ts.csv",header=T)
3  x = midcapD.ts[,2:11]
4  prices= exp(apply(x,2,cumsum))
5  options(digits=3)
6  summary(ca.jo(prices))
```

Problem 1 *How many cointegration vectors were found?*

15.5.2 Cointegration Analysis of Yields

This example is similar to Example 15.3 but uses different yield data. The data are in the `mk.zero2.csv` data set. There are 55 maturities and they are in the vector `mk.maturity`. We will use only the first 10 yields. Run the following commands in R.

```
1 library(urca)
2 mk.maturity = read.csv("mk.zero2.csv", header=T)
3 summary(ca.jo(mk.maturity[,2:11]))
```

Problem 2 *What maturities are being used? Are they short-, medium-, or long-term, or a mixture of short- and long-term maturities?*

Problem 3 *How many cointegration vectors were found? Use 1 % level tests.*

15.5.3 Cointegration Analysis of Daily Stock Prices

The CokePepsi.csv data set contains the adjusted daily closing prices of Coke and Pepsi stock from January 2007 to November 2012. Run the following commands in R.

```
1 CokePepsi = read.table("CokePepsi.csv", header=T)
2 ts.plot(CokePepsi)
```

Problem 4 *Do these two series appear cointegrated from the time series plot? Why?*

Now make a time series plot of the difference between the two prices.

```
3 ts.plot(CokePepsi[,2] - CokePepsi[,1])
```

Problem 5 *Does this difference series appear stationary? Why?*

Run the following commands to conduct Johansen's cointegration test.

```
4 library(urca)
5 summary(ca.jo(CokePepsi))
```

Problem 6 *Are these two series cointegrated? Why?*

Now consider the daily adjusted closing prices for 10 company stocks from January 2, 1987 to September 1, 2006 from the Stock_FX_Bond.csv dataset.

```
6 Stock_FX_Bond = read.csv("Stock_FX_Bond.csv", header=T)
7 adjClose = Stock_FX_Bond[,seq(from=3, to=21, by=2)]
8 ts.plot(adjClose)
9 summary(ca.jo(adjClose))
```

Problem 7 *Are these 10 stock price series cointegrated? If so, what is the rank of the cointegrating matrix, and what are the cointegrating vectors?*

Rerun the Johansen's cointegration test with lag K = 8.

```
10 summary(ca.jo(adjClose, K=8))
```

Problem 8 *Are these 10 stock price series cointegrated if Johansen's cointegration test is conducted with lag K = 8? If so, has the estimated rank of the cointegrating matrix or the cointegrating vectors changed?*

15.5.4 Simulation

In this section, you will run simulations similar to those in Sect. 2.4.2. The difference is that now the price process is mean-reverting.

Suppose a hedge fund owns a $1,000,000 position in a portfolio and used $50,000 of its own capital and $950,000 in borrowed money for the purchase. If the value of the portfolio falls below $950,000 at the end of any trading day, then the hedge fund must liquidate and repay the loan.

The portfolio was selected by cointegration analysis and its price is an AR(1) process,

$$(P_t - \mu) = \phi(P_{t-1} - \mu) + \epsilon_t,$$

where P_t is the price of the portfolio at the end of trading day t, $\mu = \$1,030,000$, $\phi = 0.99$, and the standard deviation of ϵ_t is 5000. The hedge fund knows that the price will eventually revert to $1,030,000 (assuming that the model is correct and, of course, this is a big assumption). It has decided to liquidate its position on day t if $P_t \geq \$1,020,000$. This will yield a profit of at least $20,000. However, if the price falls below $950,000, then it must liquidate and lose its entire $50,000 investment plus the difference between $950,000 and the price at liquidation.

In summary, the hedge fund will liquidate at the end of the first day such that the price is either above $1,020,000 or below $950,000. In the first case, it will achieve a profit of at least $20,000 and in the second case it will suffer a loss of at least $50,000. Presumably, the probability of a loss is small, and we will see how small by simulation.

Run a simulation experiment similar to the one in Sect. 2.4.2 to answer the following questions. Use 10,000 simulations.

Problem 9 *What is the expected profit?*

Problem 10 *What is the probability that the hedge fund will need to liquidate for a loss?*

Problem 11 *What is the expected waiting time until the portfolio is liquidated?*

Problem 12 *What is the expected yearly return on the $50,000 investment?*

15.6 Exercises

1. Show that (15.4) implies that $Y_{1,t-1} - \lambda Y_{2,t-1}$ is an AR(1) process with coefficient $1 + \phi_1 - \lambda\phi_2$.

2. In (15.2) and (15.3) there are no constants, so that $Y_{1,t} - \lambda Y_{2,t}$ is a stationary process with mean zero. Introduce constants into (15.2) and (15.3) and show how they determine the mean of $Y_{1,t} - \lambda Y_{2,t}$.
3. Verify that in Example 15.2 $Y_{1,t} - \lambda Y_{2,t}$ is stationary.
4. Suppose that $\boldsymbol{Y}_t = (Y_{1,t}, Y_{2,t})'$ is the bivariate AR(1) process in Example 15.2. Is $\boldsymbol{Y}_t$ stationary? (Hint: See Sect. 13.4.4.)

References

Alexander, C. (2001) *Market Models: A Guide to Financial Data Analysis*, Wiley, Chichester.

Alexander, C., Giblin, I., and Weddington, W. III (2001) *Cointegration and Asset Allocation: A New Hedge Fund*, ISMA Discussion Centre Discussion Papers in Finance 2001–2003.

Enders, W. (2004) *Applied Econometric Time Series*, 2nd ed., Wiley, New York.

Hamilton, J. D. (1994) *Time Series Analysis*, Princeton University Press, Princeton, NJ.

Johansen, S. (1991) Estimation and hypothesis testing of cointegration vectors in gaussian vector autoregressive models. *Econometrica*, **59**, 1551–1580.

Johansen, S. (1995) *Likelihood-Based Inference in Cointegrated Vector Autoregressive Models*, Oxford University Press, New York.

Johansen, S., and Juselius, K. (1990) Maximum likelihood estimation and inference on cointegration — With applications to the demand for money. *Oxford Bulletin of Economics and Statistics*, **52**, 2, 169–210.

Pfaff, B. (2006) *Analysis of Integrated and Cointegrated Time Series with R*, Springer, New York.

Phillips, P. C. B., and Ouliaris, S. (1990) Asymptotic properties of residual based tests for cointegration. *Econometrica*, **58**, 165–193.

Pole, A. (2007) *Statistical Arbitrage*, Wiley, Hoboken, NJ.

Vidyamurthy, G. (2004) *Pairs Trading*, Wiley, Hoboken, NJ.

16

Portfolio Selection

16.1 Trading Off Expected Return and Risk

How should we invest our wealth? Portfolio theory provides an answer to this question based upon two principles:

- we want to maximize the expected return; and
- we want to minimize the risk, which we define in this chapter to be the standard deviation of the return, though we may ultimately be concerned with the probabilities of large losses.

These goals are somewhat at odds because riskier assets generally have a higher expected return, since investors demand a reward for bearing risk. The difference between the expected return of a risky asset and the risk-free rate of return is called the *risk premium*. Without risk premiums, few investors would invest in risky assets.

Nonetheless, there are optimal compromises between expected return and risk. In this chapter we show how to maximize expected return subject to an upper bound on the risk, or to minimize the risk subject to a lower bound on the expected return. One key concept that we discuss is reduction of risk by diversifying the portfolio.

16.2 One Risky Asset and One Risk-Free Asset

We start with a simple example with one risky asset, which could be a portfolio, for example, a mutual fund. Assume that the expected return is 0.15 and the standard deviation of the return is 0.25. Assume that there is a *risk-free asset*, such as, a 90-day T-bill, and the risk-free rate is 6%, so the return on the risk-free asset is 6%, or 0.06. The standard deviation of the return on the

© Springer Science+Business Media New York 2015
D. Ruppert, D.S. Matteson, *Statistics and Data Analysis for Financial Engineering*, Springer Texts in Statistics,
DOI 10.1007/978-1-4939-2614-5_16

risk-free asset is 0 by definition of "risk-free." The rates and returns here are annual, though all that is necessary is that they be in the same time units.

We are faced with the problem of constructing an investment portfolio that we will hold for one time period, which is called the *holding period* and which could be a day, a month, a quarter, a year, 10 years, and so forth. At the end of the holding period we might want to readjust the portfolio, so for now we are only looking at returns over one time period. Suppose that a fraction w of our wealth is invested in the risky asset and the remaining fraction $1 - w$ is invested in the risk-free asset. Then the expected return is

$$E(R) = w(0.15) + (1 - w)(0.06) = 0.06 + 0.09w, \qquad (16.1)$$

the variance of the return is

$$\sigma_R^2 = w^2 (0.25)^2 + (1 - w)^2 (0)^2 = w^2(0.25)^2,$$

and the standard deviation of the return is

$$\sigma_R = 0.25 |w|. \qquad (16.2)$$

As will be discussed later, w is negative if the risky asset is sold short, so we have $|w|$ rather than w in (16.2).

To decide what proportion w of one's wealth to invest in the risky asset, one chooses either the expected return $E(R)$ one wants or the amount of risk σ_R with which one is willing to live. Once either $E(R)$ or σ_R is chosen, w can be determined.

Although σ is a measure of risk, a more direct measure of risk is actual monetary loss. In the next example, w is chosen to control the maximum size of the loss.

Example 16.1. Finding w to achieve a targeted value-at-risk

Suppose that a firm is planning to invest \$1,000,000 and has capital reserves that could cover a loss of \$150,000 but no more. Therefore, the firm would like to be certain that, if there is a loss, then it is no more than 15 %, that is, that R is greater than -0.15. Suppose that R is normally distributed. Then the only way to guarantee that R is greater than -0.15 with probability equal to 1 is to invest entirely in the risk-free asset. The firm might instead be more modest and require only that $P(R < -0.15)$ be small, for example, 0.01. Therefore, the firm should find the value of w such that

$$P(R < -0.15) = \Phi \left(\frac{-0.15 - (0.06 + 0.09\,w)}{0.25\,w} \right) = 0.01.$$

The solution is

$$w = \frac{-0.21}{0.25\,\Phi^{-1}(0.01) + 0.9} = 0.4264.$$

The value of $\Phi^{-1}(0.01)$ is calculated by `qnorm(0.01)` and is -2.33.

In Chap. 19, \$150,000 is called the value-at-risk (= VaR) and $1 - 0.01 = 0.99$ is called the confidence coefficient. What was done in this example is to find the portfolio that has a VaR of \$150,000 with 0.99 confidence.

We saw in Chap. 5 that the distributions of stock returns usually have much heavier tails than a normal distribution. In Chap. 19, VaR is estimated under more realistic assumptions, e.g., that the returns are t-distributed. □

More generally, if the expected returns on the risky and risk-free assets are μ_1 and μ_f and if the standard deviation of the risky asset is σ_1, then the expected return on the portfolio is $w\mu_1 + (1 - w)\mu_f$ while the standard deviation of the portfolio's return is $|w|\,\sigma_1$.

This model is simple but not as useless as it might seem at first. As discussed later, finding an optimal portfolio can be achieved in two steps:

1. finding the "optimal" portfolio of risky assets, called the "tangency portfolio," and
2. finding the appropriate mix of the risk-free asset and the tangency portfolio.

So we now know how to do the second step. What we still need to learn is how find the tangency portfolio.

16.2.1 Estimating $E(R)$ and σ_R

The value of the risk-free rate, μ_f, will be known since Treasury bill rates are published in sources providing financial information.

What should we use as the values of $E(R)$ and σ_R? If returns on the asset are assumed to be stationary, then we can take a time series of past returns and use the sample mean and standard deviation. Whether the stationarity assumption is realistic is always debatable. If we think that $E(R)$ and σ_R will be different from the past, we could subjectively adjust these estimates upward or downward according to our opinions, but we must live with the consequences if our opinions prove to be incorrect. Also, the sample mean and standard deviation are not particularly accurate and could be replaced by estimates from a factor model such as the CAPM or the Fama-French model; see Chaps. 17, 18, and 20.

Another question is how long a time series to use, that is, how far back in time one should gather data. A long series, say 10 or 20 years, will give much less variable estimates. However, if the series is not stationary but rather has slowly drifting parameters, then a shorter series (maybe 1 or 2 years) will be more representative of the future. Almost every time series of returns is nearly stationary over short enough time periods.

Even if the time series is stationary, it is likely to exhibit volatility clustering. In that case one might use a GARCH estimate of the conditional standard deviation of the return over the holding period. See Chap. 14.

16.3 Two Risky Assets

16.3.1 Risk Versus Expected Return

The mathematics of mixing risky assets is most easily understood when there are only two risky assets. This is where we start.

Suppose the two risky assets have returns R_1 and R_2 and that we mix them in proportions w and $1 - w$, respectively. The return on the portfolio is $R_p = wR_1 + (1 - w)R_2$. The expected return on the portfolio is $E(R_P) = w\mu_1 + (1 - w)\mu_2$. Let ρ_{12} be the correlation between the returns on the two risky assets. The variance of the return on the portfolio is

$$\sigma_R^2 = w^2\sigma_1^2 + (1 - w)^2\sigma_2^2 + 2w(1 - w)\rho_{12}\,\sigma_1\sigma_2. \tag{16.3}$$

Note that $\rho_{12}\sigma_1\sigma_2 = \sigma_{R_1,R_2}$.

Example 16.2. The expectation and variance of the return on a portfolio with two risky assets

Suppose that $\mu_1 = 0.14$, $\mu_2 = 0.08$, $\sigma_1 = 0.2$, $\sigma_2 = 0.15$, and $\rho_{12} = 0$. Then

$$E(R_P) = 0.08 + 0.06w,$$

and because $\rho_{12} = 0$ in this example,

$$\sigma_{R_P}^2 = (0.2)^2\, w^2 + (0.15)^2\, (1 - w)^2.$$

Using differential calculus, one can easily show that the portfolio with the minimum risk is $w = 0.045/0.125 = 0.36$. For this portfolio $E(R_P) = 0.08 + (0.06)(0.36) = 0.1016$ and $\sigma_{R_P} = \sqrt{(0.2)^2(0.36)^2 + (0.15)^2(0.64)^2} = 0.12$.

The somewhat parabolic curve[1] in Fig. 16.1 is the locus of values of $(\sigma_R, E(R))$ when $0 \le w \le 1$. The leftmost point on this locus achieves the minimum value of the risk and is called the *minimum variance portfolio*. The points on this locus that have an expected return at least as large as the minimum variance portfolio are called the *efficient frontier*. Portfolios on the efficient frontier are called *efficient portfolios* or, more precisely, *mean-variance efficient portfolios*.[2] The points labeled R_1 and R_2 correspond to $w = 1$ and $w = 0$, respectively. The other features of this figure are explained in Sect. 16.4.   ☐

---

[1] In fact, the curve would be parabolic if σ_R^2 were plotted on the x-axis instead of σ_R.

[2] When a risk-free asset is available, then the efficient portfolios are no longer those on the efficient frontier but rather are characterized by Result 16.1 ahead.

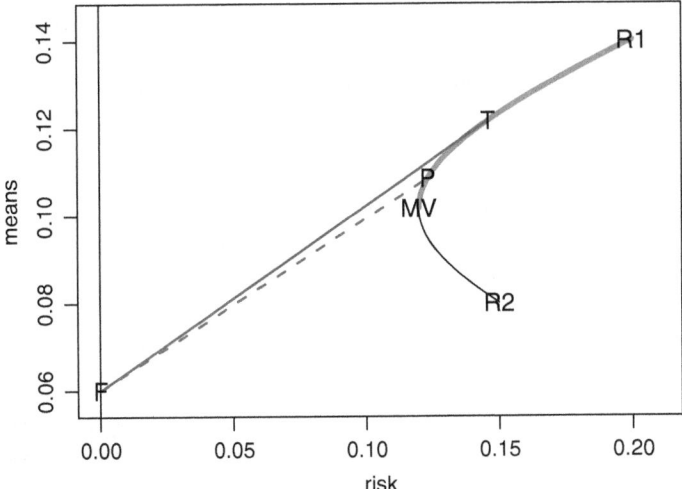

Fig. 16.1. *Expected return versus risk for Example 16.2. F = risk-free asset. T = tangency portfolio. R1 is the first risky asset. R2 is the second risky asset. MV is the minimum variance portfolio. The efficient frontier is the red curve. All points on the curve connecting R2 and R1 are attainable with $0 \leq w \leq 1$, but the ones on the black curve are suboptimal. P is a typical portfolio on the efficient frontier.*

In practice, the mean and standard deviations of the returns can be estimated as discussed in Sect. 16.2.1 and the correlation coefficient can be estimated by the sample correlation coefficient. Alternatively, in Chaps. 18 and 20 factor models are used to estimate expected returns and the covariance matrix of returns.

16.4 Combining Two Risky Assets with a Risk-Free Asset

Our ultimate goal is to find optimal portfolios combining many risky assets with a risk-free asset. However, many of the concepts needed for this task can be first understood most easily when there are only two risky assets.

16.4.1 Tangency Portfolio with Two Risky Assets

As mentioned in Sect. 16.3.1, each point on the efficient frontier in Fig. 16.1 is $(\sigma_{R_P}, E(R_p))$ for some value of w between 0 and 1. If we fix w, then we have a fixed portfolio of the two risky assets. Now let us mix that portfolio of risky assets with the risk-free asset. The point F in Fig. 16.1 gives $(\sigma_{R_P}, E(R))$ for the risk-free asset; of course, $\sigma_{R_P} = 0$ at F. The possible values of $(\sigma_{R_P}, E(R_p))$ for a portfolio consisting of the fixed portfolio of two risky assets and the risk-free asset is a line connecting the point F with a point on the efficient frontier, for example, the dashed purple line.

Notice that the solid blue line connecting F with the point labeled T lies above the dashed purple line connecting F and the typical portfolio. This means that for any value of σ_{R_P}, the solid blue line gives a higher expected return than the dashed purple line. The slope of each line is called its *Sharpe's ratio*, named after William Sharpe, whom we will meet again in Chap. 17. If $E(R_P)$ and σ_{R_P} are the expected return and standard deviation of the return on a portfolio and μ_f is the risk-free rate, then

$$\frac{E(R_P) - \mu_f}{\sigma_{R_P}} \tag{16.4}$$

is Sharpe's ratio of the portfolio. Sharpe's ratio can be thought of as a "reward-to-risk" ratio. It is the ratio of the reward quantified by the excess expected return[3] to the risk as measured by the standard deviation.

A line with a larger slope gives a higher expected return for a given level of risk, so the larger Sharpe's ratio, the better regardless of what level of risk one is willing to accept. The point T on the efficient frontier is the portfolio with the highest Sharpe's ratio. It is the optimal portfolio for the purpose of mixing with the risk-free asset. This portfolio is called the *tangency portfolio* since its line is tangent to the efficient frontier.

Result 16.1 *The optimal or efficient portfolios mix the tangency portfolio with the risk-free asset. Each efficient portfolio has two properties:*

- *it has a higher expected return than any other portfolio with the same or smaller risk, and*
- *it has a smaller risk than any other portfolio with the same or higher expected return.*

Thus we can only improve (reduce) the risk of an efficient portfolio by accepting a worse (smaller) expected return, and we can only improve (increase) the expected return of an efficient portfolio by accepting worse (higher) risk.

Note that all efficient portfolios use the same mix of the two risky assets, namely, the tangency portfolio. Only the proportion allocated to the tangency portfolio and the proportion allocated to the risk-free asset vary.

Given the importance of the tangency portfolio, you may be wondering "how do we find it?" Again, let μ_1, μ_2, and μ_f be the expected returns on the two risky assets and the return on the risk-free asset. Let σ_1 and σ_2 be the standard deviations of the returns on the two risky assets and let ρ_{12} be the correlation between the returns on the risky assets.

---

[3] Here "excess" means in excess of the risk-free rate.

Define $V_1 = \mu_1 - \mu_f$ and $V_2 = \mu_2 - \mu_f$, the excess expected returns. Then the tangency portfolio uses weight

$$w_T = \frac{V_1\sigma_2^2 - V_2\rho_{12}\,\sigma_1\sigma_2}{V_1\sigma_2^2 + V_2\sigma_1^2 - (V_1 + V_2)\rho_{12}\,\sigma_1\sigma_2} \tag{16.5}$$

for the first risky asset and weight $(1 - w_T)$ for the second.

Let R_T, $E(R_T)$, and σ_T be the return, expected return, and standard deviation of the return on the tangency portfolio. Then $E(R_T)$ and σ_T can be found by first finding w_T using (16.5) and then using the formulas

$$E(R_T) = w_T\mu_1 + (1 - w_T)\mu_2$$

and

$$\sigma_T = \sqrt{w_T^2\sigma_1^2 + (1 - w_T)^2\sigma_2^2 + 2w_T(1 - w_T)\rho_{12}\sigma_1\sigma_2}\,.$$

Example 16.3. The tangency portfolio with two risky assets

Suppose as before that $\mu_1 = 0.14$, $\mu_2 = 0.08$, $\sigma_1 = 0.2$, $\sigma_2 = 0.15$, and $\rho_{12} = 0$. Suppose as well that $\mu_f = 0.06$. Then $V_1 = 0.14 - 0.06 = 0.08$ and $V_2 = 0.08 - 0.06 = 0.02$. Plugging these values into formula (16.5), we get $w_T = 0.693$ and $1 - w_t = 0.307$. Therefore,

$$E(R_T) = (0.693)(0.14) + (0.307)(0.08) = 0.122,$$

and

$$\sigma_T = \sqrt{(0.693)^2(0.2)^2 + (0.307)^2(0.15)^2} = 0.146.$$

□

16.4.2 Combining the Tangency Portfolio with the Risk-Free Asset

Let R_p be the return on the portfolio that allocates a fraction ω of the investment to the tangency portfolio and $1 - \omega$ to the risk-free asset. Then $R_p = \omega R_T + (1 - \omega)\mu_f = \mu_f + \omega(R_T - R_f)$, so that

$$E(R_p) = \mu_f + \omega\{E(R_T) - \mu_f\} \quad \text{and} \quad \sigma_{R_p} = \omega\sigma_T.$$

Example 16.4. (Continuation of Example 16.2 and 16.3)

In this example, we will find the optimal investment with $\sigma_{R_p} = 0.05$.

The maximum expected return with $\sigma_{R_p} = 0.05$ mixes the tangency portfolio and the risk-free asset such that $\sigma_{R_p} = 0.05$. Since $\sigma_T = 0.146$, we have that $0.05 = \sigma_{R_p} = w\sigma_T = 0.146\,w$, so that $w = 0.05/0.146 = 0.343$ and $1 - w = 0.657$.

So 65.7 % of the portfolio should be in the risk-free asset, and 34.3 % should be in the tangency portfolio. Thus $(0.343)(69.3\%) = 23.7\%$ should be in the first risky asset and $(0.343)(30.7\%) = 10.5\%$ should be in the second risky asset. The total is not quite 100 % because of rounding. $\square$

Example 16.5. (Continuation of Examples 16.2–16.4)

Now suppose that you want a 10 % expected return. In this example we will compare

- the best portfolio of only risky assets, and
- The best portfolio of the risky assets and the risk-free asset.

The best portfolio of only risky assets uses w solving $0.1 = w(0.14) + (1-w)$ (0.08), which implies that $w = 1/3$. This is the *only* portfolio of risky assets with $E(R_p) = 0.1$, so by default it is best. Then

$$\sigma_{R_P} = \sqrt{w^2(0.2)^2 + (1 - w)^2(0.15)^2} = \sqrt{(1/9)(0.2)^2 + 4/9(0.15)^2} = 0.120.$$

The best portfolio of the two risky assets and the risk-free asset can be found as follows. First, $0.1 = E(R) = \mu_f + w\{E(R_T) - \mu_f\} = 0.06 + 0.062\,w = 0.06 + 0.425\,\sigma_R$, since $\sigma_{R_P} = w\sigma_T$ or $w = \sigma_{R_P}/\sigma_T = \sigma_{R_P}/0.146$. This implies that $\sigma_{R_P} = 0.04/0.425 = 0.094$ and $w = 0.04/0.062 = 0.645$. So combining the risk-free asset with the two risky assets reduces σ_{R_P} from 0.120 to 0.094 while maintaining $E(R_p)$ at 0.1. The reduction in risk is $(0.120 - 0.094)/0.094 = 28\%$, which is substantial. $\square$

16.4.3 Effect of ρ_{12}

Positive correlation between the two risky assets increases risk. With positive correlation, the two assets tend to move together which increases the volatility of the portfolio. Conversely, negative correlation is beneficial since it decreases risk. If the assets are negatively correlated, a negative return of one tends to occur with a positive return of the other so the volatility of the portfolio decreases. Figure 16.2 shows the efficient frontier and tangency portfolio when $\mu_1 = 0.14$, $\mu_2 = 0.09$, $\sigma_1 = 0.2$, $\sigma_2 = 0.15$, and $\mu_f = 0.03$. The value of ρ_{12} is varied from 0.5 to -0.7. Notice that Sharpe's ratio of the tangency portfolio returns increases as ρ_{12} decreases. This means that when ρ_{12} is small, then efficient portfolios have less risk for a given expected return compared to when ρ_{12} is large.

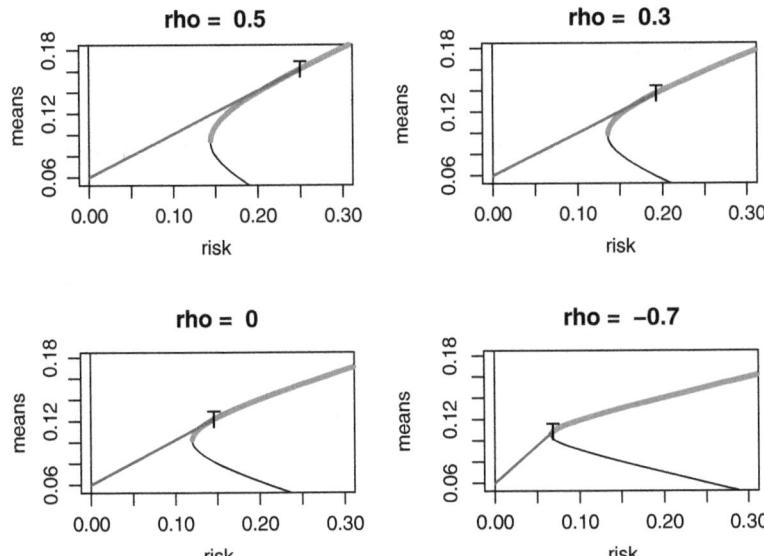

Fig. 16.2. *Efficient frontier (red) and tangency portfolio (T) when $\mu_1 = 0.14$, $\mu_2 = 0.09$, $\sigma_1 = 0.2$, $\sigma_2 = 0.15$, and $\mu_f = 0.03$. The value of ρ_{12} is varied from 0.5 to -0.7.*

16.5 Selling Short

Often some of the weights in an efficient portfolio are negative. A negative weight on an asset means that this asset is sold short. *Selling short* is a way to profit if a stock price goes *down*. To sell a stock short, one sells the stock without owning it. The stock must be borrowed from a broker or another customer of the broker. At a later point in time, one buys the stock and gives it back to the lender. This closes the short position.

Suppose a stock is selling at \$25/share and you sell 100 shares short. This gives you \$2500. If the stock goes down to \$17/share, you can buy the 100 shares for \$1700 and close out your short position. You made a profit of \$800 (ignoring transaction costs) because the stock went down 8 points. If the stock had gone up, then you would have had a loss.

Suppose now that you have \$100 and there are two risky assets. With your money you could buy \$150 worth of risky asset 1 and sell \$50 short of risky asset 2. The net cost would be exactly \$100. If R_1 and R_2 are the returns on risky assets 1 and 2, then the return on your portfolio would be

$$\frac{3}{2}R_1 + \left(-\frac{1}{2}\right)R_2.$$

Your portfolio weights are $w_1 = 3/2$ and $w_2 = -1/2$. Thus, you hope that risky asset 1 rises in price and risky asset 2 falls in price. Here, again, we have ignored transaction costs.

If one sells a stock short, one is said to have a *short position* in that stock, and owning the stock is called a *long position*.

16.6 Risk-Efficient Portfolios with N Risky Assets

In this section, we use quadratic programming to find efficient portfolios with an arbitrary number of assets. An advantage of quadratic programming is that it allows one to impose constraints such as limiting short sales. With no constraints on the allocation vector $\boldsymbol{w}$, analytic formulas for the tangency portfolio can be derived using Lagrange multipliers, but this approach does not generalize to constrained $\boldsymbol{w}$.

Assume that we have N risky assets and that the return on the ith risky asset is R_i and has expected value μ_i. Define

$$\boldsymbol{R} = \begin{pmatrix} R_1 \\ \vdots \\ R_N \end{pmatrix}$$

to be the random vector of returns,

$$E(\boldsymbol{R}) = \boldsymbol{\mu} = \begin{pmatrix} \mu_1 \\ \vdots \\ \mu_N \end{pmatrix},$$

and $\boldsymbol{\Sigma}$ to be the covariance matrix of $\boldsymbol{R}$.

Let

$$\boldsymbol{w} = \begin{pmatrix} w_1 \\ \vdots \\ w_N \end{pmatrix}$$

be a vector of portfolio weights so that $w_1 + \cdots + w_N = \mathbf{1}^\mathsf{T}\boldsymbol{w} = 1$, where

$$\mathbf{1} = \begin{pmatrix} 1 \\ \vdots \\ 1 \end{pmatrix}$$

is a column of N ones. The expected return on the portfolio is

$$\sum_{i=1}^{N} w_i \mu_i = \boldsymbol{\omega}^\mathsf{T} \boldsymbol{\mu}. \tag{16.6}$$

Suppose there is a target value, μ_P, of the expected return on the portfolio. When $N = 2$, the target expected returns is achieved by only one portfolio and its w_1-value solves $\mu_P = w_1\mu_1 + w_2\mu_2 = \mu_2 + w_1(\mu_1 - \mu_2)$. For $N \geq 3$, there will be an infinite number of portfolios achieving the target μ_P. The one with the smallest variance is called the "efficient" portfolio. Our goal is to find the efficient portfolio.

The variance of the return on the portfolio with weights w is

$$w^{\mathsf{T}} \Sigma w. \tag{16.7}$$

Thus, given a target μ_P, the efficient portfolio minimizes (16.7) subject to

$$w^{\mathsf{T}} \mu = \mu_P \tag{16.8}$$

and

$$w^{\mathsf{T}} 1 = 1. \tag{16.9}$$

Quadratic programming is used to minimize a quadratic objective function subject to linear constraints. In applications to portfolio optimization, the objective function is the variance of the portfolio return. The objective function is a function of N variables, such as the weights of N assets, that are denoted by an $N \times 1$ vector x. Suppose that the quadratic objective function to be minimized is

$$\frac{1}{2} x^{\mathsf{T}} D x - d^{\mathsf{T}} x, \tag{16.10}$$

where D is an $N \times N$ matrix and d is an $N \times 1$ vector. The factor of $1/2$ is not essential but is used here to keep our notation consistent with R. There are two types of linear constraints on x, inequality and equality constraints. The linear inequality constraints are

$$A_{\text{neq}}^{\mathsf{T}} x \geq b_{\text{neq}}, \tag{16.11}$$

where A_{neq} is an $m \times N$ matrix, b_{neq} is an $m \times 1$ vector, and m is the number of inequality constraints. The equality constraints are

$$A_{\text{eq}}^{\mathsf{T}} x = b_{\text{eq}}, \tag{16.12}$$

where A_{eq} is an $n \times N$ matrix, b_{eq} is an $n \times 1$ vector, and n is the number of equality constraints. Quadratic programming minimizes the quadratic objective function (16.10) subject to linear inequality constraints (16.11) and linear equality constraints (16.12).

To apply quadratic programming to find an efficient portfolio, we use $x = w$, $D = 2\Sigma$, and d equal to an $N \times 1$ vector of zeros so that (16.10) is $w^{\mathsf{T}} \Sigma w$, the return variance of the portfolio. There are two equality constraints, one that the weights sum to 1 and the other that the portfolio return is a specified target μ_P. Therefore, we define

$$A_{\text{eq}}^{\mathsf{T}} = \begin{pmatrix} 1^{\mathsf{T}} \\ \mu^{\mathsf{T}} \end{pmatrix}$$

and

$$b_{eq} = \begin{pmatrix} 1 \\ \mu_P \end{pmatrix},$$

so that (16.12) becomes

$$\begin{pmatrix} \mathbf{1}^\mathsf{T} w \\ \mu^\mathsf{T} w \end{pmatrix} = \begin{pmatrix} 1 \\ \mu_P \end{pmatrix},$$

which is the same as constraints (16.8) and (16.9). So far, inequality constraints have not been used.

Investors often wish to impose additional inequality constraints. If an investor cannot or does not wish to sell short, then the constraint

$$w \geq 0$$

can be used. Here $\mathbf{0}$ is a vector of N zeros. In this case A_{neq} is the $N \times N$ identical matrix and $b_{neq} = \mathbf{0}$.

To avoid concentrating the portfolio in just one or a few stocks, an investor may wish to constrain the portfolio so that no w_i exceeds a bound λ, for example, $\lambda = 1/4$ means that no more than $1/4$ of the portfolio can be in any single stock. In this case, $w \leq \lambda \mathbf{1}$ or equivalently $-w \geq -\lambda \mathbf{1}$, so that A_{neq} is minus the $N \times N$ identity matrix and $b_{neq} = -\lambda \mathbf{1}$. One can combine these constraints with those that prohibit short selling.

To find the efficient frontier, one uses a grid of values of μ_P and finds the corresponding efficient portfolios. For each portfolio, σ_P^2, which is the minimized value of the objective function, can be calculated. Then one can find the minimum variance portfolio by finding the portfolio with the smallest value of the σ_P^2. The efficient frontier is the set of efficient portfolios with expected return above the expected return of the minimum variance portfolio. One can also compute Sharpe's ratio for each portfolio on the efficient frontier and the tangency portfolio is the one maximizing Sharpe's ratio.

Example 16.6. Finding the efficient frontier, tangency portfolio, and minimum variance portfolio using quadratic programming

The following R program uses the returns on three stocks, GE, IBM, and Mobil, in the CRSPday data set in the Ecdat package. The function solve.QP() in the quadprog package is used for quadratic programming. solve.QP() combines A_{eq}^T and A_{neq}^T into a single matrix Amat by stacking A_{eq}^T on top of A_{neq}^T. The parameter meq is the number of rows of A_{eq}^T. b_{eq} and b_{neq} are handled analogously. In this example, there are no inequality constraints, so A_{neq}^T and b_{neq} are not needed, but they are used in the next example.

The efficient portfolio is found for each of 300 target values of μ_P between 0.05 and 0.14. For each portfolio, Sharpe's ratio is found at line 28 and the logical vector ind at line 29 indicates which portfolio is the tangency portfolio maximizing Sharpe's ratio. Similarly, ind2 at line 34 indicates the minimum variance portfolio. Also, ind3 at line 36 indicates the points on the efficient frontier. It is assumed that the risk-free rate is 1.3 %/year; see line 26.

```
1  llibrary(Ecdat)
2  library(quadprog)
3  data(CRSPday)
4  R = 100*CRSPday[ ,4:6]   # convert to percentages
5  mean_vect = apply(R, 2 ,mean)
6  cov_mat = cov(R)
7  sd_vect = sqrt(diag(cov_mat))
8  Amat = cbind(rep(1, 3), mean_vect) # set the constraints matrix
9  muP = seq(0.05, 0.14, length = 300)  # target portfolio means
10 # for the expect portfolio return
11 sdP = muP # set up storage for std dev's of portfolio returns
12 weights = matrix(0, nrow = 300, ncol = 3) # storage for weights
13 for (i in 1:length(muP))  # find the optimal portfolios
14 {
15   bvec = c(1, muP[i])  # constraint vector
16   result =
17      solve.QP(Dmat = 2 * cov_mat, dvec = rep(0, 3),
18        Amat = Amat, bvec = bvec, meq = 2)
19   sdP[i] = sqrt(result$value)
20   weights[i,] = result$solution
21 }
22 pdf("quad_prog_plot.pdf", width = 6, height = 5)
23 plot(sdP, muP, type = "l", xlim = c(0, 2.5),
24    ylim = c(0, 0.15), lty = 3)  # plot efficient frontier (and
25              # inefficient portfolios below the min var portfolio)
26 mufree = 1.3 / 253 # input value of risk-free interest rate
27 points(0, mufree, cex = 4, pch = "*") # show risk-free asset
28 sharpe = (muP - mufree) / sdP # compute Sharpe's ratios
29 ind = (sharpe == max(sharpe)) # Find maximum Sharpe's ratio
30 weights[ind, ] #  print the weights of the tangency portfolio
31 lines(c(0, 2), mufree + c(0, 2) * (muP[ind] - mufree) / sdP[ind],
32    lwd = 4, lty = 1, col = "blue")  # show line of optimal portfolios
33 points(sdP[ind], muP[ind], cex = 4, pch = "*") #  tangency portfolio
34 ind2 = (sdP == min(sdP)) # find minimum variance portfolio
35 points(sdP[ind2], muP[ind2], cex = 2, pch = "+") # min var portfolio
36 ind3 = (muP > muP[ind2])
37 lines(sdP[ind3], muP[ind3], type = "l", xlim = c(0, 0.25),
38    ylim = c(0, 0.3), lwd = 3, col = "red")  # plot efficient frontier
39 text(sd_vect[1], mean_vect[1], "GE", cex = 1.15)
40 text(sd_vect[2], mean_vect[2], "IBM", cex = 1.15)
41 text(sd_vect[3], mean_vect[3], "Mobil", cex = 1.15)
42 graphics.off()
```

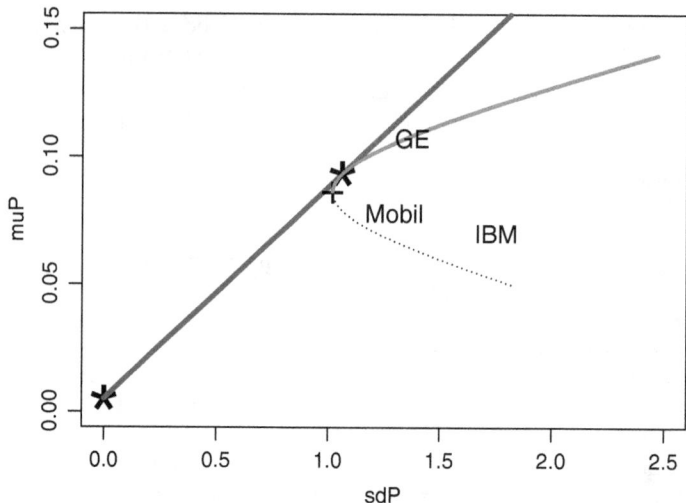

Fig. 16.3. *Efficient frontier (solid), line of efficient portfolios (dashed) connecting the risk-free asset and tangency portfolio (asterisks), and the minimum variance portfolio (plus) with three stocks (GE, IBM, and Mobil). The three stocks are also shown on reward-risk space.*

The plot produced by this program is Fig. 16.3. The program prints the weights of the tangency portfolio, which are

```
> weights[ind,] # Find tangency portfolio
[1] 0.5512 0.0844 0.3645
```

☐

Example 16.7. Finding the efficient frontier, tangency portfolio, and minimum variance portfolio with no short selling using quadratic programming

In this example, Example 16.6 is modified so that short sales are not allowed. Only three lines of code need to be changed. When short sales are prohibited, the target expected return on the portfolio must lie between the smallest and largest expected returns on the stocks. To prevent numerical errors, the target expected returns will start 0.0001 above the smallest expected stock return and end 0.0001 below the largest expected stock return. This is enforced by the following change:

```
muP = seq(min(mean_vect) + 0.0001, max(mean_vect) - 0.0001,
    length = 300)
```

To enforce no short sales, an A_{neq} matrix is needed and is set equal to a 3×3 identity matrix:

```
Amat = cbind(rep(1, 3), mean_vect, diag(1, nrow = 3))
```

Also, b_{neq} is set equal to a three-dimensional vector of zeros:

```
bvec = c(1, muP[i], rep(0, 3))
```

The new plot is shown in Fig. 16.4. Since the tangency portfolio in Example 16.6 had all weights positive, the tangency portfolio is unchanged by the prohibition of short sales. The efficient frontier is changed since without short sales, it is impossible to have expected returns greater than the expected return of GE, the stock with the highest expected return. In contrast, when short sales are allowed, there is no upper bound on the expected return (or on the risk). In Fig. 16.4 the red curve is the entire efficient frontier, but in Fig. 16.3 the efficient frontier is the red curve extended to $(+\infty, +\infty)$.   □

16.7 Resampling and Efficient Portfolios

The theory of portfolio optimization assumes that the expected returns and the covariance matrix of the returns is known. In practice, one must replace these quantities with estimates as in the previous examples. However, the effects of estimation error, especially with smaller values of N, can result in portfolios that only appear efficient. This problem will be investigated in this section using the bootstrap to quantify the effects of estimation error.

Example 16.8. The global asset allocation problem

One application of optimal portfolio selection is allocation of capital to different market segments. For example, Michaud (1998) discusses a global asset allocation problem where capital must be allocated to "U.S. stocks and government/corporate bonds, euros, and the Canadian, French, German, Japanese, and U.K. equity markets." Here we look at a similar example where we allocate capital to the equity markets of 10 different countries. Monthly returns for these markets were calculated from MSCI Hong Kong, MSCI Singapore, MSCI Brazil, MSCI Argentina, MSCI UK, MSCI Germany, MSCI Canada, MSCI France, MSCI Japan, and the S&P 500. "MSCI" means "Morgan Stanley Capital Index." The data are from January 1988 to January 2002, inclusive, so there are 169 months of data.

Assume that we want to find the tangency portfolio that maximizes Sharpe's ratio. The tangency portfolio was estimated using sample means and the sample covariance as in Example 16.6, and its Sharpe's ratio is estimated to be 0.3681. However, we should suspect that 0.3681 must be an overestimate since this portfolio only maximizes Sharpe's ratio using estimated parameters,

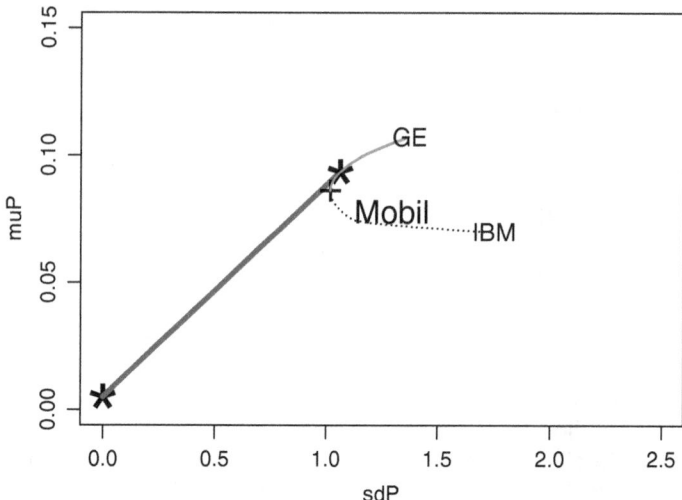

Fig. 16.4. *Efficient frontier (solid), line of efficient portfolios (dashed) connecting the risk-free asset and tangency portfolio (asterisks), and the minimum variance portfolio (plus) with three stocks (GE, IBM, and Mobil) with short sales prohibited.*

not the true means and covariance matrix. To evaluate the possible amount of overestimation, one can use the bootstrap. As discussed in Chap. 6, in the bootstrap simulation experiment, the sample is the "true population" so that the sample mean and covariance matrix are the "true parameters," and the resamples mimic the sampling process. Actual Sharpe's ratios are calculated with the sample means and covariance matrix, while estimated Sharpe's ratio use the means and covariance matrix of the resamples.

First, 250 resamples were taken and for each the tangency portfolio was estimated. Resampling was done by sampling rows of the data matrix as discussed in Sect. 7.11. For each of the 250 tangency portfolios estimated from the resamples, the actual and estimated Sharpe's ratios were calculated. Boxplots of the 250 actual and 250 estimated Sharpe's ratios of the estimated tangency portfolios are in Fig. 16.5a. "Estimated" means calculated from the resample and "true" means calculated from the sample. In this figure, there is a dashed horizontal line at height 0.3681, the actual Sharpe's ratio of the true tangency portfolio. One can see that all 250 estimated tangency portfolios have actual Sharpe's ratios below this value, as they must since the actual Sharpe's ratio is maximized by the true tangency portfolio, not the estimated tangency portfolios.

From the boxplot on the right-hand side of (a), one can see that the estimated Sharpe's ratios overestimate not only the actual Sharpe's ratios of the estimated tangency portfolios but also the somewhat larger (and unattainable) actual Sharpe's ratio of the true (but unknowable) tangency portfolio.   □

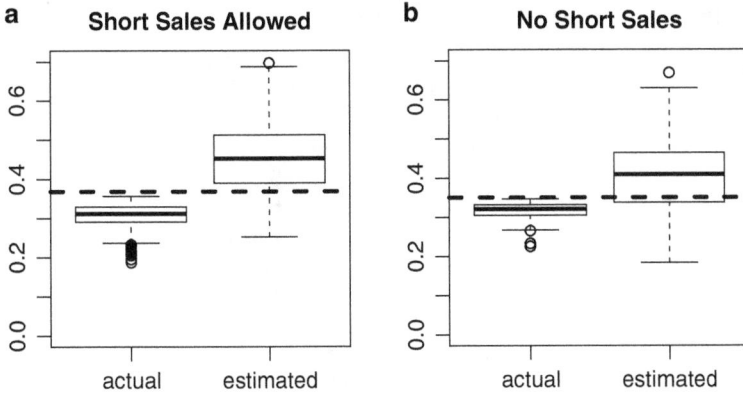

Fig. 16.5. *Bootstrapping estimation of the tangency portfolio and its Sharpe's ratio. (a) Short sales allowed. The left-hand boxplot is of the actual Sharpe's ratios of the estimated tangency portfolios for 250 resamples. The right-hand boxplot contains the estimated Sharpe's ratios for these portfolios. The horizontal dashed line indicates Sharpe's ratio of the true tangency portfolio. (b) Same as (a) but with short sales not allowed.*

There are several ways to alleviate the problems caused by estimation error when attempting to find a tangency portfolio. One can try to find more accurate estimators; the factor models of Chap. 18 and Bayes estimators of Chap. 20 (see especially Example 20.12) do this. Another possibility is to restrict short sales.

Portfolios with short sales aggressively attempt to maximize Sharpe's ratio by selling short those stocks with the smallest estimated mean returns and having large long positions in those stocks with the highest estimated mean returns. The weakness with this approach is that it is particularly sensitive to estimation error. Unfortunately, expected returns are estimated with relatively large uncertainty. This problem can be seen in Table 16.1, which has 95 % confidence intervals for the mean returns. The percentile method is used for the confidence intervals, so the endpoints are the 2.5 and 97.5 bootstrap percentiles. Notice for Singapore and Japan, the confidence intervals include both positive and negative values. In the table, the returns are expressed as percentage returns.

Example 16.9. The global asset allocation problem: short sales prohibited

This example repeats the bootstrap experimentation of Example 16.8 with short sales prohibited by using inequality constraints such as in Example 16.7. With short sales not allowed, the actual Sharpe's ratio of the true tangency portfolio is 0.3503, which is only slightly less than when short sales are allowed.

Table 16.1. *95 % percentile-method bootstrap confidence intervals for the mean returns of the 10 countries.*

Country	2.5 %	97.5 %
Hong Kong	0.186	2.709
Singapore	−0.229	2.003
Brazil	0.232	5.136
Argentina	0.196	6.548
UK	0.071	1.530
Germany	0.120	1.769
Canada	0.062	1.580
France	0.243	2.028
Japan	−0.884	0.874
U.S.	0.636	1.690

Boxplots of actual and apparent Sharpe's ratios are in Fig. 16.5b. Comparing Fig. 16.5a and b, one sees that prohibiting short sales has two beneficial effects—Sharpe's ratios actually achieved are slightly higher with no short sales allowed compared to having no constraints on short sales. In fact, the mean of the 250 actual Sharpe's ratios is 0.3060 with short sales allowed and 0.3169 with short sales prohibited. Moreover, the overestimation of Sharpe's ratio is reduced by prohibiting short sales—the mean apparent Sharpe's ratio is 0.4524 [with estimation error $(0.4524 - 0.3681) = 0.0843$] with short sales allowed but only 0.4038 [with estimation error $(0.4038 - 0.3503) = 0.0535$] with short sales prohibited. However, these effects, though positive, are only modest and do not entirely solve the problem of overestimation of Sharpe's ratio. □

Example 16.10. The global asset allocation problem: Shrinkage estimation and short sales prohibited

In Example 16.9, we saw that prohibiting short sales can increase Sharpe's ratio of the estimated tangency portfolio, but the improvement is only modest. Further improvement requires more accurate estimation of the mean vector or the covariance matrix of the returns.

This example investigates possible improvements from shrinking the 10 estimated means toward each other. Specifically, if $\overline{Y}_i$ is the sample mean of the ith country, $\overline{Y} = (\sum_{i=1}^{10} \overline{Y}_i)/10$ is the grand mean (mean of the means), and α is a tuning parameter between 0 and 1, then the estimated mean return for the ith country is

$$\widehat{\mu}_i = \alpha \overline{Y}_i + (1 - \alpha)\overline{Y}. \tag{16.13}$$

The purpose of shrinkage is to reduce the variance of the estimator, though the reduced variance comes at the expense of some bias. Since it is the mean of

10 means, $\overline{Y}$ is much less variable than any of $\overline{Y}_1, \ldots, \overline{Y}_{10}$. Therefore, $\mathrm{Var}(\widehat{\mu}_i)$ decreases as α is decreased toward 0. However,

$$E(\widehat{\mu}_i) = \alpha\mu_i + \frac{1-\alpha}{10}\sum_{i=1}^{10}\mu_i \qquad (16.14)$$

so that, for any $\alpha \neq 1$, $\widehat{\mu}_i$ is biased, except under the very likely circumstance that $\mu_1 = \cdots = \mu_{10}$. The parameter α controls the bias–variance tradeoff. In this example, $\alpha = 1/2$ will be used for illustration and short sales will not be allowed.

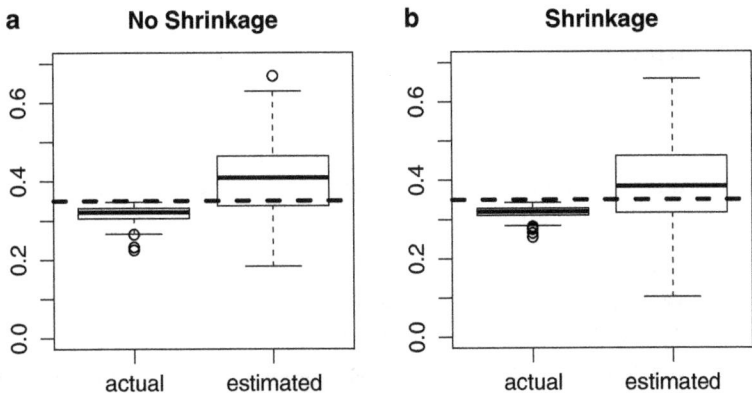

Fig. 16.6. *Bootstrapping estimation of the tangency portfolio and its Sharpe's ratio. Short sales not allowed. (a) No shrinkage. The left-hand boxplot is of the actual Sharpe's ratios of the estimated tangency portfolios for 250 resamples. The right-hand boxplot contains the estimated Sharpe's ratios for these portfolios. The horizontal dashed line indicates Sharpe's ratio of the true tangency portfolio. (b) Same as (a) but with shrinkage.*

Figure 16.6 compares the performance of shrinkage versus no shrinkage. Panel (a) contains the boxplots that we saw in panel (b) of Fig. 16.5 where $\alpha = 1$. Panel (b) has the boxplots when the tangency portfolio is estimated using $\alpha = 1/2$. Compared to panel (a), in panel (b) the actual Sharpe's ratios are somewhat closer to the dashed line indicating Sharpe's ratio of the true tangency portfolio; the means of the actual Sharpe's ratios are 0.317 and 0.318 with and without shrinkage, respectively. These values should be compared with the Sharpe ratio of the true (but unknown) tangency portfolio of 0.34.

Moreover, the estimated Sharpe's ratios in (b) are smaller and closer to the true Sharpe's ratios, so there is less overoptimization—shrinkage has helped in two ways. The mean estimated Sharpe's ratios are 0.390 and 0.404 with and without shrinkage.

The next step might be selection of α to optimize performance of shrinkage estimation. Doing this need not be difficult, since different values of α can be compared by bootstrapping. □

There are other methods for improving the estimation of the mean vector and estimation of the covariance matrix can be improved as well, for example, by using the factor models in Chap. 18 or Bayesian estimation as in Chap. 20. Moreover, one need not focus on the tangency portfolio but could, for example, estimate the minimum variance portfolio. Whatever the focus of estimation, the bootstrap can be used to compare various strategies for improving the estimation of the optimal portfolio.

16.8 Utility

Economists generally do not model economic decisions in terms of the mean and variance of the return but rather by using a *utility function*. The utility of an amount X of money is said to be $U(X)$ where the utility function U generally has the properties:

1. $U(0) = 0$;
2. U is strictly increasing;
3. the first derivative $U'(X)$ is strictly decreasing.

Assumption 1 is not necessary but is reasonable and states that the utility of 0 dollars is 0. Assumption 2 merely states that more money is better than less. Assumption 3 implies that the more money we have the less we value an extra dollar and is called *risk aversion*. Assumption 3 implies that we would decline a bet that pays $\pm\Delta$ with equal probabilities. In fact, Assumption 3 implies that we would decline any bet with a payoff that is symmetrically distributed about 0, because the expected utility of our wealth would be reduced if we accepted the bet. Mathematically, Assumption 3 implies that U is strictly concave. If the second derivative U'' exists then Assumption 3 is equivalent to the assumption that $U''(X) < 0$ for all X.

It is assumed that a rational person will make investment decisions so as to maximize

$$E\{U(X)\} = E[U\{X_0(1 + R)\}] \qquad (16.15)$$

where X is that person's final wealth, X_0 is the person's initial wealth, and R is the return from the investments. In economics this is almost a part of the definition of a rational person, with another component of the definition being that a rational person will update probabilities using Bayes' law (see Chap. 20). Each individual is assumed to have his or her own utility function and two different rational people may make different decisions because they have different utility functions.

How different are mean-variance efficient portfolios and portfolios that maximize expected utility? In the case that returns are normally distributed, this question can be answered.

Result 16.2 *If returns on all portfolios are normally distributed and if U satisfies Assumptions 3, then the portfolio that maximizes expected utility is on the efficient frontier.*

So, if one chose a portfolio to maximize expected utility, then a mean-variance efficient portfolio would be selected. Exactly which portfolio on the efficient frontier one chooses would depend on one's utility function.

Proof of Result 16.2

This result can be proven by proving the following fact: if R_1 and R_2 are normally distributed with the same means and with standard deviations σ_1 and σ_2 such that $\sigma_1 < \sigma_2$, then $E\{U(R_1)\} > E\{U(R_2)\}$. We will show that this follows from Jensen's inequality which states that if U is concave function and X is any random variable, then $E\{U(X)\} \leq U\{E(X)\}$. The inequality is strict if U is strictly convex and X is nondegenerate.[4]

Let $X = R_1 + e$ where e is independent of R_1 and normally distributed with mean 0 and variance $\sigma_2^2 - \sigma_1^2$. Then X has the same distribution as R_2, e is nondegenerate, and, using the law of iterated expectations and then Jensen's inequality and Assumption 3, we have

$$E\{U(R_2)\} = E\{U(X)\} = E[E\{U(X)|R_1\}] < E[\{U\{E(X|R_1)\}] = E\{U(R_1)\}, \tag{16.16}$$

since $E(X|R_1) = R_1$.   □

The assumption in Result 16.2 that the returns are normally distributed can be weakened to the more realistic assumption that the vector of returns on the assets is a multivariate scale mixture, e.g., has a multivariate t-distribution. To prove this extension, one conditions on the mixing variable so that the returns have a conditional multivariate normal distribution. Then (16.16) holds conditionally for all values of the mixing variable and therefore holds unconditionally.

A common class of utility functions is

$$U(x; \lambda) = 1 - \exp(-\lambda x), \tag{16.17}$$

where $\lambda > 0$ determines the amount of risk aversion. Note that $U'(x; \lambda) = \lambda \exp(\lambda x)$ and $U''(x; \lambda) = -\lambda^2 \exp(-\lambda x)$ (differentiation is with respect to x).

---

[4] Jensen's inequality is usually stated for convex functions with the inequality reversed. If U is concave then $-U$ is convex so that the two forms of Jensen's inequality are equivalent. A random variable X is degenerate if there is a constant a such that $P(X = a) = 1$. Otherwise, it is nondegenerate.

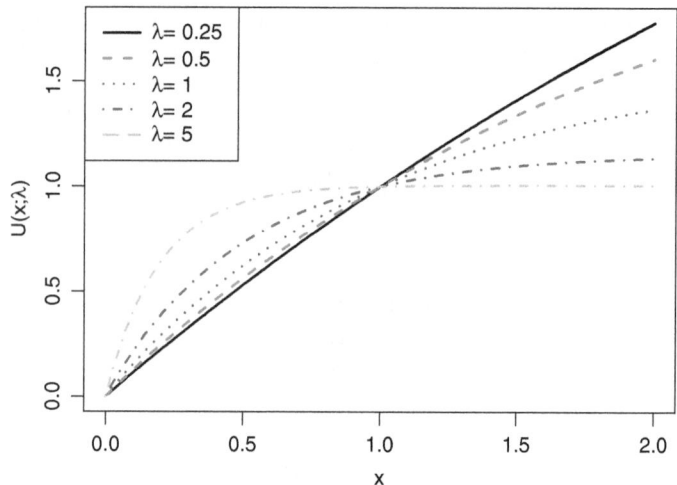

Fig. 16.7. *The utility functions* $\widetilde{U}(x;\lambda) = U(x;\lambda)/U(0,\lambda)$ *where* $U(x;\lambda) = 1 - \exp(-\lambda x)$.

Thus, $U''(x;\lambda)$ is negative for all x so Assumption 3 is met; it is easy to see that Assumptions 1 and 2 also hold. As $x \to \infty$, $U(x:\lambda) \to 1$.

Multiplying a utility function by a positive constant will not affect which decision maximizes utility and can standardize utility functions to make them more comparable. In Fig. 16.7, $\widetilde{U}(x;\lambda) := U(x;\lambda)/U(0;\lambda)$ is plotted for $\lambda = 0.25$, 0.5, 1, 2 and 5. Since $\widetilde{U}(1;\lambda) = 1$ for all λ, these utility functions have been standardized so that the utility corresponding to a return of 0 is always 1. Stated differently, a return equal to 0 has the same utility for all degrees of risk aversion.

Adding a constant to the utility function does not effect the optimal decision, so one could work with the slightly simpler utility function $-\exp(-\lambda x)$ instead of $U(x;\lambda)$ or $\widetilde{U}(x;\lambda)$.

When the utility function is given by (16.17) and R is normally distributed, then (16.15) becomes

$$1 - \exp\left[-\lambda X_0\{1 + E(R)\} + (\lambda X_0)^2 \frac{\text{var}(R)}{2}\right] \tag{16.18}$$

by properties of the lognormal distribution; see Appendix A.9.4. For given values of λ and X_0, the expected utility is maximized by maximizing

$$E(R) - (\lambda X_0) \frac{\text{var}(R)}{2}. \tag{16.19}$$

Therefore, using the notation of Sect. 16.6, one selects the allocation vector $\boldsymbol{w}$ of the portfolio to maximize

$$\boldsymbol{w}^\mathsf{T}\boldsymbol{\mu} - (\lambda X_0)\frac{\boldsymbol{w}^\mathsf{T}\boldsymbol{\Sigma}\boldsymbol{w}}{2} \tag{16.20}$$

subject to $\boldsymbol{w}^\mathsf{T}\boldsymbol{1} = 1$.

Maximizing (16.20) subject to linear constraints is a quadratic programming problem. As $\lambda \to 0$, the expected return and standard deviation of the return converge to ∞. Conversely, as $\lambda \to \infty$, the solution converges to the minimum variance portfolio. Therefore, as λ is varied from ∞ to 0, one finds all of the portfolios on the efficient frontier from left to right. This behavior is illustrated in the next example; see Fig. 16.8.

Example 16.11. Finding portfolios the maximize expected utility

We will use stock price data in the file Stock_Bond.csv. This data set was discussed in Sect. 2.4.1. For simplicity of notation, we subsume X_0 into λ. The R code below solves the quadratic program (16.20) for 250 values of $\log(\lambda)$ equally spaced from 2 to 8; this range was selected by trial-and-error.

```
1  library(quadprog)
2  dat = read.csv("Stock_Bond.csv")
3  y = dat[, c(3, 5, 7, 9, 11, 13, 15, 17, 19, 21)]
4  n = dim(y)[1]
5  m = dim(y)[2] - 1
6  r = y[-1,] / y[-n,] - 1
7  mean_vect = as.matrix(colMeans(r))
8  cov_mat = cov(r)
9  nlambda = 250
10 loglambda_vect = seq(2, 8, length = nlambda)
11 w_matrix = matrix(nrow = nlambda, ncol = 10)
12 mu_vect = matrix(nrow = nlambda, ncol = 1)
13 sd_vect = mu_vect
14 ExUtil_vect = mu_vect
15 conv_vect = mu_vect
16 for (i in 1:nlambda)
17 {
18    lambda = exp(loglambda_vect[i])
19    opt = solve.QP(Dmat = as.matrix(lambda^2 * cov_mat),
20       dvec = lambda * mean_vect, Amat = as.matrix(rep(1,10)),
21       bvec = 1, meq = 1)
22    w = opt$solution
23    mu_vect[i] = w %*% mean_vect
24    sd_vect[i] = sqrt(w %*% cov_mat %*% w)
25    w_matrix[i,] = w
26    ExUtil_vect[i] = opt$value
27 }
```

Next, the expected return and the standard deviation of the return are plotted against λ and then the efficient frontier is drawn by plotting the expect return again the standard deviation of the return. The plots are in Fig. 16.8.   □

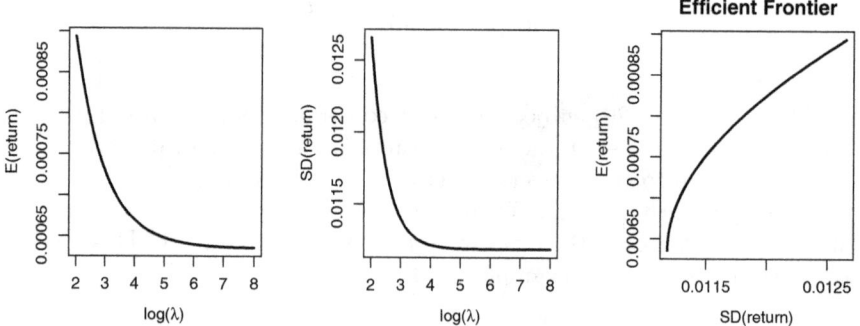

Fig. 16.8. *The expected portfolio return versus* $\log(\lambda$ *(left), the standard deviation of the return versus* $\log(\lambda)$ *(center) and the efficient frontier (right).*

16.9 Bibliographic Notes

Markowitz (1952) was the original paper on portfolio theory and was expanded into the book Markowitz (1959). Bodie and Merton (2000) provide an elementary introduction to portfolio selection theory. Bodie, Kane, and Marcus (1999) and Sharpe, Alexander, and Bailey (1999) give a more comprehensive treatment. See also Merton (1972). Formula (16.5) is derived in Example 5.10 of Ruppert (2004).

Jobson and Korkie (1980) and Britten-Jones (1999) discuss the statistical issue of estimating the efficient frontier; see the latter for additional recent references. Britten-Jones (1999) shows that the tangency portfolio can be estimated by regression analysis and hypotheses about the tangency portfolio can be tested by regression F-tests. Jagannathan and Ma (2003) discuss how imposing constraints such as no short sales can reduce risk.

16.10 R Lab

16.10.1 Efficient Equity Portfolios

This section uses daily stock prices in the data set Stock_Bond.csv that is posted on the book's website and in which any variable whose name ends with "AC" is an adjusted closing price. As the name suggests, these prices have been adjusted for dividends and stock splits, so that returns can be calculated without further adjustments. Run the following code which will read the data, compute the returns for six stocks, create a scatterplot matrix of these returns, and compute the mean vector, covariance matrix, and vector of standard deviations of the returns. Note that returns will be percentages.

```
dat = read.csv("Stock_Bond.csv", header = T)
prices = cbind(dat$GM_AC, dat$F_AC, dat$CAT_AC, dat$UTX_AC,
    dat$MRK_AC, dat$IBM_AC)
n = dim(prices)[1]
returns =  100 * (prices[2:n, ] / prices[1:(n-1), ] - 1)
pairs(returns)
mean_vect = colMeans(returns)
cov_mat = cov(returns)
sd_vect = sqrt(diag(cov_mat))
```

Problem 1 *Write an R program to find the efficient frontier, the tangency portfolio, and the minimum variance portfolio, and plot on "reward-risk space" the location of each of the six stocks, the efficient frontier, the tangency portfolio, and the line of efficient portfolios. Use the constraints that $-0.1 \leq w_j \leq 0.5$ for each stock. The first constraint limits short sales but does not rule them out completely. The second constraint prohibits more than 50 % of the investment in any single stock. Assume that the annual risk-free rate is 3 % and convert this to a daily rate by dividing by 365, since interest is earned on trading as well as nontrading days.*

Problem 2 *If an investor wants an efficient portfolio with an expected daily return of 0.07 %, how should the investor allocate his or her capital to the six stocks and to the risk-free asset? Assume that the investor wishes to use the tangency portfolio computed with the constraints $-0.1 \leq w_j \leq 0.5$, not the unconstrained tangency portfolio.*

Problem 3 *Does this data set include Black Monday?*

16.10.2 Efficient Portfolios with Apple, Exxon-Mobil, Target, and McDonald's Stock

This section constructs portfolios with stocks from four companies: Apple Inc. (AAPL), Exxon-Mobil (XOM), Target Corp. (TGT), and McDonalds (MCD). Run the following code to get 2013 returns in terms of percentage for each of the 4 companies:

```
dat = read.csv("FourStocks_Daily2013.csv", header = TRUE)
head(dat)
prices = dat[,-1]
n = dim(prices)[1]
returns = 100*(prices[-1,] / prices[-n,] - 1)
```

Problem 4 *Write an R program to plot the efficient frontier and to find the allocation weight vector w corresponding to the tangency portfolio. Use the sample mean vector and sample covariance matrix of the returns to estimate μ and Σ. Assume that the annual risk free rate is 1.3 %. Use the constraints that no w_j can be less than -0.5 or greater than 0.5. Let μ_P range from 0.045 to 0.06 %. Report both the Sharpe's Ratio and w for the tangency portfolio.*

Problem 5 *Write an R program to minimize*

$$w^{\mathsf{T}}\mu - \lambda\frac{w^{\mathsf{T}}\Sigma w}{2} \tag{16.21}$$

over w, subject to $w^{\mathsf{T}}1 = 1$, for each λ on a log-spaced grid. Plot the expected return and standard deviation of the return for the portfolios found this way and show that the curve coincides with the efficient frontier found in Problem 4. Select the range of the grid of log-λ values by trial and error to cover an interesting range of the efficient frontier. What value of λ yields a portfolio with $\mu_P = 0.046$? What value of λ yields to the tangency portfolio? What value of λ yields to the minimum variance portfolio?

16.10.3 Finding the Set of Possible Expected Returns

In Sect. 16.6 when we found the efficient frontier by quadratic programming, it was necessary to set up a grid of possible values of the expected returns on the portfolios. When there are no constraints on the allocation vector w except that its elements sum to 1, any expected return is feasible.[5] We saw in Example 16.7, that if short sales are prohibited by the constraints $0 \leq w_i \leq 1$ for all i, the the feasible expected portfolio returns lie between the smallest and largest expected returns on the individual assets.

When more complex constraints are placed on the w_i, the set of feasible expected portfolio returns can be found by linear programming. In this section, we use the same data as used in Sect. 16.10.1. We will impose the constraints that $w_i \leq \mathtt{B1}$ and $-\mathtt{B2} \leq w_i$ for all i.

The function $\mathtt{solveLP()}$ in the $\mathtt{linprog}$ package minimizes (or maximizes) over N-dimensional variable x the objection function $c^{\mathsf{T}}x$ subject to $Ax \leq b$ and $x \geq 0$. Here c is an $N \times 1$ constant vector, A is an $N \times k$ constant matrix, and b is a $k \times 1$ constant vector for some integers N and k. Also, 0 is a k-dimensional zero vector.

Since $x \geq 1$ we cannot let w be x unless we are prohibiting short sale. When short sales are allowed, we can instead let w equal $x_1 - x_2$ and $x^{\mathsf{T}} = (x_1^{\mathsf{T}}, x_2^{\mathsf{T}})$. Then the constraints are that each element of x_1 is at most $\mathtt{B1}$ and each element of x_2 is at most $\mathtt{B2}$. The objective function $w^{\mathsf{T}}\mu$ is equal to $(\mu^{\mathsf{T}}, -\mu^{\mathsf{T}})x$.

---

[5] "Feasible" means that there exists a vector w achieving that expected return.

The constraints in $Ax \leq b$ can be a mixture of equality and inequality constraints. The argument const.dir specifies the directions of the constraints; see line 17. In the program below, there are $2M + 1$ constraints where M is the number of assets. The first M constraints are that $w_i \leq$ B1 for all i, the next M constraints are that $-$B2 $\leq w_i$ for all i, and the last constraint is that $w^T 1 = 1$.

The function solveLP() is used twice, once at lines 18 and 19 to find the smallest feasible expected portfolio return and then at lines 20 and 21 to find the largest possible expected return.

```
1  dat = read.csv("Stock_Bond.csv", header = T)
2  prices = cbind(dat$GM_AC, dat$F_AC, dat$CAT_AC, dat$UTX_AC,
3      dat$MRK_AC, dat$IBM_AC)
4  n = dim(prices)[1]
5  returns =  100 * (prices[2:n, ] / prices[1:(n-1), ] - 1)
6  mean_vect = colMeans(returns)
7  M = length(mean_vect)
8  B1 = 0.3
9  B2 = 0.1
10 library(linprog)
11 AmatLP1 = cbind(diag(1, nrow = M), matrix(0, nrow = M, ncol = M))
12 AmatLP2 = cbind(matrix(0, nrow = M, ncol = M), diag(1, nrow = M))
13 AmatLP3 = c(rep(1, M), rep(-1, M))
14 AmatLP = rbind(AmatLP1, AmatLP2, AmatLP3)
15 bvecLP = c(rep(B1, M), rep(B2, M), 1)
16 cLP =  c(mean_vect, -mean_vect)
17 const.dir = c(rep("<=", 2 * M), "=")
18 resultLP_min = solveLP(cvec = cLP, bvec = bvecLP, Amat = AmatLP,
19     lpSolve=T, const.dir = const.dir, maximum = FALSE)
20 resultLP_max = solveLP(cvec = cLP, bvec = bvecLP,
21     Amat = AmatLP, lpSolve = TRUE, maximum = TRUE)
```

Problem 6 *What is the set of feasible expected portfolio returns when $-0.1 \leq w_i \leq 0.3$ for all i? What allocation vector w achieve the smallest possible expected portfolio return? What allocation vector w achieve the largest possible expected portfolio return?*

Problem 7 *Would it be possible to use* B1$= 0.15$ *and* B2 $= 0.15$? *Explain your answer.*

16.11 Exercises

1. Suppose that there are two risky assets, A and B, with expected returns equal to 2.3 % and 4.5 %, respectively. Suppose that the standard deviations of the returns are $\sqrt{6}$ % and $\sqrt{11}$ % and that the returns on the assets have a correlation of 0.17.

(a) What portfolio of A and B achieves a 3 % rate of expected return?

(b) What portfolios of A and B achieve a $\sqrt{5.5}$ % standard deviation of return? Among these, which has the largest expected return?

2. Suppose there are two risky assets, C and D, the tangency portfolio is 65 % C and 35 % D, and the expected return and standard deviation of the return on the tangency portfolio are 5 % and 7 %, respectively. Suppose also that the risk-free rate of return is 1.5 %. If you want the standard deviation of your return to be 5 %, what proportions of your capital should be in the risk-free asset, asset C, and asset D?

3. (a) Suppose that stock A shares sell at $75 and stock B shares at $115. A portfolio has 300 shares of stock A and 100 of stock B. What are the weights w and $1 - w$ of stocks A and B in this portfolio?

(b) More generally, if a portfolio has N stocks, if the price per share of the jth stock is P_j, and if the portfolio has n_j shares of stock j, then find a formula for w_j as a function of $n_1, \ldots, n_N$ and $P_1, \ldots, P_N$.

4. Let $\mathcal{R}_P$ be a return of some type on a portfolio and let $\mathcal{R}_1, \ldots, \mathcal{R}_N$ be the same type of returns on the assets in this portfolio. Is

$$\mathcal{R}_P = w_1 \mathcal{R}_1 + \cdots + w_N \mathcal{R}_N$$

true if $\mathcal{R}_P$ is a net return? Is this equation true if $\mathcal{R}_P$ is a gross return? Is it true if $\mathcal{R}_P$ is a log return? Justify your answers.

5. Suppose one has a sample of monthly log returns on two stocks with sample means of 0.0032 and 0.0074, sample variances of 0.017 and 0.025, and a sample covariance of 0.0059. For purposes of resampling, consider these to be the "true population values." A bootstrap resample has sample means of 0.0047 and 0.0065, sample variances of 0.0125 and 0.023, and a sample covariance of 0.0058.

(a) Using the resample, estimate the efficient portfolio of these two stocks that has an expected return of 0.005; that is, give the two portfolio weights.

(b) What is the estimated variance of the return of the portfolio in part (a) using the resample variances and covariances?

(c) What are the actual expected return and variance of return for the portfolio in (a) when calculated with the true population values (e.g., with using the original sample means, variances, and covariance)?

6. Stocks 1 and 2 are selling for $100 and $125, respectively. You own 200 shares of stock 1 and 100 shares of stock 2. The weekly returns on these stocks have means of 0.001 and 0.0015, respectively, and standard deviations of 0.03 and 0.04, respectively. Their weekly returns have a correlation of 0.35. Find the correlation matrix of the weekly returns on the two stocks and the mean and standard deviation of the weekly returns on the portfolio.

References

Bodie, Z., and Merton, R. C. (2000) *Finance*, Prentice-Hall, Upper Saddle River, NJ.

Bodie, Z., Kane, A., and Marcus, A. (1999) *Investments*, 4th ed., Irwin/McGraw-Hill, Boston.

Britten-Jones, M. (1999) The sampling error in estimates of mean-variance efficient portfolio weights. *Journal of Finance*, **54**, 655–671.

Jagannathan, R. and Ma, T. (2003) Risk reduction in large portfolios: Why imposing the wrong constraints helps. *Journal of Finance*, **58**, 1651–1683.

Jobson, J. D., and Korkie, B. (1980) Estimation for Markowitz efficient portfolios. *Journal of the American Statistical Association*, **75**, 544–554.

Markowitz, H. (1952) Portfolio Selection. *Journal of Finance*, **7**, 77–91.

Markowitz, H. (1959) *Portfolio Selection: Efficient Diversification of Investment*, Wiley, New York.

Merton, R. C. (1972) An analytic derivation of the efficient portfolio frontier. *Journal of Financial and Quantitative Analysis*, **7**, 1851–1872.

Michaud, R. O. (1998) *Efficient Asset Management: A Practical Guide to Stock Portfolio Optimization and Asset Allocation*, Harvard Business School Press, Boston.

Ruppert, D. (2004) *Statistics and Finance: An Introduction*, Springer, New York.

Sharpe, W. F., Alexander, G. J., and Bailey, J. V. (1999) *Investments*, 6th ed., Prentice-Hall, Upper Saddle River, NJ.

17

The Capital Asset Pricing Model

17.1 Introduction to the CAPM

The *CAPM (capital asset pricing model)* has a variety of uses. It provides a theoretical justification for the widespread practice of passive investing by holding *index funds*.[1] The CAPM can provide estimates of expected rates of return on individual investments and can establish "fair" rates of return on invested capital in regulated firms or in firms working on a cost-plus basis.[2]

The CAPM starts with the question, what would be the risk premiums on securities if the following assumptions were true?

1. The market prices are "in equilibrium." In particular, for each asset, supply equals demand.
2. Everyone has the same forecasts of expected returns and risks.
3. All investors choose portfolios optimally according to the principles of efficient diversification discussed in Chap. 16. This implies that everyone holds a tangency portfolio of risky assets as well as the risk-free asset.
4. The market rewards people for assuming unavoidable risk, but there is no reward for needless risks due to inefficient portfolio selection. Therefore, the risk premium on a single security is not due to its "standalone" risk, but rather to its contribution to the risk of the tangency portfolio. The various components of risk are discussed in Sect. 17.4.

---

[1] An index fund holds the same portfolio as some index. For example, an S&P 500 index fund holds all 500 stocks on the S&P 500 in the same proportions as in the index. Some funds do not replicate an index exactly, but are designed to track the index, for instance, by being cointegrated with the index.

[2] See Bodie and Merton (2000).

© Springer Science+Business Media New York 2015
D. Ruppert, D.S. Matteson, *Statistics and Data Analysis for Financial Engineering*, Springer Texts in Statistics,
DOI 10.1007/978-1-4939-2614-5_17

Assumption 3 implies that the market portfolio is equal to the tangency portfolio. Therefore, a broad index fund that mimics the market portfolio can be used as an approximation to the tangency portfolio.

The validity of the CAPM can only be guaranteed if all of these assumptions are true, and certainly no one believes that any of them are exactly true. Assumption 3 is at best an idealization. Moreover, some of the conclusions of the CAPM are contradicted by the behavior of financial markets; see Sect. 18.4.1 for an example. Despite its shortcomings, the CAPM is widely used in finance and it is essential for a student of finance to understand the CAPM. Many of its concepts such as the beta of an asset and systematic and diversifiable risks are of great importance, and the CAPM has been generalized to the widely used factor models introduced in Chap. 18.

17.2 The Capital Market Line (CML)

The *capital market line* (CML) relates the excess expected return on an efficient portfolio to its risk. *Excess expected return* is the expected return minus the risk-free rate and is also called the risk premium. The CML is

$$\mu_R = \mu_f + \frac{\mu_M - \mu_f}{\sigma_M}\sigma_R, \tag{17.1}$$

where R is the return on a given efficient portfolio (mixture of the market portfolio [= tangency portfolio] and the risk-free asset), $\mu_R = E(R)$, μ_f is the risk-free rate, R_M is the return on the market portfolio, $\mu_M = E(R_M)$, σ_M is the standard deviation of R_M, and σ_R is the standard deviation of R. The risk premium of R is $\mu_R - \mu_f$ and the risk premium of the market portfolio is $\mu_M - \mu_f$.

In (17.1) μ_f, μ_M, and σ_M are constant. What varies are σ_R and μ_R. These vary as we change the efficient portfolio R. Think of the CML as showing how μ_R depends on σ_R.

The slope of the CML is, of course,

$$\frac{\mu_M - \mu_f}{\sigma_M},$$

which can be interpreted as the ratio of the risk premium to the standard deviation of the market portfolio. This is Sharpe's famous "reward-to-risk ratio," which is widely used in finance. Equation (17.1) can be rewritten as

$$\frac{\mu_R - \mu_f}{\sigma_R} = \frac{\mu_M - \mu_f}{\sigma_M},$$

which says that the reward-to-risk ratio for any efficient portfolio equals that ratio for the market portfolio—all efficient portfolios have the same Sharpe's ratio as the market portfolio.

Example 17.1. The CML

Suppose that the risk-free rate of interest is $\mu_f = 0.06$, the expected return on the market portfolio is $\mu_M = 0.15$, and the risk of the market portfolio is $\sigma_M = 0.22$. Then the slope of the CML is $(0.15 - 0.06)/0.22 = 9/22$. The CML of this example is illustrated in Fig. 17.1. □

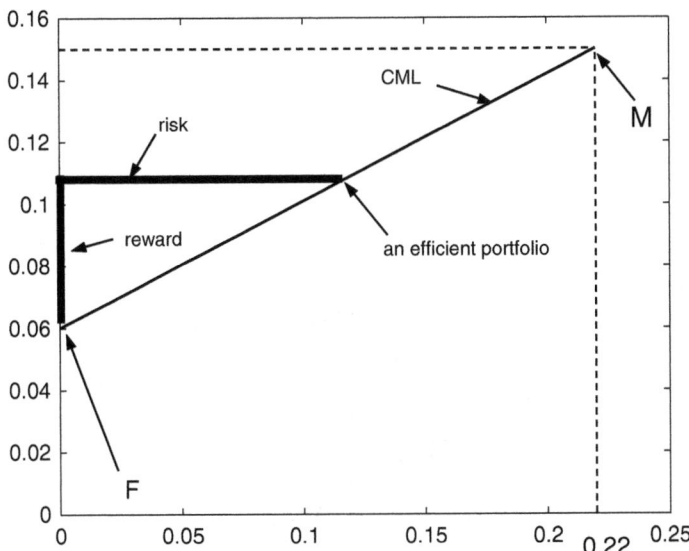

Fig. 17.1. *CML when* $\mu_f = 0.06$, $\mu_M = 0.15$, *and* $\sigma_M = 0.22$. *All efficient portfolios are on the line connecting the risk-free asset (F) and the market portfolio (M). Therefore, the reward-to-risk ratio is the same for all efficient portfolios, including the market portfolio. This fact is illustrated by the thick lines, whose lengths are the risk and reward for a typical efficient portfolio.*

The CML is easy to derive. Consider an efficient portfolio that allocates a proportion w of its assets to the market portfolio and $(1 - w)$ to the risk-free asset. Then

$$R = wR_M + (1 - w)\mu_f = \mu_f + w(R_M - \mu_f). \qquad (17.2)$$

Therefore, taking expectations in (17.2),

$$\mu_R = \mu_f + w(\mu_M - \mu_f). \qquad (17.3)$$

Also, from (17.2),

$$\sigma_R = w\sigma_M, \qquad (17.4)$$

or

$$w = \frac{\sigma_R}{\sigma_M}. \qquad (17.5)$$

Substituting (17.5) into (17.3) gives the CML.

The CAPM says that the optimal way to invest is to

1. decide on the risk σ_R that you can tolerate, $0 \leq \sigma_R \leq \sigma_M$[3];
2. calculate $w = \sigma_R/\sigma_M$;
3. invest w proportion of your investment in a market index fund, that is, a fund that tracks the market as a whole;
4. invest $1 - w$ proportion of your investment in risk-free Treasury bills, or a money-market fund.

Alternatively,

1. choose the reward $\mu_R - \mu_f$ that you want; the only constraint is that $\mu_f \leq \mu_R \leq \mu_M$ so that $0 \leq w \leq 1$[4];
2. calculate

$$w = \frac{\mu_R - \mu_f}{\mu_M - \mu_f};$$

3. do steps 3 and 4 as above.

Instead of specifying the expected return or standard deviation of return, as in Example 16.1 one can find the portfolio with the highest expected return subject to a guarantee that with confidence $1 - \alpha$ the maximum loss is below a prescribed bound M determined, say, by a firm's capital reserves. If the firm invests an amount C, then for the loss to be greater than M the return must be less than $-M/C$. If we assume that the return is normally distributed, then by (A.11), (17.3), and (17.4),

$$P\left(R < -\frac{M}{C}\right) = \Phi\left(\frac{-M/C - \{\mu_f + w(\mu_M - \mu_f)\}}{w\sigma_M}\right). \tag{17.6}$$

Thus, we solve the following equation for w:

$$\Phi^{-1}(\alpha) = \frac{-M/C - \{\mu_f + w(\mu_M - \mu_f)\}}{w\sigma_M}.$$

One can view $w = \sigma_R/\sigma_M$ as an index of the risk aversion of the investor. The smaller the value of w the more risk-averse the investor. If an investor has w equal to 0, then that investor is 100 % in risk-free assets. Similarly, an investor with $w = 1$ is totally invested in the tangency portfolio of risky assets.[5]

---

[3] In fact, $\sigma_R > \sigma_M$ is possible by borrowing money to buy risky assets on margin.

[4] This constraint can be relaxed if one is permitted to buy assets on margin.

[5] An investor with $w > 1$ is buying the market portfolio on margin, that is, borrowing money to buy the market portfolio.

17.3 Betas and the Security Market Line

The *security market line* (SML) relates the excess return on an asset to the slope of its regression on the market portfolio. The SML differs from the CML in that the SML applies to all assets while the CML applies only to efficient portfolios.

Suppose that there are many securities indexed by j. Define

$$\sigma_{jM} = \text{covariance between the returns on the } j\text{th security}$$
$$\text{and the market portfolio.}$$

Also, define

$$\beta_j = \frac{\sigma_{jM}}{\sigma_M^2}. \tag{17.7}$$

It follows from the theory of best linear prediction in Sect. 11.9.1 that β_j is the slope of the best linear predictor of the jth security's returns using returns of the market portfolio as the predictor variable. This fact follows from equation (11.37) for the slope of a best linear prediction equation. In fact, the best linear predictor of R_j based on R_M is

$$\widehat{R}_j = \beta_{0,j} + \beta_j R_M, \tag{17.8}$$

where β_j in (17.8) is the same as in (17.7). Also, $\beta_{0,j}$ is the intercept that can be calculated by taking expectations in (17.8) and solving to obtain $\beta_{0,j} = E(R_j) - \beta_j E(R_M)$.

Another way to appreciate the significance of β_j uses linear regression. As discussed in Sect. 11.9, linear regression is a method for estimating the coefficients of the best linear predictor based upon data. To apply linear regression, suppose that we have a bivariate time series $(R_{j,t}, R_{M,t})_{t=1}^n$ of returns on the jth asset and the market portfolio. Then, the estimated slope of the linear regression of $R_{j,t}$ on $R_{M,t}$ is

$$\hat{\beta}_j = \frac{\sum_{t=1}^n (R_{j,t} - \overline{R}_j)(R_{M,t} - \overline{R}_M)}{\sum_{t=1}^n (R_{M,t} - \overline{R}_M)^2}, \tag{17.9}$$

which, after multiplying the numerator and denominator by the same factor n^{-1}, becomes an estimate of σ_{jM} divided by an estimate of σ_M^2 and therefore by (17.7) an estimate of β_j.

Let μ_j be the expected return on the jth security. Then $\mu_j - \mu_f$ is the *risk premium* (or *reward for risk* or *excess expected return*) for that security. Using the CAPM, it can be shown that

$$\mu_j - \mu_f = \beta_j(\mu_M - \mu_f). \tag{17.10}$$

This equation, which is called the security market line (SML), is derived in Sect. 17.5.2. In (17.10) β_j is a variable in the linear equation, not the slope;

more precisely, μ_j is a linear function of β_j with slope $\mu_M - \mu_f$. This point is worth remembering. Otherwise, there could be some confusion since β_j was defined earlier as a slope of a regression model. In other words, β_j is a slope in one context but is the independent variable in the different context of the SML. One can estimate β_j using (17.9) and then plug this estimate into (17.10).

The SML says that the risk premium of the jth asset is the product of its beta (β_j) and the risk premium of the market portfolio ($\mu_M - \mu_f$). Therefore, β_j measures both the riskiness of the jth asset and the reward for assuming that riskiness. Consequently, β_j is a measure of how "aggressive" the jth asset is. By definition, the beta for the market portfolio is 1; i.e., $\beta_M = 1$. This suggest the rules-of-thumb

$$\beta_j > 1 \;\Rightarrow\; \text{"aggressive,"}$$
$$\beta_j = 1 \;\Rightarrow\; \text{"average risk,"}$$
$$\beta_j < 1 \;\Rightarrow\; \text{"not aggressive."}$$

Figure 17.2 illustrates the SML and an asset J that is not on the SML. This asset contradicts the CAPM, because according to the CAPM all assets are on the SML so no such asset exists.

Consider what would happen if an asset like J did exist. Investors would not want to buy it because, since it is below the SML, its risk premium is too low for the risk given by its beta. They would invest less in J and more in other securities. Therefore, the price of J would decline and *after* this decline its expected return would increase. After that increase, the asset J would be on the SML, or so the theory predicts.

17.3.1 Examples of Betas

Table 17.1 has some "five-year betas" taken from the Salomon, Smith, Barney website between February 27 and March 5, 2001. The beta for the S&P 500 is given as 1.00; why?

17.3.2 Comparison of the CML with the SML

The CML applies only to the return R of an efficient portfolio. It can be arranged so as to relate the excess expected return of that portfolio to the excess expected return of the market portfolio:

$$\mu_R - \mu_f = \left(\frac{\sigma_R}{\sigma_M} \right) (\mu_M - \mu_f). \tag{17.11}$$

The SML applies to *any* asset and like the CML relates its excess expected return to the excess expected return of the market portfolio:

$$\mu_j - \mu_f = \beta_j (\mu_M - \mu_f). \tag{17.12}$$

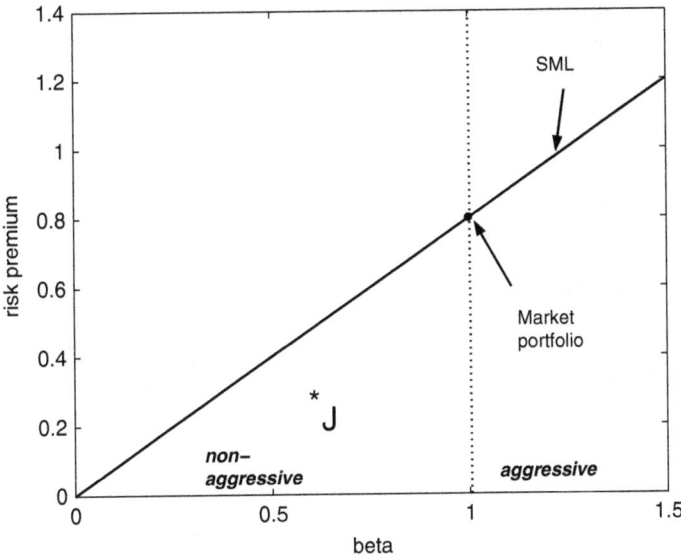

Fig. 17.2. *Security market line (SML) showing that the risk premium of an asset is a linear function of the asset's beta. J is a security not on the line and a contradiction to the CAPM. Theory predicts that the price of J decreases until J is on the SML. The vertical dotted line separates the nonaggressive and aggressive regions.*

If we take an efficient portfolio and consider it as an asset, then μ_R and μ_j both denote the expected return on that portfolio/asset. Both (17.11) and (17.12) hold so that

$$\frac{\sigma_R}{\sigma_M} = \beta_R.$$

17.4 The Security Characteristic Line

Let R_{jt} be the return at time t on the jth asset. Similarly, let $R_{M,t}$ and $\mu_{f,t}$ be the return on the market portfolio and the risk-free return at time t. The *security characteristic line* (sometimes shortened to the characteristic line) is a regression model:

$$R_{j,t} = \mu_{f,t} + \beta_j(R_{M,t} - \mu_{f,t}) + \epsilon_{j,t}, \tag{17.13}$$

where $\epsilon_{j,t}$ is $N(0, \sigma_{\epsilon,j}^2)$. It is often assumed that the $\epsilon_{j,t}$s are uncorrelated across assets, that is, that $\epsilon_{j,t}$ is uncorrelated with $\epsilon_{j',t}$ for $j \neq j'$. This assumption has important ramifications for risk reduction by diversification; see Sect. 17.4.1.

Table 17.1. Selected stocks and in which industries they are. Betas are given for each stock (Stock's β) and its industry (Ind's β). Betas taken from the Salomon, Smith, Barney website between February 27 and March 5, 2001.

Stock (symbol)	Industry	Stock's β	Ind's β
Celanese (CZ)	Synthetics	0.13	0.86
General Mills (GIS)	Food—major diversif	0.29	0.39
Kellogg (K)	Food—major, diversif	0.30	0.39
Proctor & Gamble (PG)	Cleaning prod	0.35	0.40
Exxon-Mobil (XOM)	Oil/gas	0.39	0.56
7-Eleven (SE)	Grocery stores	0.55	0.38
Merck (Mrk)	Major drug manuf	0.56	0.62
McDonalds (MCD)	Restaurants	0.71	0.63
McGraw-Hill (MHP)	Pub—books	0.87	0.77
Ford (F)	Auto	0.89	1.00
Aetna (AET)	Health care plans	1.11	0.98
General Motors (GM)	Major auto manuf	1.11	1.09
AT&T (T)	Long dist carrier	1.19	1.34
General Electric (GE)	Conglomerates	1.22	0.99
Genentech (DNA)	Biotech	1.43	0.69
Microsoft (MSFT)	Software applic.	1.77	1.72
Cree (Cree)	Semicond equip	2.16	2.30
Amazon (AMZN)	Net soft & serv	2.99	2.46
Doubleclick (Dclk)	Net soft & serv	4.06	2.46

Let $\mu_{j,t} = E(R_{j,t})$ and $\mu_{M,t} = E(R_{M,t})$. Taking expectations in (17.13) we get,

$$\mu_{j,t} = \mu_{f,t} + \beta_j(\mu_{M,t} - \mu_{f,t}),$$

which is equation (17.10), the SML, though in (17.10) it is not shown explicitly that the expected returns can depend on t. The SML gives us information about expected returns, but not about the variance of the returns. For the latter we need the characteristic line. The characteristic line is said to be a *return-generating process* since it gives us a probability model of the returns, not just a model of their expected values.

An analogy to the distinction between the SML and characteristic line is this. The regression line $E(Y|X) = \beta_0 + \beta_1 X$ gives the expected value of Y given X but not the conditional probability distribution of Y given X. The regression model

$$Y_t = \beta_0 + \beta_1 X_t + \epsilon_t \quad \text{and} \quad \epsilon_t \sim N(0, \sigma^2)$$

does give us this conditional probability distribution.

The characteristic line implies that

$$\sigma_j^2 = \beta_j^2 \sigma_M^2 + \sigma_{\epsilon,j}^2,$$

that

$$\sigma_{jj'} = \beta_j \beta_{j'} \sigma_M^2 \tag{17.14}$$

for $j \neq j'$, and that

$$\sigma_{Mj} = \beta_j \sigma_M^2.$$

For (17.14) to hold, $\epsilon_{j,t}$ and $\epsilon_{j',t}$ must be uncorrelated. The total risk of the jth asset is

$$\sigma_j = \sqrt{\beta_j^2 \sigma_M^2 + \sigma_{\epsilon,j}^2}.$$

The squared risk has two components: $\beta_j^2 \sigma_M^2$ is called the *market* or *systematic component of risk* and $\sigma_{\epsilon,j}^2$ is called the *unique, nonmarket,* or *unsystematic component of risk.*

17.4.1 Reducing Unique Risk by Diversification

The market component of risk cannot be reduced by diversification, but the unique component can be reduced or even eliminated by sufficient diversification.

Suppose that there are N assets with returns $R_{1,t}, \ldots, R_{N,t}$ for holding period t. If we form a portfolio with weights $w_1, \ldots, w_N$, then the return of the portfolio is

$$R_{P,t} = w_1 R_{1,t} + \cdots + w_N R_{N,t}.$$

Let $R_{M,t}$ be the return on the market portfolio. According to the characteristic line model $R_{j,t} = \mu_{f,t} + \beta_j(R_{M,t} - \mu_{f,t}) + \epsilon_{j,t}$, so that

$$R_{P,t} = \mu_{f,t} + \left(\sum_{j=1}^{N} \beta_j w_j \right) (R_{M,t} - \mu_{f,t}) + \sum_{j=1}^{N} w_j \epsilon_{j,t}.$$

Therefore, the portfolio beta is

$$\beta_P = \sum_{j=1}^{N} w_j \beta_j,$$

and the "epsilon" for the portfolio is

$$\epsilon_{P,t} = \sum_{j=1}^{N} w_j \epsilon_{j,t}.$$

We now assume that $\epsilon_{1,t}, \ldots, \epsilon_{N,t}$ are uncorrelated. Therefore, by equation (7.11),

$$\sigma_{\epsilon,P}^2 = \sum_{j=1}^{N} w_j^2 \sigma_{\epsilon,j}^2.$$

Example 17.2. Reduction in risk by diversification

Suppose the assets in the portfolio are equally weighted; that is, $w_j = 1/N$ for all j. Then

$$\beta_P = \frac{\sum_{j=1}^{N} \beta_j}{N},$$

and

$$\sigma_{\epsilon,P}^2 = \frac{N^{-1} \sum_{j=1}^{N} \sigma_{\epsilon,j}^2}{N} = \frac{\overline{\sigma}_\epsilon^2}{N},$$

where $\overline{\sigma}_\epsilon^2$ is the average of the $\sigma_{\epsilon,j}^2$.

As an illustration, if we assume the simple case where $\sigma_{\epsilon,j}^2$ is a constant, say σ_ϵ^2, for all j, then

$$\sigma_{\epsilon,P} = \frac{\sigma_\epsilon}{\sqrt{N}}. \tag{17.15}$$

For example, suppose that σ_ϵ is 5 %. If $N = 20$, then by (17.15) $\sigma_{\epsilon,P}$ is 1.12 %. If $N = 100$, then $\sigma_{\epsilon,P}$ is 0.5 %. There are approximately 1600 stocks on the NYSE; if $N = 1600$, then $\sigma_{\epsilon,P} = 0.125$ %, a remarkable reduction from 5 %. □

17.4.2 Are the Assumptions Sensible?

A key assumption that allows nonmarket risk to be removed by diversification is that $\epsilon_{1,t}, \ldots, \epsilon_{N,t}$ are uncorrelated. This assumption implies that *all* correlation among the cross-section[6] of asset returns is due to a single cause and that cause is measured by the market index. For this reason, the characteristic line is a "single-factor" or "single-index" model with $R_{M,t}$ being the "factor."

This assumption of uncorrelated ϵ_{jt} would not be valid if, for example, two energy stocks are correlated over and beyond their correlation due to the market index. In this case, unique risk could not be eliminated by holding a large portfolio of all energy stocks. However, if there are many market sectors and the sectors are uncorrelated, then one could eliminate nonmarket risk by diversifying across all sectors. All that is needed is to treat the sectors themselves as the underlying assets and then apply the CAPM theory.

Correlation among the stocks in a market sector can be modeled using a factor model; see Chap. 18.

17.5 Some More Portfolio Theory

In this section we use portfolio theory to show that $\sigma_{j,M}$ quantifies the contribution of the jth asset to the risk of the market portfolio. Also, we derive the SML.

---

[6] "Cross-section" of returns means returns across assets within a *single* holding period.

17.5.1 Contributions to the Market Portfolio's Risk

Suppose that the market consists of N risky assets and that $w_{1,M}, \ldots, w_{N,M}$ are the weights of these assets in the market portfolio. Then

$$R_{M,t} = \sum_{i=1}^{N} w_{i,M} R_{i,t},$$

which implies that the covariance between the return on the jth asset and the return on the market portfolio is

$$\sigma_{j,M} = \text{Cov}\left(R_{j,t}, \sum_{i=1}^{N} w_{i,M} R_{i,t}\right) = \sum_{i=1}^{N} w_{i,M} \sigma_{i,j}. \qquad (17.16)$$

Therefore,

$$\sigma_M^2 = \sum_{j=1}^{N} \sum_{i=1}^{N} w_{j,M} w_{i,M} \sigma_{i,j} = \sum_{j=1}^{N} w_{j,M} \left(\sum_{i=1}^{N} w_{i,M} \sigma_{i,j}\right) = \sum_{j=1}^{N} w_{j,M} \sigma_{j,M}. \qquad (17.17)$$

Equation (17.17) shows that the contribution of the jth asset to the risk of the market portfolio is $w_{j,M} \sigma_{j,M}$, where $w_{j,M}$ is the weight of the jth asset in the market portfolio and $\sigma_{j,M}$ is the covariance between the return on the jth asset and the return on the market portfolio.

17.5.2 Derivation of the SML

The derivation of the SML is a nice application of portfolio theory, calculus, and geometric reasoning. It is based on a clever idea of putting together a portfolio with two assets, the market portfolio and the ith risky asset, and then looking at the locus in reward-risk space as the portfolio weight assigned to the ith risky asset varies.

Consider a portfolio P with weight w_i given to the ith risky asset and weight $(1 - w_i)$ given to the market (tangency) portfolio. The return on this portfolio is

$$R_{P,t} = w_i R_{i,t} + (1 - w_i) R_{M,t}.$$

The expected return is

$$\mu_P = w_i \mu_i + (1 - w_i) \mu_M, \qquad (17.18)$$

and the risk is

$$\sigma_P = \sqrt{w_i^2 \sigma_i^2 + (1 - w_i)^2 \sigma_M^2 + 2w_i(1 - w_i)\sigma_{i,M}}. \qquad (17.19)$$

As we vary w_i, we get the locus of points on (σ, μ) space that is shown as a dashed curve in Fig. 17.3, which uses the same returns as in Fig. 16.3 and Mobil stock as asset i.

It is easy to see geometrically that the derivative of this locus of points evaluated at the tangency portfolio (which is the point where $w_i = 0$) is equal to the slope of the CML. We can calculate this derivative and equate it to the slope of the CML to see what we get. We will see that the result is the SML.

We have from (17.18)

$$\frac{d\mu_P}{dw_i} = \mu_i - \mu_M,$$

and from (17.19) that

$$\frac{d\sigma_P}{dw_i} = \frac{1}{2}\sigma_P^{-1}\left\{2w_i\sigma_i^2 - 2(1 - w_i)\sigma_M^2 + 2(1 - 2w_i)\sigma_{i,M}\right\}.$$

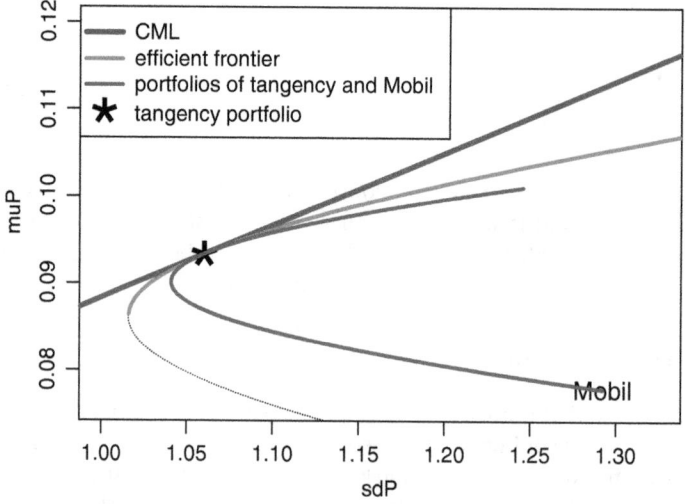

Fig. 17.3. *Derivation of the SML. The purple curve is the locus of portfolios combining Mobil stock and the tangency portfolio (asterisk). The purple curve is to the right of the efficient frontier (red) and intersects the efficient frontier at the tangency portfolio. Therefore, the derivative of the purple curve at the tangency portfolio is equal to the slope of the CML (blue), since the purple curve is tangent to the CML at the tangency portfolio.*

Therefore,

$$\frac{d\mu_P}{d\sigma_P} = \frac{d\mu_P/dw_i}{d\sigma_P/dw_i} = \frac{(\mu_i - \mu_M)\sigma_P}{w_i\sigma_i^2 - \sigma_M^2 + w_i\sigma_M^2 + \sigma_{i,M} - 2w_i\sigma_{i,M}}.$$

Next,

$$\left.\frac{d\mu_P}{d\sigma_P}\right|_{w_i=0} = \frac{(\mu_i - \mu_M)\sigma_M}{\sigma_{i,M} - \sigma_M^2}.$$

Recall that $w_i = 0$ is the tangency portfolio, the point in Fig. 17.3 where the dashed locus is tangent to the CML. Therefore,

$$\frac{d\,\mu_P}{d\,\sigma_P}\bigg|_{w_i=0}$$

must equal the slope of the CML, which is $(\mu_M - \mu_f)/\sigma_M$. Therefore,

$$\frac{(\mu_i - \mu_M)\sigma_M}{\sigma_{i,M} - \sigma_M^2} = \frac{\mu_M - \mu_f}{\sigma_M},$$

which, after some algebra, gives us

$$\mu_i - \mu_f = \frac{\sigma_{i,M}}{\sigma_M^2}(\mu_M - \mu_f) = \beta_i(\mu_M - \mu_f),$$

which is the SML given in equation (17.10).

17.6 Estimation of Beta and Testing the CAPM

17.6.1 Estimation Using Regression

Recall the security characteristic line

$$R_{j,t} = \mu_{f,t} + \beta_j(R_{M,t} - \mu_{f,t}) + \epsilon_{j,t}. \tag{17.20}$$

Let $R_{j,t}^* = R_{j,t} - \mu_{f,t}$ be the excess return on the jth security and let $R_{M,t}^* = R_{M,t} - \mu_{f,t}$, be the excess return on the market portfolio. Then (17.20) can be written as

$$R_{j,t}^* = \beta_j R_{M,t}^* + \epsilon_{j,t}. \tag{17.21}$$

Equation (17.21) is a regression model without an intercept and with β_j as the slope. A more elaborate model is

$$R_{j,t}^* = \alpha_j + \beta_j R_{M,t}^* + \epsilon_{j,t}, \tag{17.22}$$

which includes an intercept. The CAPM says that $\alpha_j = 0$ but by allowing $\alpha_j \neq 0$, we recognize the possibility of mispricing.

Given time series $R_{j,t}$, $R_{M,t}$, and $\mu_{f,t}$ for $t = 1, \ldots, n$, we can calculate $R_{j,t}^*$ and $R_{M,t}^*$ and regress $R_{j,t}^*$ on $R_{M,t}^*$ to estimate α_j, β_j, and $\sigma_{\epsilon,j}^2$. By testing the null hypothesis that $\alpha_j = 0$, we are testing whether the jth asset is mispriced according to the CAPM.

As discussed in Sect. 9.2.2, when fitting model (17.21) or (17.22) one should use daily data if available, rather than weekly or monthly data. A more difficult question to answer is how long a time series to use. Longer time series give more data, of course, but models (17.21) and (17.22) assume that β_j is constant and this might not be true over a long time period.

Example 17.3. Estimation of α and β for Microsoft

As an example, daily closing prices on Microsoft and the S&P 500 index from November 1, 1993, to April 3, 2003, were used. The S&P 500 was taken as the market price. Three-month T-bill rates were used as the risk-free returns.[7] The excess returns are the returns minus the T-bill rates. The code is

```
dat = read.csv("capm.csv", header = TRUE)
attach(dat)
n = dim(dat)[1]
EX_R_sp500 = Close.sp500[2:n] / Close.sp500[1:(n-1)]
    - 1  - Close.tbill[2:n] / (100 * 253)
EX_R_msft = Close.msft[2:n] / Close.msft[1:(n-1)]
    - 1  - Close.tbill[2:n] / (100 * 253)
fit = lm(EX_R_msft ~ EX_R_sp500)
options(digits = 3)
summary(fit)
```

and the output is

```
Call:
lm(formula = EX_R_msft ~ EX_R_sp500)

Coefficients:
            Estimate Std. Error t value Pr(>|t|)
(Intercept) 0.000914   0.000409    2.23   0.026 *
EX_R_sp500  1.247978   0.035425   35.23   <2e-16 ***
---

Residual standard error: 0.0199 on 2360 degrees of freedom
Multiple R-squared:  0.345,Adjusted R-squared:  0.344
F-statistic: 1.24e+03 on 1 and 2360 DF,  p-value: <2e-16
```

For Microsoft, we find that $\hat{\beta} = 1.25$ and $\hat{\alpha} = 0.0009$. The estimate of α is very small and, although the p-value for α is 0.026, we can conclude that for practical purposes, α is essentially 0. This is another example of an effect being statistically significant according to a test of the hypothesis of no effect but not practically significant. Very small effects are often statistically significant when the sample size is large. In this example, we have nearly 10 years of daily data and the sample size is quite large for a hypothesis testing problem, 2363.

The estimate of σ_ϵ is the root MSE which equals 0.0199. Notice that the R^2 (R-sq) value for the regression is 34.5 %. The interpretation of R^2 is the percent of the variance in the excess returns on Microsoft that is due to excess returns on the market. In other words, 34.5 % of the squared risk is due to

---

[7] Interest rates are return rates. Thus, we use the T-bill rates themselves as the risk-free returns. One does *not* take logs and difference the T-bill rates as if they were prices. However, the T-bill rates were divided by 100 to convert from a percentage and then by 253 to convert to a daily rate.

systematic or market risk $(\beta_j^2 \sigma_M^2)$. The remaining 65.5 % is due to unique or nonmarket risk (σ_ϵ^2).

If we assume that $\alpha = 0$, then we can refit the model using a no-intercept model. The code for fitting the model is changed to

```
fit_NoInt = lm(EX_R_msft ~ EX_R_sp500 - 1)
options(digits = 3)
summary(fit_NoInt)
```

Notice the "−1" in the formula. The "1" represents the intercept so "−1" indicates that the intercept is removed. The output changes to

```
Call:
lm(formula = EX_R_msft ~ EX_R_sp500 - 1)

Coefficients:
          Estimate Std. Error t value Pr(>|t|)
EX_R_sp500   1.2491     0.0355    35.2   <2e-16 ***
---
Residual standard error: 0.0199 on 2361 degrees of freedom
Multiple R-squared:  0.345,Adjusted R-squared:  0.344
F-statistic: 1.24e+03 on 1 and 2361 DF,  p-value: <2e-16
```

With no intercept $\hat{\beta}$, $\hat{\sigma}_\epsilon$ and R^2 are nearly the same as before—forcing a nearly zero intercept to be exactly zero has little effect.　　　　　　　□

17.6.2 Testing the CAPM

Testing that α equals 0 tests only one of the conclusions of the CAPM. Accepting this null hypothesis only means that the CAPM has passed one test, not that we should now accept it as true.[8] To fully test the CAPM, its other conclusions should also be tested. The factor models in Sect. 18.3 have been used to test the CAPM and fairly strong evidence against the CAPM has been found. Fortunately, these factor models do provide a generalization of the CAPM that is likely to be useful for financial decision making.

Often, as an alternative to regression using excess returns, the returns on the asset are regressed on the returns on the market. When this is done, an intercept model should be used. In the Microsoft data when using returns instead of excess returns, the estimate of beta changed hardly at all.

17.6.3 Interpretation of Alpha

If α is nonzero, then the security is mispriced, at least according to the CAPM. If $\alpha > 0$ then the security is underpriced; the returns are too large on average.

---

[8] In fact, acceptance of a null hypothesis should never be interpreted as proof that the null hypothesis is true.

This is an indication of an asset worth purchasing. Of course, one must be careful. If we reject the null hypothesis that $\alpha = 0$, all we have done is to show that the security was mispriced *in the past*.

Warning: If we use returns rather than excess returns, then the intercept of the regression equation does *not* estimate α, so one cannot test whether α is zero by testing the intercept.

17.7 Using the CAPM in Portfolio Analysis

Suppose we have estimated beta and σ_ϵ^2 for each asset in a portfolio and also estimated σ_M^2 and μ_M for the market. Then, since μ_f is also known, we can compute the expectations, variances, and covariances of all asset returns by the formulas

$$\mu_j = \mu_f + \beta_j(\mu_M - \mu_f),$$
$$\sigma_j^2 = \beta_j^2 \sigma_M^2 + \sigma_{\epsilon,j}^2,$$
$$\sigma_{jj'} = \beta_j \beta_{j'} \sigma_M^2 \text{ for } j \neq j'.$$

There is a noteworthy danger here: These estimates depend heavily on the validity of the CAPM assumptions. Any or all of the quantities beta, σ_ϵ^2, σ_M^2, μ_M, and μ_f could depend on time t. However, it is generally assumed that the betas and σ_ϵ^2s of the assets as well as σ_M^2 and μ_M of the market are independent of t so that these parameters can be estimated assuming stationarity of the time series of returns.

17.8 Bibliographic Notes

The CAPM was developed by Sharpe (1964), Lintner (1965a,b), and Mossin (1966). Introductions to the CAPM can be found in Bodie, Kane, and Marcus (1999), Bodie and Merton (2000), and Sharpe, Alexander, and Bailey (1999). I first learned about the CAPM from these three textbooks. Campbell, Lo, and MacKinlay (1997) discuss empirical testing of the CAPM. The derivation of the SML in Sect. 17.5.2 was adapted from Sharpe, Alexander, and Bailey (1999). Discussion of factor models can be found in Sharpe, Alexander, and Bailey (1999), Bodie, Kane, and Marcus (1999), and Campbell, Lo, and MacKinlay (1997).

17.9 R Lab

In this lab, you will fit model (17.20). The S&P 500 index will be a proxy for the market portfolio and the 90-day Treasury rate will serve as the risk-free rate.

This lab uses the data set `Stock_Bond_2004_to_2006.csv`, which is available on the book's website. This data set contains a subset of the data in the data set `Stock_Bond.csv` used elsewhere.

The R commands needed to fit model (17.20) will be given in small groups so that they can be explained better. First run the following commands to read the data, extract the prices, and find the number of observations:

```
dat = read.csv("Stock_Bond_2004_to_2006.csv", header = TRUE)
prices = dat[ , c(5, 7, 9, 11, 13, 15, 17, 24)]
n = dim(prices)[1]
```

Next, run these commands to convert the risk-free rate to a daily rate, compute net returns, extract the Treasury rate, and compute excess returns for the market and for seven stocks. The risk-free rate is given as a percentage so the returns are also computed as percentages.

```
dat2 =  as.matrix(cbind(dat[(2:n), 3] / 365,
    100 * (prices[2:n,] / prices[1:(n-1), ] - 1)))
names(dat2)[1] = "treasury"
risk_free = dat2[,1]
ExRet = dat2[ ,2:9] - risk_free
market = ExRet[ ,8]
stockExRet = ExRet[ ,1:7]
```

Now fit model (17.20) to each stock, compute the residuals, look at a scatterplot matrix of the residuals, and extract the estimated betas.

```
fit_reg = lm(stockExRet ~ market)
summary(fit_reg)
res = residuals(fit_reg)
pairs(res)
options(digits = 3)
betas = fit_reg$coeff[2, ]
```

Problem 1 *Would you reject the null hypothesis that alpha is zero for any of the seven stocks? Why or why not?*

Problem 2 *Use model (17.20) to estimate the expected excess return for all seven stocks. Compare these results to using the sample means of the excess returns to estimate these parameters. Assume for the remainder of this lab that all alphas are zero. (Note: Because of this assumption, one might consider reestimating the betas and the residuals with a no-intercept model. However, since the estimated alphas were close to zero, forcing the alphas to be exactly zero will not change the estimates of the betas or the residuals by much. Therefore, for simplicity, do not reestimate.)*

Problem 3 *Compute the correlation matrix of the residuals. Do any of the residual correlations seem large? Could you suggest a reason why the large correlations might be large? (Information about the companies in this data set is available at Yahoo Finance and other Internet sites.)*

Problem 4 *Use model (17.20) to estimate the covariance matrix of the excess returns for the seven companies.*

Problem 5 *What percentage of the excess return variance for UTX is due to the market?*

Problem 6 *An analyst predicts that the expected excess return on the market next year will be 4 %. Assume that the betas estimated here using data from 2004–2006 are suitable as estimates of next year's betas. Estimate the expected excess returns for the seven stocks for next year.*

17.9.1 Zero-beta Portfolios

A portfolio with beta $= 0$ is neutral to market risk and bounding the absolute weights of the portfolio reduces the portfolio's unique risk. In the next problem, you will find a low-risk portfolio with a large alpha. The data in this section have been simulated and are only for illustration. Estimation of the alphas of stock is difficult, especially the prediction of future values of alphas.

Problem 7 *The file* `AlphaBeta.csv` *contains alphas and betas on 50 stocks. Use linear programming to find the portfolio containing these stocks that has the maximum possible alpha subject to the portfolio's beta being equal to zero and weights satisfying* $-0.25 \leq w_i \leq 0.25$ *for all* $i = 1, \dots, 50$. *What are the 50 weights of your portfolio? What is its alpha?*
   Hint: This is a linear programming problem. Use the function `solveLP()`. *See Sect. 16.10.3.*

Problem 8 *If you attempt to find a zero-beta portfolio with* $-0.25 \leq w_i \leq 0.25$ *with a smaller number of stock, you will find that there is no solution. (If you like, try this with the first 20 stocks.) Discuss why is there no solution.*

17.10 Exercises

1. What is the beta of a portfolio if $E(R_P) = 16\%$, $\mu_f = 5.5\%$, and $E(R_M) = 11\%$?

2. Suppose that the risk-free rate of interest is 0.03 and the expected rate of return on the market portfolio is 0.14. The standard deviation of the market portfolio is 0.12.

   (a) According to the CAPM, what is the efficient way to invest with an expected rate of return of 0.11?

   (b) What is the risk (standard deviation) of the portfolio in part (a)?

3. Suppose that the risk-free interest rate is 0.023, that the expected return on the market portfolio is $\mu_M = 0.10$, and that the volatility of the market portfolio is $\sigma_M = 0.12$.

   (a) What is the expected return on an efficient portfolio with $\sigma_R = 0.05$?

   (b) Stock A returns have a covariance of 0.004 with market returns. What is the beta of Stock A?

   (c) Stock B has beta equal to 1.5 and $\sigma_\epsilon = 0.08$. Stock C has beta equal to 1.8 and $\sigma_\epsilon = 0.10$.

      i. What is the expected return of a portfolio that is one-half Stock B and one-half Stock C?

      ii. What is the volatility of a portfolio that is one-half Stock B and one-half Stock C? Assume that the ϵs of Stocks B and C are independent.

4. Show that equation (17.16) follows from equation (7.8).

5. True or false: The CAPM implies that investors demand a higher return to hold more volatile securities. Explain your answer.

6. Suppose that the riskless rate of return is 4 % and the expected market return is 12 %. The standard deviation of the market return is 11 %. Suppose as well that the covariance of the return on Stock A with the market return is 165 %².[9]

   (a) What is the beta of Stock A?

   (b) What is the expected return on Stock A?

   (c) If the variance of the return on Stock A is 220 %², what percentage of this variance is due to market risk?

7. Suppose there are three risky assets with the following betas and $\sigma_{\epsilon_j}^2$.

j	β_j	$\sigma_{\epsilon_j}^2$
1	0.9	0.010
2	1.1	0.015
3	0.6	0.011

   Suppose also that the variance of $R_{Mt} - \mu_{ft}$ is 0.014.

   (a) What is the beta of an equally weighted portfolio of these three assets?

   (b) What is the variance of the excess return on the equally weighted portfolio?

   (c) What proportion of the total risk of asset 1 is due to market risk?

---

[9] If returns are expressed in units of percent, then the units of variances and co-variances are percent-squared. A variance of 165 %² equals 165/10,000.

8. Suppose there are two risky assets, call them C and D. The tangency portfolio is 60 % C and 40 % D. The expected yearly returns are 4 % and 6 % for assets C and D. The standard deviations of the yearly returns are 10 % and 18 % for C and D and the correlation between the returns on C and D is 0.5. The risk-free yearly rate is 1.2 %.

(a) What is the expected yearly return on the tangency portfolio?

(b) What is the standard deviation of the yearly return on the tangency portfolio?

(c) If you want an efficient portfolio with a standard deviation of the yearly return equal to 3 %, what proportion of your equity should be in the risk-free asset? If there is more than one solution, use the portfolio with the higher expected yearly return.

(d) If you want an efficient portfolio with an expected yearly return equal to 7 %, what proportions of your equity should be in asset C, asset D, and the risk-free asset?

9. What is the beta of a portfolio if the expected return on the portfolio is $E(R_P) = 15\,\%$, the risk-free rate is $\mu_f = 6\,\%$, and the expected return on the market is $E(R_M) = 12\,\%$? Make the usual CAPM assumptions including that the portfolio alpha is zero.

10. Suppose that the risk-free rate of interest is 0.07 and the expected rate of return on the market portfolio is 0.14. The standard deviation of the market portfolio is 0.12.

(a) According to the CAPM, what is the efficient way to invest with an expected rate of return of 0.11?

(b) What is the risk (standard deviation) of the portfolio in part (a)?

11. Suppose there are three risky assets with the following betas and $\sigma_{\epsilon_j}^2$ when regressed on the market portfolio.

j	β_j	$\sigma_{\epsilon_j}^2$
1	0.7	0.010
2	0.8	0.025
3	0.6	0.012

Assume ϵ_1, ϵ_2, and ϵ_3 are uncorrelated. Suppose also that the variance of $R_M - \mu_f$ is 0.02.

(a) What is the beta of an equally weighted portfolio of these three assets?

(b) What is the variance of the excess return on the equally weighted portfolio?

(c) What proportion of the total risk of asset 1 is due to market risk?

12. As an analyst, you have constructed 2 possible portfolios. Both portfolios have the same beta and expected return, but portfolio 1 was constructed with only technology companies whereas portfolio 2 was constructed using technology, healthcare, energy, consumer products, and metals and mining companies. Should you be impartial to which portfolio you invest in? Explain why or why not.

References

Bodie, Z., and Merton, R. C. (2000) *Finance*, Prentice-Hall, Upper Saddle River, NJ.

Bodie, Z., Kane, A., and Marcus, A. (1999) *Investments*, 4th ed., Irwin/McGraw-Hill, Boston.

Campbell, J. Y., Lo, A. W., and MacKinlay, A. C. (1997) *The Econometrics of Financial Markets*, Princeton University Press, Princeton, NJ.

Lintner, J. (1965a) The valuation of risky assets and the selection of risky investments in stock portfolios and capital budgets. *Review of Economics and Statistics*, **47**, 13–37.

Lintner, J. (1965b) Security prices, risk, and maximal gains from diversification. *Journal of Finance*, **20**, 587–615.

Mossin, J. (1966) Equilibrium in capital markets. *Econometrica*, **34**, 768–783.

Sharpe, W. F. (1964) Capital asset prices: A theory of market equilibrium under conditions of risk. *Journal of Finance*, **19**, 425–442.

Sharpe, W. F., Alexander, G. J., and Bailey, J. V. (1999) *Investments*, 6th ed., Prentice-Hall, Upper Saddle River, NJ.

18

Factor Models and Principal Components

18.1 Dimension Reduction

High-dimensional data can be challenging to analyze. They are difficult to visualize, need extensive computer resources, and often require special statistical methodology. Fortunately, in many practical applications, high-dimensional data have most of their variation in a lower-dimensional space that can be found using *dimension reduction techniques*. There are many methods designed for dimension reduction, and in this chapter we will study two closely related techniques, *factor analysis* and *principal components analysis*, often called *PCA*.

PCA finds structure in the covariance or correlation matrix and uses this structure to locate low-dimensional subspaces containing most of the variation in the data.

Factor analysis explains returns with a smaller number of fundamental variables called *factors* or *risk factors*. Factor analysis models can be classified by the types of variables used as factors, macroeconomic or fundamental, and by the estimation technique, time series regression, cross-sectional regression, or statistical factor analysis.

18.2 Principal Components Analysis

PCA starts with a sample $\boldsymbol{Y}_i = (Y_{i,1}, \ldots, Y_{i,d})$, $i = 1, \ldots, n$, of d-dimensional random vectors with mean vector $\boldsymbol{\mu}$ and covariance matrix $\boldsymbol{\Sigma}$. One goal of PCA is finding "structure" in $\boldsymbol{\Sigma}$.

We will start with a simple example that illustrates the main idea. Suppose that $\boldsymbol{Y}_i = \boldsymbol{\mu} + W_i \boldsymbol{o}$, where $W_1, \ldots, W_n$ are i.i.d. mean-zero random variables and $\boldsymbol{o}$ is some fixed vector, which can be taken to have norm 1. The $\boldsymbol{Y}_i$ lie on

© Springer Science+Business Media New York 2015
D. Ruppert, D.S. Matteson, *Statistics and Data Analysis for Financial Engineering*, Springer Texts in Statistics,
DOI 10.1007/978-1-4939-2614-5_18

the line that passes through μ and is in the direction given by o, so that all variation among the mean-centered vectors $Y_i - \mu$ is in the one-dimensional space spanned by o. Also, the covariance matrix of Y_i is

$$\Sigma = E\{W_i^2 oo^\mathsf{T}\} = \sigma_W^2 oo^\mathsf{T}.$$

The vector o is called the first principal axis of Σ and is the only eigenvector of Σ with a nonzero eigenvalue, so o can be estimated by an eigen-decomposition (Appendix A.20) of the estimated covariance matrix.

A slightly more realistic situation is where $Y_i = \mu + W_i o + \epsilon_i$, where ϵ_i is a random vector uncorrelated with W_i and having a "small" covariance matrix. Then most of the variation among the $Y_i - \mu$ vectors is in the space spanned by o, but there is small variation in other directions due to ϵ_i. Having looked at some simple special cases, we now turn to the general case.

PCA can be applied to either the sample covariance matrix or the correlation matrix. We will use Σ to represent whichever matrix is chosen. The correlation matrix is, of course, the covariance matrix of the standardized variables, so the choice between the two matrices is really a decision whether or not to standardize the variables before PCA. This issue will be addressed later. Even if the data have not been standardized, to keep notation simple, we assume that the mean $\overline{Y}$ has been subtracted from each Y_i. By (A.50),

$$\Sigma = O \operatorname{diag}(\lambda_1, \ldots, \lambda_d) O^\mathsf{T}, \tag{18.1}$$

where O is an orthogonal matrix whose columns $o_1, \ldots, o_d$ are the eigenvectors of Σ and $\lambda_1 > \ldots > \lambda_d$ are the corresponding eigenvalues. The columns of O have been arranged so that the eigenvalues are ordered from largest to smallest. This is not essential, but it is convenient. We also assume no ties among the eigenvalues, which almost certainly will be true in actual applications.

A *normed linear combination* of Y_i (either standardized or not) is of the form $\alpha^\mathsf{T} Y_i = \sum_{j=1}^p \alpha_j Y_{i,j}$, where $\|\alpha\| = \sqrt{\sum_{j=1}^p \alpha_i^2} = 1$. The first principal component is the normed linear combination with the greatest variance. The variation in the direction α, where α is any fixed vector with norm 1, is

$$\operatorname{Var}(\alpha^\mathsf{T} Y_i) = \alpha^\mathsf{T} \Sigma \alpha. \tag{18.2}$$

The first principal component maximizes (18.2) over α. The maximizer is $\alpha = o_1$, the eigenvector corresponding to the largest eigenvalue, and is called the first principal axis. The projections $o_1^\mathsf{T} Y_i$, $i = 1, \ldots, n$, onto this vector are called the first principal component or principal component scores. Requiring that the norm of α be fixed is essential, because otherwise (18.2) is unbounded and there is no maximizer.

After the first principal component has been found, one searches for the direction of maximum variation perpendicular to the first principal axis (eigenvector). This means maximizing (18.2) subject to $\|\alpha\| = 1$ and $\alpha^\mathsf{T} o_1 = 0$.

The maximizer, called the second principal axis, is o_2, and the second principal component is the set of projections $o_2^T Y_i$, $i = 1, \ldots, n$, onto this axis. The reader can probably see where we are going. The third principal component maximizes (18.2) subject to $\|\alpha\| = 1$, $\alpha^T o_1 = 0$, and $\alpha^T o_2 = 0$ and is $o_3^T Y_i$, and so forth, so that $o_1, \ldots, o_d$ are the principal axes and the set of projections $o_j^T Y_i$, $i = 1, \ldots, n$, onto the jth eigenvector is the jth principal component. Moreover,

$$\lambda_i = o_i^T \Sigma o_i$$

is the variance of the ith principal component, $\lambda_i / (\lambda_1 + \cdots + \lambda_d)$ is the proportion of the variance due to this principal component, and $(\lambda_1 + \cdots + \lambda_i) / (\lambda_1 + \cdots + \lambda_d)$ is the proportion of the variance due to the first i principal components. The principal components are mutually uncorrelated since for $j \neq k$ we have

$$\mathrm{Cov}(o_j^T Y_i, o_k^T Y_i) = o_j^T \Sigma o_k = 0$$

by (A.52).

Let

$$Y = \begin{pmatrix} Y_1^T \\ \vdots \\ Y_n^T \end{pmatrix}$$

be the original data and let

$$S = \begin{pmatrix} o_1^T Y_1 & \cdots & o_d^T Y_1 \\ \vdots & \ddots & \vdots \\ o_1^T Y_n & \cdots & o_d^T Y_n \end{pmatrix}$$

be the matrix of principal components. Then

$$S = YO.$$

Postmultiplication of Y by O to obtain S is an orthogonal rotation of the data. For this reason, the eigenvectors are sometimes called the *rotations*, e.g., in output from R's pca() function.

In many applications, the first few principal components, such as, the first three to five, account for almost all of the variation, and, for most purposes, one can work solely with these principal components and discard the rest. This can be a sizable reduction in dimension. See Example 18.2 for an illustration.

So far, we have left unanswered the question of how one should decide between working with the original or the standardized variables. If the components of Y_i are comparable, e.g., are all daily returns on equities or all are yields on bonds, then working with the original variables should cause no problems. However, if the variables are not comparable, e.g., one is an unemployment rate and another is the GDP in dollars, then some variables may be many orders of magnitude larger than the others. In such cases, the large variables could completely dominate the PCA, so that the first principal component is in the direction of the variable with the largest standard deviation. To eliminate this problem, one should standardize the variables.

Example 18.1. PCA with unstandardized and standardized variables

As a simple illustration of the difference between using standardized and unstandardized variables, suppose there are two variables $(d = 2)$ with a correlation of 0.9. Then the correlation matrix is

$$\begin{pmatrix} 1 & 0.9 \\ 0.9 & 1 \end{pmatrix}$$

with normalized eigenvectors $(0.71, 0.71)$ and $(-0.71, 0.71)$[1] and eigenvalues 1.9 and 0.1. Most of the variation is in the direction $(1, 1)$, which is consistent with the high correlation between the two variables.

However, suppose that the first variable has variance 1,000,000 and the second has variance 1. The covariance matrix is

$$\begin{pmatrix} 1,000,000 & 900 \\ 900 & 1 \end{pmatrix},$$

which has eigenvectors, after rounding, equal to $(1.0000, 0.0009)$ and $(-0.0009, 1)$ and eigenvalues 1,000,000 and 0.19. The first variable dominates the principal components analysis based on the covariance matrix. This principal components analysis does correctly show that almost all of the variation is in the first variable, but this is true only with the original units. Suppose that variable 1 had been in dollars and is now converted to millions of dollars. Then its variance is equal to 10^{-6}, so that the principal components analysis using the covariance matrix will now show most of the variation to be due to variable 2. In contrast, principal components analysis based on the correlation matrix does not change as the variables' units change.   □

Example 18.2. Principal components analysis of yield curves

This example uses yields on Treasury bonds at 11 maturities, $T = 1, 3$, and 6 months and 1, 2, 3, 5, 7, 10, 20, and 30 years. Daily yields were taken from a U.S. Treasury website for the time period January 2, 1990, to October 31, 2008, A subset of these data was used in Example 15.1. The yield curves are shown in Fig. 18.1a for three different dates. Notice that the yield curves can have a variety of shapes. In this example, we will use PCA to study how the curves change from day to day.

To analyze daily changes in yields, all 11 time series were differenced. Daily yields were missing from some values of T because, for example to quote the

---

[1] The normalized eigenvalues are determined only up to sign so they could multiplied by -1 to become $(-0.71, -0.71)$ and $(0.71, -0.71)$.

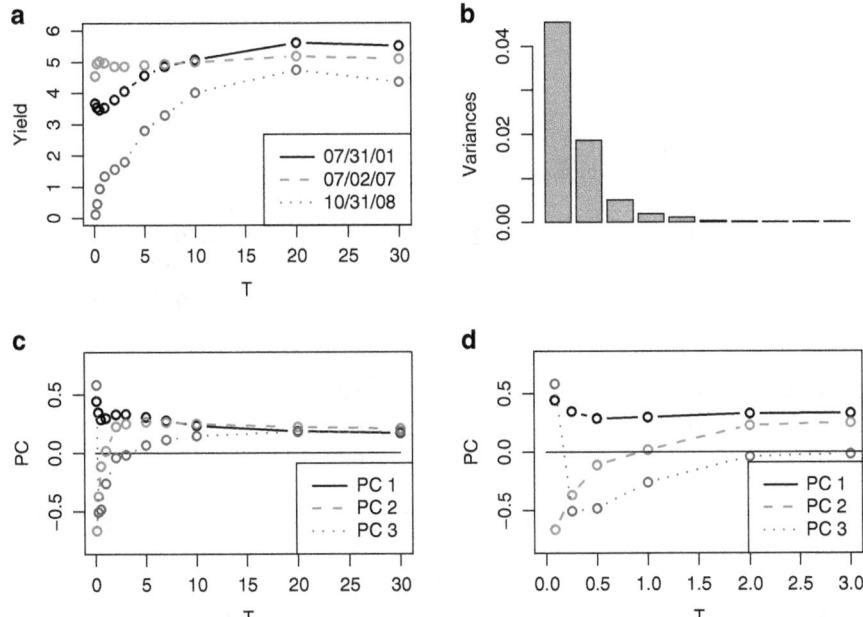

Fig. 18.1. *(a) Treasury yields on three dates. (b) Scree plot for the changes in Treasury yields. Note that the first three principal components have most of the variation, and the first five have virtually all of it. (c) The first three eigenvectors for changes in the Treasury yields. (d) The first three eigenvectors for changes in the Treasury yields in the range $0 \leq T \leq 3$.*

website, "Treasury discontinued the 20-year constant maturity series at the end of calendar year 1986 and reinstated that series on October 1, 1993." Differencing caused a few additional days to have missing values. In the analysis, all days with missing values of the differenced data were omitted. This left 819 days of data starting on July 31, 2001, when the one-month series started and ending on October 31, 2008, with the exclusion of the period February 19, 2002 to February 2, 2006 when the 30-year Treasury was discontinued. One could use much longer series by not including the one-month and 30-year series.

The covariance matrix, not the correlation matrix, was used, because in this example the variables are comparable and in the same units.

First, we will look at the 11 eigenvalues using R's function prcomp(). The code is:

```
datNoOmit = read.table("treasury_yields.txt", header = TRUE)
diffdatNoOmit = diff(as.matrix(datNoOmit[ , 2:12]))
dat = na.omit(datNoOmit)
diffdat = na.omit(diffdatNoOmit)
n = dim(diffdat)[1]
```

```
options(digits = 5)
pca = prcomp(diffdat)
summary(pca)
```

The results are:

```
Importance of components:
                        PC1  PC2   PC3    PC4    PC5     PC6
Standard deviation     0.21 0.14 0.071  0.045  0.033  0.0173
Proportion of Variance 0.62 0.25 0.070  0.028  0.015  0.0041
Cumulative Proportion  0.62 0.88 0.946  0.974  0.989  0.9932

PC7    PC8    PC9    PC10    PC11
0.0140 0.0108 0.0092 0.00789 0.00610
0.0027 0.0016 0.0012 0.00085 0.00051
0.9959 0.9975 0.9986 0.99949 1.00000
```

The first row gives the values of $\sqrt{\lambda_i}$, the second row the values of $\lambda_i/(\lambda_1 + \cdots + \lambda_d)$, and the third row the values of $(\lambda_1 + \cdots + \lambda_i)/(\lambda_1 + \cdots + \lambda_d)$ for $i = 1, \ldots, 11$. One can see, for example, that the standard deviation of the first principal component is 0.21 and represents 62 % of the total variance. Also, the first three principal components have 94.6 % of the variation, and this increases to 97.4 % for the first four principal components and to 98.9 % for the first five. The variances (the squares of the first row) are plotted in Fig. 18.1b. This type of plot is called a "scree plot" since it looks like scree, fallen rocks that have accumulated at the base of a mountain.

We will concentrate on the first three principal components since approximately 95 % of the variation in the changes in yields is in the space they span. The eigenvectors, labeled "PC," are plotted in Fig. 18.1c and d, the latter showing detail in the range $T \le 3$. The eigenvectors have interesting interpretations. The first, o_1, has all positive values.[2] A change in this direction either increases all yields or decreases all yields, and by roughly the same amounts. One could call such changes "parallel shifts" of the yield curve, though they are only approximately parallel. These shifts are shown in Fig. 18.2a, where the mean yield curve is shown as a solid black line, the mean plus o_1 is a dashed red line, and the mean minus o_1 is a dashed blue line. Only the range $T \le 7$ is shown, since the curves change less after this point. Since the standard deviation of the first principal component is only 0.21, a ± 1 shift in a single day is huge and is used only for better graphical presentation.

---

[2] As mentioned previously, the eigenvectors are determined only up to a sign reversal, since multiplication by -1 would not change the spanned space or the norm. Thus, we could instead say the eigenvector has only negative values, but this would not change the interpretation.

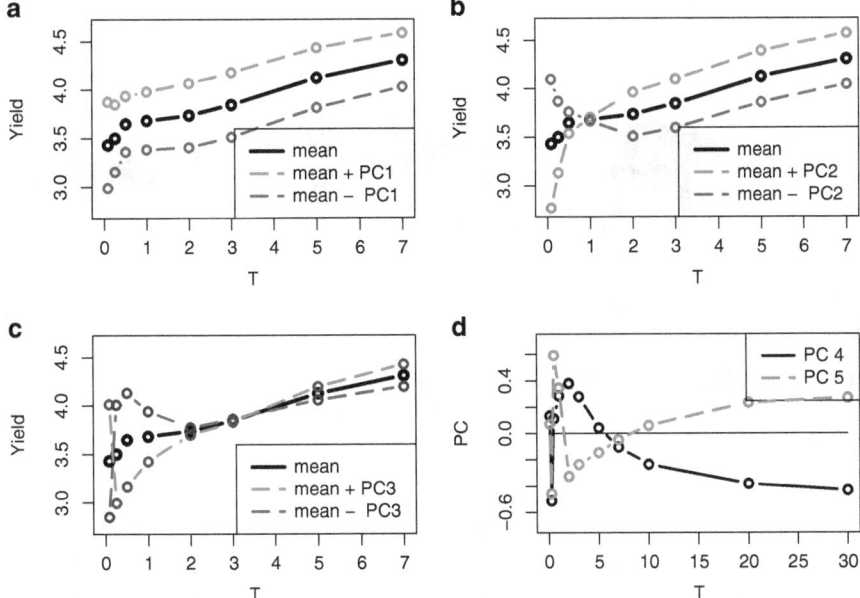

Fig. 18.2. *(a) The mean yield curve plus and minus the first eigenvector. (b) The mean yield curve plus and minus the second eigenvector. (c) The mean yield curve plus and minus the third eigenvector. (d) The fourth and fifth eigenvectors for changes in the Treasury yields.*

The graph of o_2 is everywhere decreasing[3] and changes in this direction either increase or decrease the slope of the yield curve. The result is that a graph of the mean plus or minus PC2 will cross the graph of the mean curve at approximately $T = 1$, where o_2 equals zero; see Fig. 18.2b.

The graph of o_3 is first decreasing and then increasing, and the changes in this direction either increase or decrease the convexity of the yield curve. The result is that a graph of the mean plus or minus PC3 will cross the graph of the mean curve twice; see Fig. 18.2c. It is worth repeating a point just made in connection with PC1, since it is even more important here. The standard deviations in the directions of PC2 and PC3 are only 0.14 and 0.071, respectively, so observed changes in these directions will be much smaller than those shown in Fig. 18.2b and c. Moreover, parallel shifts will be larger than changes in slope, which will be larger than changes in convexity.

Figure 18.2d plots the fourth and fifth eigenvectors. The patterns in their graphs are complex and do not have easy interpretations. Fortunately, the variation in the space they span is too small to be of much importance.

---

[3] The graph would, of course, be everywhere increasing if o_2 were multiplied by -1.

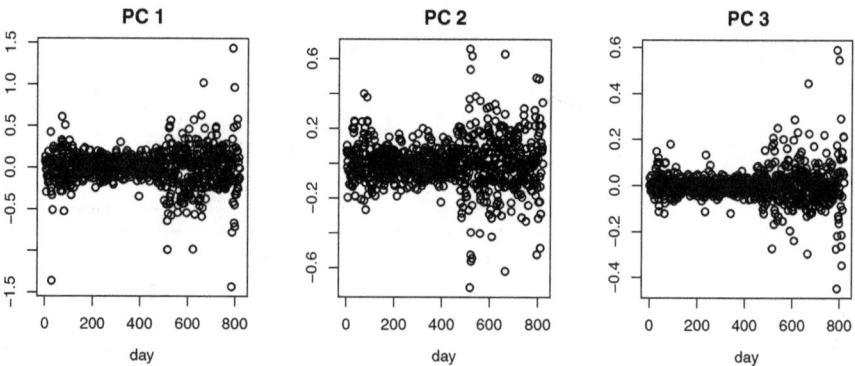

Fig. 18.3. *Time series plots of the first three principal components of the Treasury yields. There are 819 days of data, but they are not consecutive because of missing data; see text.*

A bond portfolio manager would be interested in the behavior of the yield changes over time. Time series analysis based on the changes in the 11 yields could be useful, but a better approach would be to use the first three principal components. Their time series and auto- and cross-correlation plots are shown in Figs. 18.3 and 18.4, respectively. The latter shows moderate short-term auto-correlations which could be modeled with an ARMA process, though the correlation is small enough that it might be ignored. Notice that the lag-0 cross-correlations are zero; this is not a coincidence but rather is due to the way the principal components are defined. They are defined to be uncorrelated with each other, so their lag-0 correlations are exactly zero. Cross-correlations at nonzero lags are not zero, but in this example they are small. The practical implication is that parallel shifts, changes in slopes, and changes in convexity are nearly uncorrelated and could be analyzed separately. The time series plots show substantial volatility clustering which could be modeled using the GARCH models of Chap. 14.   □

Example 18.3. Principal components analysis of equity funds

This example uses the data set `equityFunds.csv`. The variables are daily returns from January 1, 2002 to May 31, 2007 on eight equity funds: EASTEU, LATAM, CHINA, INDIA, ENERGY, MINING, GOLD, and WATER. The following code was run:

```
equityFunds = read.csv("equityFunds.csv")
pcaEq = prcomp(equityFunds[ , 2:9])
summary(pcaEq)
```

The results in this example are below and are different than those for the changes in yields, because in this example the variation is less concentrated in the first few principal components. For example, the first three principal

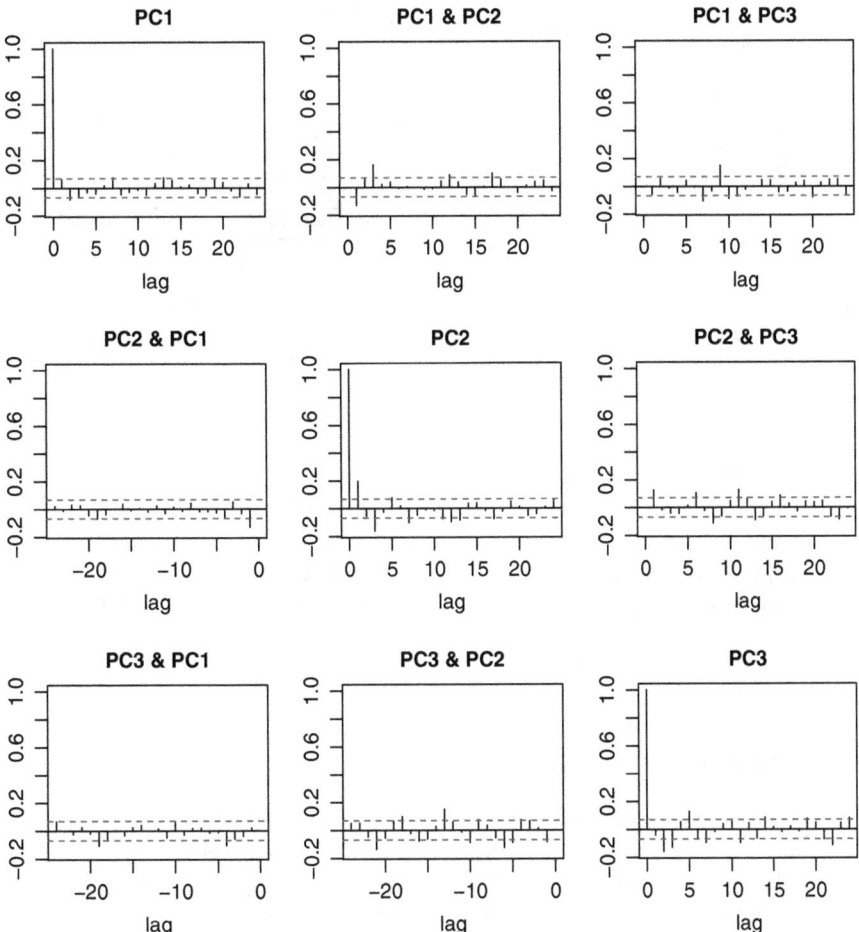

Fig. 18.4. *Sample auto- and cross-correlations of the first three principal components of the Treasury yields.*

components have only 75 % of the variance, compared to 95 % for the yield changes. For the equity funds, one needs six principal components to get 95 %. A scree plot is shown in Fig. 18.5a.

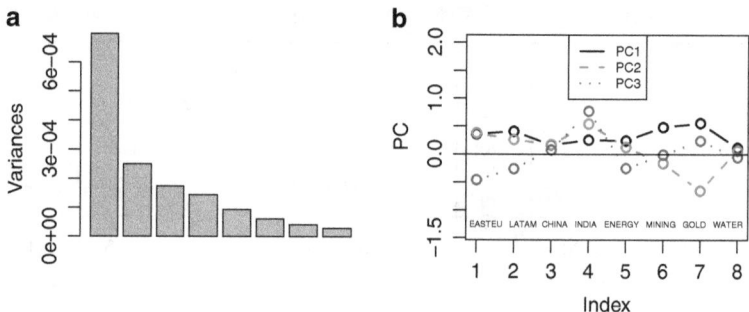

Fig. 18.5. *(a) Scree plot for the Equity Funds example. (b) The first three eigenvectors for the Equity Funds example.*

```
Importance of components:
                        PC1    PC2    PC3    PC4    PC5
Standard deviation     0.026  0.016  0.013  0.012  0.0097
Proportion of Variance 0.467  0.168  0.117  0.097  0.0627
Cumulative Proportion  0.467  0.635  0.751  0.848  0.9107

   PC6     PC7     PC8
 0.0079  0.0065  0.0055
 0.0413  0.0280  0.0201
 0.9520  0.9799  1.0000
```

The first three eigenvectors are plotted in Fig. 18.5b. The first eigenvector has only positive values, and returns in this direction are either positive for all of the funds or negative for all of them. The second eigenvector is negative for mining and gold (funds 6 and 7) and positive for the other funds. Variation along this eigenvector has mining and gold moving in the opposite direction of the other funds. Gold and mining stock moving counter to the rest of the stock market is a common occurrence and, in fact, these types of stock often have negative betas, so it is not surprising that the second principal component has 17 % of the variation. The third principal component is less easy to interpret, but its loading on India (fund 4) is higher than on the other funds, which might indicate that there is something different about Indian equities.   □

Example 18.4. Principal components analysis of the Dow Jones 30

As a further example, we will use returns on the 30 stocks on the Dow Jones average. The data are in the data set **DowJone30.csv** and cover the period from January 2, 1991 to January 2, 2002 The first five principal components have over 97 % of the variation:

```
Importance of components:
                        PC1     PC2     PC3     PC4     PC5
Standard deviation      88.53 24.967  13.44  10.602 8.2165
Proportion of Variance  0.87   0.069   0.02  0.012 0.0075
Cumulative Proportion   0.87   0.934   0.95  0.967 0.9743
```

In contrast to the analysis of the equity funds where six principal components were needed to obtain 95 % of the variance, here the first three principal components have over 95 % of the variance. Why are the Dow Jones stocks behaving differently compared to the equity funds? The Dow Jones stocks are similar to each other since they are all large companies in the United States. Thus, we can expect that their returns will be highly correlated with each other and a few principal components will explain most of the variation.   □

18.3 Factor Models

A factor model for excess equity returns is

$$R_{j,t} = \beta_{0,j} + \beta_{1,j} F_{1,t} + \cdots + \beta_{p,j} F_{p,t} + \epsilon_{j,t}, \tag{18.3}$$

where $R_{j,t}$ is either the return or the excess return on the jth asset at time t, $F_{1,t}, \ldots, F_{p,t}$ are variables, called *factors* or *risk factors*, that represent the "state of the financial markets and world economy" at time t, and $\epsilon_{1,t}, \ldots, \epsilon_{n,t}$ are uncorrelated, mean-zero random variables called the *unique risks* of the individual stocks. The assumption that unique risks are uncorrelated means that all cross-correlation between the returns is due to the factors. Notice that the factors do not depend on j since they are common to all returns. The parameter $\beta_{i,j}$ is called a factor loading and specifies the sensitivity of the jth return to the ith factor. Depending on the type of factor model, either the loadings, the factors, or both the factors and the loadings are unknown and must be estimated.

The CAPM is a factor model where $p = 1$ and $F_{1,t}$ is the excess return on the market portfolio. In the CAPM, the market risk factor is the only source of risk besides the unique risk of each asset. Because the market risk factor is the only risk that any two assets share, it is the sole source of correlation between asset returns. Factor models generalize the CAPM by allowing more factors than simply the market risk and the unique risk of each asset. A *factor* can be any variable thought to affect asset returns. Examples of factors include:

1. returns on the market portfolio;
2. growth rate of the GDP;
3. interest rate on short term Treasury bills or changes in this rate;
4. inflation rate or changes in this rate;
5. interest rate spreads, for example, the difference between long-term Treasury bonds and long-term corporate bonds;

6. return on some portfolio of stocks, for example, all U.S. stocks or all stocks with a high ratio of book equity to market equity — this ratio is called BE/ME in Fama and French (1992, 1995, 1996);
7. the difference between the returns on two portfolios, for example, the difference between returns on stocks with high BE/ME values and stocks with low BE/ME values.

With enough factors, most, and perhaps all, commonalities between assets should be accounted for in the model. Then the $\epsilon_{j,t}$ should represent factors truly unique to the individual assets and therefore should be uncorrelated across j (across assets), as is being assumed.

Factor models that use macroeconomic variables such as 1–5 as factors are called *macroeconomic factor models*. *Fundamental factor models* use observable asset characteristics (fundamentals) such as 6 and 7 as factors. Both types of factor models can be fit by time series regression, the topic of the next section. Fundamental factor models can also be fit by cross-sectional regression, as explained in Sect. 18.5.

18.4 Fitting Factor Models by Time Series Regression

Equation (18.3) is a regression model. If j is fixed, then it is a univariate multiple regression model, "univariate" because there is one response (the return on the jth asset) and "multiple" since there can be several predictor variables (the factors). If we combine these models across j, then we have a multivariate regression model, that is, a regression model with more than one response. Multivariate regression is used when fitting a set of returns to factors.

As discussed in Sect. 17.6, when fitting time series regression models, one should use data at the highest sampling frequency available, which is often daily or weekly, though only monthly data were available for the next example.

Example 18.5. A macroeconomic factor model

The efficient market hypothesis implies that stock prices change because of new information. Although there is considerable debate about the extent to which markets are efficient, one still can expect that stock returns will be influenced by unpredictable changes in macroeconomic variables. Accordingly, the factors in a macroeconomic model are not the macroeconomic variables themselves, but rather the residuals when changes in the macroeconomic variables are predicted from past data by a time series model, such as, a multivariate AR model.

In this example, we look at a subset of a case study that has been presented by other authors; see the bibliographical notes in Sect. 18.7. The macroeconomic variables in this example are changes in the logs of CPI (Consumer

Price Index) and IP (Industrial Production). The changes in these series have been analyzed before in Examples 12.10, 12.11, and 13.10 and in that last example a bivariate AR model was fit. It was found that the AR(5) model minimized AIC, but the AR(1) had an AIC value nearly as small as the AR(5) model.

In this example, we will use the residuals from the AR(5) model as the factors. Monthly returns on nine stocks were taken from the `berndtInvest.csv` data set. The returns are from January 1978 to December 1987. The CPI and IP series from July 1977 to December 1987 were used, but the month of July 1977 was lost through differencing. This left enough data (the five months August 1977 to December 1977) for forecasting CPI and IP beginning January 1978 when the return series started.

R^2 and the slopes for the regressions of the stock returns on the CPI residuals and the IP residuals are plotted in Fig. 18.6 for each of the 9 stocks. Note that the R^2-values are very small, so the macroeconomic factors have little explanatory power. The problem of low explanatory power is common with macroeconomic factor models and has been noticed by other authors. For this reason, fundamental factor models are more widely used than macroeconomic models. $\square$

18.4.1 Fama and French Three-Factor Model

Fama and French (1995) have developed a fundamental factor model with three risk factors, the first being the excess return of the market portfolio, which is the sole factor in the CAPM. The second risk factor, which is called small minus big (SMB), is the difference in returns on a portfolio of small stocks and a portfolio of large stocks. Here "small" and "big" refer to the size of the *market value*, which is the share price times the number of shares outstanding. The third factor, HML (high minus low), is the difference in returns on a portfolio of high book-to-market value (BE/ME) stocks and a portfolio of low BE/ME stocks. *Book value* is the net worth of the firm according to its accounting balance sheet. Fama and French argue that most pricing anomalies that are inconsistent with the CAPM disappear in the three-factor model. Their model of the return on the jth asset for the tth holding period is

$$R_{j,t} - \mu_{f,t} = \beta_{0,j} + \beta_{1,j}(R_{M,t} - \mu_{f,t}) + \beta_{2,j}\mathrm{SMB}_t + \beta_{3,j}\mathrm{HML}_t + \epsilon_{j,t},$$

where SMB_t and HML_t are the values of SMB and HML and $\mu_{f,t}$ is the risk-free rate for the tth holding period. Returns on portfolios have little autocorrelation, so the returns themselves, rather than residuals from a time series model, can be used.

Notice that this model does *not* use the size or the BE/ME ratio of the jth asset to explain returns. The coefficients $\beta_{2,j}$ and $\beta_{3,j}$ are the loading of the jth asset on SMB and HML. These loadings may, but need not, be

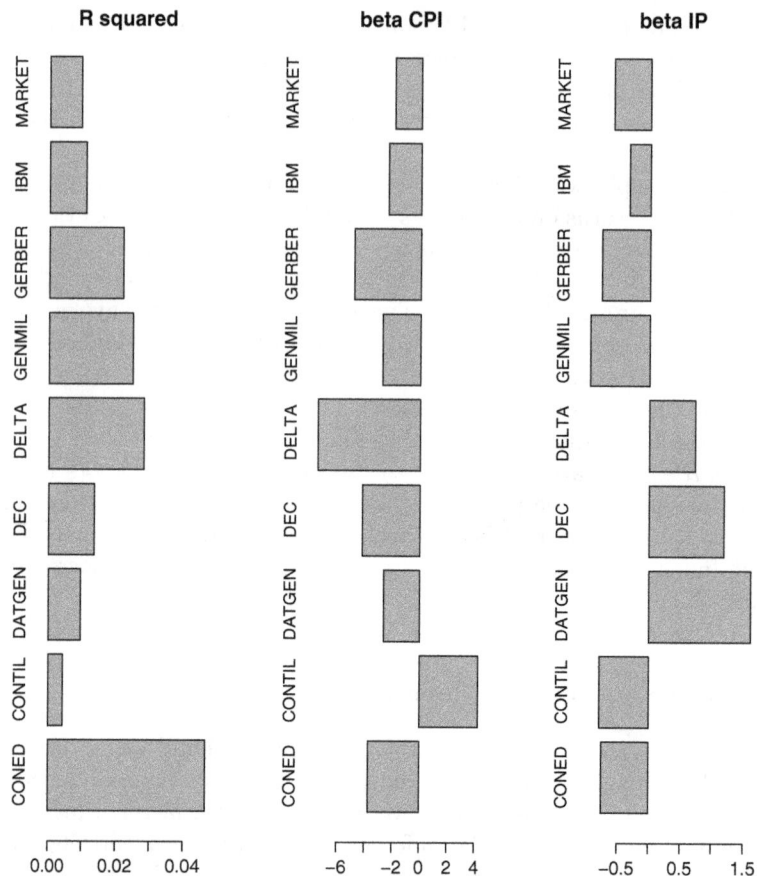

Fig. 18.6. R^2 *and slopes of regressions of stock returns on CPI residuals and IP residuals.*

related to the size and to the BE/ME ratio of the jth asset. In any event, the loadings are estimated by regression, not by measuring the size or BE/ME of the jth asset. If the loading $\beta_{2,j}$ of the jth asset on SMB is high, that might be because the jth asset is small or it might be because that asset is large but, in terms of returns, behaves similarly to small assets.

For emphasis, it is mentioned again that the factors SMB_t and HML_t do not depend on j since they are differences between returns on two fixed portfolios, not variables that are measured on the jth asset. This is true in general of the factors and loadings in model (18.3), not just the Fama–French model—only the loadings, that is, the parameters $\beta_{k,j}$, depend on the asset j. The factors are macroeconomic variables, linear combinations of returns on portfolios, or other variables that depend only on the financial markets and the economy as a whole.

There are many reasons why book and market values may differ. Book value is determined by accounting methods that do not necessarily reflect market values. Also, a stock might have a low book-to-market value because investors expect a high return on equity, which increases its market value relative to its book value. Conversely, a high book-to-market value could indicate a firm that is in trouble, which decreases its market value. A low market value relative to the book value is an indication of a stock's "cheapness," and stocks with a high market-to-book value are considered *growth stocks* for which investors are willing to pay a premium because of the promise of higher future earnings. Stocks with a low market-to-book value are called *value stocks* and investing in them is called *value investing*.

SMB and HML are the returns on portfolio that are long on one group of stocks and short on another. Such portfolios are called *hedge portfolios* since they are hedged, though perhaps not perfectly, against changes in the overall market.

Example 18.6. Fitting the Fama–French model to GE, IBM, and Mobil

This example uses two data sets. The first is CRSPmon in R's Ecdat package. This is similar to the CRSPday data set used in previous examples except that the returns are now monthly rather than daily. There are returns on three equities, GE, IBM, and Mobil, as well as on the CRSP average, though we will not use the last one here. The returns are from January 1969 to December 1998. The second data set is the Fama–French factors and was taken from the website of Prof. Kenneth French.

Figure 18.7 is a scatterplot matrix of the GE, IBM, and Mobil excess returns and the factors. Focusing on GE, we see that, as would be expected, GE excess returns are highly correlated with the excess market returns. The GE returns are negatively related with the factor HML which would indicate that GE behaves as a growth stock, since it moves in the same direction as low BE/ME stocks and in the opposite direction of high BE/ME stocks. However, this is a false impression caused by the lack of adjustment for associations between GE excess returns and the other factors. Regression analysis will be used soon to address this problem. The two Fama–French factors are not quite hedge portfolios since SMB is positively and HML negatively related to the excess market return. However, these associations are far weaker than that between the excess returns on the stocks and the market excess returns. Moreover, SMB and HML have little association between each other, so multicollinearity is not a problem.

The three excess equity returns were regressed on the three factors using the lm() function in R. The code is:

```
FF_data = read.table("FamaFrench_mon_69_98.txt", header = TRUE)
attach(FF_data)
library("Ecdat")
```

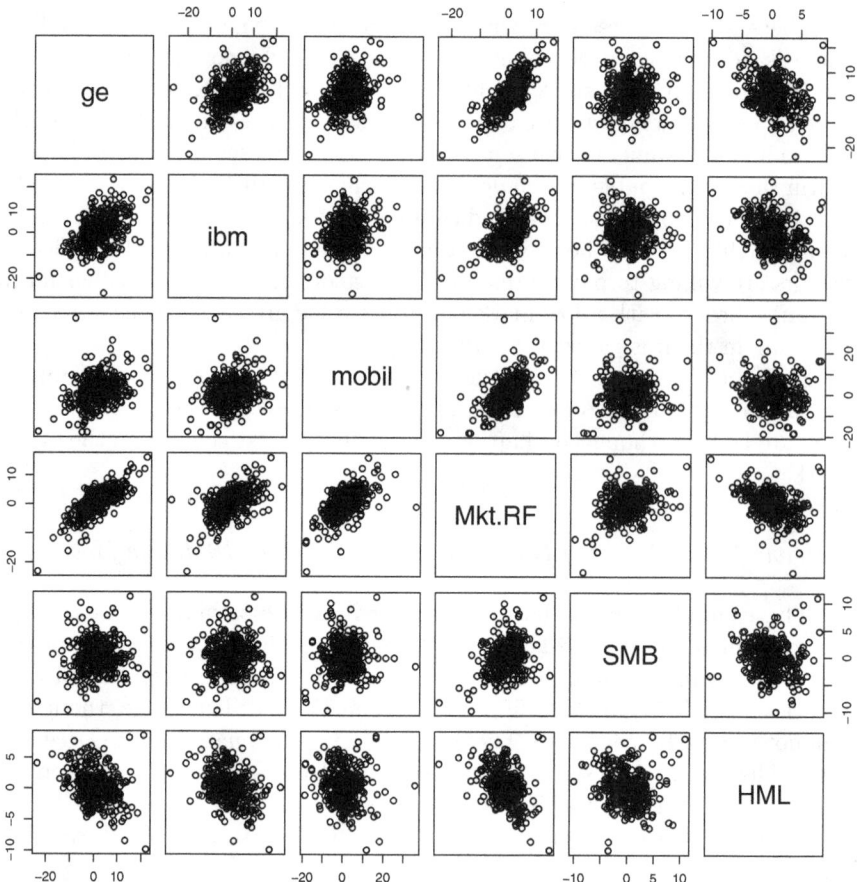

Fig. 18.7. *Scatterplot matrix of the excess returns on GE, IBM, and Mobil and the three factors in the Fama–French model.* `Mkt.RF` *is the return on the market portfolio minus the risk-free rate.*

```
library("robust")
data(CRSPmon)
ge = 100*CRSPmon[,1] - RF
ibm = 100*CRSPmon[,2] - RF
mobil = 100*CRSPmon[,3] - RF
stocks = cbind(ge, ibm, mobil)
fit = lm(cbind(ge, ibm, mobil) ~ Mkt.RF + SMB + HML)
fit
```

and the estimated coefficients are

```
Call:
lm(formula = cbind(ge, ibm, mobil) ~ Mkt.RF + SMB + HML)
```

```
Coefficients:
                ge       ibm      mobil
(Intercept)    0.3443    0.1460    0.1635
Mkt.RF         1.1407    0.8114    0.9867
SMB           -0.3719   -0.3125   -0.3753
HML            0.0095   -0.2983    0.3725
```

The coefficients of HML indicate that GE and Mobil are value stocks and IBM is a growth stock. Notice that GE now has a positive relationship with HML, not the negative relationship seen in Fig. 18.7, although its coefficient is close to 0. GE seems to be somewhere in between being a growth stock and a value stock.

All three equity returns have negative relationships with SMB, so, not surprisingly, they behave like large stocks.

Recall that one important assumption of the factor model is that the $\epsilon_{j,t}$ in (18.3) are uncorrelated. Violation of this assumption, that is, cross-correlations between $\epsilon_{j,t}$ and $\epsilon_{j',t}$, $j \neq j'$, will create biases when the factor model is used to estimate correlations between the equity returns, a topic explained in the next section. Lack of cross-correlation is not an assumption of the multivariate regression model and does not cause bias in the estimation of the regression coefficients or the variances of the $\epsilon_{j,t}$. The biases arise only when estimating covariances between the equity returns.

To check for cross-correlations, we will use the residuals from the multivariate regression. Their sample correlation matrix is

```
> cor(fit$residuals)
              ge      ibm   mobil
ge         1.000   0.071  -0.25
ibm        0.071   1.000  -0.10
mobil     -0.254  -0.102   1.00
```

The correlation between GE and Mobil is rather far from zero and is worth checking. A 95 % confidence interval for the residual correlations between GE excess returns and Mobil excess returns does not include 0, so a test would reject the null hypotheses that the true correlation is 0. The other correlations are not significantly different from 0. Because of the large negative GE–Mobil correlation, we should be careful about using the Fama–French model for estimation of the covariance matrix of the equity returns. As always, it is good practice to look at scatterplot matrices as well as correlations, since scatterplots may be outliers or nonlinear relationships affecting the correlations. Figure 18.8 contains a scatterplot matrix of the residuals. One sees that there are few outliers. although none of the outliers is really extreme, it seems worthwhile to compute robust correlations estimates and to compare them with the ordinary sample correlation matrix. Robust estimates were found using the function covRob() in R's robust package. What was found is that the robust estimates are all closer to zero than the nonrobust estimates, but the robust correlation estimate for GE and Mobil is still a large negative value.

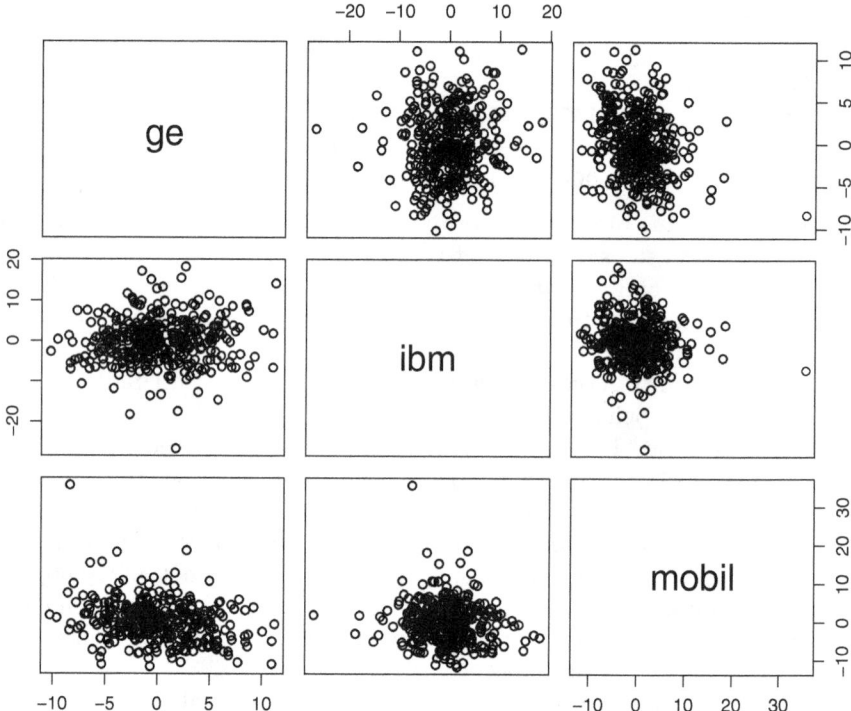

Fig. 18.8. *Scatterplot matrix of the residuals for GE, IBM, and Mobil from the Fama–French model.*

```
Call:
covRob(data = fit$residuals, corr = T)

Robust Estimate of Correlation:
          ge     ibm    mobil
ge     1.000  0.0360 -0.2479
ibm    0.036  1.0000 -0.0687
mobil -0.248 -0.0687  1.0000
```

This example is atypical of real applications because, for illustration purposes, the number of returns has been kept low, only three, whereas in portfolio management the number of returns will be larger and might be in the hundreds.   □

18.4.2 Estimating Expectations and Covariances of Asset Returns

Section 17.7 discussed how the CAPM can simplify the estimation of expectations and covariances of asset returns. However, using the CAPM for this purpose can be dangerous since the estimates depend on the validity of the CAPM. Fortunately, it is also possible to estimate return expectations and covariances using a more realistic factor model instead of the CAPM.

We start with two factors for simplicity. From (18.3), now with $p = 2$, we have

$$R_{j,t} = \beta_{0,j} + \beta_{1,j}F_{1,t} + \beta_{2,j}F_{2,t} + \epsilon_{j,t}. \tag{18.4}$$

It follows from (18.4) that

$$E(R_{j,t}) = \beta_{0,j} + \beta_{1,j}E(F_{1,t}) + \beta_{2,j}E(F_{2,t}) \tag{18.5}$$

and

$$\mathrm{Var}(R_{j,t}) = \beta_{1,j}^2 \mathrm{Var}(F_1) + \beta_{2,j}^2 \mathrm{Var}(F_2) + 2\beta_{1,j}\beta_{2,j}\mathrm{Cov}(F_1, F_2) + \sigma_{\epsilon,j}^2.$$

Also, because $R_{j,t}$ and $R_{j',t}$ are two linear combinations of the risk factors, it follows from (7.8) that for any $j \neq j'$,

$$\begin{aligned} \mathrm{Cov}(R_{j,t}, R_{j',t}) = {} & \beta_{1,j}\beta_{1,j'}\mathrm{Var}(F_1) + \beta_{2,j}\beta_{2,j'}\mathrm{Var}(F_2) \\ & + (\beta_{1,j}\beta_{2,j'} + \beta_{1,j'}\beta_{2,j})\mathrm{Cov}(F_1, F_2). \end{aligned} \tag{18.6}$$

More generally, let

$$\boldsymbol{F}_t^\mathsf{T} = (F_{1,t}, \ldots, F_{p,t}) \tag{18.7}$$

be the vector of p factors at time t and suppose that $\boldsymbol{\Sigma}_F$ is the $p \times p$ covariance matrix of $\boldsymbol{F}_t$. Define the vector of intercepts

$$\boldsymbol{\beta}_0^\mathsf{T} = (\beta_{0,1}, \ldots, \beta_{0,n})$$

and the matrix of loadings

$$\boldsymbol{\beta} = \begin{pmatrix} \beta_{1,1} & \cdots & \beta_{1,j} & \cdots & \beta_{1,n} \\ \vdots & \ddots & \vdots & \ddots & \vdots \\ \beta_{p,1} & \cdots & \beta_{p,j} & \cdots & \beta_{p,n} \end{pmatrix}.$$

Also, define

$$\boldsymbol{\epsilon}^\mathsf{T} = (\epsilon_{1,t}, \ldots, \epsilon_{n,t}) \tag{18.8}$$

and let $\boldsymbol{\Sigma_\epsilon}$ be the $n \times n$ diagonal covariance matrix of $\boldsymbol{\epsilon}$:

$$\boldsymbol{\Sigma_\epsilon} = \begin{pmatrix} \sigma_{\epsilon,1}^2 & \cdots & 0 & \cdots & 0 \\ \vdots & \ddots & \vdots & \ddots & \vdots \\ 0 & \cdots & \sigma_{\epsilon,j}^2 & \cdots & 0 \\ \vdots & \ddots & \vdots & \ddots & \vdots \\ 0 & \cdots & 0 & \cdots & \sigma_{\epsilon,n}^2 \end{pmatrix}.$$

Finally, let

$$R_t^{\mathsf{T}} = (R_{1,t} \ldots, R_{n,t}) \tag{18.9}$$

be the vector of all returns at time t. Model (18.3) then can be reexpressed in matrix notation as

$$R_t = \beta_0 + \beta^{\mathsf{T}} F_t + \epsilon_t. \tag{18.10}$$

Therefore, the $n \times n$ covariance matrix of R_t is

$$\Sigma_R = \beta^{\mathsf{T}} \Sigma_F \beta + \Sigma_\epsilon. \tag{18.11}$$

In particular, if $\beta_j = (\beta_{1,j} \quad \cdots \quad \beta_{p,j})^{\mathsf{T}}$ is the jth column of β, then the variance of the jth return is

$$\mathrm{Var}(R_j) = \beta_j^{\mathsf{T}} \Sigma_F \beta_j + \sigma_{\epsilon_j}^2, \tag{18.12}$$

and the covariance between the jth and j'th returns is

$$\mathrm{Cov}(R_j, R_j') = \beta_j^{\mathsf{T}} \Sigma_F \beta_{j'}. \tag{18.13}$$

To use (18.11), (18.12) or (18.13), one needs estimates of β, Σ_F, and Σ_ϵ. The regression coefficients are used to estimate β, the sample covariance of the factors can be used to estimate Σ_F, and $\widehat{\Sigma}_\epsilon$ can be the diagonal matrix of the mean residual sum of squared errors from the regressions; see equation (9.13).

Why estimate Σ_R via a factor model instead of simply using the sample covariance matrix? One reason is estimation accuracy. This is another example of bias–variance tradeoff. The sample covariance matrix is unbiased, but it contains $n(n + 1)/2$ estimates, one for each covariance and each variance. Each of these parameters is estimated with error and when this many errors accumulate, the result can be a sizable loss of precision. In contrast, the factor model requires estimates of $n \times p$ parameters in β, $p(p+1)/2$ parameters in Σ_F, and n parameters in the diagonal matrix Σ_ϵ, for a total of $np+n+p(p+1)/2$ parameters. Typically, n, the number of returns, is large but p, the number of factors, is much smaller, so $np+n+p(p+1)/2$ is much smaller than $n(n+1)/2$. For example, suppose there are 200 returns and 5 factors. Then $n(n + 1)/2 = 20{,}100$ but $np + n + p(p + 1)/2$ is only 1,215. The downside of the factor model is that there will be bias in the estimate of Σ_R if the factor model is misspecified, especially if Σ_ϵ is not diagonal as the factor model assumes.

Another advantage of the factor model is expediency. Having fewer parameters to estimate is one convenience and another is ease of updating. Suppose a portfolio manager has implemented a factor model for n equities and now needs to add another equity. If the manager uses the sample covariance matrix, then the n sample covariances between the new return time series and the old ones must be computed. This requires that all n of the old time series be available. In comparison, with a factor model, the portfolio manager needs only to regress the new return time series on the factors. Only the p factor time series need to be available.

Example 18.7. Estimating the covariance matrix of GE, IBM, and Mobil excess returns

This example continues Example 18.6. Recall that the number of returns has been kept artificially low, since with more returns it would not have been possible to display the results. Therefore, this example merely illustrates the calculations and is not a typical application of factor modeling.

The estimate of Σ_F is the sample covariance matrix of the factors:

```
         Mkt.RF     SMB      HML
Mkt.RF  21.1507  4.2326  -5.1045
SMB      4.2326  8.1811  -1.0760
HML     -5.1045 -1.0760   7.1797
```

The estimate of β is the matrix of regression coefficients (without the intercepts):

```
        Mkt.RF       SMB       HML
ge     1.14071  -0.37193   0.009503
ibm    0.81145  -0.31250  -0.298302
mobil  0.98672  -0.37530   0.372520
```

The estimate of Σ_ϵ is the diagonal matrix of residual error MS values:

```
         [,1]    [,2]    [,3]
[1,]  16.077   0.000   0.000
[2,]   0.000  31.263   0.000
[3,]   0.000   0.000  27.432
```

Therefore, the estimate of $\beta^T \Sigma_F \beta$ is

```
         ge     ibm    mobil
ge     24.960 19.303 19.544
ibm    19.303 15.488 14.467
mobil  19.544 14.467 16.155
```

and the estimate of $\beta^T \Sigma_F \beta + \Sigma_\epsilon$ is

```
         ge     ibm    mobil
ge     41.036 19.303 19.544
ibm    19.303 46.752 14.467
mobil  19.544 14.467 43.587
```

For comparison, the sample covariance matrix of the equity returns is

```
         ge     ibm    mobil
ge     40.902 20.878 14.255
ibm    20.878 46.491 11.518
mobil  14.255 11.518 43.357
```

The largest difference between the estimate of $\boldsymbol{\beta}^\mathsf{T} \boldsymbol{\Sigma}_F \boldsymbol{\beta} + \boldsymbol{\Sigma}_\epsilon$ and the sample covariance matrix is in the covariance between the excess returns on GE and Mobil. The reason for this large discrepancy is that the factor model assumes a zero residual correlation between these two variables, but, as we learned earlier, the data show a negative correlation of -0.25.

The code for the calculations in this example continues the code in Example 18.6. The addition code is:

```
sigF = as.matrix(var(cbind(Mkt.RF, SMB, HML)))
bbeta = as.matrix(fit$coef)
bbeta = t( bbeta[-1, ])
n = dim(CRSPmon)[1]
sigeps = (n - 1) / (n - 4) * as.matrix((var(as.matrix(fit$resid))))
sigeps = diag(as.matrix(sigeps))
sigeps = diag(sigeps, nrow = 3)
cov_equities = bbeta %*% sigF %*% t(bbeta) + sigeps
options(digits = 5)
sigF
bbeta
sigeps
bbeta %*% sigF %*% t(bbeta)
cov_equities
cov(stocks)
```

□

18.5 Cross-Sectional Factor Models

Models of the form (18.3) are *time series factor models*. They use time series data, one single asset at a time, to estimate the loadings.

As just discussed, time series factor models do not make use of variables such as dividend yields, book-to-market value, or other variables specific to the jth firm. An alternative is a *cross-sectional factor model*, which is a regression model using data from many assets but from only a single holding period. For example, suppose that R_j, $(B/M)_j$, and D_j are the return, book-to-market value, and dividend yield for the jth asset for some fixed time t. Since t is fixed, it will not be made explicit in the notation. Then a possible cross-sectional factor model is

$$R_j = \beta_0 + \beta_1 (B/M)_j + \beta_2 D_j + \epsilon_j.$$

The parameters β_1 and β_2 are unknown values at time t of a book-to-market value risk factor and a dividend yield risk factor. These values are estimated by regression.

There are two fundamental differences between time series factor models and cross-sectional factor models. The first is that with a time series factor model one estimates parameters, one asset at a time, using multiple holding periods, while in a cross-sectional model one estimates parameters, one single holding period at a time, using multiple assets. The other major difference is that in a time series factor model, the factors are directly measured and the loadings are the unknown parameters to be estimated by regression. In a cross-sectional factor model the opposite is true; the loadings are directly measured and the factor values are estimated by regression.

Example 18.8. An industry cross-sectional factor model

This example uses the `berndtInvest.csv` used in Example 18.5. This data set has monthly returns on 15 stocks over 10 years, 1978 to 1987. The 15 stocks were classified into three industries, "Tech," "Oil," and "Other," as follows:

```
       tech oil other
CITCRP  0   0    1
CONED   0   0    1
CONTIL  0   1    0
DATGEN  1   0    0
DEC     1   0    0
DELTA   0   1    0
GENMIL  0   0    1
GERBER  0   0    1
IBM     1   0    0
MOBIL   0   1    0
PANAM   0   1    0
PSNH    0   0    1
TANDY   1   0    0
TEXACO  0   1    0
WEYER   0   0    1
```

We used the indicator variables of "tech" and "oil" as loadings and fit the model

$$R_j = \beta_0 + \beta_1 \text{tech}_j + \beta_2 \text{oil}_j + \epsilon_j, \tag{18.14}$$

where R_j is the return on the jth stock, tech_j equals 1 if the jth stock is a technology stock and equals 0 otherwise, and oil_j is defined similarly. Model (18.14) was fit separately for each of the 120 months. The estimates $\widehat{\beta}_0$, $\widehat{\beta}_1$, and $\widehat{\beta}_3$ for a month were the values of the three factors for that month. The loadings were the known values of tech_j and oil_j.

Factor 1, the values of $\widehat{\beta}_0$, can be viewed as an overall market factor, since it affects all 15 returns. Factors 2 and 3 are the technology and oil factors. For example, if the value of factor 2 is positive in any given month, then Tech stocks have better-than-market returns that month. Figure 18.9 contains time series plots of the three factor series, and Fig. 18.10 shows their auto- and cross-correlation functions. The largest cross-correlation is between the tech and oil factors at lag 0, which indicates that above- (below-) market returns for technology stocks are associated with above (below) market returns for oil stocks.

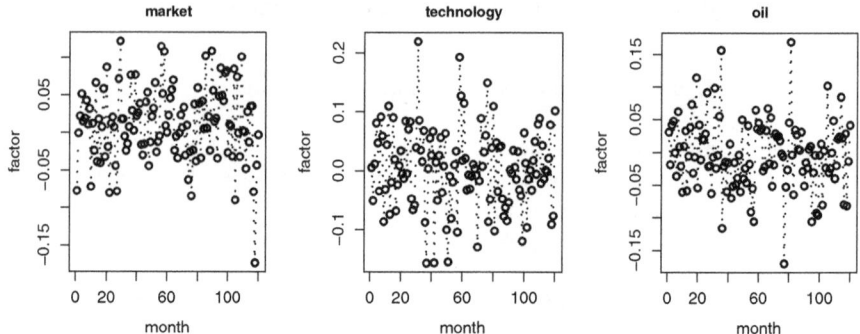

Fig. 18.9. *Time series plots of the estimated values of the three factors in the cross-sectional factor model.*

The standard deviations of the three factors are

```
market    tech    oil
 0.049   0.069   0.053
```

There are other ways of defining the factors. For example, Zivot and Wang (2006) use the model

$$R_j = \beta_1 \mathtt{tech}_j + \beta_2 \mathtt{oil}_j + \beta_3 \mathtt{other}_j + \epsilon_j, \tag{18.15}$$

with no intercept but with $\mathtt{other}_j$ as a third variable. With this model, there is no market factor but instead factors for all three industries. The model with an intercept but without $\mathtt{other}$ is equivalent to the model with $\mathtt{other}$ in place of the intercept, in the sense that the two models produce the same fitted values. □

Cross-sectional factor models are sometimes called BARRA models after BARRA, Inc., a company that has been developing cross-sectional factor models and marketing the output of their models to financial managers.

18.6 Statistical Factor Models

In a statistical factor model, neither the factor values nor the loadings are directly observable. All that is available is the sample $\boldsymbol{Y}_1, \dots, \boldsymbol{Y}_n$ or, perhaps, only the sample covariance matrix. This is the same type of data available

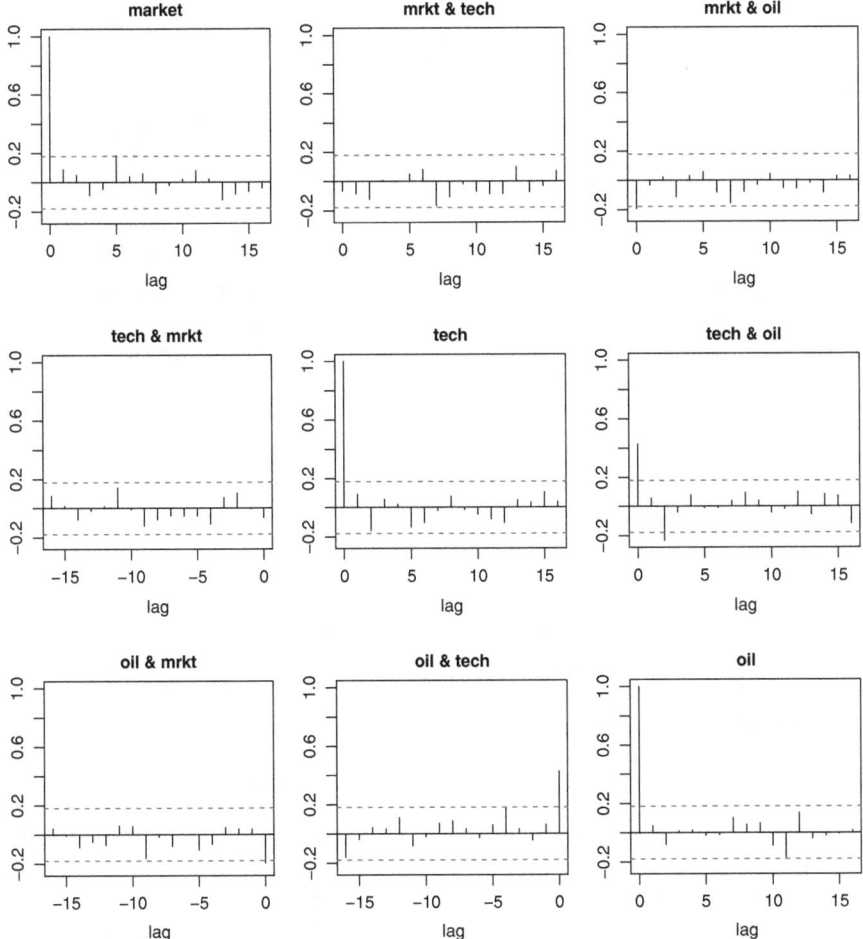

Fig. 18.10. *Auto- and cross-correlation plots of the estimated three factors in the cross-sectional factor model. Series 1–3 are the market, tech, and oil factors, respectively.*

for PCA and we will see that statistical factor analysis and PCA have some common characteristics. As with PCA, one can work with either the standardized or unstandardized variables. R's `factanal()` function automatically standardizes the variables.

We start with the multifactor model in matrix notation (18.10) and the return covariance matrix (18.11) which for convenience will be repeated as

$$\boldsymbol{R}_t = \boldsymbol{\beta}_0 + \boldsymbol{\beta}^\mathsf{T} \boldsymbol{F}_t + \boldsymbol{\epsilon}_t. \tag{18.16}$$

and

$$\boldsymbol{\Sigma}_R = \boldsymbol{\beta}^\mathsf{T} \boldsymbol{\Sigma}_F \boldsymbol{\beta} + \boldsymbol{\Sigma}_\epsilon. \tag{18.17}$$

Here $\boldsymbol{\beta}^\mathsf{T}$ is $d \times p$ where d is the dimension of R_t and p is the number of factors.

The only component of (18.17) that can be estimated directly from the data is $\boldsymbol{\Sigma}_R$. One can use this estimate to find estimates of $\boldsymbol{\beta}$, $\boldsymbol{\Sigma}_F$, and $\boldsymbol{\Sigma}_\epsilon$. However, it is too much to ask that all three of these matrices be identified from $\boldsymbol{\Sigma}_R$ alone. Here is the problem: Let $\boldsymbol{A}$ be any $p \times p$ invertible matrix. Then the returns vector $\boldsymbol{R}_t$ in (18.16) is unchanged if $\boldsymbol{\beta}^\mathsf{T}$ is replaced by $\boldsymbol{\beta}^\mathsf{T} \boldsymbol{A}^{-1}$ and $\boldsymbol{F}_t$ is replaced by $\boldsymbol{A}\boldsymbol{F}_t$. Therefore, the returns only determine $\boldsymbol{\beta}$ and $\boldsymbol{F}_t$ up to a nonsingular linear transformation, and consequently a set of constraints is needed to identify the parameters. The usual constraints are the factors are uncorrelated and standardized, so that

$$\boldsymbol{\Sigma}_F = \boldsymbol{I}, \tag{18.18}$$

where $\boldsymbol{I}$ is the $p \times p$ identity matrix. With these constraints, (18.17) simplifies to the statistical factor model

$$\boldsymbol{\Sigma}_R = \boldsymbol{\beta}^\mathsf{T} \boldsymbol{\beta} + \boldsymbol{\Sigma}_\epsilon. \tag{18.19}$$

However, even with this simplification, $\boldsymbol{\beta}$ is only determined up to a rotation, that is, by multiplication by an orthogonal matrix. To appreciate why this is so, let $\boldsymbol{P}$ be any orthogonal matrix, so that $\boldsymbol{P}^\mathsf{T} = \boldsymbol{P}^{-1}$. Then (18.19) is unchanged if $\boldsymbol{\beta}$ is replaced by $\boldsymbol{P}\boldsymbol{\beta}$ since

$$(\boldsymbol{P}\boldsymbol{\beta})^\mathsf{T}(\boldsymbol{P}\boldsymbol{\beta}) = \boldsymbol{\beta}^\mathsf{T} \boldsymbol{P}^\mathsf{T} \boldsymbol{P}\boldsymbol{\beta} = \boldsymbol{\beta}^\mathsf{T} \boldsymbol{P}^{-1} \boldsymbol{P}\boldsymbol{\beta} = \boldsymbol{\beta}^\mathsf{T} \boldsymbol{\beta}.$$

Therefore, to determine $\boldsymbol{\beta}$ a further set of constraints is needed. One possible set of constraints is that $\boldsymbol{\beta}\boldsymbol{\Sigma}_\epsilon^{-1}\boldsymbol{\beta}^\mathsf{T}$ is diagonal (Mardia et al., 1979, p. 258). Output from R's function `factanal()` satisfies this constraint when the argument `rotation` is set to `"none"`. If $\boldsymbol{\beta}$ is rotated as discussed in Sect. 18.6.1, then this constraint no longer holds.

If the main purpose of the statistical factor model is to estimate $\boldsymbol{\Sigma}_R$ by (18.19), then the choice of constraint is irrelevant since all constraints lead to the same product $\boldsymbol{\beta}^\mathsf{T}\boldsymbol{\beta}$. In particular, rotation of $\boldsymbol{\beta}$ does not change the estimate of $\boldsymbol{\Sigma}_R$.

It is helpful to compare three estimates of $\boldsymbol{\Sigma}_R$. The sample covariance matrix has full rank (rank $= d$) provided that $n > d$ as will be assumed here. Instead of the sample covariance matrix, one can perform PCA and estimate $\boldsymbol{\Sigma}_R$ by the sample covariance matrix of the first $p < d$ principal components. Then

$$\widehat{\boldsymbol{\Sigma}}_R = \boldsymbol{O}^\mathsf{T}\boldsymbol{O}.$$

where $\boldsymbol{O}^\mathsf{T}$ is the $d \times p$ matrix whose columns are the first d principal axes (eigenvectors) and the rank of $\widehat{\boldsymbol{\Sigma}}_R$ is only p so less than full rank. In contrast, (18.19) provides a full-rank estimate of $\boldsymbol{\Sigma}_R$ but with a simple structure, the sum of a rank p matrix and a diagonal matrix.

Example 18.9. Factor analysis of equity funds

This example continues the analysis of the equity funds data set that was used in Example 18.3 to illustrate PCA. The code for fitting a 4-factor model ($p = 4$) using factanal() is:

```
equityFunds = read.csv("equityFunds.csv")
fa_none = factanal(equityFunds[ , 2:9], 4, rotation = "none")
print(fa_none,cutoff = 0.1)
```

Here we specify no rotations. The output is:

```
> factanal(equityFunds[,2:9],4,rotation="none")

Call:
factanal(x = equityFunds[, 2:9], factors = 4,
        rotation = "none")

Uniquenesses:
EASTEU  LATAM  CHINA  INDIA ENERGY MINING   GOLD  WATER
 0.735  0.368  0.683  0.015  0.005  0.129  0.005  0.778

Loadings:
        Factor1 Factor2 Factor3 Factor4
EASTEU   0.387   0.169   0.293
LATAM    0.511   0.167   0.579
CHINA    0.310   0.298   0.362
INDIA    0.281   0.951
ENERGY   0.784                   0.614
MINING   0.786           0.425  -0.258
GOLD     0.798                  -0.596
WATER    0.340           0.298   0.109

                Factor1 Factor2 Factor3 Factor4
SS loadings       2.57    1.07    0.82    0.82
Proportion Var    0.32    0.13    0.10    0.10
Cumulative Var    0.32    0.46    0.56    0.66

Test of the hypothesis that 4 factors are sufficient.
The chi square statistic is 17 on 2 degrees of freedom.
The p-value is 2e-04
```

The "loadings" are the estimates $\widehat{\beta}^{\mathsf{T}}$. Since there are eight funds and four factors, the loadings are in an 8×4 matrix fa_none$loadings. The output above gives the sums of squares of the eight loadings for each factor. The Proportion Var row contains the SS loadings divided by 8, where 8 is the sum of the variances of the eight variables, since each variable has been standardized to have variance equal to 1.

By convention, any loading with an absolute value less than the parameter cutoff is not printed, and the default value of cutoff is 0.1.

Because all its loadings have the same sign, the first factor is an overall index of the eight funds. The second factor has large loadings on the four regional funds (EASTEU, LATAM, CHINA, INDIA) and small loadings on the four industry section funds (ENERGY, MINING, GOLD, WATER). The four regions are all emerging markets, so the second factor might be interpreted as an emerging markets factor. The fourth factor is a contrast of MINING and GOLD with ENERGY and WATER, and mimics a hedge portfolio that is long on ENERGY and WATER and short on GOLD and MINING. The third factor is less interpretable. The uniquenesses are the diagonal elements of the estimate $\widehat{\Sigma}_\epsilon$.

The output gives a p-value for testing the null hypothesis that there are at most four factors. The p-value is small, indicating that the null hypothesis should be rejected. However, four is that maximum number of factors that can be used by factanal() when there are only eight returns. Should we be concerned that we are not using enough factors? Recall the important distinction between statistical and practical significance that has been emphasized elsewhere in this book. One way to assess practical significance is to see how well the factor model can reproduce the sample correlation matrix. Since factanal() standardizes the variables, the factor model estimate of the correlation matrix is the estimate of the covariance matrix, that, using (18.19), is

$$\widehat{\beta}^\mathsf{T}\widehat{\beta} + \widehat{\Sigma}_\epsilon. \tag{18.20}$$

The code to calculate this estimate is

```
B_none = fa_none$loadings[ , ]
BB_none = B_none %*% t(B_none)
D_none = diag(fa_none$unique)
Sigma_R_hat = BB_none + D_none
```

Here B_none is $\widehat{\beta}^\mathsf{T}$ with no rotation, BB_none equals $\widehat{\beta}^\mathsf{T}\widehat{\beta}$ and D_none equals $\widehat{\Sigma}_\epsilon$.

The difference between this estimate and the sample correlation matrix is a 8×8 matrix. We would like all of its entries to be close to 0. Unfortunately, they are not as small as we would like. There are various ways to check if a matrix this size is "small." The smallest entry is -0.063 and the largest is 0.03. These are reasonably large discrepancies between correlation matrices. Also, the eigenvalues of the difference are

```
-7.5e-02 -6.0e-03 -3.4e-15 -2.0e-15
-1.3e-15  3.0e-15  7.7e-03  7.3e-02
```

Another way to check for smallness of the difference between the two estimates is to look at the estimates of the variance of an equally weighted portfolio (of the standardized returns), which is

$$w^\mathsf{T} \Sigma_R w,$$

where $w^\mathsf{T} = (1/8, \ldots, 1/8)$. These estimates are 0.37 and 0.42 using the factor model and the sample correlation matrix, respectively. The absolute difference, 0.06, is relatively large compared to either of the estimates. It is unclear whether this difference is due to a more parsimonious and accurate fit by the factor model (good) or due to bias from a lack of fit by the factor model (not good).   □

18.6.1 Varimax Rotation of the Factors

As discussed earlier, the estimate of the covariance matrix is unchanged if the loadings β are rotated by multiplication by an orthogonal matrix. Rotation might increase the interpretability of the loadings. In some applications, it is desirable for each loading to be either close to 0 or large, so that a variable will load only on a few factors, or even on only one factor. *Varimax* rotation attempts to make each loading either small or large by maximizing the sum of the variances of the squared loadings. Varimax rotation is the default with R's factanal() function, but this can be changed as in Example 18.9 where no rotation was used. In finance, having variables loading on only one or a few factors is not that important, and may even be undesirable, so varimax rotation may not advantageous.

We repeat again for emphasis that the estimate of Σ_R is not changed by rotation. The uniquenesses are also unchanged. Only the loadings change.

Example 18.10. Factor analysis of equity funds: Varimax rotation

The statistical factor analysis in Example 18.9 is repeated here but now with varimax rotation.

```
Call:
factanal(x = equityFunds[, 2:9], factors = 4,
        rotation = "varimax")

Uniquenesses:
EASTEU  LATAM   CHINA  INDIA ENERGY MINING   GOLD  WATER
 0.735  0.368   0.683  0.015  0.005  0.129  0.005  0.778

Loadings:
        Factor1 Factor2 Factor3 Factor4
EASTEU 0.436    0.175   0.148   0.148
LATAM  0.748    0.174           0.180
CHINA  0.494            0.247
INDIA  0.243            0.959
ENERGY 0.327    0.118           0.934
```

```
MINING 0.655   0.637              0.168
GOLD   0.202   0.971
WATER  0.418                      0.188


               Factor1 Factor2 Factor3 Factor4
SS loadings       1.80    1.45    1.03    1.00
Proportion Var    0.23    0.18    0.13    0.12
Cumulative Var    0.23    0.41    0.54    0.66
```

```
Test of the hypothesis that 4 factors are sufficient.
The chi square statistic is 17 on 2 degrees of freedom.
The p-value is 2e-04
```

The most notable change compared to the nonrotated loadings is that now all loadings with an absolute value above 0.1 are positive. Therefore, the factors all represent long positions, whereas before some were more like hedge portfolios. However, the rotated factors seem less interpretable compared to the unrotated factors, so a financial analyst might prefer the unrotated factors. □

18.7 Bibliographic Notes

The Fama–French three-factor model was introduced by Fama and French (1993) and discussed further in Fama and French (1995, 1996). Connor (1995) compares the three types of factor models and finds that macroeconomic factor models have less explanatory power than other factor models. Example 18.5 was adopted from Zivot and Wang (2006). Sharpe, Alexander, and Bailey (1999) has a brief description of the BARRA, Inc. factor model. The yields.txt data set is from the Rsafd package distributed by Professor René Carmona.

18.8 R Lab

18.8.1 PCA

In the first section of this lab, you will do a principal components analysis of daily yield data in the file yields.txt. R has functions, which we will use later, that automate PCA, but it is easy to do PCA "from scratch" and it is instructive to do this. First load the data and, to get a feel for what yield curves look like, plot the yield curves on days 1, 101, 201, 301, ..., 1101. There are 1352 yield curves in the data, so you will see a representative sample of them. The yield curves change slowly, which is why one should look at yield curves that are spaced rather far (100 days) apart.

```
yieldDat = read.table("yields.txt", header = T)
maturity = c((0:5), 5.5, 6.5, 7.5, 8.5, 9.5)
pairs(yieldDat)
par(mfrow = c(4,3))
for (i in 0:11)
{
plot(maturity, yieldDat[100 * i + 1, ], type = "b")
}
```

Next compute the eigenvalues and eigenvectors of the sample covariance matrix, print the results, and plot the eigenvalues as a scree plot.

```
eig = eigen(cov(yieldDat))
eig$values
eig$vectors
par(mfrow = c(1, 1))
barplot(eig$values)
```

The following R code plots the first four eigenvectors.

```
par(mfrow=c(2, 2))
plot(eig$vector[ , 1], ylim = c(-0.7, 0.7), type = "b")
abline(h = 0)
plot(eig$vector[ , 2], ylim = c(-0.7, 0.7), type = "b")
abline(h = 0)
plot(eig$vector[ , 3], ylim = c(-0.7, 0.7), type = "b")
abline(h = 0)
plot(eig$vector[ , 4], ylim = c(-0.7, 0.7), type = "b")
abline(h = 0)
```

Problem 1 *It is generally recommended that PCA be applied to time series that are stationary. Plot the first column of* yieldDat. *(You can look at other columns as well. You will see that they are fairly similar.) Does the plot appear stationary? Why or why not? Include your plot with your work.*

Another way to check for stationarity is to run the augmented Dickey–Fuller test. You can do that with the following code:

```
library("tseries")
adf.test(yieldDat[ , 1])
```

Problem 2 *Based on the augmented Dickey–Fuller test, do you think the first column of* yieldDat *is stationary? Why or why not?*

Run the following code to compute changes in the yield curves. Notice the use of [-1,] to delete the first row and similarly the use of [-n,].

```
n=dim(yieldDat)[1]
delta_yield = yieldDat[-1, ] - yieldDat[-n, ]
```

Plot the first column of `delta_yield` and run the augmented Dickey–Fuller test to check for stationarity.

Problem 3 *Do you think the first column of* `delta_yield` *is stationary? Why or why not?*

Run the following code to perform a PCA using the function `princomp()`, which is similar, although not identical, to `prcomp()`. By default, `princomp()` does a PCA on the covariance matrix, though there is an option to use the correlation matrix instead. We will use the covariance matrix. The second line of the code will print the names of the components in the object that is returned by `princomp()`. As you can see, the `names` function can be useful for learning just what is being returned. You can also get this information by typing `?princomp`.

```
pca_del = princomp(delta_yield)
names(pca_del)
summary(pca_del)
par(mfrow = c(1, 1)
plot(pca_del)
```

Problem 4 *(a) The output from* **names** *includes the following:*

```
[1] "sdev"  "loadings" "center"  "scores"
```

Describe each of these components in mathematical terms. To answer this part of the question, you can print and plot the components to see what they contain and use R's help for further information.
(b) What are the first two eigenvalues of the covariance matrix?
(c) What is the eigenvector corresponding to the largest eigenvalue?
(d) Suppose you wish to "explain" at least 95 % of the variation in the changes in the yield curves. Then how many principal components should you use?

18.8.2 Fitting Factor Models by Time Series Regression

In this section, we will start with the one-factor CAPM model of Chap. 17 and then extend this model to the three-factor Fama–French model. We will use the data set `Stock_Bond_2004_to_2005.csv` on the book's website, which contains stock prices and other financial time series for the years 2004 and 2005. Data on the Fama–French factors are available at Prof. Kenneth French's website

```
http://mba.tuck.dartmouth.edu/pages/faculty/ken.french/
data_library.html#Research
```

where RF is the risk-free rate and Mkt.RF, SMB, and HML are the Fama–French factors.

Go to Prof. French's website and get the daily values of RF, Mkt.RF, SMB, and HML for the years 2004–2005. It is assumed here that you've put the data in a text file FamaFrenchDaily.txt. Returns on this website are expressed as percentages.

Now fit the CAPM to the four stocks using the lm command. This code fits a linear regression model separately to the four responses. In each case, the independent variable is Mkt.RF.

```
#  Uses daily data 2004-2005

stocks = read.csv("Stock_Bond_2004_to_2005.csv",header=T)
attach(stocks)
stocks_subset = as.data.frame(cbind(GM_AC,  F_AC, UTX_AC, MRK_AC))
stocks_diff = as.data.frame(100 * apply(log(stocks_subset),
    2, diff) - FF_data$RF)
names(stocks_diff) = c("GM", "Ford", "UTX", "Merck")

FF_data = read.table("FamaFrenchDaily.txt", header = TRUE)
FF_data = FF_data[-1, ] # delete first row since stocks_diff
                        # lost a row due to differencing

fit1 = lm(as.matrix(stocks_diff) ~ FF_data$Mkt.RF)
summary(fit1)
```

Problem 5 *The CAPM predicts that all four intercepts will be zero. For each stock, using $\alpha = 0.025$, can you accept the null hypothesis that its intercept is zero? Why or why not? Include the p-values with your work.*

Problem 6 *The CAPM also predicts that the four sets of residuals will be uncorrelated. What is the correlation matrix of the residuals? Give a 95 % confidence interval for each of the six correlations. Can you accept the hypothesis that all six correlations are zero?*

Problem 7 *Regardless of your answer to Problem 6, assume for now that the residuals are uncorrelated. Then use the CAPM to estimate the covariance matrix of the excess returns on the four stocks. Compare this estimate with the sample covariance matrix of the excess returns. Do you see any large discrepancies between the two estimates of the covariance matrix?*

Next, you will fit the Fama–French three-factor model. Run the following R code, which is much like the previous code except that the regression model has two additional predictor variables, SMB and HML.

```
fit2 = lm(as.matrix(stocks_diff) ~ FF_data$Mkt.RF +
    FF_data$SMB + FF_data$HML)
summary(fit2)
```

Problem 8 *The CAPM predicts that for each stock, the slope (beta) for SMB and HML will be zero. Explain why the CAPM makes this prediction. Do you accept this null hypothesis? Why or why not?*

Problem 9 *If the Fama–French model explains all covariances between the returns, then the correlation matrix of the residuals should be diagonal. What is the estimated correlations matrix? Would you accept the hypothesis that the correlations are all zero?*

Problem 10 *Which model, CAPM or Fama–French, has the smaller value of AIC? Which has the smaller value of BIC? What do you conclude from this?*

Problem 11 *What is the covariance matrix of the three Fama–French factors?*

Problem 12 *In this problem, Stocks 1 and 2 are two stocks, not necessarily in the Stock_FX_Bond_2004_to_2005.csv data set. Suppose that Stock 1 has betas of 0.5, 0.4, and −0.1 with respect to the three factors in the Fama–French model and a residual variance of 23.0. Suppose also that Stock 2 has betas of 0.6, 0.15, and 0.7 with respect to the three factors and a residual variance of 37.0. Regardless of your answer to Problem 9, when doing this problem, assume that the three factors do account for all covariances.*

(a) *Use the Fama–French model to estimate the variance of the excess return on Stock 1.*
(b) *Use the Fama–French model to estimate the variance of the excess return on Stock 2.*
(c) *Use the Fama–French model to estimate the covariance between the excess returns on Stock 1 and Stock 2.*

18.8.3 Statistical Factor Models

This section applies statistical factor analysis to the log returns of 10 stocks in the data set Stock_FX_Bond.csv. The data set contains adjusted closing

(AC) prices of the stocks, as well as daily volumes and other information that we will not use here.

The following R code will read the data, compute the log returns, and fit a two-factor model. Note that `factanal` works with the correlation matrix or, equivalently, with standardized variables.

```
dat = read.csv("Stock_FX_Bond.csv")
stocks_ac = dat[ , c(3, 5, 7, 9, 11, 13, 15, 17)]
n = length(stocks_ac[ , 1])
stocks_returns = log(stocks_ac[-1, ] / stocks_ac[-n, ])
fact = factanal(stocks_returns, factors = 2, rotation = "none")
print(fact)
```

Loadings less than the parameter `cutoff` are not printed. The default value of `cutoff` is 0.1, but you can change it as in "`print(fact,cutoff = 0.01)`" or "`print(fact, cutoff = 0)`".

Problem 13 *What are the factor loadings? What are the variances of the unique risks for Ford and General Motors?*

Problem 14 *Does the likelihood ratio test suggest that two factors are enough? If not, what is the minimum number of factors that seems sufficient?*

The following code will extract the loadings and uniquenesses.

```
loadings = matrix(as.numeric(loadings(fact)), ncol = 2)
unique = as.numeric(fact$unique)
```

Problem 15 *Regardless of your answer to Problem 14, use the two-factor model to estimate the correlation of the log returns for Ford and IBM.*

18.9 Exercises

1. The file `yields2009.csv` on this book's website contains daily Treasury yields for 2009. Perform a principal components analysis on changes in the yields. Describe your findings. How many principal components are needed to capture 98 % of the variability?
2. Perform a statistical factor analysis of the returns in the data set `mid-capD.ts` on the book's website. How many factors did you select? Use (18.20) to estimate the covariance matrix of the returns.
3. Verify equation (18.6).
4. Compute the eigenvectors in Example 18.3 and offer an interpretation of the first two eigenvectors.

References

Connor, G. (1995) The three types of factor models: a comparison of their explanatory power. *Financial Analysts Journal.* 42–46.

Fama, E. F., and French, K. R. (1992) The cross-section of expected stock returns. *Journal of Finance*, **47**, 427–465.

Fama, E. F., and French, K. R. (1993) Common risk factors in the returns on stocks and bonds. *Journal of Financial Economics*, **33**, 3–56.

Fama, E. F., and French, K. R. (1995) Size and book-to-market factors in earnings and returns. *Journal of Finance*, **50**, 131–155.

Fama, E. F., and French, K. R. (1996) Multifactor explanations of asset pricing anomalies. *Journal of Finance*, **51**, 55–84.

Mardia, K. V., Kent, J. T., and Bibby, J. M. (1979) *Multivariate Analysis*, Academic Press, London.

Sharpe, W. F., Alexander, G. J., and Bailey, J. V. (1999) *Investments*, 6th ed., Prentice-Hall, Upper Saddle River, NJ.

Zivot, E., and Wang, J. (2006) *Modeling Financial Time Series with S-PLUS*, 2nd ed., Springer, New York.

19

Risk Management

19.1 The Need for Risk Management

The financial world has always been risky, and financial innovations such as the development of derivatives markets and the packaging of mortgages have now made risk management more important than ever, but also more difficult.

There are many different types of risk. *Market risk* is due to changes in prices. *Credit risk* is the danger that a counterparty does not meet contractual obligations, for example, that interest or principal on a bond is not paid. *Liquidity risk* is the potential extra cost of liquidating a position because buyers are difficult to locate. *Operational risk* is due to fraud, mismanagement, human errors, and similar problems.

Early attempts to measure risk such as duration analysis, discussed in Sect. 3.8.1 and used to estimate the market risk of fixed income securities, were somewhat primitive and of only limited applicability. In contrast, value-at-risk (VaR) and expected shortfall (ES) are widely used because they can be applied to all types of risks and securities, including complex portfolios.

VaR uses two parameters, the time horizon and the confidence level, which are denoted by T and $1 - \alpha$, respectively. Given these, the VaR is a bound such that the loss over the horizon is less than this bound with probability equal to the confidence coefficient. For example, if the horizon is one week, the confidence coefficient is 99 % (so $\alpha = 0.01$), and the VaR is $5 million, then there is only a 1 % chance of a loss exceeding $5 million over the next week. We sometimes use the notation VaR(α) or Var(α, T) to indicate the dependence of VaR on α or on both α and the horizon T. Usually, VaR(α) is used with T being understood.

If $\mathcal{L}$ is the loss over the holding period T, then VaR(α) is the αth upper quantile of $\mathcal{L}$. Equivalently, if $\mathcal{R} = -\mathcal{L}$ is the revenue, then VaR(α) is minus the αth quantile of $\mathcal{R}$. For continuous loss distributions, VaR(α) solves

© Springer Science+Business Media New York 2015
D. Ruppert, D.S. Matteson, *Statistics and Data Analysis for Financial Engineering*, Springer Texts in Statistics,
DOI 10.1007/978-1-4939-2614-5_19

$$P\{\mathcal{L} > \text{VaR}(\alpha)\} = P\{\mathcal{L} \geq \text{VaR}(\alpha)\} = \alpha, \tag{19.1}$$

and for any loss distribution, continuous or not,

$$\text{VaR}(\alpha) = \inf\{x : P(\mathcal{L} > x) \leq \alpha\}. \tag{19.2}$$

As will be discussed later, VaR has a serious deficiency—it discourages diversification—and for this reason it is being replaced by newer risk measures. One of these newer risk measures is the expected loss given that the loss exceeds VaR, which is called by a variety of names: *expected shortfall*, the *expected loss given a tail event*, *tail loss*, and *shortfall*. The name *expected shortfall* and the abbreviation ES will be used here.

For any loss distribution, continuous or not,

$$\text{ES}(\alpha) = \frac{\int_0^\alpha \text{VaR}(u)\,du}{\alpha}, \tag{19.3}$$

which is the average of VaR(u) over all u that are less than or equal to α. If $\mathcal{L}$ has a continuous distribution,

$$\text{ES}(\alpha) = E\{\mathcal{L} \mid \mathcal{L} > \text{VaR}(\alpha)\} = E\{\mathcal{L} \mid \mathcal{L} \geq \text{VaR}(\alpha)\}. \tag{19.4}$$

Example 19.1. VaR with a normally distributed loss

Suppose that the yearly return on a stock is normally distributed with mean 0.04 and standard deviation 0.18. If one purchases \$100,000 worth of this stock, what is the VaR with T equal to one year?

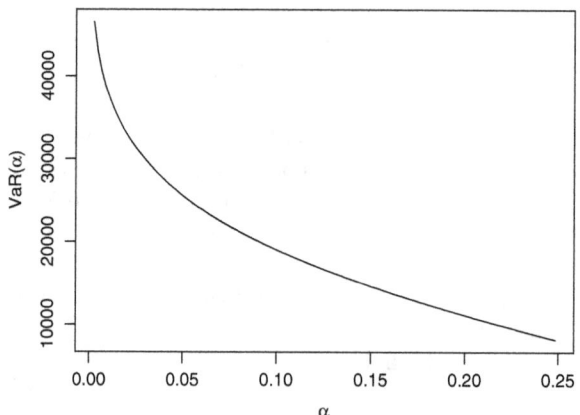

Fig. 19.1. VaR(α) *for* $0.025 < \alpha < 0.25$ *when the loss distribution is normally distributed with mean* -4000 *and standard deviation* 18,000.

To answer this question, we use the fact that the loss distribution is normal with mean -4000 and standard deviation 18,000, with all units in dollars. Therefore, VaR is

$$-4000 + 18{,}000 z_\alpha,$$

where z_α is the α-upper quantile of the standard normal distribution. $\mathrm{VaR}(\alpha)$ is plotted as a function of α in Fig. 19.1. VaR depends heavily on α and in this figure ranges from 46,527 when α is 0.025 to 8,226 when α is 0.25.   □

In applications, risk measures will rarely, if ever, be known exactly as in these simple examples. Instead, risk measures are estimated, and estimation error is another source of uncertainty. This uncertainty can be quantified using a confidence interval for the risk measure. We turn next to these topics.

19.2 Estimating VaR and ES with One Asset

To illustrate the techniques for estimating VaR and ES, we begin with the simple case of a single asset. In this section, these risk measures are estimated using historic data to estimate the distribution of returns. We make the assumption that returns are stationary, at least over the historic period we use. This is usually a reasonable assumption. We will also assume that the returns are independent. Independence is a much less reasonable assumption because of volatility clustering, and later we will remove this assumption by using GARCH models.

Two cases are considered, first without and then with the assumption of a parametric model for the return distribution.

19.2.1 Nonparametric Estimation of VaR and ES

We start with *nonparametric* estimates of VaR and ES, meaning that the loss distribution is not assumed to be in a parametric family such as the normal or t-distributions.

Suppose that we want a confidence coefficient of $1-\alpha$ for the risk measures. Therefore, we estimate the α-quantile of the return distribution, which is the α-upper quantile of the loss distribution. In the nonparametric method, this quantile is estimated as the α-quantile of a sample of historic returns, which we will call $\hat{q}(\alpha)$. If S is the size of the current position, then the nonparametric estimate of VaR is

$$\widehat{\mathrm{VaR}}^{\,\mathrm{np}}(\alpha) = -S \times \hat{q}(\alpha),$$

with the minus sign converting revenue (return times initial investment) to a loss. In this chapter, superscripts and subscripts will sometimes be placed on VaR and ES to provide information. Here, the superscript "np" means "nonparametrically estimated."

To estimate ES, let $R_1, \ldots, R_n$ be the historic returns and define $\mathcal{L}_i = -S \times R_i$. Then

$$\widehat{\text{ES}}^{\text{np}}(\alpha) = \frac{\sum_{i=1}^{n} \mathcal{L}_i \, I\{\mathcal{L}_i > \widehat{\text{VaR}}(\alpha)\}}{\sum_{i=1}^{n} I\{\mathcal{L}_i > \widehat{\text{VaR}}(\alpha)\}} = -S \times \frac{\sum_{i=1}^{n} R_i \, I\{R_i < \widehat{q}(\alpha)\}}{\sum_{i=1}^{n} I\{R_i < \widehat{q}(\alpha)\}}, \quad (19.5)$$

which is the average of all $\mathcal{L}_i$ exceeding $\widehat{\text{VaR}}^{\text{np}}(\alpha)$. Here $I\{\mathcal{L}_i > \widehat{\text{VaR}}^{\text{np}}(\alpha)\}$ is the indicator that $\mathcal{L}_i$ exceeds $\widehat{\text{VaR}}^{\text{np}}(\alpha)$, and similarly for $I\{R_i < \widehat{q}(\alpha)\}$.

Example 19.2. Nonparametric VaR and ES for a position in an S&P 500 index fund

As a simple example, suppose that you hold a $20,000 position in an S&P 500 index fund, so your returns are those of this index, and that you want a 24-h VaR. We estimate this VaR using the 1,000 daily returns on the S&P 500 for the period ending in April 1991. These log returns are a subset of the data set SP500 in R's Ecdat package. The full time series is plotted in Fig. 4.1. Black Monday, with a log return of -0.23, occurs near the beginning of the shortened time series used in this example.

Suppose you want 95 % confidence. The 0.05 quantile of the returns computed by R's quantile() function is -0.0169. In other words, a daily return of -0.0169 or less occurred only 5 % of the time in the historic data, so we estimate that there is a 5 % chance of a return of that size occurring during the next 24 h. A return of -0.0169 on a $20,000 investment yields a revenue of $-\$337.5$, and therefore the estimated $\widehat{\text{VaR}}(0.05, 24 \text{ h})$ is $337.43.

ES(0.05) is obtained by averaging all returns below -0.0169 and multiplying this average by $-20,000$. The result is $\widehat{\text{ES}}^{\text{np}}(0.05) = \$619.3$. The code for this example is below.

```
1 data(SP500, package="Ecdat")
2 n = 2783
3 SPreturn = SP500$r500[(n - 999):n]
4 year = 1981 + (1:n) * (1991.25 - 1981) / n
5 year = year[(n - 999):n]
6 alpha = 0.05
7 q = as.numeric(quantile(SPreturn, alpha))
8 VaR_nonp = -20000 * q
9 IEVaR = (SPreturn < q)
10 sum(IEVaR)
11 ES_nonp = -20000 * sum(SPreturn * IEVaR) / sum(IEVaR)
12 options(digits = 5)
13 VaR_nonp
14 ES_nonp
```

□

19.2.2 Parametric Estimation of VaR and ES

Parametric estimation of VaR and ES has a number of advantages. For example, parametric estimation allows the use of GARCH models to adapt the risk measures to the current estimate of volatility. Also, risk measures can be easily computed for a portfolio of stocks if we assume that their returns have a joint parametric distribution, such as a multivariate t-distribution. Nonparametric estimation using sample quantiles works best when the sample size and α are reasonably large. With smaller sample sizes or smaller values of α, it is preferable to use parametric estimation. In this section, we look at parametric estimation of VaR and ES when there is a single asset.

Let $F(y|\boldsymbol{\theta})$ be a parametric family of distributions used to model the return distribution and suppose that $\widehat{\boldsymbol{\theta}}$ is an estimate of $\boldsymbol{\theta}$, such as, the MLE computed from historic returns. Then $F^{-1}(\alpha|\widehat{\boldsymbol{\theta}})$ is an estimate of the α-quantile of the return distribution and

$$\widehat{\text{VaR}}^{\text{par}}(\alpha) = -S \times F^{-1}(\alpha|\widehat{\boldsymbol{\theta}}) \tag{19.6}$$

is a parametric estimate of VaR(α). As before, S is the size of the current position.

Let $f(y|\boldsymbol{\theta})$ be the density of $F(y|\boldsymbol{\theta})$. Then the estimate of expected shortfall is

$$\widehat{\text{ES}}^{\text{par}}(\alpha) = -\frac{S}{\alpha} \times \int_{-\infty}^{F^{-1}(\alpha|\widehat{\boldsymbol{\theta}})} x f(x|\widehat{\boldsymbol{\theta}}) \, dx. \tag{19.7}$$

The superscript "par" denotes "parametrically estimated." Computing this integral is not always easy, but in the important cases of normal and t-distributions there are convenient formulas.

Suppose the return has a t-distribution with mean equal to μ, scale parameter equal to λ, and tail index[1] ν. Let f_ν and F_ν be, respectively, the t-density and t-distribution function with ν degrees of freedom. The expected shortfall is

$$\widehat{\text{ES}}^{t}(\alpha) = S \times \left\{ -\mu + \lambda \left(\frac{f_\nu\{F_\nu^{-1}(\alpha)\}}{\alpha} \left[\frac{\nu + \{F_\nu^{-1}(\alpha)\}^2}{\nu - 1} \right] \right) \right\}. \tag{19.8}$$

The formula for normal loss distributions is obtained by a direct calculation or letting $\nu \to \infty$ in (19.8). The result is

$$\text{ES}^{\text{norm}}(\alpha) = S \times \left\{ -\mu + \sigma \left(\frac{\phi\{\Phi^{-1}(\alpha)\}}{\alpha} \right) \right\}, \tag{19.9}$$

where μ and σ are the mean and standard deviation of the returns and ϕ and Φ are the standard normal density and CDF. The superscripts "t" and

---

[1] The tail index parameter for the t-distribution is also commonly referred to as the degrees-of-freedom parameter by its association with the theory of linear regression, and some R functions use the abbreviations df or nu.

"norm" denote estimates assuming a t-distributed return and normal return, respectively.

Parametric estimation with one asset is illustrated in the next example.

Example 19.3. Parametric VaR and ES for a position in an S&P 500 index fund

This example uses the same data set as in Example 19.2 so that parametric and nonparametric estimates can be compared. We will assume that the returns are i.i.d. with a t-distribution. Under this assumption, VaR is

$$\widehat{\text{VaR}}^t(\alpha) = -S \times \{\widehat{\mu} + q_{\alpha,t}(\widehat{\nu})\widehat{\lambda}\}, \tag{19.10}$$

where $\widehat{\mu}$, $\widehat{\lambda}$, and $\widehat{\nu}$ are the estimated mean, scale parameter, and tail index of a sample of returns. Also, $q_{\alpha,t}(\widehat{\nu})$ is the α-quantile of the t-distribution with tail index $\widehat{\nu}$, so that $\{\widehat{\mu} + q_{\alpha,t}(\widehat{\nu})\widehat{\lambda}\}$ is the αth quantile of the fitted distribution.

The t-distribution was fit using R's `fitdistr()` function and the estimates were $\widehat{\mu} = 0.000689$, $\widehat{\lambda} = 0.007164$, and $\widehat{\nu} = 2.984$. For later reference, the estimated standard deviation is $\widehat{\sigma} = \widehat{\lambda}\sqrt{\widehat{\nu}/(\widehat{\nu}-2)} = 0.01248$.

The 0.05-quantile of the t-distribution with tail index 2.984 is -2.3586. Therefore, by (19.6),

$$\widehat{\text{VaR}}^t(0.05) = -20000 \times \{0.000689 - (2.3586)(0.007164)\} = \$324.17.$$

Notice that the nonparametric estimate, $\widehat{\text{VaR}}^{np}(0.05) = \$337.55$, is similar to, but somewhat larger than the parametric estimate, \$324.17.

The parametric estimate of $\text{ES}^t(0.05)$ is \$543.81 and is found by substituting $S = 20,000$, $\alpha = 0.05$, $\widehat{\mu} = 0.000689$, $\widehat{\lambda} = 0.007164$, and $\widehat{\nu} = 2.984$ into (19.8). The parametric estimate of $\text{ES}^t(0.05)$ is noticeably shorter than the nonparametric. The reason the two estimates differ is that the extreme left tail of the returns, roughly the smallest 10 of 1,000 returns, is heavier than the tail of a t-distribution with 2.984 degrees of freedom; see the t-plot in Fig. 19.2. The code for this example is below.

```
15  data(SP500, package="Ecdat")
16  n = 2783
17  SPreturn = SP500$r500[(n - 999):n]
18  year = 1981 + (1:n) * (1991.25 - 1981) / n
19  year = year[(n - 999):n]
20  alpha = 0.05
21  library(MASS)
22  fitt = fitdistr(SPreturn, "t")
23  param = as.numeric(fitt$estimate)
24  mean = param[1]
25  df = param[3]
26  sd = param[2] * sqrt((df) / (df - 2))
```

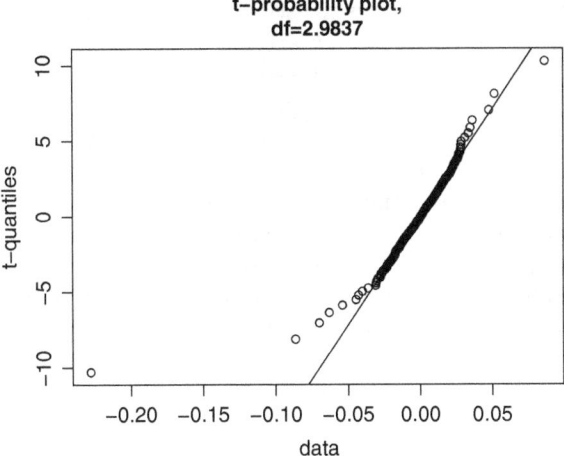

Fig. 19.2. *t-plot of the S&P 500 returns used in Examples 19.2 and 19.3. The deviations from linearity in the tails, especially the left tail, indicate that the t-distribution does not fit the data in the extreme tails. The reference line goes through the first and third quartiles. The t-quantiles use 2.98 degrees of freedom, the MLE. The deviation in the left tail of the data from the t-distribution explains why the parametric estimate of ES is smaller than the nonparametric estimate.*

```
27  lambda = param[2]
28  qalpha = qt(alpha, df = df)
29  VaR_par = -20000 * (mean + lambda * qalpha)
30  es1 = dt(qalpha, df = df) / (alpha)
31  es2 = (df + qalpha^2) / (df - 1)
32  es3 = -mean + lambda * es1 * es2
33  ES_par = 20000*es3
34  VaR_par
35  ES_par
```

□

19.3 Bootstrap Confidence Intervals for VaR and ES

The estimates of VaR and ES are precisely that, just estimates. If we had used a different sample of historic data, then we would have gotten different estimates of these risk measures. We just calculated VaR and ES values to five significant digits, but do we really have that much precision? The reader has probably guessed (correctly) that we do not, but how much precision do we have? How can we learn the true precision of the estimates? Fortunately, a confidence interval for VaR or ES is rather easily obtained by bootstrapping. Any of the confidence interval procedures in Sect. 6.3 can be used. We will see that even with 1,000 returns to estimate VaR and ES, these risk measures are estimated with considerable uncertainty.

For now, we will assume an i.i.d. sample of historic returns and use model-free resampling. In Sect. 19.4 we will allow for dependencies, for instance, GARCH effects, in the data and we will use model-based resampling.

Suppose we have a large number, B, of resamples of the returns data. Then a VaR(α) or ES(α) estimate is computed from each resample and for the original sample. The confidence interval can be based upon either a parametric or nonparametric estimator of VaR(α) or ES(α). Suppose that we want the confidence coefficient of the interval to be $1 - \gamma$. The interval's confidence coefficient should not be confused with the confidence coefficient of VaR, which we denote by $1 - \alpha$. The $\gamma/2$-lower and -upper quantiles of the bootstrap estimates of VaR(α) and ES(α) are the limits of the basic percentile method confidence intervals.

It is worthwhile to restate the meanings of α and γ, since it is easy to confuse these two confidence coefficients, but they need to be distinguished since they have rather different interpretations. VaR(α) is defined so that the probability of a loss being greater than VaR(α) is α. On the other hand, γ is the confidence coefficient for the confidence interval for VaR(α) and ES(α). If many confidence intervals are constructed, then approximately γ of them do not contain the true risk measure. Thus, α is about the loss from the investment while γ is about the confidence interval being correct. An alternative way to view the difference between α and γ is that VaR(α) and ES(α) are measuring risk due to uncertainty about future losses, assuming perfect knowledge of the loss distribution, while the confidence intervals tell us the uncertainty of these risk measures due to imperfect knowledge of the loss distribution.

Example 19.4. Bootstrap confidence intervals for VaR and ES for a position in an S&P 500 index fund

In this example, we continue Examples 19.2 and 19.3 and find an approximate confidence interval for VaR(α) and ES(α). We use $\alpha = 0.05$ as before and $\gamma = 0.1$. $B = 5,000$ resamples were taken.

The basic percentile confidence intervals for VaR(0.05) were (297, 352) and (301, 346) using nonparametric and parametric estimators of VaR(0.05), respectively. For ES(0.05), the corresponding basic percentile confidence intervals were (487, 803) and (433, 605). We see that there is considerable uncertainty in the risk measures, especially for ES(0.05) and especially using nonparametric estimation.

When the first edition was written, the bootstrap computation took 33.3 minutes using an R program and a 2.13 GHz Pentium[TM] processor running under Windows[TM]. The computations took this long because the optimization step to find the MLE for parametric estimation is moderately expensive in computational time, at least if it is repeated 5,000 times. However, the same computation took only 5.23 minutes on a 2.9 GHz MacBook Pro when the second edition was being written in 2014. Nonetheless, more computationally expensive estimators could easily take one-half hour or more to bootstrap even on a fast computer.

Waiting over a half an hour for the confidence interval may not be an attractive proposition. However, a reasonable measure of precision can be obtained with far fewer bootstrap repetitions. One might use only 50 repetitions, which would take less than a minute. This is not enough resamples to use basic percentile bootstrap confidence intervals, but instead one can use the normal approximation bootstrap confidence interval, (6.4). As an example, the normal approximation interval for the nonparametric estimate of VaR(0.05) is (301, 361) using only the first 50 bootstrap resamples. This interval gives the same general impression of accuracy as the above basic percentile method interval, (297, 352), that uses all 5,000 resamples.

The normal approximation interval assumes that $\widehat{\text{VaR}}(0.05)$ is approximately normally distributed. This assumption is justified by the central limit theorem for sample quantiles (Sect. 4.3.1) and the fact that $\widehat{\text{VaR}}(0.05)$ is a multiple of a sample quantile. The normal approximation does *not* require that the returns are normally distributed. In fact, we are modeling them as t-distributed when computing the parametric estimates.   □

19.4 Estimating VaR and ES Using ARMA+GARCH Models

As we have seen in Chaps. 12 and 14, daily equity returns typically have a small amount of autocorrelation and a greater amount of volatility clustering. When calculating risk measures, the autocorrelation can be ignored if it is small enough, but the volatility clustering is less ignorable. In this section, we use ARMA+GARCH models so that VaR(α) and ES(α) can adjust to periods of high or low volatility.

Assume that we have n returns, $R_1, \ldots, R_n$ and we need to estimate VaR and ES for the next return R_{n+1}. Let $\widehat{\mu}_{n+1|n}$ and $\widehat{\sigma}_{n+1|n}$ be the estimated conditional mean and variance of tomorrow's return R_{n+1}, conditional on the current information set, which in this context is simply $\{R_1, \ldots, R_n\}$. We will also assume that R_{n+1} has a conditional t-distribution with tail index ν. After fitting an ARMA+GARCH model, we have estimates of $\widehat{\nu}$, $\widehat{\mu}_{n+1|n}$, and $\widehat{\sigma}_{n+1|n}$. The estimated conditional scale parameter is

$$\widehat{\lambda}_{n+1|n} = \sqrt{(\widehat{\nu} - 2)/\widehat{\nu}}\ \widehat{\sigma}_{n+1|n}. \qquad (19.11)$$

VaR and ES are estimated as in Sect. 19.2.2 but with $\widehat{\mu}$ and $\widehat{\lambda}$ replaced by $\widehat{\mu}_{n+1|n}$ and $\widehat{\lambda}_{n+1|n}$.

Example 19.5. VaR and ES for a position in an S&P 500 index fund using a GARCH(1,1) model

An AR(1)+GARCH(1,1) model was fit to the log returns on the S&P 500. The AR(1) coefficient was small and not significantly different from 0, so a GARCH(1,1) was used for estimation of VaR and ES. The GARCH(1,1) fit is

```
36  library(rugarch)
37  garch.t = ugarchspec(mean.model=list(armaOrder=c(0,0)),
38                        variance.model=list(garchOrder=c(1,1)),
39                        distribution.model="std")
40  sp.garch.t = ugarchfit(data=SPreturn, spec=garch.t)
41  show(sp.garch.t)
```

```
    Optimal Parameters
    ------------------------------------

            Estimate   Std. Error    t value  Pr(>|t|)
    mu      0.000714     0.000264    2.70872  0.006754
    omega   0.000003     0.000004    0.79083  0.429046
    alpha1  0.032459     0.019439    1.66979  0.094961
    beta1   0.939176     0.009296  101.02598  0.000000
    shape   4.417464     0.560553    7.88054  0.000000
```

```
42  pred = ugarchforecast(sp.garch.t, data=SPreturn, n.ahead=1) ; pred
```

```
            Series      Sigma
    T+1  0.0007144  0.009478
```

```
43  alpha = 0.05
44  nu = as.numeric(coef(sp.garch.t)[5])
45  q = qstd(alpha, mean=fitted(pred), sd=sigma(pred), nu=nu)
46  VaR = -20000*q ; VaR
```

```
    T+1              276.7298
```

```
47  lambda = sigma(pred)/sqrt( (nu)/(nu-2) )
48  qalpha = qt(alpha, df=nu)
49  es1 = dt(qalpha, df=nu)/(alpha)
50  es2 = (nu + qalpha^2) / (nu - 1)
51  es3 = -mean + lambda*es1*es2
52  ES_par = 20000*es3 ; ES_par
```

```
    T+1              413.6518
```

The conditional mean and standard deviation of the next return were estimated to be 0.00071 and 0.00950. For the estimation of VaR and ES, the next return was assumed to have a t-distribution with these values for the mean and standard deviation and tail index 4.417. The estimate of VaR was $276.73 and the estimate of ES was $413.65. The VaR and ES estimates using the GARCH model are considerably smaller than the parametric estimates

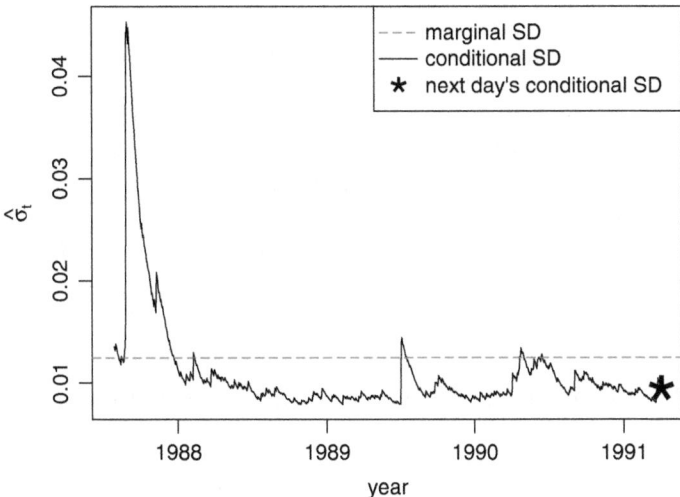

Fig. 19.3. *Estimated conditional standard deviation of the daily S&P 500 index returns based on a GARCH(1,1) model. The asterisk is a forecast of the next day's conditional standard deviation from the end of the return series, and the height of the dashed horizontal line is an estimate of the marginal (unconditional) standard deviation.*

in Example 19.2 ($323.42 and $543.81), because the conditional standard deviation used here (0.00950) is smaller than the marginal standard deviation (0.01248) used in Example 19.2; see Fig. 19.3, where the dashed horizontal line's height is the marginal standard deviation and the conditional standard deviation of the next day's return is indicated by a large asterisk. The marginal standard deviation is inflated by periods of higher volatility such as in October 1987 (near Black Monday) on the left-hand side of Fig. 19.3.   □

19.5 Estimating VaR and ES for a Portfolio of Assets

When VaR is estimated for a portfolio of assets rather than a single asset, parametric estimation based on the assumption of multivariate normal or t-distributed returns is very convenient, because the portfolio's return will have a univariate normal or t-distributed return. The portfolio theory and factor models developed in Chaps. 16 and 18 can be used to estimate the mean and variance of the portfolio's return.

Estimating VaR becomes complex when the portfolio contains stocks, bonds, options, foreign exchange positions, and other assets. However, when a portfolio contains only stocks, then VaR is relatively straightforward to estimate, and we will restrict attention to this case—see Sect. 19.10 for discussion of the literature covering more complex cases.

With a portfolio of stocks, means, variances, and covariances of returns could be estimated directly from a sample of returns as discussed in Chap. 16 or using a factor model as discussed in Sect. 18.4.2. They can also be estimated using the multivariate time series models discussed in Chaps. 13 and 14. Once these estimates are available, they can be plugged into Eqs. (16.6) and (16.7) to obtain estimates of the expected value and variance of the return on the portfolio, which are denoted by $\widehat{\mu}_P$ and $\widehat{\sigma}_P^2$. Then, analogous to (19.10), VaR can be estimated, assuming normally distributed returns on the portfolio (denoted with a subscript "P"), by

$$\widehat{\mathrm{VaR}}_P^{\mathrm{norm}}(\alpha) = -S \times \{\widehat{\mu}_P + \Phi^{-1}(\alpha)\widehat{\sigma}_P\}, \qquad (19.12)$$

where S is the initial value of the portfolio. Moreover, using (19.9), the estimated expected shortfall is

$$\widehat{\mathrm{ES}}_P^{\mathrm{norm}}(\alpha) = S \times \left\{ -\widehat{\mu}_P + \widehat{\sigma}_P \left(\frac{\phi\{\Phi^{-1}(\alpha)\}}{\alpha} \right) \right\}. \qquad (19.13)$$

If the stock returns have a joint t-distribution, then the returns on the portfolio have a univariate t-distribution with the same tail index, and VaR and ES for the portfolio can be calculated using formulas in Sect. 19.2.2. If the returns on the portfolio have a t-distribution with mean μ_P, scale parameter λ_P, and tail index ν, then the estimated VaR is

$$\widehat{\mathrm{VaR}}_P^t(\alpha) = -S \times \{\widehat{\mu}_P + F_\nu^{-1}(\alpha)\widehat{\lambda}_P\}, \qquad (19.14)$$

and the estimated expected shortfall is

$$\widehat{\mathrm{ES}}_P^t(\alpha) = S \times \left\{ -\widehat{\mu}_P + \widehat{\lambda}_P \left(\frac{f_{\widehat{\nu}}\{F_{\widehat{\nu}}^{-1}(\alpha)\}}{\alpha} \left[\frac{\widehat{\nu} + \{F_{\widehat{\nu}}^{-1}(\alpha)\}^2}{\widehat{\nu} - 1} \right] \right) \right\}. \qquad (19.15)$$

Example 19.6. VaR and ES for portfolios of the three stocks in the CRSPday *data set*

This example uses the data set CRSPday used earlier in Examples 7.1 and 7.4. There are four variables—returns on GE, IBM, Mobil, and the CRSP index and we found in Example 7.4 that their returns can be modeled as having a multivariate t-distribution with tail index 5.94. In this example, we will only consider the returns on the three stocks. The t-distribution parameters were reestimated without the CRSP index and $\widehat{\nu}$ changed slightly to 5.81. The estimated mean was

$$\widehat{\mu} = (\,0.000858 \quad 0.000325 \quad 0.000616\,)^\mathsf{T}$$

and the estimated covariance matrix was

$$\widehat{\Sigma} = \begin{pmatrix} 1.27e-04 & 5.04e-05 & 3.57e-05 \\ 5.04e-05 & 1.81e-04 & 2.40e-05 \\ 3.57e-05 & 2.40e-05 & 1.15e-04 \end{pmatrix}.$$

For an equally weighted portfolio with $w = (1/3 \quad 1/3 \quad 1/3)^\mathsf{T}$, the mean return for the portfolio is estimated to be

$$\widehat{\mu}_P = \widehat{\mu}^\mathsf{T} w = 0.0006$$

and the standard deviation of the portfolio's return is estimated as

$$\widehat{\sigma}_P = \sqrt{w^\mathsf{T} \widehat{\Sigma} w} = 0.00846.$$

The return on the portfolio has a t-distribution with this mean and standard deviation, and the same tail index as the multivariate t-distribution of the three stock returns. The scale parameter, using $\widehat{\nu} = 5.81$, is

$$\widehat{\lambda}_P = \sqrt{(\widehat{\nu}-2)/\widehat{\nu}} \times 0.00846 = 0.00685.$$

Therefore,

$$\widehat{\mathrm{VaR}}^t(0.05) = -S \times \{\widehat{\mu}_P + \widehat{\lambda}_P\,\widehat{q}_{0.05,t}(\widehat{\nu})\} = S \times 0.0128,$$

so, for example, with $S = \$20,000$, $\widehat{\mathrm{VaR}}^t(0.05) = \$256$.
    The estimated ES using (19.8) and $S = \$20,000$ is

$$\widehat{\mathrm{ES}}^t(0.05) = S \times \left\{ -\widehat{\mu}_P + \widehat{\lambda}_P \left(\frac{f_{\widehat{\nu}}\{\widehat{q}_{0.05,t}(\widehat{\nu})\}}{\alpha} \left[\frac{\widehat{\nu} + \{\widehat{q}_{0.05,t}(\widehat{\nu})\}^2}{\widehat{\nu}-1} \right] \right) \right\} = \$363.$$

$\square$

19.6 Estimation of VaR Assuming Polynomial Tails

There is an interesting compromise between using a totally nonparametric estimator of VaR as in Sect. 19.2.1 and a parametric estimator as in Sect. 19.2.2. The nonparametric estimator is feasible for large α, but not for small α. For example, if the sample had 1,000 returns, then reasonably accurate estimation of the 0.05-quantile is feasible, but not estimation of the 0.0005-quantile. Parametric estimation can estimate VaR for any value of α, but is sensitive to misspecification of the tail when α is small. Therefore, a methodology intermediary between totally nonparametric and parametric estimation is attractive.
    The approach used in this section assumes that the return density has a polynomial left tail, or equivalently that the loss density has a polynomial right tail. Under this assumption, it is possible to use a nonparametric estimate of

VaR(α_0) for a *large* value of α_0 to obtain estimates of VaR(α_1) for *small* values of α_1. It is assumed here that VaR(α_1) and VaR(α_0) have the same horizon T.

Because the return density is assumed to have a polynomial left tail, the return density f satisfies

$$f(y) \sim Ay^{-(a+1)}, \text{ as } y \to -\infty, \tag{19.16}$$

where $A > 0$ is a constant, $a > 0$ is the tail index, and "$\sim$" means that the ratio of the left-hand to right-hand sides converges to 1. Therefore,

$$P(R \leq y) \sim \int_{-\infty}^{y} f(u)\,du = \frac{A}{a}y^{-a}, \text{ as } y \to -\infty, \tag{19.17}$$

and if $y_0 > 0$ and $y_1 > 0$, then

$$\frac{P(R < -y_0)}{P(R < -y_1)} \approx \left(\frac{y_0}{y_1}\right)^{-a}. \tag{19.18}$$

Now suppose that $y_0 = $ VaR(α_1) and $y_1 = $ VaR(α_0), where $0 < \alpha_1 < \alpha_0$ and, for simplicity and without loss of generality, we use $S = 1$ in the following calculation. Then (19.18) becomes

$$\frac{\alpha_1}{\alpha_0} = \frac{P\{R < -\text{VaR}(\alpha_1)\}}{P\{R < -\text{VaR}(\alpha_0)\}} \approx \left(\frac{\text{VaR}(\alpha_1)}{\text{VaR}(\alpha_0)}\right)^{-a} \tag{19.19}$$

or

$$\frac{\text{VaR}(\alpha_1)}{\text{VaR}(\alpha_0)} \approx \left(\frac{\alpha_0}{\alpha_1}\right)^{1/a},$$

so, now dropping the subscript "1" of α_1 and writing the approximate equality as exact, we have

$$\text{VaR}(\alpha) = \text{VaR}(\alpha_0) \left(\frac{\alpha_0}{\alpha}\right)^{1/a}. \tag{19.20}$$

Equation (19.20) becomes an estimate of VaR(α) when VaR(α_0) is replaced by a nonparametric estimate and the tail index a is replaced by one of the estimates discussed soon in Sect. 19.6.1. Notice another advantage of (19.20), that it provides an estimate of VaR(α) not just for a single value of α but for all values. This is useful if one wants to compute and compare VaR(α) for a variety of values of α, as is illustrated in Example 19.7 ahead. The value of α_0 must be large enough that VaR(α_0) can be accurately estimated, but α can be any value less than α_0.

A model combining parametric and nonparametric components is called *semiparametric*, so estimator (19.20) is semiparametric because the tail index is specified by a parameter, but otherwise the distribution is unspecified.

To find a formula for ES, we will assume further that for some $c < 0$, the returns density satisfies

$$f(y) = A|y|^{-(a+1)}, \quad y \leq c, \tag{19.21}$$

so that we have equality in (19.16) for $y \leq c$. Then, for any $d \leq c$,

$$P(R \leq d) = \int_{-\infty}^{d} A|y|^{-(a+1)} \, dy = \frac{A}{a} |d|^{-a}, \tag{19.22}$$

and the conditional density of R given that $R \leq d$ is

$$f(y|R \leq d) = \frac{Ay^{-(a+1)}}{P(R \leq d)} = a|d|^a|y|^{-(a+1)}. \tag{19.23}$$

It follows from (19.23) that for $a > 1$,

$$E\left(|R| \mid R \leq d\right) = a|d|^a \int_{-\infty}^{d} |y|^{-a} dy = \frac{a}{a-1} |d|. \tag{19.24}$$

(For $a \leq 1$, this expectation is $+\infty$.) If we let $d = -\text{VaR}(\alpha)$, then we see that

$$\text{ES}(\alpha) = \frac{a}{a-1} \text{VaR}(\alpha) = \frac{1}{1-a^{-1}} \text{VaR}(\alpha), \text{ if } a > 1. \tag{19.25}$$

Formula (19.25) enables one to estimate $\text{ES}(\alpha)$ using an estimate of $\text{VaR}(\alpha)$ and an estimate of a.

19.6.1 Estimating the Tail Index

In this section, we estimate the tail index assuming a polynomial left tail. Two estimators will be introduced, the regression estimator and the Hill estimator.

Regression Estimator of the Tail Index

It follows from (19.17) that

$$\log\{P(R \leq -y)\} = \log(L) - a \log(y), \tag{19.26}$$

where $L = A/a$.

If $R_{(1)}, \ldots, R_{(n)}$ are the order statistics of the returns, then the number of observed returns less than or equal to $R_{(k)}$ is k, so we estimate $\log\{P(R \leq R_{(k)})\}$ to be $\log(k/n)$. Then, from (19.26), we have

$$\log(k/n) \approx \log(L) - a \log(-R_{(k)}) \tag{19.27}$$

or, rearranging (19.27),

$$\log(-R_{(k)}) \approx (1/a) \log(L) - (1/a) \log(k/n). \tag{19.28}$$

The approximation (19.28) is expected to be accurate only if $-R_{(k)}$ is large, which means k is small, perhaps only 5 %, 10 %, or 20 % of the sample size n. If we plot the points $\{[\log(k/n), \log(-R_{(k)})] : k = 1, \ldots, m\}$ for m equal to a small percentage of n, say 10 %, then we should see these points fall on roughly a straight line. Moreover, if we fit the straight-line model (19.28) to these points by least squares, then the estimated slope, call it $\widehat{\beta}_1$, estimates $-1/a$. Therefore, we will call $-1/\widehat{\beta}_1$ the *regression estimator of the tail index*.

Hill Estimator

The Hill estimator of the left tail index a of the return density f uses all data less than a constant c, where c is sufficiently small such that

$$f(y) = A|y|^{-(a+1)} \qquad (19.29)$$

is assumed to be true for $y < c$. The choice of c is crucial and will be discussed below. Let $Y_{(1)}, \ldots, Y_{(n)}$ be order statistics of the returns and $n(c)$ be the number of Y_1 less than or equal to c. By (19.23), the conditional density of Y_i given that $Y_i \leq c$ is

$$a|c|^a |y|^{-(a+1)}. \qquad (19.30)$$

Therefore, the likelihood for $Y_{(1)}, \ldots, Y_{(n(c))}$ is

$$L(a) = \left(\frac{a|c|^a}{|Y_1|^{a+1}} \right) \left(\frac{a|c|^a}{|Y_2|^{a+1}} \right) \cdots \left(\frac{a|c|^a}{|Y_{n(c)}|^{a+1}} \right),$$

and the log-likelihood is

$$\log\{L(a)\} = \sum_{i=1}^{n(c)} \left\{ \log(a) + a\log(|c|) - (a+1)\log(|Y_{(i)}|) \right\}. \qquad (19.31)$$

Differentiating the right-hand side of (19.31) with respect to a and setting the derivative equal to 0 gives the equation

$$\frac{n(c)}{a} = \sum_{i=1}^{n(c)} \log\left(Y_{(i)}/c \right).$$

Therefore, the MLE of a, which is called the *Hill estimator*, is

$$\widehat{a}^{\text{Hill}}(c) = \frac{n(c)}{\sum_{i=1}^{n(c)} \log\left(Y_{(i)}/c \right)}. \qquad (19.32)$$

Note that $Y_{(i)} \leq c < 0$, so that $Y_{(i)}/c$ is positive.

How should c be chosen? Usually c is equal to one of $Y_1, \ldots, Y_n$ so that $c = Y_{(n(c))}$, and therefore choosing c means choosing $n(c)$. The choice involves a bias–variance tradeoff. If $n(c)$ is too large, then $f(y) = A|y|^{-(a+1)}$ will not hold for all values of $y \leq c$, causing bias. If $n(c)$ is too small, then there will be too few Y_i below c and $\widehat{a}^{\text{Hill}}(c)$ will be highly variable and unstable because it uses too few data. However, we can hope that there is a range of values of $n(c)$ where $\widehat{a}^{\text{Hill}}(c)$ is reasonably constant because it is neither too biased nor too variable.

A *Hill plot* is a plot of $\widehat{a}^{\text{Hill}}(c)$ versus $n(c)$ and is used to find this range of values of $n(c)$. In a Hill plot, one looks for a range of $n(c)$ where the estimator is nearly constant and then chooses $n(c)$ in this range.

Example 19.7. Estimating the left tail index of the daily S&P 500 index returns

This example uses the 1,000 daily S&P 500 index returns used in Examples 19.2 and 19.3. First, the regression estimator of the tail index was calculated. The values $\{[\log(k/n), \log(-R_{(k)})] : k = 1, \ldots, m\}$ were plotted for $m = $ 50, 100, 200, and 300 to find the largest value of m giving a roughly linear plot, of which $m = 100$ was selected. The plotted points and the least-squares lines can be seen in Fig. 19.4. The slope of the line with $m = 100$ was -0.506, so a was estimated to be $1/0.506 = 1.975$.

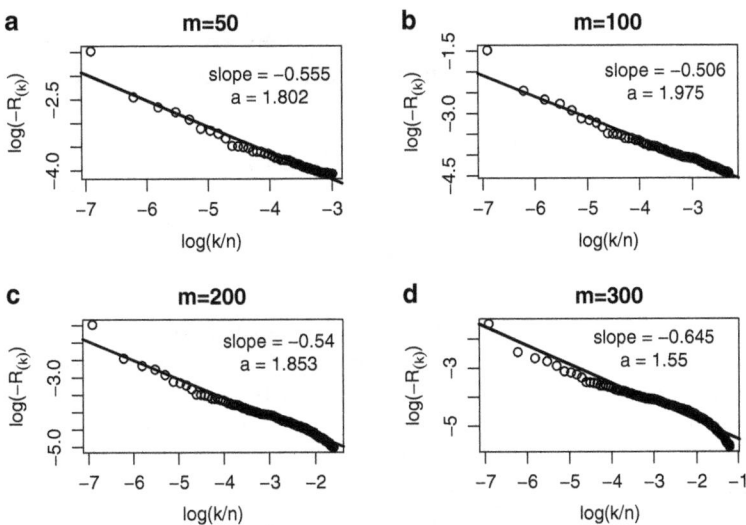

Fig. 19.4. *Plots for estimating the left tail index of the S&P 500 returns by regression. The "slope" is the least-squares slope estimate and "a" is $-1/slope$.*

Suppose we have invested $20,000 in an S&P 500 index fund. We will use $\alpha_0 = 0.1$. VaR(0.1, 24 h) is estimated to be $-\$20,000$ times the 0.1-quantile of the 1,000 returns. The sample quantile is -0.0117, so $\widehat{\text{VaR}}^{np}(0.1, 24\text{ h}) = \$234$. Using (19.20) and $a = 1.975$ (i.e., $1/a = 0.506$), we have

$$\widehat{\text{VaR}}(\alpha) = 234 \left(\frac{0.1}{\alpha}\right)^{0.506}. \tag{19.33}$$

The black curve in Fig. 19.5 is a plot of $\widehat{\text{VaR}}(\alpha)$ for $0.0025 \le \alpha \le 0.25$ using (19.33) and the regression estimator of a. The red curve is the same plot but with the Hill estimator of a, which is 2.2—see below. The blue curve is VaR(α) estimated assuming t-distributed returns as discussed in Sect. 19.2.2, and the purple curve is estimated assuming normally distributed returns. The

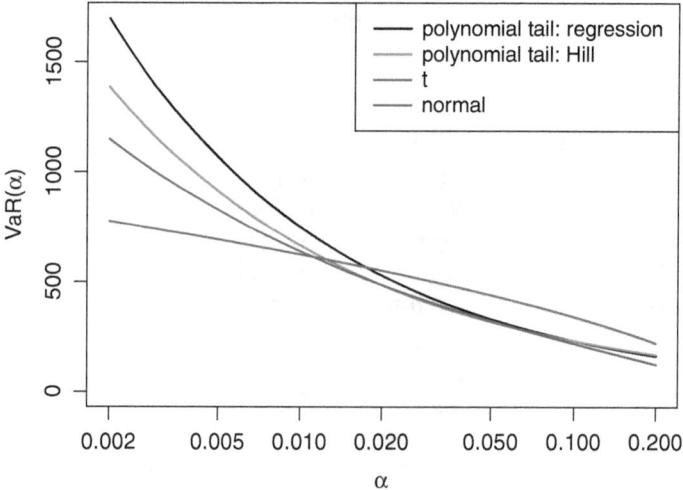

Fig. 19.5. *Estimation of VaR(α) using formula (19.33) and the regression estimator of the tail index (black line), using formula (19.33) and the Hill estimator of the tail index (red line), assuming t-distributed returns (blue line), and assuming normally distributed returns (purple line). Note the log-scale on the x-axis.*

return distribution has much heavier tails than a normal distribution, and the latter curve is included only to show the effect of model misspecification. The parametric estimates based on the t-distribution are similar to the estimates assuming a polynomial tail except when α is very small. The difference between the two estimates for small α ($\alpha < 0.01$) is to be expected because the polynomial tail with tail index 1.975 or 2.2 is heavier than the tail of the t-distribution with $\nu = a = 2.984$. The estimate based on the t-distribution is somewhat biased since it assumes a symmetric density and uses data in the right, as well as the left, tails to estimate the left tail; the problem with this is that the right tail is lighter than the left tail. If α is in the range 0.01 to 0.2, then $\widehat{\text{VaR}}(\alpha)$ is relatively insensitive to the choice of model, except for the poorly fitting normal model. This is a good reason for preferring $\alpha \geq 0.01$.

It follows from (19.25) using the regression estimate $\widehat{a} = 1.975$ that

$$\widehat{\text{ES}}(\alpha) = \frac{1.975}{0.975} \, \widehat{\text{VaR}}(\alpha) = 2.026 \, \widehat{\text{VaR}}(\alpha). \tag{19.34}$$

The Hill estimator of a was also implemented. Figure 19.6 contains Hill plots, that is, plots of the Hill estimate $\widehat{a}_{\text{Hill}}(c)$ versus $n(c)$. In panel (a), $n(c)$ ranges from 25 to 250. There seems to be a region of stability when $n(c)$ is between 25 and 120, which is shown in panel (b). In panel (b), we see a region of even greater stability when $n(c)$ is between 60 and 100. Panel (c) zooms in on this region. We see in panel (c) that the Hill estimator is close to 2.2 when $n(c)$ is between 60 and 100, and we will take 2.2 as the Hill estimate. Thus, the Hill estimate is similar to the regression estimate (1.975) of the tail index.

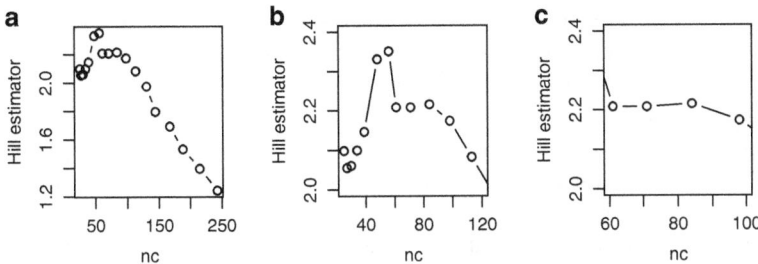

Fig. 19.6. *Estimation of the tail index by applying a Hill plot to the daily returns on the S&P 500 index for 1,000 consecutive trading days ending on March 4, 2003. (a) Full range of n_c. (b) Zoom in to n_c between 25 and 120. (c) Zoom in further to n_c between 60 and 100.*

The advantage of the regression estimate is that one can use the linearity of the plots of $\{[\log(k/n), \log(-R_{(k)})] : k = 1, \ldots, m\}$ for different m to guide the choice of m, which is analogous to $n(c)$. A linear plot indicates a polynomial tail. In contrast, the Hill plot checks for the stability of the estimator and does not give a direct assessment whether or not the tail is polynomial.   □

19.7 Pareto Distributions

The Pareto distribution with location parameter $c > 0$ and shape parameter $a > 0$ has density

$$f(y|a, c) = \begin{cases} ac^a \, y^{-(a+1)}, & y > c, \\ 0, & \text{otherwise.} \end{cases} \qquad (19.35)$$

The expectation is $ac/(a - 1)$ if $a > 1$, and $+\infty$ otherwise. The Pareto distribution has a polynomial tail and, in fact, a polynomial tail is often called a Pareto tail.

Equation (19.30) states that the loss, conditional on being above $|c|$, has a Pareto distribution. A property of the Pareto distribution that was exploited before [see (19.23)] is that if Y has a Pareto distribution with parameters a and c and if $d > c$, then the conditional distribution of Y, given that $Y > d$, is Pareto with parameters a and d.

19.8 Choosing the Horizon and Confidence Level

The choice of horizon and confidence coefficient are somewhat interdependent, and also depend on the eventual use of the VaR estimate. For shorter horizons such as one day, a large α (small confidence coefficient $= 1 - \alpha$) would result in frequent losses exceeding VaR. For example, $\alpha = 0.05$ would result in a loss exceeding VaR approximately once per month since there are slightly more than 20 trading days in a month. Therefore, we might wish to use smaller values of α with a shorter horizon.

One should be wary, however, of using extremely small values of α, such as, values less than 0.01. When α is very small, then VaR and, especially, ES are impossible to estimate accurately and are very sensitive to assumptions about the left tail of the return distribution. As we have seen, it is useful to create bootstrap confidence intervals to indicate the amount of precision in the VaR and ES estimates. It is also important to compare estimates based on different tail assumptions as in Fig. 19.5, for example, where the three estimates of VaR are increasingly dissimilar as α decreases below 0.01.

There is, of course, no need to restrict attention to only one horizon or confidence coefficient. When VaR is estimated parametrically and i.i.d. normally distributed returns are assumed, then it is easy to reestimate VaR with different horizons. Suppose that $\widehat{\mu}_P^{1\,\text{day}}$ and $\widehat{\sigma}_P^{1\,\text{day}}$ are the estimated mean and standard deviation of the return for one day. Assuming only that returns are i.i.d., the mean and standard deviation for M days are

$$\widehat{\mu}_P^{\text{M days}} = M\widehat{\mu}_P^{1\,\text{day}} \tag{19.36}$$

and

$$\widehat{\sigma}_P^{\text{M days}} = \sqrt{M}\widehat{\sigma}_P^{1\,\text{day}}. \tag{19.37}$$

Therefore, if one assumes further that the returns are normally distributed, then the VaR for M days is

$$\text{VaR}_P^{\text{M days}} = -S \times \left\{ M\widehat{\mu}_P^{1\,\text{day}} + \sqrt{M}\Phi^{-1}(\alpha)\widehat{\sigma}_P^{1\,\text{day}} \right\}, \tag{19.38}$$

where S is the size of the initial investment. The power of Eq. (19.38) is, for example, that it allows one to change from a daily to a weekly horizon without reestimating the mean and standard deviation with weekly instead of daily returns. Instead, one simply uses (19.38) with $M = 5$. The danger in using (19.38) is that it assumes normally distributed returns and no autocorrelation or GARCH effects (volatility clustering) of the daily returns. If there is positive autocorrelation, then (19.38) underestimates the M-day VaR. If there are GARCH effects, then (19.38) gives VaR based on the marginal distribution, but one should be using VaR based on the conditional distribution given the current information set.

If the returns are not normally distributed, then there is no simple analog to (19.38). For example, if the daily returns are i.i.d. but t-distributed then one cannot simply replace the normal quantile $\Phi^{-1}(\alpha)$ in (19.38) by a t-quantile. The problem is that the sum of i.i.d. t-distributed random variables is not itself t-distributed. Therefore, if the daily returns are t-distributed then the sum of M daily returns is not t-distributed. However, for large values of M and i.i.d. returns, the sum of M independent returns will be close to normally distributed by the central limit theorem, so (19.38) could be used for large M even if the returns are not normally distributed.

19.9 VaR and Diversification

A serious problem with VaR is that it may *discourage* diversification. This problem was studied by Artzner, Delbaen, Eber, and Heath (1997, 1999), who ask the question, what properties can reasonably be required of a risk measure? They list four properties that any risk measure should have, and they call a risk measure *coherent* if it has all of them.

One property among the four that is very desirable is *subadditivity*. Let $\Re(P)$ be a risk measure of a portfolio P, for example, VaR or ES. Then $\Re$ is said to be subadditive, if for any two portfolios P_1 and P_2, $\Re(P_1 + P_2) \leq \Re(P_1) + \Re(P_2)$. Subadditivity says that the risk for the combination of two portfolios is at most the sum of their individual risks, which implies that diversification reduces risk or at least does not increase risk. For example, if a bank has two traders, then the risk of them combined is less than or equal to the sum of their individual risks if a subadditive risk measure is used. Subadditivity extends to more than two portfolios, so if $\Re$ is subadditive, then for m portfolios, $P_1, \ldots, P_m$,

$$\Re(P_1 + \cdots + P_m) \leq \Re(P_1) + \cdots + \Re(P_m).$$

Suppose a firm has 100 traders and monitors the risk of each trader's portfolio. If the firm uses a subadditive risk measure, then it can be sure that the total risk of the 100 traders is at most the sum of the 100 individual risks. Whenever this sum is acceptable, there is no need to compute the risk measure for the entire firm. If the risk measure used by the firm is not subadditive, then there is no such guarantee.

Unfortunately, as the following example shows, VaR is *not* subadditive and therefore is incoherent. ES is subadditive, which is a strong reason for preferring ES to VaR.

Example 19.8. An example where VaR is not subadditive

This simple example has been designed to illustrate that VaR is not subadditive and can discourage diversification. A company is selling par \$1,000 bonds with a maturity of one year that pay a simple interest of 5 % so that the bond pays \$50 at the end of one year if the company does not default. If the bank defaults, then the entire \$1,000 is lost. The probability of no default is 0.96. To make the loss distribution continuous, we will assume that the loss is $N(-50, 1)$ with probability 0.96 and $N(1000, 1)$ with probability 0.04. The main purpose of making the loss distribution continuous is to simplify calculations. However, the loss would be continuous, for example, if the portfolio contained both the bond and some stocks. Suppose that there is a second company selling bonds with exactly the same loss distribution and that the two companies are independent.

Consider two portfolios. Portfolio 1 buys two bonds from the first company and portfolio 2 buys one bond from each of the two companies. Both portfolios have the same expected loss, but the second is more diversified. Let $\Phi(x|\mu, \sigma^2)$ be the normal CDF with mean μ and variance σ^2. For portfolio 1, the loss CDF is

$$0.04\,\Phi(x|2000, 4) + 0.96\,\Phi(x|-100, 4),$$

while for portfolio 2, by independence of the two companies, the loss distribution CDF is

$$0.04^2\,\Phi(x|2000, 2) + 2(0.96)(0.04)\,\Phi(x|950, 2) + 0.96^2\,\Phi(x|-100, 2).$$

We should expect the second portfolio to seem less risky, but VaR(0.05) indicates the opposite. Specifically, VaR(0.05) is -95.38 and 949.53 for portfolios 1 and 2, respectively. Notice that a negative VaR means a negative loss (positive revenue). Therefore, portfolio 1 is much less risky than portfolio 2, as measured by VaR(0.05). For each portfolio, VaR(0.05) is shown in Fig. 19.7 as the loss at which the CDF crosses the horizontal dashed line at 0.95.

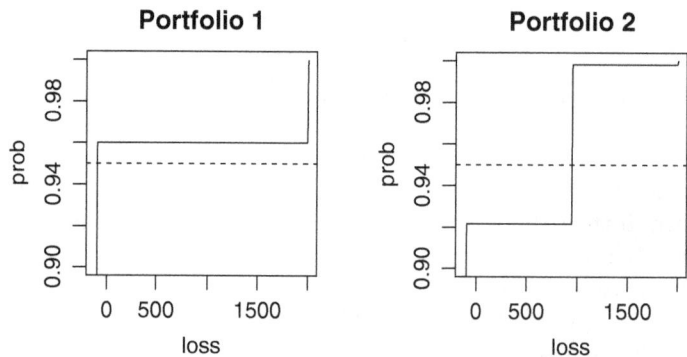

Fig. 19.7. *Example where VaR discourages diversification. Plots of the CDF of the loss distribution. VaR(0.05) is the loss at which the CDF crosses the horizontal dashed line at 0.95.*

Notice as well that which portfolio has the highest value of VaR(α) depends heavily on the values of α. When α is below the default probability, 0.04, portfolio 1 is more risky than portfolio 2.   □

Although VaR is often considered the industry standard for risk management, Artzner, Delbaen, Eber, and Heath (1997) make an interesting observation. They note that when setting margin requirements, an exchange should use a subadditive risk measure so that the aggregate risk due to all customers is guaranteed to be smaller than the sum of the individual risks. Apparently, no organized exchanges use quantiles of loss distributions to set margin requirements. Thus, exchanges may be aware of the shortcomings of VaR, and VaR is not the standard for measuring risk within exchanges.

19.10 Bibliographic Notes

Risk management is an enormous subject and we have only touched upon a few aspects, focusing on statistical methods for estimating risk. We have not considered portfolios with bonds, foreign exchange positions, interest rate derivatives, or credit derivatives. We also have not considered risks other than market risk or how VaR and ES can be used for risk management. To cover risk management thoroughly requires at least a book-length treatment of that subject. Fortunately, excellent books exist, for example, Dowd (1998), Crouhy, Galai, and Mark (2001), Jorion (2001), and McNeil, Frey, and Embrechts (2005). The last has a strong emphasis on statistical techniques, and is recommended for further reading along the lines of this chapter. Generalized Pareto distributions were not covered here but are discussed in McNeil, Frey, and Embrechts.

Alexander (2001), Hull (2003), and Gourieroux and Jasiak (2001) have chapters on VaR and risk management. The semiparametric method of estimation based on the assumption of a polynomial tail and Eq. (19.20) are from Gourieroux and Jasiak (2001). Drees, de Haan, and Resnick (2000) and Resnick (2001) are good introductions to Hill plots.

19.11 R Lab

19.11.1 Univariate VaR and ES

In this section we will compare VaR and ES parametric (unconditional) estimates with those from using ARMA+GARCH (conditional) models. Consider the daily returns for Coca-Cola stock from January 2007 to November 2012.

```
1 CokePepsi = read.table("CokePepsi.csv", header=T)
2 price = CokePepsi[,1]
3 returns = diff(price)/lag(price)[-1]
4 ts.plot(returns)
```

First, assume that the returns are iid and follow a t-distribution. Run the following commands to get parameter estimates in R.

```
5  S = 4000
6  alpha = 0.05
7  library(MASS)
8  res = fitdistr(returns,'t')
9  mu = res$estimate['m']
10 lambda = res$estimate['s']
11 nu = res$estimate['df']
12 qt(alpha, df=nu)
13 dt(qt(alpha, df=nu), df=nu)
```

Problem 1 *What quantities are being computed in the last two lines above?*

Problem 2 *For an investment of $4,000, what are estimates of $VaR^t(0.05)$ and $ES^t(0.05)$?*

Now, fit a ARMA(0,0)+GARCH(1,1) model to the returns and calculate one step forecasts.

```
14 library(fGarch) # for qstd() function
15 library(rugarch)
16 garch.t = ugarchspec(mean.model=list(armaOrder=c(0,0)),
17                      variance.model=list(garchOrder=c(1,1)),
18                      distribution.model = "std")
19 KO.garch.t = ugarchfit(data=returns, spec=garch.t)
20 show(KO.garch.t)
21 plot(KO.garch.t, which = 2)
22 pred = ugarchforecast(KO.garch.t, data=returns, n.ahead=1) ; pred
23 fitted(pred) ; sigma(pred)
24 nu = as.numeric(coef(KO.garch.t)[5])
25 q = qstd(alpha, mean = fitted(pred), sd = sigma(pred), nu = nu) ; q
26 sigma(pred)/sqrt( (nu)/(nu-2) )
27 qt(alpha, df=nu)
28 dt(qt(alpha, df=nu), df=nu)
```

Problem 3 *Carefully express the fitted ARMA(0,0)+GARCH(1,1) model in mathematical notation.*

Problem 4 *What are the one-step ahead predictions of the conditional mean and conditional standard deviation?*

Problem 5 *Again, for an investment of $4,000, what are estimates of VaR^t (0.05) and $ES^t(0.05)$ for the next day based on the fitted ARMA+GARCH model?*

19.11.2 VaR Using a Multivariate-*t* Model

Run the following code to create a data set of returns on two stocks, DATGEN and DEC.

```
1 library(mnormt)
2 berndtInvest = read.csv("berndtInvest.csv")
3 Berndt = berndtInvest[,5:6]
4 names(Berndt)
```

Problem 6 *Fit a multivariate-t model to the returns in* **Berndt**; *see Sect. 7.13.3 for an example of fitting such a model. What are the estimates of the mean vector, tail index, and scale matrix? Include your R code and output with your work.*

Problem 7

(a) What is the distribution of the return on a $100,000 portfolio that is 30 % invested in DATGEN and 70 % invested in DEC? Include your R code and output with your work.

(b) Find $VaR^t(0.05)$ and $ES^t(0.05)$ for this portfolio.

Problem 8 Use the model-free bootstrap to find a basic percentile bootstrap confidence interval for VaR(0.05) for the portfolio in Problem 7. Use a 90 % confidence coefficient for the confidence interval. Use 250 bootstrap resamples. This amount of resampling is not enough for a highly accurate confidence interval, but will give a reasonably good indication of the uncertainty in the estimate of VaR(0.05), which is all that is really needed.

Also, plot kernel density estimates of the bootstrap distribution of the tail index and $VaR^t(0.05)$. Do the densities appear Gaussian or skewed? Use a normality test to check if they are Gaussian.

Problem 9 This problem uses the variable DEC. Estimate the left tail index using the Hill estimator. Use a Hill plot to select n_c. What is your choice of n_c?

19.12 Exercises

1. This exercise uses daily BMW returns in the bmwRet data set on the book's website. For this exercise, assume that the returns are i.i.d., even though there may be some autocorrelation and volatility clustering is likely. Suppose a portfolio holds $1,000 in BMW stock (and nothing else).
   (a) Compute nonparametric estimates of VaR(0.01, 24 h) and ES(0.01, 24 h).
   (b) Compute parametric estimates of VaR(0.01, 24 h) and ES(0.01, 24 h) assuming that the returns are normally distributed.
   (c) Compute parametric estimates of VaR(0.01, 24 h) and ES(0.01, 24 h) assuming that the returns are t-distributed.
   (d) Compare the estimates in (a), (b), and (c). Which do you feel are most realistic?
2. Assume that the loss distribution has a polynomial tail and an estimate of a is 3.1. If VaR(0.05) = $252, what is VaR(0.005)?
3. Find a source of stock price data on the Internet and obtain daily prices for a stock of your choice over the last 1,000 days.
   (a) Assuming that the loss distribution is t, find the parametric estimate of VaR(0.025, 24 h).
   (b) Find the nonparametric estimate of VaR(0.025, 24 h).
   (c) Use a t-plot to decide if the normality assumption is reasonable.

(d) Estimate the tail index assuming a polynomial tail and then use the estimate of VaR(0.025, 24 h) from part (a) to estimate VaR(0.0025, 24 h).

4. This exercise uses daily Microsoft price data in the `msft.dat` data set on the book's website. Use the closing prices to compute daily returns. Assume that the returns are i.i.d., even though there may be some autocorrelation and volatility clustering is likely. Suppose a portfolio holds $1,000 in Microsoft stock (and nothing else). Use the model-free bootstrap to find 95 % confidence intervals for parametric estimates of VaR(0.005, 24 h) and ES(0.005, 24 h) assuming that the returns are t-distributed.

5. Suppose the risk measure $\mathfrak{R}$ is VaR(α) for some α. Let P_1 and P_2 be two portfolios whose returns have a joint normal distribution with means μ_1 and μ_2, standard deviations σ_1 and σ_2, and correlation ρ. Suppose the initial investments are S_1 and S_2. Show that $\mathfrak{R}(P_1 + P_2) \leq \mathfrak{R}(P_1) + \mathfrak{R}(P_2)$ under joint normality.[2]

6. This problem uses daily stock price data in the file `Stock_Bond.csv` on the book's website. In this exercise, use only the first 500 prices on each stock. The following R code reads the data and extracts the first 500 prices for five stocks. "AC" in the variables' names means "adjusted closing" price.

```
dat = read.csv("Stock_Bond.csv", header = T)
prices = as.matrix(dat[1:500, c(3, 5, 7, 9, 11)])
```

(a) What are the sample mean vector and sample covariance matrix of the 499 returns on these stocks?

(b) How many shares of each stock should one buy to invest $50 million in an equally weighted portfolio? Use the prices at the end of the series, e.g., `prices[,500]`.

(c) What is the one-day VaR(0.1) for this equally weighted portfolio? Use a parametric VaR assuming normality.

(d) What is the five-day Var(0.1) for this portfolio? Use a parametric VaR assuming normality. You can assume that the daily returns are uncorrelated.

References

Alexander, C. (2001) *Market Models: A Guide to Financial Data Analysis*, Wiley, Chichester.

Artzner, P., Delbaen, F., Eber, J.-M., and Heath, D. (1997) Thinking coherently. *RISK*, **10**, 68–71.

---

[2] This result shows that VaR is subadditive on a set of portfolios whose returns have a joint normal distribution, as might be true for portfolios containing only stocks. However, portfolios containing derivatives or bonds with nonzero probabilities of default generally do not have normally distributed returns.

Artzner, P., Delbaen, F., Eber, J.-M., and Heath, D. (1999) Coherent measures of risk. *Mathematical Finance*, **9**, 203–238.

Crouhy, M., Galai, D., and Mark, R. (2001) *Risk Management*, McGraw-Hill, New York.

Drees, H., de Haan, L., and Resnick, S. (2000) How to make a Hill plot, *Annals of Statistics*, **28**, 254–274.

Dowd, K. (1998) *Beyond Value At Risk*, Wiley, Chichester.

Gourieroux, C., and Jasiak, J. (2001) *Financial Econometrics*, Princeton University Press, Princeton, NJ.

Hull, J. C. (2003) *Options, Futures, and Other Derivatives*, 5th ed., Prentice-Hall, Upper Saddle River, NJ.

Jorion, P. (2001) *Value At Risk*, McGraw-Hill, New York.

McNeil, A. J., Frey, R., and Embrechts, P. (2005) *Quantitative Risk Management*, Princeton University Press, Princeton, NJ.

Resnick, S. I. (2001) *Modeling Data Networks*, School of Operations Research and Industrial Engineering, Cornell University, Technical Report #1345.

20

Bayesian Data Analysis and MCMC

20.1 Introduction

Bayesian statistics is based up a philosophy different from that of other methods of statistical inference. In Bayesian statistics all unknowns, and in particular unknown parameters, are considered to be random variables and their probability distributions specify our beliefs about their likely values. Estimation, model selection, and uncertainty analysis are implemented by using Bayes's theorem to update our beliefs as new data are observed.

Non-Bayesians distinguish between two types of unknowns, parameters and latent variables. To a non-Bayesian, parameters are fixed quantities without probability distributions while latent variables are random unknowns with probability distributions. For example, to a non-Bayesian, the mean μ, the moving average coefficients $\theta_1, \ldots, \theta_q$, and the white noise variance σ_ϵ^2 of an MA(q) process are fixed parameters while the unobserved white noise process itself consists of latent variables. In contrast, to a Bayesian, the parameters and the white noise process are both unknown random quantities. Since this chapter takes a Bayesian perspective, there is no need to distinguish between the parameters and latent variables, since they can now be treated in the same way. Instead, we will let $\boldsymbol{\theta}$ denote the vector of all unknowns and call it the "parameter vector." In the context of time series forecasting, for example, $\boldsymbol{\theta}$ could include both the unobserved white noise and the future values of the series being forecast.

A hallmark of Bayesian statistics is that one *must* start by specifying prior beliefs about the values of the parameters. Many statisticians have been reluctant to use Bayesian analysis since the need to start with prior beliefs seems too subjective. Consequently, there have been heated debates between Bayesian and non-Bayesian statisticians over the philosophical basis of statistics. However, much of mainstream statistical thought now supports the more pragmatic notion that we should use whatever works satisfactorily.

© Springer Science+Business Media New York 2015   581
D. Ruppert, D.S. Matteson, *Statistics and Data Analysis for Financial Engineering*, Springer Texts in Statistics,
DOI 10.1007/978-1-4939-2614-5_20

If one has little prior knowledge about a parameter, this lack of knowledge can be accommodated by using a so-called noninformative prior that provides very little information about the parameter relative to the information supplied by the data. In practice, Bayesian and non-Bayesian analyses of data usually arrive at similar conclusions when the Bayesian analysis uses only weak prior information so that knowledge of the parameters comes predominately from the data.

Moreover, in finance and many other areas of application, analysts often have substantial prior information and are willing to use it. In business and finance, there is no imperative to strive for objectivity as there is in scientific study. The need to specify a prior can be viewed as a strength, not a weakness, of the Bayesian view of statistics, since it forces the analyst to think carefully about how much and what kind of prior knowledge is available.

There has been a tremendous increase in the use of Bayesian statistics over the past few decades, because the Bayesian philosophy is becoming more widely accepted and because Bayesian estimators have become much easier to compute. In fact, Bayesian techniques often are the most satisfactory way to compute estimates for complex models. We have heard one researcher say "I am not a Bayesian but I use Bayesian methods" and undoubtedly others would agree.

For an overview of this chapter, assume we are interested in a parameter vector $\boldsymbol{\theta}$. A Bayesian analysis starts with a *prior* probability distribution for $\boldsymbol{\theta}$ that summarizes all prior knowledge about $\boldsymbol{\theta}$; "prior" means before the data are observed. The likelihood is defined in the same way in a non-Bayesian analysis, but in Bayesian statistics the likelihood has a different interpretation—the likelihood is the conditional distribution of the data given $\boldsymbol{\theta}$. The key step in Bayesian inference is the use of Bayes's theorem to combine the prior knowledge about $\boldsymbol{\theta}$ with the information in the data. This is done by computing the conditional distribution of $\boldsymbol{\theta}$ given the data. This distribution is called the *posterior distribution*. In many, if not most, practical problems, it is impossible to compute the posterior analytically and numerical methods are used instead. A very successful class of numerical Bayesian methods is Markov chain Monte Carlo (MCMC), which simulates a Markov chain in such a way that the stationary distribution of the chain is the posterior distribution of the parameters. The simulated data from the chain are used to compute Bayes estimates and perform uncertainty analysis.

20.2 Bayes's Theorem

Bayes's theorem applies to both discrete events and to continuously distributed random variables. We will start with the case of discrete events. The continuous case is covered in Sect. 20.3.

Suppose that $B_1, \ldots, B_K$ is a partition of the sample space $\mathcal{S}$ (the set of all possible outcomes). By "partition" is meant that $B_i \cap B_j = \emptyset$ if $i \neq j$ and $B_1 \cup B_2 \cup \cdots \cup B_K = \mathcal{S}$. For any set A, we have that

$$A = (A \cap B_1) \cup \cdots \cup (A \cap B_K),$$

and therefore, since $B_1, \ldots, B_K$ are disjoint,

$$P(A) = P(A \cap B_1) + \cdots + P(A \cap B_K). \tag{20.1}$$

It follows from (20.1) and the definition of conditional probability that

$$P(B_j | A) = \frac{P(A|B_j)P(B_j)}{P(A)} = \frac{P(A|B_j)P(B_j)}{P(A|B_1)P(B_1) + \cdots + P(A|B_K)P(B_K)}. \tag{20.2}$$

Equation (20.2) is called *Bayes's theorem*, and is also known as Bayes's rule or Bayes's law. Bayes's theorem is a simple, almost trivial, mathematical result, but its implications are profound. The importance of Bayes's theorem comes from its use for updating probabilities. Here is an example, one that is far too simple to be realistic but that illustrates how Bayes's theorem can be applied.

Example 20.1. Bayes's theorem in a discrete case

Suppose that our prior knowledge about a stock indicates that the probability θ that the price will rise on any given day is either 0.4 or 0.6. Based upon past data, say from similar stocks, we believe that θ is equally likely to be 0.4 or 0.6. Thus, we have the *prior* probabilities

$$P(\theta = 0.4) = 0.5 \quad \text{and} \quad P(\theta = 0.6) = 0.5.$$

We observe the stock for five consecutive days and its price rises on all five days. Assume that the price changes are independent across days, so that the probability that the price rises on each of five consecutive days is θ^5. Given this information, we may suspect that θ is 0.6, not 0.4. Therefore, the probability that θ is 0.6, given five consecutive price increases, should be greater than the prior probability of 0.5, but how much greater? As notation, let A be the event that the prices rises on five consecutive days. Then, using Bayes's theorem, we have

$$\begin{aligned}
P(\theta = 0.6 | A) &= \frac{P(A|\theta = 0.6)P(\theta = 0.6)}{P(A|\theta = 0.6)P(\theta = 0.6) + P(A|\theta = 0.4)P(\theta = 0.4)} \\
&= \frac{(0.6)^5(0.5)}{(0.6)^5(0.5) + (0.4)^5(0.5)} \\
&= \frac{(0.6)^5}{(0.6)^5 + (0.4)^5} = \frac{0.07776}{0.07776 + 0.01024} = 0.8836.
\end{aligned}$$

Thus, our probability that θ is 0.6 was 0.5 before we observed five consecutive price increases but is 0.8836 after observing this event. Probabilities before observing data are called the *prior probabilities* and the probabilities conditional on observed data are called the *posterior probabilities*, so the prior probability that θ equals 0.6 is 0.5 and the posterior probability is 0.8836. □

Bayes's theorem is extremely important because it tells us exactly how to update our beliefs in light of new information. Revising beliefs after receiving additional information is something that humans do poorly without the help of mathematics.[1] There is a human tendency to put either too little or too much emphasis on new information, but this problem can be mitigated by using Bayes's theorem for guidance.

20.3 Prior and Posterior Distributions

We now assume that $\boldsymbol{\theta}$ is a continuously distributed parameter vector. The *prior distribution* with density $\pi(\boldsymbol{\theta})$ expresses our beliefs about $\boldsymbol{\theta}$ prior to observing data. The likelihood function is interpreted as the conditional density of the data $\boldsymbol{Y}$ given $\boldsymbol{\theta}$ and written as $f(\boldsymbol{y}|\boldsymbol{\theta})$. Using Eq. (A.19), the joint density of $\boldsymbol{\theta}$ and $\boldsymbol{Y}$ is the product of the prior and the likelihood; that is,

$$f(\boldsymbol{y},\boldsymbol{\theta}) = \pi(\boldsymbol{\theta})f(\boldsymbol{y}|\boldsymbol{\theta}). \tag{20.3}$$

The marginal density of $\boldsymbol{Y}$ is found by integrating $\boldsymbol{\theta}$ out of the joint density so that

$$f(\boldsymbol{y}) = \int \pi(\boldsymbol{\theta})f(\boldsymbol{y}|\boldsymbol{\theta})d\boldsymbol{\theta}, \tag{20.4}$$

and the conditional density of $\boldsymbol{\theta}$ given $\boldsymbol{Y}$ is

$$\pi(\boldsymbol{\theta}|\boldsymbol{Y}) = \frac{\pi(\boldsymbol{\theta})f(\boldsymbol{Y}|\boldsymbol{\theta})}{f(\boldsymbol{Y})} = \frac{\pi(\boldsymbol{\theta})f(\boldsymbol{Y}|\boldsymbol{\theta})}{\int \pi(\boldsymbol{\theta})f(\boldsymbol{Y}|\boldsymbol{\theta})d\boldsymbol{\theta}}. \tag{20.5}$$

Equation (20.5) is another form of Bayes's theorem. The density on the left-hand side of (20.5) is called the *posterior density* and gives the probability distribution of $\boldsymbol{\theta}$ after observing the data $\boldsymbol{Y}$.

Notice our use of π to denote densities of $\boldsymbol{\theta}$, so that $\pi(\boldsymbol{\theta})$ is the prior density and $\pi(\boldsymbol{\theta}|\boldsymbol{Y})$ is the posterior density. In contrast, f is used to denote densities of the data, so that $f(\boldsymbol{y})$ is the marginal density of the data and $f(\boldsymbol{y}|\boldsymbol{\theta})$ is the conditional density given $\boldsymbol{\theta}$.

Bayesian estimation and uncertainty analysis are based upon the posterior. The most common Bayes estimators are the mode and the mean of the

---

[1] See Edwards (1982) .

posterior density. The mode is called the *maximum a posteriori estimator*, or *MAP estimator*. The mean of the posterior is

$$E(\boldsymbol{\theta}|\boldsymbol{Y}) = \int \boldsymbol{\theta}\,\pi(\boldsymbol{\theta}|\boldsymbol{Y})d\boldsymbol{\theta} = \frac{\int \boldsymbol{\theta}\,\pi(\boldsymbol{\theta})f(\boldsymbol{Y}|\boldsymbol{\theta})d\boldsymbol{\theta}}{\int \pi(\boldsymbol{\theta})f(\boldsymbol{Y}|\boldsymbol{\theta})d\boldsymbol{\theta}} \tag{20.6}$$

and is also called the posterior expectation.

Example 20.2. Updating the prior beliefs about the probability that a stock price will increase

We continue Example 20.1 but change the simple, but unrealistic, prior that said that θ was either 0.4 or 0.6 to a more plausible prior where θ could be any value in the interval $[0, 1]$, but with values near $1/2$ more likely. Specifically, we use a Beta(2,2) prior so that

$$\pi(\theta) = 6\theta(1 - \theta), \quad 0 < \theta < 1.$$

Let Y be the number of times the stock price increases on five consecutive days. Then Y is Binomial(n, θ) and the density of Y is

$$f(y|\theta) = \binom{5}{y} \theta^y (1 - \theta)^{5-y}, \quad y = 0, 1, \ldots, 5.$$

Since we observed that $Y = 5$, $f(Y|\theta) = f(5|\theta) = \theta^5$ and the posterior density is

$$\pi(\theta|5) = \frac{6\,\theta(1 - \theta)\theta^5}{\int 6\,\theta(1 - \theta)\theta^5 d\theta} = 56\,\theta^6(1 - \theta),$$

which is a Beta(7,2) density.

The prior and posterior densities are shown in Fig. 20.1. The posterior density is shifted towards the right compared to the prior because five consecutive days saw increased prices. The 0.05 lower and upper quantiles of the posterior distribution are 0.529 and 0.953, respectively, and are shown on the plot. Thus, there is 90 % posterior probability that θ is between 0.529 and 0.953. For this reason, the interval $[0.529, 0.953]$ is called a 90 % *posterior interval* and provides us with the set of likely values of θ. Posterior intervals are Bayesian analogs of confidence intervals and are discussed further in Sect. 20.6. Posterior intervals are also called *credible intervals*.

The posterior expectation is

$$\int_0^1 \theta\,\pi(\theta|5)d\theta = \int_0^1 56\,\theta^7(1 - \theta)d\theta = \frac{56}{72} = 0.778. \tag{20.7}$$

The MAP estimate is $6/7 = 0.857$ and its location is shown by a dotted vertical line in Fig. 20.1.

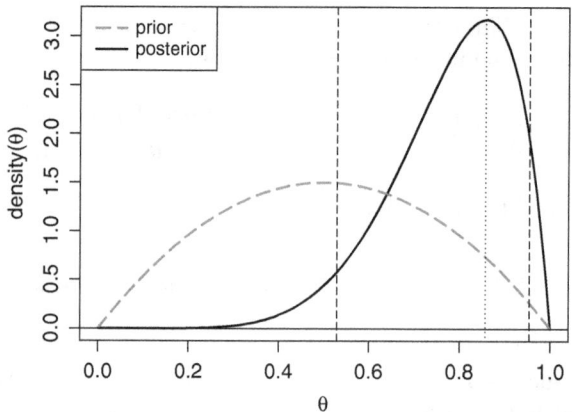

Fig. 20.1. *Prior and posterior densities in Example 20.2. The dashed vertical lines are at the lower and upper 0.05-quantiles of the posterior, so they mark off a 90 % equal-tailed posterior interval. The dotted vertical line shows the location of the posterior mode at $\theta = 6/7 = 0.857$.*

The posterior CDF is

$$\int_0^\theta \pi(x|5)dx = \int_0^\theta 56x^6(1-x)dx = 56\left(\frac{\theta^7}{7} - \frac{\theta^8}{8}\right), \quad 0 \le \theta \le 1.$$

□

20.4 Conjugate Priors

In Example 20.2, the prior and the posterior were both beta distributions. This is an example of a family of conjugate priors. A family of distributions is called a *conjugate prior family* for a statistical model (or, equivalently, for the likelihood) if the posterior is in this family whenever the prior is in the family. Conjugate families are convenient because they make calculation of the posterior straightforward. All one needs to do is to update the parameters in the prior. To see how this is done, we will generalize Example 20.2.

Example 20.3. Computing the posterior density of the probability that a stock price will increase—general case of a conjugate prior

Suppose now that the prior for θ is Beta(α, β) so that the prior density is

$$\pi(\theta) = K_1\theta^{\alpha-1}(1-\theta)^{\beta-1}, \tag{20.8}$$

where K_1 is a constant. As we will see, knowing the exact value of K_1 is not important, but from (A.14) we know that $K_1 = \frac{\Gamma(\alpha+\beta)}{\Gamma(\alpha)\Gamma(\beta)}$. The parameters in a prior density must be known, so here α and β are chosen by the data analyst in accordance with the prior knowledge about the value of θ. The choice of these parameters will be discussed later.

Suppose that the stock price is observed on n days and increases on Y days (and does not increase on $n - Y$ days). Then the likelihood is

$$f(Y|\theta) = K_2\theta^Y(1-\theta)^{n-Y}, \tag{20.9}$$

where $K_2 = \binom{n}{Y}$ is another constant. The joint density of θ and Y is

$$\pi(\theta)f(Y|\theta) = K_3\theta^{\alpha+Y-1}(1-\theta)^{\beta+n-Y-1}, \tag{20.10}$$

where $K_3 = K_1K_2$. Then, the posterior density is

$$\pi(\theta|Y) = \frac{\pi(\theta)f(Y|\theta)}{\int_0^1 \pi(\theta)f(Y|\theta)d\theta} = K_4\theta^{\alpha+Y-1}(1-\theta)^{\beta+n-Y-1}. \tag{20.11}$$

where

$$K_4 = \frac{1}{\int_0^1 \theta^{\alpha+Y-1}(1-\theta)^{\beta+n-Y-1}d\theta}. \tag{20.12}$$

The posterior distribution is Beta$(\alpha + Y, \beta + n - Y)$.

We did not need to keep track of the values of $K_1, \ldots, K_4$. Since (20.11) is proportional to a Beta$(\alpha + Y, \beta + n - Y)$ density and since all densities integrate to 1, we can deduce that the constant of proportionality is 1 and the posterior is Beta$(\alpha + Y, \beta + n - Y)$. It follows from (A.14) that

$$K_4 = \frac{\Gamma(\alpha + \beta + n)}{\Gamma(\alpha + Y)\Gamma(\beta + n - Y)}.$$

It is worth noticing how easily the posterior can be found. One simply updates the prior parameters α and β to $\alpha + Y$ and $\beta + n - Y$, respectively.

Using the results in Appendix A.9.7 about the mean and variance of beta distributions, the mean of the posterior is

$$E(\theta|Y) = \frac{\alpha + Y}{\alpha + \beta + n} \tag{20.13}$$

and the posterior variance is

$$\begin{aligned}
\text{var}(\theta|Y) &= \frac{(\alpha + Y)(\beta + n - Y)}{(\alpha + \beta + n)^2(\alpha + \beta + n + 1)} \\
&= \frac{E(\theta|Y)\{1 - E(\theta|Y)\}}{(\alpha + \beta + n + 1)}. \tag{20.14}
\end{aligned}$$

For values of α and β that are small relative to Y and n, $E(\theta|Y)$ is approximately equal to the MLE, which is Y/n. If we had little prior knowledge

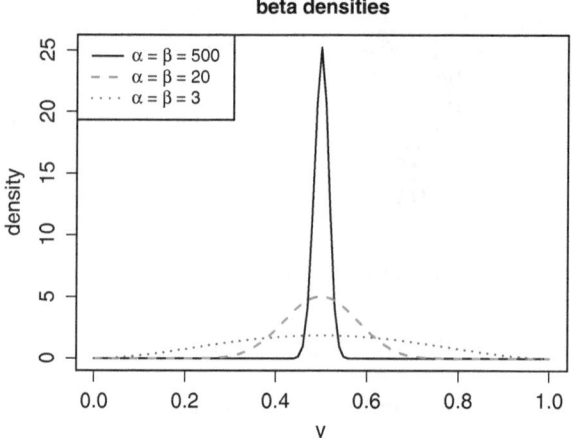

Fig. 20.2. *Examples of beta probability densities with $\alpha = \beta$.*

of θ, we might take both α and β close to 0. However, since θ is the probability of a positive daily return on a stock, we might be reasonably certain that θ is close to 1/2. In that case, choosing $\alpha = \beta$ and both fairly large (so that the prior precision is large) makes sense. One could plot several beta densities with $\alpha = \beta$ and decide which seem reasonable choices of the prior. For example, Fig. 20.2 contains plots of beta densities with $\alpha = \beta = 3, 20$, and 500. When 500 is the common value of α and β, then the prior is quite concentrated about 1/2. This prior could be used by someone who is rather sure that θ is close to 1/2. Someone with less certainty might instead prefer to use $\alpha = \beta = 20$, which has almost all of the prior probability between 0.3 and 0.6. The choice $\alpha = \beta = 3$ leads to a very diffuse prior and would be chosen if one had very little prior knowledge of θ and wanted to "let the data speak for themselves."

The posterior mean in (20.13) has an interesting interpretation. Suppose that we had prior information from a previous sample of size $\alpha + \beta$ and in that sample the stock price increased α times. If we combined the two samples, then the total sample size would be $\alpha + \beta + n$, the number of days with a price increase would be $\alpha + Y$, and the MLE of θ would be $(\alpha + Y)/(\alpha + \beta + n)$, the posterior mean given by (20.13). We can think of the prior as having as much information as would be given by a prior sample of size $\alpha + \beta$ and $\alpha/(\alpha + \beta)$ can be interpreted as the MLE of θ from that sample. Therefore, the three priors in Fig. 20.2 can be viewed as having as much information as samples of sizes 6, 40, and 1000. For a fixed value of $E(\theta|Y)$, we see from (20.14) that the posterior variance of θ becomes smaller as α, β, or n increases; this makes sense since n is the sample size and $\alpha + \beta$ quantifies the amount of information in the prior.

Since it is not necessary to keep track of constants, we could have omitted them from the previous calculations and, for example, written (20.8) as

$$\pi(\theta) \propto \theta^{\alpha-1}(1-\theta)^{\beta-1}. \tag{20.15}$$

In the following examples, we will omit constants in this manner.   □

Example 20.4. Posterior distribution when estimating the mean of a normal population with known variance

Suppose $Y_1, \ldots, Y_n$ are i.i.d. $N(\mu, \sigma^2)$ and σ^2 is known. The unrealistic assumption that σ^2 is known is made so that we can start simple and will be removed later.

The conjugate prior for μ is the family of normal distributions. To show this, assume that the prior on μ is $N(\mu_0, \sigma_0^2)$ for known values of μ_0 and σ_0^2. We learned in Example 20.3 that it is not necessary to keep track of quantities that do not depend on the unknown parameters (but could depend on the data or known parameters), so we will keep track only of terms that depend on μ.

Simple algebra shows that the likelihood is

$$f(Y_1, \ldots, Y_n|\mu) = \prod_{i=1}^{n}\left[\frac{1}{\sqrt{2\pi}\sigma}\exp\left\{-\frac{1}{2\sigma^2}(Y_i - \mu)^2\right\}\right]$$

$$\propto \exp\left\{-\frac{1}{2\sigma^2}\left(-2n\overline{Y}\mu + n\mu^2\right)\right\}. \tag{20.16}$$

The prior density is

$$\pi(\mu) = \frac{1}{\sqrt{2\pi}\sigma_0}\exp\left\{-\frac{1}{2\sigma_0^2}(\mu - \mu_0)^2\right\} \propto \exp\left\{-\frac{1}{2\sigma_0^2}(-2\mu\mu_0 + \mu^2)\right\}. \tag{20.17}$$

A *precision* is the reciprocal of a variance, and we let $\tau = 1/\sigma^2$ denote the population precision. Multiplying (20.16) and (20.17), we can see that the posterior density is

$$\pi(\mu|Y_1, \ldots, Y_n) \propto \exp\left\{\left(\frac{n\overline{Y}}{\sigma^2} + \frac{\mu_0}{\sigma_0^2}\right)\mu - \left(\frac{n}{2\sigma^2} + \frac{1}{2\sigma_0^2}\right)\mu^2\right\}$$

$$= \exp\left\{(\tau_{\overline{Y}}\overline{Y} + \tau_0\mu_0)\mu - \frac{1}{2}(\tau_{\overline{Y}} + \tau_0)\mu^2\right\}, \tag{20.18}$$

where $\tau_{\overline{Y}} = n\tau = n/\sigma^2$ and $\tau_0 = 1/\sigma_0^2$, so that $\tau_{\overline{Y}}$ is the precision of $\overline{Y}$ and τ_0 is the precision of the prior distribution.

One can see that $\log\{\pi(\mu|Y_1, \ldots, Y_n)\}$ is a quadratic function of μ, so $\pi(\mu|Y_1, \ldots, Y_n)$ is a normal density. Therefore, to find the posterior distribution we need only compute the posterior mean and variance. The posterior mean is the value of μ that maximizes the posterior density, that is, the posterior mode, so to calculate the posterior mean, we solve

$$0 = \frac{\partial}{\partial \mu} \log\{\pi(\mu|Y_1,\ldots,Y_n)\} \tag{20.19}$$

and find that the mean is

$$E(\mu|Y_1,\ldots,Y_n) = \frac{\tau_{\overline{Y}}\overline{Y} + \tau_0\mu_0}{\tau_{\overline{Y}} + \tau_0} = \frac{\frac{n\overline{Y}}{\sigma^2} + \frac{\mu_0}{\sigma_0^2}}{\frac{n}{\sigma^2} + \frac{1}{\sigma_0^2}}. \tag{20.20}$$

We can see from (A.10) that the precision of a normal density $f(y)$ is -2 times the coefficient of y^2 in $\log\{f(y)\}$. Therefore, the posterior precision is -2 times the coefficient of μ^2 in (20.18). Consequently, the posterior precision is $\tau_{\overline{Y}} + \tau_0 = n/\sigma^2 + 1/\sigma_0^2$, and the posterior variance is

$$\mathrm{Var}(\mu|Y_1\ldots,Y_n) = \frac{1}{\frac{n}{\sigma^2} + \frac{1}{\sigma_0^2}}. \tag{20.21}$$

In summary, the posterior distribution is

$$N\left(\frac{\frac{n\overline{Y}}{\sigma^2} + \frac{\mu_0}{\sigma_0^2}}{\frac{n}{\sigma^2} + \frac{1}{\sigma_0^2}}, \frac{1}{\frac{n}{\sigma^2} + \frac{1}{\sigma_0^2}}\right) = N\left(\frac{\tau_{\overline{Y}}\overline{Y} + \tau_0\mu_0}{\tau_{\overline{Y}} + \tau_0}, \frac{1}{\tau_{\overline{Y}} + \tau_0}\right). \tag{20.22}$$

We can see that the posterior precision $(\tau_{\overline{Y}} + \tau_0)$ is the sum of the precision of $\overline{Y}$ and the precision of the prior; this makes sense since the posterior combines the information in the data with the information in the prior.

Notice that as $n \to \infty$, the posterior precision $\tau_{\overline{Y}}$ converges to ∞ and the posterior distribution is approximately

$$N(\overline{Y}, \sigma^2/n). \tag{20.23}$$

What this result tells us is that as the amount of data increases, the effect of the prior becomes negligible. The posterior density also converges to (20.23) as $\sigma_0 \to \infty$ with n fixed, that is, as the prior becomes negligible because the prior precision decreases to zero.

A common Bayes estimator is the posterior mean given by the right-hand side of (20.20). Many statisticians are neither committed Bayesians nor committed non-Bayesians and like to look at estimators from both perspectives. A non-Bayesian would analyze the posterior mean by examining its bias, variance, and mean-squared error. We will see that, in general, the Bayes estimator is biased but is less variable than $\overline{Y}$, and the tradeoff between bias and variance is controlled by the choice of the prior.

To simplify notation, let $\widehat{\mu}$ denote the posterior mean. Then

$$\widehat{\mu} = \delta\overline{Y} + (1-\delta)\mu_0, \tag{20.24}$$

where $\delta = \tau_{\overline{Y}}/(\tau_{\overline{Y}} + \tau_0)$, and $E(\widehat{\mu}|\mu) = \delta\mu + (1-\delta)\mu_0$, so the bias of $\widehat{\mu}$ is $\{E(\widehat{\mu}|\mu) - \mu\} = (\delta - 1)(\mu - \mu_0)$ and $\widehat{\mu}$ is biased unless $\delta = 1$ or $\mu_0 = \mu$.

We will have $\delta = 1$ only in the limit as the prior precision τ_0 converges to 0 and $\mu_0 = \mu$ means that the prior mean is exactly equal to the true parameter, but of course this beneficial situation cannot be arranged since μ is not known.

The variance of $\widehat{\mu}$ is

$$\mathrm{Var}(\widehat{\mu}|\mu) = \frac{\delta^2 \sigma^2}{n},$$

which is less than $\mathrm{Var}(\overline{Y}) = \sigma^2/n$, except in the extreme case where $\delta = 1$. We see that smaller values of δ lead to more bias but smaller variance. The best bias–variance tradeoff minimizes the mean square error of $\widehat{\mu}$, which is

$$\mathrm{MSE}(\widehat{\mu}) = \mathrm{BIAS}^2(\widehat{\mu}) + \mathrm{Var}(\widehat{\mu}) = (\delta - 1)^2(\mu - \mu_0)^2 + \frac{\delta^2 \sigma^2}{n}. \qquad (20.25)$$

It is best, of course, to have $\mu_0 = \mu$, but this is not possible since μ is unknown. What is known is $\delta = \tau_{\overline{Y}}/(\tau_{\overline{Y}} + \tau_0)$ and δ can be controlled by the choice of τ_0.

Figure 20.3 shows the MSE as a function of $\delta \in (0, 1)$ for three values of $\mu - \mu_0$, which is called the "prior bias" since it is the difference between the true value of the parameter and the prior mean. In this figure $\sigma^2/n = 1/2$. For each of the two larger values of the prior bias, there is a range of values of δ where the Bayes estimator has a smaller MSE than $\overline{Y}$, but if δ is below this range, then the Bayes estimator has a larger MSE than $\overline{Y}$ and the range of "good" δ-values decreases as the prior bias increases. If the prior bias is large and δ is too small, then the MSE of the Bayes estimator can be quite large since it converges to the squared prior bias as $\delta \to 0$; see (20.25) or Fig. 20.3. This result shows the need either to have a good prior guess of μ or to keep the prior precision small so that δ is large. However, when δ is large, then the Bayes estimator cannot improve much over $\overline{Y}$ and, in fact, converges to $\overline{Y}$ as $\delta \to 1$.

In summary, it can be challenging to choose a prior that offers a substantial improvement over $\overline{Y}$. One way to do this is to combine several related

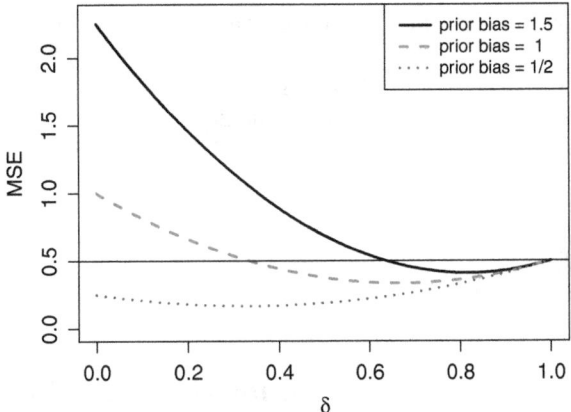

Fig. 20.3. *MSE versus δ for three values of "prior bias" $= \mu - \mu_0$ when $\sigma^2/n = 1/2$. The horizontal line represents the MSE of the maximum likelihood estimator $(\overline{Y})$.*

estimation problems using a hierarchical prior; see Sect. 20.8. When it is not possible to combine related problems and there is no other way to get information about μ, then the prudent data analyst will forgo the attempt to improve upon the MLE and instead will choose a small value for the prior precision τ_0.   □

Example 20.5. Posterior distribution when estimating a normal precision

Now suppose that $Y_1, \ldots, Y_n$ are i.i.d. with a known mean μ and an unknown variance σ^2 and precision $\tau = 1/\sigma^2$. We will show that the conjugate priors for τ are the gamma distributions and we will find the posterior distribution of τ. Define $s^2 = n^{-1} \sum_{i=1}^n (Y_i - \mu)^2$, which is the MLE of σ^2.

Simple algebra shows that the likelihood is

$$f(Y_1, \ldots, Y_n | \tau) \propto \exp\left(-\frac{1}{2} n \tau s^2\right) \tau^{n/2}. \tag{20.26}$$

Let the prior distribution be the gamma distribution with shape parameter α and scale parameter b which has density

$$\pi(\tau) = \frac{\tau^{\alpha-1}}{\Gamma(\alpha)b^\alpha} \exp(-\tau/b) \propto \tau^{\alpha-1} \exp(-\tau/b). \tag{20.27}$$

Multiplying (20.26) and (20.27), we see that the posterior density for τ is

$$\pi(\tau|Y_1, \ldots, Y_n) \propto \tau^{n/2+\alpha-1} \exp\{-(ns^2/2 + b^{-1})\tau\}, \tag{20.28}$$

which shows that the posterior distribution is gamma with shape parameter $n/2 + \alpha$ and scale parameter $(ns^2/2 + b^{-1})^{-1}$; that is,

$$\pi(\tau|Y_1, \ldots, Y_n) = \text{Gamma}\left\{n/2 + \alpha, \left(ns^2/2 + b^{-1}\right)^{-1}\right\}. \tag{20.29}$$

We have shown that gamma distributions are conjugate for a normal precision parameter.

The expected value of a gamma distribution is the product of the shape and scale parameters, so the posterior mean of τ is

$$E(\tau|Y_1, \ldots, Y_n) = \frac{\frac{n}{2} + \alpha}{\frac{ns^2}{2} + b^{-1}}.$$

Notice that $E(\tau|Y_1, \ldots, Y_n)$ converges to s^{-2} as $n \to \infty$, which is not surprising since the MLE of σ^2 is s^2, so that the MLE of τ is s^{-2}.   □

20.5 Central Limit Theorem for the Posterior

For large sample sizes, the posterior distribution obeys a central limit theorem that can be roughly stated as follows:

Result 20.6. *Under suitable assumptions and for large enough sample sizes, the posterior distribution of θ is approximately normal with mean equal to the true value of θ and with variance equal to the inverse of the Fisher information matrix.*

This result is also known as the *Bernstein–von Mises Theorem.* See Sect. 20.14 for references to a precise statement of the theorem.

This theorem is an important result for several reasons. First, a comparison with Result 5.1 shows that the Bayes estimator and the MLE have the same large-sample distributions. In particular, we see that for large sample sizes, the effect of the prior becomes negligible, because the asymptotic distribution does not depend on the prior. Moreover, the theorem shows a connection between confidence and posterior intervals that is discussed in the next section.

One of the assumptions of this theorem is that the prior remains fixed as the sample size increases, so that eventually nearly all of the information comes from the data. The more informative the prior, the larger the sample size needed for the posterior distribution to approach its asymptotic limit.

20.6 Posterior Intervals

Bayesian posterior intervals were mentioned in Example 20.2 and will now be discussed in more depth.

Posterior intervals have a different probabilistic interpretation than confidence intervals. The theory of confidence intervals views the parameter as fixed and the interval as random because it is based on a random sample. Thus, when we say "the probability that the confidence interval will include the true parameter is ...," it is the probability distribution of the interval, not the parameter, that is being considered. Moreover, the probability expresses the likelihood *before* the data are collected about what will happen after the data are collected. For example, if we use 95 % confidence, then the probability is 0.95 that we will obtain a sample whose interval covers the parameter. After the data have been collected and the interval is known, a non-Bayesian will say that either the interval covers the parameter or it does not, so the probability that the interval covers the parameter is either 1 or 0, though, of course, we do not know which value is the actual probability.

In the Bayesian theory of posterior intervals, the opposite is true. The sample is considered fixed since we use posterior probabilities, that is, probabilities conditional on the data. Therefore, the posterior interval is considered a fixed quantity. But in Bayesian statistics, parameters are treated as random. Therefore, when a Bayesian says "the probability that the posterior interval will include the true parameter is ...," the probability distribution being considered is the posterior distribution of the parameter. The random quantity is the parameter, the interval is fixed, and the probability is after the data have been collected.

Despite these substantial philosophical differences between confidence and posterior intervals, in many examples where both a confidence interval and a posterior interval have been constructed, one finds that they are nearly equal. This is especially common when the prior is relatively noninformative compared to the data, for example, in Example 20.3 if $\alpha + \beta$ is much smaller than n.

There are solid theoretical reasons based on central limit theorems why confidence and posterior intervals are nearly equal for large sample sizes. By Result 20.6 (the central limit theorem for the posterior), a large-sample posterior interval for the ith component of $\boldsymbol{\theta}$ is

$$E(\theta_i|\boldsymbol{Y}) \pm z_{\alpha/2}\sqrt{\operatorname{var}(\theta_i|\boldsymbol{Y})}. \qquad (20.30)$$

By Results 5.1 and 7.6 (the univariate and multivariate central limit theorems for the MLE), the large-sample confidence interval (5.20) based on the MLE and the large-sample posterior interval (20.30) will approach each other as the sample size increases. Therefore, practically-minded non-Bayesian data analysts are often happy to use a posterior interval and interpret it as a large-sample approximation to a confidence interval. Except in simple problems, all confidence intervals are based on large-sample approximations. This is true for confidence intervals that use profile likelihood, the central limit theorem for the MLE and Fisher information, or the bootstrap, in other words, for all of the major methods for constructing confidence intervals.

There are two major types of posterior intervals, highest probability and equal-tails. Let $\psi = \psi(\boldsymbol{\theta})$ be a scalar function of the parameter vector $\boldsymbol{\theta}$ and let $\pi(\psi|\boldsymbol{Y})$ be the posterior density of ψ. A highest-probability interval is of the form $\{\psi : \pi(\psi|\boldsymbol{Y}) > k\}$ for some constant k. As k increases from 0 to ∞, the posterior probability of this interval decreases from 1 to 0, and k is chosen so that the probability is $1 - \alpha$. If $\pi(\psi|\boldsymbol{Y})$ has multiple modes, then the set $\{\psi : \pi(\psi|\boldsymbol{Y}) > k\}$ might not be an interval and in that case it should be called a posterior set or posterior region rather than a posterior interval. In any case, this region has the interpretation of being the smallest set with $1 - \alpha$ posterior probability. When the highest-posterior region is an interval, it can be found by computing all intervals that range from the α_1-lower quantile of $\pi(\psi|\boldsymbol{Y})$ to the α_2-upper quantile of $\pi(\psi|\boldsymbol{Y})$, where $\alpha_1 + \alpha_2 = \alpha$, and the using the shortest of these intervals.

The equal-tails posterior interval has lower and upper limits equal to the lower and upper $\alpha/2$-quantiles of $\pi(\psi|\boldsymbol{Y})$. The two types of intervals coincide when $\pi(\psi|\boldsymbol{Y})$ is symmetric and unimodal, which will be at least approximately true for large samples by the central limit theorem for the posterior.

Posterior intervals of either type are easy to compute when using Monte Carlo methods; see Sect. 20.7.3.

Example 20.7. Posterior interval for a normal mean when the variance is known

This example continues Example 20.4. By (20.20) and (20.21), a $(1 - \alpha)$ 100% posterior interval for μ is

$$\frac{\tau_{\overline{Y}}\overline{Y} + \tau_0\mu_0}{\tau_{\overline{Y}} + \tau_0} \pm z_{\alpha/2}\sqrt{\frac{1}{\frac{n}{\sigma^2} + \frac{1}{\sigma_0^2}}}, \tag{20.31}$$

where $z_{\alpha/2}$ is the $\alpha/2$-upper quantile of the standard normal distribution.

If either $n \to \infty$ or $\sigma_0 \to \infty$, then the information in the prior becomes negligible relative to the information in the data because $\tau_{\overline{Y}}/\tau_0 \to \infty$, and the posterior interval converges to

$$\overline{Y} \pm z_{\alpha/2}\frac{\sigma}{\sqrt{n}},$$

which is the usual non-Bayesian confidence interval. $\square$

20.7 Markov Chain Monte Carlo

Although the Bayesian calculations in the simple examples of the last few sections were straightforward, this is generally not true for problems of practical interest. Frequently, the integral in the denominator of posterior density (20.5) is impossible to calculate analytically. The same is true of the integral in the numerator of the posterior mean given by (20.6). Because of computational difficulties, until approximately 1990 Bayesian data analysis was much less widely used than now. Fortunately, Monte Carlo simulation methods for approximating posterior densities and expectations have been developed. They have been a tremendous advance and not only have they made Bayesian methods practical, but also they have led to the solution of applied problems that heretofore could not be tackled.

The most widely applicable Monte Carlo method for Bayesian analysis simulates a Markov chain whose stationary distribution is the posterior. The sample from this chain is used for Bayesian inference. This technique is called *Markov chain Monte Carlo*, or *MCMC*. The BUGS language implements MCMC. There are three widely used versions of BUGS, OpenBUGS, WinBUGS, and JAGS. Most BUGS programs will run on all three versions, but there are exceptions. JAGS is in one way the most versatile of the three versions since it is the only one that will run under MacOS.[2]

This section is an introduction to MCMC and BUGS. First, we discuss Gibbs sampling, the simplest type of MCMC. Gibbs sampling works well when it is applicable, but it is applicable only to limited set of problems. Next, the Metropolis–Hastings algorithm is discussed. Metropolis–Hastings is applicable to nearly every type of Bayesian analysis. BUGS is a sophisticated program that is able to select an MCMC algorithm that is suitable for a particular model.

---

[2] OpenBUGS will run on a Mac under WINE.

20.7.1 Gibbs Sampling

Gibbs sampling is the simplest MCMC method. Suppose that the parameter vector $\boldsymbol{\theta}$ can be partitioned into M subvectors so that

$$\boldsymbol{\theta} = \begin{pmatrix} \boldsymbol{\theta}_1 \\ \vdots \\ \boldsymbol{\theta}_M \end{pmatrix}.$$

Let $[\boldsymbol{\theta}_j | \boldsymbol{Y}, \boldsymbol{\theta}_k, k \neq j]$ be the conditional distribution of $\boldsymbol{\theta}_j$ given the data $\boldsymbol{Y}$ and the values of the other subvectors; $[\boldsymbol{\theta}_j | \boldsymbol{Y}, \boldsymbol{\theta}_k, k \neq j]$ is called the *full conditional distribution* of $\boldsymbol{\theta}_j$. Gibbs sampling is feasible if one can sample from each of the full conditionals.

Gibbs sampling creates a Markov chain that repeatedly samples the subvectors $\boldsymbol{\theta}_1, \ldots, \boldsymbol{\theta}_M$ in the following manner. The chain starts with an arbitrary starting value $\boldsymbol{\theta}^{(0)}$ for the parameter vector $\boldsymbol{\theta}$. Then the subvector $\boldsymbol{\theta}_1^{(1)}$ is sampled from the full conditional $[\boldsymbol{\theta}_1 | \boldsymbol{Y}, \boldsymbol{\theta}_k, k \neq 1]$ with each of the remaining subvectors $\boldsymbol{\theta}_k, k \neq 1$, set at its current value which is $\boldsymbol{\theta}_k^{(0)}$. Next $\boldsymbol{\theta}_2^{(1)}$ is sampled from $[\boldsymbol{\theta}_2 | \boldsymbol{Y}, \boldsymbol{\theta}_k, k \neq 2]$ with $\boldsymbol{\theta}_k, k \neq 2$, set at its current value, which is $\boldsymbol{\theta}_k^{(1)}$ for $k = 1$ and $\boldsymbol{\theta}_k^{(0)}$ for $k \geq 2$. One continues it this way until each of $\boldsymbol{\theta}_1, \ldots, \boldsymbol{\theta}_M$ has been updated and one has $\boldsymbol{\theta}^{(1)} = (\boldsymbol{\theta}_1^{(1)}, \ldots, \boldsymbol{\theta}_M^{(1)})^{\mathsf{T}}$.

Then $\boldsymbol{\theta}^{(2)}$ is found starting at $\boldsymbol{\theta}^{(1)}$ in the same way that $\boldsymbol{\theta}^{(1)}$ was obtained starting at $\boldsymbol{\theta}^{(0)}$. Continuing in this way, we obtain the sequence $\boldsymbol{\theta}^{(1)}, \ldots, \boldsymbol{\theta}^{(N)}$ that is a Markov chain with the remarkable property that its stationary distribution is the posterior distribution of $\boldsymbol{\theta}$. Moreover, regardless of the starting value $\boldsymbol{\theta}^{(0)}$, the chain will converge to the stationary distribution. After convergence to the stationary distribution, the Markov chain samples the posterior distribution and the MCMC sample is used to compute posterior expectations, quantiles, and other characteristics of the posterior distribution.

Since the Gibbs sample does not start in the stationary distribution, the first N_0 iterations are discarded as a burn-in period for an appropriately chosen value of N_0. We will assume that this has been done and $\boldsymbol{\theta}^{(1)}, \ldots, \boldsymbol{\theta}^{(N)}$ is the sample from the chain after the burn-in period. In Sect. 20.7.5, methods for choosing N_0 are discussed.

Example 20.8. Gibbs sampling for a normal mean and precision

In Example 20.7, we found the posterior for a normal mean when the precision is known, and in Example 20.5, we found the posterior for a normal precision when the mean is known. These two results specify the two full conditionals and allow one to apply Gibbs sampling to the problem of estimating a normal mean and precision when both are unknown. The idea is simple. A starting value $\tau^{(0)}$ for τ is selected. The starting value might be the MLE, for example. However, there are advantages to using multiple chains with random starting values that are *overdispersed*, meaning that their probability distribution is more scattered than that posterior distribution; see Sect. 20.7.5.

Then, treating τ as known and equal to $\tau^{(0)}$, $\mu^{(1)}$ is drawn randomly from its Gaussian full conditional posterior distribution given in (20.22). Note: the starting value $\tau^{(0)}$ for the population precision τ should not be confused with the precision τ_0 in the prior for μ; $\tau^{(0)}$ is used only once, to start the Gibbs sampling algorithm; after burn-in, the Gibbs sample will not depend on the actual value of $\tau^{(0)}$. In contrast, τ_0 is fixed and is part of the posterior so the Gibbs sample should and will depend on τ_0.

After $\mu^{(1)}$ has been sampled, μ is treated as known and equal to $\mu^{(1)}$ and $\tau^{(1)}$ is drawn from the full conditional (20.29). Gibbs sampling continues in this way, alternatively between sampling μ and τ from their full conditionals. ☐

20.7.2 Other Markov Chain Monte Carlo Samplers

It is often difficult or impossible to sample directly from the full conditionals of the posterior and then Gibbs sampling is infeasible. Fortunately, there is a large variety of other sampling algorithms that can be used when Gibbs sampling cannot be used. Programming Monte Carlo algorithms "from scratch" is beyond the scope of this book but is explained in the references in Sect. 20.14. The BUGS language discussed in Sect. 20.7.4 allows analysts to use MCMC without the time-consuming and error-prone process of programming the details.

20.7.3 Analysis of MCMC Output

The analysis of MCMC output typically examines scalar-valued functions of the parameter vector $\boldsymbol{\theta}$. The analysis should be performed on each scalar quantity of interest. Let $\psi = \psi(\boldsymbol{\theta})$ be one such function. Suppose $\boldsymbol{\theta}_1, \ldots, \boldsymbol{\theta}_N$ is an MCMC sample from the posterior distribution of $\boldsymbol{\theta}$, either from a single Markov chain or from combining multiple chains, and define $\psi_i = \psi(\boldsymbol{\theta}_i)$. We will assume that the burn-in period and the chain lengths are sufficient so that $\psi_1, \ldots, \psi_N$ is a representative sample from the posterior distribution of ψ. Methods for diagnosing convergence and adequacy of the Monte Carlo sample size are explained in Sect. 20.7.5.

The MCMC sample mean $\overline{\psi} = N^{-1} \sum_{i=1}^{N} \psi_i$ estimates the posterior expectation $E(\psi|\boldsymbol{Y})$, which is the most common Bayes estimator. The MCMC sample standard deviation $s_\psi = \left\{ (N-1)^{-1} \sum_{i=1}^{N} (\psi_i - \overline{\psi})^2 \right\}^{1/2}$ estimates the posterior standard deviation of ψ and will be called the *Bayesian standard error*. If the sample size of the data is sufficiently large, then the posterior distribution will be approximately normal by Result 20.6 and an approximate $(1-\alpha)$ posterior interval for ψ is

$$\overline{\psi} \pm z_{\alpha/2} \, s_\psi. \tag{20.32}$$

Interval (20.32) is an MCMC approximation to (20.30).

However, one need not use this normal approximation to find posterior intervals. If $L(\alpha_1)$ is the α_1-lower sample quantile and $U(\alpha_2)$ is the α_2-upper sample quantile of $\psi_1, \ldots, \psi_N$, then $[L(\alpha_1), U(\alpha_2)]$ is a $1 - (\alpha_1 + \alpha_2)$ posterior interval. For an equal-tailed posterior interval, one uses $\alpha_1 = \alpha_2 = \alpha/2$. For a highest-posterior density interval, one chooses α_1 and α_2 on a fine grid such that $\alpha_1 + \alpha_2 = \alpha$ and $U(\alpha_2) - L(\alpha_1)$ is minimized. One should check that the posterior density of ψ is unimodal using a kernel density estimate. If there are several modes and sufficiently deep troughs between them, then highest-posterior density posterior region could be a union of intervals, not a single interval. However, even in this somewhat unusual case, $[L(\alpha_1), U(\alpha_2)]$ might still be used as the shortest $1 - \alpha$ posterior *interval*.

Kernel density estimates can be used to visualize the shapes of the posterior densities. As an example, see Fig. 20.4 discussed in Example 20.9 ahead. Most automatic bandwidth selectors for kernel density estimation are based on the assumption of an independent sample. When applied to MCMC output, they might undersmooth. If the `density()` function in R is used, one might correct this undersmoothing by using a value of the `adjust` parameter greater than the default value of 1. However, Fig. 20.4 uses the default value and the amount of smoothing seems adequate; this could be due to the large Monte Carlo sample size, $N = 10{,}000$.

20.7.4 JAGS

JAGS is a implementation of the BUGS (Bayesian analysis Using Gibbs Sampling) program that can be run from Windows, Mac OS, or Linux. JAGS can be used as a standalone program or it can be called from within R using the `rjags` package.

Example 20.9. Using JAGS to fit a t-distribution to returns

In this example, a *t*-distribution will be fit to S&P 500 returns using JAGS called from R. The BUGS program below is in the file `univt.bug`. The program will run under any of OpenBUGS, WinBUGS, or JAGS. In this example, JAGS will be used.

```
 1  model{
 2  for(i in 1:N)
 3  {
 4     r[i] ~ dt(mu, tau, k)
 5  }
 6  mu ~ dnorm(0.0, 1.0E-6)
 7  tau ~ dgamma(0.1, 0.01)
 8  k ~ dunif(2, 50)
 9  sigma2 <- (k / (k - 2)) / tau
10  sigma <- sqrt(sigma2)
11  }
```

In BUGS, dnorm(mu,tau) is the normal distribution with mean equal to mu and precision equal to tau. Also, dt(mu, tau, k) is the t-distribution with mean equal to mu, degrees of freedom equal to k, and inverse scale parameter equal to the square root of tau (so tau is proportional to, rather than equal to, the precision and the constant of proportionality is $\sqrt{(k-2)/k}$). In the BUGS program, the "for loop" (lines 3–5) specifies the likelihood and lines 6–8 specify the priors for mu, tau, and k. Line 9 computes the variance from tau and line 10 computes the standard deviation.

The R program is:

```
1  library(rjags)
2  library("Ecdat")
3  data(SP500)
4  r = SP500$r500
5  N = length(r)
6  data = list(r = r, N = N)
7  inits = function(){list(mu = rnorm(1, mean = mean(r),
8     sd = 2 * sd(r)), tau = runif(1, 0.2/var(r), 2/var(r)),
9     k = runif(1, 2.5, 10))}
10 t1 = proc.time()
11 univt.mcmc <- jags.model("univt.bug", data = data, inits = inits,
12    n.chains = 3, n.adapt = 1000, quiet = FALSE)
13 nthin = 20
14 univt.coda = coda.samples(univt.mcmc, c("mu", "k", "sigma"),
15    100*nthin, thin = nthin)
16 summary(univt.coda, digits = 2)
17 t2 = proc.time()
18 (t2 - t1) / 60
19 pdf("basic_plot.pdf", width = 4, height = 7)  ## Figure 20.4
20 par(mfrow = c(4, 2))
21 plot(univt.coda, auto.layout = F)   ## Figure 20.4
22 graphics.off()
23 gelman.diag(univt.coda)
24 effectiveSize(univt.coda)
25 pdf("gelman_plot.pdf", width = 6, height = 6)  ## Figure 20.6
26 gelman.plot(univt.coda)
27 graphics.off()
28 dic.samples(univt.mcmc, 100*nthin, thin = nthin, type = "pD:)
```

Line 6 creates a data list that is given to JAGS and lines 7–9 creates a function inits() that generates starting values for each chain. The function jags.model() at lines 11–12 creates an object univt.mcmc of class jags containing a graphical model description of the model specified in the file univt.bug. This object is one of the arguments of coda.samples() at lines 14–15. The function coda.samples() produces an object univt.coda of class mcmc.list containing MCMC output. Objects of this class can be used as input to functions in the coda package such as gelman.diag(), effectiveSize(), gelman.plot(), and summary(). Line 18 prints the computation time in minutes. The computation time for this example was about 6 minutes, but this number is, of course, hardware dependent.

The output from line 16 is:

```
> summary(univt.coda, digits = 2)

Iterations = 3020:5000
Thinning interval = 20
Number of chains = 3
Sample size per chain = 100

1. Empirical mean and standard deviation for each variable,
   plus standard error of the mean:

            Mean        SD  Naive SE Time-series SE
k       6.0451630 0.5443919 3.143e-02      3.137e-02
mu      0.0005129 0.0001850 1.068e-05      1.071e-05
sigma   0.0103017 0.0002078 1.200e-05      1.146e-05

2. Quantiles for each variable:

           2.5%       25%       50%       75%     97.5%
k     5.1407126 5.6780998 6.0380664 6.4039149 7.1849194
mu    0.0001289 0.0003821 0.0005046 0.0006191 0.0008763
sigma 0.0099304 0.0101560 0.0102910 0.0104427 0.0107435
```

Figure 20.4 produced at line 21 contains trace plots and kernel density estimates for mu, k, and sigma arranged alphabetically. The trace plots are simply time series plots of the three chains. The interpretation of trace plots is discussed in Sect. 20.7.5. The Gelman plot produced by line 26 is shown later as Fig. 20.6.

The diagnostics from lines 23–28 will discuss briefly here and described in more detail later. The Gelman diagnostics produced by line 23 are:

```
> gelman.diag(univt.coda)
Potential scale reduction factors:

        Point est. Upper C.I.
k            1.00       1.01
mu           1.00       1.01
sigma        1.02       1.06

Multivariate psrf

1.02
```

The effective sample sizes calculated at line 24 are:

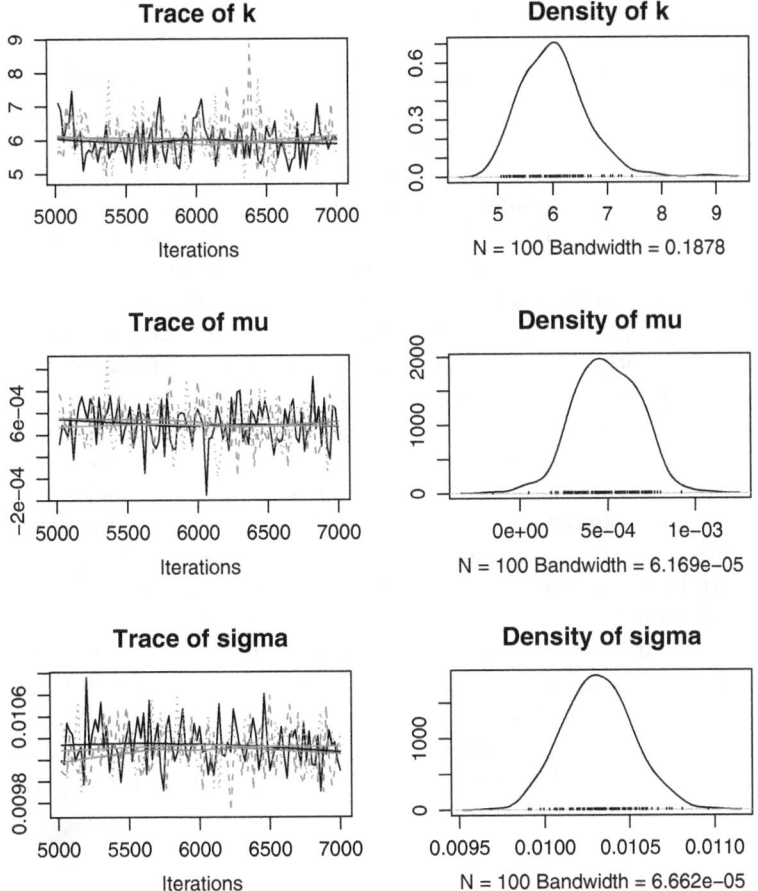

Fig. 20.4. *Trace plots and kernel density estimates in Example 20.9.*

```
> effectiveSize(univt.coda)
       k        mu      sigma
366.8874 300.0000 300.0000
```

Gelman diagnostics and effective sample sizes are discussed soon in Sect. 20.7.5.
DIC and p_D, which are discussed in Sect. 20.7.6 and produced at line 28, are
DIC $= -18,062$ and $p_D = 2.664$:

```
> dic.samples(univt.mcmc, 100*nthin, thin = nthin, type = "pD")
  |**************************************************| 100%
Mean deviance:  -18065
penalty 2.664
Penalized deviance: -18062
```

20.7.5 Monitoring MCMC Convergence and Mixing

The length N_0 of the burn-in period must be sufficiently large that the Markov chain has converged to the stationary distribution by the end of burn-in. The length N of the chain after burn-in must be large enough that moments, quantiles, and other quantities computed from the MCMC sample are accurate estimates of the corresponding characteristics of posterior. Markov chains are dependent sequences and the chains used in MCMC typically have positive autocorrelation. Because of the autocorrelation, to achieve accurate estimates Markov chain samples must be larger, often far larger, than would be necessary with independent sampling. A chain that moves about the posterior slowly is said to mix poorly. The slower the mixing of the chain, the larger the sample size needed for accurate estimation.

In principle, one long Markov chain is all that is needed to sample the posterior. However, if several chains are generated, then one can compare them to decide if the burn-in period N_0 and chain length N are sufficiently large. If the amount of between-chain variation in the chain means is large relative to the within-chain variation, then the chains are mixing poorly. Consequently, diagnostics for convergence and mixing can be based on between- and within-chain variation.

Between-chain variability will be artificially low if the chains have similar starting values. For this reason, it is recommended that the starting values be randomly sampled from a distribution with greater dispersion than the posterior. For example, one might use a Gaussian or t-distribution with mean equal to the MLE and covariance matrix equal to k times the inverse Fisher information for some $k > 1$, e.g., $k = 1.5$ or 2.

Example 20.10. Good mixing and poor mixing

Excellent and poor mixing are contrasted in Fig. 20.5. The model is linear regression with two predictor variables and i.i.d. Gaussian noise. There are two simulated data sets. In the first data set the predictors are highly correlated (sample correlation $= 0.996$). Trace plots for this data set are in the top row. In the second data set the predictors are independent and the trace plots are in the middle row. Except for this difference in the amount of collinearity, the two data sets have the same distributions. In both of these cases, there are three chains and for each chain there is a burn-in period of $N_0 = 100$ iterations and then 1000 iterations that are retained. In each row, trace plots, are shown for the three regression coefficients (intercept and two slopes) and for the residual precision (inverse variance).

In each case the three chains were started at randomly chosen initial values. The probability distribution was centered at the least-squares and "overdispersed" relative to the posterior distribution. Specifically, the regression coefficients have a Gaussian starting value distribution centered at the least-squares estimate and with covariance matrix 1.5 times the covariance matrix of the least-squares estimator. The noise variance had a starting distribution that was uniformly distributed between 0.25 and 4 times the least-squares estimate [e.g., $\widehat{\sigma}_\epsilon^2$ in (9.16)] of the noise variance. By using overdispersed starting values, one can discover how quickly the chains move from their starting values to the stationary distribution. The chains for the regression coefficients move very quickly in the middle row but slowly in the top row. The residual precision is unaffected by collinearity and moves quickly even in the high collinearity case.

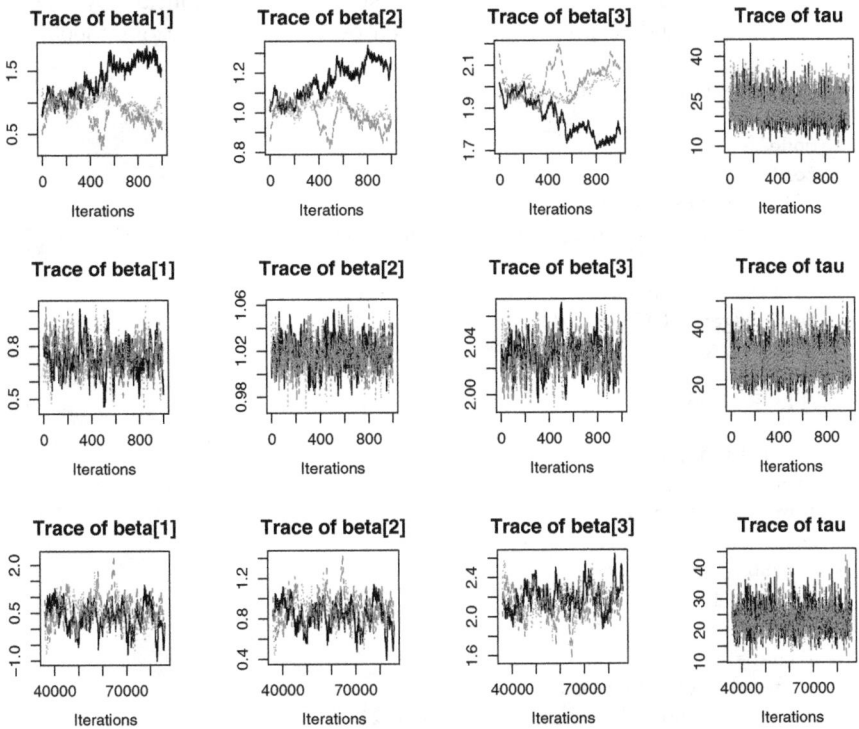

Fig. 20.5. *MCMC analysis of a linear regression model with two predictor variables. Simulated data. Trace plots of the regression coefficients and residual precision for three chains. The burn-in period was 100 and the chain lengths are 1000. The trace plots contain the MCMC output after the burn-in period. The intercept is beta[1] and the slopes are beta[2] and beta[3].* **Top row:** *The two predictors are highly correlated and the strong collinearity is causing poor mixing of the regression coefficients. Notice that the chains have not converged to the stationary distribution by the start of the sampling period and that the between-chain variation is large.* **Middle row:** *The burn-in period was 100 and the chain lengths are 1000 as in the top row. The two predictors are independent and there is very good mixing because there is no collinearity. Notice that the chains have converged to the stationary distribution by the start of the sampling period and there is little between-chain variation.* **Bottom row:** *Same data set as the top row but with a burn-in period of 5000 and chain lengths of 30,000. The chains have been thinned so that only every 10th iteration is retained.*

One solution to poor mixing is to increase the burn-in period and the chain lengths. The bottom row uses the same data set as in the top row but with a longer burn-in (5000 iterations) and longer chains (30,000 iterations). The chains have been thinned so that only every 10th iteration is retained. Thinning can speed the analysis of the MCMC output by reducing the Monte Carlo sample size and can improve the appearance of trace plots—a trace plot of 3 chains of 30,000 iterations each would be almost completely filled in. The chains appear to have converged to the stationary distribution by the end of the burn-in and to mix reasonably well over 30,000 iterations (3000 after thinning).

The BUGS code for this model is:

```
1  model{
2  for(i in 1:N){
3  y[i] ~ dnorm(mu[i],tau)
4  mu[i] <- x[i,1]*beta[1] + x[i,2]*beta[2] +  x[i,3]*beta[3]
5  }
6  for(i in 1:3)beta[i] ~ dnorm(0,.00001)
7  tau ~ dgamma(0.01,0.01)
8  }
```

The R code is:

```
1  library(rjags)
2  library(coda)
3  library(mvtnorm)
4  set.seed(90201)
5  N = 50
6  beta1 = 1
7  beta2 = 2
8  alpha = 1
9  x1 = rnorm(N, mean = 3, sd = 2)
10  x2 = x1 + rnorm(N, mean = 3, sd = 0.2)
11  x = cbind(rep(1, N), x1, x2)
12  y = alpha + beta1 * x1 + beta2 * x2 + rnorm(N, mean = 0, sd = 0.2)
13  data = list(y = y, x = x, N = N)
14  summ = summary(lm(y ~ x1 + x2))
15  betahat = as.numeric(summ$coeff)[1:3]
16  covbetahat = summ$sigma^2 * solve(t(x) %*% x)
17  inits = function(){list(beta = as.numeric(rmvnorm(n = 1,
18     mean = betahat, sigma=1.5 * covbetahat)),
19     tau=runif(1, 1/(4 * summ$sigma^2), 4 / summ$sigma^2))}
20  regr <- jags.model("lin_reg_vect.bug", data = data, inits = inits,
21     n.chains = 3, n.adapt = 1000, quiet = FALSE)
22  regr.coda = coda.samples(regr, c("beta", "tau"), 1000, thin = 1)
23  regr.coda.largeN = coda.samples(regr, c("beta", "tau"),
24     50000, thin = 100)
25  #####   no collinearity  #####
26  set.seed(90201)
```

```
27 x1 = rnorm(N, mean = 3, sd = 2)
28 x2 = rnorm(N,mean = 3, sd = 2) + rnorm(N, mean = 3, sd = 0.2)
29 x = cbind(rep(1, N), x1, x2)
30 y = alpha + beta1 * x1 + beta2 * x2 + rnorm(N, mean = 0, sd = 0.2)
31 data = list(y = y, x = x, N = N)
32 summ = summary(lm(y ~ x1 + x2))
33 betahat = as.numeric( summ$coeff )[1:3]
34 covbetahat = summ$sigma^2 * solve(t(x) %*% x)
35 inits=function(){list(beta = as.numeric(rmvnorm(n = 1,
36    mean = betahat, sigma = 1.5 * covbetahat)) ,
37    tau=runif(1, 1 / (4 * summ$sigma^2), 4 / summ$sigma^2))}
38 regr.noco <- jags.model("lin_reg_vect.bug", data = data,
39    inits = inits, n.chains = 3, n.adapt = 1000, quiet = FALSE)
40 regr.coda.noco = coda.samples(regr.noco, c("beta", "tau"),
41    1000, thin = 1)
42 pdf("linRegMCMC.pdf", width = 7, height = 6)
43 par(mfrow=c(3,4))
44 traceplot(regr.coda)
45 traceplot(regr.coda.noco)
46 traceplot(regr.coda.largeN)
47 graphics.off()
```

□

We now introduce two widely used diagnostics, Rhat and n.eff. Suppose one samples M chains, each of length N after burn-in. Let $\theta_{i,j}$ be the ith iterate from the jth chain and let $\psi_{i,j} = \psi(\theta_{i,j})$ for some scalar-valued function ψ. For example, to extract the kth parameter, one would use $\psi(x) = x_k$, or ψ might compute the standard deviation or the variance from the precision. We also use ψ to denote the estimand $\psi(\theta)$.

Let

$$\overline{\psi}_{.,j} = N^{-1} \sum_{i=1}^{N} \psi_{i,j} \tag{20.33}$$

be the mean of the jth chain and let

$$\overline{\psi}_{.,.} = M^{-1} \sum_{j=1}^{M} \overline{\psi}_{.,j}. \tag{20.34}$$

$\overline{\psi}_{.,.}$ is the average of the chain means and is the Monte Carlo approximation to $E(\psi|Y)$. Then define

$$B = \frac{N}{M-1} \sum_{j=1}^{M} \left(\overline{\psi}_{.,j} - \overline{\psi}_{.,.}\right)^2. \tag{20.35}$$

B/N is the sample variance of the chain means. Define

$$s_j^2 = (N-1)^{-1} \sum_{i=1}^{N} \left(\psi_{i,j} - \overline{\psi}_{.,j} \right)^2, \tag{20.36}$$

the variance of the jth chain, and define

$$W = M^{-1} \sum_{j=1}^{M} s_j^2. \tag{20.37}$$

W is the pooled within-chain variance. The two variances, B and W, are combined into

$$\widehat{\text{var}}^+(\psi|\boldsymbol{Y}) = \frac{N-1}{N} W + \frac{1}{N} B, \tag{20.38}$$

where, as before, $\boldsymbol{Y}$ is the data.

To assess convergence, one can use

$$\widehat{R} = \sqrt{\frac{\widehat{\text{var}}^+(\psi|\boldsymbol{Y})}{W}}. \tag{20.39}$$

$\widehat{R}$ is called the "potential scale reduction factor" in output produced by the function gelman.diag(); see the output in Example 20.9. $\widehat{R}$ is also called the "shrink factor" or sometime the "Gelman shrink factor."

When the chains have not yet reached the stationary distribution, the numerator $\widehat{\text{var}}^+(\psi|\boldsymbol{Y})$ inside the radical is an upward-biased estimate of $\text{var}(\psi|\boldsymbol{Y})$ and the denominator W is a downward-biased estimator of this quantity. Both biases converge to 0 as the burn-in period and Monte Carlo sample size increase. Therefore, larger values of $\widehat{R}$ indicate nonconvergence. If $\widehat{R}$ is approximately equal to 1, say at most 1.1, then the chains are considered to have converged to the stationary distribution and $\widehat{\text{var}}^+(\psi|\boldsymbol{Y})$ can be used as an estimate of $\text{var}(\psi|\boldsymbol{Y})$. A larger value of $\widehat{R}$ is an indication that a longer burn-in period is needed. A small value of $\widehat{R}$ is evidence that the burn-in period is adequate, but we need another diagnostic, the effective sample size, to know if the sampling period was long enough.

The *effective sample size* of the chain is

$$N_{\text{eff}} = MN \frac{\widehat{\text{var}}^+(\psi|\boldsymbol{Y})}{B}. \tag{20.40}$$

The interpretation of N_{eff} is that the Markov chain can estimate the posterior expectation of ψ with approximately the same precision as would be obtained from an independent sample from the posterior of size N_{eff}. (Of course, it is usually impossible to actually obtain an independent sample, which is why MCMC is used.)

N_{eff} is derived by comparing the variance of $\overline{\psi}_{.,.}$ from Markov chain sampling with the variance of $\overline{\psi}_{.,.}$ under hypothetical independent sampling. Since

$\overline{\psi}_{.,.}$ is the average of the means of M independent chains and since B/N is the sample variance of these M chain means,

$$M^{-1}\frac{B}{N} \tag{20.41}$$

estimates the Monte Carlo variance of $\overline{\psi}_{.,.}$. Suppose instead of sampling M chains, each of length N, one could take an independent sample of size N^* from the posterior. The Monte Carlo variance of the mean of this sample would be

$$\frac{\mathrm{var}(\psi|\boldsymbol{Y})}{N^*},$$

which can be estimated by

$$\frac{\widehat{\mathrm{var}}^+(\psi|\boldsymbol{Y})}{N^*}. \tag{20.42}$$

By definition N_{eff} is the value of N^* that makes (20.41) equal to (20.42) and therefore N^* is given by (20.40). Because B/N is the sample variance of M chains and because M is typically quite small, often between 2 and 5, B has considerable Monte Carlo variability. Therefore, N_{eff} is at best a crude estimate of the effective sample size.

$\widehat{R}$ and N_{eff} are computed by the functions `gelman.diag()` and `effective-Size()` in the `coda` package. The function `gelman.plot()` in the `coda` package plots the $\widehat{R}$ evaluated at various times along the simulation. The documentation for this function notes that "A potential problem with gelman.diag is that it may mis-diagnose convergence if the shrink factor happens to be close to 1 by chance. By calculating the shrink factor at several points in time, gelman.plot shows if the shrink factor has really converged, or whether it is still fluctuating." Figure 20.6 shows the Gelman plot from Example 20.9. The dashed red line is an upper 97.5 % confidence limit. Although $\widehat{R}$ varies during the simulations, it appears to have converged to values close to 1 by the end of the simulations.

How large should N_{eff} be? Of course, larger means better Monte Carlo accuracy, but larger values of N_{eff} require more or longer chains, so we do not want N_{eff} to be unnecessarily large. The effect of N_{eff} on estimation error can be seen by decomposing the estimation error $\psi - \overline{\psi}_{.,.}$ into two parts, which will be called E_1 and E_2:

$$\psi - \overline{\psi}_{.,.} = \{\psi - E(\psi|\boldsymbol{Y})\} + \{E(\psi|\boldsymbol{Y}) - \overline{\psi}_{.,.}\} = E_1 + E_2. \tag{20.43}$$

If $E\{\psi|\boldsymbol{Y}\}$ could be computed exactly so that it, not $\overline{\psi}_{.,.}$, would be the estimator of ψ, then E_1 would be the only error. E_2 is the error due to the Monte Carlo approximation of $E\{\psi|\boldsymbol{Y}\}$ by $\overline{\psi}_{.,.}$. The two errors E_1 and E_2 are uncorrelated, so

$$\mathrm{var}\{(\psi - \overline{\psi}_{.,.})|\boldsymbol{Y}\} = \mathrm{var}(E_1|\boldsymbol{Y}) + \mathrm{var}(E_2|\boldsymbol{Y})$$

$$= \text{var}(\psi|\boldsymbol{Y}) + \frac{\text{var}(\psi|\boldsymbol{Y})}{N_{\text{eff}}}$$

$$= \text{var}(\psi|\boldsymbol{Y}) \left(1 + \frac{1}{N_{\text{eff}}}\right)$$

by the definitions of $\text{var}(\psi|\boldsymbol{Y})$ and N_{eff} and using the approximation $\widehat{\text{var}}^+$ $(\psi|\boldsymbol{Y}) \approx \text{var}(\psi|\boldsymbol{Y})$. Using the Taylor series approximation $\sqrt{1+\delta} \approx 1 + \delta/2$ for small values of δ, we see that

$$\sqrt{\text{var}\{(\psi - \overline{\psi}_{.,.})|\boldsymbol{Y}\}} \approx \sqrt{\text{var}(\psi|\boldsymbol{Y})} \left(1 + \frac{1}{2N_{\text{eff}}}\right). \qquad (20.44)$$

Recall that $\sqrt{\text{var}\{(\psi - \overline{\psi}_{.,.})|\boldsymbol{Y}\}}$ is the "Bayesian standard error." If $N_{\text{eff}} \geq 50$, then we see from (20.44) that the standard error is inflated by Monte Carlo error by at most 1 %. Thus, one might use the rule-of-thumb that N_{eff} should be at least 50. Remember, however, that N_{eff} is estimated only crudely because the number of chains is small. Thus, we might want to have N_{eff} at least 100 to provide some leeway for error in the estimation of N_{eff}.

The value of N_{eff} can vary between different choices of ψ. Recall that the values of N_{eff} from Example 20.9 were:

```
> effectiveSize(univt.coda)
       k       mu    sigma
366.8874 300.0000 300.0000
```

In this example, N_{eff} is 300 for mu and sigma and only slightly larger for k. (In more complex models, much greater variation in N_{eff} is common.) In this example, 300 is the Monte Carlo sample size since there are three chains, each of length 100 after burn-in and thinning.[3] Therefore, in this simple example, MCMC sampling is as effective as independent sampling; this is not a typical case.

For convenience, $\widehat{R}$ values in Example 20.9 are listed again:

```
> gelman.diag(univt.coda)
Potential scale reduction factors:

      Point est. Upper C.I.
k           1.00       1.01
mu          1.00       1.01
sigma       1.02       1.06

Multivariate psrf

1.02
```

---

[3] The effective sample size can be larger than the actual sample size if there is negative correlation, but negative correlation is unlikely. It is more likely that some of the effective sample sizes exceed the actual sample sizes due to random variation, i.e., estimation error.

One can see that $\widehat{R}$ is at most 1.02 for all of the parameters that were monitored, which is another indication that the amount of MCMC sampling was sufficient. Even the 95 % upper confidence limits are satisfactory, at most 1.06.

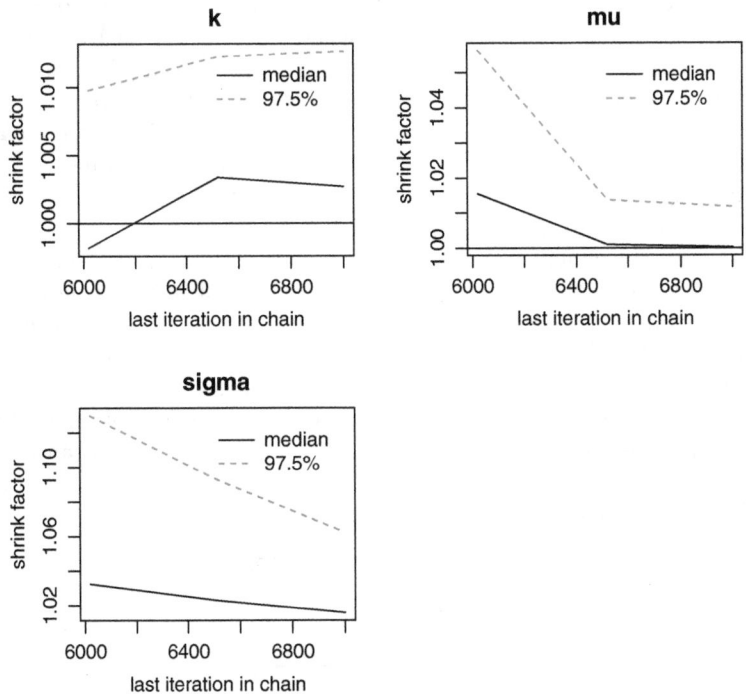

Fig. 20.6. *Gelman plot in Example 20.9.*

20.7.6 DIC and p_D for Model Comparisons

In this section, we introduce two widely used statistics, DIC and p_D. DIC is used to compare several models for the same data set and is a Bayesian analog of AIC. p_D is a Bayesian analog to the number of parameters in the model.

Recall from Sect. 5.12 that the deviance, denoted now by $D(\boldsymbol{Y}, \boldsymbol{\theta})$, is minus twice the log-likelihood, and AIC defined by (5.29) is

$$\text{AIC} = D(\boldsymbol{Y}, \boldsymbol{\theta}_{\text{ML}}) + 2p, \tag{20.45}$$

where $\widehat{\boldsymbol{\theta}}_{\text{ML}}$ is the MLE and p is the dimension of $\boldsymbol{\theta}$. A Bayesian analog of the MLE is the posterior mean, the usual Bayes estimator, which can be estimated by MCMC.

We need a Bayesian analog of p, the number of parameters. It may seem strange at first that we do not simply use p itself as in a non-Bayesian analysis. After all, the number of parameters has not changed just because we now have a prior and are using Bayesian estimation. However, the prior information used in a Bayesian analysis somewhat constrains the estimated parameters, which makes the *effective* number of parameters less than p. To appreciate why this is true, consider an example where there are d returns on equities that are believed to be similar. Assume the returns have a multivariate normal distribution. Let's focus on the d expected returns, call them $\mu_1, \ldots, \mu_d$. To a non-Bayesian, there are two ways to model $\mu_1, \ldots, \mu_d$. The first is to assume that they are all equal, say to μ, and then there is only one parameter to model the means. The other possibility is to assume that the expected returns are not equal so that there are d parameters.

A Bayesian can achieve a compromise between these two extreme by specifying a prior such that $\mu_1, \ldots, \mu_d$ are similar but not identical. For example, we could assume that they are i.i.d. $N(\mu, \sigma_\mu^2)$, and σ_μ^2 would specify the degree of similarity. It is important to appreciate that σ_μ^2 can be estimated from the data, that is, there is no need to specify in advance the degree of similarity between the means. The result of using such prior information is that the *effective* number of parameters to specify $\mu_1, \ldots, \mu_d$ is greater than 1 but less than d.

The effective number of parameters is defined as

$$p_D = \widehat{D}_{\text{avg}} - D(\boldsymbol{Y}, \overline{\boldsymbol{\theta}}), \tag{20.46}$$

where

$$\overline{\boldsymbol{\theta}} = (NM)^{-1} \sum_{j=1}^{M} \sum_{i=1}^{N} \boldsymbol{\theta}_{i,j}$$

is the average of the MCMC sample of $\boldsymbol{\theta}_{i,j}$ and estimates the posterior expectation of θ, and

$$\widehat{D}_{\text{avg}} = (NM)^{-1} \sum_{j=1}^{M} \sum_{i=1}^{N} D(\boldsymbol{Y}, \boldsymbol{\theta}_{i,j})$$

is an MCMC estimate of

$$D_{\text{avg}} = E\{D(\boldsymbol{Y}, \boldsymbol{\theta}) | \boldsymbol{Y}\}. \tag{20.47}$$

In analogy with (20.45), DIC is defined as

$$\text{DIC} = D(\boldsymbol{Y}, \overline{\boldsymbol{\theta}}) + 2p_D.$$

By (20.46), we have as well that

$$\text{DIC} = \widehat{D}_{\text{avg}} + p_D.$$

The function `dic.samples()` in the `rjags` package reports $\widehat{D}_{\text{avg}}$ as the "mean deviance," p_D as the "penalty," and DIC as the "penalized deviance." See the output in Example 20.9.

As the following example illustrates, p_D is primarily a measure of the posterior variability of $\boldsymbol{\theta}$, which increases as p increases or the amount of prior information about $\boldsymbol{\theta}$ decreases relative to the information in the sample.

Example 20.11. p_D when estimating a normal mean with known precision

Suppose that $\boldsymbol{Y} = (Y_1, \ldots, Y_n)$ are i.i.d. $N(\mu, 1)$, so $\boldsymbol{\theta} = \mu$ in this example. Then the log-likelihood is

$$
\log\{L(\mu)\} = -\frac{1}{2}\sum_{i=1}^{n}(Y_i - \mu)^2 - \frac{n}{2}\log(2\pi)
$$

$$
= -\frac{1}{2}\left\{\sum_{i=1}^{n}(Y_i - \overline{Y})^2 + n(\overline{Y} - \mu)^2\right\} - \frac{n}{2}\log(2\pi),
$$

and so

$$
D(\boldsymbol{Y}, \mu) = \sum_{i=1}^{n}(Y_i - \overline{Y})^2 + n(\overline{Y} - \mu)^2 + n\log(2\pi). \tag{20.48}
$$

When p_D is computed, quantities not depending on μ cancel with the subtraction in (20.46). Therefore, for the purpose of computing p_D, we can use

$$
D(\boldsymbol{Y}, \mu) = n(\overline{Y} - \mu)^2. \tag{20.49}
$$

Then

$$
D\{\boldsymbol{Y}, E(\mu|\boldsymbol{Y})\} = \{\overline{Y} - E(\mu|\boldsymbol{Y})\}^2, \tag{20.50}
$$

and

$$
\begin{aligned}
D_{\text{avg}} &= n\, E\{(\overline{Y} - \mu)^2 | \boldsymbol{Y}\} \\
&= n\left(\{\overline{Y} - E(\mu|\boldsymbol{Y})\}^2 + E\left[\{E(\mu|\boldsymbol{Y}) - \mu\}^2 | \boldsymbol{Y}\right]\right) \\
&= n\left[\{\overline{Y} - E(\mu|\boldsymbol{Y})\}^2 + \text{Var}(\mu|\boldsymbol{Y})\right] \\
&= D\{\boldsymbol{Y}, E(\mu|\boldsymbol{Y})\} + n\text{Var}(\mu|\boldsymbol{Y}), \tag{20.51}
\end{aligned}
$$

because $\{\overline{Y} - E(\mu|\boldsymbol{Y})\}$ and $\{E(\mu|\boldsymbol{Y}) - \mu\}$ are conditionally uncorrelated given $\boldsymbol{Y}$. Therefore,

$$
\begin{aligned}
p_D &= \widehat{D}_{\text{avg}} - D\{\boldsymbol{Y}, E(\mu|\boldsymbol{Y})\} \\
&\approx D_{\text{avg}} - D\{\boldsymbol{Y}, E(\mu|\boldsymbol{Y})\} = n\text{Var}(\mu|\boldsymbol{Y}) = \frac{n}{n + \tau_0}, \tag{20.52}
\end{aligned}
$$

where the last equality uses (20.21) and τ_0 is the prior precision for μ. The approximation ("$\approx$") in (20.52) becomes equality as the Monte Carlo sample size N increases to ∞.

As $\tau_0 \to 0$, the amount of prior information becomes negligible and the right-hand side of (20.52) converges to $p = 1$. Conversely, as $\tau_0 \to \infty$, the amount of prior information increases without bound and the right-hand side of (20.52) converges to 0. This is an example of a general phenomenon—more prior information means less effective parameters.   □

Generally, $p_D \approx p$ when p is small and there is little prior information. In other cases, such as when d means are modeled as coming from a common normal distribution, p_D could be considerably less than p—see Example 20.12.

When comparing models using DIC, smaller is better, though, like AIC and BIC, DIC should never be used blindly. Often subject-matter considerations or model simplicity will lead an analyst to select a model other than the one minimizing DIC.

The function dic.sample() returns both DIC and p_D, as can be seen in the output from Example 20.9 which was:

```
> dic.samples(univt.mcmc, 100*nthin, thin = nthin, type = "pD")
 |**************************************************| 100%
Mean deviance:  -18065
penalty 2.664
Penalized deviance: -18062
```

20.8 Hierarchical Priors

A common situation is having a number of parameters that are believed to have similar, but not identical, values. For example, the expected returns on several equities might be thought similar. In such cases, it can be useful to pool information about the parameters to improve the specification of the prior, because the use of good prior information will improve the accuracy of the estimation. A effective method for pooling information is a Bayesian analysis with a so-called "hierarchical prior" that allows one to shrink the estimates toward each other or toward some other target. An example of the latter would be shrinking the sample covariance matrix of returns toward an estimate from the CAPM or another factor model. This type of shrinkage would achieve a tradeoff between the high variability of the sample covariance matrix and the potential bias of the covariance matrix estimator from a factor model when the factor model does not fit perfectly.

As before, let the likelihood be $f(\boldsymbol{y}|\boldsymbol{\theta})$. The likelihood is the first layer (or stage) in the hierarchy. So far in this chapter, the prior density of $\boldsymbol{\theta}$, which is the second layer, has been $\pi(\boldsymbol{\theta}|\boldsymbol{\gamma})$, where the parameter vector $\boldsymbol{\gamma}$ in the prior has a known value, say $\boldsymbol{\gamma}_0$. For example, in Example 20.3 the prior had a beta distribution with both parameters fixed.

In a *hierarchical* or multistage prior, γ is unknown and has its own prior $\pi(\gamma|\delta)$ (the third layer). Typically, δ has a known value, though one can add further layers to the hierarchy by making δ unknown with its own prior, and so forth.

It is probably easiest to understand hierarchical priors using examples.

Example 20.12. Estimating expected returns on midcap stocks

This example uses the `midcapD.ts` dataset. This data set contains 500 daily returns on 20 midcap stocks and on the market and was used in Example 5.2.

The data set will be divided into the "training data," which contains the first 100 days of returns and the "test" data containing the last 400 days of returns. Only the training data will be used for estimation. The test data will be used to compare the estimates from the training data. The test data sample size was chosen intentionally to be relatively large so that we can consider the mean returns from the test data to be the "true" expected returns on the 20 stocks, though, of course, this is only an approximation. The "true" expected returns will be estimated using the training data.

We will compare three possible estimators of the true expected returns.

(a) sample means (the 20 mean returns on the midcap stocks for the first 100 days);
(b) pooled estimation (total shrinkage where every expected return has the same estimate);
(c) Bayes estimation with a hierarchical prior (shrinkage).

Method (a) is the "usual" non-Bayesian estimator where each expected return is estimated by the sample mean of that stock. In method (b), every expected return has the same estimate, which is the "mean of means," that is, the average of the 20 means from (a). Bayes shrinkage, which will be explained in this example, shrinks the 20 individual means toward the mean of means using a hierarchical prior. Bayesian shrinkage is a compromise between (a) and (b). Shrinkage was also used in Example 16.10 though in that example the amount of shrinkage was chosen arbitrarily because Bayesian methods had not yet been introduced.

Let $R_{i,t}$ be the tth daily return on i stock expressed as a percentage. For Bayesian shrinkage, the first layer will be the simple model

$$R_{i,t} = \mu_i + \epsilon_{i,t},$$

where $\epsilon_{i,t}$ are i.i.d. $N(0, \sigma_\epsilon^2)$. This model has several unrealistic aspects: (a) the assumption that the standard deviation of $\epsilon_{i,t}$ does not depend on i; (b) the assumption that $\epsilon_{i,t}$ and $\epsilon_{i',t}$ are independent (we know that there will be cross-sectional correlations); (c) the assumption that there are no GARCH effects; (d) the assumption that the $\epsilon_{i,t}$ are normally distributed rather than having heavy tails. Nonetheless, for the purpose of estimating expected returns, this model should be adequate. Remember, "all models are wrong but some models are useful," and, of course, what is "useful" depends on the objectives of the analysis.

The hierarchy prior has second layer

$$\mu_i \sim \text{ i.i.d. } N(\alpha, \sigma_\mu^2).$$

The assumption here is that the expected returns for the 20 midcap stocks have been sampled from a large population of expected returns, perhaps of all midcap stocks or even a larger population. The mean of that population is α and the standard deviation is σ_μ.

If we used a non-hierarchical prior, then we would need to specify values of α and σ_μ. This is exactly what was done in Example 20.4, except in that example σ_ϵ^2 also was known. We probably have a rough idea of the values of α and σ_μ, but it is unlikely that we have precise information about them, and we saw in Example 20.4 that a rather accurate specification of the prior is needed for the Bayes estimator to improve upon the sample means. In fact, the Bayes estimator can easily be inferior to the sample means if the prior is poorly chosen.

The third layer will be a prior on α and σ_μ and will let us use the data to estimate these parameters. It is important to appreciate why we can estimate α and σ_μ in this example, but they could not be estimated in Example 20.4. The reason is that we now have 20 expected returns (the μ_i) that are distributed with the same mean α and standard deviation σ_μ. In contrast, in Example 20.4 there is only a single μ and so it not possible to estimate the mean and variance of the population from which this μ was sampled.

Because there is now a substantial amount of information in the data about α, σ_ϵ^2, and σ_μ^2, we could use fairly noninformative priors for them to "let the data speak for themselves."

The BUGS program for this example is:

```
1 model{
2 for (i in 1:n)
3 {
4   for (j in 1:m)
5   {
6     x1[i,j] ~ dnorm(mu[j], tau_eps)
7   }
8 }
9 for (j in 1:m)
10 {
11   mu[j] ~ dnorm(alpha, tau_mu)
12 }
13 alpha ~ dnorm(0.0, 1.0E-3)
14 tau_eps ~ dgamma(0.1, 0.01)
15 tau_mu ~ dgamma(0.1, 0.01)
```

```
16 sigma_eps <- 1 / sqrt(tau_eps)
17 sigma_mu <- 1 / sqrt(tau_mu)
18 }
```

The R code for this example is below:

```
1  library(rjags)
2  dat = read.csv("midcapD.ts.csv")
3  market = 100 * as.matrix(dat[, 22])
4  x = 100 * as.matrix(dat[, -c(1, 22)])
5  m = 20
6  k = 100
7  x1 = x[1:k, ]
8  x2 = x[(k+1):500, ]
9  mu1 = apply(x1, 2, mean)
10 mu2 = apply(x2, 2, mean)
11 means = apply(x1, 2, mean)
12 sd2 = apply(x1, 2, sd)
13 tau_mu = 1 / mean(sd2^2)
14 tau_eps = 1 / sd(means)^2
15 n = k
16 data = list(x1 = x1, n = n, m = m)
17 inits.midCap = function(){list(alpha = 0.001, mu = means,
18     tau_eps = tau_eps, tau_mu = tau_mu)}
19 midCap <- jags.model("midCap.bug", data = data, inits = inits.midCap,
20     n.chains = 3, n.adapt = 1000, quiet = FALSE)
21 nthin = 20
22 midCap.coda = coda.samples(midCap, c("mu", "tau_mu", "tau_eps",
23     "alpha", "sigma_mu", "sigma_eps"), 500 * nthin, thin = nthin)
24 summ.midCap = summary(midCap.coda)
25 summ.midCap
26 post.means = summ.midCap[[1]][2:21, 1]
27 pdf("midcap.pdf", width = 6, height = 3.75)
28 par(mfrow = c(1, 2))
29 plot(c(rep(1, m), rep(2, m)), c(mu1, mu2),
30     xlab = "estimate                    target",ylab = "mean",
31     main = "sample means",
32     ylim = c(-0.3, 0.7), axes = FALSE)
33 axis(2)
34 axis(1, labels = FALSE, tick = TRUE, lwd.tick = 0)
35 for (i in 1:m){lines(1:2, c(mu1[i], mu2[i]), col = i)}
36 plot(c(rep(1, m), rep(2, m)), c(post.means, mu2),
37     xlab = "estimate                    target", ylab = "mean",
38     main = "Bayes",
39     ylim=c(-0.3, 0.7), axes = FALSE)
40 axis(2)
41 axis(1, labels = FALSE, tick = TRUE, lwd.tick = 0)
42 for (i in 1:m){lines(1:2, c(post.means[i], mu2[i]) ,col=i)}
43 graphics.off()
44 options(digits = 2)
```

```
45 sum((mu1 - mu2)^2 )
46 sum((post.means - mu2)^2)
47 sum((mean(mu1) - mu2)^2)
```

The output is below.

```
> summ.midCap

Iterations = 20:10000
Thinning interval = 20
Number of chains = 3
Sample size per chain = 500

1. Empirical mean and standard deviation for each variable,
   plus standard error of the mean:

              Mean         SD  Naive SE Time-series SE
alpha      0.08730  0.102433 2.645e-03       3.101e-03
mu[1]      0.11121  0.171456 4.427e-03       4.546e-03
mu[2]      0.12128  0.169230 4.369e-03       4.892e-03
mu[3]      0.07871  0.170849 4.411e-03       4.702e-03

(edited to save space)

mu[19]     0.05082  0.175466 4.531e-03       4.422e-03
mu[20]     0.03997  0.184614 4.767e-03       4.873e-03
sigma_eps  4.30691  0.067810 1.751e-03       1.629e-03
sigma_mu   0.14970  0.067054 1.731e-03       1.671e-03
tau_eps    0.05395  0.001699 4.386e-05       4.074e-05
tau_mu    75.70128 68.715669 1.774e+00       1.738e+00

2. Quantiles for each variable:

              2.5%       25%      50%      75%     97.5%
alpha     -0.11465  0.017800  0.08600  0.15560   0.29111
mu[1]     -0.21768 -0.001040  0.10910  0.21864   0.43408
mu[2]     -0.21107  0.018025  0.11704  0.22382   0.47727
mu[3]     -0.27262 -0.032547  0.08392  0.19634   0.40900

(edited to save space)

mu[19]    -0.30441 -0.059642  0.05612  0.16521   0.38525
mu[20]    -0.34436 -0.069959  0.04462  0.16479   0.37048
sigma_eps  4.17808  4.261352  4.30714  4.35226   4.43733
sigma_mu   0.06155  0.100600  0.13853  0.18225   0.31293
tau_eps    0.05079  0.052792  0.05390  0.05507   0.05729
tau_mu    10.21311 30.107752 52.10738 98.81013 263.94725
```

The posterior means of σ_μ and σ_ϵ are $0.150\,\%$ and $4.31\,\%$, respectively (the returns are as percentages). If we look at precisions instead of standard deviations, then we find that the posterior means of τ_μ and τ_ϵ are 75.7 and 0.0540. Using the notation of (20.24), in the present example $\tau_{\overline{Y}}$ is $100\tau_\epsilon = 5.4$ and $\tau_0 = \tau_\mu = 75.7$. Therefore, δ in (20.24) is $5.4/(5.4 + 75.7) = 0.067$. Recall that δ close to 0 (far from 1) results in substantial shrinkage, so δ equal to 0.064 causes a great amount of shrinkage of the sample means toward the mean of means, as can be seen in Fig. 20.7.

To compare the estimators, we use the sum of squared errors (SSE) defined as

$$\text{SSE} = \sum_{i=1}^{20} (\widehat{\mu}_i - \mu_i)^2,$$

where μ_i is the ith "true" mean from the test data and $\widehat{\mu}_i$ is an estimate from the training data. The values of the SSE are found in Table 20.1. The SSE for the sample means is about 11 (1.9/0.18) times larger than for the Bayes estimate. Clearly, shrinkage is very successful in this example.

Interestingly, complete shrinkage to the pooled mean is even better than Bayesian shrinkage. Bayesian shrinkage attempts to estimate the optimal amount of shrinkage, but, of course, it cannot do this perfectly. Although complete shrinkage is better than Bayesian shrinkage in this example, complete shrinkage is, in general, dangerous since it will have a large SSE in examples where the true means differ more than in this case. If one has a strong prior belief that the true means are very similar, one should use this

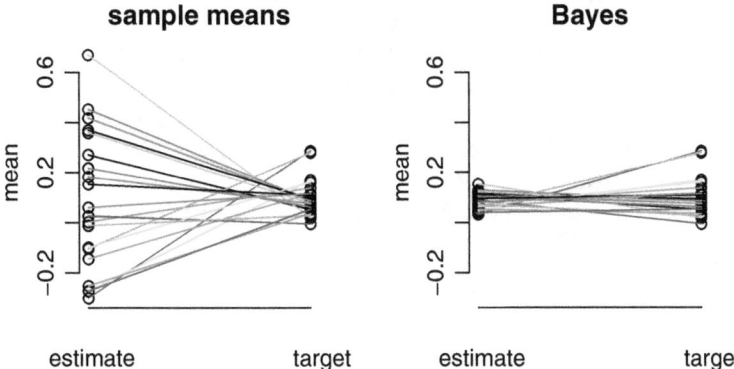

Fig. 20.7. *Estimation of the average returns for 20 midcap stocks. "Target" is the quantity being estimated, specifically the average return over 400 days of test data. "Estimate" is an estimate based on the 100 previous days of training data. On the left, the estimates are the 20 individual sample means. On the right, the estimates are the sample means shrunk toward their mean. In each panel, the estimate and target for each stock are connected by a line. On the left, the sample means of the training data are so variable that the stocks with smaller (larger) means in the training data often have larger (smaller) means in the test data. The Bayes estimates on the right are much closer to the targets.*

belief when specifying a prior for σ_μ. Instead of using a noninformative prior as in this example, one would use a prior more concentrated near 0.   □

Table 20.1. Sum of squared errors (SSE) for three estimators of the expected returns of 20 midcap stocks.

Estimate	SSE
(a) sample means	1.9
(b) pooled mean	0.12
(c) Bayes	0.18

20.9 Bayesian Estimation of a Covariance Matrix

In this section, we assume that $Y_1, \ldots, Y_n$ is an i.i.d. sample from a d-dimensional $N(\mu, \Sigma)$ distribution or a d-dimensional $t_\nu(\mu, \Lambda)$ distribution. We will focus on estimation of the covariance matrix Σ of a multivariate normal distribution or the scale matrix Λ of a multivariate t-distribution. The *precision matrix* is defined as Σ^{-1} or Λ^{-1} for the Gaussian and t-distributions, respectively. This definition is analogous to the univariate case where the precision is defined as the reciprocal of the variance or squared scale parameter.

We will start with Gaussian distributions.

20.9.1 Estimating a Multivariate Gaussian Covariance Matrix

In the multivariate Gaussian case, the conjugate prior for the precision matrix Σ^{-1} is the Wishart distribution. The Wishart distribution, denoted by Wishart(ν, A), has a univariate parameter ν called the degrees of freedom and a matrix parameter A that can be any nonsingular covariance matrix. There is a simple definition of the Wishart(ν, A) distribution when ν is an integer. Let $Z_i, \ldots, Z_n$ be i.i.d. $N(\mu, A)$. In this case, the distribution of

$$\sum_{i=1}^{n} (Z_i - \mu)(Z_i - \mu)^\mathsf{T}$$

is Wishart(n, A). Also, the distribution of

$$\sum_{i=1}^{n} (Z_i - \overline{Z})(Z_i - \overline{Z})^\mathsf{T} \tag{20.53}$$

is Wishart$(n - 1, A)$. Because the sum in (20.53) is $n - 1$ times the sample covariance matrix, the Wishart distribution is important for inference about the covariance matrix of a Gaussian distribution.

The density of a Wishart$(\nu, \boldsymbol{A})$ distribution for any positive value of ν is

$$f(\boldsymbol{W}) = C(\nu, d) \, |\boldsymbol{A}|^{-\nu/2} \, |\boldsymbol{W}|^{(\nu-d-1)/2} \exp\left\{-\frac{1}{2} \text{tr}(\boldsymbol{A}^{-1}\boldsymbol{W})\right\} \qquad (20.54)$$

with normalizing constant

$$C(\nu, d) = \left\{2^{\nu d/2} \, \pi^{d(d-1)/4} \prod_{i=1}^{d} \Gamma\left(\frac{\nu+1-i}{2}\right)\right\}^{-1}.$$

The argument $\boldsymbol{W}$ is a nonsingular covariance matrix. The expected value is $E(\boldsymbol{W}) = \nu\boldsymbol{A}$. In the univariate case $(d = 1)$, the Wishart distribution is a gamma distribution.

If $\boldsymbol{W}$ is Wishart$(\nu, \boldsymbol{A})$ distributed, then the distribution of $\boldsymbol{W}^{-1}$ is called the inverse Wishart distribution with parameters ν and $\boldsymbol{A}^{-1}$ and denoted Inv-Wishart$(\nu, \boldsymbol{A}^{-1})$.

Let $\boldsymbol{Y} = (\boldsymbol{Y}_1, \ldots, \boldsymbol{Y}_n)$ denote the data. To derive the full conditional for the precision matrix $\boldsymbol{\Sigma}^{-1}$, assume that $\boldsymbol{\mu}$ is known. We know from (7.15) that the likelihood is

$$f(\boldsymbol{Y}|\boldsymbol{\Sigma}^{-1}) = \prod_{i=1}^{n} \left[\frac{1}{(2\pi)^{d/2}|\boldsymbol{\Sigma}|^{1/2}} \exp\left\{-\frac{1}{2}(\boldsymbol{Y}_i - \boldsymbol{\mu})^{\mathsf{T}} \boldsymbol{\Sigma}^{-1}(\boldsymbol{Y}_i - \boldsymbol{\mu})\right\}\right].$$

After some simplification,

$$f(\boldsymbol{Y}|\boldsymbol{\Sigma}^{-1}) \propto |\boldsymbol{\Sigma}^{-1}|^{n/2} \exp\left\{-\frac{1}{2}\sum_{i=1}^{n}(\boldsymbol{Y}_i - \boldsymbol{\mu})^{\mathsf{T}} \boldsymbol{\Sigma}^{-1}(\boldsymbol{Y}_i - \boldsymbol{\mu})\right\}.$$

Define

$$\boldsymbol{S} = \sum_{i=1}^{n}(\boldsymbol{Y}_i - \boldsymbol{\mu})(\boldsymbol{Y}_i - \boldsymbol{\mu})^{\mathsf{T}}.$$

Next

$$\sum_{i=1}^{n}(\boldsymbol{Y}_i - \boldsymbol{\mu})^{\mathsf{T}} \boldsymbol{\Sigma}^{-1}(\boldsymbol{Y}_i - \boldsymbol{\mu}) = \text{tr}\left\{\sum_{i=1}^{n}(\boldsymbol{Y}_i - \boldsymbol{\mu})^{\mathsf{T}} \boldsymbol{\Sigma}^{-1}(\boldsymbol{Y}_i - \boldsymbol{\mu})\right\} = \text{tr}(\boldsymbol{\Sigma}^{-1}\boldsymbol{S}).$$
$$(20.55)$$

The first equality in (20.55) is the trivial result that a scalar is also a 1×1 matrix and equal to its trace. The second equality uses the result that $\text{tr}(\boldsymbol{A}\boldsymbol{B}) = \text{tr}(\boldsymbol{B}\boldsymbol{A})$ for any matrices $\boldsymbol{B}$ and $\boldsymbol{A}$ such that the products $\boldsymbol{B}\boldsymbol{A}$ and $\boldsymbol{A}\boldsymbol{B}$ are defined. If follows that

$$f(\boldsymbol{Y}|\boldsymbol{\Sigma}^{-1}) \propto |\boldsymbol{\Sigma}^{-1}|^{n/2} \exp\left\{-\frac{1}{2}\text{tr}(\boldsymbol{\Sigma}^{-1}\boldsymbol{S})\right\}. \qquad (20.56)$$

Suppose that the prior on the precision matrix $\boldsymbol{\Sigma}^{-1}$ is Wishart$(\nu_0, \boldsymbol{\Sigma}_0^{-1})$. Then the prior density is

$$\pi(\boldsymbol{\Sigma}^{-1}) \propto |\boldsymbol{\Sigma}^{-1}|^{(\nu_0-d-1)/2} \exp\left\{-\frac{1}{2}\mathrm{tr}(\boldsymbol{\Sigma}^{-1}\boldsymbol{\Sigma}_0)\right\}. \tag{20.57}$$

Since the posterior density is proportional to the product of the prior density and the likelihood, it follows from (20.56) and (20.57) that the posterior density is

$$\pi(\boldsymbol{\Sigma}^{-1}|\boldsymbol{Y}) \propto |\boldsymbol{\Sigma}^{-1}|^{(n+\nu_0-d-1)/2} \exp\left[-\frac{1}{2}\mathrm{tr}\left\{\boldsymbol{\Sigma}^{-1}(\boldsymbol{S}+\boldsymbol{\Sigma}_0)\right\}\right]. \tag{20.58}$$

Therefore, the posterior distribution of $\boldsymbol{\Sigma}^{-1}$ is Wishart$\{n+\nu_0, (\boldsymbol{S}+\boldsymbol{\Sigma}_0)^{-1}\}$. The posterior expectation is

$$E(\boldsymbol{\Sigma}^{-1}|\boldsymbol{Y}) = (n+\nu_0)\left\{(\boldsymbol{S}+\boldsymbol{\Sigma}_0)^{-1}\right\}. \tag{20.59}$$

If ν_0 and $\boldsymbol{\Sigma}_0$ are both small, then

$$E(\boldsymbol{\Sigma}^{-1}|\boldsymbol{Y}) \approx n\boldsymbol{S}^{-1} \tag{20.60}$$

The MLE of $\boldsymbol{\Sigma}$ is $n^{-1}\boldsymbol{S}$, so the MLE of $\boldsymbol{\Sigma}^{-1}$ is $n\boldsymbol{S}^{-1}$. Therefore, for small values of ν_0 and $\boldsymbol{\Sigma}_0$, the Bayesian estimator of $\boldsymbol{\Sigma}^{-1}$ is close to the MLE.

The full conditional for $\boldsymbol{\Sigma}^{-1}$ can be combined with a model for $\boldsymbol{\mu}$ to estimate both parameters. For application to asset returns, a hierarchical prior for $\boldsymbol{\mu}$ such as in Example 20.12 might be used. In either case, an MCMC analysis would be straightforward.

20.9.2 Estimating a Multivariate-t Scale Matrix

The Wishart distribution is not a conjugate prior for the scale matrix of a multivariate t-distribution, but it can be used as the prior nonetheless, since MCMC does not require the use of conjugate priors.

Example 20.13. Estimating the correlation matrix of the CRSPday *data*

In Example 7.4, the correlation matrix of the CRSPday returns data was estimated by maximum likelihood. In this example, the MLE will be compared to a Bayes estimate and the two estimates will be found to be very similar. The BUGS program used in this example is

```
model{
for(i in 1:N)
{
  y[i,1:m] ~ dmt(mu[], tau[,], df_likelihood)
}
mu[1:m] ~ dmt(mu0[], Prec_mu[,], df_prior)
tau[1:m,1:m] ~ dwish(Prec_tau[,], df_wishart)
lambda[1:m,1:m] <- inverse(tau[,])
}
```

In the BUGS program, mu is the mean vector, tau is the precision matrix, lambda is the scale matrix of the returns. Also, dmt is the multivariate-t distribution, and dwish is the Wishart distribution.

At the time of this writing (Dec 2014), JAGS does not have a sampler that will sample the posterior from this model. Therefore, WinBUGS will be used and will be called using the bugs() function in the R2WinBUGS package. The R code is below.

```
 1 library(R2WinBUGS)
 2 library(MASS)  # need to mvrnorm
 3 library(MCMCpack) # need for rwish
 4 library(mnormt)
 5 data(CRSPday, package = "Ecdat")
 6 y = CRSPday[,4:7]
 7 N = dim(y)[1]
 8 m = dim(y)[2]
 9 mu0 = rep(0,m)
10 Prec_mu = diag(rep(1, m)) / 10000
11 Prec_tau =  diag(rep(1, m)) / 10000
12 df_wishart = 6
13 df_likelihood = 6
14 df_prior = 6
15 data = list(y = y, N = N, Prec_mu = Prec_mu,
16     Prec_tau = Prec_tau,
17     mu0 = mu0, m = m, df_likelihood = df_likelihood,
18     df_prior = df_prior, df_wishart = df_wishart)
19 inits_t_CRSP = function(){list(mu = mvrnorm(1, mu0,
20     diag(rep(1, m) / 100)),
21     tau = rwish(6, diag(rep(1, m)) / 100))}
22 library(R2WinBUGS)
23 multi_t.sim = bugs(data, inits_t_CRSP ,
24     model.file = "mult_t_CRSP.bug",
25     parameters = c("mu", "tau"), n.chains = 3,
26     n.iter = 2200, n.burnin = 200, n.thin = 2,
27     program = "WinBUGS", bugs.seed = 13, codaPkg = FALSE)
28 print(multi_t.sim, digits = 2)
29 tauhat = multi_t.sim$mean$tau
30 lambdahat = solve(tauhat)
31 sdinv = diag(1/sqrt(diag(lambdahat)))
32 cor = sdinv %*% lambdahat %*% sdinv
33 print(cor,digits=4)
```

The data list that is an input to the BUGS program contain y which is the matrix of returns, df_likelihood which is the degrees of freedom of the t-distribution in the likelihood, mu0 which is the prior mean for mu, df_prior which is the degrees of freedom in the t prior on mu, and df_wishart which is the degrees of freedom of the Wishart prior on tau.

Ideally, df_likelihood should be an unknown parameter, but WinBUGS does not allow this parameter to be estimated. Instead, we fix it at the MLE (rounded to 6) computed in Example 7.4. The need to fix this parameter at the MLE is due to limitations of WinBUGS and could, with considerably more effort, be circumvented by programming the MCMC in R or another language rather than using WinBUGS.

Note that codaPkg = FALSE was specified in the call to bugs(); this was not necessary since it is the default. When codaPkg = FALSE then bugs() returns an object of class bugs which cannot be used directly by functions in the coda package since these functions take objects of class mcmc.list. However, the function as.mcmc.list() will convert a bugs object to an mcmc.list object.

There were three chains, each of length 2000 after a burn-in of 200 and thinned to every second iteration. Thus, the total sample size was 3000 after thinning. The convergence to the stationary distribution and mixing were both quite rapid. N_{eff} was at least 1500 and $\hat{R}$ essentially 1 for all parameters, which indicate adequate burn-in and chain lengths.

```
> print(multi_t.sim, digits = 2)
Inference for Bugs model at "mult_t_CRSP.bug", fit using WinBUGS,
 3 chains, each with 2200 iterations (first 200 discarded), n.thin = 2
 n.sims = 3000 iterations saved
           mean   sd   2.5%    25%    50%    75%  97.5% Rhat n.eff
mu[1]         0    0      0      0      0      0      0    1  3000
mu[2]         0    0      0      0      0      0      0    1  3000
mu[3]         0    0      0      0      0      0      0    1  3000
mu[4]         0    0      0      0      0      0      0    1  1800
tau[1,1]  14706  473  13780  14390  14690  15020  15630    1  3000

 (edited to save space)

tau[4,4]  65197 2102  61180  63738  65190  66600  69360    1  3000
deviance -69858   61 -69980 -69900 -69860 -69820 -69730    1  3000

For each parameter, n.eff is a crude measure of effective sample size,
and Rhat is the potential scale reduction factor (at convergence, Rhat=1).

DIC info (using the rule, pD = Dbar-Dhat)
pD = 13.8 and DIC = -69843.9
DIC is an estimate of expected predictive error (lower deviance is better).
```

Since μ is close to zero, multi_t.sim needed to be printed again, this time with more digits than 2:

```
           mean      sd     2.5%     25%     50%     75%   97.5% Rhat n.eff
mu[1]  9.4e-04 2.4e-04  4.6e-04 7.7e-04 9.4e-04 1.1e-03 1.4e-03    1  3000
mu[2]  4.4e-04 2.9e-04 -1.3e-04 2.4e-04 4.6e-04 6.4e-04 1.0e-03    1  3000
mu[3]  6.9e-04 2.3e-04  2.3e-04 5.3e-04 6.9e-04 8.4e-04 1.1e-03    1  3000
mu[4]  7.7e-04 1.3e-04  5.1e-04 6.8e-04 7.7e-04 8.6e-04 1.0e-03    1  1800
```

The Bayes estimate of the precision matrix was converted to a correlation matrix at lines 29–33 of the R program. The estimated correlation matrix is below.

```
         [,1]    [,2]    [,3]    [,4]
[1,]  1.0000  0.3191  0.2841  0.6756
[2,]  0.3191  1.0000  0.1586  0.4696
[3,]  0.2841  0.1586  1.0000  0.4300
[4,]  0.6756  0.4696  0.4300  1.0000
```

In Example 7.4, the MLE of the correlation matrix was found to be

```
$cor
        [,1]    [,2]    [,3]    [,4]
[1,] 1.0000 0.3192 0.2845 0.6765
[2,] 0.3192 1.0000 0.1584 0.4698
[3,] 0.2845 0.1584 1.0000 0.4301
[4,] 0.6765 0.4698 0.4301 1.0000
```

Notice the similarity between the Bayes estimate and the MLE.   □

20.9.3 Non-Wishart Priors for the Covariate Matrix

We saw in Example 20.13 that a Wishart prior with noninformative choices
of the prior parameters more or less replicates maximum likelihood estimation. Often, however, one wishes to shrink the covariance matrix toward some
target, perhaps a estimate from a factor model. See Example 20.16.

20.10 Stochastic Volatility Models

Stochastic volatility models are an alternative to GARCH models for modeling
conditional heteroscedasticity. In the ARIMA/GARCH models of Chap. 14,
there was a single white noise process that drove both the conditional mean
and the conditional variance. In contrast, stochastic volatility models use one
white noise process to drive the conditional expectation and another to drive
the conditional variance. Therefore, stochastic volatility models are more challenging to fit because the unobserved volatility process is driven by its own
white noise process, which, of course, is also unobserved; see (20.63) below.
Bayesian analysis is particularly good at dealing with unobserved variables
and seems the best way to meet the challenge of fitting stochastic volatility
models.

We will illustrate stochastic volatility models with the model

$$Y_t = \mu + \sum_{j=1}^{k} \beta_j X_{j,t} + a_t, \tag{20.61}$$

where Y_t is an observed process, e.g., the returns on an asset, and $X_{j,t}$, $j =$
$1, \ldots, k$, is the jth covariate at time t and could be a lagged value of Y_t.
Also, a_t is a weak white noise process with conditional heteroscedasticity.
Specifically,

$$a_t = \sqrt{h_t}\epsilon_t \tag{20.62}$$

where $\log(h_t)$ follows the ARMA(p,q) process

$$\log(h_t) = \beta_0 + \sum_{j=1}^{p} \phi_j \log(h_{t-j}) + \sum_{j=1}^{q} \theta_j v_{t-j} + v_t, \tag{20.63}$$

ϵ_t and v_t are mutually independent iid white noises, and $\text{Var}(\epsilon_t) = 1$. Notice from (20.62) that $\sqrt{h_t}$ is the conditional standard deviation of Y_t. As mentioned above, none of the variables in (20.63) are observable.

Example 20.14. Fitting an ARMA(1,1) stochastic volatility model to the S&P 500 stock returns

As an illustration, model (20.61)–(20.63) will be fit to daily S&P 500 log returns from January 2011 through October 2014. Model (20.61) will be used with no covariates so that $Y_t = \mu + a_t$. An ARMA(1,1) stochastic volatility model will be used so that $\log(h_t) = \beta_0 + \phi \log(h_{t-1}) + \theta v_{t-1} + v_t$. In (20.62) ϵ_t has a t-distribution.

A BUGS program to fit the stochastic volatility model is below.

```
1  model
2  {
3    for (i in 1:N)
4    {
5      y[i] ~ dt(mu, tau[i], nu)
6    }
7    logh[1] ~ dnorm(0, 1.0E-6)
8    for(i in 2:N)
9    {
10     logh[i] ~ dnorm(beta0 + phi * logh[i-1] + theta * v[i-1], tau_v)
11     v[i] <- logh[i] - beta0 + phi * logh[i-1] + theta * v[i-1]
12   }
13   for (i in 1:N)
14   {
15     tau[i] <- exp(-logh[i])
16     h[i] <- 1/tau[i]
17   }
18   mu ~ dnorm(0.0, 1)
19   beta0 ~ dnorm(0, 0.0001)
20   phi ~ dnorm(0.4, 0.0001)
21   theta ~ dnorm(0, 0.0001)
22   tau_v ~ dgamma(0.01, 0.01)
23   v[1] ~ dnorm(0, 0.001)
24   nu ~ dunif(1,30)
25   sigma_v <- 1 / sqrt(tau_v)
26 }
```

Line 5 specifies the likelihood conditional on h_t. Line 7 gives a prior for the h_1. Lines 8–12 specify model (20.63) starting at $t = 2$ with $p = q = 1$; line 23 gives v_1 a noninformative prior to start this recursion. Lines 18–25 specify diffuse priors on μ, β_0, ϕ, θ, and σ_v.

S&P 500 prices from Jan 3, 2011 to Oct 31, 2014 are in the file S&P500_new.csv. There are 964 log returns starting on Jan 4, 2011. The following R code computes the log returns and fit the stochastic volatility model to the log returns. The MCMC took about 10 minutes.

```
1  library(rjags)
2  dat = read.csv("S&P500_new.csv")
3  prices = dat$Adj.Close
4  y = diff(log(prices))
5  #####  get initial estimates  #####
6  N = length(y)
7  logy2 = log(y^2)
8  fitar = lm(logy2[2:N] ~ logy2[1:(N - 1)])
9  beta0Init = as.numeric(fitar$coef[1])
10  phiInit = as.numeric(fitar$coef[2])
11  sfitar = summary(fitar)
12  tauInit = 1/sfitar$sigma^2
13  #####  Set up for MCMC  #####
14  N = length(y)
15  data = list(y = y, N = N)
16  inits_stochVol_ARMA11 = function(){list(mu = rnorm(1, mean = mean(y),
17    sd = sd(y) / sqrt(N)), logh = log(y^2),
18    beta0 = runif(1, beta0Init * 1.5, beta0Init/1.5),
19    phi = runif(1,phiInit / 1.5, phiInit * 1.5),
20    tau_v = runif(1, tauInit / 1.5, tauInit * 1.5),
21    theta = runif(1, -0.5, 0.5))}
22  stochVol_ARMA11 <- jags.model("stochVol_ARMA11.bug", data = data,
23    inits = inits_stochVol_ARMA11,
24    n.chains = 3, n.adapt = 1000, quiet = FALSE)
25  nthin = 20
26  stochVol_ARMA.coda = coda.samples(stochVol_ARMA11, c("mu", "beta0",
27    "phi", "theta", "tau_v", "nu", "tau"), 100 * nthin, thin = nthin)
28  summ_stochVol_ARMA11 = summary(stochVol_ARMA.coda)
29  head(summ_stochVol_ARMA11[[1]], 8)
30  tail(summ_stochVol_ARMA11[[1]], 8)
31  dic.stochVol_ARMA11 = dic.samples(stochVol_ARMA11, 100 * nthin,
32    thin = nthin, type = "pD")
33  dic.stochVol_ARMA11
```

Lines 8–12 compute rough initial values for β_0, ϕ, and σ_v by using Y_t^2 a proxy for h_t and regressing $\log(Y_t^2)$ on $\log(Y_{t-1}^2)$.

Since nearly 1000 variables are monitored, we do not want to look at the output for all of them. Instead, the MCMC output is summarized for β_0, μ, ϕ, θ, τ_v, and the first and last few h_i.

```
> head(summ_stochVol_ARMA11[[1]], 8)
                 Mean           SD     Naive SE Time-series SE
beta0  -1.327118e+01 6.986557e+00 4.033691e-01    2.956974e+00
mu      8.977085e-04 2.290611e-04 1.322485e-05    1.197846e-05
nu      2.116160e+01 5.785812e+00 3.340440e-01    3.884508e-01
phi     3.651321e-05 2.254858e-02 1.301843e-03    8.364831e-03
tau[1]  2.379568e+05 4.180067e+05 2.413363e+04    2.413332e+04
tau[2]  6.815322e+04 5.764209e+04 3.327968e+03    3.933972e+03
tau[3]  6.472217e+04 5.089951e+04 2.938685e+03    3.247533e+03
```

```
tau[4]   6.030660e+04 4.190914e+04 2.419625e+03   2.815956e+03
> tail(summ_stochVol_ARMA11[[1]], 8)
                  Mean           SD      Naive SE Time-series SE
tau[959] 1.328635e+04 7.236487e+03 4.177988e+02   4.432135e+02
tau[960] 1.468882e+04 8.124815e+03 4.690864e+02   5.062925e+02
tau[961] 1.318162e+04 7.091475e+03 4.094265e+02   3.959164e+02
tau[962] 1.514622e+04 8.918848e+03 5.149299e+02   5.163099e+02
tau[963] 1.574039e+04 1.065051e+04 6.149072e+02   6.165450e+02
tau[964] 1.403511e+04 8.598674e+03 4.964447e+02   5.021540e+02
tau_v    5.161879e+00 1.409690e+00 8.138847e-02   1.765265e-01
theta    4.819691e-01 1.219201e-02 7.039058e-04   3.262414e-03

> dic.stochVol_ARMA11
Mean deviance:  -6653
penalty 119.6
Penalized deviance: -6533
```

DIC and pD are called the "penalized deviance" and the "penalty" in the output of `dic.samples()` and in this example are -6536 and 127.3, respectively.   □

20.11 Fitting GARCH Models with MCMC

Like stochastic volatility models, GARCH models are easy to fit by MCMC. One reason for fitting a GARCH model with BUGS is that this model can then be compared using DIC with a stochastic volatility model that is also fit using BUGS.

The following BUGS program fits a GARCH(1,1) model with t-distributed noise. This program runs under JAGS but crashes with no useful error messages under OpenBUGS and WinBUGS. The model is

$$y_t = \mu + a_t$$
$$a_t = \sqrt{h_t}\epsilon_t$$
$$h_t = \alpha_0 + \alpha_1 a_{t-1}^2 + \beta_1 h_{t-1}$$
$$\epsilon_t \sim t_\nu(0, 1).$$

```
1  model{
2  for (t in 1:N)
3  {
4     y[t] ~ dt(mu, tau[t], nu)
5     a[t] <- y[t] - mu
6     tau[t] <- 1/h[t]
7  }
8  for (t in 2:N)
9  {
10     h[t] <- alpha0 + alpha1 * pow(a[t-1], 2) + beta1 * h[t-1]
```

```
11  }
12  mu ~ dnorm(0, 0.001)
13  h[1] ~ dunif(0, 0.0012)
14  alpha0 ~ dunif(0, 0.2)
15  alpha1 ~ dunif(0.00001, 0.8)
16  beta0 ~ dunif(0.00001, 0.8)
17  nu ~ dunif(1,30)
18  }
```

Example 20.15. Fitting a GARCH(1,1) model to the S&P 500 stock returns

The following R code fits a GARCH(1,1) model to the data in Example 20.14.

```
1   library(rjags)
2   dat = read.csv("S&P500_new.csv")
3   prices = dat$Adj.Close
4   y = diff(log(prices))
5   N = length(y)
6   data = list(y = y, N = N)
7   inits_garch11 = function(){list(alpha0 = runif(1, 0.001, 0.25),
8       beta1 = runif(1, 0.001, 0.25), mu = runif(1, 0.001, 0.25),
9       alpha1 = runif(1, 0.001, 0.25), nu = runif(1, 2, 10))}
10  garch11 <- jags.model("garch11.bug", data=data,
11      inits = inits_garch11,
12      n.chains = 3, n.adapt = 1000, quiet = FALSE)
13  nthin = 20
14  garch11.coda = coda.samples(garch11,c("mu", "beta1", "alpha0",
15      "alpha1", "nu", "tau"), 100*nthin, thin = nthin)
16  dic.garch11 = dic.samples(garch11, 100*nthin, thin = nthin)
17  dic.garch11
18  diffdic(dic.garch11, dic.stochVol_ARMA11)
19  summ_garch11 = summary(garch11.coda)
20  head(summ_garch11[[1]])
21  tail(summ_garch11[[1]])
```

The output is below.

```
> dic.garch11
Mean deviance:  -6508
penalty 5.737
Penalized deviance: -6502
> diffdic(dic.garch11, dic.stochVol_ARMA11)
Difference: 30.90573
Sample standard error: 15.12843
> head(summ_garch11[[1]])
               Mean          SD     Naive SE Time-series SE
alpha0 3.875413e-06 9.405668e-07 5.430365e-08   6.060383e-08
alpha1 1.287614e-01 2.170856e-02 1.253344e-03   1.465383e-03
beta1  7.677071e-01 2.621658e-02 1.513615e-03   1.870831e-03
mu     8.663882e-04 2.290809e-04 1.322599e-05   1.247927e-05
```

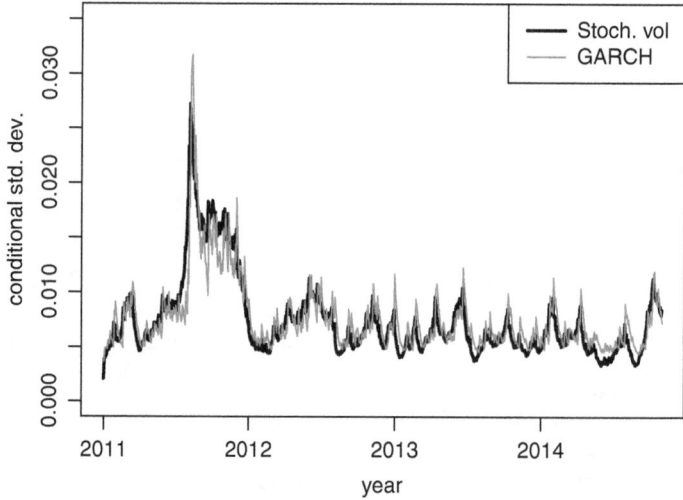

Fig. 20.8. *The conditional standard deviations of the log returns estimated by the stochastic volatility model in Example 20.14 and the GARCH(1,1) model.*

```
nu        7.570399e+00 1.990408e+00 1.149163e-01   1.077795e-01
tau[1]    7.432585e+04 1.242142e+05 7.171511e+03   7.376238e+03
> tail(summ_garch11[[1]])
                 Mean        SD   Naive SE Time-series SE
tau[959]  10765.65  1050.613   60.65714       56.26545
tau[960]  12495.45  1207.416   69.71019       64.84224
tau[961]  15145.85  1500.491   86.63091       81.57595
tau[962]  14266.32  1339.490   77.33549       72.44932
tau[963]  17144.03  1653.541   95.46724       90.31518
tau[964]  19112.62  1869.545  107.93821      107.54300
```

Line 18 of the R code uses the function `diffdic()` in the `rjags` package to compare the stochastic volatility and GARCH models by DIC. In the output from `diffdic()`, we see that DIC for the GARCH model is larger than for the stochastic volatility model; the difference between the two DIC values is 30.9 but with a standard error of 15.1. Figure 20.8 compares the estimates of the conditional standard deviations of the log returns by the two models. Figure 20.9 compares the ACF's of the squared standardized residuals[4] from the two models and also assuming that the log returns are i.i.d. Despite the large difference in DIC, the two models are about equally successful in modeling the conditional heteroscedasticity.

□

---

[4] A residual is standardized by dividing it by its conditional standard deviation.

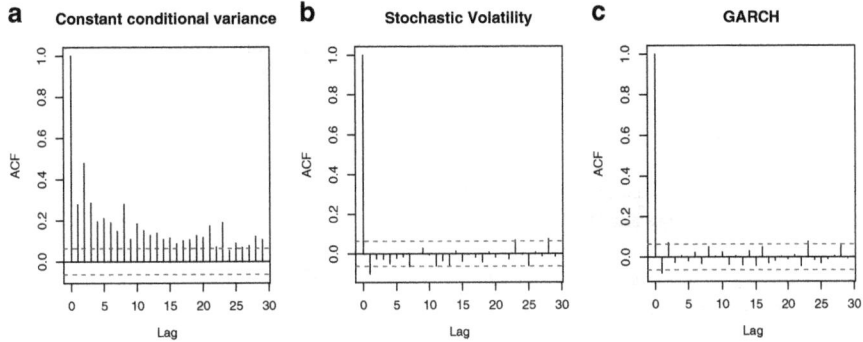

Fig. 20.9. *ACF's of the squared standardized residuals. (a) Assuming a constant conditional standard deviation. The model is $R_t = \mu + \epsilon_t$ where ϵ_t is independent white noise. (b) Assuming an ARMA(1,1) stochastic volatility model. (c) Assuming a GARCH(1,1) model.*

20.12 Fitting a Factor Model

Factor models can be fit easily using JAGS. An advantage of a Bayesian analysis of a factor model is that a hierarchial model can be used to shrink the betas towards each other.

Example 20.16. Fitting a one factor model to stock returns

In this example, the factor will be the returns on the S&P 500 so this is a Bayesian version of the CAPM. The model that will be used is

$$R_{j,t} = \beta_j R_{M,t} + \epsilon_{j,t}.$$

where, as in Chap. 17, $R_{j,t}$ is the return on the j stock at time t, $j = 1, \ldots, 10$, and $R_{M,t}$ is the return on the S&P 500 at time t. It is assumed that for $j = 1, \ldots, 10$, $\{\epsilon_{j,t}, \ t = 1, \ldots\}$ are mutually independent i.i.d. white noise processes with $\mathrm{var}(\epsilon_{j,t}) = \sigma_{\epsilon,j}^2$. For simplicity we have assumed that all the alphas are zero.

We will put a hierarchical on $\beta_1, \ldots, \beta_{10}$, specifically

$$\beta_j \sim N(\mu_\beta, \sigma_\beta^2),$$

with non-informative priors on μ_β and σ_β^2.

The BUGS program is below. At line 11, μ_β is called meanbeta and σ_β^{-2} is called taubeta. Also, tauepsilon on line 6 is σ_ϵ^{-2}.

```
1 model{
2 for (t in 1:N)
3   {
4     for (j in 1:m)
```

```
5   {
6         R[t,j] ~ dnorm(beta[j]*mkt[t], tauepsilon[j])
7       }
8     }
9   for (j in 1:m)
10  {
11    beta[j] ~ dnorm(meanbeta, taubeta)
12    tauy[j] ~ dgamma(0.1, 0.001)
13  }
14  meanbeta ~ dnorm(1, 0.000001)
15  taubeta ~ dunif(1, 100)
16  }
```

This example is a continuation of Example 16.11 in that it uses the stock price data in the file Stock_Bond.csv. Since there are nearly 5000 days of returns, one can create 20 blocks of returns, each block with 250 days except that the last block would be somewhat short of 250. Each block can be used for training (parameter estimation) and the next block for testing.

The R program below illustrates this strategy using only the first two blocks, the first as training data and the second as test data. During training, the optimal allocation vector w is estimated. Here "optimal" means maximizing the expected utility of the returns using the utility function

$$U(R; \lambda) = 1 - \exp\{-\lambda(1 + R)\}. \tag{20.64}$$

The value of λ is set equal to 3 at line 38 of the R code. A value of λ in the range 2 to 8 is reasonable since these are daily returns and typically in the range ± 0.05. Also, in (20.64) we are implicitly assuming that the initial wealth is equal to 1, since the initial wealth is not explicitly included there.

The first part of the R program fits the factor model:

```
1   library(rjags)
2   dat = read.csv("Stock_Bond.csv")
3   y = dat[, c(3, 5, 7, 9, 11, 13, 15, 17, 19, 21, 22)]
4   n = dim(y)[1]
5   m = dim(y)[2] - 1
6   r = y[-1, ] / y[-n, ] - 1
7   k1 = 250
8   k2 = k1 + 250
9   rtrain = r[1:k1, 1:m]
10  mkt_train = r[1:k1, 11]
11  rtest = r[(k1+1):k2, 1:m]
12  data = list(R = rtrain, N = k1, mkt = mkt_train, m = m)
13  inits.Capm = function(){list(beta = rep(1,m))}
14  Capm.jags <- jags.model("BayesCapm.bug.R", data = data,
15      inits = inits.Capm, n.chains = 1, n.adapt = 1000, quiet = FALSE)
16  nthin = 10
17  N = 500
18  Capm.coda = coda.samples(Capm.jags,
```

```
19      c("beta", "tauepsilon", "taubeta"),
20      N * nthin, thin = nthin)
21  MCMC_out = Capm.coda[[1]]
22  summ = as.matrix(summary(Capm.coda)[[1]][,1])
23  beta = summ[1:10]
24  taubeta = summ[11]
25  tauy = summ[12:21]
26  sigmaepsilon = tauepsilon^(-.5)
```

The next section of R code defines the utility function and finds the optimal allocation vector w by using quadratic programming as in Example 16.11. The vector w is found twice, once using the sample mean vector and covariance matrix of the training sample returns to estimate μ and Σ in Eq. (16.20) (model-free) and once using the factor model to estimate μ and Σ (CAPM).

```
27  ExUtil = function(w)
28  {
29    -1 + exp(-lambda * (1 + t(w) %*% mu) +
30    lambda^2 * t(w) %*% Omega %*% w / 2 )
31  }
32
33  mu_model_free = colMeans(rtrain)
34  Omega_model_free = cov(rtrain)
35  mu_Capm = beta * mean(mkt)
36  Omega_Capm = beta %o% beta * var(mkt_train) + diag(sigmaepsilon^2)
37
38  lambda = 3
39  library(quadprog)
40  mu = mu_model_free
41  Omega = Omega_model_free
42  opt1 = solve.QP(Dmat = as.matrix(lambda^2 * Omega),
43    dvec = lambda * mu, Amat = as.matrix(rep(1,10)),
44    bvec = 1, meq = 1)
45  w_model_free = opt1$solution
46
47  mu = mu_Capm
48  Omega = Omega_Capm
49  opt2 = solve.QP(Dmat = as.matrix(lambda^2 * Omega),
50    dvec = lambda * mu, Amat = as.matrix(rep(1,10)),
51    bvec = 1, meq = 1)
52  w_Capm = opt2$solution
```

Next, the utility of the portfolio's returns is averaged over the test data using the model-free and CAPM based estimates of the optimal portfolio. Also, the mean and standard deviations of the portfolio's returns on the test data are computed.

```
53  return_model_free = as.matrix(rtest) %*% w_model_free
54  ExUt_model_free = mean(1 - exp(-lambda * return_model_free))
55
```

```
56 return_Capm = as.matrix(rtest) %*% w_Capm
57 ExUt_Capm = mean(1 - exp(-lambda * return_Capm))
58
59 print(c(ExUt_model_free, ExUt_Capm), digits = 2)
60 print(c(mean(return_model_free), mean(return_Capm)), digits = 2)
61 print(c(sd(return_model_free), sd(return_Capm)), digits = 2)
```

The output is below. We see that the portfolio selected using the CAPM estimates outperformed the portfolio based on the sample mean vector and covariance matrix. Interestingly, the CAPM selected portfolio not only has a higher average utility, but it also has both a higher mean and a lower standard deviation of the returns compared to the model-free estimates. Usually a higher expected return comes with higher risk (larger standard deviation), but a better estimate can achieve a higher return without higher risk.

```
62 > print(c(ExUt_model_free, ExUt_Capm), digits = 2)
63 [1] -0.0179  0.0023
64 > print(c(mean(return_model_free), mean(return_Capm)), digits = 2)
65 [1] 0.00060 0.00099
66 > print(c(sd(return_model_free), sd(return_Capm)), digits = 2)
67 [1] 0.067 0.012
```

These results are based on only one block of training data and one block of test data, and by themselves they are not a convincing demonstration of the superiority of CAPM estimates. The analysis could be continued using all twenty blocks of data; see Exercise 4.

Figure 20.10 plots the allocation vectors, w, using model-free and CAPM-based estimators. The model-free w oscillates widely and has substantial short-selling. In contrast, the CAPM-based w is much closer to assigning equal weights to the 10 stocks and has minimal short selling.

Another issue is how hierarchical Bayes estimation of the CAPM compares with ordinary least-squares estimation. Figure 20.11 plots the least-squares estimates of β versus the Bayes estimates. With the exception of the largest beta, the Bayes estimates are only slightly shrunk together and are similar to the least-squares estimates. This suggest that the Bayes estimates will, at best, be only a moderate improvement over least-squares in this example. Most of the improvement is due to using the CAPM to estimate the expected returns.   □

20.13 Sampling a Stationary Process

This section provides the theory behind the statistics B, W, and $\widehat{\text{var}}^+(\psi|\boldsymbol{Y})$ used in Sect. 20.7.5 to monitor MCMC convergence and mixing.

Suppose that $Y_1, Y_2, \ldots, Y_n$ is a sample from a stationary process with mean μ and autocovariance function $\gamma(h)$. Let $\overline{Y} = n^{-1} \sum_{i=1}^{n} Y_i$ be the sample mean. Then

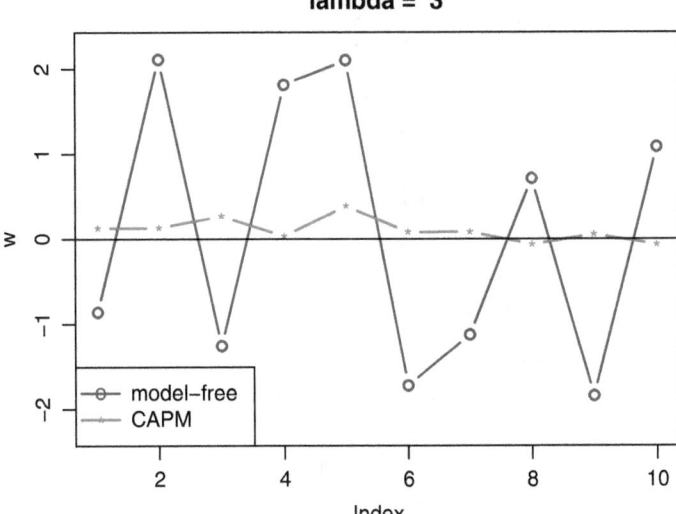

Fig. 20.10. *Allocation (or weight) vectors using the sample mean vector and co-variance matrix (model-free) and CAPM-based estimates.*

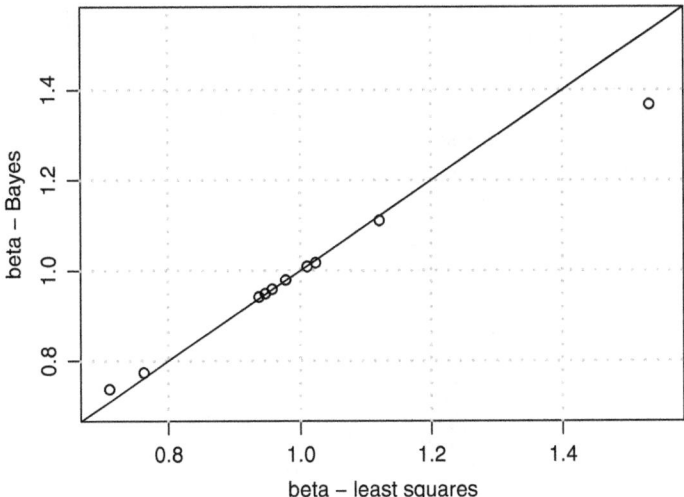

Fig. 20.11. *Plot of the Bayes estimator of β versus the least-squares estimator.*

$$\text{var}(\overline{Y}) = n^{-2} \sum_{i=1}^{n} \sum_{j=1}^{n} \text{Cov}(Y_i, Y_j)$$

$$= n^{-2} \sum_{i=1}^{n} \sum_{j=1}^{n} \gamma(i - j)$$

$$= n^{-2} \left\{ n\gamma(0) + 2 \sum_{h=1}^{n-1} \gamma(h)(n - h) \right\}$$

$$= \frac{\gamma(0)}{n} R_n, \tag{20.65}$$

where $R_n = \left\{ 1 + 2 \sum_{h=1}^{n-1} \rho(h) \left(1 - \frac{h}{n}\right) \right\}$. If $Y_1, Y_2, \ldots, Y_n$ is an uncorrelated process (white noise), then $R_n = 1$ and (20.65) agrees with (7.13).

Most stationary processes generated by MCMC have $\rho(h) \geq 0$ for all h so that R_n is inflated by the autocorrelation. The inflation can be severe. Consider the case of a stationary AR(1) process, $Y_n = \phi Y_{n_1} + \epsilon_i$. AR(1) processes often are reasonably good approximations to MCMC processes. For an AR(1) process we can approximate R_n:

$$R_n \approx \left\{ 1 + 2 \sum_{h=1}^{\infty} \rho(h) \right\} = \left\{ 2 \sum_{h=0}^{\infty} \phi^h - 1 \right\} = \left(\frac{2}{1 - \phi} - 1 \right) = \frac{1 + \phi}{1 - \phi}, \tag{20.66}$$

where we have used summation formula for geometric series (3.4) with $T = \infty$. Notice that the right-hand side of (20.66) increases without bound as $\phi \to 1$.

From the identity

$$\sum_{i=1}^{n} (Y_i - \mu)^2 = \sum_{i=1}^{n} (Y_i - \overline{Y})^2 + n(\overline{Y} - \mu)^2,$$

we obtain

$$E \left\{ \sum_{i=1}^{n} (Y_i - \overline{Y})^2 \right\} = \gamma(0)(n - R_n) \tag{20.67}$$

since $\gamma(0) = E\left\{(Y_i - \mu)^2\right\}$ and $\gamma(0)R_n = E\left\{n(\overline{Y} - \mu)^2\right\}$ by definitions. Therefore, an unbiased estimate of the process variance $\gamma(0)$ is

$$\hat{\gamma}(0) = \frac{\sum_{i=1}^{n} (Y_i - \overline{Y})^2}{n - R_n}. \tag{20.68}$$

When the process is uncorrelated so that $R_n = 1$, the right-hand side of (20.68) is the sample variance (A.7). For positively autocorrelated processes, $R_n > 1$ and the sample variance (which uses 1 in place of R_n) is biased downward.

To obtain an unbiased estimate of $\gamma(0)$, one can use

$$\frac{\sum_{i=1}^{n} (Y_i - \overline{Y})^2 + \widehat{\gamma(0)} R_n}{n}, \tag{20.69}$$

where $\widehat{\gamma(0)R_n}$ is an unbiased estimator of $\gamma(0)R_n$. There are several methods for estimating $\gamma(0)R_n$. The simplest uses several independent realizations of the process. Let $\overline{Y}_1, \ldots, \overline{Y}_M$ be the means of M independent realizations of the process and let $\overline{Y} = M^{-1} \sum_{j=1}^m \overline{Y}_j$. Then

$$\widehat{\gamma(0)R_n} = \frac{\sum_{j=1}^M (\overline{Y}_j - \overline{Y})^2}{M - 1} \qquad (20.70)$$

is an unbiased estimator of $\gamma(0)R_n$. The statistic $\widehat{\mathrm{var}}^+(\psi|\boldsymbol{Y})$ used in Sect. 20.7.5 for MCMC monitoring is a special case of (20.68) and (20.70).

20.14 Bibliographic Notes

There are many excellent books on Bayesian statistics. Gelman, Carlin, Stern, Dunson, Vehtari, and Rubin (2013) and Carlin and Louis (2008) are introductions to Bayesian statistics written at about the same mathematical level as this book. Box and Tiao (1973) is a classic work on Bayesian statistics with a wealth of examples and still worth reading despite its age. Berger (1985) is a standard reference on Bayesian analysis and decision theory. Bernardo and Smith (1994) and Robert (2007) are more recent books on Bayesian theory. Rachev, Hsu, Bagasheva, and Fabozzi (2008) covers many applications of Bayesian statistics to finance.

Albert (2007) is an excellent introduction to Bayesian computations in R. Chib and Greenberg (1995) explain how the Metropolis–Hastings algorithm works and why its stationary distribution is the posterior. Congdon (2001, 2003) covers the more recent developments in Bayesian computing with an emphasis on OpenBUGS software. There are other Bayesian Monte Carlo samplers besides MCMC, for example, importance sampling. Robert and Casella (2005) discuss these as well as MCMC. Gelman et al., (2013) have examples of Bayesian computations in R and OpenBUGS in an appendix. Lunn, Thomas, Best, and Spiegelhalter (2000) describe the design of OpenBUGS. Lunn, Jackson, Best, and Spiegelhalter (2013) is a comprehensive introduction to BUGS.

The diagnostics $\widehat{R}$ and N_{eff} are due to Gelman and Rubin (1992) though Sect. 20.7.5 uses the somewhat different notation of Gelman et al., (2013). Spiegelhalter, Best, Carlin, and van der Linde (2002) proposed DIC and p_D. The Gelman plot was introduced by Brooks and Gelman (1998). Kass et al. (1998) discuss their practical experiences with MCMC.

Bayesian modeling of yield curves models is discussed by Chib and Ergashev (2009). Bayesian time series are discussed by Albert and Chib (1993), Chib and Greenberg (1994), and Kim, Shephard, and Chib (1998); the first two papers cover ARMA process and the last covers ARCH and stochastic volatility models. There is a vast literature on the important and difficult problem of Bayesian estimation of covariance matrices with nonconjugate priors. Daniels and Kass (1999) review some of the literature in addition to providing their own suggestions.

We have not discussed empirical Bayes inference, but Carlin and Louis (2000) can be consulted for an introduction to that literature. Empirical Bayes inference uses a hierarchical prior but estimates the parameters in the lower level in a non-Bayesian manner and then, treating those parameter as known and fixed, performs a Bayesian analysis. The result is shrinkage estimation much like that achieved by a Bayesian analysis. The advantage of an empirical Bayes analysis is that it can be somewhat simpler than a fully Bayesian analysis. The disadvantage is that it underestimates uncertainty because estimated parameters in the prior are treated as if they were known. There are shrinkage estimators that are not exactly Bayesian or even empirical Bayes procedures. Ledoit and Wolf (2003) propose a shrinkage estimator for the covariance matrix of stock returns. Their shrinkage target is an estimate from a factor model, for example, the CAPM. Shrinkage estimation goes back at least to Stein (1956) and is often called Stein estimation.

The central limit theorem for the posterior is discussed by Gelman et al. (2013), Lehmann (1983), and van der Vaart (1998), in increasing order of technical level.

See Greyserman, Jones, and Strawderman (2006) for more information on portfolio selection by Bayesian methods.

20.15 R Lab

20.15.1 Fitting a t-Distribution by MCMC

In this section of the lab, you will fit the t-distribution to monthly returns on IBM using JAGS to estimate the posterior distribution by MCMC sampling.

Run the following R code to load the rjags package, input the data, and prepare the data for use by JAGS.

```
library(rjags)
data(CRSPmon, package = "Ecdat")
ibm = CRSPmon[ , 2]
r = ibm
N = length(r)
ibm_data = list(r = r, N = N)
```

Next, put the following BUGS code in a text file. I will assume that you name this file univt.bug, though you can use another name provided you make appropriate changes in the R code that follows. BUGS code is somewhat similar to, but not the same as, R code. For example, in R "dt" is the t-density, but in BUGS it is the t-distribution.

```
model{
for (t in 1:N)
{
```

```
    r[t] ~ dt(mu, tau, k)
}
mu ~ dnorm(0.0, 1.0E-6)
tau ~ dgamma(0.1, 0.01)
nu ~ dunif(2, 50)
sigma2 <- (k / (k - 2)) / tau
sigma <- sqrt(sigma2)
}
```

BUGS programs are difficult to debug, so be careful to enter the code exactly as it appears here. It has been tested and runs as written, but any error will cause problems. Our experience is that JAGS is better at providing error messages and easier to debug than WinBUGS and OpenBUGS.

The BUGS code above provides a description of the statistical model and specifies the prior distributions. The model states that the data are i.i.d. from a t-distribution. The ~ symbol assigns a distribution to a random variable so y[i] ~ dt(mu, tau, k) gives the likelihood of the data. Here mu, tau, and k are the mean, precision, and degrees of freedom, respectively, of the t-distribution. For a t-distribution, the precision is $\tau = 1/\lambda^2$ where λ is the scale parameter. Also, mu ~ dnorm(0.0, 1.0E-6) specifies the prior for the mean mu to be normal with mean 0 and precision 1.0E-6. The precision of a normal distribution is the reciprocal of its variance, so here the prior variance of μ is 1.0E6.

The symbol <- is used to assign a value (rather than a distribution) to a variable. Thus, sigma <- 1/sqrt(tau) makes sigma the scale parameter of the t-distribution of the data. In R, "=" can be used in place of "<-" for assigning a value to a variable, but this is not true in BUGS. The parameter sigma is not needed, but, by defining this variable in the BUGS program, we generate a sample from its posterior distribution.

Next, run the following R code that defines a function inits(). This function is used to generate random starting values for the chains.

```
inits = function(){list(mu = rnorm(1, 0, 0.3),
    tau = runif(1, 1, 10), k = runif(1, 1, 30))}
```

The next code uses the jags.model() and coda.samples() functions in the rjags package. Notice that the arguments specify the data, the function to create initial values of the chains, the file containing the BUGS program, the parameters to be monitored and returned, the number of chains, the number of iterations per chain, the number of iterations to discard as burn-in, the amount of thinning.

```
univ_t <- jags.model("univt.bug", data = ibm_data,
    inits = inits, n.chains = 3, n.adapt = 1000, quiet = FALSE)
nthin = 2
univ_t.coda = coda.samples(univ_t, c("mu", "tau", "k",
    "sigma"), n.iter = 500 * nthin, thin = nthin)
```

Next, print and plot the results.

```
summary(univ_t.coda)
effectiveSize(univ_t.coda)
gelman.diag(univ_t.coda)
```

Problem 1

(a) Which parameter mixes best according to N_{eff} in the output?
(b) Which parameter mixes worst according to N_{eff} in the output?
(c) Give a 95 % posterior interval for the degrees-of-freedom parameter.

Next, plot the results to check for convergence to the stationary distribution (posterior distribution) using Gelman plots and trace plots.

```
gelman.plot(univ_t.coda)
par(mfrow = c(2, 2))
traceplot(univ_t.coda)
```

Plotting the ACFs gives much insight into how well the chains are mixing. The less autocorrelation, the better. The function `autocorr.plot()` plots ACFs separately for each chain.

```
par(mfrow = c(2, 2))
autocorr.plot(univ_t.coda, auto.layout = FALSE)
```

Problem 2

(a) Which parameter mixes best and which mixes worse according to the ACF plots? Explain your answers.
(b) Find the posterior skewness and kurtosis of the degrees of freedom parameter.

The function `densityplot()` gives kernel density estimate from each chain.

```
library(lattice)
densityplot(univ_t.coda)
```

Problem 3 *Which posterior densities are most skewed?*

The kurtosis of a t-distribution is $3(\nu - 2)/(\nu - 4)$ if $\nu > 4$ and is $+\infty$ if $\nu \leq 4$. Variables in R can have infinite values: `Inf` is $+\infty$ and `-Inf` is $-\infty$, so R can handle infinite values of kurtosis if they occur.

Problem 4 *Write* R *code to compute 1500 MCMC values of the kurtosis.* *(1500 = 3*2000/2; there are 3 chains of length 2000 after burn-in and they are thinned to every 2nd iteration.)*

(a) *Find the 0.01, 0.05, 0.25, 0.5, 0.75, 0.95, and 0.99 quantiles of the posterior distribution of the kurtosis of IBM returns.*

(b) *Estimate the posterior probability that the kurtosis of the distribution of IBM returns is finite.*

(c) *Compute the 0.01, 0.05, 0.25, 0.5, 0.75, 0.95, and 0.99 quantiles of the bootstrap distribution of the sample kurtosis of IBM. Take 1000 resamples using both a model-free and a model-based bootstrap. Compare the two sets of bootstrap quantiles with the posterior quantiles in (a).*

(d) *Compare 90 % bootstrap basic percentile confidence intervals for the kurtosis with the 90 % posterior interval. Which interval is shortest? Why might it be shortest?*

20.15.2 AR Models

In this section of the lab, you will fit an AR(1) model to the changes in the log of the GDP. First, run the following code to process the data. Notice that the log-GDP time series is differenced before fitting.

```
1 library(rjags)
2 data(Tbrate, package = "Ecdat")
3 #   r = the 91-day treasury bill rate
4 #   y = the log of real GDP
5 #   pi = the inflation rate
6 del_dat = diff(Tbrate)
7 y = del_dat[,2]
8 N = length(y)
9 GDP_data=list(y = y, N = N)
```

Next create a file called `ar1.bug` containing the following WinBUGS code.

```
1 model{
2 for(i in 2:N){
3     y[i] ~ dnorm(mu + phi * (y[i-1] - mu), tau)
4 }
5 mu ~ dnorm(0, 0.00001)
6 phi ~ dnorm(0, 0.00001)
7 tau ~ dgamma(0.1, 0.0001)
8 sigma <- 1/sqrt(tau)
9 }
```

Finally, run the following code to fit an AR(1) model using **JAGS** and also using R's `arima()` function to compute the MLE, which will be compared with the Bayes estimator.

```
1 inits = function(){list(mu = rnorm(1, 0, 2 * sd(y) / sqrt(N)),
2    phi = rnorm(1, 0, 0.3), tau = runif(1, 1, 10))}
3 ar1 <- jags.model("ar1.bug", data = GDP_data, inits = inits,
4    n.chains = 3, n.adapt = 1000, quiet = FALSE)
5 nthin = 20
6 ar1.coda = coda.samples(ar1, c("mu", "phi", "sigma"),
7    n.iter = 500 * nthin, thin = nthin)
8 summary(ar1.coda, digits = 3)
9 arima(y, order = c(1, 0, 0))
```

Problem 5 *Construct time series and ACF plots of the parameters* phi *and* sigma.

(a) *Do you believe that the MCMC sample size is adequate? Why or why not? Is the burn-in iterations adequate? Why or why not? If you feel that either the number of iterations or the length of the burn-in period is inadequate, then rerun with a larger burn-in period and/or MCMC sample size.*

(b) *Compute the MLEs for this model using* arima()*. How closely do the Bayes estimates and MLEs agree? Could you explain any possible disagreement?*

(c) *The model in the* BUGS *program does not assume that the time series is in its stationary distribution. In fact, the model does not even assume that there is a stationary distribution. Explain why.*

(d) *Modify the* BUGS *program to utilize the marginal distribution of* y_1, *assuming that the process starts in its stationary distribution.*

20.15.3 MA Models

Next you will fit an MA(1) to simulated data. The function arima.sim() is used to create the data.

```
1 library(rjags)
2 set.seed(5640)
3 N = 600
4 y = arima.sim(n = N, list(ma = -0.5), sd = 0.4)
5 y = as.numeric(y) + 3
6 q = 5
7 ma.sim_data = list(y = y, N = N, q = q)
8 inits.ma = function(){list(mu = rnorm(1, mean(y), 2*sd(y)/sqrt(N)),
9    theta = rnorm(1, -0.05, 0.1), tau = runif(1, 5, 8))}
```

Put the following BUGS program in the file ma1.bug. This program not only fits the MA(1) model but also predicts q steps ahead; q is an input parameter chosen by the user and, from the viewpoint of BUGS, q is part of the data and is set equal to 5 in the code above. The predicted values will be included in the output and called ypred.

```
1  model{
2  for (i in 2:N)
3  {
4     w[i] <- y[i] - mu - theta * w[i-1]
5  }
6  w[1] ~ dnorm(0, 0.01)
7  for (i in 2:N)
8  {
9     y[i] ~ dnorm(mu + theta * w[i-1], tau)
10 }
11 mu ~ dnorm(0, 0.0001)
12 theta ~ dnorm(0, 0.0001)
13 tau ~ dgamma(0.01, 0.0001)
14 sigma <- 1/sqrt(tau)
15 for (i in 1:q)
16 {
17    ypred[i] ~ dnorm(theta * w[N + i - 1], tau)
18    w[i + N] <- ypred[i] - theta * w[N + i - 1]
19 }
20 }
```

Now run this R code.

```
1  ma1 <- jags.model("ma1.bug", data = ma.sim_data, inits = inits.ma,
2     n.chains = 3, n.adapt = 1000, quiet = FALSE)
3  nthin = 5
4  ma1.coda = coda.samples(ma1, c("mu", "theta", "sigma", "ypred"),
5     n.iter = 500 * nthin, thin = nthin)
6  summary(ma1.coda)
```

Problem 6

(a) *Do you believe that the MCMC sample size is adequate? Why or why not? If you feel it is inadequate, than rerun JAGS with a larger MCMC sample size. Is the length of the burn-in periods adequate?*

(b) *Construct time series and ACF plots of the parameters* theta, sigma, ypred[1], *and* ypred[2]. *What do the plots tell us about MCMC mixing and convergence?*

(c) *Find a 90 % posterior interval for the next observation after the observed data.*

20.15.4 ARMA Models

Create a simulated sample from an ARMA(1,1) process with the following R code.

```
set.seed(5640)
N = 600
```

```
y = arima.sim(n = N, list(ar = 0.9, ma = -0.5), sd = 0.4)
y = as.numeric(y)
```

Problem 7 *Create* BUGS *and* R *code to fit the ARMA(1,1) model to the simulated data. Monitor the result to make certain that the MCMC sample size is large enough.*

(a) *Discuss how well the chains mix and whether the Monte Carlo sample size is adequate.*
(b) *Find 99 % posterior intervals for the AR and MA parameters.*

20.16 Exercises

1. Show in Example 20.2 that the MAP estimator is 6/7.
2. Verify (20.26).
3. In the derivation of (20.51), it was stated that "$\{\overline{Y} - E(\mu|Y)\}$ and $\{E(\mu|Y) - \mu\}$ are conditionally uncorrelated given Y." Verify this statement.
4. Continue the analysis in Example 20.16. Divide the data into 20 blocks of 250 days each, except that the last block will have only 212 days. Use each of the first 19 blocks as training data with the subsequent block as test data. How does portfolio selection based on the CAPM compare with model-free estimation when averaged over the 19 pairs of training and test data sets?
5. One of the strength of fitting models by MCMC using BUGS is that a very wide range of models can be fit. As an example, in this exercise a regression model with MA(1) errors will be used. For data, use the first 1500 returns[5] on GM and on the S&P 500 index in the data set Stock_Bond.csv. Fit the model
$$R_t = \beta R_{M,t} + \epsilon_t + \theta\epsilon_{t-1}.$$
where R_t is the tth return on GM, $R_{M,t}$ is the tth return on the S&P 500, and $\epsilon_1,\ldots,\epsilon_{1500}$ are i.i.d. $N(0,\sigma_\epsilon^2)$. Use non-informative priors on β, θ, and σ_ϵ^2.
6. Expand the model in Exercise 5 so that $\epsilon_1,\ldots$ is a GARCH(1,1) process. Revise the BUGS and R code of Exercise 5 to fit this expanded model.
7. So far we have treated the sample mean vector and covariance matrix as fixed when considering the risk of a portfolio. Stated differently, estimation risk has been ignored. A methodology for taking risk due to estimation error into account was proposed by Greyserman, Jones, and Strawderman (2006). Assume that the vector of returns R is $N(\mu, \Sigma)$ distributed. Let $(\mu^{(k)}, \Sigma^{(k)})$, $k = 1,\ldots,K$, be an MCMC sample from the posterior distribution of (μ, Σ). For each k, let $R^{(k)}$ be $N(\mu^{(k)}, \Sigma^{(k)})$ distributed. Then

---

[5] JAGS had trouble when the full data set was used, probably because there are nearly 5000 latent variables. This problem is likely hardware dependent.

$\boldsymbol{R}^{(1)}, \ldots, \boldsymbol{R}^{(K)}$ is a sample from the posterior predictive distribution of $\boldsymbol{R}$ and take uncertainty about $\boldsymbol{\mu}$ and $\boldsymbol{\Sigma}$ into account and

$$K^{-1} \sum_{k=1}^{K} U\{X_0(1 + \boldsymbol{w}^\mathsf{T} \boldsymbol{R}^{(k)})\} \tag{20.71}$$

estimates the expected utility if the allocation vector is $\boldsymbol{w}$. Here, as in Sect. 16.8, X_0 is the initial wealth and U is the utility function. Continue the analysis in Example 20.16 using the CAPM model and maximize (20.71). Maximizing (20.71) is a nonlinear optimization problem, so a good starting value is essential. As a starting value, use the $\boldsymbol{w}$ found in Example 20.16 that ignores estimation error.

References

Albert, J. (2007) *Bayesian Computation with R*, Springer, New York.

Albert, J. H. and Chib, S. (1993) Bayes inference via Gibbs sampling of autoregressive time series subject to Markov mean and variance shifts, *Journal of Business & Economic Statistics*, 11, 1–15.

Berger, J. O. (1985) *Statistical Decision Theory and Bayesian Analysis* 2nd ed., Springer-Verlag, Berlin.

Bernardo, J. M., and Smith, A. F. M. (1994) *Bayesian Theory*, Wiley, Chichester.

Box, G. E. P., and Tiao, G. C. (1973) *Bayesian Inference in Statistical Analysis*, Addison-Wesley, Reading, MA.

Brooks, S. P. and Gelman, A. (1998) General Methods for Monitoring Convergence of Iterative Simulations. *Journal of Computational and Graphical Statistics*, 7, 434–455.

Carlin, B. P., and Louis, T. A. (2000) Empirical Bayes: Past, present and future. *Journal of the American Statistical Association*, **95**, 1286–1289.

Carlin, B. , and Louis, T. A. (2008) *Bayesian Methods for Data Analysis*, 3rd ed., Chapman & Hall, New York.

Chib, S., and Ergashev, B. (2009) Analysis of multifactor affine yield curve models. *Journal of the American Statistical Association*, **104**, 1324–1337.

Chib, S., and Greenberg, E. (1994) Bayes inference in regression models with ARMA(p, q) errors. *Journal of Econometrics*, **64**, 183–206.

Chib, S., and Greenberg, E. (1995) Understanding the Metropolis–Hastings algorithm. *American Statistician*, **49**, 327–335.

Congdon, P. (2001) *Bayesian Statistical Modelling*, Wiley, Chichester.

Congdon, P. (2003) *Applied Bayesian Modelling*, Wiley, Chichester.

Daniels, M. J., and Kass, R. E. (1999) Nonconjugate Bayesian estimation of covariance matrices and its use in hierarchical models. *Journal of the American Statistical Association*, **94**, 1254–1263.

Edwards, W. (1982) Conservatism in human information processing. In *Judgement Under Uncertainty: Heuristics and Biases*, D. Kahneman, P. Slovic, and A. Tversky, ed., Cambridge University Press, New York.

Gelman, A., and Rubin, D. B. (1992) Inference from iterative simulation using multiple sequence (with discussion). *Statistical Science*, **7**, 457–511.

Gelman, A., Carlin, J. B., Stern, H. S., Dunson, D. B., Vehtari, A., and Rubin, D. B. (2013) *Bayesian Data Analysis*, 3rd ed., Chapman & Hall, London.

Greyserman, A., Jones, D. H., and Strawderman, W. E. (2006) Portfolio selection using hierarchical Bayesian analysis and MCMC methods, *Journal of Banking and Finance*, 30, 669–678.

Kass, R. E., Carlin, B. P., Gelman, A., and Neal, R. (1998) Markov chain Monte Carlo in practice: A roundtable discussion. *American Statistician*, **52**, 93–100.

Kim, S., Shephard, N., and Chib, S. (1998) Stochastic volatility: likelihood inference and comparison with ARCH models. *Review of Economic Studies*, **65**, 361–393.

Ledoit, O., and Wolf, M. (2003) Improved estimation of the covariance matrix of stock returns with an application to portfolio selection. *Journal of Empirical Finance*, **10**, 603–621.

Lehmann, E. L. (1983) *Theory of Point Estimation*, Wiley, New York.

Lunn, D., Jackson, C., Best, N., Thomas, A., and Spiegelhalter, D. (2013) *The BUGS Book*, Chapman & Hall.

Lunn, D. J., Thomas, A., Best, N., and Spiegelhalter, D. (2000) OpenBUGS— A Bayesian modelling framework: Concepts, structure, and extensibility. *Statistics and Computing*, **10**, 325–337.

Rachev, S. T., Hsu, J. S. J., Bagasheva, B. S., and Fabozzi, F. J. (2008) *Bayesian Methods in Finance*, Wiley, Hoboken, NJ.

Robert, C. P. (2007) *The Bayesian Choice: From Decision-Theoretic Foundations to Computational Implementation*, 2nd ed., Springer, New York.

Robert, C. P., and Casella, G. (2005) *Monte Carlo Statistical Methods, 2nd ed.*, Springer, New York.

Spiegelhalter, D. J., Best, N. G., Carlin, B. P., and van der Linde, A. (2002) Bayesian measures of model complexity and fit. *Journal of the Royal Statistical Society, Series B, Methodological*, **64**, 583–616.

Stein, C. (1956) Inadmissibility of the usual estimator for the mean of a multivariate normal distribution. In *Proceedings of the Third Berkeley Symposium on Mathematical and Statistical Probability*, J. Neyman, ed., University of California, Berkeley, pp. 197–206, Volume 1.

van der Vaart, A. W. (1998) *Asymptotic Statistics*, Cambridge University Press, Cambridge.

21

Nonparametric Regression and Splines

21.1 Introduction

As discussed in Chap. 9, regression analysis estimates the conditional expectation of a response given predictor variables. The conditional expectation is called the regression function and is the best predictor of the response based upon the predictor variables, because it minimizes the expected squared prediction error.

There are three types of regression, linear, nonlinear parametric, and nonparametric. *Linear regression* assumes that the regression function is a linear function of the parameters and estimates the intercept and slopes (regression coefficients). *Nonlinear parametric regression*, which was discussed in Sect. 11.2, does not assume linearity but does assume that the regression function is of a *known* parametric form, for example, the Nelson-Siegel model. In this chapter, we study *nonparametric regression*, where the form of the regression function is also nonlinear but, unlike nonlinear parametric regression, not specified by a model but rather determined from the data. Nonparametric regression is used when we know, or suspect, that the regression function is curved, but we do not have a model for the curve.

There are many techniques for nonparametric regression, but local polynomial regression and splines are the most widely used, and only these will be discussed here. Local polynomial regression and splines generally work well and, since they usually give similar estimates, it is difficult to recommend one over the over. Local polynomial estimation might be somewhat simpler to understand. Splines are used in many areas of mathematics, such as, for interpolation, and so it is worthwhile to be familiar with them. Also, splines are useful as components in complex models. The R lab at the end of this chapter gives an example.

© Springer Science+Business Media New York 2015

D. Ruppert, D.S. Matteson, *Statistics and Data Analysis for Financial Engineering*, Springer Texts in Statistics,

DOI 10.1007/978-1-4939-2614-5_21

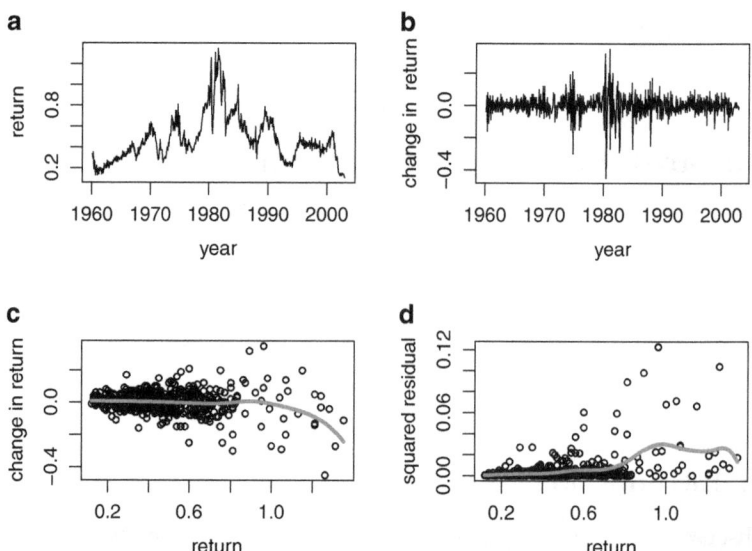

Fig. 21.1. *Risk-free monthly returns. The returns are 1/12th the yearly rate. (a) Time series plot of the returns. (b) Time series plot of the changes in the returns. (c) Plot of changes in returns against lagged returns with a local linear estimate of the drift. (d) Plot of squared residuals against lagged returns with a local linear estimate of the squared diffusion coefficient.*

Models for the evolution of short-term interest rates are important in finance, for example, because they are needed for the pricing of interest rate derivatives. Figure 21.1 contains plots of the monthly risk-free returns[1] in the Capm data set in R's Ecdat package. This data set has been used for various purposes in several previous chapters. Here we will use it to illustrate nonparametric regression. Panels (a) and (b) are time series plots of the returns and the changes in the returns.

A common model for changes in short-term interest rates is

$$\Delta r_t = \mu(r_{t-1}) + \sigma(r_{t-1})\epsilon_t, \qquad (21.1)$$

where r_t is the rate at time t, $\Delta r_t = r_t - r_{t-1}$, $\mu(\cdot)$ is the drift function, $\sigma(\cdot)$ is the volatility function, also called the diffusion function, and ϵ_t is $N(0,1)$ noise. Many different parametric models have been proposed for $\mu(\cdot)$ and $\sigma(\cdot)$, for example, by Merton (1973), Vasicek (1977), Cox, Ingersoll, and Ross (1985), Yau and Kohn (2003), and Chan et al. (1992). The simplest model, due to Merton (1973), is that $\mu(\cdot)$ and $\sigma(\cdot)$ are constant. Chan et al. (1992) assume that $\mu(r) = \beta(r - \alpha)$ and $\sigma(r) = \theta r^\gamma$, where $\alpha > 0$, $\beta < 0$, $\theta > 0$, and γ are unknown parameters—this process reverts to a mean equal to α. Chan et al.'s model was used as an example of nonlinear regression in Sect. 11.12.1.

---

[1] The risk-free rate is called the risk-free return in the Capm package.

The approach of Yau and Kohn (2003) that is used here is to model both $\mu(\cdot)$ and $\sigma(\cdot)$ nonparametrically. Doing this allows one to check which parametric models, if any, fit the data and to have a nonparametric alternative if none of the parametric models fits well.

The solid red curves in Fig. 21.1c and d are estimates of $\mu(\cdot)$ and $\sigma^2(\cdot)$ by a nonparametric regression method *local linear regression*, a special case of *local polynomial regression*. By (21.1), $E(\Delta r_t) = \mu(r_{t-1})$ and $\mathrm{Var}(\Delta r_t) = \sigma^2(r_{t-1})$, so $\widehat{\mu}(\cdot)$ is obtained by regressing Δr_t on r_{t-1} and $\widehat{\sigma}^2(\cdot)$ by regressing $\{\Delta r_t - \widehat{\mu}(r_{t-1})\}^2$ on r_{t-1}. The latter is an example of estimating a conditional variance; see Sect. 14.2.

The code to produce Fig. 21.1 is below. The local linear regression estimates were produced by the function `locpoly()` in the **KernSmooth** package. This function computes the fitted function on a grid of 401 equally spaced points; this is sufficient for plotting the fitted function as in lines 21 and 24 but not for computing the residuals. Instead, to compute squared residuals at line 11, the `spline()` function is used at line 10 to interpolate the fit from the 401-point grid to the values of the explanatory variable.

```
 1  library(Ecdat)
 2  library(KernSmooth)
 3  data(Capm)
 4  attach(Capm)
 5  n = length(rf)
 6  year = seq(1960.125, 2003, length = n)
 7  diffrf = diff(Capm$rf)
 8  rf_lag = rf[1:(n-1)]
 9  ll_mu <- locpoly(rf_lag, diffrf, bandwidth = dpill(rf_lag, diffrf))
10  muhat = spline(ll_mu$x, ll_mu$y, xout = rf_lag)$y
11  epsilon_sqr = (diffrf-muhat)^2
12  ll_sig <- locpoly(rf_lag, epsilon_sqr,
13      bandwidth = dpill(rf_lag, epsilon_sqr) )
14  pdf("riskfree01.pdf", width = 6, height = 5)
15  par(mfrow=c(2, 2))
16  plot(year, rf, ylab = "return", main = "(a)", type = "l" )
17  plot(year[2:n], diffrf, ylab = "change in  return", main = "(b)",
18      type = "l", xlab = "year")
19  plot(rf_lag, diffrf, ylab = "change in return", xlab = "return",
20      main="(c)",type="p",cex=.7)
21  lines(ll_mu$x, ll_mu$y, lwd = 4, col = "red")
22  plot(rf_lag, (diffrf - muhat)^2, xlab = "return",
23      ylab = "squared residual", main = "(d)", cex = 0.7)
24  lines(ll_sig$x, ll_sig$y, lwd = 4, col = "red")
25  graphics.off()
```

21.2 Local Polynomial Regression

Local polynomial regression is based on the principle that a smooth function can be approximated locally by a low-degree polynomial. Suppose we have a sample (X_i, Y_i), $i = 1, \ldots, n$, and $E(Y|X = x) = \mu(x)$ for a smooth function μ. The function μ will be estimated on a grid of x-values, $x_1, \ldots, x_M$. These could, but need not, be the same values $X_1, \ldots, X_n$, as where we observe Y.

The estimation is done one point at a time on the grid $x_1, \ldots, x_M$. To estimate μ at x_ℓ, one fits a pth-degree polynomial using only (X_i, Y_i) with X_i near x_ℓ. This is done using weights determined by a kernel function K. K is a probability density function symmetric about 0 and such that $K(x)$ decreases as $|x|$ increases, for instance, a normal density with mean 0. We have seen kernels used for density estimation in Sect. 4.2.

The regression function at x_ℓ is estimated by kernel-weighted least squares, which minimizes

$$\sum_{i=1}^{n} \left[Y_i - \{\beta_0 + \beta_1(X_i - x_\ell) + \cdots + \beta_p(X_i - x_\ell)^p\} \right]^2 K\{(X_i - x_\ell)/h\} \quad (21.2)$$

and then $\widehat{\mu}(x_\ell) = \widehat{\beta_0}$ since the regression model $\beta_0 + \beta_1(x - x_\ell) + \cdots + \beta_p(x - x_\ell)^p$ equals β_0 at $x = x_\ell$. The weights $K\{(X_i - x_\ell)/h\}$ decrease as $|X_i - x_\ell|$ increases, so only the data near x_ℓ are used. The parameter h is called the bandwidth and determines how much data are used for estimation; the larger the value of h, the more data used.

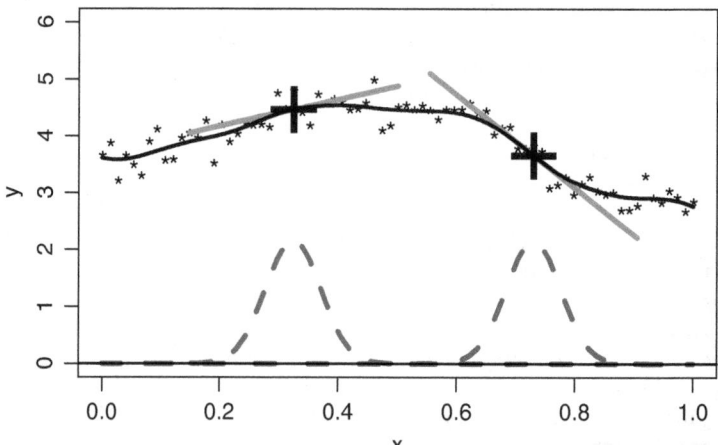

Fig. 21.2. *Local linear fit (solid curve) to 75 data points (asterisks) with bandwidth chosen by the direct plug-in method. The regression function μ is estimated at each of the 75 points and the estimates are connected to create the solid curve. Estimation at $x_{25} = 0.32$ and $x_{55} = 0.72$ is illustrated by the kernels (dashed curves), the linear fits (solid lines), and the fitted points (large +).*

Local linear estimation, where $p = 1$, is illustrated in Fig. 21.2. The kernel functions are shown as blue dashed curves at two points, $x_{25} = 0.32$ and $x_{75} = 0.72$. Above each kernel, the local linear fit is shown as a red line and the large "+" is placed at $\{x, \widehat{\mu}(x)\}$. The black curve $\widehat{\mu}$ is obtained by finding local fits on a grid of 75 x_ℓ-values and plotting $\{x_\ell, \widehat{\mu}(x_\ell)\}$ for all x_ℓ on this grid. For example, the curve in Fig. 21.2 used the R function `locpoly()` in R's `KernSmooth` package and has a grid of 401 equally spaced x-values (the default). Often the grid is simply the observed X-values, $X_1, \ldots, X_n$.

The bandwidth h is called a "smoothing parameter" because it determines the smoothness of $\widehat{\mu}$. A larger value of h gives a smoother curve. The choice of h is important. If h is too large, then the polynomial approximation may be poor and the estimate of $\mu(x)$ will be badly biased. Conversely, if h is too small, then too few data are used and the estimate of μ will be too variable. A good choice of the bandwidth minimizes the mean squared error of the estimator, which is the variance plus the squared bias. Both the squared bias and variance of the estimator are unknown and must be estimated, or at least their sum must be estimated. Automatic bandwidth selection, which either directly or indirectly estimates and attempts to minimize the mean-squared error, has been an area of intense research and a number of data-based bandwidth selectors are available. The curve in Fig. 21.2 used the bandwidth chosen by the popular direct plug-in (dpi) bandwidth selector of Ruppert, Sheather and Wand (1995). The dpi selector estimates the mean integrated squared error (MISE) of $\widehat{\mu}$, which is

$$
E\left[\int_{\min(X_i)}^{\max(X_i)} \{\mu(x) - \widehat{\mu}(x)\}^2 \, dx \right],
\tag{21.3}
$$

and finds the bandwidth that minimizes the estimated MISE.

Nonparametric regression estimators are also called *smoothers* because they smooth out the noise in the data. Using a bandwidth that is too small causes *overfitting*, which is *undersmoothing*. Conversely, a bandwidth that is too large will result in *underfitting*, which is *oversmoothing*—see Sect. 4.2 for further discussion of under- and oversmoothing in the context of kernel density estimation.

Figure 21.3 illustrates the effect of varying the bandwidth. The solid black curve uses the dpi bandwidth, the dashed red curve uses three times the dpi bandwidth, and the dotted-and-dashed blue curve uses one-third the dpi bandwidth. The dashed red curve is too smooth to follow the data closely, that is, it underfits, while the dotted-and-dashed blue curve is wiggly because it is tracking random noise in the data, that is, it overfits. In this example, the data were simulated, so the true regression function, $\mu(x) = 3.6 + 0.1x + \sin(5\,x^{1.5})$, is known and it is possible to calculate the average squared error, $\sum_{i=1}^{n}\{\widehat{\mu}(X_i) - \mu(X_i)\}^2$, for each bandwidth. The average squared errors are 1.34 and 2.27 times larger using 3*dpi and dpi/3, respectively, compared to using dpi.

Besides the dpi bandwidth selector, the bandwidth can also be chosen by minimizing either the AIC or GCV (generalized cross-validation) criterion. The definition of AIC for a parametric model uses the number of parameters in the model, but local polynomial estimation is not parametric, so one cannot count parameters. Nonetheless, it is possible to define the "effective number of parameters" and this is done in Sect. 21.3.1. GCV is defined in Sect. 21.3.2.

Example 21.1. Local polynomial estimation of forward rates.

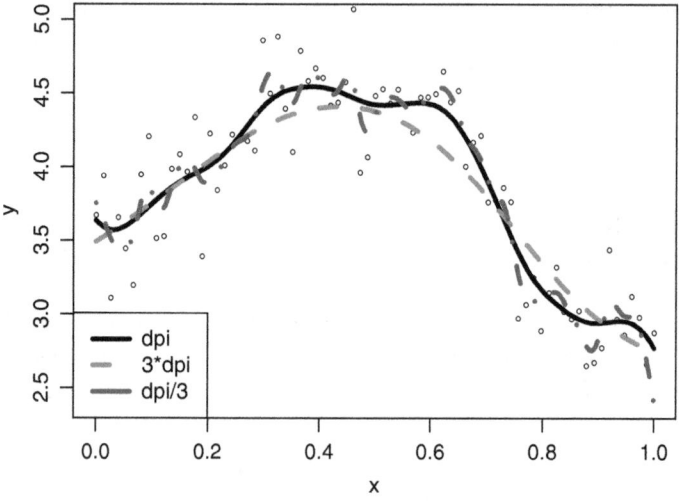

Fig. 21.3. *Local linear estimators with three bandwidths: dpi (direct plug-in), which gives an appropriate amount of smoothing; three times the dpi, which oversmooths (underfits); and one-third the dpi, which undersmooths (overfits). Simulated data.*

The R code in this section computes two estimates of the forward rate function, a parametric estimate based on the Nelson-Siegel model and a nonparametric estimates using local polynomial smoothing. The program is split into several chucks of code with discussion between the chucks.

Line 1 reads in prices of STRIPS on Dec 31, 1995. This data set was used in Sect. 11.3. There are 29 1/4 years of quarterly maturities T, for a total of 117 (= 29.25 * 4) prices.

The price of a zero-coupon bond as a percentage of par is price = $100 \exp\left\{-\int_0^T f(s)ds\right\}$ so that $-\log(\text{price}) = \int_0^T f(s)ds - \log(100)$. Thus, the integrated forward rate, called Int_F, is defined on line 7 to be $-\log(\text{price}) + \log(100)$. Lines 4–7 use the order() function to order the maturities T from smallest to largest and to order the prices accordingly. Ordering the data by T is needed when the plotted points are connected by lines. If they were not ordered by T, then the plot could look like a spider web.

```
1 dat = read.table("strips_dec95.txt", header = T)
2 T = dat$T
3 n = length(T)
4 ord = order(T)
5 T = T[ord]
6 price = dat$price[ord]
7 Int_F = - log(price) + log(100)
```

In lines 8–9, the function locfit() in the locfit package estimates the first derivative of Int_F by local cubic fitting to produce an estimate of the forward rate (since Int_F estimates the integrated forward rate). Note that deriv = 1 specifies estimation of the first derivative and deg = 3 specifies using cubic polynomial fitting. The function locfit() is similar to locpoly() used in Sect. 21.1 to create the curves in Fig. 21.1. We could have used locpoly() again here but wanted to illustrate both local polynomial regression functions.

```
8 library(locfit)
9 fit_loc_Int_F = locfit(Int_F ~ T, deriv = 1, deg = 3)
```

The function Nelson-Siegel() in lines 10–18 returns a list containing the forward rate, called f, and the integrated forward rate, called int_f. Lines 19–24 estimate the parameters of the Nelson-Siegel model by fitting the Nelson-Siegel integrated forward rate to Int_F. On line 21, Yhat is the integrated forward rate because "[[2]]" is the second element of the list that is output by NelsonSiegel().

```
10 NelsonSiegel = function(theta){
11    ####  f = forward rate and int_f = intergrated forward rate  ####
12    f = theta[1] + (theta[2] + theta[3] * T) * exp(-theta[4] * T)
13    int_f = theta[1] * T - theta[2] / theta[4]
14    * (exp(-theta[4] * T) - 1) -
15    theta[3] * (T * exp(-theta[4] * T) / theta[4] +
16    (exp(-theta[4] * T) - 1) / theta[4]^2)
17      list("f" = f, "inf_t" = int_f)
18 }
19 fit_NS = optim(c(0.05, 0.001, 0.001, 0.08),
20 fn = function(theta){
21    Yhat = NelsonSiegel(theta)[[2]]
22    sum((Int_F - Yhat)^2)},
23    control = list(maxit=30000, reltol = 1e-10))
24 NS_yhat = NelsonSiegel(fit_NS$par)[[1]]
```

Figure 21.4 is produced by lines 25–36. Notice that the Nelson-Siegel fit (in blue) tends to overestimate the forward rate when $5 < T < 15$ and $T > 25$ and to underestimate when $T < 3$ and $15 < T < 25$. In comparison, the local polynomial estimates show no bias.

```
25 pdf("strips02.pdf",width=6,height=5)
26 par(mfrow=c(1,1))
```

```
27 plot(fit_loc_Int_F, ylim = c(.025,.075), ylab = "Forward rate",
28    col = "red", lwd = 3)
29 lines(fit_loc_Int_F, col = "red", lwd=3) # to widen to linewidth = 3
30 lines(T, NS_yhat, col = "blue", lwd = 3, lty = 4)
31 points(T[2:n], diff(Int_F) / diff(T))
32 legend("bottomleft", c("local cubic",
33    "Nelson-Siegel", "Empirical"),
34    lty=c(1, 4,NA),pch=c(NA, NA, "o"),
35    col=c("red", "blue", "black"), lwd = c(3, 3, NA))
36 graphics.off()
```

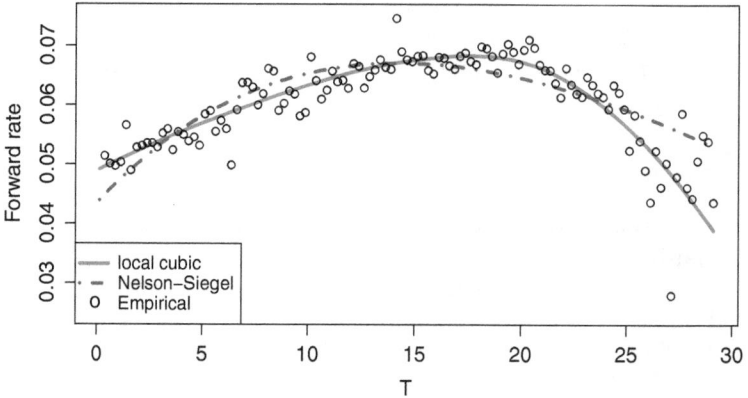

Fig. 21.4. Local cubic (nonparametric), Nelson-Siegel (parametric), and empirical estimates of the forward rate curve. Notice that the nonparametric estimates are much closer than the parametric estimates to the empirical estimates.

□

21.2.1 Lowess and Loess

Loess and its earlier version lowess are local polynomial smoothers with spatially varying bandwidths controlled by a parameter called *span*. Span is the fraction of the data used for estimation at each point. The bandwidth, call it $h(x, \text{span})$, for estimation at a point x is adjusted so that whenever span ≤ 1 then $K\{(X_i - x)/h(x, \text{span})\}$ is nonzero for span $\times$ 100 % of the X_i.

If span $= 1$, then all of the data are used for estimation at each point, but the data farthest from X_i get small weights. Because of these small weights, for small data sets, a lowess (or loess) smooth with a span of 1 might not be smooth enough. To solve this problem, span is defined for values greater than 1 by

$$h(x, \text{span}) = \text{span} \times h(x, 1).$$

As span increases beyond 1, the weights $K\{(X_i - x)/h(x, \text{span})\}$ become more and more equal. As span $\to \infty$, the weights converge to a constant, $K(0)$, and the lowess (or loess) fit converges to a polynomial regression fit.

21.3 Linear Smoothers

Local polynomial regression as well as penalized spline regression—to be covered soon—are examples of linear smoothers. A linear smoother has an $n \times n$ smoother matrix $\boldsymbol{H}$, which does not depend on $\boldsymbol{Y}$, such that

$$\widehat{\boldsymbol{Y}} = \boldsymbol{HY}, \tag{21.4}$$

where $\boldsymbol{Y} = (Y_1, \ldots, Y_n)^\mathsf{T}$ is the vector of responses and $\widehat{\boldsymbol{Y}} = (\widehat{Y}_1, \ldots, \widehat{Y}_n)^\mathsf{T}$ is the vector of fitted values. Equation (21.4) can be written as

$$\widehat{Y}_i = \sum_{j=1}^{n} H_{ij} Y_j, \quad i = 1, \ldots, n. \tag{21.5}$$

The smoother matrix will depend on a smoothing parameter, which for local polynomial regression is the bandwidth. We will let λ denote the smoothing parameter and denote the smoother matrix by $\boldsymbol{H}(\lambda)$. The smoother matrix is an analog of the hat matrix of linear regression and is, itself, often called a hat matrix.

21.3.1 The Smoother Matrix and the Effective Degrees of Freedom

In a parametric model, the number of parameters quantifies the ability of the model to fit the data. In nonparametric estimation, the potential to fit (and overfit) can be quantified by the *effective number of parameters* or the *effective degrees of freedom of the fit*. Conceptually, the effective number of parameters is similar to the Bayesian p_D in Sect. 20.7.6.

By (21.5), the hat diagonal $H(\lambda)_{ii}$ gives the *leverage* or *self-influence* of the Y_i since it is the weight given to Y_i when calculating $\widehat{Y}_i$. A large value of $H(\lambda)_{ii}$ means a high potential for overfitting. The effective number of parameters is the sum of the leverages:

$$p_{\text{eff}} = \sum_{i=1}^{n} H(\lambda)_{ii} = \text{tr}\{\boldsymbol{H}(\lambda)\}. \tag{21.6}$$

If p_{eff} is too small (too large), then the data are underfit (overfit).

The residual mean sum of squares is

$$\sum_{i=1}^{n} (Y_i - \widehat{Y}_i)^2 = \|\boldsymbol{Y} - \widehat{\boldsymbol{Y}}\|^2 = \|\{\boldsymbol{I} - \boldsymbol{H}(\lambda)\}\boldsymbol{Y}\|^2, \tag{21.7}$$

where $\boldsymbol{I}$ is the $n \times n$ identity matrix. The noise variance is estimated by

$$\widehat{\sigma}(\lambda)^2 = \frac{\|\{\boldsymbol{I} - \boldsymbol{H}(\lambda)\}\boldsymbol{Y}\|^2}{n - p_{\text{eff}}}, \tag{21.8}$$

which is a direct analog of (9.16).

21.3.2 AIC, CV, and GCV

For linear regression models, AIC is

$$\text{AIC} = n \log(\widehat{\sigma}^2) + 2(1 + p),$$

where $1 + p$ is the number of parameters (intercept plus p slopes). For a linear smoother, AIC uses p_{eff} in place of $p + 1$, so that

$$\text{AIC}(\lambda) = n \log\{\widehat{\sigma}^2(\lambda)\} + 2\,p_{\text{eff}}.$$

We can then select λ by minimizing AIC.

The cross-validation or CV statistic is

$$\text{CV}(\lambda) = \sum_{i=1}^{n} \{Y_i - \widehat{Y}_{-i}(\lambda)\}^2$$

where, to prevent overfitting, $\widehat{Y}_{-1}(\lambda)$ is the ith fitted value computed with the ith observation deleted. One, of course, should choose a λ with a small value of $\text{CV}(\lambda)$.

The generalized cross-validation statistic (GCV) is

$$\text{GCV}(\lambda) = \frac{\|\boldsymbol{Y} - \widehat{\boldsymbol{Y}}(\lambda)\|^2}{(n - p_{\text{eff}})^2}. \tag{21.9}$$

$\text{GCV}(\lambda)$ can be computed more quickly than $\text{CV}(\lambda)$ and $\text{GCV}(\lambda)$ is a good approximation to $\text{CV}(\lambda)/n^2$ so minimizing GCV is another way to choose λ.

AIC and GCV can both be computed very quickly and usually give essentially the same amount of smoothing. In fact, it has been shown theoretically that both criteria should give similar estimates. Therefore, it does not matter much which is used, but GCV is more commonly used than AIC in nonparametric regression.

21.4 Polynomial Splines

The use of polynomial splines in nonparametric regression, as well as many other areas of mathematics, is based on the same principle as local polynomial regression—a smooth function can be accurately approximated locally by a low-degree polynomial. A pth-degree polynomial spline is constructed by piecing together pth-degree polynomials, so that they join together at specified locations called *knots*. The polynomials are spliced together, so that the spline has $p - 1$ continuous derivatives. The pth derivative of the spline is constant between knots and can jump at the knots.

21.4.1 Linear Splines with One Knot

We start simple, a linear spline with one knot. Figure 21.5a illustrates such a spline. This spline is defined as

$$f(x) = \begin{cases} 0.5 + 0.2x, & x < 2, \\ -0.5 + 0.7x, & x \geq 2. \end{cases}$$

Because $0.5 + 0.2x = 0.9 = -0.5 + 0.7x$ when $x = 2$, the two linear components are equal at the point $x = 2$, so that they join together there.

The point $x = 2$ where the spline switches from one linear function to the other is called a *knot*. A linear spline with a knot at the point t can be constructed as follows. The spline is defined to be $s(x) = a + bx$ for $x < t$ and $s(x) = c + dx$ for $x > t$. The parameters a, b, c, and d can be chosen arbitrarily except that they must satisfy the equality constraint

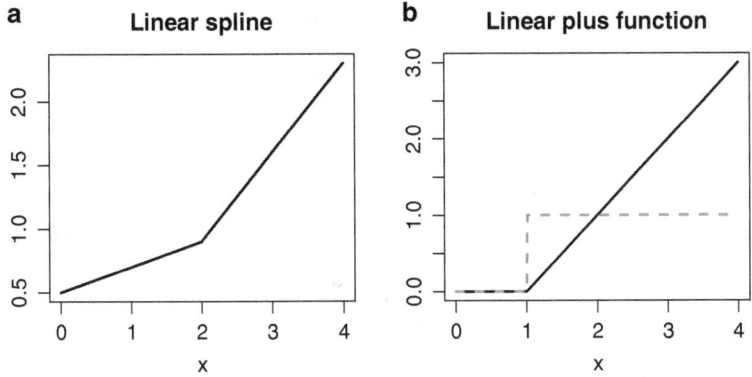

Fig. 21.5. *(a) Example of a linear spline with a knot at 2. (b) The linear plus function $(x - 1)_+$ with a knot at 1 (black) and its first derivative (red).*

$$a + bt = c + dt, \tag{21.10}$$

which assures us that the two lines join together at $x = t$. Solving for c in (21.10), we get $c = a + (b - d)t$. Substituting this expression for c into the definition of $s(x)$ and doing some rearranging, we have

$$s(x) = \begin{cases} a + bx, & x < t, \\ a + bx + (d - b)(x - t), & x \geq t. \end{cases} \tag{21.11}$$

Recall the definition that for any number y,

$$(y)_+ = \begin{cases} 0, & y < 0, \\ y, & y \geq 0. \end{cases}$$

By this definition,

$$(x - t)_+ = \begin{cases} 0, & x < t, \\ x - t, & x \geq t. \end{cases}$$

We call $(x - t)_+$ a linear *plus function* with a knot at t. It is also called a truncated line, though we will stick with "plus function." The spline $s(x)$ in (21.11) can be written using this plus function:

$$s(x) = a + bx + (d - b)(x - t)_+.$$

The plus function simplifies the problem of keeping the spline continuous at t. Figure 21.5b illustrates a linear plus function with a knot at 1 and its first derivative. Notice that

$$\frac{d}{dx}(x - t)_+ = \begin{cases} 0, & x < t, \\ 1, & x > t. \end{cases}$$

21.4.2 Linear Splines with Many Knots

Plus functions are very convenient when defining linear splines with more than one knot, because plus functions automatically join the component linear functions together so that the spline is continuous. For example, suppose we want a linear spline to have K knots, $t_1 < \cdots < t_K$, for the spline to equal $s(x) = \beta_0 + \beta_1 x$ for $x < t_1$, and for the first derivative of the spline to jump by the amount b_k at knot t_k, for $k = 1, \ldots, K$. Then the spline can be constructed from linear plus functions, one for each knot:

$$s(x) = \beta_0 + \beta_1 x + b_1(x - t_1)_+ + b_2(x - t_2)_+ + \cdots + b_K(x - t_K)_+.$$

Because the plus functions are continuous, the spline is the sum of continuous functions and is therefore continuous itself.

21.4.3 Quadratic Splines

A linear spline is continuous but has "kinks" at its knots, where its first derivative jumps. If we want a function without these kinks, we cannot use a linear spline. A quadratic spline is a function obtained by piecing together quadratic polynomials. More precisely, $s(x)$ is a quadratic spline with knots $t_1 < \cdots < t_K$ if $s(x)$ equals one quadratic polynomial to the left of t_1 and equals a second quadratic polynomial between t_1 and t_2, and so on. The quadratic polynomials are pieced together, so that the spline is continuous and, to guarantee no kinks, its first derivative is also continuous. Figure 21.6a shows a quadratic spline with a knot at 1. Notice that the function does not have a kink at the knot but changes from convex to concave there.

As with linear splines, continuity can be enforced by using plus functions. Define the quadratic plus function

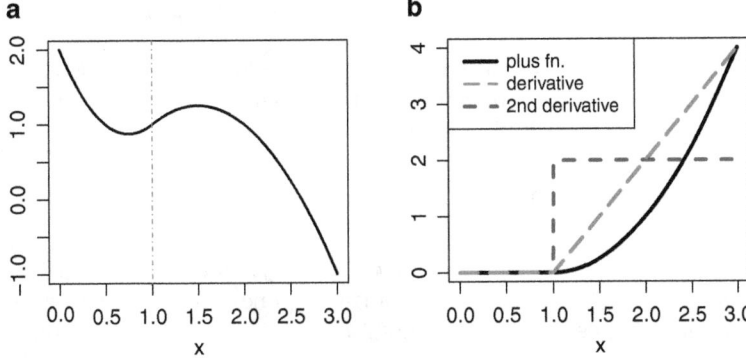

Fig. 21.6. *(a) Quadratic spline with a knot at 1. The dotted red vertical line marks the knot's location. (b) The quadratic plus function $(x-1)^2_+$ with a knot at 1 (black) and its first (red) and second (blue) derivatives.*

$$(x-t)^2_+ = \begin{cases} 0, & x < t, \\ (x-t)^2, & x \geq t. \end{cases}$$

Notice that $(x-t)^2_+$ equals $\{(x-t)_+\}^2$, not $\{(x-t)^2\}_+ = (x-t)^2$.

Figure 21.6b shows a quadratic plus function and its first and second derivatives. One can see that for $x > t$

$$\frac{d}{dx}(x-t)^2_+ = 2(x-t)_+$$

and

$$\frac{d^2}{dx^2}(x-t)^2_+ = 2(x-t)^0_+,$$

where $(x-t)^0_+ = \{(x-t)_+\}^0$, so that $(x-t)^0_+$ is the 0th-degree plus function

$$(x-t)^0_+ = \begin{cases} 0, & x < t, \\ 1, & x \geq t. \end{cases}$$

Therefore, the second derivative of $(x-t)^2_+$ jumps from 0 to 2 at the knot t.

A quadratic spline with knots $t_1 < \cdots < t_K$ can be written as

$$s(x) = \beta_0 + \beta_1 x + \beta_2 x^2 + b_1(x-t_1)^2_+ + b_2(x-t_2)^2_+ + \cdots + b_K(x-t_K)^2_+.$$

The second derivative of s jumps by the amount $2b_k$ at knot t_k for $k = 1, \ldots, K$.

21.4.4 pth Degree Splines

The way to define a general pth-degree spline with knots $t_1 < \cdots < t_K$ should now be obvious:

$$s(x) = \beta_0 + \beta_1 x + \cdots + \beta_p x^p + b_1(x - t_1)_+^p + \cdots + b_K(x - t_K)_+^p, \quad (21.12)$$

where, as we have seen for the specific case of $p = 2$, $(x-t)_+^p$ equals $\{(x-t)_+\}^p$. The first $p-1$ derivatives of s are continuous while the pth derivative takes a jump equal to $p!\,b_k$ at the kth knot.

21.4.5 Other Spline Bases

Given a degree p and knots $\kappa_1, \ldots, \kappa_K$, the polynomials $1, x, \ldots, x^p$ and plus functions $(x - \kappa_1)_+^p, \ldots, (x - \kappa_K)_+^p$ form a spline basis. What this means is that any pth degree spline with knots $\kappa_1, \ldots, \kappa_K$ is a linear combination of these basis functions. The basis of polynomials and plus functions is simple to understand, but is known to be numerically unstable if the number of knots is large. For this reason, other bases are often used for numerical computations. The B-spline basis is particular popular. It is assumed here that the reader will not be programming spline estimators from scratch but rather will be using spline software. Therefore, B-splines and other bases will not be covered here, but see Sect. 21.6 for further reading.

21.5 Penalized Splines

Because a pth degree spline with K knots has $1 + p + K$ parameters, an ordinary least-squares fit will usually overfit the data unless both p and K are kept small, for instance, $1 + p + K \leq 6$. (There is nothing especial about the number 6 and it is just being used as a rule of thumb. Any number between 5 and 10 would be equally good.) An example is the quadratic spline with one knot (so $1 + p + K = 4$) used as a forward-rate curve in Example 11.3. However, a spline with p and K both small is essentially a parametric model. To have the flexibility of a nonparametric model, that is, a wide range of potential values of p_{eff}, we need to have K large and find another way to avoid overfitting. Penalized least-squares estimation does this.

Let $\mu(x; \beta) = \boldsymbol{B}(x)^{\mathsf{T}} \boldsymbol{\beta}$ be a spline, where $\boldsymbol{\beta}$ is a vector of coefficients and $\boldsymbol{B}(x) = (B_1(x), \ldots, B_{1+p+K}(x))^{\mathsf{T}}$ is a spline basis. For example, $\boldsymbol{B}(x) = (1, x, \ldots, x_p, (x - \kappa_1)_+^p, \ldots, (x - \kappa_K)_+^p)$ if we use model (21.12). A penalized least-squares estimator minimizes over $\boldsymbol{\beta}$ the penalized sum of squares

$$\sum_{i=1}^{n} \{Y_i - \mu(X_i; \beta)\}^2 + \lambda \boldsymbol{\beta}^{\mathsf{T}} \boldsymbol{D} \boldsymbol{\beta}, \quad (21.13)$$

where $\boldsymbol{D}$ is a positive semidefinite matrix and $\lambda > 0$ is a penalty parameter.

A common choice of $\boldsymbol{D}$ has the i, jth element equal to

$$\int_a^b B_i^{(2)}(x) B_j^{(2)}(x) dx \quad (21.14)$$

for some $a < b$, such as, $a = \min(X_i)$ and $b = \max(X_i)$. Here $B_i^{(2)}(x)$ is the second derivative of $B_i(x)$. With this D,

$$\lambda \boldsymbol{\beta}^{\mathsf{T}} \boldsymbol{D} \boldsymbol{\beta} = \lambda \int_a^b \left\{ \mu^{(2)}(x; \boldsymbol{\beta}) \right\}^2 dx, \qquad (21.15)$$

Since $\mu^{(2)}(x)$ is the amount of curvature of μ at x, this choice of D penalizes wiggly functions and, if λ is chosen appropriately, prevents overfitting. If $\lambda = 0$, then there is no penalization and the effective number of parameters is $1 + p + K$. With this D, in the limit as $\lambda \to \infty$, any curvature at all receives an infinite penalty, so the estimator converges to a linear polynomial fit and the effective number of parameters converges to 2. Any value of p_{eff} between 2 and $1 + p + K$ is achievable by the some value of λ between the extremes of 0 and ∞.

Let X be the $n \times (1 + p + K)$ matrix with i, jth element $B_j(X_i)$ and let $Y = (Y_1, \ldots, Y_n)^{\mathsf{T}}$. The penalized least-squares estimate is

$$\widehat{\boldsymbol{\beta}}(\lambda) = \left(\boldsymbol{X}^{\mathsf{T}} \boldsymbol{X} + \lambda \boldsymbol{D} \right)^{-1} \boldsymbol{X}^{\mathsf{T}} \boldsymbol{Y}, \qquad (21.16)$$

which is obtained by setting the gradient of (21.13) equal to zero and solving. The fitted values are

$$\widehat{\boldsymbol{Y}}(\lambda) = \boldsymbol{X}\widehat{\boldsymbol{\beta}}(\lambda) = \left\{ \boldsymbol{X}(\boldsymbol{X}^{\mathsf{T}}\boldsymbol{X} + \lambda\boldsymbol{D})^{-1}\boldsymbol{X}^{\mathsf{T}} \right\} \boldsymbol{Y} = \boldsymbol{H}(\lambda)\,\boldsymbol{Y}, \qquad (21.17)$$

where $\boldsymbol{H}(\lambda) = \left\{ \boldsymbol{X}(\boldsymbol{X}^{\mathsf{T}}\boldsymbol{X} + \lambda\boldsymbol{D})^{-1}\boldsymbol{X}^{\mathsf{T}} \right\}$ is the smoother matrix.

21.5.1 Cubic Smoothing Splines

A very widely used nonparametric regression estimator is the cubic smoothing spline. This estimator uses a knot at each unique value of $\{X_1, \ldots, X_n\}$ and the second-derivative penalty in (21.15). Using this many knots is not really necessary, and a variation on the cubic smoothing spline also uses penalty (21.15) but fewer knots. The knots could be equally-spaced or at selected quantiles of $\{X_1, \ldots, X_n\}$.

The function `smooth.spline()` in R fits a cubic smoothing spline if the argument `all.knots` is `TRUE` or if $n < 50$. If `all.knots = FALSE` and $n > 49$, then it uses less than the full set of knots.

21.5.2 Selecting the Amount of Penalization

The penalty parameter λ determines the amount of smoothing and can be chosen by AIC or GCV. Another popular method for choosing λ is REML (restricted maximum likelihood). REML is based on a so-called mixed model, where some of the spline coefficients are random variables. A description of mixed models and REML is beyond the scope of this book, but the interested reader may consult the references in Sect. 21.6.

Example 21.2. Estimating the drift and volatility for the evolution of the risk-free returns

In this example, we return to estimating the drift and squared volatility functions for the evolution of the risk-free returns. Three estimators will be used: local linear, local quadratic, and a penalized spline. The R code is:

```
1  library(Ecdat)
2  library(KernSmooth)
3  library(locfit)
4  library(mgcv)
5  data(Capm)
6  attach(Capm)
7  n = length(rf)
8  year = seq(1960.125,2003,length=n)
9  diffrf=diff(Capm$rf)
10 rf_lag = rf[1:(n-1)]
11 log_rf_lag = log(rf_lag)
12 ll_mu <- locpoly(rf_lag, diffrf, bandwidth = dpill(rf_lag,diffrf))
13 muhat = spline(ll_mu$x, ll_mu$y, xout = rf_lag)$y
14 epsilon_sqr = (diffrf - muhat)^2
15 ll_sig <- locpoly(rf_lag, epsilon_sqr,
16    bandwidth = dpill(rf_lag, epsilon_sqr) )
17 gam_mu = gam(diffrf ~ s(rf_lag, bs = "cr"), method = "REML")
18 epsilon_sqr = (diffrf-gam_mu$fit)^2
19 gam_sig = gam(epsilon_sqr ~ s(rf_lag, bs = "cr"), method = "REML")
20 locfit_mu = locfit(diffrf ~ rf_lag)
21 epsilon_sqr = (diffrf - fitted(locfit_mu))^2
22 locfit_sig = locfit(epsilon_sqr ~ rf_lag)
23 std_res = (diffrf - fitted(locfit_mu)) / sqrt(fitted(locfit_sig))
24 min(rf_lag[(gam_mu$fit < 0)])
25 orrf = order(rf_lag)
26 pdf("riskfree02.pdf", width = 8, height = 4)
27 par(mfrow=c(1, 2))
28 plot(rf_lag[orrf], gam_mu$fit[orrf], type = "l", lwd = 3, lty = 1,
29    xlab = "lagged rate", ylab = "change in rate", main = "(a)")
30 lines(ll_mu$x,ll_mu$y, lwd = 3, lty = 2, col = "red")
31 lines(locfit_mu, lwd = 3, lty = 3, col = "blue")
32 legend(0.1, -0.05, c("spline", "local linear", "local quadratic"),
33    lty = c(1, 2, 3), cex = 0.85, lwd = 3,
34    col = c("black", "red", "blue"))
35 rug(rf_lag)
36 abline(h = 0, lwd = 2)
37 plot(rf_lag[orrf], gam_sig$fit[orrf], type="l", lwd = 3, lty = 1,
38    ylim = c(0, 0.03), xlab = "lagged rate",
39    ylab = "squared residual", main = "(b)")
40 lines(ll_sig$x,ll_sig$y, lwd = 3, lty = 2, col = "red")
41 lines(locfit_sig, lwd = 3, lty = 3, col = "blue")
42 abline(h = 0, lwd = 2)
```

```
43  legend("topleft", c("spline", "local linear", "local quadratic"),
44     lty = c(1, 2, 3), cex = 0.85, lwd = 3,
45     col = c("black", "red", "blue"))
46  rug(rf_lag)
47  graphics.off()
```

The first estimator, local linear, is computed at line 12 using the function locpoly() in R's KernSmooth package. The dpi plug-in bandwidth selector is computed using the function dpill() in this package.[2]

In the R code, the changes in the risk-free returns (diffrf) are regressed on the lagged returns (rf_lag) to estimate the drift. The local linear estimator is computed on an equally-spaced grid, and to compute residuals the function spline() is used at line 13 to interpolate the fit to the observed values of rf_lag. Finally, the squared residuals (epsilon_sqr) computed at line 14 are regressed at lines 15–16 on the lagged returns to estimate the squared volatility function. The estimated drift function is in the object ll_mu and the estimated squared volatility function is in ll_sig.

The penalized spline estimator is computed at line 17 by the gam() function in the mgcv package. The specification bs = "cr" requests a cubic spline fit with penalty (21.15). The REML method is used to select the amount of smoothing.

The local quadratic estimator is computed at line 20 with the function locfit() in R's locfit package. Spline interpolation is not necessary here, since with locfit() the fitted values can be computed with the fitted function.

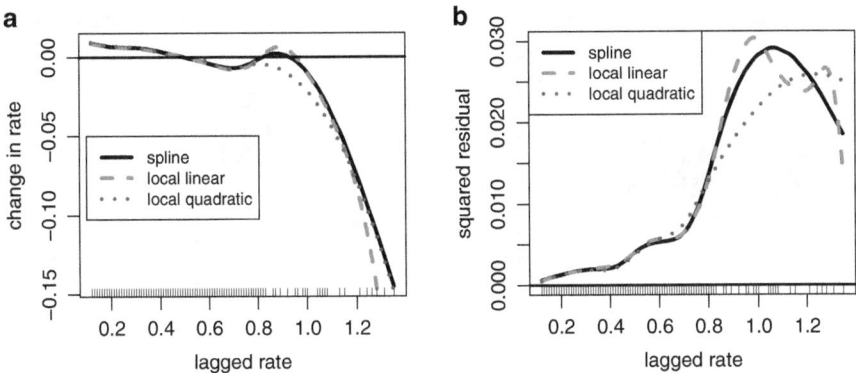

Fig. 21.7. *Risk-free monthly returns. (a) Estimates of the drift function. (b) Estimates of the squared volatility function.*

---

[2] "dpill" means "direct plug-in, local linear."

All three estimated drift functions are shown in Fig. 21.7a and the squared volatility function estimates are in Fig. 21.7b.

The drift functions have a general decreasing trend and are negative to the right of 0.51 (approximately), except that the estimates have humps around 0.9–1.0 and the spline and local linear estimates are slightly positive at this hump. It is likely that the hump is due to random variation, which increases as one moves from left to right (see Fig. 21.1). If we use the local quadratic fit, then the estimated drift is positive to the left of 0.51 and negative to the right of 0.51. The drift will cause reversion to a mean of 0.51, which is an annual rate of $6.12\% = (12)(0.51)\%$. The Chan et al. (1992) drift function, $\mu(r) = \beta(r - \alpha)$, is also mean-reverting, but linear. In contrast, the local quadratic estimated drift function in Fig. 21.7 is nonlinear and shows much faster reversion to the mean when the rate is high.

The squared volatility estimates show that volatility increases with the rate, at least to a point. For very high rates, the estimated volatility function becomes decreasing. There is not enough data with extremely high rates to tell if this phenomenon is "real" or due to random estimation error. The extremely high rates occurred only for the brief period in the early 1980s; see Fig. 21.1a.

The standardized residuals $\{\Delta r_t - \widehat{\mu}(r_{t-1})\}/\widehat{\sigma}(r_{t-1})$ show negative serial correlation and GARCH-type volatility clustering; see Fig. 21.8. Neither of these is surprising. Negative lag-1 autocorrelation is common in a differenced series and volatility clustering is certainly to be expected in any financial time series. This case study could be continued by fitting an ARMA/GARCH model to the standardized residuals.   □

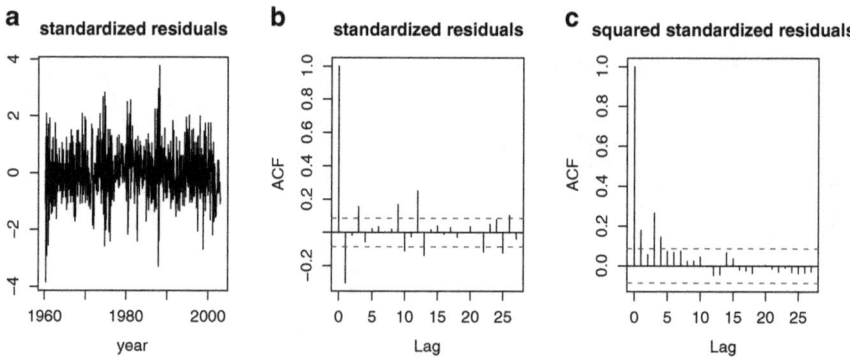

Fig. 21.8. *Risk-free monthly returns. Residual analysis. (a) Time series plot of standardized residuals. (b) ACF of standardized residuals. (c) ACF of squared standardized residuals.*

Example 21.3. Spline estimation of a forward rate

This example used the STRIPS data that were already analyzed in Example 11.3. In that example, an unpenalized spline was fit to the bond prices by nonlinear regression, and smoothness was controlled by using only knot.

The function `gam()` in the `mgcv` is a powerful tool that can fit a wide variety of spline models with penalties. In this example, `gam()` is used to fit a cubic spline to the empirical forward rates that are defined at the beginning of Sect. 11.3. The code is below.

```
1  dat = read.table("strips_dec95.txt", header = T)
2  T = dat$T
3  n = length(T)
4  ord = order(T)
5  T = T[ord]
6  price = dat$price[ord]
7  Int_F = - log(price) + log(100)
8  emp_forward = diff(Int_F)/diff(T)
9  library(mgcv)
10 X = T[-1]
11 fit_gam = gam(emp_forward ~ s(X, bs = "cr"))
12 pred_gam = predict(fit_gam, as.data.frame(X)   )
13 pdf("forward_spline.pdf", width = 6, height = 5)
14 plot(X, emp_forward, xlab = "maturity", ylab = "forward rate")
15 lines(X, pred_gam, col = "red", lwd = 2)
16 graphics.off()
```

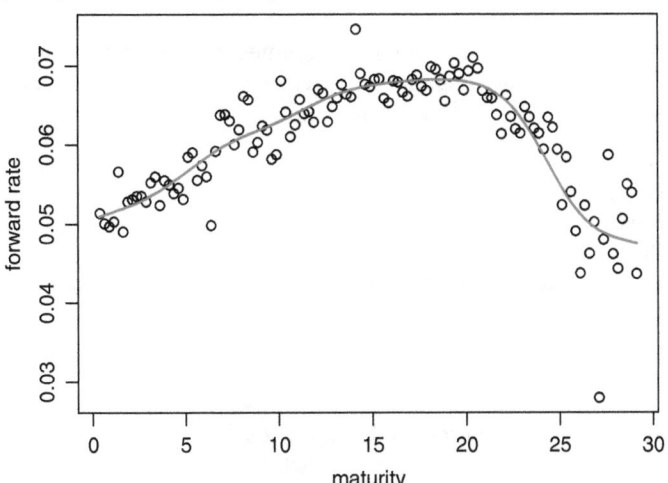

Fig. 21.9. *Empirical forward rates (circle) with a spline fit.*

Figure 21.9 shows the empirical forward rates and the cubic spline fit to them. Notice that the spline goes through the empirical forward rates with little sign of bias and yet is smooth.   □

21.6 Bibliographic Notes

Ruppert, Wand, and Carroll (2003) and Wood (2006) offer comprehensive introductions to nonparametric and semiparametric modeling and their applications. Wand and Jones (1995) and Fan and Gijbels (1996) are good sources of information about local polynomial regression. REML is discussed in detail by Ruppert, Wand, and Carroll (2003) and Wood (2006) . Wasserman (2006) is an interesting modern synthesis of nonparametric estimation. Wood (2006) is a good introduction to his mgcv package.

21.7 R Lab

21.7.1 Additive Model for Wages, Education, and Experience

This section uses the Current Population Survey data in the CPS1988 data set introduced in Sect. 10.4.1. We will fit spline effects for both predictors, education and experience. This is easily done with the gam() function in the mgcv package. The model being fit is

$$\log(\text{wage}) = \beta_0 + s_1(\text{education}) + s_2(\text{experience}) + \beta_1\text{ethnicity} + \epsilon_i,$$

where β_0 is the intercept, s_1 and s_2 are splines, ethnicity is 0 for Caucasians and 1 for African Americans, and ϵ_i is white noise. To fit this model, print its summary, and plot the estimates of s_1 and s_2, run:

```
library(AER)
library(mgcv)
data(CPS1988)
attach(CPS1988)
fitGam = gam(log(wage)~s(education)+s(experience)+ethnicity)
summary(fitGam)
par(mfrow=c(1,2))
plot(fitGam)
```

Problem 1 *What are the estimates of β_0 and β_1?*

Problem 2 *Describe the shapes of s_1 and s_2.*

21.7.2 An Extended CKLS Model for the Short Rate

In this section, we use splines to extend the CKLS model in Sect. 11.12 by
letting the drift parameters a and θ vary with time so that

$$\mu(t, r) = a(t)\{\theta(t) - r\}. \tag{21.18}$$

One could also let the volatility parameters σ and γ vary as well with t, but,
for simplicity, we will not do that here. We will fit this model with $a(t)$ being
linear in time and $\theta(t)$ being a piecewise linear spline. [Letting both $a(t)$ and
$\theta(t)$ be splines can lead to unstable estimates, so we will restrict $a(t)$ to be
linear.] First, read in the data, and then create the knots and the truncated
line basis functions.

```
#  CKLS, extended
library(Ecdat)
data(Irates)
r1 = Irates[,1]
n = length(r1)
lag_r1 = lag(r1)[-n]
delta_r1 = diff(r1)
n = length(lag_r1)
knots = seq(from=1950,to=1985,length=10)
t = seq(from=1946,to =1991+2/12,length=n)
X1 = outer(t,knots,FUN="-")
X2 = X1 * (X1>0)
X3 = cbind(rep(1,n), (t - 1946),X2)
m2 = dim(X3)[2]
m = m2 - 1
```

Problem 3 *How many knots are being used here? What does the* outer()
function do here? What is done by the statement X2 = X1 * (X1>0)*?
Describe what is in the variable* X3.

Now fit the CKLS model with time-varying drift.

```
nlmod_CKLS_ext = nls(delta_r1 ~ X3[,1:2]%*%a *
   (X3%*%theta-lag_r1),
   start=list(theta = c(10,rep(0,m)),
   a=c(.01,0)),control=list(maxiter=200))
AIC(nlmod_CKLS_ext)
param4 = summary(nlmod_CKLS_ext)$parameters[,1]
par(mfrow=c(1,3))
plot(t,X3%*%param4[1:m2],ylim=c(0,16),ylab="rate",
   main="(a)",col="red",type="l",lwd=2)
lines(t,lag_r1)
legend("topleft",c("theta(t)","lagged rate"),lwd=c(2,1),
   col=c("red","black"))
```

```
plot(t,X3[,1:2]%*%param4[(m2+1):(m2+2)],ylab="a(t)",
    col="red",type="l",lwd=2,main="(b)")

res_sq = residuals(nlmod_CKLS_ext)^2
nlmod_CKLS_ext_res <- nls(res_sq ~ A*lag_r1^B,
    start=list(A=.2,B=1/2) )

plot(lag_r1,sqrt(res_sq),pch=5,ylim=c(0,6),ylab="",main="(c)")
lines(lag_r1,sqrt(fitted(nlmod_CKLS_ext_res)),
    lw=3,col="red",type="l")
legend("topleft",c("abs res","volatility fn"),lty=c(NA,1),
    pch=c(5,NA),col=c("black","red"),lwd=1:2)
```

Problem 4 *Explain why* X3[,1:2]%*%a *is a linear function but* X3%*%theta *is a spline.*

Problem 5 *What is the interpretation of a time-varying* θ *? Note that in panel (a),* θ *seems to track the interest rate. Does this make sense? Why or why not?*

Problem 6 *Would you accept or reject the null hypothesis that* $a(t)$ *is constant, that is, that the slope of the linear function* $a(t)$ *is zero? Justify your answer.*

21.8 Exercises

1. A linear spline $s(t)$ has knots at 1, 2, and 3. Also, $s(0) = 1$, $s(1) = 1.3$, $s(2) = 5.5$, $s(4) = 6$, and $s(5) = 6$.
   (a) What is $s(0.5)$?
   (b) What is $s(3)$?
   (c) What is $\int_2^4 s(t) \, dt$?
2. Suppose that (21.1) holds with $\mu(r) = 0.1(0.035 - r)$ and $\sigma(r) = 2.3r$.
   (a) What is the expected value of r_t given that $r_{t-1} = 0.04$?
   (b) What is the variance of r_t given that $r_{t-1} = 0.02$?
3. Let the spline $s(x)$ be defined as

$$s(x) = (x)_+ - 3(x - 1)_+ + (x - 2)_+.$$

   (a) Is $s(x)$ either a probability density function (pdf) or a cumulative distribution function (cdf)? Explain your answer.
   (b) If X is a random variable and s is its pdf or cdf [whichever is the correct answer in (a)], then what is the 90th percentile of X?
4. Let s be the spline

$$s(x) = 1 + 0.65x + x^2 + (x - 1)_+^2 + 0.6(x - 2)_+^2.$$

   (a) What are $s(1.5)$ and $s'(1.5)$?
   (b) What is $s''(2.2)$?

References

Chan, K. C., Karolyi, G. A., Longstaff, F. A., and Sanders, A. B. (1992) An empirical comparison of alternative models of the short-term interest rate. *Journal of Finance*, **47**, 1209–1227.

Cox, J. C., Ingersoll, J. E., and Ross, S. A. (1985) A theory of the term structure of interest rates. *Econometrica*, **53**, 385–407.

Fan, J., and Gijbels, I. (1996) *Local Polynomial Modelling and Its Applications*, Chapman & Hall, London.

Merton, R. C. (1973) Theory of rational option pricing. *Bell Journal of Economics and Management Science*, **4**, 141–183.

Ruppert, D., Sheather, S., and Wand, M. P. (1995) An effective bandwidth selector for local least squares kernel regression, *Journal of the American Statistical Association*, **90**, 1257–1270.

Ruppert, D., Wand, M. P., and Carroll, R. J. (2003) *Semiparametric Regression*, Cambridge University Press, Cambridge.

Vasicek, O. A. (1977) An equilibrium characterization of the term structure. *Journal of Financial Economics*, **5**, 177–188.

Wand, M. P., and Jones, M. C. (1995) *Kernel Smoothing*, Chapman & Hall, London.

Wasserman, L. (2006) *All of Nonparametric Statistics*, Springer, New York.

Wood, S. (2006) *Generalized Additive Models: An Introduction with R*, Chapman & Hall, Boca Raton, FL.

Yau, P., and Kohn, R. (2003) Estimation and variable selection in nonparametric heteroskedastic regression. *Statistics and Computing*, **13**, 191–208.

A

Facts from Probability, Statistics, and Algebra

A.1 Introduction

It is assumed that the reader is already familiar with the basics of probability, statistics, matrix algebra, and other mathematical topics needed in this book, and so the goal of this appendix is merely to provide a quick review and cover some more advanced topics that may not be familiar.

A.2 Probability Distributions

A.2.1 Cumulative Distribution Functions

The *cumulative distribution function (CDF)* of Y is defined as

$$F_Y(y) = P\{Y \leq y\}.$$

If Y has a PDF f_Y, then

$$F_Y(y) = \int_{-\infty}^{y} f_Y(u)\, du.$$

Many CDFs and PDFs can be calculated by computer software packages, for instance, `pnorm()`, `pt()`, and `pbinom()` in R calculate, respectively, the CDF of a normal, t, and binomial random variable. Similiarly, `dnorm()`, `dt()`, and `dbinom()` calculate the PDFs of these distributions.

© Springer Science+Business Media New York 2015   669
D. Ruppert, D.S. Matteson, *Statistics and Data Analysis for Financial Engineering*, Springer Texts in Statistics,
DOI 10.1007/978-1-4939-2614-5

A.2.2 Quantiles and Percentiles

If the CDF $F(y)$ of a random variable Y is continuous and strictly increasing, then it has an inverse function F^{-1}. For each q between 0 and 1, $F^{-1}(q)$ is called the q-*quantile* or $100q$th percentile.

The median is the 50 % percentile or 0.5-quantile. The 25 % and 75 % percentiles (0.25- and 0.75-quantiles) are called the first and third quartiles and the median is the second quartile. The three quartiles divide the range of a continuous random variable into four groups of equal probability. Similarly, the 20 %, 40 %, 60 %, and 80 % percentiles are called quintiles and the 10 %, 20 %, ..., 90 % percentiles are called deciles.

For any CDF F, invertible or not, the *pseudo-inverse* is defined as

$$F^-(x) = \inf\{y : F(y) \geq x\}.$$

Here "inf" is the infinum or greatest lower bound of a set; see Appendix A.5. For any q between 0 and 1, the qth quantile will defined as $F^-(q)$. If F is invertible, then $F^{-1} = F^-$, so this definition of quantile agree with the one for invertible CDFs. F^- is often called the *quantile function*.

Sometimes a $(1 - \alpha)$-quantile is called an α-upper quantile, to emphasize the amount of probability above the quantile. In analogy, a quantile might also be referred to as lower quantile.

Quantiles are said to "respect transformations" in the following sense. If Y is a random variable whose q-quantile equals y_q, if g is a strictly increasing function, and if $X = g(Y)$, then $g(y_q)$ is the q-quantile of X; see (A.5).

A.2.3 Symmetry and Modes

A probability density function (PDF) f is said to be *symmetric* about μ if $f(\mu - y) = f(\mu + y)$ for all y. A *mode* of a PDF is a local maximum, that is a value y such that for some $\epsilon > 0$, $f(y) > f(x)$ if $y - \epsilon < x < y$ or $y < x < y + \epsilon$. A PDF with one mode is called *unimodal*, with two modes *bimodal*, and with two or more modes *multimodal*.

A.2.4 Support of a Distribution

The support of a *discrete* distribution is the set of all y that have a positive probability. More generally, a point y is in the support of a distribution if, for every $\epsilon > 0$, the interval $(y - \epsilon, y + \epsilon)$ has positive probability. For example, the support of a normal distribution is $(-\infty, \infty)$, the support of a gamma or log-normal distribution is $[0, \infty)$, and the support of a binomial(n, p) distribution is $\{0, 1, 2, \ldots, n\}$ provided $p \neq 0, 1$.[1]

---

[1] It is assumed that most readers are already familiar with the normal, gamma, log-normal, and binomial distributions. However, these distributions will be discussed in some detail later.

A.3 When Do Expected Values and Variances Exist?

The expected value of a random variable could be infinite or not exist at all. Also, a random variable need not have a well-defined and finite variance. To appreciate these facts, let Y be a random variable with density f_Y. The expectation of Y is

$$\int_{-\infty}^{\infty} y f_Y(y) dy$$

provided that this integral is defined. If

$$\int_{-\infty}^{0} y f_Y(y) dy = -\infty \text{ and } \int_{0}^{\infty} y f_Y(y) dy = \infty, \qquad (A.1)$$

then the expectation is, formally, $-\infty + \infty$, which is not defined, so the expectation does not exist. If integrals in (A.1) are both finite, then $E(Y)$ exists and equals the sum of these two integrals. The expectation can exist but be infinite, because if

$$\int_{-\infty}^{0} y f_Y(y) dy = -\infty \text{ and } \int_{0}^{\infty} y f_Y(y) dy < \infty,$$

then $E(Y) = -\infty$, and if

$$\int_{-\infty}^{0} y f_Y(y) dy > -\infty \text{ and } \int_{0}^{\infty} y f_Y(y) dy = \infty,$$

then $E(Y) = \infty$.

If $E(Y)$ is not defined or is infinite, then the variance that involves $E(Y)$ cannot be defined either. If $E(Y)$ is defined and finite, then the variance is also defined. The variance is finite if $E(Y^2) < \infty$; otherwise the variance is infinite.

The nonexistence of finite expected values and variances is of importance for modeling financial markets data, because, for example, the popular GARCH models discussed in Chap. 14 need not have finite expected values and variances. Also, t-distributions that, as demonstrated in Chap. 5, can provide good fits to equity returns may have nonexistent means or variances.

One could argue that any variable Y derived from financial markets will be bounded, that is, that there is a constant $M < \infty$ such that $P(|Y| \leq M) = 1$. In this case, the integrals in (A.1) are both finite, in fact at most M, and $E(Y)$ exists and is finite. Also, $E(Y^2) \leq M^2$, so the variance of Y is finite. So should we worry at all about the mathematically niceties of whether expected values and variances exist and are finite? The answer is that we should. A random variable might be bounded in absolute value by a very large constant M and yet, if M is large enough, behave much like a random variable that does not have an expected value or has an expected value that is infinite or has a finite expected value but an infinite variance. This can be seen in the simulations of GARCH processes. Results from computer simulations are bounded by the maximum size of a number in the computer. Yet these simulations behave as if the variance were infinite.

A.4 Monotonic Functions

The function g is increasing if $g(x_1) \leq g(x_2)$ whenever $x_1 < x_2$ and strictly increasing if $g(x_1) < g(x_2)$ whenever $x_1 < x_2$. Decreasing and strictly decreasing are defined similarly, and g is (strictly) monotonic if it is either (strictly) increasing or (strictly) decreasing.

A.5 The Minimum, Maximum, Infinum, and Supremum of a Set

The minimum and maximum of a set are its smallest and largest values, if these exists. For example, if $A = \{x : 0 \leq x \leq 1\}$, then the minimum and maximum of A are 0 and 1. However, not all sets have a minimum or a maximum, for example, $B = \{x : 0 < x < 1\}$ has neither a minimum nor a maximum. Every set as an infinum (or inf) and a supremum (or sup). The inf of a set C is the largest number that is less than or equal to all elements of C. Similarly, the sup of C is the smallest number that is greater than or equal to every element of C. The set B just defined has an inf of 0 and a sup of 1. The following notation is standard: $\min(C)$ and $\max(C)$ are the minimum and maximum of C, if these exist, and $\inf(C)$ and $\sup(C)$ are the infinum and supremum.

A.6 Functions of Random Variables

Suppose that X is a random variable with PDF $f_X(x)$ and $Y = g(X)$ for g a strictly increasing function. Since g is strictly increasing, it has an inverse, which we denote by h. Then Y is also a random variable and its CDF is

$$F_Y(y) = P(Y \leq y) = P\{g(X) \leq y\} = P\{X \leq h(y)\} = F_X\{h(y)\}. \quad (A.2)$$

Differentiating (A.2), we find the PDF of Y:

$$f_Y(y) = f_X\{h(y)\}h'(y). \quad (A.3)$$

Applying a similar argument to the case, where g is strictly decreasing, one can show that whenever g is strictly monotonic, then

$$f_Y(y) = f_X\{h(y)\}|h'(y)|. \quad (A.4)$$

Also from (A.2), when g is strictly increasing, then

$$F_Y^{-1}(p) = g\{F_X^{-1}(p)\}, \qquad (A.5)$$

so that the pth quantile of Y is found by applying g to the pth quantile of X. When g is strictly decreasing, then it maps the pth quantile of X to the $(1 - p)$th quantile of Y.

Result A.1 *Suppose that $Y = a + bX$ for some constants a and $b \neq 0$. Let $g(x) = a + bx$, so that the inverse of g is $h(y) = (y - a)/b$ and $h'(y) = 1/b$. Then*

$$
\begin{aligned}
F_Y(y) &= F_X\{b^{-1}(y - a)\}, \quad b > 0, \\
&= 1 - F_X\{b^{-1}(y - a)\}, \quad b < 0, \\
f_Y(y) &= |b|^{-1} f_X\{b^{-1}(y - a)\},
\end{aligned}
$$

and

$$
\begin{aligned}
F_Y^{-1}(p) &= a + bF_X^{-1}(p), \quad b > 0 \\
&= a + bF_X^{-1}(1 - p), \quad b < 0.
\end{aligned}
$$

A.7 Random Samples

We say that $\{Y_1, \ldots, Y_n\}$ is a *random sample* from a probability distribution if they each have that probability distribution and are independent. In this case, we also say that they are *independent and identically distributed* or simply i.i.d. The probability distribution is often called the population and its expected value, variance, CDF, and quantiles are called the *population mean*, *population variance*, *population CDF*, and *population quantiles*. It is worth mentioning that the population is, in effect, infinite. There is a statistical theory of sampling, usually without replacement, from finite populations, but sampling of this type will not concern us here. Even in cases where the population is finite, such as, when sampling house prices, the population is usually large enough, so that it can be treated as infinite.

If $Y_1, \ldots, Y_n$ is a sample from an unknown probability distribution, then the population mean can be estimated by the *sample mean*

$$\overline{Y} = n^{-1} \sum_{i=1}^{n} Y_i, \qquad (A.6)$$

and the population variance can be estimated by the *sample variance*

$$s_Y^2 = \frac{\sum_{i=1}^{n}(Y_i - \overline{Y})^2}{n - 1}. \qquad (A.7)$$

The reason for the denominator of $n-1$ rather than n is discussed in Sect. 5.9. The *sample standard deviation* is s_Y, the square root of s_Y^2.

A.8 The Binomial Distribution

Suppose that we conduct n experiments for some fixed (nonrandom) integer n. On each experiment there are two possible outcomes called "success" and "failure"; the probability of a success is p, and the probability of a failure is $q = 1 - p$. It is assumed that p and q are the same for all n experiments. Let Y be the total number of successes, so that Y will equal $0, 1, 2, \ldots$, or n. If the experiments are independent, then

$$P(Y = k) = \binom{n}{k} p^k q^{n-k} \quad \text{for } k = 0, 1, 2, \ldots, n,$$

where

$$\binom{n}{k} = \frac{n!}{k!(n-k)!}.$$

The distribution of Y is called the *binomial distribution* and denoted Binomial(n, p). The expected value of Y is np and its variance is npq. The Binomial($1, p$) distribution is also called the Bernoulli distribution and its density is

$$P(Y = y) = p^y (1 - p)^{1-y}, \quad y = 0, 1. \tag{A.8}$$

Notice that p^y is equal to either p (when $y = 1$) or 1 (when $y = 0$), and similarly for $(1 - p)^{1-y}$.

The functions pbinom(), dbinom(), qbinom(), and rbinom() compute binomial CDFs, pdfs, quantiles, and random numbers, respectively. For example,

```
> pbinom(3,6,0.5)
[1] 0.65625
```

shows that the probability of 3 or less heads in 6 tosses of a fair coin is 0.65625.

A.9 Some Common Continuous Distributions

A.9.1 Uniform Distributions

The uniform distribution on the interval (a, b) is denoted by Uniform(a, b) and has PDF equal to $1/(b - a)$ on (a, b) and equal to 0 outside this interval. It is easy to check that if Y is Uniform(a, b), then its expectation is

$$E(Y) = \frac{1}{b-a} \int_a^b Y \, dY = \frac{a+b}{2},$$

which is the midpoint of the interval. Also,

$$E(Y^2) = \frac{1}{b-a} \int_a^b Y^2 \, dY = \frac{Y^3|_a^b}{3(b-a)} = \frac{b^2 + ab + a^2}{3}.$$

Therefore,

$$\sigma_Y^2 = E(Y^2) - \{E(Y)\}^2 = \frac{b^2 + ab + a^2}{3} - \left(\frac{a+b}{2}\right)^2 = \frac{(b-a)^2}{12}.$$

Reparameterization means replacing the parameters of a distribution by an equivalent set. The uniform distribution can be reparameterized by using $\mu = (a+b)/2$ and $\sigma = (b-a)/\sqrt{12}$ as the parameters. Then μ is a location parameter and σ is the scale parameter. Which parameterization of a distribution is used depends upon which aspects of the distribution one wishes to emphasize. The parameterization (a, b) of the uniform specifies its endpoints while the parameterization (μ, σ) gives the mean and standard deviation. One is free to move back and forth between two or more parameterizations, using whichever is most useful in a given context. The uniform distribution does not have a shape parameter since the shape of its density is always rectangular.

The functions `punif()`, `dunif()`, `qunif()`, and `runif()` compute uniform CDFs, pdfs, quantiles, and random numbers, respectively. For example,

```
> runif(3,0,5)
[1] 1.799252 4.003232 3.978002
```

are three random numbers uniformly distributed between 0 and 5.

A.9.2 Transformation by the CDF and Inverse CDF

If Y has a continuous CDF F, then $F(Y)$ has a Uniform(0,1) distribution. $F(Y)$ is often called the *probability transformation* of Y. This fact is easy to see if F is strictly increasing, since then F^{-1} exists, so that

$$P\{F(Y) \leq y\} = P\{Y \leq F^{-1}(y)\} = F\{F^{-1}(y)\} = y. \tag{A.9}$$

The result holds even if F is not strictly increasing, but the proof is slightly more complicated. It is only necessary that F be continuous.

If U is Uniform(0,1) and F is a CDF, then $Y = F^-(U)$ has F as its CDF. Here F^- is the pseudo-inverse of F. This can be proved easily when F is continuous and strictly increasing, since then $F^{-1} = F^-$ and

$$P(Y \leq y) = P\{F^{-1}(U) \leq y\} = P\{Y \leq F(y)\} = F(y).$$

In fact, the result holds for any CDF F, but it is more difficult to prove in the general case. $F^-(U)$ is often called the *quantile transformation* since F^- is the quantile function.

A.9.3 Normal Distributions

The *standard normal distribution* has the familiar bell-shaped density

$$\phi(y) = \frac{1}{\sqrt{2\pi}} \exp\left(-y^2/2\right), \quad -\infty < y < \infty.$$

The standard normal has mean 0 and variance 1. If Z is standard normal, then the distribution of $\mu + \sigma Z$ is called the *normal distribution with mean μ and variance σ^2* and denoted by $N(\mu, \sigma^2)$. By Result A.1, the $N(\mu, \sigma^2)$ density is

$$\frac{1}{\sigma}\phi\left(\frac{y-\mu}{\sigma}\right) = \frac{1}{\sqrt{2\pi}\sigma} \exp\left\{-\frac{(y-\mu)^2}{2\sigma^2}\right\}. \qquad (A.10)$$

The parameter μ is a location parameter and σ is a scale parameter. The normal distribution does not have a shape parameter since its density is always the same bell-shaped curve.[2] The standard normal CDF is

$$\Phi(y) = \int_{-\infty}^{y} \phi(u)du.$$

Φ can be evaluated using software such as R's pnorm function. If Y is $N(\mu, \sigma^2)$, then since $Y = \mu + \sigma Z$, where Z is standard normal, by Result A.1,

$$F_Y(y) = \Phi\{(y-\mu)/\sigma\}. \qquad (A.11)$$

Normal distribution are also called Gaussian distributions after the great German mathematician Carl Friedrich Gauss.

Normal Quantiles

The q-quantile of the $N(0, 1)$ distribution is $\Phi^{-1}(q)$ and, more generally, the q-quantile of an $N(\mu, \sigma^2)$ distribution is $\mu + \sigma\Phi^{-1}(q)$. The α-upper quantile of Φ, that is, $\Phi^{-1}(1-\alpha)$, is denoted by z_α. As shown later, z_α is widely used for confidence intervals.

For example, $z_{0.1}$ and $z_{0.01}$ are 1.282 and 2.326, respectively, as can be seen in the following R output:

```
> round(qnorm(c(0.1, 0.01), lower.tail = FALSE), 3)
[1] 1.282 2.326
```

A.9.4 The Lognormal Distribution

If Z is distributed $N(\mu, \sigma^2)$, then $Y = \exp(Z)$ is said to have a Lognormal(μ, σ^2) distribution. In other words, Y is *lognormal* if its logarithm is normally

---

[2] In contrast, a t-density is also a bell curve, but the exact shape of the bell depends on a shape parameter, the degrees of freedom which is a tail index.

lognormal densities

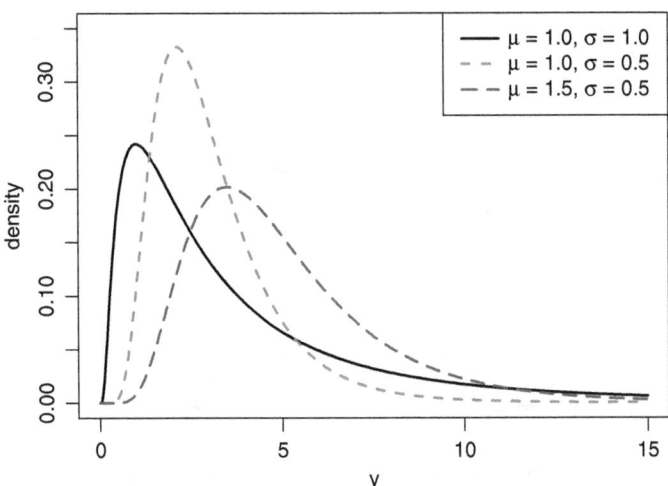

Fig. A.1. *Examples of lognormal probability densities. Here μ and σ are the log-mean and log-standard deviation, that is, the mean and standard deviation of the logarithm of the lognormal random variable.*

distributed. We will call μ the log-mean and σ the log-standard deviation. Also, σ^2 will be called the log-variance.

The median of Y is $\exp(\mu)$ and the expected value of Y is $\exp(\mu + \sigma^2/2)$.[3] The expectation is larger than the median because the lognormal distribution is right skewed, and the skewness is more extreme with larger values of σ. Skewness is discussed further in Sect. 5.4. The probability density functions of several lognormal distributions are shown in Fig. A.1.

The log-mean μ is a scale parameter and the log-standard deviation σ is a shape parameter. The lognormal distribution does not have a location parameter since its support is fixed to start at 0.

Use the functions `plnorm()`, `dlnorm()`, `qlnorm()`, and `rlnorm()` for the lognormal distribution. For example,

```
> options(digits = 3)
> dlnorm(0.5, meanlog = 1, sdlog = 2)
[1] 0.279
```

computes the lognormal density at 0.5 when the log-mean is 1 and the log-standard deviation is 2.

---

[3] It is important to remember that if Y is lognormal(μ, σ), then μ is the expected value of $\log(Y)$, not of Y.

A.9.5 Exponential and Double-Exponential Distributions

The *exponential distribution* with scale parameter $\theta > 0$, which we denote by Exponential(θ), has CDF

$$F(y) = 1 - e^{-y/\theta}, \quad y > 0.$$

The Exponential(θ) distribution has PDF

$$f(y) = \frac{e^{-y/\theta}}{\theta}, \tag{A.12}$$

expected value θ, and standard deviation θ. The inverse CDF is

$$F^{-1}(y) = -\theta \log(1 - y), \quad 0 < y < 1.$$

Use the functions `pexp()`, `dexp()`, `qexp()`, and `rexp()` for the exponential distribution.

The *double-exponential* or *Laplace distribution* with mean μ and scale parameter θ has PDF

$$f(y) = \frac{e^{-|y-\mu|/\theta}}{2\theta}. \tag{A.13}$$

If Y has a double-exponential distribution with mean μ, then $|Y - \mu|$ has an exponential distribution. A double-exponential distribution has a standard deviation of $\sqrt{2}\theta$. The mean μ is a location parameter and θ is a scale parameter.

A.9.6 Gamma and Inverse-Gamma Distributions

The *gamma distribution* with scale parameter $b > 0$ and shape parameter $\alpha > 0$ has density

$$\frac{y^{\alpha-1}}{\Gamma(\alpha)b^{\alpha}} \exp(-y/b),$$

where Γ is the gamma function defined in Sect. 5.5.2. The mean, variance, and skewness coefficient of this distribution are $b\alpha$, $b^2\alpha$, and $2\alpha^{-1/2}$, respectively. Figure A.2 shows gamma densities with shape parameters equal to 0.75, 3/2, and 7/2 and each with a mean equal to 1.

The gamma distribution is often parameterized using $\beta = 1/b$, so that the density is

$$\frac{\beta^{\alpha}y^{\alpha-1}}{\Gamma(\alpha)} \exp(-\beta y).$$

With this form of the parameterization, β is an *inverse-scale parameter* and the mean and variance are α/β and α/β^2. Also, β is often called the rate parameter, e.g., in R.

Use the functions `pgamma()`, `dgamma()`, `qgamma()`, and `rgamma()` for the gamma distribution. For example, the median of the gamma distribution with $\alpha = 2$ and $\beta = 3$ can be computed in two equivalent ways:

gamma densities

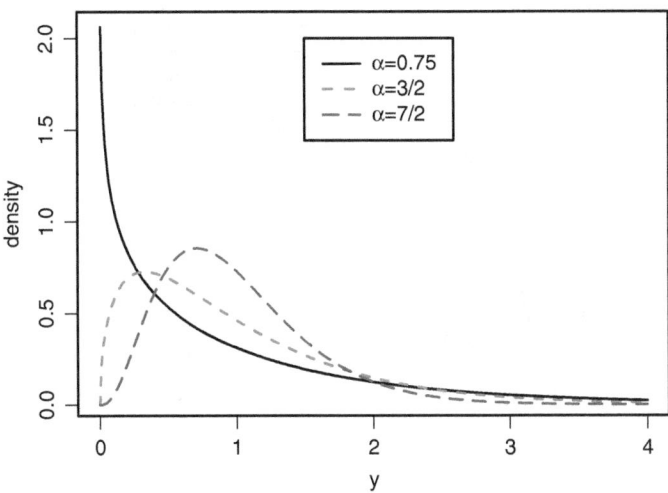

Fig. A.2. *Examples of gamma probability densities with differing shape parameters. In each case, the scale parameter has been chosen so that the expectation is 1.*

```
> qgamma(0.5, shape = 2, rate = 3)
[1] 0.559
> qgamma(0.5, shape = 2, scale = 1/3)
[1] 0.559
```

If X has a gamma distribution with inverse-scale parameter β and shape parameter α, then we say that $1/X$ has an *inverse-gamma distribution* with scale parameter β and shape parameter α. The mean of this distribution is $\beta/(\alpha - 1)$ provided $\alpha > 1$ and the variance is $\beta^2/\{(\alpha - 1)^2(\alpha - 2)\}$ provided that $\alpha > 2$.

A.9.7 Beta Distributions

The beta distribution with shape parameters $\alpha > 0$ and $\beta > 0$ has density

$$\frac{\Gamma(\alpha + \beta)}{\Gamma(\alpha)\Gamma(\beta)}\, y^{\alpha-1}(1 - y)^{\beta-1}, \quad 0 < y < 1. \tag{A.14}$$

The mean and variance are $\alpha/(\alpha + \beta)$ and $(\alpha\beta)/\{(\alpha + \beta)^2(\alpha + \beta + 1)\}$, and if $\alpha > 1$ and $\beta > 1$, then the mode is $(\alpha - 1)/(\alpha + \beta - 2)$.

Figure A.3 shows beta densities for several choices of shape parameters. A beta density is right-skewed, symmetric about $1/2$, or left-skewed depending on whether $\alpha < \beta$, $\alpha = \beta$, or $\alpha > \beta$.

Use the functions pbeta(), dbeta(), qbeta(), and rbeta() for the beta distribution. For example, the code below created Fig. A.3.

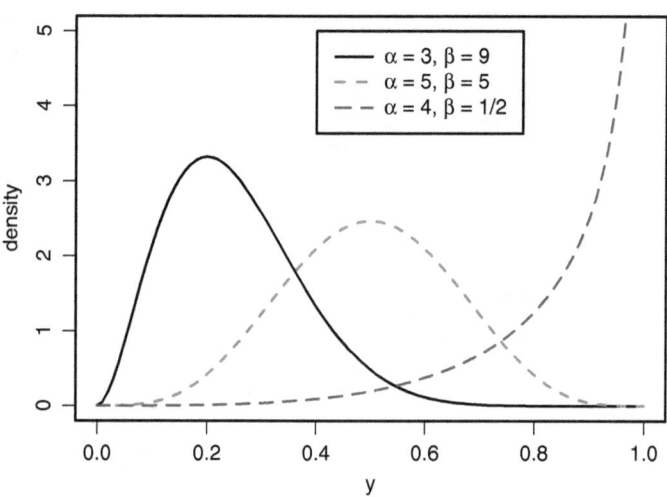

Fig. A.3. *Examples of beta probability densities with differing shape parameters.*

```
pdf("beta_densities.pdf", width = 6, height = 5)  ##  Figure A.3
par(lwd = 2)
x = seq(0, 1, 0.01)
plot(x, dbeta(x, 3, 9), type = "l", lty = 1, xlab = "y",
     ylab = "density", main = "beta densities", ylim = c(0, 5))
lines(x, dbeta(x, 5, 5), type = "l", lty = 2, col = "red")
lines(x, dbeta(x, 4, 1/2), type = "l", lty = 5, col = "blue")
legend(0.4, 5, c(
  expression(paste(alpha," = 3, ",beta," = 9")) ,
  expression(paste(alpha," = 5, ",beta," = 5")),
  expression(paste(alpha," = 4, ",beta," = 1/2"))),
     lty = c(1, 2,5 ), col = c("black", "red", "blue"), lwd = 2)
graphics.off()
```

A.9.8 Pareto Distributions

A random variable X has a Pareto distribution, named after the Swiss economics professor Vilfredo Pareto (1848–1923), if its CDF for some $a > 0$

$$F(x) = 1 - \left(\frac{c}{x}\right)^a, \quad x > c, \tag{A.15}$$

where $c > 0$ is the minimum possible value of X.

The PDF of the distribution in (A.15) is

$$f(x) = \frac{ac^a}{x^{a+1}}, \quad x > c, \tag{A.16}$$

so a Pareto distribution has polynomial tails and a is the *tail index*. It is also called the *Pareto constant*.

A.10 Sampling a Normal Distribution

A common situation is that we have a random sample from a normal distribution and we wish to have confidence intervals for the mean and variance or test hypotheses about these parameters. Then, the following distributions are very important, since they are the basis for many commonly used confidence intervals and tests.

A.10.1 Chi-Squared Distributions

Suppose that $Z_1, \ldots, Z_n$ are i.i.d. $N(0,1)$. Then, the distribution of $Z_1^2 + \cdots + Z_n^2$ is called the *chi-squared distribution* with n *degrees of freedom*. This distribution has an expected value of n and a variance of $2n$. The α-upper quantile of this distribution is denoted by $\chi_{\alpha,n}^2$ and is used in tests and confidence intervals about variances; see Appendix A.10.1 for the latter. Also, as discussed in Sect. 5.11, $\chi_{\alpha,n}^2$ is used in likelihood ratio testing. As an example, $\chi_{0.05,10}^2$ is 18.31 and can be computed in two ways:

```
> qchisq(0.05, 10, lower.tail = FALSE)
[1] 18.31
> qchisq(0.95, 10)
[1] 18.31
```

So far, the degrees-of-freedom parameter has been an integer-valued, but this can be generalized. The chi-squared distribution with ν degrees of freedom is equal to the gamma distribution with scale parameter equal to 2 and shape parameter equal to $\nu/2$. Thus, since the shape parameter of a gamma distribution can be any positive value, the chi-squared distribution can be defined for any positive value of ν as the gamma distribution with scale and shape parameters equal to 2 and $\nu/2$, respectively.

A.10.2 F-Distributions

If U and W are independent and chi-squared-distributed with n_1 and n_2 degrees of freedom, respectively, then the distribution of

$$\frac{U/n_1}{W/n_2}$$

is called the F-distribution with n_1 and n_2 degrees of freedom. The α-upper quantile of this distribution is denoted by F_{α,n_1,n_2}. F_{α,n_1,n_2} is used as a critical value for F-tests in regression. For example, $F_{0.95,3,7}$ is 4.347:

```
> qf(0.95, 3, 7)
[1] 4.347
> qf(0.05, 3, 7, lower.tail = FALSE)
[1] 4.347
```

The degrees-of-freedom parameters of the chi-square, t-, and F-distributions are shape parameters.

A.11 Law of Large Numbers and the Central Limit Theorem for the Sample Mean

Suppose that $\overline{Y}_n$ is the mean of an i.i.d. sample $Y_1, \ldots, Y_n$. We assume that their common expected value $E(Y_1)$ exists and is finite and call it μ. The *law of large numbers* states that

$$P(\overline{Y}_n \to \mu \text{ as } n \to \infty) = 1.$$

Thus, the sample mean will be close to the population mean for large enough sample sizes. However, even more is true. The famous *central limit theorem* (CLT) states that if the common variance σ^2 of $Y_1, \ldots, Y_n$ is finite, then the probability distribution of $\overline{Y}_n$ gets closer to a normal distribution as n converges to ∞. More precisely, the CLT states that

$$P\{\sqrt{n}(\overline{Y}_n - \mu) \le y\} \to \Phi(y/\sigma) \text{ as } n \to \infty \text{ for all } y. \quad (A.17)$$

Stated differently, for large n, $\overline{Y}$ is approximately $N(\mu, \sigma^2/n)$.

Students often misremember or misunderstand the CLT. A common misconception is that a large *population* is approximately normally distributed. The CLT says nothing about the distribution of a population; it is only a statement about the distribution of a sample mean. Also, the CLT does not assume that the population is large; it is the size of the sample that is converging to infinity. Assuming that the sampling is with replacement, the population could be quite small, in fact, with only two elements.

When the variance of $Y_1, \ldots, Y_n$ is infinite, then the limit distribution of $\overline{Y}_n$ may still exist but will be a nonnormal stable distribution.

Although the CLT was first discovered for the sample mean, other estimators are now known to also have approximate normal distributions for large sample sizes. In particular, there are central limit theorems for the maximum likelihood estimators of Sect. 5.9 and the least-squares estimators discussed in Chap. 9. This is very important, since most estimators we use will be maximum likelihood estimators or least-squares estimators. So, if we have a reasonably large sample, we can assume that these estimators have an approximately normal distribution and the normal distribution can be used for testing and constructing confidence intervals.

A.12 Bivariate Distributions

Let $f_{Y_1, Y_2}(y_1, y_2)$ be the joint density of a pair of random variables (Y_1, Y_2). Then, the *marginal density* of Y_1 is obtained by "integrating out" Y_2:

$$f_{Y_1}(y_1) = \int f_{Y_1, Y_2}(y_1, y_2) \, dy_2,$$

and similarly $f_{Y_2}(y_2) = \int f_{Y_1, Y_2}(y_1, y) \, dy_1$.

The *conditional density* of Y_2 given Y_1 is

$$f_{Y_2|Y_1}(y_2|y_1) = \frac{f_{Y_1,Y_2}(y_1,y_2)}{f_{Y_1}(y_1)}. \tag{A.18}$$

Equation (A.18) can be rearranged to give the joint density of Y_1 and Y_2 as the product of a marginal density and a conditional density:

$$f_{Y_1,Y_2}(y_1,y_2) = f_{Y_1}(y_1)f_{Y_2|Y_1}(y_2|y_1) = f_{Y_2}(y_2)f_{Y_1|Y_2}(y_1|y_2). \tag{A.19}$$

The *conditional expectation* of Y_2 given Y_1 is just the expectation calculated using $f_{Y_2|Y_1}(y_2|y_1)$:

$$E(Y_2|Y_1 = y_1) = \int y_2 f_{Y_2|Y_1}(y_2|y_1)dy_2,$$

which is, of course, a function of y_1. The conditional variance of Y_2 given Y_1 is

$$\mathrm{Var}(Y_2|Y_1 = y_1) = \int \{y_2 - E(Y_2|Y_1 = y_1)\}^2 f_{Y_2|Y_1}(y_2|y_1)\, dy_2.$$

A formula that is important elsewhere in this book is

$$f_{Y_1,\ldots,Y_n}(y_1,\ldots,y_n) = f_{Y_1}(y_1)f_{Y_2|Y_1}(y_2|y_1) \cdots$$
$$f_{Y_n|Y_1,\ldots,Y_{n-1}}(y_n|y_1,\ldots,y_{n-1}), \tag{A.20}$$

which follows from repeated use of (A.19).

The marginal mean and variance are related to the conditional mean and variance by

$$E(Y) = E\{E(Y|X)\} \tag{A.21}$$

and

$$\mathrm{Var}(Y) = E\{\mathrm{Var}(Y|X)\} + \mathrm{Var}\{E(Y|X)\}. \tag{A.22}$$

Result (A.21) has various names, especially the *law of iterated expectations* and the *tower rule*.

Another useful formula is that if Z is a function of X, then

$$E(ZY|X) = ZE(Y|X). \tag{A.23}$$

The idea here is that, given X, Z is constant and can be factored outside the conditional expectation.

A.13 Correlation and Covariance

Expectations and variances summarize the individual behavior of random variables. If we have two random variables, X and Y, then it is convenient to have some way to summarize their joint behavior—correlation and covariance do this.

The *covariance* between two random variables X and Y is

$$\text{Cov}(X,Y) = \sigma_{XY} = E\Big[\{X - E(X)\}\{Y - E(Y)\}\Big].$$

The two notations $\text{Cov}(X,Y)$ and σ_{XY} will be used interchangeably. If (X,Y) is continuously distributed, then using (A.36), we have

$$\sigma_{XY} = \int \{x - E(X)\}\{y - E(Y)\} f_{XY}(x,y)\, dx\, dy.$$

The following are useful formulas:

$$\sigma_{XY} = E(XY) - E(X)E(Y), \tag{A.24}$$
$$\sigma_{XY} = E[\{X - E(X)\}Y], \tag{A.25}$$
$$\sigma_{XY} = E[\{Y - E(Y)\}X], \tag{A.26}$$
$$\sigma_{XY} = E(XY) \text{ if } E(X) = 0 \text{ or } E(Y) = 0. \tag{A.27}$$

The covariance between two variables measures the linear association between them, but it is also affected by their variability; all else equal, random variables with larger standard deviations have a larger covariance. Correlation is covariance after this size effect has been removed, so that correlation is a pure measure of how closely two random variables are related, or more precisely, linearly related. The *Pearson correlation coefficient* between X and Y is

$$\text{Corr}(X,Y) = \rho_{XY} = \sigma_{XY}/\sigma_X\,\sigma_Y. \tag{A.28}$$

The Pearson correlation coefficient is sometimes called simply the correlation coefficient, though there are other types of correlation coefficients; see Sect. 8.5.

Given a bivariate sample $\{(X_i, Y_i)\}_{i=1}^n$, the sample covariance, denoted by s_{XY} or $\hat{\sigma}_{XY}$, is

$$s_{XY} = \hat{\sigma}_{XY} = (n-1)^{-1} \sum_{i=1}^n (X_i - \overline{X})(Y_i - \overline{Y}), \tag{A.29}$$

where $\overline{X}$ and $\overline{Y}$ are the sample means. Often the factor $(n-1)^{-1}$ is replaced by n^{-1}, but this change has little effect relative to the random variation in $\hat{\sigma}_{XY}$. The *sample correlation* is

$$\hat{\rho}_{XY} = r_{XY} = \frac{s_{XY}}{s_X s_Y}, \tag{A.30}$$

where s_X and s_Y are the sample standard deviations.

To provide the reader with a sense of what particular values of a correlation coefficient imply about the relationship between two random variables, Fig. A.4 shows scatterplots and the sample correlation coefficients for nine bivariate random samples. A *scatterplot* is just a plot of a bivariate sample, $\{(X_i, Y_i)\}_{i=1}^n$. Each plot also contain the *linear* least-squares fit (Chap. 9) to illustrate the linear relationship between y and x. Notice that

- an absolute correlation of 0.25 or less is weak—see panels (a) and (b);
- an absolute correlation of 0.5 is only moderately strong—see (c);
- an absolute correlation of 0.9 is strong—see (d);
- an absolute correlation of 1 implies an exact linear relationship—see (e) and (h);
- a strong nonlinear relationship may or may not imply a high correlation—see (f) and (g);
- positive correlations imply an increasing relationship (as X increases, Y increases on average)—see (b)–(e) and (g);
- negative correlations imply a decreasing relationship (as X increases, Y decreases on average)—see (h) and (i).

If the correlation between two random variables is equal to 0, then we say that they are *uncorrelated*.

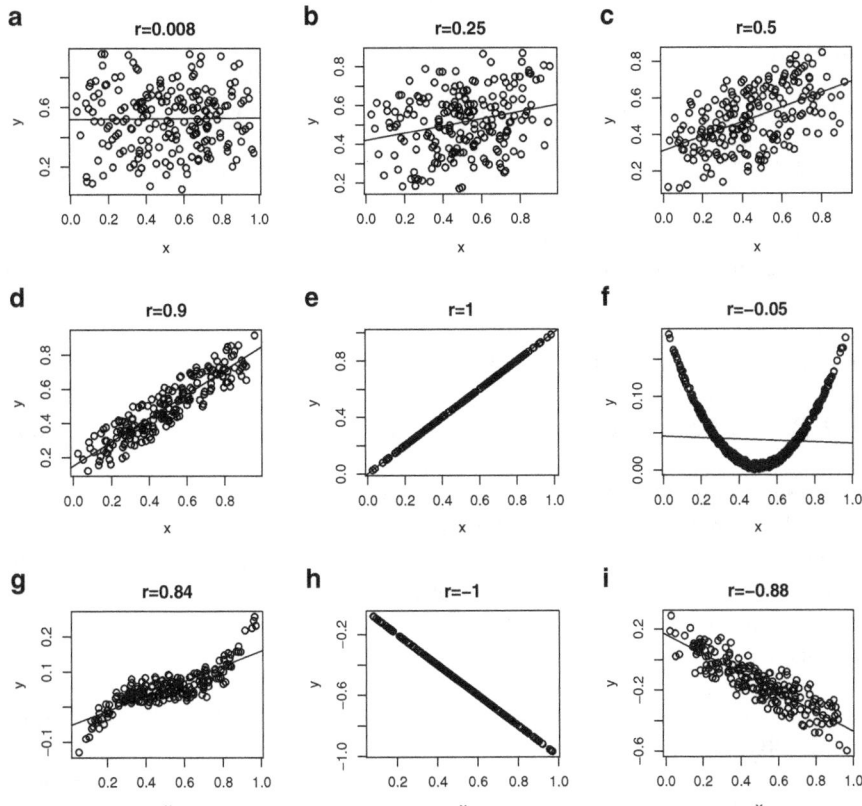

Fig. A.4. *Sample correlation coefficients for nine random samples. Each plot also contains the linear regression line of y on x.*

If X and Y are independent, then for all functions g and h,

$$E\{g(X)h(Y)\} = E\{g(X)\}E\{h(Y)\}. \tag{A.31}$$

This fact can be used to prove that if X and Y are independent, then $\sigma_{XY} = 0$, so the variables are uncorrelated. The opposite is not true. For example, if X is uniformly distributed on $[-1, 1]$ and $Y = X^2$, then a simple calculation shows that $\sigma_{XY} = 0$, but the two random variables are not independent. The key point here is that Y is related to X, in fact, completely determined by X, but the relationship is highly nonlinear and correlation measures linear association.

Another example of random variables that are uncorrelated but dependent is the bivariate t-distribution. For this distribution, the two variates are dependent even when their correlation is 0; see Sect. 7.6.

If $E(Y|X) = 0$, then Y and X are uncorrelated, since

$$E(Y) = E\{E(Y|X)\} = 0 \tag{A.32}$$

by the law of iterated expectations, and then

$$\text{Cov}(Y, X) = E(YX) = E\{E(YX|X)\} = E\{XE(Y|X)\} = 0 \tag{A.33}$$

by (A.27), a second application of the law of iterated expectations, (A.23) with $Z = X$, and (A.32).

Result (A.22) has an important interpretation. If X is known and one needs to predict Y, then $E(Y|X)$ is the best predictor in that it minimizes the expected squared prediction error. If the best predictor is used, then the prediction error is $Y - E(Y|X)$ and $E\{Y - E(Y|X)\}^2$ is the expected squared prediction error. From the law of iterated expectations, that latter is

$$E\{Y - E(Y|X)\}^2 = E\left(E[\{Y - E(Y|X)\}^2|X]\right) = E\{\text{Var}(Y|X)\}, \tag{A.34}$$

the first summand on the right-hand side of (A.22). Also, $\text{Var}\{E(Y|X)\}$, the second summand there, is the variability of the best predictor and a measure of how well $E(Y|X)$ can track Y—the more $E(Y|X)$ can vary, the better it can track Y. Therefore, the sum of the tracking ability and the expected squared prediction error is the constant $\text{Var}(Y)$—increasing the tracking ability decreases the expected squared prediction error.

Some insight can be gained by looking at the worst and best cases. The worst case is when X is independent of Y. Then, $E(Y|X) = E(Y)$, the tracking ability is $\text{Var}\{E(Y|X)\} = 0$, and the expected squared prediction takes on its maximum value, $\text{Var}(Y)$. The best case is when Y is a function of X, say $y = g(X)$ for some g. Then, $E(Y|X) = g(X) = Y$, the prediction error is 0, and the tracking ability is $\text{Var}(Y)$, its maximum possible value.

A.13.1 Normal Distributions: Conditional Expectations and Variance

The calculation of conditional expectations and variances can be difficult for some probability distributions, but it is quite easy for a pair (Y_1, Y_2) that has a bivariate normal distribution.

For a bivariate normal pair, the conditional expectation of Y_2 given Y_1 equals the best linear predictor[4] of Y_2 given Y_1:

$$E(Y_2 | Y_1 = y_1) = E(Y_2) + \frac{\sigma_{Y_1, Y_2}}{\sigma_{Y_1}^2}\{y_1 - E(Y_1)\}.$$

Therefore, for normal random variables, best linear prediction is the same as best prediction. Also, the conditional variance of Y_2 given Y_1 is the expected squared prediction error:

$$\text{Var}(Y_2 | Y_1 = y_1) = \sigma_{Y_2}^2(1 - \rho_{Y_1, Y_2}^2). \tag{A.35}$$

In general, $\text{Var}(Y_2 | Y_1 = y_1)$ is a function of y_1 but we see in (A.35) that for the special case of a bivariate normal distribution, $\text{Var}(Y_2 | Y_1 = y_1)$ is constant, that is, independent of y_1.

A.14 Multivariate Distributions

Multivariate distributions generalized the bivariate distributions of Appendix A.12. A *random vector* is a vector whose elements are random variable. A random vector of continuously distributed random variables, $\boldsymbol{Y} = (Y_1, \ldots, Y_d)$, has a *multivariate probability density function* $f_{Y_1, \ldots, Y_d}(y_1, \ldots, y_d)$ if

$$P\{(Y_1, \ldots, Y_d) \in A\} = \int \int_A f_{Y_1, \ldots, Y_d}(y_1, \ldots, y_d)\, dy_1 \cdots dy_d$$

for all sets $A \subset \Re^p$.

The PDF of Y_j is obtained by integrating the other variates out of $f_{Y_1, \ldots, Y_d}$:

$$f_{Y_j}(y_j)$$

$$= \int_{y_1} \cdots \int_{y_{j-1}} \int_{y_{j+1}} \cdots \int_{y_d} f_{Y_1, \ldots, Y_d}(y_1, \ldots, y_d)\, dy_1 \cdots dy_{j-1} dy_{j+1} \cdots dy_d.$$

Similarly, the PDF of any subset of $(Y_1, \ldots, Y_d)$ is obtained by integrating the other variables out of $f_{Y_1, \ldots, Y_d}(y_1, \ldots, y_d)$.

The expectation of a function g of $Y_1, \ldots, Y_d$ is given by the formula

$$E\{g(Y_1, \ldots, Y_d)\} = \int_{y_1} \cdots \int_{y_d} g(y_1, \ldots, y_d) f_{Y_1, \ldots, Y_d}(y_1, \ldots, y_d)\, dy_1 \cdots dy_d. \tag{A.36}$$

---

[4] See Sect. 11.9.

If $Y_1, \ldots, Y_d$ are discrete, then their joint probability distribution specifies $P\{Y_1 = x_1, \ldots, Y_d = y_d\}$ for all values of $y_1, \ldots, y_d$. If $Y_1, \ldots, Y_d$ are discrete and independent, then

$$P\{Y_1 = y_1, \ldots, Y_d = y_d\} = P\{Y_1 = y_1\} \cdots P\{Y_d = y_d\}. \tag{A.37}$$

The joint CDF of $Y_1, \ldots, Y_d$, whether they are continuous or discrete, is

$$F_{Y_1, \ldots, Y_d}(x_1, \ldots, y_d) = P(Y_1 \leq y_1, \ldots, Y_d \leq y_d).$$

Suppose there is a sample of size n of d-dimensional random vectors, $\{Y_i = (Y_{i,1}, \ldots, Y_{i,d}) : i = 1, \ldots, n\}$. Then the empirical CDF is

$$F_n(y_1, \ldots, y_d) = \frac{\sum_{i=1}^{n} I\{Y_{i,j} \leq y_j, \text{ for } j = 1, \ldots, d\}}{n}. \tag{A.38}$$

A.14.1 Conditional Densities

The conditional density of $Y_1, \ldots, Y_q$ given $Y_{q+1} \ldots, Y_d$, where $1 \leq q < d$, is

$$f_{Y_1, \ldots, Y_q \mid Y_{q+1} \ldots, Y_d}(y_1, \ldots, y_q \mid y_{q+1} \ldots, y_d) = \frac{f_{Y_1, \ldots, Y_d}(y_1, \ldots, y_d)}{f_{Y_{q+1} \ldots, Y_d}(y_{q+1} \ldots, y_d)}. \tag{A.39}$$

Since $Y_1, \ldots, Y_d$ can be arranged in any order that is convenient, (A.39) provides a formula for the conditional density of any subset of the variables, given the other variables. Also, (A.39) can be rearranged to give the *multiplicative formula*

$$f_{Y_1, \ldots, Y_d}(y_1, \ldots, y_d)$$
$$= f_{Y_1, \ldots, Y_q \mid Y_{q+1} \ldots, Y_d}(y_1, \ldots, y_q \mid y_{q+1} \ldots, y_d) f_{Y_{q+1} \ldots, Y_d}(y_{q+1} \ldots, y_d). \tag{A.40}$$

Repeated use of (A.40) gives a formula that will be useful later for calculating likelihoods for dependent data

$$f_{Y_1, \ldots, Y_d}(y_1, \ldots, y_d)$$
$$= f_{Y_1}(y_1) f_{Y_2 \mid Y_1}(y_2 \mid y_1) f_{Y_3 \mid Y_1, Y_2}(y_3 \mid y_1, y_2) \cdots f_{Y_d \mid Y_1, \ldots, Y_{d-1}}(y_d \mid y_1, \ldots, y_{d-1}). \tag{A.41}$$

If $Y_1, \ldots, Y_d$ are independent, then

$$f_{Y_1, \ldots, Y_d}(y_1, \ldots, y_d) = f_{Y_1}(y_1) \cdots f_{Y_d}(y_d). \tag{A.42}$$

A.15 Stochastic Processes

A discrete-time stochastic process is a sequence of random variables $\{Y_1, Y_2, Y_3, \ldots\}$. The distribution of Y_n is called its marginal distribution. The process is said to be Markov, or Markovian, if the conditional distribution of Y_{n+1}

given $\{Y_1, Y_2, \ldots, Y_n\}$ equals the conditional distribution of Y_{n+1} given Y_n, so Y_{n+1} depends only on the previous value of the process. The AR(1) process in Sect. 12.4 is a simple example of a Markov process. A process generated by computer simulation will be Markov if only Y_n and random numbers independent of $\{Y_1, Y_2, \ldots, Y_{n-1}\}$ are used to generate Y_{n+1}. An important example is Markov chain Monte Carlo, the topic of Sect. 20.7.

A distribution π is a stationary distribution for a Markov process if, for all n, Y_{n+1} has distribution π whenever Y_n has distribution π.

Stochastic processes can also have a continuous-time parameter. Examples are Brownian motion and geometric Brownian motion, which are used, *inter alia*, to model the log-prices and prices of equities, respectively, in continuous time.

A.16 Estimation

A.16.1 Introduction

One of the major areas of statistical inference is estimation of unknown parameters, such as a population mean, from data. An estimator is defined as any function of the observed data. The key question is which of many possible estimators should be used. If θ is an unknown parameter and $\widehat{\theta}$ is an estimator, then $E(\widehat{\theta}) - \theta$ is called the *bias* and $E\{\widehat{\theta} - \theta\}^2$ is called the *mean-squared error* (MSE). One seeks estimators that are efficient, that is, having the smallest possible value of the MSE (or of some other measure of inaccuracy). It can be shown from simple algebra that the MSE is the squared bias plus the variance, that is,

$$E\{\widehat{\theta} - \theta\}^2 = \{E(\widehat{\theta}) - \theta\}^2 + \mathrm{Var}(\widehat{\theta}), \qquad (A.43)$$

so an efficient estimator will have both a small bias and a small variance. An estimator with a zero bias is called *unbiased*. However, it is not necessary to use an unbiased estimator—we only want the bias to be small, not necessarily exactly zero. One should be willing to accept a small bias if this leads to a significant reduction in variance.

The most popular methods of estimation are least squares (Sect. 9.2.1), maximum likelihood (Sects. 5.9 and 5.14), and Bayes estimation (Chap. 20).

A.16.2 Standard Errors

When a estimator is calculated from a random sample, it is a random variable, but this fact is often not appreciated by beginning students. When first exposed to statistical estimation, students tend not to think of estimators such as a sample mean as random. If we have only a single sample, then the sample mean does not *appear* random. However, if we realize that the observed sample is only one of many possible samples that could have been drawn, and that each sample has a different sample mean, then we see that the mean is in fact random.

Since an estimator is a random variable, it has an expectation and a standard deviation. We have already seen that the difference between its expectation and the parameter is called the bias. The standard deviation of an estimator is called its *standard error*. If there are unknown parameters in the formula for this standard deviation, then they can be replaced by estimates. If $\hat{\theta}$ is an estimator of θ, then $s_{\hat{\theta}}$ will denote its standard error with any unknown parameters replaced by estimates.

Example A.1. The standard error of the mean

Suppose that $Y_1, \ldots, Y_n$ are i.i.d. with mean μ and variance σ^2. Then, it follows from (7.13) that the standard deviation of $\overline{Y}$ is $\sigma/\sqrt{n}$. Thus, $\sigma/\sqrt{n}$, or when σ is unknown $s_Y/\sqrt{n}$, is called the standard error of the sample mean. That is, $s_{\overline{Y}}$ is $\sigma/\sqrt{n}$ or $s_Y/\sqrt{n}$ depending on whether or not σ is known. $\square$

A.17 Confidence Intervals

Instead of estimating an unknown parameter by a single number, it is often better to provide a range of numbers that gives a sense of the uncertainty of the estimate. Such ranges are called *interval estimates*. One type of interval estimate, the Bayesian credible interval, is introduced in Chap. 20. Another type of interval estimate is the confidence interval. A *confidence interval* is defined by the requirement that the probability that the interval will include the true parameter is a specified value called the *confidence coefficient,*, so, for example, if a large number of independent 90 % intervals are constructed, then approximately 90 % of them will contain the parameter.

A.17.1 Confidence Interval for the Mean

If $\overline{Y}$ is the mean of a sample from a normal population, then

$$\overline{Y} \pm t_{\alpha/2, n-1}\, s_{\overline{Y}} \tag{A.44}$$

is a confidence interval with $(1 - \alpha)$ confidence. This confidence interval is derived in Sect. 6.3.2. If $\alpha = 0.05$ (0.95 or 95 % confidence) and if n is reasonably large, then $t_{\alpha/2, n-1}$ is approximately 2, so $\overline{Y} \pm 2\, s_{\overline{Y}}$ is often used as an approximate 95 % confidence interval. Since $s_{\overline{Y}} = s_Y/\sqrt{n}$, the confidence can also be written as $\overline{Y} \pm 2\, s_Y/\sqrt{n}$. When n is reasonably large, say 20 or more, then $\overline{Y}$ will be approximately normally distributed by the central limit theorem, and the assumption that the population itself is normal can be dropped.

Example A.2. Confidence interval for a normal mean

Suppose we have a sample of size 25 from a normal distribution, $s_Y^2 = 2.7$, $\overline{Y} = 16.1$, and we want a 99 % confidence interval for μ. We need $t_{0.005,24}$. This quantile can be found, for example, using the R function qt and $t_{0.005,24} = 2.797$. Then, the 99 % confidence interval for μ is

$$16.1 \pm \frac{(2.797)\sqrt{2.7}}{\sqrt{25}} = 16.1 \pm 0.919 = [15.18, \ 17.02].$$

Since $n = 25$ is reasonably large, this interval should have approximately 99 % confidence even if the population is not normally distributed. The exception would be if the population was extremely heavily skewed or had very heavy tails; in such cases a sample size larger than 25 might be necessary for this confidence interval to have near 99 % coverage.

Just how large a sample is needed for $\overline{Y}$ to be nearly normally distributed depends on the population. If the population is symmetric and the tails are not extremely heavy, then approximate normality is often achieved with n around 10. For skewed populations, 30 observations may be needed, and even more in extreme cases. If the data appear to come from a highly skewed or heavy-tailed population, it might be better to assume a parametric model and compute the MLE as discussed in Chap. 5 and perhaps to use the bootstrap (Chap. 6) for finding the confidence interval.

The function t.test() computes a confidence interval for a normal mean. The output below gives a 99 % confidence interval for daily log-returns on Ford using t.test() and then using (A.44). The interval is $(-0.000417, 0.003407)$

```
> ford = read.csv("RecentFord.csv")
> returns = diff(log(ford[ , 7]))
> options(digits = 3)
> t.test(returns, conf.level = 0.99)

One Sample t-test

data:  returns
t = 2.02, df = 1256, p-value = 0.04388
alternative hypothesis: true mean is not equal to 0
99 percent confidence interval:
 -0.000417  0.003407
sample estimates:
mean of x
  0.00149

> n = length(returns)
> mean(returns) + c(-1, 1) *
    qt(0.995, n - 1) * sd(returns) / sqrt(n)
[1] -0.000417  0.003407
```

A.17.2 Confidence Intervals for the Variance and Standard Deviation

A $(1 - \alpha)$ confidence interval for the variance of a normal distribution is given by

$$\left[\frac{(n-1)s_Y^2}{\chi_{\alpha/2,n-1}^2}, \ \frac{(n-1)s_Y^2}{\chi_{1-\alpha/2,n-1}^2} \right], \tag{A.45}$$

where n is the sample size, s_Y^2 is the sample variance given by equation (A.7), and, as defined in Appendix A.10.1, $\chi_{\gamma,n-1}^2$ is the $(1-\gamma)$-quantile of the chi-square distribution with $n-1$ degrees of freedom.

Example A.3. Confidence interval for a normal standard deviation

Suppose we have a sample of size 25 from a normal distribution, $s_Y^2 = 2.7$, and we want a 90 % confidence interval for σ^2. The quantiles we need for constructing the interval are $\chi_{0.95,24}^2 = 13.848$ and $\chi_{0.05,24}^2 = 36.415$. These values can be found using software such as qchisq() in R. The 90 % confidence interval for σ^2 is

$$\left[\frac{(2.7)(24)}{36.415}, \ \frac{(2.7)(24)}{13.848} \right] = [1.78, \ 4.68].$$

Taking square roots of both endpoints, we get $1.33 < \sigma < 2.16$ as a 90 % confidence interval for the standard deviation.

As another example, confidence intervals for the variance and standard deviation of daily Ford returns are calculate below. The confidence interval for the standard deviation is (0.0253, 0.0273).

```
> ford = read.csv("RecentFord.csv")
> returns = diff(log(ford[,7]))
> n = length(returns)
> options(digits = 3)
> ci = (n - 1) * var(returns) / qchisq(c(0.025, 0.975), n - 1,
    lower.tail = FALSE)
> ci
[1] 0.000639 0.000748
> sqrt(ci)
[1] 0.0253 0.0273
```

□

Unfortunately, the assumption that the population is normally distributed cannot be dispensed with, even if the sample size is large. If a normal probability plot or test of normality (see Sect. 4.4) suggests that the population might be nonnormally distributed, then one might instead construct a confidence interval for σ using the bootstrap; see Chap. 6. Another possibility is to assume a nonnormal parametric model such as the t-model if the data are symmetric and heavy-tailed; see Example 5.3.

A.17.3 Confidence Intervals Based on Standard Errors

Many estimators are approximately unbiased and approximately normally distributed. Then, an approximate 95 % confidence interval is the estimator plus or minus twice its standard error; that is,

$$\widehat{\theta} \pm 2\, s_{\widehat{\theta}}$$

is an approximate 95 % confidence interval for θ.

A.18 Hypothesis Testing

A.18.1 Hypotheses, Types of Errors, and Rejection Regions

Statistical hypothesis testing uses data to decide whether a certain statement called the *null hypothesis* is true. The negation of the null hypothesis is called the *alternative hypothesis*. For example, suppose that $Y_1, \ldots, Y_n$ are i.i.d. $N(\mu, 1)$ and μ is unknown. The null hypothesis could be that μ is 1. Then, we write $H_0: \mu = 1$ and $H_1: \mu \neq 1$ to denote the null and alternative hypotheses.

There are two types of errors that we hope to avoid. If the null hypothesis is true but we reject it, then we are making a *type I error*. Conversely, if the null hypothesis is false and we accept it, then we are making a *type II error*.

The *rejection region* is the set of possible samples that lead us to reject H_0. For example, suppose that μ_0 is a hypothesized value of μ and the null hypothesis is $H_0: \mu = \mu_0$ and the alternative is $H_1: \mu \neq \mu_0$. One rejects H_0 if $|\overline{Y} - \mu_0|$ exceeds an appropriately chosen cutoff value c called a *critical value*. The rejection region is chosen to keep the probability of a type I error below a prespecified small value called the *level* and often denoted by α. Typical values of α used in practice are 0.01, 0.05, or 0.1. As α is made smaller, the rejection region must be made smaller. In the example, since we reject the null hypothesis when $|\overline{Y} - \mu_0|$ exceeds c, the critical value c gets larger as the α gets smaller. The value of c is easy to determine. Assuming that σ is known, c is $z_{\alpha/2}\, \sigma/\sqrt{n}$, where, as defined in Appendix A.9.3, $z_{\alpha/2}$ is the $\alpha/2$-upper quantile of the standard normal distribution. If σ is unknown, then σ is replaced by s_X and $z_{\alpha/2}$ is replaced by $t_{\alpha/2, n-1}$, where, as defined in Sect. 5.5.2, $t_{\alpha/2, n-1}$ is the $\alpha/2$-upper quantile of the t-distribution with $n-1$ degrees of freedom. The test using the t-quantile is called the *one-sample t-test*.

A.18.2 p-Values

Rather than specifying α and deciding whether to accept or reject the null hypothesis at that α, we might ask "for what values of α do we reject the null hypothesis?" The *p-value* for a sample is defined as the smallest value of α for which the null hypothesis is rejected. Stated differently, to perform the test using a given sample, we first find the p-value of that sample, and then H_0 is rejected if we decide to use α larger than the p-value and H_0 is accepted if we use α smaller than the p-value. Thus,

- a small p-value is evidence *against* the null hypothesis

while

- a large p-value shows that the *data are consistent* with the null hypothesis.

Example A.4. Interpreting p-values

If the p-value of a sample is 0.033, then we reject H_0 if we use α equal to 0.05 or 0.1, but we accept H_0 if we use $\alpha = 0.01$. $\qquad\qquad\square$

The p-value not only tells us whether the null hypothesis should be accepted or rejected, but it also tells us whether or not the decision to accept or reject H_0 is a close call. For example, if we are using $\alpha = 0.05$ and the p-value were 0.047, then we would reject H_0 but we would know the decision was close. If instead the p-value were 0.001, then we would know the decision was not so close.

When performing hypothesis tests, statistical software routinely calculates p-values. Doing this is much more convenient than asking the user to specify α, and then reporting whether the null hypothesis is accepted or rejected for that α.

A.18.3 Two-Sample t-Tests

Two-sample t-tests are used to test hypotheses about the difference between two population means. The independent-samples t-test is used when we sample independently from the two populations. Let μ_i, $\overline{Y}_i$, s_i, and n_i be the population mean, sample mean, sample standard deviation, and sample size for the ith sample, $i = 1, 2$, respectively. Let Δ_0 be a hypothesized value of $\mu_1 - \mu_2$. We assume that the two populations have the same standard deviation and estimate this parameter by the *pooled standard deviation*, which is

$$s_{\text{pool}} = \left\{ \frac{(n_1 - 1)s_1^2 + (n_2 - 1)s_2^2}{n_1 + n_2 - 2} \right\}^{1/2}. \tag{A.46}$$

The independent-samples t-statistic is

$$t = \frac{\overline{Y}_1 - \overline{Y}_2 - \Delta_0}{s_{\text{pool}}\sqrt{\frac{1}{n_1} + \frac{1}{n_2}}}.$$

If the hypotheses are H_0: $\mu_1 - \mu_2 = \Delta_0$ and H_1: $\mu_1 - \mu_2 \neq \Delta_0$, then H_0 is rejected if $|t| > t_{\alpha/2|n_1+n_2-2}$. If the hypotheses are H_0: $\mu_1 - \mu_2 \leq \Delta_0$ and H_1: $\mu_1 - \mu_2 > \Delta_0$, then H_0 is rejected if $t > t_{\alpha|n_1+n_2-2}$ and if they are H_0: $\mu_1 - \mu_2 \geq \Delta_0$ and H_1: $\mu_1 - \mu_2 < \Delta_0$, then H_0 is rejected if $t < -t_{\alpha|n_1+n_2-2}$.

Sometimes the samples are paired rather than independent. For example, suppose we wish to compare returns on small-cap versus large-cap[5] stocks and for each of n years we have the returns on a portfolio of small-cap stocks and on a portfolio of large-cap stocks. For any year, the returns on the two portfolios will be correlated, so an independent-samples test is not valid. Let $d_i = X_{i,1} - X_{i,2}$ be the difference between the observations from populations 1 and 2 for the ith pair, and let $\overline{d}$ and s_d be the sample mean and standard deviation of $d_1, \ldots, d_n$. The paired-sample t-statistics is

$$t = \frac{\overline{d} - \Delta_0}{s_d/\sqrt{n}}. \tag{A.47}$$

The rejection regions are the same as for the independent-samples t-tests except that the degrees-of-freedom parameter for the t-quantiles is $n-1$ rather than $n_1 + n_2 - 2$.

The power of a test is the probability of correctly rejecting H_0 when H_1 is true. Paired samples are often used to obtain more power. In the example of comparing small- and large-cap stocks, the returns on both portfolios will have high year-to-year variation, but the d_i will be free of this variation, so that s_d should be relatively small compared to s_1 and s_2. A small variation in the data means that $\mu_1 - \mu_2$ can be more accurately estimated and deviations of this parameter from Δ_0 are more likely to be detected.

Since $\overline{d} = \overline{Y}_1 - \overline{Y}_2$, the numerators in (A.46) and (A.47) are equal. What differs are the denominators. The denominator in (A.47) will be smaller than in (A.46) when the correlation between observations $(Y_{i,2}, Y_{i,2})$ in a pair is positive. It is the smallness of the denominator in (A.47) that gives the paired t-test increased power.

Suppose someone had a paired sample but incorrectly used the independent-samples t-test. If the correlation between $Y_{i,1}$ and $Y_{i,2}$ is zero, then the paired samples behave the same as independent samples and the effect of using the incorrect test would be small. Suppose that this correlation is positive. The result of using the incorrect test would be that if H_0 is false, then the true p-value would be overestimated and one would be less likely to reject H_0 than if the paired-sample test had been used. However, if the p-value is small, then one can be confident in rejecting H_0 because the p-value for the paired-sample test would be even smaller.[6] Unfortunately, statistical methods are often used

---

[5] The market capitalization of a stock is the product of the share price and the number of shares outstanding. If stocks are ranked based on market capitalization, then all stocks below some specified quantile would be small-cap stocks and all above another specified quantile would be large-cap.

[6] An exception would be the rare situation, where $Y_{i,1}$ and $Y_{i,2}$ are *negatively* correlated.

by researchers without a solid understanding of the underlying theory, and this can lead to misapplications. The hypothetical use just described of an incorrect test is often a reality, and it is sometimes necessary to evaluate whether the results that are reported can be trusted.

Confidence intervals can also be constructed for the difference between the two means and are

$$\overline{Y}_1 - \overline{Y}_2 \pm t_{\alpha/2|n_1+n_2-2} s_{\text{pool}} \sqrt{\frac{1}{n_1} + \frac{1}{n_2}} \tag{A.48}$$

for unpaired samples and

$$\overline{d} \pm t_{\alpha/2|n_1+n_2-2} s_d / \sqrt{n}. \tag{A.49}$$

for paired samples.

Example A.5. A Paired Two-sample t-test and Confidence Interval

In the next example, a 95 % confidence interval is created for the difference between the mean daily log-returns on Merck and Pfizer. Since the prices were taken over the same time intervals, the daily log-returns are highly correlated ($\hat{\rho} = 0.547$), so a paired test and interval were used. The confidence interval was also calculated using (A.49).

```
> prices = read.csv("Stock_Bond.csv")
> prices_Merck = prices[ , 11]
> return_Merck = diff(log(prices_Merck))
> prices_Pfizer = prices[ , 13]
> return_Pfizer = diff(log(prices_Pfizer))
> cor(return_Merck,return_Pfizer)
[1] 0.547
> t.test(return_Merck, return_Pfizer, paired = TRUE)

Paired t-test

data:  return_Merck and return_Pfizer
t = -0.406, df = 4961, p-value = 0.6849
alternative hypothesis: true difference in means is not equal to 0
95 percent confidence interval:
 -0.000584  0.000383
sample estimates:
mean of the differences
              -1e-04

> differences = return_Merck - return_Pfizer
> n = length(differences)
> mean(differences) + c(-1,1) * qt(0.025, n - 1,
    lower.tail = FALSE) * sd(differences) / sqrt(n)
[1] -0.000584  0.000383
```

□

A.18.4 Statistical Versus Practical Significance

When we reject a null hypothesis, we often say there is a *statistically significant effect*. In this context, the word "significant" is easily misconstrued. It does *not* mean that there is an effect of practical importance. For example, suppose we were testing the null hypothesis that the means of two populations are equal versus the alternative that they are unequal. Statistical significance simply means that the two sample means are sufficiently different that this difference cannot reasonably be attributed to mere chance. Statistical significance does *not* mean that the population means are so dissimilar that their difference is of any practical importance. When large samples are used, small and unimportant effects are likely to be statistically significant.

When determining practical significance, confidence intervals are more useful than tests. In the case of the comparison between two population means, it is important to construct a confidence interval and to conclude that there is an effect of practical significance only if *all* differences in that interval are large enough to be of practical importance. How large is "large enough" is *not* a statistical question but rather must be answered by a subject-matter expert. For an example, suppose a difference between the two population means that exceeds 0.2 is considered important, at least for the purpose under consideration. If a 95 % confidence interval were [0.23, 0.26], then with 95 % confidence we could conclude that there is an important difference. If instead the interval were [0.13, 0.16], then we could conclude with 95 % confidence that there is no important difference. If the confidence interval were [0.1, 0.3], then we could not state with 95 % confidence whether the difference is important or not.

A.19 Prediction

Suppose that Y is a random variable that is unknown at the present time, for example, a future change in an interest rate or stock price. Let X be a known random vector that is useful for predicting Y. For example, if Y is a future change in a stock price or a macroeconomic variable, X might be the vector of recent changes in that stock price or macroeconomic variable.

We want to find a function of X, which we will call $\widehat{Y}(X)$, that best predicts Y. By this we mean that the mean-squared error $E[\{Y-\widehat{Y}(X)\}^2]$ is made as small as possible. The function $\widehat{Y}(X)$ that minimizes the mean-squared error will be called the best predictor of Y based on X. Note that $\widehat{Y}(X)$ can be any function of X, not necessarily a linear function as in Sect. 11.9.1. The *best predictor* is theoretically simple—it is the conditional expectation of Y given X. That is, $E(Y|X)$ is the best predictor of Y in the sense of minimizing $E[\{Y - \widehat{Y}(X)\}^2]$ among *all* possible choices of $\widehat{Y}(X)$ that are arbitrary functions of X.

If Y and $\boldsymbol{X}$ are independent, then $E(Y|\boldsymbol{X}) = E(Y)$. If $\boldsymbol{X}$ were unobserved, then $E(Y)$ would be used to predict $\boldsymbol{Y}$. Thus, when Y and $\boldsymbol{X}$ are independent, the best predictor of Y is the same as if $\boldsymbol{X}$ were unknown, because $\boldsymbol{X}$ contains no information that is useful for prediction of Y.

In practice, using $E(Y|\boldsymbol{X})$ for prediction is not trivial. The problem is that $E(Y|\boldsymbol{X})$ may be difficult to estimate whereas the best linear predictor can be estimated by linear regression as described in Chap. 9. However, the newer technique of *nonparametric regression* can be used to estimate $E(Y|\boldsymbol{X})$. Nonparametric regression is discussed in Chap. 21.

A.20 Facts About Vectors and Matrices

The norm of the vector $\boldsymbol{x} = (x_1, \ldots, x_p)^{\mathsf{T}}$ is $\|\boldsymbol{x}\| = (\sum_{i=1}^{p} x_i^2)^{1/2}$.

A square matrix $\boldsymbol{A}$ is diagonal if $A_{i,j} = 0$ for all $i \neq j$. We use the notation $\mathrm{diag}(d_1, \ldots, d_p)$ for a $p \times p$ diagonal matrix $\boldsymbol{A}$ such that $A_{i,i} = d_i$.

A matrix $\boldsymbol{O}$ is orthogonal if $\boldsymbol{O}^{\mathsf{T}} = \boldsymbol{O}^{-1}$. This implies that the columns of $\boldsymbol{O}$ are mutually orthogonal (perpendicular) and that their norms are all equal to 1.

Any symmetric matrix $\boldsymbol{\Sigma}$ has an *eigenvalue-eigenvector decomposition*, eigen-decomposition for short, which is

$$\boldsymbol{\Sigma} = \boldsymbol{O}\,\mathrm{diag}(\lambda_i)\,\boldsymbol{O}^{\mathsf{T}}, \tag{A.50}$$

where $\boldsymbol{O}$ is an orthogonal matrix whose columns are the eigenvectors of $\boldsymbol{\Sigma}$ and $\lambda_1, \ldots, \lambda_p$ are the eigenvalues of $\boldsymbol{\Sigma}$. Also, if all of $\lambda_1, \ldots, \lambda_p$ are nonzero, then $\boldsymbol{\Sigma}$ is nonsingular and

$$\boldsymbol{\Sigma}^{-1} = \boldsymbol{O}\,\mathrm{diag}(1/\lambda_i)\,\boldsymbol{O}^{\mathsf{T}}.$$

Let $\boldsymbol{o}_1, \ldots, \boldsymbol{o}_p$ be the columns of $\boldsymbol{O}$. Then, since $\boldsymbol{O}$ is orthogonal,

$$\boldsymbol{o}_j^{\mathsf{T}} \boldsymbol{o}_k = 0 \tag{A.51}$$

for any $j \neq k$. Moreover,

$$\boldsymbol{o}_j^{\mathsf{T}} \boldsymbol{\Sigma} \boldsymbol{o}_k = 0 \tag{A.52}$$

for $j \neq k$. To see this, let $\boldsymbol{e}_j$ be the jth unit vector, that is, the vector with a one in the jth coordinate and zeros elsewhere. Then, $\boldsymbol{o}_j^{\mathsf{T}} \boldsymbol{O} = \boldsymbol{e}_j^{\mathsf{T}}$ and $\boldsymbol{O}^{\mathsf{T}} \boldsymbol{o}_k = \boldsymbol{e}_k$, so that for $j \neq k$,

$$\boldsymbol{o}_j^{\mathsf{T}} \boldsymbol{\Sigma} \boldsymbol{o}_k = \boldsymbol{o}_j^{\mathsf{T}} \left\{ \boldsymbol{O}\,\mathrm{diag}(\lambda_i)\,\boldsymbol{O}^{\mathsf{T}} \right\} \boldsymbol{o}_k = \lambda_j \lambda_k \boldsymbol{e}_j^{\mathsf{T}} \boldsymbol{e}_k = 0.$$

The eigenvalue-eigenvector decomposition of a covariance matrix is used in Sect. 7.8 to find the orientation of elliptically contoured densities. This decomposition can be important even if the density is not elliptically contoured and is the basis of principal components analysis (PCA).

Example A.6. An Eigendecomposition

In the next example, a 3×3 symmetric matrix `Sigma` is created and its eigenvalues and eigenvectors are computed using the function `eigen()`. The eigenvalues are in the vector `decomp$values` and the eigenvectors are in the matrix `decomp$vectors`. It is also verified that `decomp$vectors` is an orthogonal matrix.

```
> Sigma = matrix(c(1, 3, 4, 3, 6, 2, 4, 2, 8), nrow = 3,
    byrow = TRUE)
> Sigma
       [,1] [,2] [,3]
[1,]    1    3    4
[2,]    3    6    2
[3,]    4    2    8
> decomp = eigen(Sigma)
> decomp
$values
[1] 11.59  4.79 -1.37

$vectors
         [,1]     [,2]     [,3]
[1,] -0.426 -0.0479  0.903
[2,] -0.499 -0.8203 -0.279
[3,] -0.754  0.5699 -0.326

> round(decomp$vectors %*% t(decomp$vectors), 5)
     [,1] [,2] [,3]
[1,]    1    0    0
[2,]    0    1    0
[3,]    0    0    1
```

$\square$

A.21 Roots of Polynomials and Complex Numbers

The roots of polynomials play an important role in the study of ARMA processes. Let $p(x) = b_0 + b_1 x + \cdots b_p x^p$, with $b_p \neq 0$, be a pth-degree polynomial. The fundamental theorem of algebra states that $p(x)$ can be factored as

$$b_p(x - r_1)(x - r_2) \cdots (x - r_p),$$

where $r_1, \ldots, r_p$ are the roots of $p(x)$, that is, the solutions to $p(x) = 0$. The roots need not be distinct and they can be complex numbers. In R, the roots of a polynomial can be found using the function `polyroot()`.

A complex number can be written as $a + bi$, where $i = \sqrt{-1}$. The absolute value or magnitude of $a + bi$ is $\sqrt{a^2 + b^2}$. The complex plane is the set of all two-dimensional vectors (a, b), where (a, b) represents the complex number $a + bi$. The unit circle is the set of all complex number with magnitude 1.

A complex number is inside or outside the unit circle depending on whether its magnitude is less than or greater than 1.

Example A.7. Roots of a Cubic Polynomial

As an example, the roots of the cubic polynomial $1 + 2x + 3x^2 + 4x^3$ are computed below. We see that there is one real root, -0.606 and two complex roots, $-0.072 \pm 0.638i$. It is also verified that these are roots.

```
> roots = polyroot(c(1, 2, 3, 4))
> roots
[1] -0.072+0.638i -0.606-0.000i -0.072-0.638i
> fn = function(x){1 + 2 * x + 3 * x^2 + 4 * x^3}
> round(fn(roots), 5)
[1] 0+0i 0+0i 0+0i
```

A.22 Bibliographic Notes

Casella and Berger (2002) covers in greater detail most of the statistical theory in this chapter and elsewhere in the book. Wasserman (2004) is a modern introduction to statistical theory and is also recommended for further study. Alexander (2001) is a recent introduction to financial econometrics and has a chapter on covariance matrices; her technical appendices cover maximum likelihood estimation, confidence intervals, and hypothesis testing, including likelihood ratio tests. Evans, Hastings, and Peacock (1993) provides a concise reference for the basic facts about commonly used distributions in statistics. Johnson, Kotz, and Kemp (1993) discusses most of the common discrete distributions, including the binomial. Johnson, Kotz, and Balakrishnan (1994, 1995) contain a wealth of information and extensive references about the normal, lognormal, chi-square, exponential, uniform, t, F, Pareto, and many other continuous distributions. Together, these works by Johnson, Kotz, Kemp, and Balakrishnan are essentially an encyclopedia of statistical distributions.

References

Alexander, C. (2001) *Market Models: A Guide to Financial Data Analysis*, Wiley, Chichester.

Casella, G. and Berger, R. L. (2002) *Statistical Inference*, 2nd ed., Duxbury/ Thomson Learning, Pacific Grove, CA.

Evans, M., Hastings, N., and Peacock, B. (1993) *Statistical Distributions*, 2nd ed., Wiley, New York.

Gourieroux, C., and Jasiak, J. (2001) *Financial Econometrics*, Princeton University Press, Princeton, NJ.

Johnson, N. L., Kotz, S., and Balakrishnan, N. (1994) *Continuous Univariate Distributions, Vol. 1*, 2nd ed., Wiley, New York.

Johnson, N. L., Kotz, S., and Balakrishnan, N. (1995) *Continuous Univariate Distributions, Vol. 2*, 2nd ed., Wiley, New York.

Johnson, N. L., Kotz, S., and Kemp, A. W. (1993) *Discrete Univariate Distributions*, 2nd ed., Wiley, New York.

Wasserman, L. (2004) *All of Statistics*, Springer, New York.

Index

∩, xxv
∪, xxv
$\imath$, 699
ρ_{XY}, xxv, 64, 684
σ_{XY}, xxv, 684
∼, xxvi
x_+, 41

bias–variance tradeoff, 483
package in R, 664

A-C skewed distributions, 102, 117
abcnon() function in R, 147
abcpar() function in R, 147
Abramson, I., 77
absolute residual plot, 258, 276
absolute value
of a complex number, 699
ACF, *see* autocorrelation function
acf() function in R, 387
ADF test, 340
adf.test() function in R, 340, 341, 371
adjust parameter, 598
adjustment matrix (of a VECM), 457
AER package in R, 77, 243, 263, 286, 403
AIC, 109, 110, 199, 232, 350, 654
corrected, 112, 126, 342
theory behind, 126
underlying statistical theory, 126
Alexander, C., 352, 433, 443, 460, 575, 700

Alexander, G., 10, 36, 510
alpha, 507, 509
analysis of variance table, 227, 229
Anderson, D. R., 126
Anderson–Darling test, 64
ANOVA table, *see* analysis of variance table
AOV table, *see* analysis of variance table
APARCH, 421
ar() function in R, 349, 385
AR process, 325
multivariate, 384
potential need for many parameters, 326
AR(1) process, 314
checking assumptions, 319
nonstationary, 317
AR(1)+ARCH(1) process, 409
AR(p) process, 325, 330
ARCH process, 405
ARCH(1) process, 407
ARCH(p) process, 411
ARFIMA, 391
arima() function in R, 318, 325, 337, 348, 378
ARIMA model
automatic selection, 342
ARIMA process, 104, 331, 343
arima.sim() function in R, 335, 640

© Springer Science+Business Media New York 2015

D. Ruppert, D.S. Matteson, *Statistics and Data Analysis for Financial Engineering*, Springer Texts in Statistics,
DOI 10.1007/978-1-4939-2614-5

CPSIA information can be obtained
at www.ICGtesting.com
Printed in the USA
LVHW081033160722
723670LV00003B/42